AF342547

LEAD-FREE ELECTRONIC SOLDERS
A Special Issue of the Journal of Materials Science:
Materials in Electronics

LEAD-FREE ELECTRONIC SOLDERS

A Special Issue of the Journal of Materials Science: Materials in Electronics

K.N. Subramanian

Chemical Engineering & Materials Science
Michigan State University

K.N. Subramanian
Michigan State University
Chemical Engineering & Materials Science
East Lansing, MI 48824-1226

Lead-Free Electronic Solders
A Special Issue of the Journal of Materials Science: Materials in Electronics

Library of Congress Control Number: 2006935131

ISBN 0-387-48431-0   e-ISBN 0-387-48433-7
ISBN 978-0387-48431-0   e-ISBN 978-0387-48433-4

Printed on acid-free paper.

9 8 7 6 5 4 3 2 1

springer.com

Contents

Preface

K. N. Subramanian

Published online: 8 September 2006

Due to the popularity of the subject of lead-free electronic solders within the past decade, the number of researchers active in this field has increased phenomenally, and the publications resulting from their studies have exploded. These publications appear in more than 20 different scientific journals and several more technical publications. As a consequence it has become impossible to keep track of current status of understanding of a given issue. Many publications in reviewed scientific journals are written by persons active in different academic disciplines, such as material science, physics, electrical engineering, mechanical engineering, etc. Similarly, publications from industrial counterparts often times do not indicate an awareness of the academic research contributions in the area. Hence it becomes necessary to provide a comprehensive and interdisciplinary account of the current status of various issues relevant to lead-free electronic solders.

This book contains the papers that were invited for a special issue of *Journal of Materials Science: Materials in Electronics*. In this reviewed scientific journal publication there were 23 articles written by recognized authorities in the field. Because this journal may not be a regular source of scientific information for academic researchers in fields other than Materials Science and those in industry, and to provide wider awareness of the current status of lead-free electronic solders to those persons active in the area but who are not regular readers of the *Journal of Materials Science: Materials in Electronics*, these articles are being reprinted in this book.

In the last few decades, the effect of lead contamination on human health has received significant attention. Based on such concerns elimination of lead from ceramic glaze, paint, plumbing etc. has been legislated and implemented. However, till recently, solders used in electronics, based on suitability and knowledge-base developed over a long period of time, remained lead-based. Successive rapid advances in microelectronic devices in recent decades make them obsolete within a very short period after their introduction resulting in significant quantities of electronic wastes in landfills. Leaching of toxic lead from such electronic wastes can result in contamination of the human food chain causing serious health hazards. As a consequence, several European and Pacific Rim countries have passed legislations warranting elimination of lead from electronic solders by fast approaching deadlines. Global economic pressures brought on by such legislations have resulted in a flurry of research activities to find suitable lead-free substitutes for the traditional leaded electronic solders.

Inspite of the exhaustive number of studies over the past two decades, no suitable drop-in substitutes have been found for lead-based electronic solders. Among the various lead-free solder alloys considered as potential substitutes only a few, especially those based on high tin, are emerging as leading candidates. Most of these tin-based compositions are eutectics or modifications of the same. These solders are being investigated, and their performances under thermal and electrical excursions are being evaluated. In addition to such an alloying approach where the intermetallic

K. N. Subramanian (✉)
University Distinguished Faculty, Professor of Materials
Science, Department of Chemical Engineering and
Materials Science, Michigan State University, East Lansing,
MI 48824-1226, USA
e-mail: subraman@egr.msu.edu

compounds play significant roles, composite solders with intentionally incorporated inert, strongly bonded, compatible reinforcements are being developed. This latter approach provides a reinforcement that will not coarsen during service and lose its effectiveness.

Solder joints are multi-component systems that have several entities such as substrate, solder/substrate intermetallic layers, solder present in the joint, and intermetallics (and other reinforcements) within the solder, etc. Such a complex system with its constraints that will arise as a result of joint geometry, and severe anisotropy of tin, poses significant complications.

The worldwide multi-facetted research efforts to arrive at suitable solutions, especially as the deadline for implementation of lead-free electronic solders approaches, have resulted in an exhaustive number of research papers in several reviewed scientific journals. Similarly, there have been presentations in several national and international meetings of various technical societies. It is impossible for any researcher or student to be aware of all the materials that have been, and are being, published in this area. So it becomes essential to have most of the relevant and currently available information in a single source such as in a special edition of a reviewed scientific journal and/or a book.

With this goal in mind the important issues that are encountered in the lead-free electronic solder area were identified, and researchers recognized for their significant scientific contributions in those areas were invited to write articles on those topics. They were asked to address the importance of a given issue, the current status of understanding and available solutions, the problems that still need to be tackled and suggestions for potential approaches to do so. Virtually all of the invitees accepted the challenge and have provided their diligently prepared papers for this special issue.

The papers that appear in this effort are arranged in the following order. Thermodynamic and phase diagram issues are presented in the first few articles. They are followed by articles addressing lead-free solder development and processing issues. Since solder joints in modern electronic applications have to possess structural integrity in addition to being electrical connections, mechanical behavior and modeling are addressed in the following articles. Since service environments impose thermal excursions, Thermomechanical Fatigue (TMF) that results from Coefficient of Thermal Expansion (CTE) mismatches of entities present in the solder joints is discussed in the following articles. Microminiaturization of electronic devices has brought out the importance of high current density with associated issues of electro-migration (EM), and whisker growth (WG) to prominence in addition to TMF. Following the papers on EM and WG, potential implications of polymorphic transformation of body-centered tetragonal tin into the diamond cubic structure at low temperatures (known as tin pest) on solder joint reliability in low temperature service is addressed. The last two papers are from researchers with extensive industrial experience and are well known for their contributions in this area. These papers deal with lead-free solder implementation in consumer electronics, and in high-end electronic components that warrant high reliability such as main-frame computers and space applications.

These papers basically cover all aspects that are relevant to lead-free electronic solder implementation and hopefully provide an overall perspective of the current status and issues to be addressed.

Thermodynamics and phase diagrams of lead-free solder materials

H. Ipser · H. Flandorfer · Ch. Luef · C. Schmetterer ·
U. Saeed

Published online: 14 September 2006

Abstract Many of the existing and most promising lead-free solders for electronics contain tin or tin and indium as a low melting base alloy with small additions of silver and/or copper. Layers of nickel or palladium are frequently used contact materials. This makes the two quaternary systems Ag–Cu–Ni–Sn and Ag–In–Pd–Sn of considerable importance for the understanding of the processes that occur during soldering and during operation of the soldered devices. The present review gives a brief survey on experimental thermodynamic and phase diagram research in our laboratory. Thermodynamic data were obtained by calorimetric measurements, whereas phase equilibria were determined by X-ray diffraction, thermal analyses and metallographic methods (optical and electron microscopy). Enthalpies of mixing for liquid alloys are reported for the binary systems Ag–Sn, Cu–Sn, Ni–Sn, In–Sn, Pd–Sn, and Ag–Ni, the ternary systems Ag–Cu–Sn, Cu–Ni–Sn, Ag–Ni–Sn, Ag–Pd–Sn, In–Pd–Sn, and Ag–In–Sn, and the two quaternary systems themselves, i.e. Ag–Cu–Ni–Sn, and Ag–In–Pd–Sn. Enthalpies of formation are given for solid intermetallic compounds in the three systems Ag–Sn, Cu–Sn, and Ni–Sn. Phase equilibria are presented for binary Ni–Sn and ternary Ag–Ni–Sn, Ag–In–Pd and In–Pd–Sn. In addition, enthalpies of mixing of liquid alloys are also reported for the two ternary systems Bi–Cu–Sn and Bi–Sn–Zn which are of interest for Bi–Sn and Sn–Zn solders.

H. Ipser (✉) · H. Flandorfer · Ch. Luef ·
C. Schmetterer · U. Saeed
Institut für Anorganische Chemie/Materialchemie,
Universität Wien, Währingerstrasse 42, A-1090 Wien,
Austria
e-mail: herbert.ipser@univie.ac.at

1 Introduction

Joining by soldering in electronics can be considered as a three step procedure:

- Melting of the solder alloy
- Contacting and solidification
- Aging of the solder joint during operation

The melting and solidification behavior as well as the interfacial reactions occurring during soldering and during operation highly influence the soldering process and the performance and durability of solder joints. Eutectic structures and solid solutions are without doubt necessary for a good soldered joint to form but the role of the intermetallics that are formed is somewhat ambivalent. According to [1], the formation of a thin layer of intermetallic compounds is absolutely necessary, however, the formation of thick layers of intermetallic compounds is frequently detrimental to the properties of the joint. As intermetallic compounds are usually brittle, they may be the cause for cracks in a particular solder joint and, as a consequence, be responsible for failure of the entire electronic part.

For a comprehensive understanding of all these processes information on the phase equilibria in the intermetallic systems generated by soldering are indispensable, and this should be supported by the relevant information on thermodynamic properties. Lead-free solders are binary but often also ternary or higher-order alloys, and together with one or more contact materials this results in multi-component systems. Any experimental study of such multi-component systems is tedious and time consuming but the well-known CALPHAD procedure (see for example Ref. [2] and references therein) provides a possible way to calculate—or extrapolate—phase equilibria in such systems based on reliable binary and ternary data. This reduces the experimental effort considerably since only a limited number of well planned key experiments will be necessary.

Many of the existing and most promising lead-free solders for electronics contain Sn or Sn and In as a low melting base alloy with small additions of Ag and/or Cu, whereas layers of Ni or Pd are frequently used contact materials. This makes the two quaternary systems Ag–Cu–Ni–Sn and Ag–In–Pd–Sn of considerable importance for the understanding of the processes that occur during soldering and during operation of the soldered devices. Thus the major aim of our series of investigations was the determination of thermodynamic properties and phase equilibria in these systems as well as in several of the binary and ternary subsystems. This should provide the basis for CALPHAD-type optimizations as part of a Thermodynamic Database [3]. Thermodynamic measurements included enthalpies of mixing, $\Delta_{mix}H$, of liquid alloys in the quaternaries themselves and in most of the constituent ternaries. Binary liquid alloys were only investigated where literature data were contradictory or incomplete. Extended calorimetric investigations at different temperatures were performed for liquid Ag–Sn, Cu–Sn, and Ni–Sn alloys since noticeable temperature dependencies of the $\Delta_{mix}H$-values had been reported for Ag–Sn and Cu–Sn [4]. In addition, enthalpies of formation, $\Delta_f H$, of solid intermetallic compounds in the systems Ag–Sn, Cu–Sn, and Ni–Sn were determined by tin solution calorimetry.

As far as phase equilibria were concerned, literature information was available for all binary subsystems. Therefore, only a few samples were prepared to check for consistency. Especially in one case, in the Ni–Sn system, it was found that the published phase equilibria [5] were obviously not correct and needed a detailed re-investigation. The corresponding phase relations and those in the ternary systems were investigated by means of X-ray diffraction (XRD), thermal analyses (DTA and DSC) and metallography including EDX and EPMA.

Since small additions of Bi and/or Zn are frequently used to adjust an appropriate melting regime of lead-free solder alloys for different applications, an experimental determination of thermochemical data (e.g. the enthalpy of mixing) of relevant ternary systems was started. First results for the ternary systems Cu–Bi–Sn and Bi–Sn–Zn will be presented here.

In this sense, this paper is supposed to give an overview over the series of investigations of thermochemical properties and phase equilibria in our laboratory within the last few years. Table 1 gives a full list of investigated systems together with the references for those cases that have already been published.

2 Literature data

A short literature overview is given for all those systems for which experimental results are reported here that have not yet been published elsewhere. For a detailed overview in all other cases the reader is referred to the corresponding original references (see Table 1).

2.1 Binary systems

2.1.1 Ag–Sn

The system Ag–Sn, now a key system for lead free solder materials, has been extensively studied in the past for various other reasons. Several experimental data sets concerning phase equilibria and thermochemistry are available, with first publications dating back to the end of the 19th century. Thermodynamic evaluations and assessments of the phase diagram including thermochemical data were done by Karakaya and Thompson [15], Chevalier [16] and Xie and Qiao [17]. There is some evidence for a temperature dependence of the enthalpy of mixing of liquid alloys.

Table 1 List of investigated systems together with references

System	Calorimetric measurements	Phase diagram measurements
Ag–Sn	This work	Not investigated
Cu–Sn	This work	Not investigated
Ni–Sn	This work	This work
Pd–Sn	[6]	Not investigated
In–Sn	[6]	Not investigated
Bi–Cu	This work	Not investigated
Ag–Ni	[7]	Not investigated
Ag–Cu–Sn	[8]	Not investigated
Cu–Ni–Sn	[8]	Not investigated
Ag–Ni–Sn	[7]	This work
Ag–Pd–Sn	[9]	Not investigated
In–Pd–Sn	[6]	[10]
Ag–In–Pd	Not investigated	[11, 12]
Bi–Cu–Sn	This work	Not investigated
Bi–Sn–Zn	[13]	Not investigated
Ag–Cu–Ni–Sn	This work	Not investigated
Ag–In–Pd–Sn	[14]	Not investigated

2.1.2 Cu–Sn

Experimental work on this important intermetallic system (brass alloys) has started at the beginning of the 20th century. Extended experimental investigations and a presentation of a phase diagram were published by Raynor [18]. Existing phase diagram versions are mainly based on his work. Thermodynamic evaluations and assessments of the phase diagram including thermochemical data were done by Saunders and Miodownik [19] and Shim et al. [20]. Again, there is evidence for a temperature dependence of the enthalpy of mixing in this system.

2.1.3 Ni–Sn

Comprehensive calorimetric investigations of this system were presented by Haddad et al. [21]. According to these results no significant temperature dependence of the enthalpy of mixing could be observed. Thermal analyses, XRD experiments and metallographic investigations date back to the first half of the 20th century, and the current version of the phase diagram, as assessed by Nash and Nash [5], is mainly based on these data. Further thermodynamic assessments and phase diagram calculations were published by Nash et al. [22] and by Ghosh [23]. Recently, Leineweber et al. [24–26] presented detailed XRD investigations of the various low-temperature modifications of Ni_3Sn_2.

2.1.4 Bi–Cu

Several calorimetric investigations and emf measurements of Bi–Cu alloys can be found in the literature, the first one dating back to 1930. The resulting enthalpies of mixing of liquid alloys are generally endothermic but not always in good agreement with each other. There are two thermodynamic assessments of the Bi–Cu system, from Niemelä et al. [27] and Teppo et al. [28], both of the same research group. The enthalpies of mixing were calculated based on an optimized thermodynamic data set. In general, there is no indication for a significant and systematic temperature dependence of the enthalpies of mixing.

2.2 Ternary systems

2.2.1 Ag–Ni–Sn

Though being one of the constituents of the quaternary Ag–Cu–Ni–Sn key system, literature on the ternary Ag–Ni–Sn system is rather scarce. An experimental isothermal section at 240°C was established by Chen and Hsu [29], whereas calculated isothermal sections are available from Ghosh at 230°C [30] and Chen and Hsu at 240°C [29]. The Sn-rich part of the liquidus projection has been suggested by Chen et al. [31] mainly based on evaluation of primary crystallization fields. Apparently due to experimental difficulties caused by liquid demixing, they made no effort to investigate the (Ag, Ni)-rich part. So far, no ternary compound has been reported in this system.

2.2.2 Bi–Cu–Sn

To the best knowledge of the authors no experimental thermochemical data for the Bi–Cu–Sn ternary system are available from literature.

2.3 Quaternary systems

Experimental data on phase relations in the tin-rich part of Ag–Cu–Ni–Sn, based on the observation of primary crystallization, were published by Chen et al. [32] and Chang et al. [33]. Otherwise, no data are available in the literature for the two quaternary systems Ag–Cu–Ni–Sn and Ag–In–Pd–Sn.

3 Experimental procedures

A very brief and general description of the experimental techniques is given below. For detailed information on sample preparation, heat treatment and specific experimental conditions the reader is referred to the corresponding references [6, 8–14] which are also listed in Table 1 and cited with the different systems in Chapter 4.

3.1 Sample preparation

All samples were synthesized from high purity elements: Ag (shot, 99.98%, ÖGUSSA, Vienna, Austria), heated in a carbon crucible at 700°C for 10 min to remove surface impurities; Bi (pellets, 99.999%, ASARCO, South Plainfield, NJ, USA); In (rod, 99.9999%, ASARCO); Cu (wire, 99.98% Goodfellow, Cambridge, UK), treated under H_2 flow at 150°C for 2 h to remove oxide layers; Ni (sheet, 99.98% Alfa Johnson–Matthey); Pd (sponge, 99.9 %, ÖGUSSA, Vienna, Austria); Sn (rod, 99.9985%); Zn (shot, 99.999%; both Alfa Johnson–Matthey). The Zn was melted under vacuum and filtered through quartz wool for further purification.

3.2 Calorimetric measurements

Calorimetric measurements were carried out in a Calvet-type micro calorimeter (SETARAM, Lyon, France) with a thermopile consisting of more than 200 thermocouples, wire wound resistance furnace, suitable for temperatures up to 1000°C, automatic drop device for up to 30 drops, control and data evaluation with LabView and HiQ as described earlier by [34].

The experimental data were treated by a least squares fit using the Redlich–Kister formalism for the description of substitutional solutions. In the case of ternary alloy systems the so-called Redlich–Kister–Muggiano polynomial was used which takes the additional ternary interactions into account [35, 36]:

$$\Delta H_{\text{mix}}^{\text{tern}} = \sum_i \sum_{j>i} \left[x_i x_j \sum_v L_{i,j}^{(v)} \left(x_i - x_j \right)^v \right] \\ + x_i x_j x_k \left(M_{i,j,k}^{(0)} x_i + M_{i,j,k}^{(1)} x_j + M_{i,j,k}^{(2)} x_k \right) \tag{1}$$

The Redlich–Kister–Muggiano formalism can also be extended to describe quaternary systems by the sum of the six binary terms, the four ternary terms and an additional quaternary contribution (see, for example, Fiorani et al. [37]):

$$\Delta H_{\text{mix}}^{\text{quat}} = \sum_i \sum_{j>i} \left[x_i x_j \sum_v L_{i,j}^{(v)} \left(x_i - x_j \right)^v \right] \\ + \sum_{i,j,k} P_{i,j,k} x_i x_j x_k + C_{i,j,k,l} x_i x_j x_k x_l \tag{2}$$

where $P_{i,j,k} = x_i M_{i,j,k}^{(0)} + x_j M_{i,j,k}^{(1)} + x_k M_{i,j,k}^{(2)}$ is the ternary contribution and $C_{i,j,k,l}$ is the symmetric quaternary interaction parameter.

3.3 X-ray diffraction

A Guinier–Huber film camera with Cu-K$_{\alpha 1}$ radiation was used to analyze the phase composition of the samples. The powdered samples were fixed on a plastic foil, and pure Si (99,9999%) was used as an internal standard. High Temperature XRD was performed on a Bruker D8 powder diffractometer equipped with an ultra-high speed Vantec 1 detector and with Cu K$_{\alpha 1}$ radiation. The heating device was an Anton Paar HTK 1200N chamber using Ar as inert gas.

3.4 Thermal analyses

Differential Thermal Analysis (DTA) was performed on a Netzsch 404S Thermal Analyzer, equipped with Pt/Pt10%Rh (S-type) thermocouples. Pieces of the samples weighing 200 to 300 mg were used for the DTA measurements. Differential Scanning Calorimetry (DSC) measurements were done on Netzsch 404 DSC equipment using Pt/Pt10%Rh (S-type) thermocouples, with samples weighing between 100 and 150 mg. For increased sensitivity, a few Ni–Sn samples were measured in a Setaram Multi-HTC instrument equipped with a heat-flow DSC transducer using Pt6%Rh/Pt30%Rh (B-type) thermocouples.

3.5 Metallography

Samples to be examined were embedded in a mixture (1:2 volume parts) of Cu Powder and Resinar F (Wirtz/Buehler, Düsseldorf, Germany). This mixture polymerizes by application of pressure and heat yielding a solid block after cooling. Cu is added to obtain a conductive material which is necessary for further investigation by electron microscopy. After embedding the samples were polished first with SiC discs with 600 and 1000 mesh and then with Al$_2$O$_3$ (1 μm).

Metallographic investigations were done with a Zeiss Axiotech light microscope with magnifications of 50, 100, 200, 500 and 1000, equipped with a Sony DSC-S75 digital still camera switched to full zoom. If necessary the samples were etched using 1–5% Nital solution (HNO$_3$ in ethanol) in order to be able to distinguish different phases from each other.

EPMA measurements were carried out on a Cameca SX 100 electron probe using wavelength dispersive spectroscopy (WDS) for quantitative analyses and employing pure elements as standard materials. Conventional ZAF matrix correction was used to calculate the compositions from the measured X-ray intensities.

4 Results and discussion

4.1 Ag–Cu–Ni–Sn and subsystems

4.1.1 Calorimetry

4.1.1.1 Binary systems
The integral enthalpies of mixing, $\Delta_{\text{mix}}H$, of liquid binary Ag–Sn, Cu–Sn, and Ni–Sn alloys were determined at different temperatures. The results are presented in Figs. 1–3. The curve fitting procedure and all extrapolations are based on the Redlich–Kister polynomials for substitutional solutions (see Chapter 3.2.); the corresponding interaction parameters, $L^{(v)}$, are listed in Table 2.

All three systems show basically an exothermic mixing behavior with exothermic minima in the Sn-

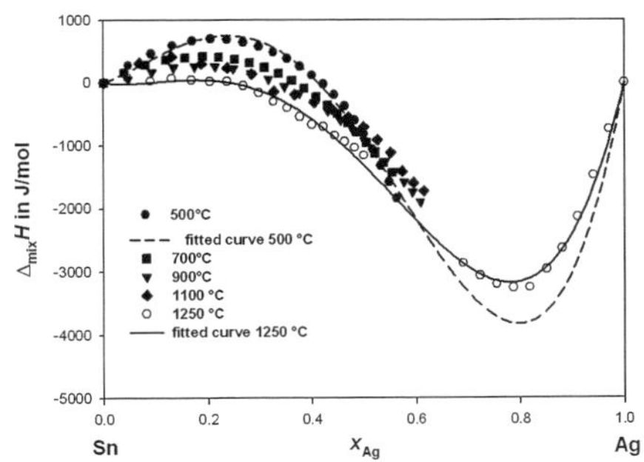

Fig. 1 Integral enthalpy of mixing of liquid Ag–Sn alloys at different temperatures; reference state: Ag(l) and Sn(l)

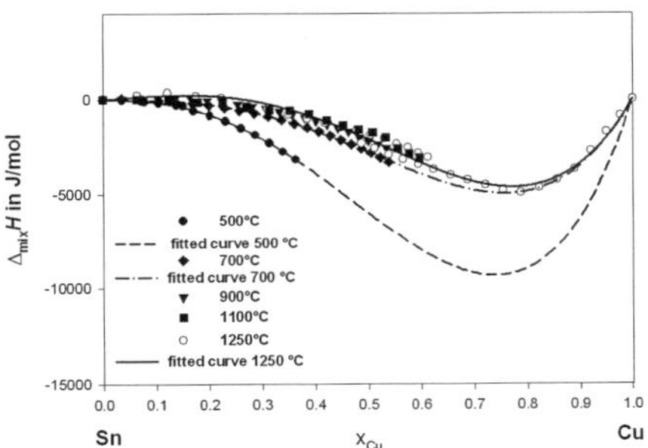

Fig. 2 Integral enthalpy of mixing of liquid Cu–Sn alloys at different temperatures; reference state: Cu(l) and Sn(l)

poor sections. Depending on temperature, the minima are at x_{Ag} = 0.75–0.8 and $\Delta_{mix}H$ = – 3000 to – 4000 J/mol for Ag–Sn, at x_{Cu} = 0.7–0.8 and $\Delta_{mix}H$ = – 4000 to – 9000 J/mol for Cu–Sn, and at x_{Ni}=0.6 and $\Delta_{mix}H$ = – 20000 J/mol for Ni–Sn. These minima correspond nicely to the compositions of the respective most stable solid compounds, Ag_3Sn, Cu_3Sn and

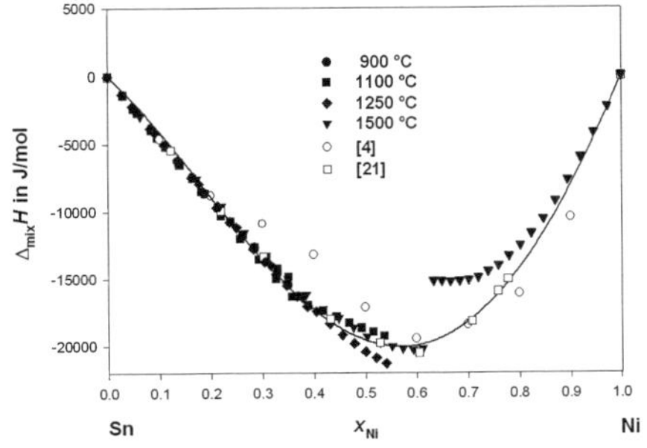

Fig. 3 Integral enthalpy of mixing of liquid Ni–Sn alloys at different temperatures compared with literature data. The solid line shows a fit based on Eq. (1); reference state: Ni(l) and Sn(l)

Ni_3Sn_2. The minimum values become more negative in the order Ag to Cu to Ni. Additionally, in the case of Ag–Sn and Cu–Sn a significant temperature dependence of these minima was found: the lower the temperature, the lower (i.e. more exothermic) the minimum value. In general, the temperature dependence is significant close to the liquidus and becomes very weak at higher temperatures. These results can be explained according to the association theory for liquid alloys (described in detail by Sommer [38]) by postulating short range ordering in the liquid alloys which is specially pronounced at lower temperatures close to solidification. The compositions of the "associates" correspond to the concentrations of the minima of the integral enthalpies of mixing.

Whereas $\Delta_{mix}H$ in Ni–Sn is exothermic over the entire concentration range, a slightly endothermic behavior can be observed in the Sn-rich part of Ag–Sn and Cu–Sn. The effect is rather weak in the case of Cu–Sn but more pronounced in Ag–Sn, having a maximum of approx. 800 J/mol at x_{Ag} = 0.3. For a more detailed view on this behavior and especially the temperature dependence of $\Delta_{mix}H$, the limiting partial enthalpies of mixing, $\Delta_{mix}H°$ for Ag, Cu and Ni, respectively, in Sn were determined at 500, 600, 700 and 800°C. All results are presented in Fig. 4, including literature values for Cu in Sn according to Deneuville et al. [39]. As one could expect, the values for Ni in Sn are strongly exothermic, with approx. –60000 J/mol at 500°C. The values show a clear temperature dependence, especially at lower temperatures. Above 800°C the positive temperature coefficient becomes smaller which explains the virtual absence of a significant temperature dependence of $\Delta_{mix}H$ at higher temperatures as seen in Fig. 3.

For Ag in Sn, the partial enthalpy values are positive, with approx. 4300 J/mol at 500°C, showing a very weak negative temperature coefficient.[1] The values for Cu in Sn are slightly positive, approx. 900 J/mol at 500°C, exhibiting a positive temperature coefficient.

The formation of clusters possessing short range order in pure liquid Sn was described by Waseda [40], based on synchrotron X-ray experiments at different temperatures. The clusters disappear at temperatures above 900°C and their structure is similar to that of tetragonal solid Sn, revealing a covalent bonding character. According to the above mentioned association theory Sn-clusters exhibiting a repulsive interaction with

---

[1] The association theory postulates a positive temperature coefficient of the enthalpy of mixing, but only for systems exhibiting exo- or endo-thermic behavior over the entire concentration range.

Table 2 Interaction parameters in the system Ag–Cu–Ni–Sn

Interaction Parameter	T [°C]	v,α	J/mol
$L^{(v)}_{Ag,Cu}$		0	17396[*]
		1	2535[*]
$L^{(v)}_{Ag,Ni}$	1500	0	51381
		1	– 20292
$L^{(v)}_{Ag,Sn}$	500	0	– 2612
		1	– 21559
		2	– 14243
	700	0	– 3719
		1	– 16793
		2	– 10165
	900	0	– 3831
		1	– 15575
		2	– 10888
$L^{(v)}_{Cu,Ni}$	1250	0	14027
		1	– 427
$L^{(v)}_{Cu,Sn}$	500	0	– 23981
		1	– 42563
		2	– 18445
	700	0	– 11093
		1	– 23767
		2	– 13720
	900	0	– 8621
		1	– 21742
		2	– 13128
	1250	0	– 9276
		1	– 24839
		2	– 13346
$L^{(v)}_{Ni,Sn}$	1250	0	– 80939
		1	– 40849
$M^{(\alpha)}_{Ag,Cu,Sn}$	500	0	– 195690
		1	43555
		2	– 27221
	700	0	– 157414
		1	– 164967
		2	– 3843
	900	0	– 127089
		1	– 123769
		2	9929
$M^{(\alpha)}_{Ag,Ni,Sn}$	1000	0	– 235299
		1	– 229480
		2	– 31033
$M^{(\alpha)}_{Cu,Ni,Sn}$	1250	0	– 155990
		1	– 274441
		2	– 19610
$C_{Ag,Cu,Ni,Sn}$	1000		– 471120

[*] Temperature independent data of different sources

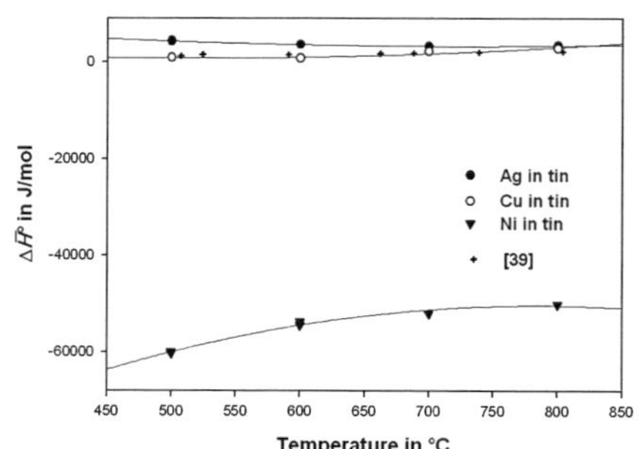

Fig. 4 Partial enthalpies of mixing at infinite dilution for Ag, Cu and Ni in liquid Sn as a function of temperature; reference state: Ag(l), Ni(l) and Cu(l)

$$\Delta_f H^{298} = x \cdot \Delta_{sol}\overline{H}^\infty(A) + y \cdot \Delta_{sol}\overline{H}^\infty(B) - \Delta_{sol}\overline{H}^\infty(A_xB_y) \tag{3}$$

where

- $\Delta_f H^{298}$ is the enthalpy of formation of the intermetallic compounds at 298 K referred to the solid components
- x, y are the mole fractions of the components A and B
- $\Delta_{sol}\overline{H}^\infty(A)$, $\Delta_{sol}\overline{H}^\infty(B)$, and $\Delta_{sol}\overline{H}^\infty(A_xB_y)$ are the limiting enthalpies of dissolution of the solid components in Sn.

All results are summarized in Table 3 including the nominal compositions, heat treatments and a comparison with available literature data.

Figure 5 shows the experimental data of the integral molar enthalpy of mixing for liquid Ag–Ni alloys at 1500°C, measured in several runs from both sides, i.e. starting from pure Ag and pure Ni. As expected for a system with an extended miscibility gap in the liquid, the enthalpy of mixing is positive for all compositions. The curves show a kink at about 4.5 at.% Ni and 1.5 at.% Ag, resp., indicating demixing of the liquid. The extrapolation of fitted experimental values gives a maximum of about 13500 J/mol at approx. 40 at% Ag. Using these results as input into a CALPHAD-type calculation of the binary Ag–Ni phase diagram, the critical temperature of the miscibility gap at 3800 K [49] comes out in good agreement with theoretical calculations of Colinet and Pasturel [50].

4.1.1.2 Ternary Systems Ag–Cu–Sn and Cu–Ni–Sn The original experimental data of the measurement series in the Ag–Cu–Sn system at 500, 700, and

free silver or copper atoms lead to the endothermic enthalpies of mixing.

In the case of additions of Ni to Sn the formation of more stable Ni–Sn associates is favorable. Thus the entire system shows an exothermic enthalpy of mixing.

The enthalpies of formation, $\Delta_f H$ of the binary intermetallic compounds Ag$_4$Sn, Ag$_3$Sn, Cu$_3$Sn (ε-phase), Cu$_3$Sn (δ-phase), Cu$_6$Sn$_5$ (η-phase), Ni$_3$Sn, Ni$_3$Sn$_2$ and Ni$_3$Sn$_4$ were obtained from tin solution calorimetry, applying the following equation:

Table 3 Enthalpies of formation of binary intermetallic compounds of Ag–Sn, Cu–Sn and Ni–Sn

Compound	Composition	Annealing temp. [°C]	$\Delta_f H^{298}$ [kJ/mol]	Literature data
Ag$_3$Sn (ε)	Ag$_{0.744}$Sn$_{0.256}$	340	−4.18(±1)	−4.5(±0.25)723 K [41]
Ag$_4$Sn (ζ)	Ag$_{0.848}$Sn$_{0.152}$	640	−2.82 (±1)	−2.5(±0.25)723 K [41] −3.6(±0.10) 723 K [42]
Cu$_3$Sn (ε)	Cu$_{0.748}$Sn$_{0.252}$	640	−8.22(±1)	−7.81(±0.2),723K [43] −7.53(±0.2), 723K [4] −7.82(±0.2), 293 K [44] −8.36, 723 K [61]
Cu$_{41}$Sn$_{11}$(δ)	Cu$_{0.80}$Sn$_{0.20}$	700	−5.68(±1)	−5.45(±0.2), 723 K [43] −5.46(±0.2), 723 K [4]
Cu$_6$Sn$_5$ (η)	Cu$_{0.548}$Sn$_{0.452}$	200	−6.11 (±1)	−7.03 (±0.05) 273 K [45]
Ni$_3$Sn (LT)	Ni$_{0.74}$Sn$_{0.26}$	1050	−24.9 (±1)	−26(±0.5), 1060 K [46] −23.4(±4), 298 K [47] −23(±0.28), 293 K [4]
Ni$_3$Sn$_2$ (HT)	Ni$_{0.577}$Sn$_{0.423}$	1050	−34.6(±1)	−31(±0.28), 293 K [4] −32(±0.5), 1060 K [46] −39(±0.2), 1023 K [48] −31.3(±4), 298 K [47]
Ni$_3$Sn$_4$	Ni$_{0.42}$Sn$_{0.58}$	600	−24(±1)	−25(±0.5), 1060 K [46] −33.7(±0.2), 1023 K [48]

900°C, resp., and in the Cu–Ni–Sn system at 1250°C can be found in Ref. [8]. As the result of a least squares fit, the ternary interaction parameters $M^{(\alpha)}_{Ag:Cu:Sn}$ and $M^{(\alpha)}_{Cu:Ni:Sn}$ were established. All binary and ternary interaction parameters can be found in Table 2.

Based on the Redlich–Kister–Muggiano polynomial (1), iso-enthalpy curves of $\Delta_{mix}H$ in the liquid state were calculated and plotted on Gibbs triangles (see Fig. 6 for Ag–Cu–Sn at 900°C and Fig. 7 for Cu–Ni–Sn at 1250°C). The corresponding analytical functions describe the enthalpy of mixing in the entire system, but it has to be noted that the iso-enthalpy graphs can only be experimentally determined at concentrations where the alloy is completely liquid at the respective temperature. All the other calculated values refer to metastable supercooled liquid alloys. In the graphic representation, the liquidus isotherms from the compilation by Villars et al. [51] were included showing all partially solid regions at the corresponding temperatures shaded in gray.

The Ag–Cu–Sn iso-enthalpy curves are somewhat more complicated than those of Cu–Ni–Sn. With decreasing temperature the enthalpy of mixing of the binary Cu–Sn system becomes significantly more negative. The minimum is situated around the binary Cu$_3$Sn compound, which is also the global minimum for the Ag–Cu–Sn system (in our fit at $x_{Cu} = 0.74$ and $\Delta_{mix}H = -9.4$ kJ/mol at 500°C, $x_{Cu} = 0.76$ and $\Delta_{mix}H = -5.0$ kJ/mol at 700°C and $x_{Cu} = 0.77$ and $\Delta_{mix}H = -4.3$ kJ/mol at 900°C). A "valley" of negative $\Delta_{mix}H$ values runs through the ternary system at around 20 to 30 at.% Sn, connecting the Cu–Sn with the Ag–Sn minimum. The maximum of the fitted surface is positive and appears at the bordering Ag–Cu

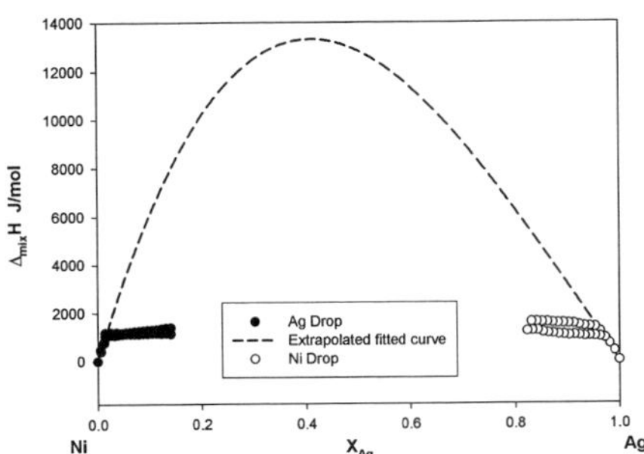

Fig. 5 Integral enthalpy of mixing of liquid Ag–Ni alloys at 1500°C; reference state: Ag (l) and Ni (l)

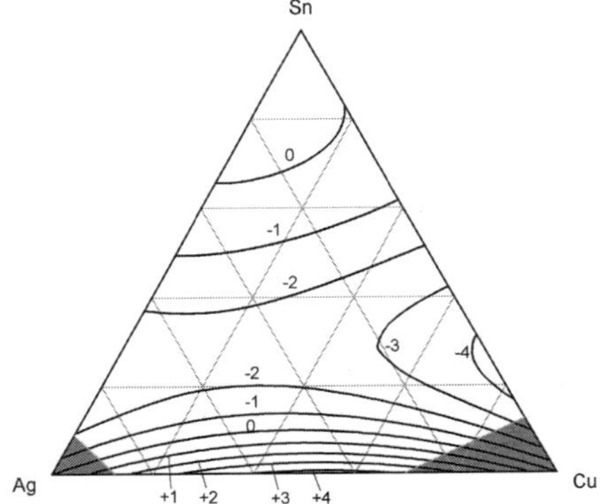

Fig. 6 Integral enthalpy of mixing of liquid Ag–Cu–Sn alloys at 900°C; reference state: Ag(l), Cu(l), Sn(l). Values are in kJ/mol

 Springer

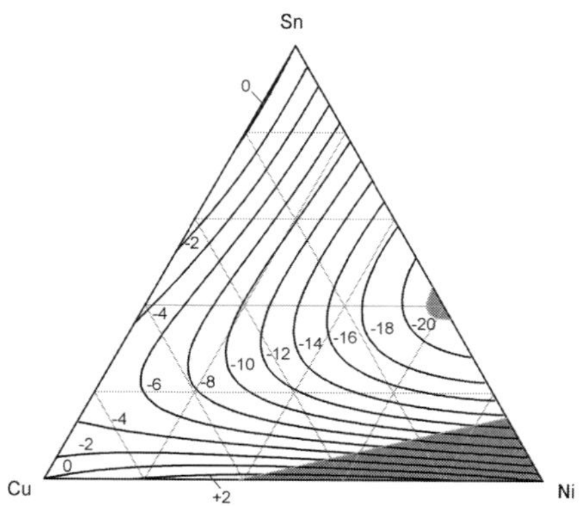

Fig. 7 Integral enthalpy of mixing of liquid Cu–Ni–Sn alloys at 1250°C; reference state: Cu(l), Ni(l), Sn(l). Values are in kJ/mol

binary system. As the temperature decreases, the liquid region in the Ag–Cu–Sn system gets significantly smaller. At 500°C only a small portion in the Sn-rich corner is still liquid, which is the region interesting for lead-free solder materials.

The enthalpy of mixing for the Cu–Ni–Sn system at 1250°C shows a global minimum around the binary Ni_3Sn_2 compound (in our fit at 0 at.% Cu and 60.8 at.% Ni, –21.4 kJ/mol). A "valley" runs through the enthalpy surface connecting the binary minima in the Ni–Sn and Cu–Sn system. Next to the binary compound Ni_3Sn_2 and in the Ni-rich corner the system is (partially) solid at this temperature. The maximum of the enthalpy surface is positive and appears along the bordering Cu–Ni binary system, far away from our region of interest for lead-free soldering. Our results for 1250°C are in good agreement with the results for 1307°C by Pool et al. [52].

The enthalpy of mixing of the ternary Ag–Cu–Sn system shows a significant dependence on temperature. The temperature dependence of the ternary interaction parameters $M^{(\alpha)}_{Ag,Cu,Sn}$ is illustrated in Table 2. To obtain a general expression for the value of these parameters as a function of the absolute temperature (in K) they were fitted using a linear polynomial. The results are as follows:

$$M^{(0)}_{Ag,Cu,Sn} = -326936 + 171.50 \cdot T$$

$$M^{(1)}_{Ag,Cu,Sn} = -365395 + 205.99 \cdot T$$

$$M^{(2)}_{Ag,Cu,Sn} = -97412 + 92.88 \cdot T$$

Note that for the temperature dependence of the parameter $M^{(1)}$ only the values for 700 and 900°C were considered since its value at 500°C deviates consider-

ably from the linear trend and would require a much more complicated equation. Therefore the equations given above should only be used to calculate the ternary interaction parameters for temperatures above 700°C, for lower temperatures the dependence might be more complicated. With the combined results it is now possible to calculate the enthalpy of mixing of liquid Ag–Cu–Sn alloys for any desired ternary concentration at different temperatures.

4.1.1.3 Ternary system Ag–Ni–Sn In the ternary Ag–Ni–Sn system the enthalpies of mixing were determined at 1000, 1220 and 1400°C. In one series of experiments, pieces of pure Ni (20 to 40 mg) were dropped into approx. 1 g of molten Ag–Sn alloys (10, 20, 30, 70, 80, and 90 at.% Sn), in another series, pure Ag was added to liquid Ni–Sn alloys (10, 20, 30, 70, 80, and 90 at.% Sn). From the composition dependence of the integral enthalpy values and from the constant values of the partial enthalpy, the concentration limits of the homogeneous liquid phase and the type of the second phase segregating from the liquid could be deduced. It was either the primary crystallization of solid Ni_3Sn_2 or the appearance of a second liquid phase due to the extension of the liquid miscibility gap in the ternary system. All these results are presented in Fig. 8.

Applying the polynomial described above, iso-enthalpy curves for liquid ternary Ag–Ni–Sn mixtures at 1000°C are presented in Fig. 9. As in the other

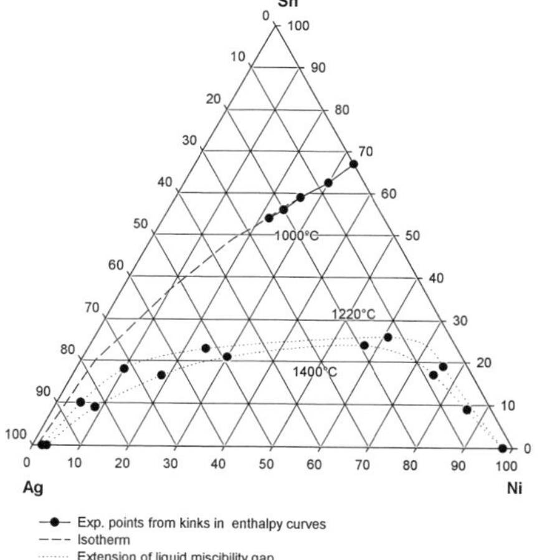

Fig. 8 Extension of the homogeneous liquid phase in the Ag–Ni–Sn system at 1000°C (liquidus), 1220 and 1400°C (liquid miscibility gap) from calorimetric measurement

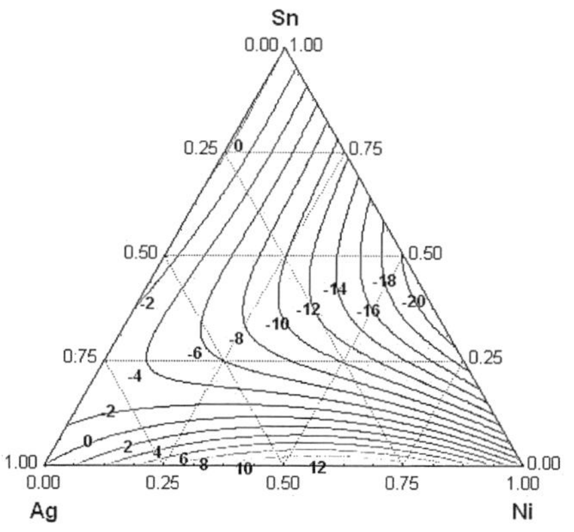

Fig. 9 Integral enthalpy of mixing of liquid Ag–Ni–Sn alloys at 1000°C; reference state: Ag(l), Ni(l), Sn(l). Values are in kJ/mol

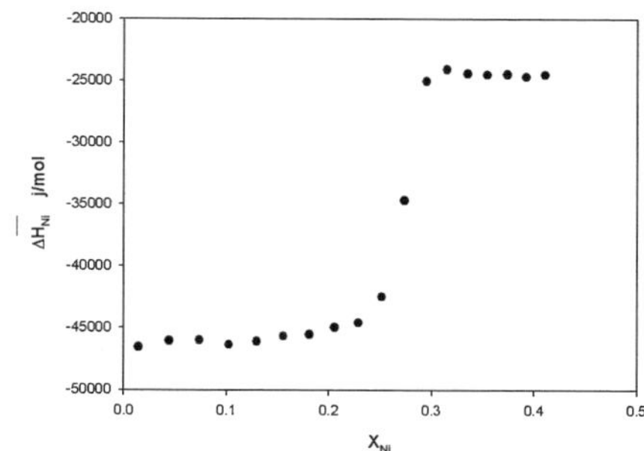

Fig. 11 Partial enthalpy of mixing of Ni in quaternary liquid Ag–Cu–Ni–Sn alloys along a section starting from $Ag_5Cu_5Sn_{90}$ at 1000°C; reference state: Ag(l), Cu(l), Ni(l), Sn(l)

ternary systems, the plot refers to the metastable liquid phase for all those compositions where the alloy is not completely liquid at 1000°C.

It can be seen that $\Delta_{mix}H$ is mostly negative except in the region close to the Ag–Ni binary (where one can find the global endothermic maximum of 13.5 kJ/mol) and in the narrow region close to the Sn-rich part of the Ag–Sn binary system. The enthalpy of mixing shows a minimum of – 20 kJ/mol corresponding to the binary minimum of Ni–Sn around 60 at.% Ni.

4.1.1.4 Quaternary System Ag–Cu–Ni–Sn In the quaternary Ag–Cu–Ni–Sn system, the enthalpies of mixing were determined at 1000°C along nine sections. Starting with ternary Ag–Cu–Sn alloys ($0.4 \leq x_{Sn} \leq 0.9$) in the crucible, 15 to 18 drops of pure nickel were added for each section. Figures 10 and 11 show,

as an example, the integral and partial liquid enthalpies of mixing along one section, starting from $Ag_5Cu_5Sn_{90}$, in the quaternary Ag–Cu–Ni–Sn system at 1000°C. It can be seen that the $\Delta_{mix}H$ curve in Fig. 10 shows a kink at approx. 28 at.% Ni which corresponds to a jump in the partial enthalpy to a constant value of approx. –25000 J/mol (Fig. 11). This is caused by the primary crystallization of a solid phase, most probably Ni_3Sn_2. Similar behavior was found in all other investigated sections.

The experimental enthalpies of mixing of the quaternary Ag–Cu–Ni–Sn alloys were fitted based on Eq. (2) (see Chapter 3.2). All the necessary binary and ternary interaction parameter can be found in Table 2. Following Luo et al. [53], the term for the ternary interactions in Ag–Cu–Ni was skipped assuming ideal mixing of the binary liquid alloys.

Usually, from a statistical point of view, the contribution of quaternary interactions to the overall enthalpy of mixing is expected to be small. Indeed, the

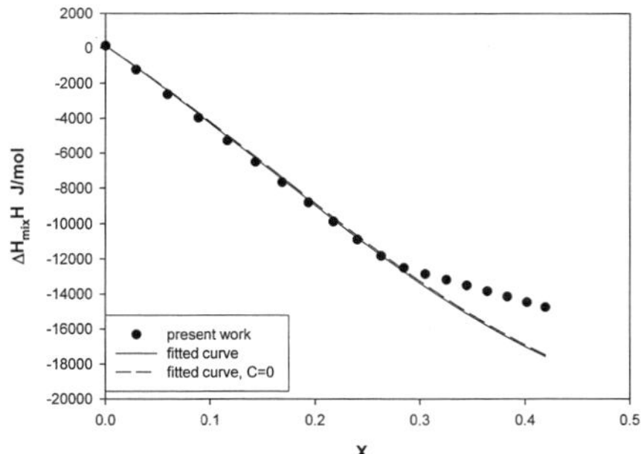

Fig. 10 Integral enthalpy of mixing of quaternary liquid Ag–Cu–Ni–Sn alloys along a section starting from ternary $Ag_5Cu_5Sn_{90}$ at 1000°C; reference state Ag(l), Cu(l), Ni(l), Sn(l)

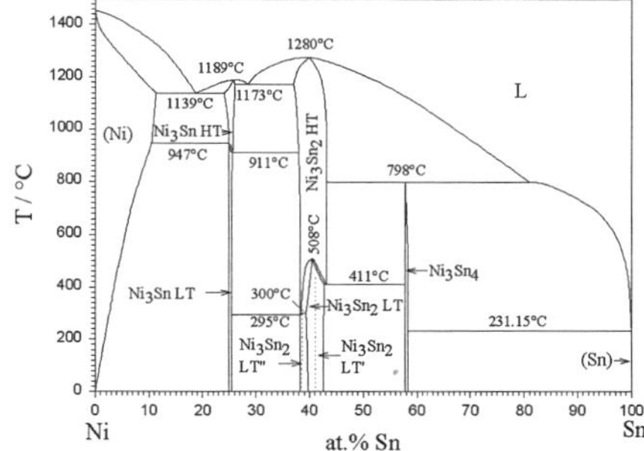

Fig. 12 Ni–Sn phase diagram revised according to the results of the present investigation, including data from Leineweber et al. [24–26]

resulting symmetric quaternary interaction parameter of $C_{Ag,Cu,Ni,Sn} = -471120$ J/mol is relatively small, considering that it has to be multiplied by a factor $x_{Ag} \cdot x_{Cu} \cdot x_{Ni} \cdot x_{Sn}$: it gives a maximum contribution of 1840 J/mol to the integral enthalpy at $x_{Ag} = x_{Cu} = x_{Ni} = x_{Sn}$. In Fig. 10 the experimental values of a quaternary section are compared with the calculated curve including $C_{Ag,Cu,Ni,Sn}$ (solid line) and without quaternary interaction (dashed line). Considering the estimated errors of these calorimetric techniques it is clear that $C_{Ag,Cu,Ni,Sn}$ can be neglected, as has been shown recently for the general case by Fiorani et al. [37].

4.1.2 Phase diagrams

4.1.2.1 Ni–Sn A new version of the Ni–Sn phase diagram was established based on thermal analyses, XRD techniques and EPMA on samples annealed at various temperatures. Homogeneity ranges of intermetallic phases and reaction temperatures were modified according to our results.

The crystal structure of the HT-Ni$_3$Sn phase was found to be cubic (BiF$_3$ type) using high temperature XRD. This phase either undergoes martensitic transformation to a β-Cu$_3$Ti structure on quenching at very high cooling rates or a massive transformation to the hexagonal LT-Ni$_3$Sn phase. The phase transition was found to comprise a peritectoid and a eutectoid reaction at 947 and 911°C, resp., which is in contrast to the available literature [5]. Furthermore, the Ni$_3$Sn$_2$ region was considerably modified: two incommensurate Ni$_3$Sn$_2$ low temperature phases (LT$'$ and LT$''$), first reported by Leineweber et al. [24–26], were included as well as the rather complicated transition to the Ni$_3$Sn$_2$ HT-phase. Figure 12 shows the modified Ni–Sn phase diagram according to our results.

4.1.2.2 Ag–Ni–Sn This ternary system is dominated by a liquid miscibility gap which extends from the binary Ag–Ni system into the ternary up to about 35 at.% Sn. No ternary intermetallic compound has been found in this system so far, and the binary phases have very limited or negligible ternary solubilities. Isothermal sections at 200, 450, 700 and 1050°C were established based on results of EPMA and XRD for more than 120 ternary samples prepared by different methods and annealed at several temperatures. As an example, the isothermal section at 450°C is shown in Fig. 13. The liquidus was estimated from thermal analysis and metallography. Certain phase fields involving the various modifications of Ni$_3$Sn$_2$ have not been clarified yet because of the complexity of the

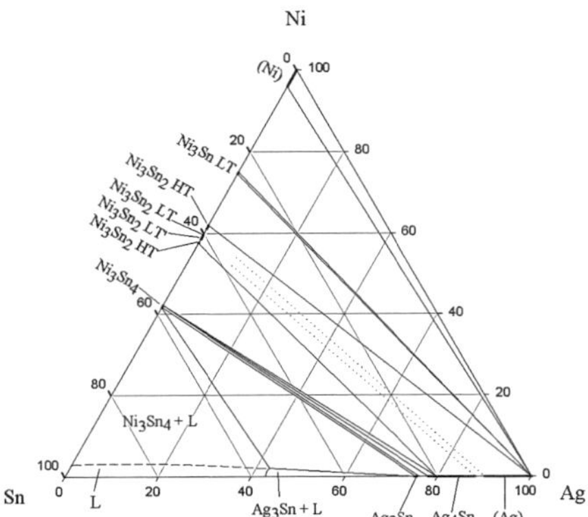

Fig. 13 Isothermal section of the Ag–Ni–Sn phase diagram at 450°C

involved phase transitions; any phase boundaries are therefore represented by dotted lines.

The tentative liquidus projection shown in Fig. 14 was deduced from results of thermal analyses. According to our experiments, the invariant reaction U2 was placed at Ag$_{64}$Ni$_1$Sn$_{35}$ and at a temperature of 570°C which is in contrast to previous work by Chen et al. [31]. All invariant reactions are listed in Table 4. The (Ag,Ni)-rich part of the liquidus projection remains still somewhat unclear because of severe experimental difficulties in that area.

4.2 Ag–In–Pd–Sn and subsystems

4.2.1 Calorimetry

The original experimental data of the measurement series in the ternary systems In–Pd–Sn and Ag–Pd–Sn,

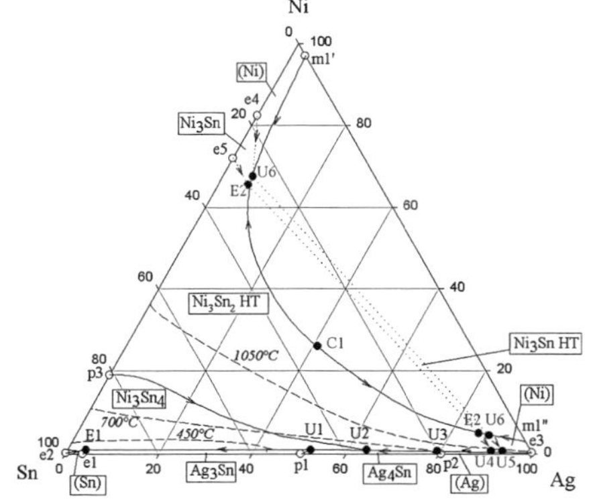

Fig. 14 Liquidus projection of Ag–Ni–Sn phase diagram; fields of primary crystallization are indicated

Table 4 Invariant reactions in the system Ag–Ni–Sn

Reaction	Type	Temperature [°C]	
		This work	Chen et al. [31]
L → (Sn) + Ni$_3$Sn$_4$ + Ag$_3$Sn	E1	220	219
L + Ag$_4$Sn → Ni$_3$Sn$_4$ + Ag$_3$Sn	U1	485	488
L + Ni$_3$Sn$_2$ → Ag$_4$Sn + Ni$_3$Sn$_4$	U2	570	516.5
L + (Ag) → Ag$_4$Sn + Ni$_3$Sn$_2$ HT	U3	725	
L2 + Ni$_3$Sn HT → Ni$_3$Sn$_2$ HT + (Ag)	U4	Unknown	
L2 + (Ni) → Ni$_3$Sn HT + (Ag)	U5	Unknown	
L1 → L2 + Ni$_3$Sn HT + Ni$_3$Sn$_2$ HT	E2	1130	
L1 + (Ni) → L2 + Ni$_3$Sn HT	U6	~1135	
Critical point	C1	1130 < T < 1270	

and the quaternary system Ag–In–Pd–Sn can be found in Refs. [6, 9, 14], respectively. All quaternary measurements were performed at 900°C. The $\Delta_{mix}H$ values of the ternary liquid alloys were again fitted employing Eq. (1). Binary and ternary interaction parameters are listed in Table 5.

The resulting iso-enthalpy plots for In–Pd–Sn and Ag–Pd–Sn are shown in Figs. 15 and 16, respectively. In the diagrams, the estimated course of the liquidus

Table 5 Interaction parameters in the system Ag–In–Pd–Sn

Interaction Parameter	T [°C]	v,α	J/mol
$L_{Ag,In}^{(v)}$	1000	0	− 15443
		1	− 12728
		2	3844
$L_{Ag,Pd}^{(v)}$	1400	0	− 19141
		1	− 15925
$L_{Ag,Sn}^{(v)}$	500	0	− 2612
		1	− 21559
		2	− 14243
	700	0	− 3719
		1	− 16793
		2	− 10165
	900	0	− 3831
		1	− 15575
		2	− 10888
$L_{In,Pd}^{(v)}$	900	0	− 202640
		1	85610
$L_{In,Sn}^{(v)}$	900	0	− 1481
		1	− 499
$L_{Pd,Sn}^{(v)}$	900	0	− 215814
		1	− 126046
$M_{Ag,In,Pd}^{(\alpha)}$	938	0	− 275878
		1	66245
		2	− 653632
$M_{Ag,In,Sn}^{(\alpha)}$	727–980	0	32696
		1	44749
		2	10393
$M_{Ag,Pd,Sn}^{(\alpha)}$	900	0	− 313084
		1	− 422417
		2	113838
$M_{In,Pd,Sn}^{(\alpha)}$	900	0	156065
		1	253787
		2	211126
$C_{Ag,In,Pd,Sn}$	900		351260

isotherm at 900°C is included, showing all regions shaded in gray where the system is not totally liquid.

It can be seen that the enthalpy of mixing in the In–Pd–Sn system is negative over the entire composition range and shows a global minimum of –57.0 kJ/mol on the binary Pd–Sn edge at 56 at.% Pd. In the Ag–Pd–Sn system the enthalpy of mixing is negative almost over the entire composition range, except for a very small region along the Sn-rich part of the binary Ag–Sn where one can find the global endothermic maximum ($\Delta_{mix}H_m$ = + 256 J/mol at 21 at.% Ag). The enthalpy of mixing shows again its minimum of − 57.0 kJ/mol on the binary Pd–Sn edge at 56 at.% Pd.

The experimental integral enthalpy of mixing data in the quaternary system Ag–In–Pd–Sn at 900°C were fitted using Eq. (2), with the necessary interaction parameters taken from Table 5. All those parameters that have been determined at a different temperature were assumed to be temperature independent. Multiplying the resulting value of $C_{Ag,Cu,Ni,Sn}$ = − 351260 J/mol by $x_{Ag} \cdot x_{In} \cdot x_{Pd} \cdot x_{Sn}$ yields a maximum quaternary contribution of 1372 J/mol at $x_{Ag} = x_{In} = x_{Pd} = x_{Sn}$.

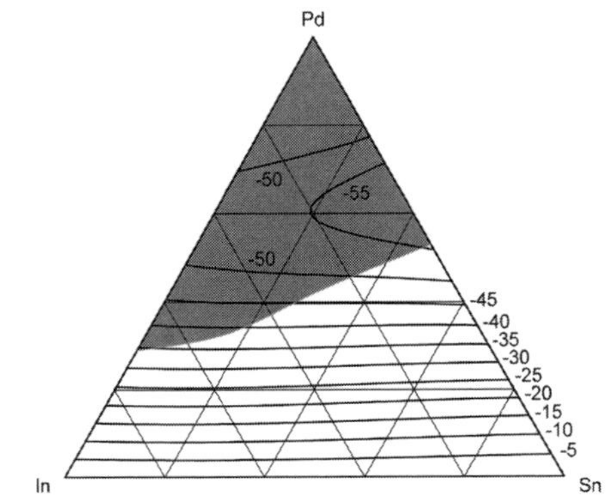

Fig. 15 Integral enthalpy of mixing of liquid In–Pd–Sn alloys at 900°C; reference state: In(l), Pd(l), Sn(l), values are in kJ/mol

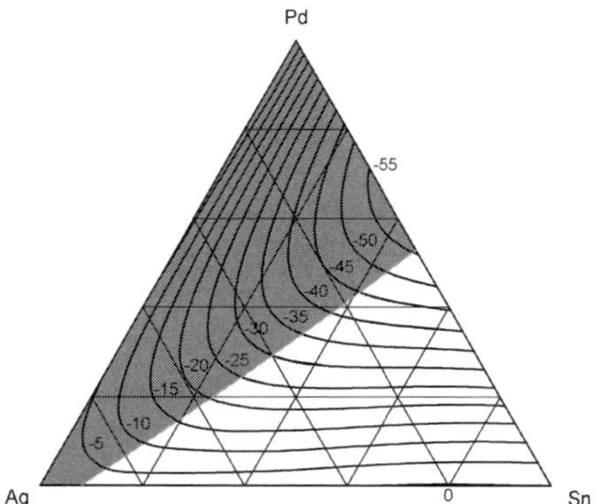

Fig. 16 Integral enthalpy of mixing of liquid Ag–Pd–Sn alloys at 900°C; reference state: Ag(l), Pd(l), Sn(l), values are in kJ/mol

This shows again that, within the experimental uncertainty, this quaternary term can be neglected as in the case of the Ag–Cu–Ni–Sn system (see above).

The enthalpies of mixing of quaternary Ag–In–Pd–Sn alloys at 900°C along one selected section (pure Ag dropped into molten $In_{40}Pd_{20}Sn_{40}$), calculated from Eq. (2), with the interaction parameters from Table 5 and with (solid line) and without (dashed line) $C_{Ag:In:Pd:Sn} = 351260$ J/mol, are shown in Fig. 17. It can be seen that the measured values are described well by both fits.

4.2.2 Phase diagrams

4.2.2.1 In–Pd–Sn Ternary In–Pd–Sn alloys were synthesized to evaluate the phase relations in that system for Pd contents up to about 60 at.%. The phase compositions of the annealed (180, 500, and 700°C) and quenched samples were characterized by XRD

and—for a number of selected samples—also by scanning electron microscopy. The original experimental data can be found in Ref. [10]. Based on binary data from Refs. [54] and [34] and the present results, isothermal sections were constructed for 180, 500, and 700°C.

As an example, the isothermal section at 500°C is presented in Fig. 18. The phase InPd, with an extended homogeneity range in the binary ($0.43 \leq x_{In} \leq 0.52$), dissolves up to 7 at.% Sn, whereas PdSn was found to dissolve significant amounts of In (approx. 7 at.%); a similar amount of In is dissolved in $Pd_{20}Sn_{13}$. The phase $PdSn_2$ shows an even larger solubility for In: Sn can be substituted by In up to 22 at.% at constant Pd contents. All samples in the composition range below the $[In_7Pd_3 + PdSn_2]$ two-phase field were found to be in equilibrium with the liquid. Our results are in general agreement with results of Kosovinc et al. [55, 56] who suggested a continuous solid solubility between the isotypic orthorhombic compounds $\alpha InPd_2$ and Pd_2Sn. Although this has not been proven unambiguously up to now it is in accord with our results.

4.2.2.2 Ag–In–Pd To obtain a first idea of the phase relationships in the Ag–In–Pd system, a CALPHAD-type calculation was employed and three isothermal sections were calculated at 200, 500, and 700°C, based on optimized versions of the binary systems only, taken from the COST 531 Thermodynamic Data Base [3]. In a next step, the three isothermal sections were investigated experimentally [11, 12] and, as an example, the

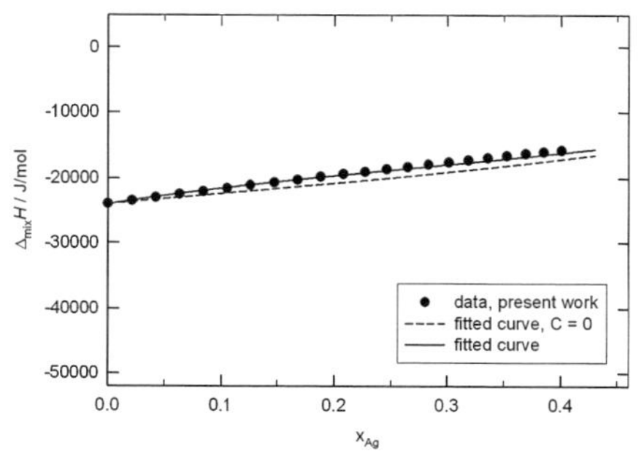

Fig. 17 Integral enthalpy of mixing of quaternary liquid Ag–In–Pd–Sn alloys along a section starting from ternary $In_{40}Pd_{20}Sn_{40}$ at 900°C; reference state: Ag(l), In(l), Pd(l), Sn(l)

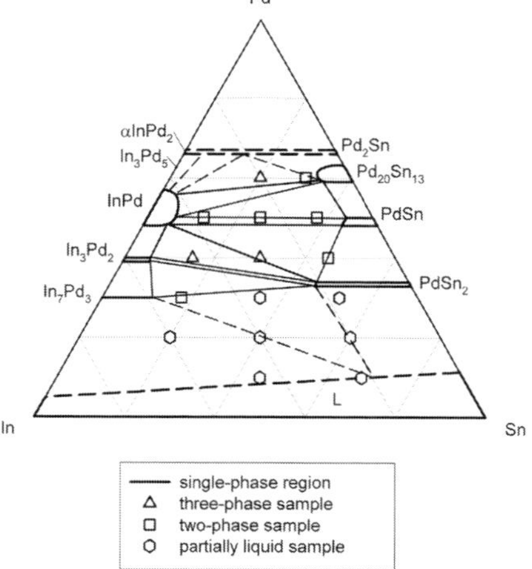

Fig. 18 Isothermal section of In–Pd–Sn phase diagram at 500°C

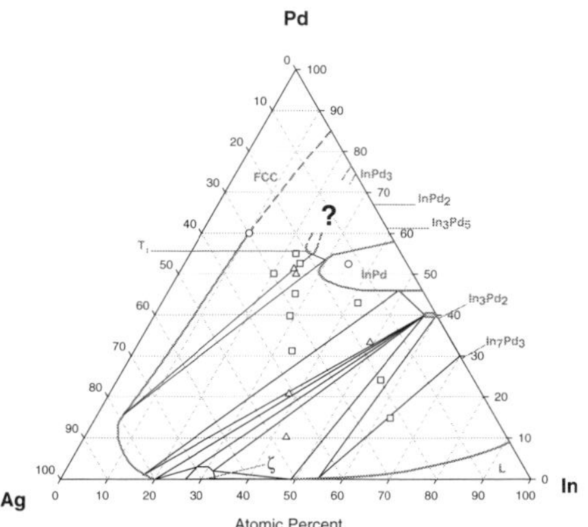

Fig. 19 Phase equilibria in the Ag–In–Pd system at 500°C based on experimental results

section at 500°C is shown in Fig. 19. As can be seen, a ternary compound (T₁) was identified with a crystal structure of the HgMn-type which is related to the structure of the binary compound InPd₃ (Al₃Ti type). This structural relationship is probably the reason that its homogeneity range, starting approximately at $Ag_{20}Pd_{55}In_{25}$, extends towards binary InPd₃, ending probably around $Ag_{10}Pd_{65}In_{25}$ although the exact position of the two-phase field is still not fully clear.

All experimental phase diagram data were taken together with experimental values for the enthalpy of mixing of liquid Ag–In–Pd alloys [14, 57], and a new CALPHAD optimization of the ternary system was performed [12]. Figure 20 shows the calculated isothermal section at 500°C which can now be compared with the experimental phase diagram in Fig. 19. It can

be seen that the agreement is good even if the calculated homogeneity range of InPd is smaller and the ternary phase T₁ is shown as a stoichiometric compound.

4.3 Bi–Cu–Sn

The partial and integral enthalpies of mixing at 800°C of liquid ternary Bi–Cu–Sn alloys were determined along nine sections in a large composition range. Additionally, binary alloys of the constituent system Bi–Cu were investigated at 800 and 1000°C.

For Bi–Cu, the $\Delta_{mix}H$ values are generally positive as expected for this simple eutectic intermetallic system with no mutual solid solubility of the constituents. The maximum value of $\Delta_{mix}H$ at 800°C is approx. 4000 J/mol at $x_{Bi} = 0.55$, whereas the measurements at 1000°C gave a maximum of $\Delta_{mix}H = 5000$ J/mol at $x_{Bi} = 0.50$. This positive temperature coefficient of $\Delta_{mix}H$ is in agreement with the association theory for liquid intermetallic alloys [38].

The integral enthalpy of mixing in the ternary system Bi–Cu–Sn is presented as an iso-enthalpy plot in Fig. 21. It was obtained by fitting our experimental results applying the Redlich–Kister–Muggiano polynomial (Eq. 1). The binary and ternary interaction parameters are listed in Table 6. The minima and maxima of the enthalpy surface correspond to the binary minima and maxima in the systems Cu–Sn and Bi–Cu, resp.

4.4 Bi–Sn–Zn

The original experimental data of the measurement series in the Bi–Sn–Zn system can be found in Luef

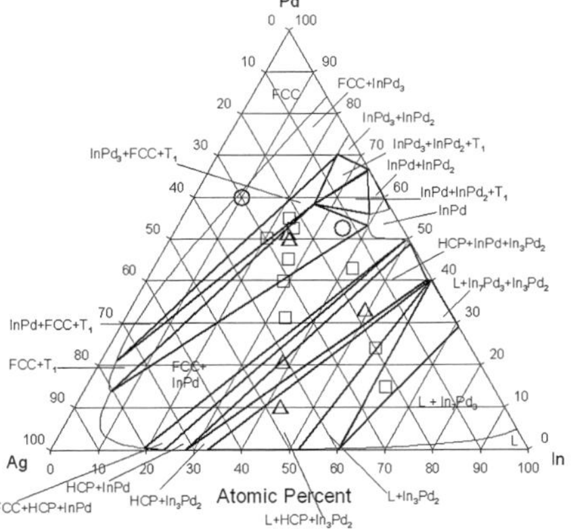

Fig. 20 Phase equilibria in the Ag–In–Pd system at 500°C from a CALPHAD-type calculation and comparison with experimental results: ○, single phase; □, two-phase; △, three-phase

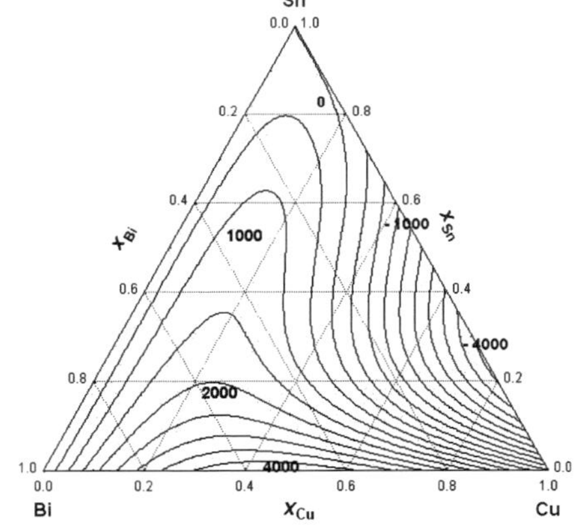

Fig. 21 Integral enthalpy of mixing of liquid Bi–Cu–Sn alloys at 800°C; reference state: Bi(l), Cu(l), Sn(l), values are in J/mol

Table 6 Interaction parameters in the systems Bi–Cu–Sn and Bi–Sn–Zn

Interaction Parameter	T [°C]	v,α	J/mol
$L^{(v)}_{Bi,Sn}$	450	0	442
		1	− 298
$L^{(v)}_{Bi,Cu}$	800	0	17810
		1	3331
$L^{(v)}_{Cu,Sn}$	800	0	− 9857
		1	− 22755
		2	− 13423
$L^{(v)}_{Bi,Zn}$	470–570	0	17782
		1	− 4775
$L^{(v)}_{Sn,Zn}$	422–577	0	12558
		1	− 5623
		2	4149
$M^{(\alpha)}_{Bi,Cu,Sn}$	800	0	− 275878
		1	66245
		2	− 653632
$M^{(\alpha)}_{Bi,Sn,Zn}$	500	0	− 15489
		1	− 10785
		2	− 12528

et al. [13]. All measurements were performed at 500°C. Again, the ternary experimental $\Delta_{mix}H$ values of the liquid alloys were fitted by means of Eq. (1), and the obtained binary and ternary interaction parameters are listed in Table 6. Figure 22 shows a corresponding iso-enthalpy plot of the integral enthalpy of mixing. The enthalpy of mixing is positive (i.e. endothermic) over the entire composition range. The global maximum of the Bi–Sn–Zn system can be found in the binary Bi–Zn bordering system ($\Delta_{mix}H_m$ = 4523 J/mol at 44 at.% Bi).

Liquidus temperatures and the temperature of the ternary eutectic reaction $E(L \rightleftharpoons (Bi) + (Sn) + (Zn))$ were determined by DSC measurements of 32 samples at constant Zn contents of 3, 5, and 7 at.%, annealed at 150°C. Three different heating rates (10, 5, and 1 K/ min) were employed and the thermal effects were linearly extrapolated to zero heating rate. The reaction E was detected in all 32 samples at a temperature of 135.0 ± 0.5°C. This is 5°C higher than the value reported by Muzaffar [58] and also higher than indicated in the recent assessments by Malakhov et al. [59] (126.1°C) and Moelans et al. [60] (131.2°C). From the course of the liquidus temperatures in the three sections it can be estimated that the composition of the liquid in the ternary eutectic reaction is approx. 41 at.% Bi, 57 at.% Sn, and 2 at.% Zn. This is shifted slightly towards lower Zn and higher Sn contents as compared to the values calculated in the two assessments [59, 60].

Acknowledgements The authors acknowledge the financial support of the Austrian "Fonds zur Förderung der wissenschaftlichen Forschung (FWF)", projects No. P-15620, P-16495 and P-17346. The financial support of the Hochschuljubilaeumsstiftung der Stadt Wien (Project No. H-812/2005) is also gratefully acknowledged. This research is a contribution to the European COST Action 531 on "Lead-free Solder Materials".

References

1. A. Rahn, in *The Basics of Soldering* (John Wiley &Sons Inc., 1993), p. 1
2. Y.A. Chang, S. Chen, F. Zhang, X. Yan, F. Xie, R. Schmid-Fetzer, W.A. Oates, Progr. Mater. Sci. **49**, 313 (2004)
3. A.T. Dinsdale, A. Watson, A. Kroupa, A. Zemanova, J. Vrestal, J. Vizdal, COST 531 Thermodynamic Database, Version 2.0 (2006) (http://www.slihot.co.uk/COST531/ td_database.htm)
4. R. Hultgren, P.D. Desai, D. Hawkins, M. Gleiser, K. Kelley, in "Selected Values of the thermodynamic properties of binary alloys" (AMS Metals Park, Ohio, 1971)
5. P. Nash, A. Nash, Bull. Alloy Phase Diagrams **6**, 350 (1985)
6. C. Luef, H. Flandorfer, H. Ipser, Thermochim. Acta **417**, 47 (2004)
7. U. Saeed, H. Flandorfer, H. Ipser, J. Mater. Res. **21**, 1294 (2006)
8. C. Luef, H. Flandorfer, H. Ipser, Z. Metallkde. **95**, 151 (2004)
9. C. Luef, A. Paul, H. Flandorfer, A. Kodentsov, H. Ipser, J. Alloys Comp. **391**, 67 (2005)
10. C. Luef, H. Flandorfer, A. Paul, A. Kodentsov, H. Ipser, Intermetallics **13**, 1207 (2005)
11. A. Zemanova, A. Kroupa, J. Vrestal, O. Semenova, K. Chandrasekaran, K.W. Richter, H. Ipser, Monatsh. Chem. **136**, 1931 (2005)
12. A. Zemanova, O. Semenova, A. Kroupa, J. Vrestal, K. Chandrasekaran, K.W. Richter, H. Ipser, Intermetallics, **14**, (2006), in press
13. C. Luef, A. Paul, J. Vizdal, A. Kroupa, A. Kodentsov, H. Ipser, Monatsh. Chem. **137**, 381 (2006)
14. C. Luef, H. Flandorfer, H. Ipser, Metall. Mater. Trans. A **36A**, 1273 (2005)

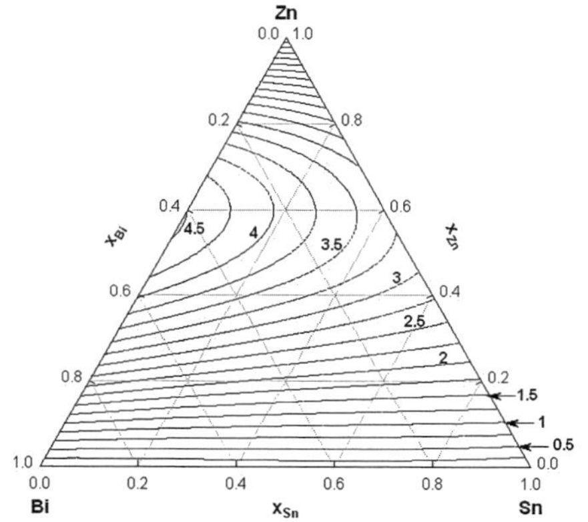

Fig. 22 Integral enthalpy of mixing curves of liquid Bi–Sn–Zn alloys at 500°C; ref. state: Bi(l), Sn(l) and Zn(l); values are in kJ/mol

15. I. Karakaya, W.T. Thompson, Bull. Alloy Phase Diagrams **8**, 340 (1987)
16. P.-Y. Chevalier, Thermochim. Acta **136**, 45 (1988)
17. Y. Xie, Z. Qiao, J. Phase Equil. **17**, 208 (1996)
18. G.-V. Raynor, in *Annotated Equilibrium Diagram Series, No. 2* (The Institute of Metals, London, 1944)
19. N. Saunders, A.P. Miodownik, Bull. Alloy Phase Diagrams **11**, 278 (1990)
20. J.-H. Shim, C.-S. Oh, Z. Metallkde. **87**, 205 (1996)
21. R. Haddad, M. Gaune-Escard, J.-P. Bros, A. Ranninger-Havlicek, E. Hayer, K.L. Komarek, J. Alloys Comp. **247**, 82 (1997)
22. P. Nash, H. Choo, R.B. Schwarz, J. Mat. Sci. **33**, 4949 (1998)
23. G. Ghosh, Metall. Mater. Trans. **30A**, 1481 (1999)
24. A. Leineweber, M. Ellner, E.J. Mittemeijer, J. Solid State Chem. **159**, 191 (2001)
25. A. Leineweber, O. Oeckler, U. Zachwieja, J. Solid State Chem. **177**, 936 (2004)
26. A. Leineweber, J. Solid State Chem. **177**, 1197 (2004)
27. J. Niemelä, G. Effenberg, K. Hack, P. Spencer, CALPHAD **10**, 77 (1986)
28. O. Teppo, J. Niemelä, P. Taskinen, Thermochim. Acta **173**, 137 (1990)
29. S.-W. Chen, H.-F. Hsu, Acta Materialia **52**, 2541 (2004)
30. G. Ghosh, J. Electron. Mater. **29**, 1182 (2000)
31. S.-W. Chen, H.-F. Hsu, Ch.-W. Lin, J. Mater. Res. **19**, 2267 (2004)
32. S.-W. Chen, C.-A. Chang, J. Electr. Mat. **33**, 1071 (2004)
33. C.-A. Chang, S.-W. Chen, C.-N. Chiu, Y.-C. Huang, J. Electr. Mat. **34**, 1135 (2005)
34. H. Flandorfer, J. Alloys Comp. **336**, 176 (2002)
35. I. Ansara, N. Dupin, in *COST 507 Thermochemical database for light metal alloys*, vol. 2 (European Commission DG XII, Luxembourg, 1998) p. 1
36. N. Saunders, A.P. Miodownik, in "CALPHAD (Calculation of Phase Diagrams): A Comprehensive Guide" (Pergamon Press, Oxford, UK, 1998)
37. J.M. Fiorani, C. Naguet, J. Hertz, A. Bourkba, L. Bouirden, Z. Metallkde. **88**, 711 (1997)
38. F. Sommer, Z. Metallkde. **73**, 72 (1982)
39. J.-F. Deneuville, C. Chatillon-Colinet, J.-C. Mathieu, E. Bonnier, J. Chim. Phys. **73**, 273 (1976)
40. Y. Waseda, in *The structure of non-crystalline materials* (McGraw-Hill Inc., 1980), p. 56
41. O.J. Kleppa. J. Phys. Chem. **60**, 852 (1956)
42. G.H. Lauri, A.W.H. Morris, J.N. Pratt, Trans. Met. Soc. **236**, 1390 (1966)
43. O.J. Kleppa, Acta Metall. **3**, 255 (1955)
44. J.B. Cohen, J.S Leach, M.B Bever, J. Met. **6**, 1257 (1954)
45. A. Gangulee, G.C. Das, M.B. Bever, Metall. Trans. **4**, 2063 (1973)
46. B. Predel, W. Vogelbein, Thermochim. Acta **30**, 201 (1979)
47. F. Korber, W. Oelsen, Mitt. K. Wilhem Inst. Eisenforsch. Düsseldorf **19**, 209 (1937)
48. H. Dannoehl, H.L. Lukas, Z. Metallkde. **65**, 642 (1974)
49. A. Zemanova, J. Vrestal, A. Kroupa, Research in progress Masaryk University Bruno (2005)
50. C. Colinet, A. Pasturel, Z. Metallkde. **89**, 863 (1998)
51. P. Villars, A. Prince, H. Okamoto, in *Handbook of Ternary Alloy Phase Diagrams* (ASM International, Metals Park, Ohio, 1995)
52. M.J. Pool, I. Arpshofen, B. Predel, E. Schultheiss, Z. Metallkde. **70**, 656 (1979)
53. H.-T. Luo, S. W. Chen, J. Mater. Sci. **31**, 5059 (1996)
54. T.B. Massalski, J.L. Murray, L.H. Bennett, H. Baker, in "Binary Alloy Phase Diagrams" (ASM, Materials Park, Ohio, 1990)
55. I. Kosovinc, M. El-Boragy, K. Schubert, Metall **26**, 917 (1972)
56. I. Kosovinc, T. Grgasovic, Rud.-Metal. Zborn. **1**, 71 (1972)
57. E. Hayer, Calorimetrie Analyse Thermique **26**, 262 (1995)
58. S.D. Muzaffar, J. Chem. Soc. **123**, 2341 (1923)
59. D.V. Malakhov, X.J. Liu, I. Ohnuma, K. Ishida, J. Phase Equil. **21**, 514 (2000)
60. N. Moelans, K.C. Hari Kumar, P. Wollants, J. Alloys Comp. **360**, 98 (2003)
61. W. Biltz, W. Wagner, H. Pieper, W. Holverscheit, Z. Anorg. Chem. **134**, 25 (1924)

Phase diagrams of Pb-free solders and their related materials systems

**Sinn-Wen Chen · Chao-Hong Wang ·
Shih-Kang Lin · Chen-Nan Chiu**

Published online: 1 September 2006

Abstract Replacing Pb–Sn with Pb-free solders is one of the most important issues in the electronic industry. Melting, dissolution, solidification and interfacial reactions are encountered in the soldering processes. Phase diagrams contain equilibrium phase information and are important for the understanding and prediction of phase transformation and reactive phase formation at the solder joints. This study reviews the available phase diagrams of the promising Pb-free solders, and their related materials systems. The solders are Sn–Ag, Sn–Cu, Sn–Ag–Cu, Sn–Zn, Sn–Bi, Sn–In and Sn–Sb. The materials systems are the solders with the Ag, Au, Cu, Ni substrates, such as Sn–Ag–Au, Sn–Ag–Ni, Sn–Cu–Au, and Sn–Cu–Ni ternary systems. For the Pb-free solders and their related ternary and quaternary systems, preliminary phase equilibria information is available; however, complete and reliable phase diagrams over the entire compositional and temperature ranges of soldering interests are lacking.

1 Introduction

Soldering is the most important joining technology in the electronic industries [1]. Solders are heated up and melt first during the soldering processes. The molten solders wet the substrates, and liquid (solder)/solid (substrate) contacts are formed. The joints are then cooled down, solders solidify, and solder joints are formed. Since melting is required in the soldering processes, solders are usually low melting point alloys. Pb–Sn alloys are the most popular solders and their properties are extensively investigated. However, due to health and environmental concerns, the European Union has made resolution prohibiting Pb usage in the electronic products [2], and the electronic industries are now shifting to Pb-free solders [3–5].

Soldering is conducted at the completely molten state of solders, while the solder products are in use with the solders at their solid phases. The primary materials characteristics of solders are their liquidus and solidus temperatures which define the temperature boundaries of completely molten and completely solid, respectively. Solders melt and wet the substrates at temperatures higher than their melting points, and there are dissolution of substrates and interfacial reactions between solders and substrates [1, 4–6]. The liquid phase then transforms to various solid phases when the joints are cooled down. The different kinds and relative amounts of the solid phases formed during solidification are important for the solder joint properties.

Since solders are low melting point alloys, the products' operation temperatures are relatively high temperatures and diffusion is significant for most solders. In addition, good wetting is required for good solders, thus interfacial reactions with substrates are usually significant not only at the liquid/solid contacts, but at the solid/solid contacts at the operation temperatures as well. Interfacial reactions at the solder joints are the key reliability factor of electronic products. Understanding and controlling of the interfacial reactions are important, because the interfacial

S.-W. Chen (✉) · C.-H. Wang · S.-K. Lin · C.-N. Chiu
Department of Chemical Engineering, National Tsing Hua University, Hsin-chu, Taiwan
e-mail: swchen@che.nthu.edu.tw

reactions are more complicated with Pb-free solders and with the emerging flip chip technologies which are with more pronounced electromigration effects.

Solders are usually binary and ternary alloys. When they are in contact with substrates, the materials systems are then ternary, quaternary or even of higher orders. As mentioned above, phase transformation and reactive phase transformation occurs at the solder joints. Phase diagrams contain the basic phase equilibria and phase transformation information, such as the solidus and liquidus temperatures of alloys. Besides, the liquidus projection can be used for the illustration of solidification path. Isothermal section is crucial for the understanding of interfacial reactions. Phase diagrams of Pb-free solders and their related materials systems are important for the applications of Pb-free solders.

Among various kinds of Pb-free solders, Sn–Ag, Sn–Cu, Sn–Ag–Cu, Sn–Zn-(based), Sn–Bi, Sn–In and Sn–Sb alloys are most promising. Ag, Au, Cu and Ni are the most common substrates. This study reviews the research status of the phase diagrams of these promising solders and the materials systems including solders and substrates, such as Sn–Ag–Au, Sn–Ag–Ni, Sn–Cu–Au and Sn–Cu–Ni. Since interfacial reactions are important for soldering applications, and they also provide the phase formation information, interfacial reactions literatures related directly with phase formations are also included in this study, especially for the materials system with no available phase equilibria information.

2 Sn–Ag–Cu (Sn–Ag and Sn–Cu)

Eutectic and near eutectic Sn–Ag–Cu (SAC) alloys are recommended by JEIDA and NEMI [7, 8]. Sn–0.7wt%Cu (Sn–1.3at%Cu) and Sn–3.5wt%Ag (Sn–3.8at%Ag) are also promising Pb-free solders. The most important Pb-free solders are at the Sn-rich corner of the Sn–Ag–Cu ternary system composed of the binary Sn–Cu eutectic, binary Sn–Ag eutectic and the ternary Sn–Ag–Cu eutectic.

As shown in Fig. 1 [9] of the Sn–Ag binary system, there are two intermetallic compounds, ζ-Ag$_4$Sn and ε-Ag$_3$Sn, two peritectic reactions, and one eutectic reaction, liquid = Ag$_3$Sn + Sn [9]. The eutectic composition is at Sn–3.5wt%Ag and its melting temperature is at 221°C. Kattner and Boettinger [10] have thermodynamically modeled this system. They used solution models for Ag, Sn, liquid, and the ζ phase, and they assumed the ε-Ag$_3$Sn to be a line compound. The calculated phase diagrams are in good agreement with

the experimental determinations, and the temperature and liquid composition of the eutectic are at 220.9°C and Sn–3.87at%Ag, respectively.

The binary Sn–Cu system is a complicated binary system. As shown in Fig. 2, there are seven intermetallic compounds (β, γ, ζ, δ, ε-Cu$_3$Sn, η-Cu$_6$Sn$_5$ and η'-Cu$_6$Sn$_5$), and 13 invariant reactions [11]. For soldering application purposes, phase equilibria at the Sn-rich corner are more important, and there are 3 intermetallic compounds, ε-Cu$_3$Sn, η-Cu$_6$Sn$_5$ and η'-Cu$_6$Sn$_5$, one eutectic, liquid = Sn + η – Cu$_6$Sn$_5$, and one possible eutectoid, η-Cu$_6$Sn$_5$ = Sn + η' – Cu$_6$Sn$_5$. The liquid composition and the temperature of the eutectic are at Sn–0.7wt%Cu and 227°C. Shim et al. [12] and Boettinger et al. [13] have thermodynamically modeled the binary Sn–Cu system. The ζ, δ, ε-Cu$_3$Sn, η-Cu$_6$Sn$_5$ and η'-Cu$_6$Sn$_5$ are assumed to be line compounds. Solution model and two-sublattices model are used for the β and γ phases, respectively. The calculated phase diagrams are in good agreement with the experimental determinations.

Gebhardt and Petzow [14] and Yen and Chen [15] experimentally determined the isothermal sections of the phase equilibria of the ternary Sn–Ag–Cu system. No ternary compounds have been found. As shown in Fig. 3a of the 240°C isothermal section, the ε-Cu$_3$Sn phase has tie-lines with all the solid phases, and the ternary solubilities of the binary compounds are limited. Gebhardt and Petzow [14] proposed a class II reaction, L + η-Cu$_6$Sn$_5$ = ε-Ag$_3$Sn + Sn, at the Sn-rich corner at 225°C. The experimental results of Miller et al. [16] indicated a different result and the invariant reaction should be a class I reaction L = η-Cu$_6$Sn$_5$ + ε-Ag$_3$Sn + Sn at 216.8°C. The Sn–Ag–Cu liquidus projection is shown in Fig. 3b. More recent phase equilibria studies all agree that the invariant reaction at the Sn-rich corner is a class I reaction [15, 17–19]. However, the exact composition and temperature of the class I reaction, the ternary eutectic varied. Loomans and Fine [17] reported the eutectic composition to be at Sn–3.5wt%Ag–0.9wt%Cu (Sn–3.81at%Ag–1.66at%Cu). The liquid composition and temperature of the ternary eutectic determined by Moon et al. [18] are Sn–3.5wt%Ag–0.9wt%Cu and 217.2°C, and are Sn–3.66wt%Ag–0.91wt%Cu (Sn–3.98at%Ag–1.68at%Cu) and 216.3°C from their calculation. The ternary eutectic is at Sn–3.24wt%Ag-0.57wt%Cu and 217.7°C as determined by Ohnuma et al. [19].

The materials systems consisting of the solders and popular substrates, Au, Ag, Cu, Ni, are important for soldering applications as well. Massalski and Pops [20] and Evans and Prince [21] experimentally determined the phase equilibria of the Sn–Ag–Au ternary system.

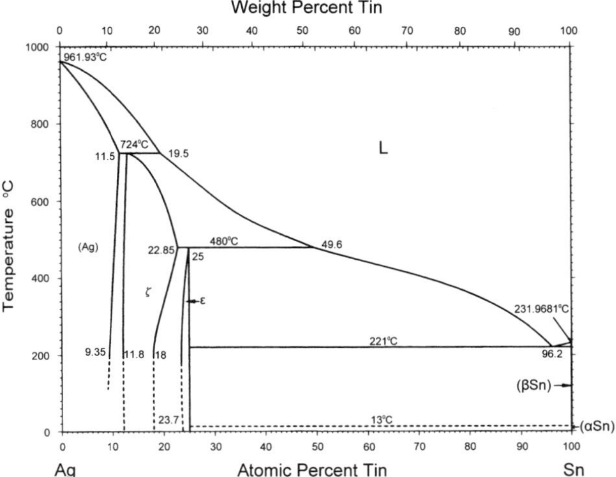

Fig. 1 The Sn–Ag binary phase diagram [9]

Figure 4a is the phase equilibria isothermal section of the ternary Sn–Ag–Au system at 200°C proposed based on phase equilibria and interfacial reaction results [20–22]. No ternary compounds have been found. Continuous solid solutions are formed between Ag and Au, and the ζ phase in the Au–Sn and Ag–Sn systems. The ε-Ag$_3$Sn phase has about 10at% Au solubility, while the ternary solubilities of other Au–Sn binary compounds are negligible. Figure 4b is the liquidus projection of the Sn–Ag–Au system [21]. At the Sn-rich corner, there is class I reaction, ternary eutectic liquid = ε-Ag$_3$Sn + Sn + η-AuSn$_4$, at 206°C. Chen and Yen [22] examined the interfacial reactions in the Ag–Sn/Au couples. Three binary phases, δ-AuSn, ε-AuSn$_2$, and η-AuSn$_4$, were formed in all the couples. The results are in agreement with the phase diagrams studies that all the compounds do not have detectable ternary solubility.

Interfacial reactions between Sn–3.5wt%Ag and Cu substrates have been studied by various groups [23–27]. In agreement with the Sn–Ag–Cu phase diagrams [15, 17–19], no ternary compound is formed, and the reaction products are binary ε-Cu$_3$Sn and η-Cu$_6$Sn$_5$ phase in the temperature range between 70°C and 360°C. It needs mentioning that although an order-disorder transformation of the Cu$_6$Sn$_5$ phase reaction is observed [11–13], the effect of this transformation has not been discussed in either the ternary phase

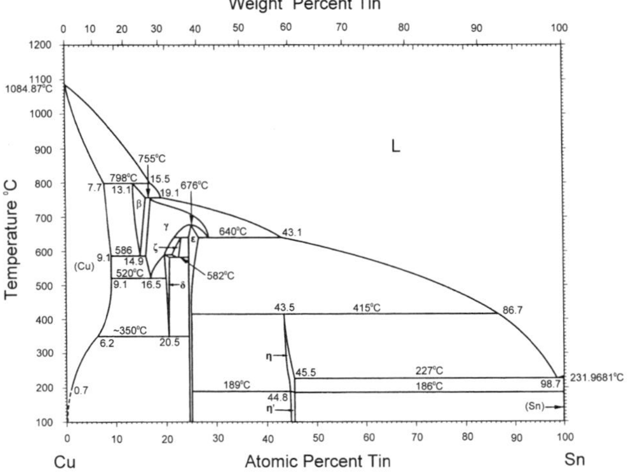

Fig. 2 The Sn–Cu binary phase diagram [11]

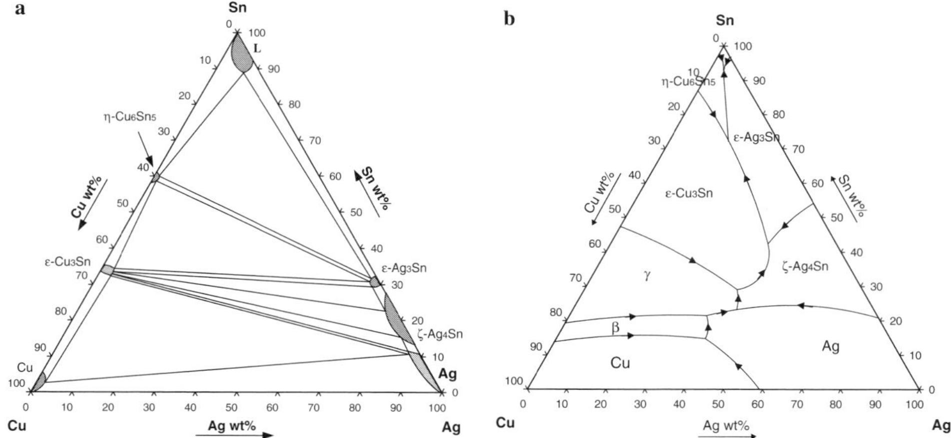

Fig. 3 (**a**) The 240°C isothermal section of the Sn–Ag–Cu ternary system [15]. (**b**) Liquidus projection of the Sn–Ag–Cu ternary system [14]

equilibria studies [14–19] or in the interfacial reaction studies [23–27].

Hsu and Chen [28] and Chen et al. [29] experimentally determined the isothermal sections and liquidus projections of the Sn–Ag–Ni ternary system. The phase diagrams of the ternary system were also calculated using the binary thermodynamic models without introducing any ternary interaction parameters [28]. Figure 5a, b are the 240°C isothermal section and the liquidus projection of the Sn–Ag–Ni ternary system, respectively. No ternary compounds have been found. The Ni_3Sn_2 phase is a very stable phase, and is in equilibrium with Ag, ζ-Ag_4Sn, ε-Ag_3Sn, Ni_3Sn, and Ni_3Sn_4 phases. It is also observed that all the ternary solubilities of the binary compounds are negligible. The liquid miscibility gap in the Ag–Ni binary system extends into the ternary system as shown in Fig. 8 of the liquidus projection. It is also shown in Fig. 5b that the class I reaction, ternary eutectic liquid = Sn + Ag_3Sn + Ni_3Sn_4, exists at the Sn-rich corner at 212.3°C. The interfacial reactions between Sn–3.5wt%Ag and Ni have been examined [28, 30–32]. In agreement with the phase diagram study [28], no ternary compound is formed in the Sn–Ag/Ni couples,

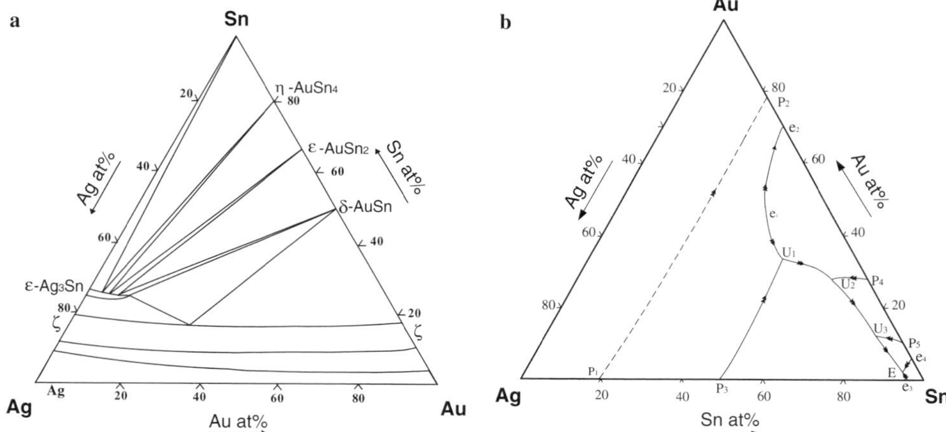

Fig. 4 (**a**) The 200°C isothermal section of the Sn–Ag–Au ternary system [22]. (**b**) Liquidus projection of the Sn–Ag–Au ternary system [21]

 Springer

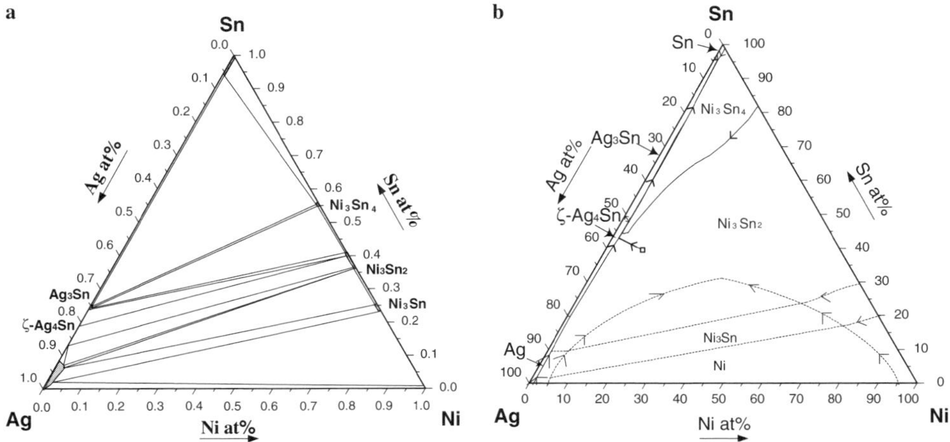

Fig. 5 (**a**) The 240°C isothermal section of the Sn–Ag–Ni ternary system [28]. (**b**) Liquidus projection of the Sn–Ag–Ni ternary system [29]

and Ni_3Sn_4 phase is the primary interfacial reaction product with negligible Ag solubility. Interfacial reactions with Ni are frequently examined because Ni is a popular substrate. However, it is worthy noting that Ni(P) and Ni–7wt%V substrates are often used in the electronic products besides pure Ni, and it has been found the interfacial reactions at the solder/Ni and solder/Ni–7wt%V are different [33].

The 360°C isothermal section of the Sn–Cu–Au phase diagram was determined by Karlsen et al. [34] as shown in Fig. 6. Continuous solid solutions are formed between Cu and Au and between the δ-AuSn and η-Cu_6Sn_5 phases. Three ternary intermetallic compounds, designated as A, B, and C, were found. The compositional ranges are $Sn_{20}Cu_{38}Au_{42}$–$Sn_{20}Cu_{33}Au_{47}$, $Sn_{20}Cu_{69}Au_{11}$–$Sn_{20}Cu_{42}Au_{38}$, and $Sn_{333}Cu_{337}Au_{330}$–$Sn_{333}Cu_{297}Au_{370}$, respectively. It can be noticed that all the homogeneity ranges of the binary compounds and the ternary compounds are all in the directions parallel to Au–Cu side, indicating the Au and Cu atoms are easier to interchange atomic sites with each other. Peplinski and Zakel [35] investigated the Sn–Cu–Au alloys by X-ray powder diffraction, and they reported the existence of a ternary compound $Sn_4Cu_8Au_8$ with compositional ranging from $Sn_4Cu_9Au_7$ to $Sn_4Cu_7Au_9$. This phase is in agreement with the A phase reported by Karlsen et al. [34]. Luciano et al. [36] studied the pseudobinary phase diagram, and they found the B phase, a ternary compound, in the results by Karlsen et al. is the ordered AuCu III′ phase. Although Au surface is frequently encountered in electronic products, and the formation of the $AuSn_4$ phase causes

various problems [37], no interfacial reactions studies of the Sn–Cu alloy with pure Au substrate are available.

Lin et al. [38], Wang and Chen [39] and Chen et al. [40] experimentally determined the phase equilibria of the Sn–Cu–Ni ternary system. Figure 7a is the 240°C Sn–Cu–Ni phase equilibria isothermal section. No ternary compounds have been found. Cu and Ni form a continuous (Cu,Ni) solid solution, and Cu_3Sn and Ni_3Sn formed a continuous $(Cu,Ni)_3Sn$ solid solution [38]. It is very distinguishable in the Sn–Cu–Ni system that the binary compounds are with very extensive solubility, but only along the direction parallel to the Cu–Ni side. However, Murakami and Kachi [41] reported a phase

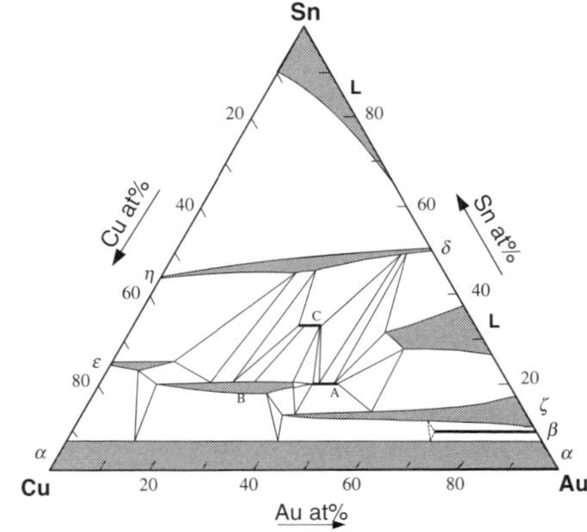

Fig. 6 The 360°C isothermal section of the Sn–Cu–Au ternary system [34]

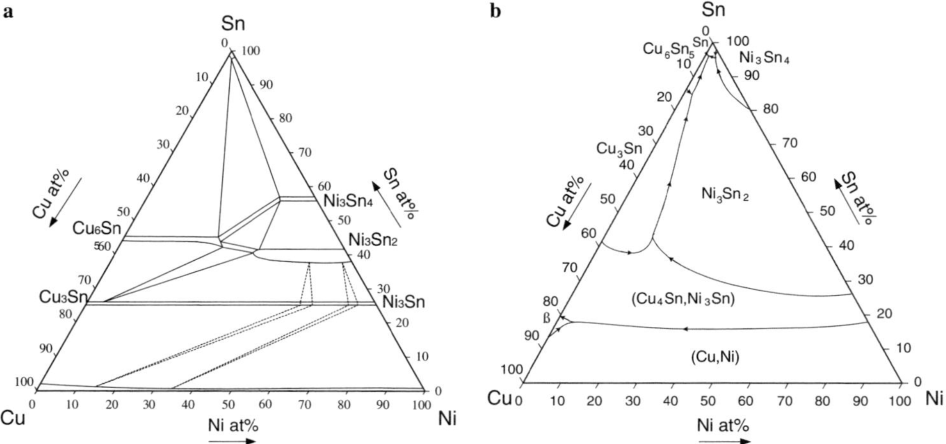

Fig. 7 (**a**) The 240°C isothermal section of the Sn–Cu–Ni ternary system [38]. (**b**) Liquidus projection of the Sn–Cu–Ni ternary system [38]

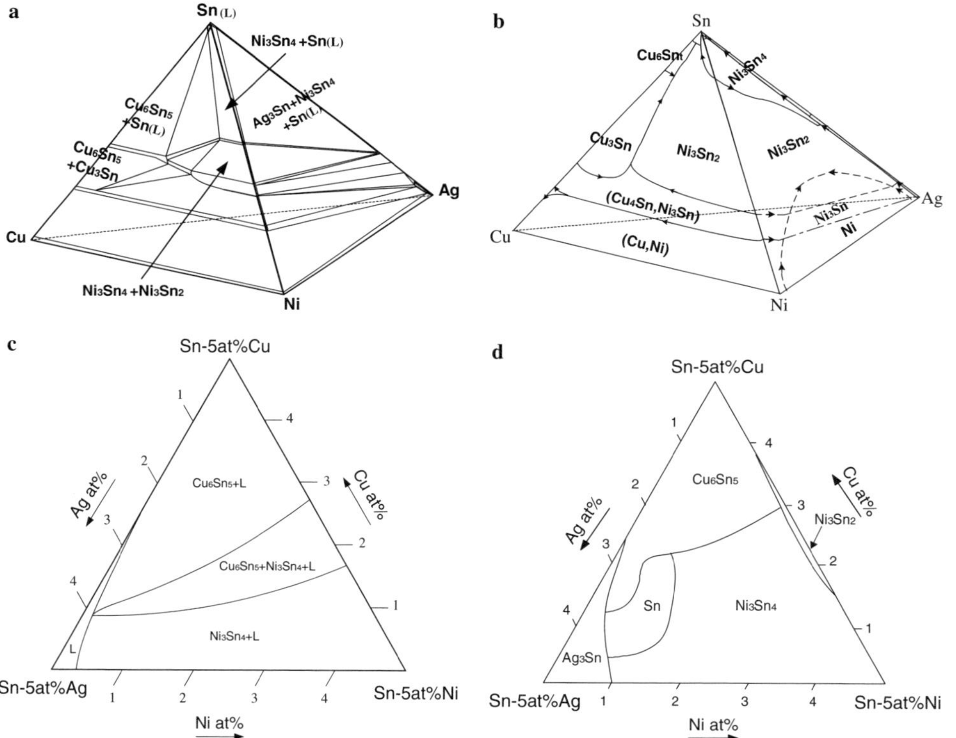

Fig. 8 (**a**) The 250°C isothermal tetrahedron of the Sn–Ag–Cu–Ni system [48]. (**b**) Liquidus projection tetrahedron of the Sn–Ag–Cu–Ni system [49]. (**c**) The 95at%Sn isoplethal section of the

250°C isothermal phase equilibria tetrahedron of the Sn–Ag–Cu–Ni system [50]. (**d**) The 95 at%Sn isoplethal section of the liquidus projection of the Sn–Ag–Cu–Ni system [49]

 Springer

transformation of this $(Cu,Ni)_3Sn$ solid solution. A continuous solid solution cannot exist if there is a phase transformation. Lin et al. [38] marked dashed lines on the 240°C phase diagrams regarding the inconsistency. Oberndorff determined the isothermal section at 235°C [42], and they reported the formation of a ternary compound with an average composition $Sn_{44}Cu_{27}Ni_{29}$. Fig. 7b is the liquidus projection of the Sn–Cu–Ni system. At the Sn-rich corner which is of primary soldering interests, there is a class I reaction, a ternary eutectic liquid = $Sn + Ni_3Sn_2 + Cu_6Sn_5$ and a class II reaction, liquid + $Ni_3Sn_4 = Sn + Ni_3Sn_2$.

Miettinen [43] is the only work that has thermodynamically modeled the Sn–Cu–Ni system but only at the Cu–Ni side. A good description of the thermodynamic models at the Sn-rich corner which is interesting to the soldering application is still lacking. The interfacial reactions between Sn–Cu/Ni have been examined by various investigators [40, 44–47]. No ternary compounds have been observed. It is found that the interfacial products are sensitive to Cu contents. When the Cu is less than 0.4 wt%, Ni_3Sn_4 phase formed, and the product is the Cu_6Sn_5 phase when the Cu content is higher. A peculiar interfacial reaction phenomenon has been observed in the Cu/Sn/Ni type couples [40], and the Cu_6Sn_5 phase is formed on both sides of the Sn solder. As shown in Fig. 7a, the Cu_6Sn_5 phase has very wide compositional homogeneity ranges, and thus the Cu_6Sn_5 phase can be stabilized with addition of Ni. The very significant Ni and Cu mutual solubilities in the Cu_6Sn_5 phase and the Ni_3Sn_4 phase result in complicated cross effects, and is the primary reason for the complicated interfacial reactions in the Sn–Cu/Ni couples. It is worth mentioning that owing to the very large mutual solubilities, some researchers use $(Cu,Ni)_6Sn_5$ and $(Cu,Ni)_3Sn_4$ for the Cu_6Sn_5 and Ni_3Sn_4 phases.

Chen et al. [48–50] determined the phase relationship of Sn–Ag–Cu–Ni ternary system. A tetrahedron is used for the description of the isothermal phase equilibria and for the liquidus projection as shown in Fig. 8a, b. No ternary and quaternary compounds have been found. Figure 8c,d are the 95at%Sn isoplethal sections of the 250°C phase equilibria and the liquidus projection, respectively. Since the Cu_6Sn_5 phase and the Ni_3Sn_4 phases contain negligible Ag, while the Ag_3Sn phase contains negligible Cu and Ni, the Cu_6Sn_5–Ni_3Sn_4 edge of the four-phase tetrahedron is very close to the Sn–Cu–Ni side of the Sn–Ag–Cu–Ni tetrahedron, and the Ag_3Sn corner is nearly on the Sn–Ag–Ni side. The compositional phase regimes consisting Ag phases are very small which indicates the phase relationship in most compositional part is similar to that of Sn–Cu–Ni.

Various investigators have studied the Sn–Ag–Cu/Ni interfacial reactions [51–55]. In agreement with the phase diagram studies, the interfacial reactions are similar to those in the Sn–Cu/Ni couples [40, 44–47], and the Ag does not actively participate in the interfacial reactions. The reaction products are sensitive to Cu contents, and are either Cu_6Sn_5 or Ni_3Sn_4 phases. Since Sn–Ag–Cu alloy is the most promising Pb-free solder, phase diagram of the quaternary Sn–Ag–Cu–Ni system is important for Pb-free soldering. For the higher order materials systems, it is very difficult and not practical to experimentally determine the phase diagram through the whole temperature and compositional ranges. The CALPHAD (CALculation of PHAse Diagrams) [56] approach combines thermochemistry and phase equilibria. With the CALPHAD approach, the phase diagram of the quaternary Sn–Ag–Cu–Ni system can be calculated with the thermodynamic models of their constituent ternary systems and possible interaction parameters. The interaction parameters can be determined with limited experimental quaternary results [48–50]. Thermodynamic models for the Sn–Ag–Cu, Sn–Ag–Ni, and Ag–Cu–Ni are available [18, 19, 28, 57], and the only important part that is missing is a good thermodynamic description of the Sn–Cu–Ni ternary system.

3 Sn–Zn

Although Sn–Ag–Cu alloys are the most promising Pb-free solders, their melting points are much higher than that of the conventional Pb–Sn eutectic which is at 183°C. Various efforts have been carried out to develop Pb-free solders of more suitable melting points. The melting point of Sn–Zn eutectic alloy is at 198.5°C [58] and is closer to that of the eutectic Pb–Sn. Besides, Sn–Zn eutectic has good mechanical properties and low cost, and is likely a good candidate [59–66]. It has been recognized that poor wetting properties and serious oxidation problems are the obstacles for industrial applications of the pure binary Sn–Zn eutectic alloys. Improvements are achieved with introducing alloying elements and the results indicate the Sn–Zn-based alloys are promising low melting point solders [60, 63, 66].

As shown in Fig. 9, the binary Sn–Zn system is a simple eutectic system. The eutectic is at 198.5°C and Sn–8.8wt%Zn (Sn–14.9at%Zn), and the mutual solubilities of Sn and Zn are negligible [58]. Thermodynamic assessments of the Sn–Zn binary system and phase diagrams are calculated by Lee [67] and Ohtani et al. [68]. The results by Lee [67] are satisfactory, and

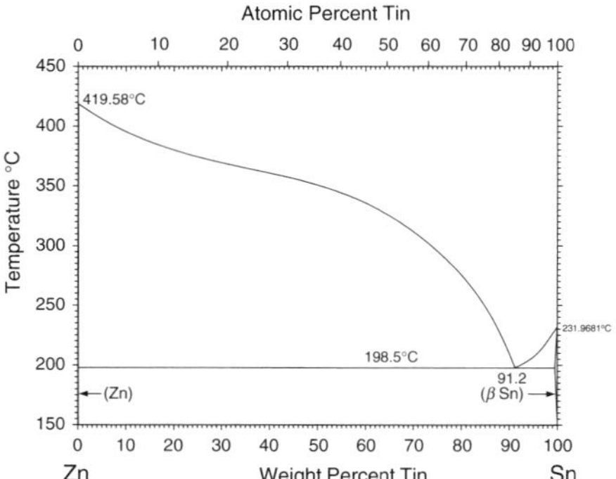

Fig. 9 The Sn–Zn binary phase diagram [58]

the calculated eutectic is at 198.25°C and Sn–14.8at%Zn.

Ohtani et al. [68] carried out thermal analysis and experimental phase equilibria studies of the Sn–Zn–Ag ternary system. They also developed thermodynamic models and calculated the isothermal sections at 420 and 190°C and the liquidus projection. Figure 10a is the 190°C isothermal section calculated by Ohtani et al. [68]. No ternary compounds have been found, and ternary solubilities of all the binary compounds are very limited. The calculated Sn-rich corner of the Sn–Zn–Ag liquidus projection is shown in Fig. 10b. The calculated ternary invariant reaction of the lowest reaction temperature is a class II reaction, liquid + ε-$AgZn_3$ = Sn + Zn, at 193.7°C. However, the binary Sn–Zn eutectic temperature calculated by Ohtani et al.

[68] is at 181°C (as label on the figure, or at 191°C as read directly from the figure) and is too low comparing to the experimental value. The class II invariant reaction might be a class I reaction if the binary eutectic liquid = Sn + Zn is at 198.5°C. No interfacial reaction studies between Sn–Zn alloys and Ag substrate are available. Song and Lin [63] studied the solidification behavior of the Sn–8.87wt%Zn–1.5wt%Ag (Sn–14.99at%Zn–1.53at%Ag) alloy. They have found the existence of γ-Ag_5Zn_8 and ε-$AgZn_3$, and they also conclude that the intermetallic compound solidified first. These observations are in agreement with sequence predicted from the liquidus projection. However, the reaction temperatures are different. More experimental efforts and thermodynamic re-assessment are needed for the Sn–Zn–Ag system.

No phase diagram is available for the Sn–Zn–Au ternary system. Binary Au–Zn compounds are found from the interfacial reactions between Sn–Zn solder and Au finish [64, 65], and to date no ternary Sn–Zn–Au compounds are reported. Lee et al. calculated the phase diagram of Sn–Zn–Cu [69] with very limited experimental results. Chou and Chen [70] experimentally determined the isothermal sections of phase equilibria at 210, 230 and 250°C. Figure 11 is the Sn–Zn–Cu 230°C isothermal section. No ternary compounds have been found. However, in contrast to the calculated results [69], the experimental results indicate that Sn phase has tie-lines with the η-Cu_6Sn_5 phase and all the binary Cu–Zn compounds at 230 and 250°C. Thermodynamic re-assessment of the Sn–Zn–Cu ternary system is needed, so that accurate phase diagrams

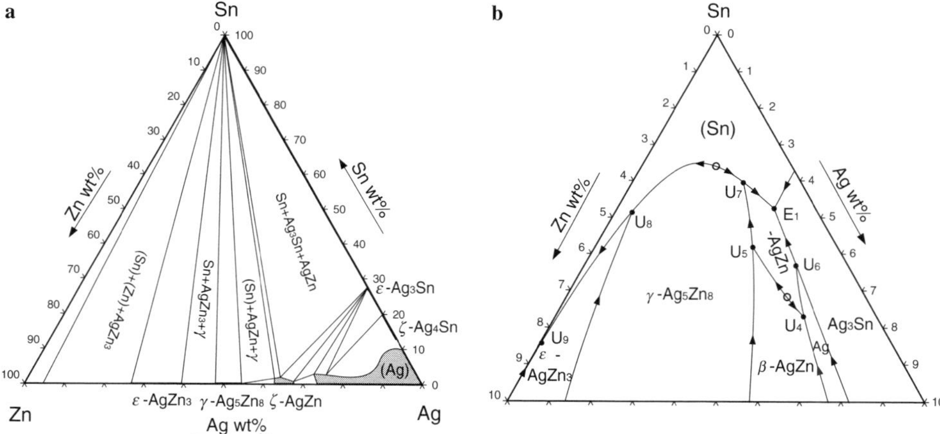

Fig. 10 (**a**) The 190°C isothermal section of the Sn–Zn–Ag ternary system [68]. (**b**) Liquidus projection of the Sn–Zn–Ag ternary system at the Sn-rich corner [68]

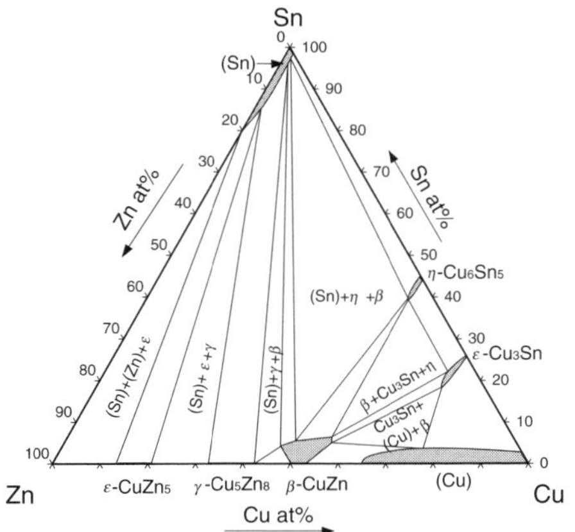

Fig. 11 The 230°C isothermal section of the Sn–Zn–Cu ternary system [70]

can be calculated. Various groups have studied the interfacial reactions Sn–Zn alloys reactions with Cu substrate [59, 61, 62, 64–66, 71]. Even though the Zn content is only 9 wt%, the interfacial reactions products are the Cu–Zn binary compounds with γ-Cu₅Zn₈ is the most dominating reaction phase.

No phase diagrams are available for the Sn–Zn–Ni ternary system. γ-Ni₅Zn₂₁ compound is the primary interfacial reaction products in the Sn–Zn/Ni couples substrates [61, 64–66, 71]. However, with 41 h reaction at 280°C, besides the γ-Ni₅Zn₂₁ phase, a layer of Sn–30.9at%Ni–16.3at%Zn composition is formed and it is likely to be the δ-Ni₃Sn₄ phase with 16.3 at%Zn [71]. A similar double layer formation is found in the Sn–8wt%Zn–3wt%Bi/Ni couple reacted at 325°C, and be-

sides the γ-Ni₅Zn₂₁ phase, a layer of Sn–35at%Ni–22at%Zn is formed [72]. Phase diagram study of the Sn–Zn–Ni system is needed to verify whether the compound is a thermodynamically stable or meta-stable phase and whether it is a binary or ternary compound.

4 Sn–Bi

Sn–Bi alloy is another low melting-point Pb-free solder. As shown Fig. 12, similar to the Sn–Pb system, the binary Sn–Bi system is a simple eutectic system without any intermetallic compound [10, 73, 74]. The eutectic is at 139°C and Sn–57wt%Bi (Sn–42.95at%Bi). Kattner and Boettinger [10] and Lee et al. [74] have developed thermodynamic models to calculate the Sn–Bi phase diagram, and the calculated results are in good agreement with the experimental determinations.

Various groups have determined the phase diagrams of the the Sn–Bi–Ag system [10, 75, 76]. No ternary compounds have been found, and all the binary compounds have negligible ternary solubilities. Figure 13a, b are the 230°C phase equilibria isothermal section and liquidus projection, respectively [10]. The liquidus projections obtained by three groups most recently [10, 75, 76] are similar. All of the results suggest there are one class I reaction, liquid = Sn + Bi+ε-Ag₃Sn, and two class II reactions. The temperatures of the class I reaction determined by the three groups are similar as well, and are 136.5, 138.4 and 139.2°C, by Kattner and Boettinger [10], Hassam et al. [75] and Ohtani et al. [76], respectively. However, the temperatures of the class II reaction, liquid+ζ = ε-Ag₃Sn + Bi, are very different at 238.6, 261.8, and 262.5°C, respectively.

Fig. 12 The Sn–Bi binary phase diagram [73]

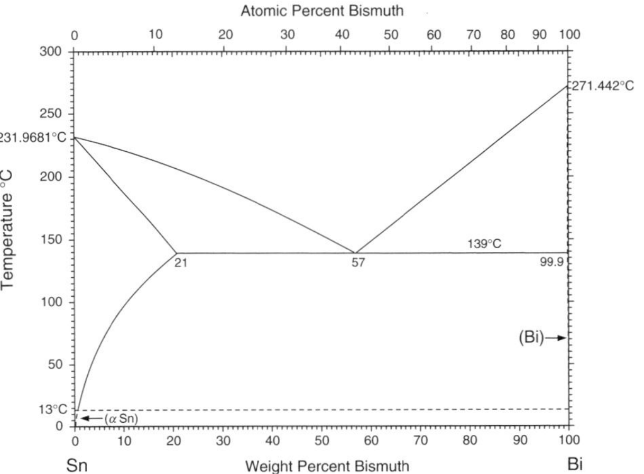

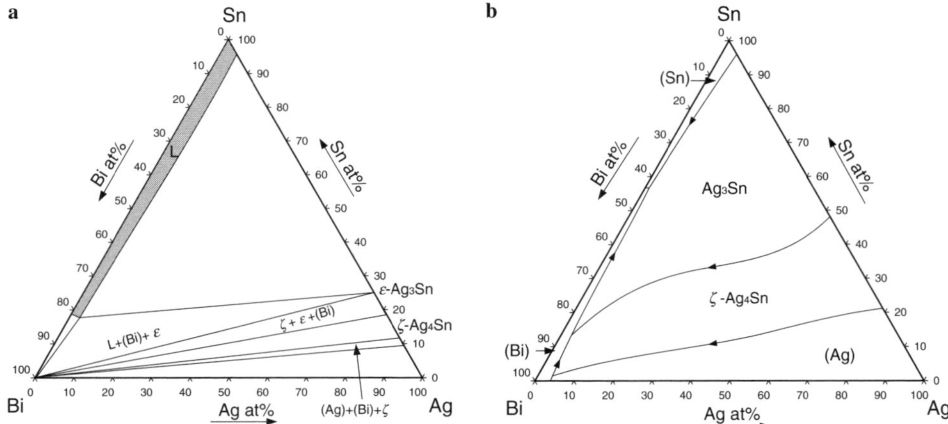

Fig. 13 (**a**) The 230°C isothermal section of the Sn–Bi–Ag ternary system [10]. (**b**) Liquidus projection of the Sn–Bi–Ag ternary system [10]

Partial isothermal sections of the Sn–Bi–Au ternary system at 139, 160, 200 and 240°C and the liquidus projection at the Sn-rich corner in the Sn–Bi–AuSn compositional regime have been determined [77]. As shown in Fig. 14a of the 200°C isothermal section, no ternary compounds have been found in this temperature and composition regime. The Bi solubility in the binary compounds, AuSn, AuSn$_2$ and AuSn$_4$ are negligible. Figure 14b is the partial liquidus projection, and there is a class I reaction, liquid = Sn + Bi + AuSn$_4$, at the Sn-rich corner. Interfacial reactions between Sn–Bi and Au are examined in a new fluxless bonding process in air using Sn–Bi with Au cap [78]. In

agreement with the phase diagrams, the intermetallic compounds are Au–Sn binary compounds with almost no detectable Bi [78].

Lee et al. [69] calculated the phase diagrams of the Sn–Bi–Cu ternary system with the thermodynamic models developed from extension of those of the constituent binary systems without ternary interaction parameter. Doi et al. [79] experimentally examined the phase boundaries with DSC and developed thermodynamic models with a ternary interaction parameter of the liquid phase for the Sn–Bi–Cu system. No ternary compounds are found, and all the binary compounds are with negligible solubility [79]. Figure 15a, b

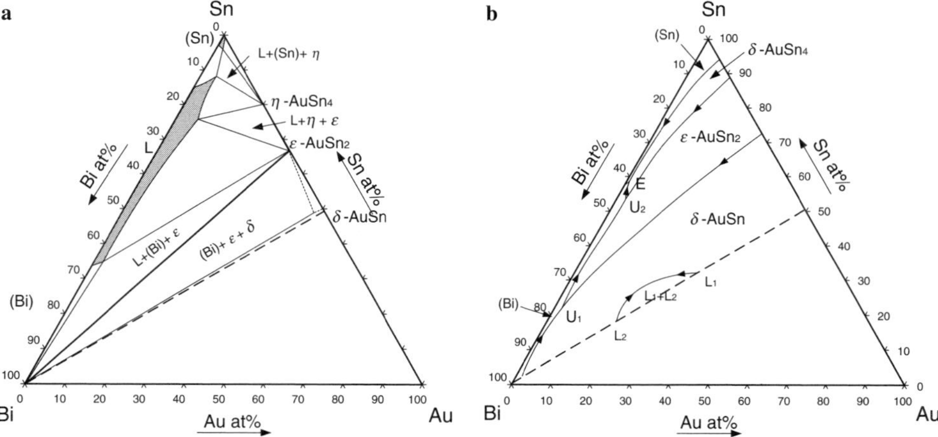

Fig. 14 (**a**) A partial 200°C isothermal section of the Sn–Bi–Au ternary system [77]. (**b**) A partial liquidus projection of the Sn–Bi–Au ternary system [77]

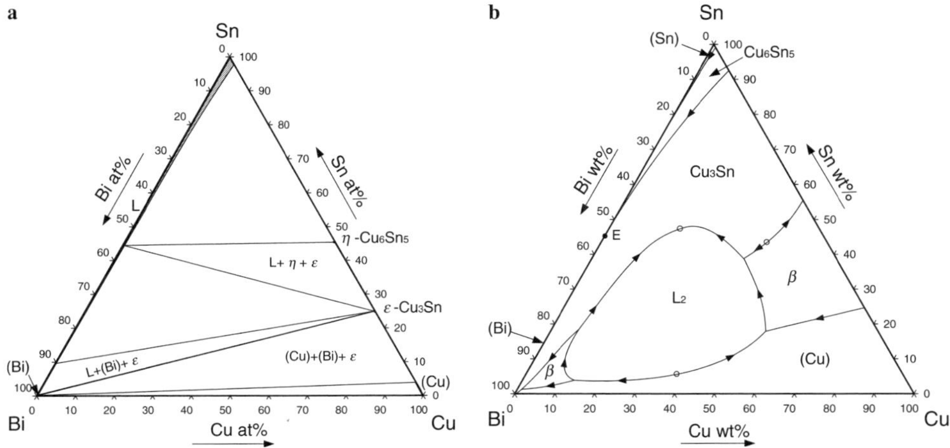

Fig. 15 (**a**) The 250°C isothermal section of the Sn–Bi–Cu ternary system [69]. (**b**) The liquidus projection of the Sn–Bi–Cu ternary system [79]

are the 250°C isothermal section and the liquidus projection [79]. A class I reaction, L = Sn + Bi + ε is at 140.1°C and Sn–54.6wt%Bi–0.014wt%Cu (Sn–40.6at%Bi–0.034at%Cu). Lee et al. [69] predicts the interfacial reaction products between Cu and Sn–58wt%Bi to be the η-Cu$_6$Sn$_5$ phase, which is in agreement with the experimental determinations by Vianco et al. [80] and Yoon et al. [81].

Figure 16 is a 300°C phase equilibria isothermal section of the Sn–Bi–Ni system proposed by Lee et al. [82] based on phase diagrams of the three binary constituent systems and very limited phase equilibria measurement. There have been no experimental phase

equilibria measurements available except the work of Lee et al. [82]. No ternary compounds have been found and the ternary solubilities of the binary compounds are negligible. Ni$_3$Sn$_2$ phase has tie-lines with all the other compounds and the liquid phase. The interfacial reaction product in the Sn–58wt%Bi/Ni couples is the Ni$_3$Sn$_4$ phase [81–83]. Lee et al. [82] studied the Sn–Bi/Ni interfacial reactions at 300°C with Sn–Bi alloys of various compositions, and they found that the NiBi$_3$ phase would form when the Bi content was higher than 97.5 wt% (95.7 at%).

5 Sn–Zn–Bi

Sn–Zn–Bi alloys are also possible candidates as Pb-free solders [84–86]. Malakhov et al. [84] carried out thermodynamic assessment and calculated phase diagrams of the Sn–Zn–Bi system. Figure 17a, b are the 170°C isothermal section and liquidus projection, respectively. There are no intermetallic compounds in this system. Both Sn–Zn and Bi–Zn are simple eutectic systems without binary compounds, and the Bi–Zn system has a monotectic and a eutectic reaction [85]. As shown in Fig. 17b, the liquid miscibility gap of the binary Bi–Zn system extends into the ternary system. The class I reaction, liquid = Bi + Sn + Zn, is at 130°C and at Sn–54.54wt%Bi–2.71wt%Zn (Sn–39.39at%Bi–6.25at%Zn). As mentioned previously, the Sn–8wt%Zn–3wt%Bi/Ni couple reacted at 325°C has been examined. Besides the γ-Ni$_5$Zn$_{21}$ phase, a layer of Sn–35at%Ni–22at%Zn was formed as well [72].

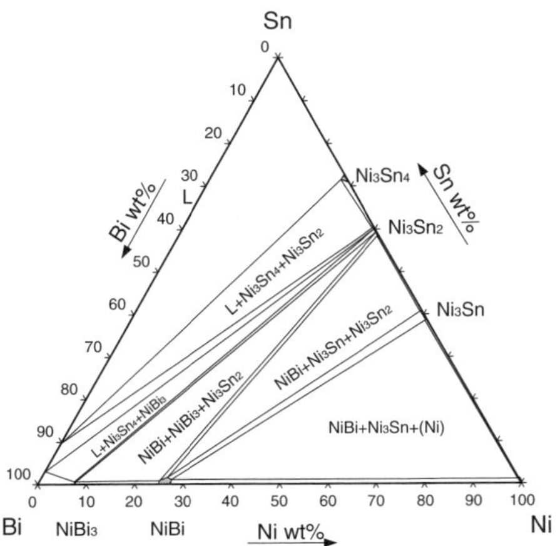

Fig. 16 The 300°C isothermal section of the Sn–Bi–Ni ternary system [82]

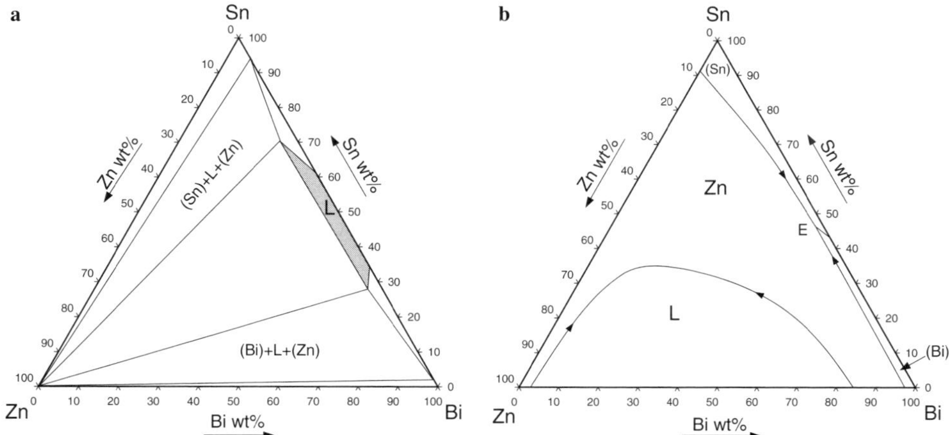

Fig. 17 (**a**) The 170°C isothermal section of the Sn–Bi–Zn ternary system [84]. (**b**) The liquidus projection of the Sn–Bi–Zn ternary system [84]

6 Sn–In

Sn–In alloys have low melting temperatures, high ductility, low substrate dissolution, and good fatigue properties [4, 87–90]. Even though In is expensive, Sn–In alloys are used in various occasions, especially when low soldering temperature is required. Okamoto [91] had a review of the available experimental phase equilibria. Lee et al. thermodynamically modeled the binary Sn–In system and calculated the phase diagram [74]. Figure 18 is the Sn–In binary phase diagram. Both of the intermetallic compounds, β and γ, are with large compositional homogeneity ranges. Solution models

were used for the descriptions of these two compounds [74]. The eutectic temperature and composition assessed by Okamoto [91] and Lee et al. [74] are 120°C, Sn–51.7wt%In (Sn–52.5at%In) and 118°C, Sn–52.5at%In, respectively.

Korhonen and Kivilahti [92] and Liu et al. [93] studied the Sn–In–Ag ternary system both experimentally and by thermodynamic modeling. Most recently, Vassilev et al. [94] experimentally determined the Sn–In–Ag equilibrium phase relationship. Isothermal sections at temperatures varied from 113 and 400°C are reported [92–94]. As shown in Fig. 19a of the 180°C isothermal section, no ternary compounds have been

Fig. 18 The Sn–In binary phase diagram [91]

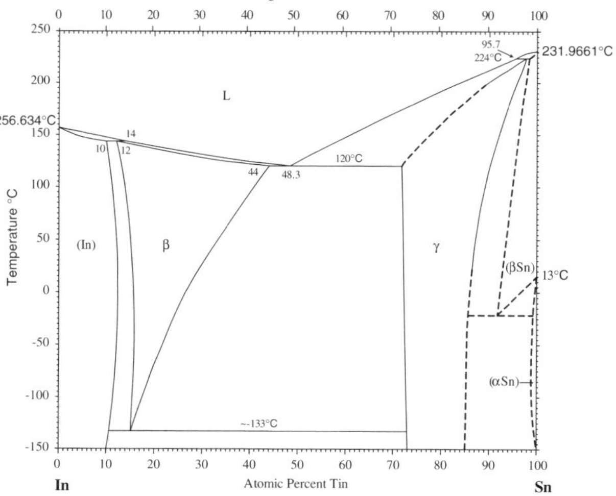

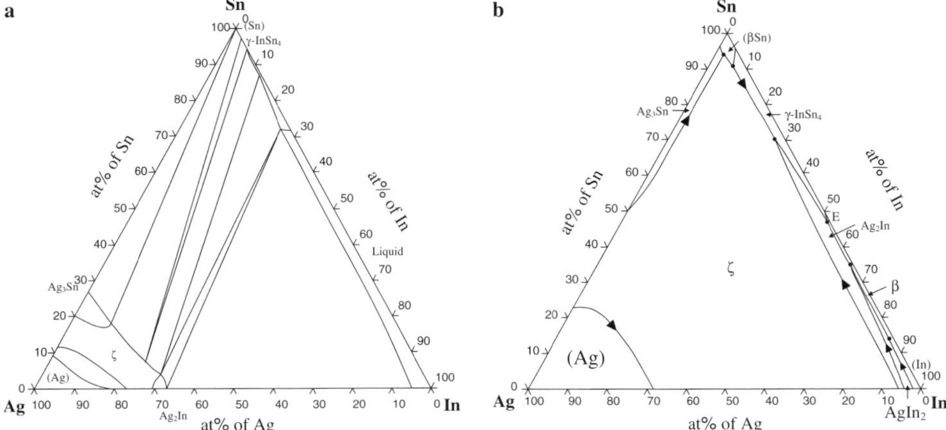

Fig. 19 (**a**) The 180°C isothermal section of the Sn–In–Ag ternary system [93]. (**b**) The liquidus projection of the Sn–In–Ag ternary system [93]

found. A continuous solid solution is formed between the ζ phase in the Ag–Sn and Ag–In systems. However, except for the solid solution, all the other binary compounds do not have significant ternary solubilities. Figure 19b is the calculated liquidus projection. The calculated invariant reaction with the lowest reaction temperature is the class I reaction, liquid = Sn + γ–InSn$_4$ + Ag$_2$In, at 114°C and Sn–52.2wt%In–0.9wt%Ag (Sn–52.98at%In–0.97at%Ag) [93]. The results are in agreement with experimental determinations. Liu and Chuang [95] and Cheng et al. [96] determined the interfacial reactions between the Sn–In alloys and Ag

substrate. In agreement with phase diagram studies, they reported the formation of the Ag$_2$In and AgIn$_2$ but no ternary compound. Since AgIn$_2$ phase is not stable at most of the reactions temperatures, formation of the AgIn$_2$ phase is likely by precipitation from soldering instead of by interfacial reactions.

Phase diagrams of the Sn–In–Au ternary system has been experimentally determined and assessed [97]. A ternary compound, Au$_4$In$_3$Sn$_3$, with Pt$_2$Sn$_3$ has been found. Liu et al. [98] developed thermodynamic models to calculate the phase diagrams at 27, 227, 427°C [98]. As shown in Fig. 20 of the calculated 227°C

Fig. 20 The 227°C isothermal section of the Sn–In–Au ternary system [98]

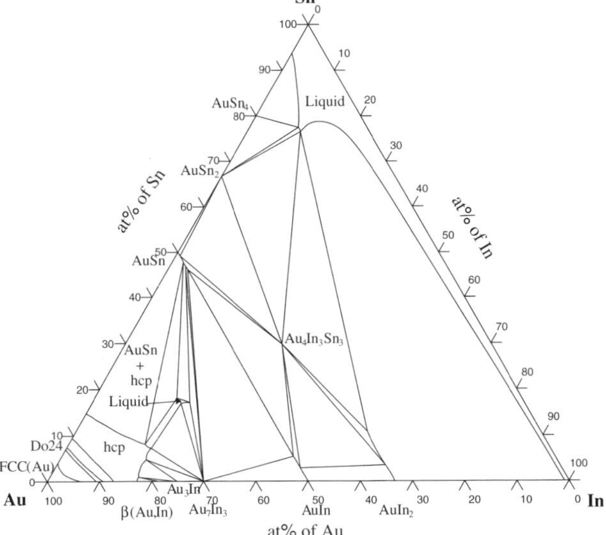

isothermal section, two continuous solid solutions are formed at the Au-rich corner, β-Au$_{10}$Sn and α_1-Au$_{88}$In$_{12}$ and the ζ phase in Au–In and in Au–Sn systems. Liu et al. [98] also indicated that the Au$_4$In$_3$Sn$_3$ was stable in the temperature range from 82 to 428°C, and it melted congruently. Chuang et al. [99] examined the Sn–49wt%In solder ball reactions with the Au/Ni/Cu UBM. They found the interfacial reaction products were AuIn$_2$/AuIn when the temperature was below 170°C, an it transformed into AuIn$_2$ when the temperature was higher than 220°C. The reaction Au–In compounds have significant Ni solubility. Kim and Tu [100] examined the interfacial reactions of 77.2Sn–20In–2.8Ag/Au at 200°C. An AuIn$_2$ type Au$_3$In$_5$Sn compound was formed, and after 60 s, the phase decomposed and an Au$_2$Sn$_7$In phase was formed. Since both compounds are not found in the Sn–In–Au phase diagrams, they might be meta-stable phases.

Köster et al. [101] determined the phase equilibria of the Sn–In–Cu ternary system at the Cu-rich side, but the phase equilibria close to the In–Sn side which is of primary soldering interests is lacking. At 400°C, two ternary compounds and a continuous solid solution formed between the η-Cu$_2$In and the η-Cu$_6$Sn$_5$ phases are observed. Liu et al. [102] determined the phase equilibria of the Sn–In–Cu system both by thermodynamic calculation and experimental determination. They have found only one ternary compound, Cu$_{16}$In$_3$Sn at higher temperature, and a ternary compound Cu$_2$In$_3$Sn at 110°C. Liu et al. [102] also calculated the isothermal sections at various temperatures using thermodynamic models developed by them. Lin et al. [103] experimentally determined the phase dia-

grams. Figure 21a is the 250°C isothermal section of the Sn–In–Cu system [103]. They confirmed the formation of the continuous solid solutions; however, they do not observe the formation of ternary compounds. Significant solubilities are found for the ε-Cu$_3$Sn with In and for the Cu$_7$In$_3$ with Sn. Figure 21b is its liquidus projection determined by Lin et al. [103]. The result is similar to that calculated by Liu et al. [102]. However, the Sn–Cu phase diagram calculated by Liu et al. is different from those by Shim et al. [12] and Boettinger et al. [13]. The Sn–Cu phase diagram by Liu et al. does not have the peritectic reaction, liquid + β = γ, thus the liquidus trough originated from this binary peritectic is missing in the Sn–In–Cu liquidus projection proposed by Liu et al.[102].

Romig et al. [104] studied the interfacial reactions of the Sn–50wt%In/Cu couples at 70 and 90°C, and Cu$_2$(Sn,In) and Cu$_2$In$_3$Sn were found. Vianco et al. [105] examined the Sn–50wt%In/Cu couples at temperatures varied from 55 to 100°C, and they found the formation of Cu$_{26}$Sn$_{13}$In$_8$ and Cu$_{17}$Sn$_9$In$_{24}$. It is presumed these two phases are the η-Cu$_6$Sn$_5$ with high In solubilities, and CuIn$_2$ with high Sn solubility. Chuang et al. [106] studied the Sn–51wt%In/Cu couple at 150–400°C, and they found ε-Cu$_3$(In,Sn) and η-Cu$_6$(In,Sn)$_5$. Sommadossi et al. [107] studied the interfacial reactions of the Sn–52at%In/Cu couples. They found the formation of η-Cu$_3$Sn at 180°C, and ε-Cu$_3$Sn and η-Cu$_6$Sn$_5$ at 290°C. The compositions of these phases are Sn–27at%In–57at%Cu and Sn–10at%In–77at%Cu, respectively. Kim and Jung [108] studied the Sn–52at%In/Cu interfacial reactions at 70–100°C, and they found the formation of the Cu(In,Sn)$_2$

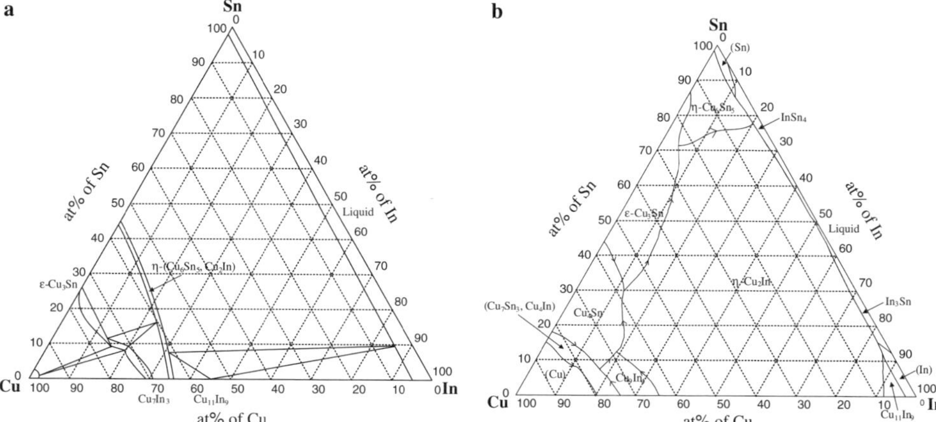

Fig. 21 (a) The 250°C isothermal section of the Sn–In–Cu ternary system [103]. (b) The liquidus projection of the Sn–In–Cu ternary system [103]

and $Cu_6(In,Sn)_5$ phases. The reaction products can be well interpreted with the isothermal section as shown in Fig. 21a except for the $CuIn_2$ phase. The $CuIn_2$ phase is likely a metastable phase since it is not found in the binary Cu–In system either [102].

Burkhardt and Schubert [109], Bhargava and Schubert [110], and Shadangi et al. [111] determined the phase relationship of the Sn–In–Ni ternary system at higher temperature and at the Ni-rich corner. Burkhardt and Schubert [109] reported a continuous solid solution formed by Ni_3Sn and Ni_3In. Bhargava and Schubert [110] indicated the existence of the ternary compound Ni_6InSn_5, and one more solid solution formed by Ni_3In_2 and Ni_3Sn_2. Huang and Chen [112] examined the phase diagram at lower temperature at 240°C. As shown in Fig. 22 of the 240°C Sn–In–Ni isothermal section, they found the continuous solid solution between Ni_3In and Ni_3Sn. They also found most of the binary compounds have significant solubilities but did not find the ternary compound. Huang and Chen [112] examined the Sn–In/Ni interfacial reactions at 160 and 240°C, and they found Ni_3Sn_4 is the primary reaction products and the $Ni_{28}In_{72}$ phase is formed only in the couples prepared with In-rich alloys. Kim and Jung [113] examined the intermetallic compound layer growth between In and Sn–52at%In with Ni/Cu substrate at 70–120°C. They found the formation of $In_{27}Ni_{10}$ and $Ni_3(In,Sn)_4$ for indium and the Sn–In alloy, respectively. The very extensive solubility found in the Ni_3Sn_4 phase by Kim and Jung [112] is in agreement with the phase diagram determination [112]. Chuang et al. [114] studied the interfacial reactions between the Sn–20wt%In–2.8wt%Ag with Ni substrate, and they also found the formation of Ni_3Sn_4

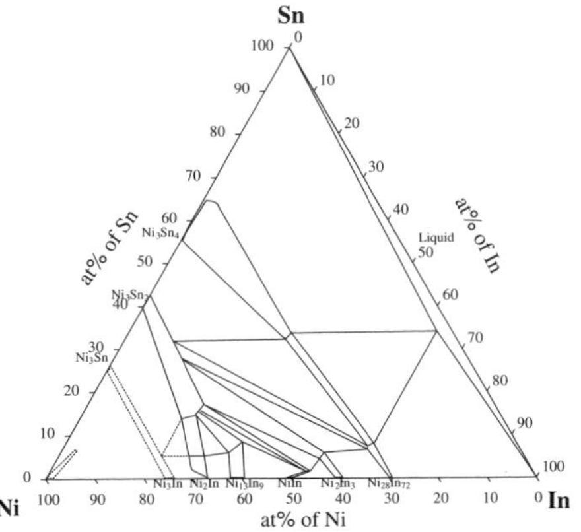

Fig. 22 The 240°C isothermal section of the Sn–In–Ni ternary system [112]

phase. However, no significant In solubility in the Ni_3Sn_4 phase was found.

7 Sn–Sb

Because of the requirements of step soldering, solders of different melting points are needed. Although there are very intensive studies of Pb-free solders for replacing the low-melting-point eutectic Sn–Pb, very few efforts are found for the replacement of high temperature Sn–95wt%Pb and Sn–90wt%Pb solders. Sn–Sb and Sn–Sb based alloys are promising Pb-free solders. This is especially true of Sn–5wt%Sb (Sn–4.53at%Sb), which is used as high temperature solder [115–117].

Predel and Schwermann [118] carried out phase equilibria study of the Sn–Sb system, and proposed the diagram as shown in Fig. 23 based on their experimental determinations and the phase equilibria results in the literatures. There are three peritectic and one eutectoid reactions. The temperature of the peritectic reaction at the Sn side, liquid + Sn_3Sb_2 = Sn, is at 250°C. The Sn_3Sb_2 compound has negligible compositional homogeneity and is stable only in a short temperature range between 242°C and 324°C. Jonsson and Agren [119] analyzed the experimental thermodynamic and phase equilibria data and developed thermodynamic models. The intermetallic β phase has large compositional homogeneity range and is described by using two-sublattices models. The Sn_3Sb_2 is treated as a line compound. The calculated Sn–Sb phase diagram using the developed models is in good agreement with the experimental determinations as shown in Fig. 23.

Oh et al. [120] and Moser et al. [121] thermodynamically assessed the ternary Sn–Sb–Ag system, and calculated the isothermal sections, isoplethal sections, and liquidus projection. Figure 24a is the calculated 100°C isothermal section of the Sn–Sb–Ag phase equilibria [121]. No ternary compounds have been found. Two continuous solid solutions are formed, and they are the ζ-$Ag_4(Sb,Sn)$ and the ε-$Ag_3(Sb,Sn)$. Figure 24b is the calculated liquidus projection. The calculation result is in agreement with the experimentally determined liquidus projection by Masson and Kirkpatrick [122]. The invariant reaction at the Sn corner with the lowest reaction temperature is a class II reaction, liquid + Sn_3Sb_2 = ε-$Ag_3(Sb,Sn)$ + Sn. The reaction temperatures and composition of the liquid are 235°C and Sn–6at%Sb–5at%Sb and 231.5°C and Sn–6.2at%Sb–3.4at%Sb as calculated by Oh et al. [120] and Moser et al. [121], respectively.

Fig. 23 The Sn–Sb binary phase diagram [118]

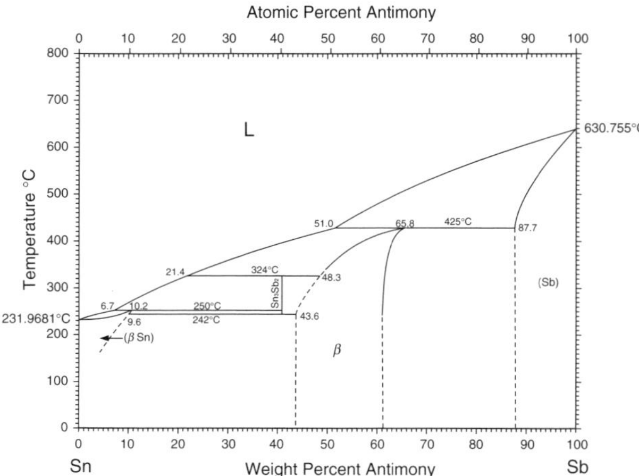

A thermodynamic assessment of the Sn–Sb–Au ternary system was conducted by Kim et al. [123] and they also calculated its isothermal sections, isoplethal sections and liquidus projection. Figure 25a, b are the isothermal section at 200°C and the liquid projection, respectively. No ternary compounds have been observed. The $AuSb_2$ is with about 15at%Sn solubility; however, the Sb solubility in $AuSn_2$ is negligible. As shown in Fig. 25b, the invariant reaction at the Sn-rich corner with the lowest reaction temperature is a class II reaction, liquid + Sb_2Sn_3 = Sn + $AuSn_4$ at 491°C.

There are no phase equilibria studies of the Sn–Sb–Cu system. Jang et al. [116] and Takaku et al. [124] examined interfacial reactions of the Sn–Sb solders on Cu substrates. They found the formation of ε-Cu_3Sn and η-Cu_6Sn_5 phases. Jang et al. [116] also found the formation of Sn_3Sb_2 phase in the Sn–15wt%Sb (Sn–14.68at%Sb) couples. No ternary compounds were reported in both studies [116, 124]. Available phase diagrams of the Sn–Sb–Ni ternary system are all at the Ni-rich corner and their temperatures are at 450°C and 500°C [125]. No phase equilibria information of the Sn–Sb–Ni ternary system is at the temperature and compositional ranges of soldering interests, and no interfacial reactions studies of the Sn–Sb/Ni have been found either.

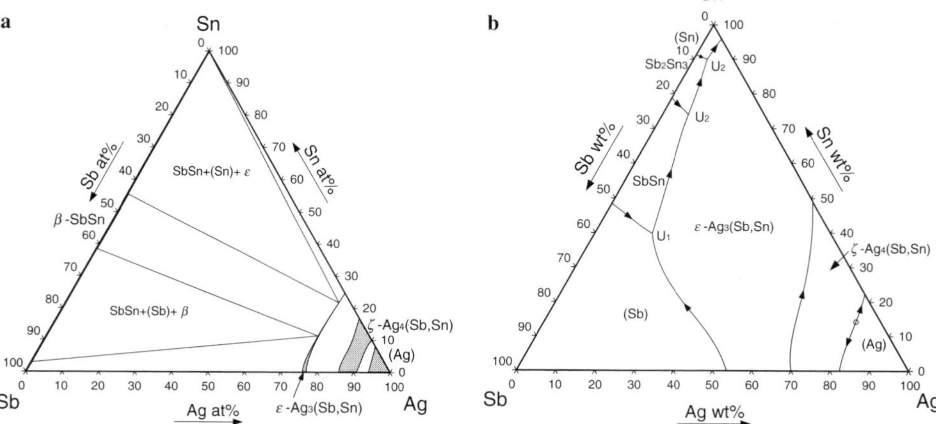

Fig. 24 (**a**) The 100°C isothermal section of the Sn–Sb–Ag ternary system [121]. (**b**) The liquidus projection of the Sn–Sb–Ag ternary system [121]

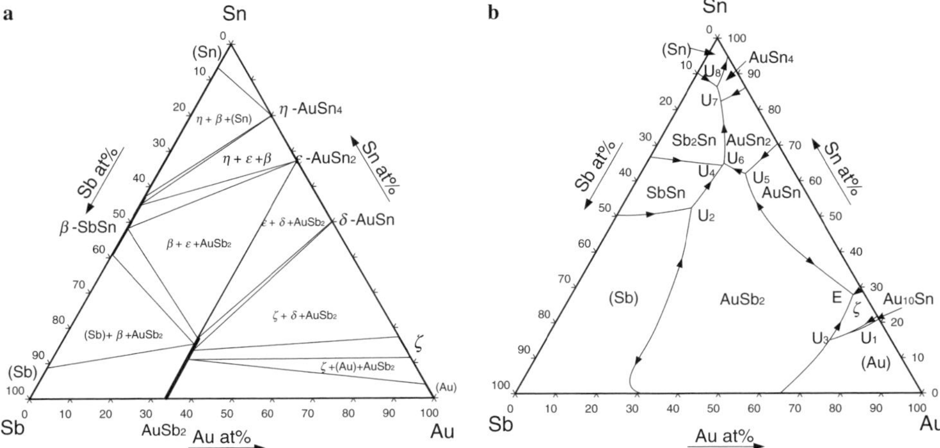

Fig. 25 (**a**) The 200°C isothermal section of the Sn–Sb–Au ternary system [123]. (**b**) The liquidus projection of the Sn–Sb–Au ternary system [123]

8 Conclusions

Phase diagrams are the roadmaps for the exploration of materials technology. Bird's-eye views of the phase diagrams of Pb-free solders and their related materials systems are mostly available. However, complete and accurate phase diagrams are lacking for almost all the ternary and quaternary systems. Experimental determinations of the multi-component phase diagrams over the entire compositional and temperature ranges of soldering interests are extremely difficult if not impossible. The CALPHAD approach which combines thermodynamic models of constituent binary systems and limited phase equilibria data of ternary and quaternary systems is suitable for the phase diagram determinations of higher order materials systems, but the calculated phase diagrams cannot be accurate without the input of reliable experimental values at some critical points. Only after the first-step calculation, the critical points can be pointed out and experimental efforts can be carried out with focus. Close cooperation and interaction between phase diagram calculation and experimental determination are crucial for the determinations of Pb-free solders and their related materials systems.

Acknowledgement The authors acknowledge the financial support of National Science Council. (NSC94-2214-E-007-006).

References

1. R.J. Klein Wassink, in *Soldering in Electronics*, 2nd edn. (Electrochemical Publications, Isle of Man, 1989)
2. Official Journal of the European Union, p. L 37/19-L 37/23, 13.2.2003
3. J. Bath, C. Handwerker, E. Bradley, Circ. Assembly **5**, 31 (2000)
4. M. Abtew, G. Selvaduray, Mater. Sci. Eng. R **27**, 95 (2000)
5. T. Laurila, V. Vuorinen, J.K. Kivilahti, Mater. Sci. Eng. R **49**, 1 (2005)
6. D.R. Frear, W.B. Jones, K.R. Kinsman, in *Solder Mechanics: A State of the Art Assessment* (Minerals, Metals & Materials Society, Warrendale, Pa, 1991)
7. "Roadmap 2002 for Commercialization of Lead-free Solder", Lead-Free Soldering Roadmap Committee, Japan Electronics and Information Technology industries Association (JEITA) (2002)
8. "Lead-free Assembly Projects", National Electronics Manufacturing Initiatives(NEMI) (1999)
9. I. Karakaya, W.T. Thompson, Bull. Alloy Phase Diagrams **8**, 340 (1987)
10. U.R. Kattner, W.J. Boettinger, J. Electron. Mater. **23**, 603 (1994)
11. N. Saunders, A.P. Miodownik, Bull. Alloy Phase Diagrams **11**, 278 (1990)
12. J.H. Shim, C.S. Oh, B.J. Lee, D.N. Lee, Z. Metallkd. **87**, 205 (1996)
13. W.J. Boettinger, C.A. Handwerker, U.R. Kattner, in *The Mechanics of Solder Alloy Wetting and Spreading* (Van Nostrand Reinhold, New York, 1993), p. 103
14. E. Gebhardt, G. Petzow, Z. Metallkd. **50**, 597 (1959)
15. Y.W. Yen, S.-W. Chen, J. Mater. Res. **19**, 2298 (2004)
16. C.M. Miller, I.E. Anderson, J.F. Smith, J. Electron. Mater. **23**, 595 (1994)
17. M.E. Loomans, M.E. Fine, Metall. Trans. A **31**, 1155 (2000)
18. K.W. Moon, W.J. Boettinger, U.R. Kattner, F.S. Biancaniello, C.A. Handwereker, J. Electron. Mater. **29**, 1122 (2000)
19. I. Ohnuma, M. Miyashita, K. Anzai, X.J. Liu, H. Ohtani, R. Kainuma, K. Ishida, J. Electron. Mater. **29**, 1137 (2000)
20. T.B. Massalski, H. Pops, Acta Metall. **18**, 961 (1970)
21. D.S. Evans, A. Prince, Met. Sci. **8**, 286 (1974)
22. S.-W. Chen, Y.W. Yen, J. Electron. Mater. **30**, 1133 (2001)
23. J. London, D.W. Ashall, Brazing Solder. **11**, 49 (1986)

24. P.T. Vianco, K.L. Erickson, P.L. Hopkins, J. Electron. Mater. **23**, 721 (1994)
25. S.-W. Chen, Y.W. Yen, J. Electron. Mater. **28**, 1203 (1999)
26. S. Choi, T.R. Bieler, J.P. Lucas, K.N. Subramanian, J. Electron. Mater. **28**, 1209 (1999)
27. T.Y. Lee, W.J. Choi, K.N. Tu, J.W. Jang, S.M. Kuo, J.K. Lin, D.R. Frear, K. Zeng, J.K. Kivilahti, J. Mater. Res. **17**, 291 (2002)
28. H.F. Hsu, S.-W. Chen, Acta Mater. **52**, 2541 (2004)
29. S.-W. Chen, H.F. Hsu, C.W. Lin, J. Mater. Res. **19**, 2262 (2004)
30. H.D. Blair, T.Y. Pan, J.M. Nicholson, in *Proceedings of the 48th Electronic Components and Technology Conference*, 1998 p. 259
31. C.M. Chen, S.-W. Chen, J. Appl. Phys. **90**, 1208 (2001)
32. C.E. Ho, R.Y. Tsai, Y.L. Lin, C.R. Kao, J. Electron. Mater. **31**, 584 (2002)
33. C.C. Chen, S.-W. Chen, C.Y. Kao, J. Electron. Mater. **35**, 922 (2006)
34. O.B. Karlsen, A. Kjekshus, E. Roest, Acta Chem. Scand. **46**, 147 (1992)
35. B. Peplinski, E. Zakel, Mater. Sci. Forum **166–169**, 443 (1994)
36. C.N.C. Luciano, K.I. Udoh, M. Nakagawa, S. Matsuya, M. Ohta, J. Alloys Comp. **337**, 289 (2002)
37. C.E. Ho, Y.M. Chen, C.R. Kao, J. Electron. Mater. **28**, 1231 (1999)
38. C.H. Lin, S.-W. Chen, C.H. Wang, J. Electron. Mater. **31**, 907 (2002)
39. C.H. Wang, S.-W. Chen, Metall. Trans. A **34**, 2281 (2003)
40. S.-W. Chen, S.H. Wu, S.W. Lee, J. Electron. Mater. **32**, 1188 (2003)
41. Y. Murakami, S. Kachi, Trans. JIM. **24**, 9 (1983)
42. P. Oberndorff, PhD Thesis, Technical University of Eindhoven, Eindhoven, Nerthelands (2001)
43. J. Miettinen, CALPHAD **27**, 309 (2003)
44. J.K. Lin, A. De Silva, D. Frear, Y. Guo, J.W. Jang, L. Li, D. Mitchell, B. Yeung, C. Zhang, in *Proceedings of the 51st Electronic Components and Technology Conference*, 2001 p. 455
45. W.T. Chen, C.E. Ho, C.R. Kao, J. Mater. Res. **17**, 263 (2002)
46. S.-W. Chen, S.W. Lee, M.C. Yip, J. Electron. Mater. **32**, 1284 (2003)
47. C.H. Wang, S.-W. Chen, Acta Mater. **54**, 247 (2006)
48. S.-W. Chen, C.A. Chang, J. Electron. Mater. **33**, 1071 (2004)
49. C.A. Chang, S.-W. Chen, C.N. Chiu, Y.C. Huang, J. Electron. Mater. **34**, 1135 (2005)
50. C.N. Chiu, Y.C. Huang, A.R. Zi, S.-W. Chen, Mater. Trans. **46**, 2426 (2005)
51. A. Zribi, A. Clark, L. Zavalij, P. Borgesen, E.J. Cotts, J. Electron. Mater. **30**, 1157 (2001)
52. K. Zeng, V. Vuorinen, J.K. Kivilahti, IEEE Trans. Electron. Packag. Manufact. **25**, 162 (2002)
53. K.S. Kim, S.H. Huh, K. Suganuma, J. Alloys Comp. **352**, 226 (2003)
54. M.D. Cheng, S.Y. Chang, S.F. Yen, T.H. Chuang, J. Electron. Mater. **33**, 171 (2004)
55. W.C. Luo, C.E. Ho, J.Y. Tsai, Y.L. Lin, C.R. Kao, Mater. Sci. Eng. A **396**, 385 (2005)
56. U.R. Kattner, JOM **49**(12), 14 (1997)
57. H.T. Luo, S.-W. Chen, J. Mater. Sci. **31**, 5059 (1996)
58. Z. Moser, J. Dutkiewicz, W. Gasior, J. Salawa, Bull. Alloy Phase Diagrams **4**, 330 (1985)
59. K. Suganuma, K. Niihara, T. Shoutoku, Y. Nakamura, J. Mater. Res. **13**, 2859 (1998)
60. K.-L. Lin, T.-P. Liu, Oxid. Met. **50**, 255 (1998)
61. Y.C. Chan, M.Y. Chiu, T.H. Chuang, Z. Metallkd. **93**, 95 (2002)
62. I. Shohji, T. Nakamura, F. Mori, S. Fujiuchi, Mater. Trans. **43**, 1797 (2002)
63. J.-M. Song, K.-L. Lin, J. Mater. Res. **18**, 2060 (2003)
64. K.-S. Kim, J.-M. Yang, C.-H. Yu, I.-O. Jung, H.-H. Kim, J. Alloys Compd. **379**, 314 (2004)
65. M. Date, K.N. Tu, T. Shoji, M. Fujiyoshi, K. Sato, J. Mater. Res. **19**, 2887 (2004)
66. C.-W. Huang, K.-L. Lin, J. Mater. Res. **19**, 3560 (2004)
67. B.-J. Lee, CALPHAD **20**, 471 (1996)
68. H. Ohtani, M. Miyashita, K. Ishida, J. Jpn. Inst. Met. **63**, 685 (1999)
69. B.-J. Lee, N.M. Hwang, H.M. Lee, Acta Mater. **45**, 1867 (1997)
70. C.-Y. Chou, S.-W. Chen, Acta Mater. **54**, 2393 (2006)
71. C.-Y. Chou, S.-W. Chen, Y. S. Chang, J. Mater. Res. **21**, 1849 (2006)
72. M.Y. Chiu, S.S. Wang, T.H. Chuang, J. Electron. Mater. **31**, 494 (2002)
73. H. Okamoto, in *Binary Alloy Phase Diagram*, 2nd edn. (ASM, Metals Park, Ohio, 1990), p. 794
74. B.-J. Lee, C.-S. Oh, J.-H. Shim, J. Electron. Mater. **25**, 983 (1996)
75. S. Hassam, E. Dichi, B. Legendre, J. Alloys Compd. **268**, 199 (1998)
76. H. Ohtani, I. Satoh, M. Miyashita, K. Ishida, Mater. Trans. **42**, 722 (2001)
77. A. Prince, G.V. Raynor, D.S. Evans, in *Phase Diagrams of Ternary Gold Alloys* (Institute of Metals, London, 1990) p. 168
78. D. Kim, C.C. Lee, in *Proceedings of the 53rd Electronic Components and Technology Conference*, New Orleans, USA, May 2003, p. 668
79. K. Doi, H. Ohtani, M. Hasebe, Mater. Trans. **45**, 380 (2004)
80. P.T. Vianco, A.C. Kilgo, R. Grant, J. Electron. Mater. **24**, 1493 (1995)
81. J.-W. Yoon, C.-B. Lee, S.-B. Jung, Mater. Trans. **43**, 1821 (2002)
82. J.-I. Lee, S.-W. Chen, H.-Y. Chang, C.-M. Chen, J. Electron. Mater. **32**, 117 (2003)
83. W.H. Tao, C. Chen, C.E. Ho, W.T. Chen, C.R. Kao, Chem. Mater. **13**, 1051 (2001)
84. D.V. Malakhov, X.J. Liu, I. Ohnuma, K. Ishida, J. Phase Equilib. **21**, 514 (2000)
85. H. Paul, Solder. Surf. Mount Technol. **11**, 46 (1999)
86. N. Moelans, K.C. Hari, P. Wollants, J. Alloys Compd. **360**, 98 (2003)
87. Z. Mei, J.W. Morris Jr., J. Electron. Mater. **21**, 401 (1992)
88. K. Shimizu, T. Nakanishi, K. Karasawa, K. Hashimoto, K. Niwa, J. Electron. Mater. **24**, 39 (1995)
89. N.C. Lee, Solder. Surf. Mount Technol. **9**, 65 (1997)
90. A.T. Kosilov, G.L. Polner, I.U. Smurov, V.V. Zenin, Sci. Technol. Weld. Joining **9**, 169 (2004)
91. H. Okamoto, in *Phase Diagrams of Indium Alloys and Their Engineering Application*, ed. by C.E.T. White, H. Okamoto (ASM international, Materials Park, OH, 1992), pp. 255–257
92. T.M. Korhonen, J.K. Kivilahti, J. Electron. Mater. **27**, 149 (1998)
93. X.J. Liu, Y. Inohana, Y. Takaku, I. Ohnuma, R. Kainuma, K. Ishida, Z. Moser, W. Gasior, J. Pstrus, J. Electron. Mater. **31**, 1139 (2002)
94. G.P. Vassilev, E.S. Dobrev, J.-C. Tedenac, J. Alloys Compd. **399**, 118 (2005)

95. Y.M. Liu, T.H. Chuang, J. Electron. Mater. **29**, 1328 (2000)
96. M.D. Cheng, S.S. Wang, T.H. Chuang, J. Electron. Mater. **31**, 171 (2002)
97. A. Prince, G.V. Raynor, D.S. Evans, in *Phase Diagrams of Ternary Gold Alloys*, ed. by A. Prince, V. Raynor, D.S. Evans (The Institute of Metals, London, 1990), pp. 300–302
98. H.S. Liu, C. Liu, K. Ishida, Z.P. Jin, J. Electron. Mater. **32**, 1290 (2003)
99. T.H. Chuang, S.Y. Chang, L.C. Tsao, W.P. Weng, H.M. Wu, J. Electron. Mater. **32**, 195 (2003)
100. P.G. Kim, K.N. Tu, Mater. Chem. Phys. **53**, 165 (1998)
101. W. Köster, T. Gödecke, D. Heine, Z. Metallkd. Bd. 63, H. 12 (1972) 802
102. X.J. Liu, H.S. Liu, I. Ohnuma, R. Kainuma, K. Ishida, S. Itabashi, K. Kameda, K. Yamaguchi, J. Electron. Mater. **30**, 1093 (2001)
103. S.-K. Lin, T.-Y. Chung, S.-W. Chen, Y.-W. Yen, unpublished results, (2006)
104. A.D. Romig Jr., F.G. Yost, P.F. Hlava, in *Microbeam Analysis-1984* ed. by A.D. Romig Jr., J.I. Goldstein (San Francisco Press, 1984), p. 87
105. P.T. Vianco, P.F. Hlava, A.C. Kilgo, J. Electron. Mater. **23**, 583 (1994)
106. T.H. Chuang, C.L. Yu, S.Y. Chang, S.S. Wang, J. Electron. Mater. **31**, 640 (2002)
107. S. Sommadossi, W. Gust, E.J. Mitteneijer, Mater. Chem. Phys. **77**, 924 (2003)
108. D.-G. Kim, S.-B. Jung, J. Alloys Compd. **386**, 1515 (2005)
109. W. Burkhardt, K. Schubert, Z. Metallkd **50**, 442 (1959)
110. M.K. Bhargava, K. Schubert, Z. Metallkd. **67**, 318 (1976)
111. S.K. Shadangi, M. Singh, S.C. Panda, S. Bhan, Indian J. Technol. **24**, 105 (1986)
112. C.-Y. Huang, S.-W. Chen, J. Electron. Mater. **31**, 152 (2002)
113. D.-G. Kim, S.-B. Jung, J. Electron. Mater. **33**, 1561 (2004)
114. T.H. Chuang, K.W. Huang, W.H. Lin, J. Electron. Mater. **33**, 374 (2004)
115. R.K. Mahidhara, S.M.L. Sastry, I. Turlik, K.K. Murty, Scripta Metall. Mater. **31**, 1145 (1994)
116. J.W. Jang, P.G. Kim, K.N. Tu, M. Lee, J. Mater. Res. **14**, 3895 (1999)
117. A.R. Geranmayeh, R. Mahmudi, J. Mater. Sci. **40**, 3361 (2005)
118. B. Predel, W. Schwermann, J. Inst. Met. **99**, 169 (1971)
119. B. Jonsson, J. Agren, Mater. Sci. Technol. **2**, 913 (1986)
120. C.-S. Oh, J.-H. Shim, B.-J. Lee, D.N. Lee, J. Alloys Compd. **238**, 155 (1996)
121. Z. Moser, W. Gasior, J. Pstrus, S. Ishihara, X.J. Liu, I. Ohnuma, R. Kainuma, K. Ishida, Mater. Trans. **45**, 652 (2004)
122. D.B. Masson, B.K. Kirkpatrick, J. Electro. Mater. **15**, 349 (1986)
123. J.H. Kim, S.W. Jeong, H.M. Lee, J. Electron. Mater. **31**, 557 (2002)
124. Y. Takaku, X.J. Liu, I. Ohnuma, R. Kainuma, K. Ishida, Mater. Trans. **45**, 646 (2004)
125. P. Villars, A. Prince, H. Okamoto, in *Handbook of Ternary Alloy Phase Diagram* (ASM, Metals Park, Ohio, 1995), p. 13010

The effects of suppressed beta tin nucleation on the microstructural evolution of lead-free solder joints

D. Swenson

Published online: 8 September 2006
© Springer Science+Business Media, LLC 2006

Abstract Most lead-free solders comprise tin (Sn) as the majority component, and nominally pure β-Sn is the majority phase in the microstructure of these solders. It is well established that nucleation of β-Sn from Sn-base liquid alloys is generally difficult. Delays in the onset of β-Sn formation have a profound effect upon the microstructural development of solidified Sn-base alloys. Utilizing stable and metastable phase diagrams, along with solidification principles, the effects of inhibited β-Sn nucleation on microstructural development are discussed, employing the widely studied Sn–Ag–Cu (SAC) alloy as a model system. This analysis shows that the main effect of suppressed β-Sn nucleation on near-eutectic SAC solders is to increase the number and/or volume fraction of primary or primary-like microconstituents, while simultaneously decreasing the volume fraction of eutectic microconstituent. General strategies are outlined for avoiding unwanted microconstituent development in these materials, including the use of metastable phase diagrams for selecting alloy compositions, employment of inoculants to promote β-Sn nucleation, and utilization of high cooling rates to limit solid phase growth. Finally, areas for future research on the development of inoculated Sn-base solder alloys are outlined.

1 Introduction

In the electronics industry's ongoing search for a standard lead-free solder formulation, tin-base alloys stand at the forefront of all candidates. This position is attributable to tin's attractive combination of economic advantages (wide availability, relatively low cost, environmental benignity, a long history of use in solders, and availability of fluxes) and physical properties (low melting temperature, high electrical conductivity, and good wettability of common transition metals and alloys) [1–4].

Most Sn-base lead-free solders contain only minor amounts of alloying additions. For example, in the widely studied Sn–Ag–Cu (SAC) alloy, more than 95% of the solder is Sn, as measured by weight [5]. Owing to this high Sn content, and to the generally low solid solubilities of most elements in solid Sn [6], the majority phase in most Sn-base solder joints is body centered tetragonal Sn, or β-Sn. Moreover, this phase may be present in several microconstituents simultaneously, ranging from β-Sn dendrites to eutectic or eutectic-like microconstituents. As such, the overall microstructure of a Sn-base solder is highly dependent upon the details of the formation of the β-Sn phase from the liquid.

It is noteworthy, therefore, that the kinetics of β-Sn formation from most Sn-base liquid metal alloys are unique among common metal alloys, in that heterogeneous β-Sn nucleation is unusually difficult (see, for example, [7–9]). In turn, owing to the ubiquity of the β-Sn phase in Sn-base solders, the inhibition of β-Sn nucleation from the liquid results in the development of nonequilibrium, and often non-uniform, microconstituents within the microstructures of these solders.

D. Swenson (✉)
Department of Materials Science and Engineering,
Michigan Technological University, 1400 Townsend Drive,
Houghton 49931 MI, USA
e-mail: dswenson@mtu.edu

In this paper, the microstructural effects of β-Sn nucleation difficulties are outlined, utilizing the SAC alloy as a model system. Experimentally observed microstructural anomalies in SAC solder joints are rationalized using solidification principles and metastable portions of the Sn–Ag–Cu phase diagram. Subsequently, strategies for overcoming anomalous microconstituent development are addressed, and finally, suggestions are made as to areas for future research on inoculant development for improved SAC-based lead free solder formulations.

2 The nucleation behavior of β-Sn

Owing to its low melting temperature ($T_m = 232°C$, or 505 K) and low equilibrium vapor pressure (5.78×10^{-21} Pa at T_m), Sn was among the first elements employed in fundamental studies of the solidification of metals. The results of these early experiments [10, 11] were used to extend the original theories of liquid phase nucleation from the vapor [12, 13] to the liquid–solid transformation [14, 15]. In these original investigations, it was shown that liquid Sn may undergo substantial undercooling prior to β-Sn nucleation. Maximum undercoolings of 107 K and 120 K were reported by Vonnegut [10] and Pound and La Mer [11], respectively, corresponding to $\Delta T^{-}_{max} = 0.22T_m$. It was generally assumed that these maximum observable undercoolings represented conditions of homogeneous nucleation. However, more recent work by Perepezko [16] has demonstrated substantially greater undercoolings prior to Sn solidification (in the vicinity of 190 K, or $\Delta T^{-}_{max} = 0.37T_m$). Perepezko employed improved sample preparation techniques over those utilized in the earlier studies cited above, and his results indicated that in all earlier investigations, heterogeneous rather than homogeneous nucleation had been responsible for solidification.

In all of these studies of Sn solidification [10, 11, 16], special effort was made to avoid heterogeneous nucleation by methods such as microdroplet dispersal (whereby some droplets are unlikely to contain potent heterogeneous nucleant impurities) and surface encapsulation (in which the potency of surface nucleation is reduced by replacing the liquid surface with a liquid–liquid, liquid–solid or liquid–amorphous phase interface). Nevertheless, even in typical laboratory and industrial settings, where special precautions are absent, liquid Sn and Sn-base alloys may easily be undercooled by 15–40 K prior to solidification [7–9]. This degree of undercooling (up to $\Delta T^{-} \sim 0.08T_m$) is well beyond the $\Delta T^{-} \sim 0.02T_m$ typically observed for heterogeneous nucleation in liquid metals [17].

The author is aware of no study that has focused on rationalizing the relative difficulty of β-Sn nucleation. However, the fact that such difficulties exist suggests that common impurities, such as native oxide, are not particularly potent heterogeneous nucleants for β-Sn. (Indeed, it is noteworthy that both Vonnegut and Pound and La Mer, previously cited [10, 11], deliberately oxidized the surfaces of their Sn droplets prior to solidification experiments so that the oxide would prevent their coalescence during dilatometry characterization.) Moreover, common alloying additions to Sn-base solders, such as Ag, Cu, Ni, Bi, etc., form oxides that are less thermodynamically stable than SnO_2, and would therefore be unlikely to form oxides in the presence of excess Sn and small amounts of oxygen [18].

Additionally, given that substantial liquid undercooling has been observed in Sn-base alloys containing primary metallic phases, such as Ag_3Sn and Cu_6Sn_5 in the case of SAC alloys, many crystalline metals and intermetallic compounds are also relatively ineffective for catalyzing β-Sn nucleation. Finally, in the case of microelectronics solder joints, the small dimensions of the joints may reduce the likelihood of the presence of a potent heterogeneous nucleation site within the joint, since the probability of a heterogeneous nucleation event is generally expected to scale with the surface or interfacial area of material to be transformed [19].

3 Solidification of Sn-base solders in the absence of the β-Sn phase

Whatever the specific reason for the suppression of β-Sn nucleation in Sn-bearing liquid alloys, these nucleation difficulties have a substantial effect upon the microstructural evolution of Sn-base solders. In the present paper, the SAC alloy system has been employed to illustrate these microstructural phenomena. Many of the following points have been discussed elsewhere in the literature with regard to SAC solders (see, for example, [8, 9, 20–25]), but to date they have not been presented together and in detail to give an overall picture of the myriad microstructural effects of β-Sn nucleation difficulties manifested in SAC solders.

3.1 Equilibrium solidification in the Sn–Ag–Cu (SAC) alloy system

The SAC alloy system [2] is the most widely studied lead-free solder for use in microelectronics applications.

SAC solders are located in the Sn-rich corner of the Sn–Ag–Cu phase diagram, and lie within the composition range Sn–(2.0–4.0 wt.%Ag)–(0.5–1.0 wt.%Cu) [5]. This composition range was chosen based on the existence of a ternary eutectic reaction: L → β-Sn + Ag$_3$Sn + Cu$_6$Sn$_5$ that is located in this compositional vicinity [2]. According to Loomans and Fine [26] and Moon et al. [8], this ternary eutectic reaction occurs at the composition Sn–3.5wt.%Ag–0.9wt.%Cu and at a temperature of 217°C (490 K).

Figure 1a shows a portion of the Sn–Ag–Cu phase diagram's liquidus projection, as calculated by Moon et al. [8] via phase diagram optimization, based on experimental ternary phase diagram data and extrapolation of thermodynamic models of the constituent binary systems. For near-eutectic alloys, seven different microconstituents may develop during equilibrium solidification, depending upon the overall alloy composition and the specific topology of the phase diagram. In the Sn–Ag–Cu system, these microconstituents are: (i) primary Ag$_3$Sn, (ii) primary Cu$_6$Sn$_5$, (iii) primary β-Sn, (iv) monovariant Ag$_3$Sn + β-Sn, (v) monovariant Cu$_6$Sn$_5$ + β-Sn, (vi) monovariant Ag$_3$Sn + Cu$_6$Sn$_5$, and (vii) eutectic β-Sn + Ag$_3$Sn + Cu$_6$Sn$_5$. The compositional regions of formation of each of these microconstituents are labeled in Fig. 1a. Additionally, isotherms are provided that show the temperature of the surface at each overall composition.

A liquidus projection corresponds to intersecting liquidus surfaces, a liquidus surface being the boundary between the liquid phase and a two phase liquid plus solid region. Because this portion of the Sn–Ag–Cu phase diagram contains three solid phases, there are three liquidus surfaces here. Each surface intersects the other two surfaces individually along lines referred to as liquidus valleys (in analogy to topographical maps) and all three surfaces intersect at a single point corresponding to the eutectic point.

Underneath each point on the liquidus surface is a tie-line connecting the surface to a solid phase at a specific composition (a two phase, or bivariant solid + liquid equilibrium). Underneath each point on a liquidus valley is a tie-triangle connecting the valley to specific compositions of two different solid phases (a three phase, or monovariant solid 1 + solid 2 + liquid equilibrium). Underneath the eutectic point is a two-dimensional tie-tetrahedron (the projection of one vertex and three edges of a tetrahedron onto its opposite face) connecting the eutectic point to specific compositions of all three solid phases (a four phase, or invariant solid 1 + solid 2 + solid 3 + liquid equilibrium).

During cooling under equilibrium conditions, solidification will commence when the alloy reaches the temperature at which its overall composition intersects the system's liquidus projection. Once this occurs, one of the solid microconstituents described above will form, shifting the liquid to a new composition and a lower temperature. This process continues, with the composition of the remaining liquid continually changing, its volume fraction continually decreasing, and its temperature continually decreasing as well. As the liquid composition and temperature change, the microconstituents that form will also generally change. Eventually, the composition of the remaining liquid will correspond to the eutectic composition at the eutectic temperature. Upon further cooling, this last liquid present will solidify as ternary eutectic microconstituent.

In Fig. 1b, the liquid composition of an alloy of the overall compositon Sn–5.0wt.%Ag–0.5wt.%Cu at every point during solidification has been mapped onto the Sn–Ag–Cu liquidus projection. Such a mapping of liquid composition is referred to as a solidification path. (This alloy composition was chosen for illustrative purposes only. It resides outside the composition range of interest for lead-free solder applications.) Note that the solidification path intersects three microconstituent regions: primary Ag$_3$Sn from approximately 243°C to 218°C), monovariant Ag$_3$Sn + β-Sn (from just under 218°C to 217°C) and eutectic β-Sn + Ag$_3$Sn + Cu$_6$Sn$_5$ (at 217°C, the ternary eutectic temperature). This indicates that these three microconstituents would form from an alloy of this overall composition when solidified under equilibrium conditions, and demonstrates the utility of the liquidus projection.

Two complicating factors should be mentioned here. First, unlike the case of a binary phase diagram, where the solidification path directly follows liquidus lines, the solidification path is generally not evident on a liquidus projection diagram, since no liquid–solid equilibrium tie-lines are shown (indeed, tie-lines to solid phases are "covered up" by the liquidus surfaces). The solidification path may, however, be calculated from a thermodynamic model of the system, such as that of Moon et al. [8], or read from a series of detailed ternary phase diagram isotherms.

Second, in actual solidification, even under nominally equilibrium conditions, liquids with compositions near the liquidus valleys and near the eutectic point may solidify directly as monovariant or eutectic microconstituents. In other words, during actual solidification the liquidus valleys are effectively regions rather than lines, and the eutectic point is a region rather than a point. This is due to kinetic factors rather than thermodynamic factors; coupled or cooperative growth of two or three phases is often faster than

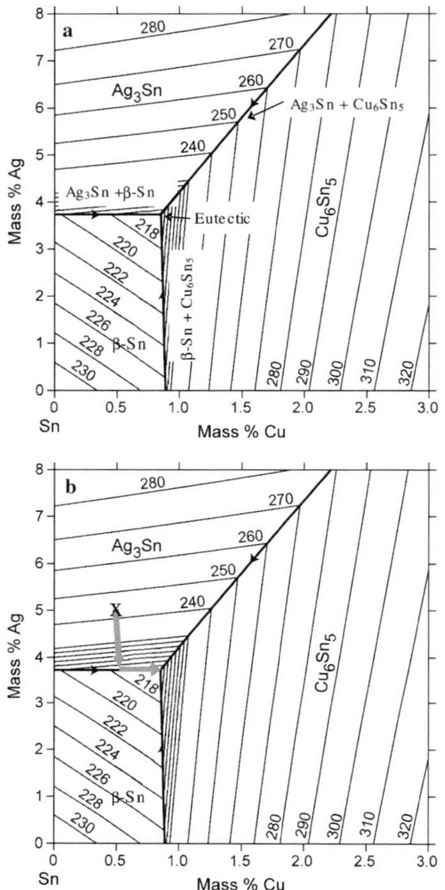

Fig. 1 (**a**) Sn-rich corner of the Sn–Ag–Cu liquidus projection, after Moon et al. [8], showing the seven different regions of microconstituent formation. (**b**) The equilibrium solidification path of an alloy of the composition Sn–5.0wt.%Ag–0.5wt.%Cu. This path traverses portions of the Ag₃Sn liquidus surface and the Ag₃Sn + β-Sn monovariant liquidus valley before terminating at the Ag₃Sn + Cu₆Sn₅ + β-Sn ternary eutectic point

work to ternary systems [28] and related theoretical treatments by Himemiya and Umeda [29] for further information.

Commonly observed morphologies of microconstituents present in SAC alloys include Ag₃Sn plates, hexagonal, hollow Cu₆Sn₅ needles and nonfaceted β-Sn dendrites for the primary phases and in some instances the monovariant microconstituents, and a β-Sn matrix embedded with rodlike Ag₃Sn and/or Cu₆Sn₅ particles for the eutectic microconstituent and in other instances, the monovariant microconstituents (see, for example, [2, 8, 9, 20, 26]). Monovariant, two phase microconstituents are often eutectic-like in appearance, indicating coupled growth. However, kinetic factors inhibit coupled growth when both phases show strong faceting tendencies, as would be expected in the case of monovariant Ag₃Sn + Cu₆Sn₅ [30]; if growth is not coupled, monovariant microconstituents may resemble primary phases.

3.2 Nonequlibrium solidification in the Sn–Ag–Cu (SAC) alloy system

For a near-eutectic SAC alloy, the example shown in Fig. 1b and the discussion from the previous section suggest that up to three microconstituents may form during equilibrium solidification: one primary microconstituent, one monovariant two phase microconstituent, and the ternary eutectic microconstituent. Moreover, these microconstituents would not be present in equal amounts; the lion's share of the volume fraction would be eutectic microconstituent, with substantially lesser amounts of primary and monovariant microconstituents.

(If growth of the phases within the monovariant microconstituent were uncoupled, and the solidification rate were moderate, the phases within the monovariant microconstituent may grow to substantial dimensions and become separated by liquid that ultimately transforms to eutectic microconstituent [28–30]. Under these conditions, the monovariant microconstituent might strongly resemble two primary phases, one of which must be the same as the true primary microconstituent. Whatever its morphology, however, the volume fraction of monovariant microconstituent is not affected, and would be very small for near-eutectic alloys.)

Figure 2 shows an optical micrograph of a SAC solder ball of the overall composition Sn–3.4 wt.% Ag–0.9wt.%Cu that has been melted and cooled at a rate of 0.25 K/s until it became solidified. The microconstituents present appear to be primary Ag₃Sn (large plates), primary Cu₆Sn₅ (large needles), primary β-Sn

growth that involves back-diffusion of solute atoms into the liquid, because the orderly configuration of solid phases minimizes diffusion distances. Moreover, coupled growth of faceting and nonfaceting solid phases is favored for off-monovariant and off-eutectic alloys, owing to unequal undercoolings required to drive growth of faceted and unfaceted phases. This latter phenomenon have been recognized for some time, and was first clearly demonstrated by Mollard and Flemings [27] for off-eutectic binary alloys. The reader is directed to a more recent extension of this early

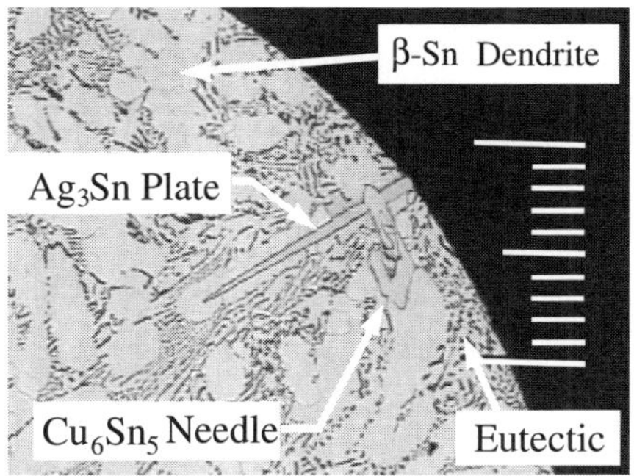

Fig. 2 Optical micrograph of a Sn–3.4wt.%Ag–0.9wt.%Cu solder solidified at a rate of 0.25 K/s, showing the simultaneous presence of at least four microconstituents: Ag₃Sn plates, primary-Cu₆Sn₅ needles, β-Sn dendrites and ternary eutectic. (Scale marker equals 100 microns.) This microstructure cannot be rationalized based on principles of equilibrium solidification

(dendrites) and at least one multiphase microconstituent, which could be a combination of monovariant and eutectic microconstituents but is most likely primarily eutectic microconstituent. This microstructure is noteworthy, because it is inconsistent with the solidification principles described in Sect. 3.1. As was the case in Fig. 1b, one would here expect the final microstructure to consist primarily of ternary eutectic β-Sn + Ag₃Sn + Cu₆Sn₅, with small amounts of primary Ag₃Sn and monovariant Ag₃Sn + β-Sn microconstituents. Instead, the majority microconstituent is clearly primary β-Sn, in amounts far too great to represent uncoupled growth of monovariant β-Sn + Ag₃Sn. Moreover, neither primary nor uncoupled monovariant Cu₆Sn₅ could form from an alloy of this overall composition, since both such regions lie outside of its likely solidification path.

The observed microstructure may be rationalized fully, however, if one assumes suppression of β-Sn nucleation to a temperature substantially below the ternary eutectic temperature. In order to elucidate this statement, let us first consider the effect of suppressed β-Sn nucleation on phase equilibria in the Sn–Ag–Cu system. Under such a circumstance, metastable equilibrium conditions will prevail, whereby the total Gibbs energy of the system is minimized at each temperature-composition pair in the absence of the β-Sn phase. Another way to describe this situation is to imagine "removing" the β-Sn phase (and with it any multiphase region that includes β-Sn) from the Sn–Ag–Cu phase diagram, and then redrawing the diagram as though the β-Sn phase did not exist. Figure 3a shows schematically the topology of the Sn-rich corner of the Sn–Ag–Cu liquidus projection in the absence of the β-Sn phase.

(*NOTE:* In principle, one may use the thermodynamic model of the Sn–Ag–Cu system established by Moon et al. [8] to calculate this metastable phase diagram, although this has not been done in the construction of Fig. 3a.)

Upon comparing Figs. 3a to 1a, one notices the following differences: *i.)* there is no longer a region of primary β-Sn formation; *ii.)* there are no longer Ag₃Sn + β-Sn and Cu₆Sn₅ + β-Sn liquidus valleys; and *iii.)* there is no longer a ternary eutectic point. The Ag₃Sn and Cu₆Sn₅ liquidus surfaces occupy the compositional space previously held by the primary β-Sn

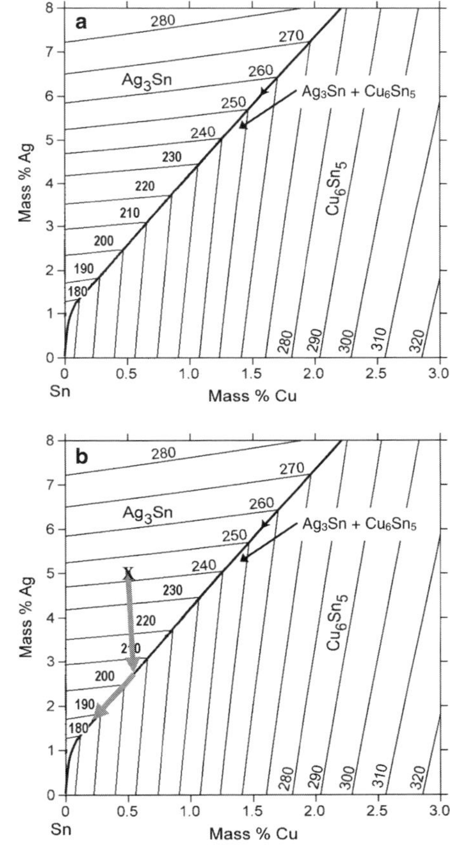

Fig. 3 (a) Metastable Sn–Ag–Cu liquidus projection, in the absence of the β-Sn phase, showing the three regions of microconstituent formation. **(b)** The metastable solidification path of an alloy of the composition Sn–5.0wt.%Ag–0.5wt.%Cu. This path traverses stable and metastable portions of the Ag₃Sn liquidus surface, and a portion of the metastable Ag₃Sn + Cu₆Sn₅ monovariant liquidus valley before terminating at 187°C, the temperature at which β-Sn nucleation occurs

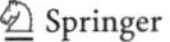

region, and the $Ag_3Sn + Cu_6Sn_5$ liquidus valley extends downward toward the Sn corner of the diagram.

In Fig. 3b, the solidification path of an alloy of the composition Sn–5.0wt.%Ag–0.5wt.%Cu is superimposed on the metastable phase diagram of Fig. 3a. This alloy composition intersects the Ag_3Sn liquidus surface at a temperature of about 243°C. Below this temperature, primary Ag_3Sn will form, with the liquid becoming progressively more Sn and Cu-rich. Now, however, the temperature range of primary Ag_3Sn formation is greatly extended over that of the equilibrium phase diagram, shown in Fig. 1b, since the β-Sn + Ag_3Sn liquidus valley no longer exists. At a temperature of approximately 205°C, the liquid composition will enter the Ag_3Sn–Cu_6Sn_5 liquidus valley. Note that this temperature is below the equilibrium eutectic temperature of the system (217°C), and that under equilibrium conditions an alloy of the given compostion have solidified prior to this point (see Fig. 1b). As the temperature continues to decrease, Ag_3Sn and Cu_6Sn_5 will form simultaneously, and as discussed previously it is likely that their growth is uncoupled, owing to their strong faceting tendencies, which would make them resemble primary Ag_3Sn and primary Cu_6Sn_5 microconstituents.

Formation of monovariant $Ag_3Sn + Cu_6Sn_5$ will continue until the onset of β-Sn nucleation. Once β-Sn nucleation occurs, the metastable phase diagram is no longer applicable, and one must revert to the equilibrium phase diagram of Fig. 1a for guidance. Let us say in this example that β-Sn nucleation occurs at an undercooling of 30 K, corresponding to a temperature of 187°C. At this temperature, Fig. 3b indicates that the remaining liquid will have a composition of about Sn–1.5wt.%Ag–0.25wt.%Cu and will be at a temperature that is approximately 39 K below the equilibrium β-Sn liquidus surface. Hence, the remaining liquid is in a state referred to as constitutional supercooling, meaning that its composition and temperature place it inside a two phase solid + liquid region of the phase diagram. A constitutionally supercooled liquid will solidify via dendrite formation, since a planar solid–liquid interface is unstable [31]. Moreover, at such a high liquid undercooling the growth rate of β-Sn is expected to be extremely rapid [32]. Rapid growth of β-Sn dendrites will have two additional effects. First, since the β-Sn phase is nearly pure Sn, most remaining Ag and Cu will be rejected into the liquid at the dendrite/liquid interface, increasing the Ag and Cu content of any remaining liquid substantially. Secondly, the latent heat of β-Sn solidification will be released, raising the temperature of the remaining liquid. Together, these effects begin to ameliorate the conditions of constitutional supercooling.

Eventually, the remaining liquid will be sufficiently Ag- and Cu-rich and at a high enough temperature for the formation of monovariant and/or ternary microconstituents. Owing to the Sn-rich nature of the undercooled liquid, if monovariant microconstituents were to form they would likely comprise $Ag_3Sn + \beta$-Sn or $Cu_6Sn_5 + \beta$-Sn. Both monovariant reactions consist of a faceting phase and a nonfaceting phase, and hence coupled, eutectic-like growth would be most likely [30].

3.2.1 Enhanced primary phase formation due to the suppression of β-Sn nucleation: intermetallic phases

It is important to note that because the compositions of Ag_3Sn, Cu_6Sn_5 and β-Sn are nearly independent of temperature [6, 8], the volume fractions of these three *phases* are unaffected by nonequilibrium solidification caused by the suppression of β-Sn nucleation. Rather, what the suppression of β-Sn nucleation affects is the volume fractions of *microconstituents* present. Specifically, the effect of suppressed β-Sn nucleation is the redistribution of these three phases among the microconstituents present; to wit, the volume fractions of primary phases (or primary-like microconstituents, such as monovariant, uncoupled $Ag_3Sn + Cu_6Sn_5$) are dramatically increased, whereas the volume fraction of eutectic microconstituent undergoes a corresponding decrease. In the example shown in Fig. 2, Cu_6Sn_5 that should have been present only as part of the eutectic microconstituent, has formed as a primary-like phase. The total volume fractions of primary or primary-like Ag_3Sn and in particular of β-Sn are much greater than would otherwise be the case, and again their presence as primary phases comes at the expense of formation of eutectic microconstituent.

The increase in volume fraction of primary intermetallic phases may have deleterious effects on the mechanical properties of SAC solder joints, potentially affecting such important factors as thermal fatigue life. A dramatic example of this has been discussed by Lewis et al. [20], Henderson et al. [21], and Kim et al. [33], who found that in many instances, particularly under conditions of slow to moderate cooling, primary or primary-like intermetallic microconstituents may grow to very large dimensions. This is particularly true for the Ag_3Sn phase. Henderson et al. [21] found that in some SAC solder joints, primary Ag_3Sn particles grew to be so large that they subtended the entire joint. An example of such a large Ag_3Sn plate, as observed by Buckmaster et al. [22], is depicted in Fig. 4. In principle, these large plates could have deleterious effects upon the thermal fatigue resistance of SAC solder joints, since the interface between the particle and the remaining solder provides a direct crack propagation route across the entire joint.

Similarly, large Cu_6Sn_5 needles have been observed in SAC solder joints. These needles, if oriented unfavorably within a joint, might also have disadvantageous effects upon joint properties such as thermal fatigue resistance. However, this possibility has not been investigated. Finally, the numerous primary β-Sn dendrites have their own unique issues, as is discussed in the following section.

3.2.2 Enhanced primary phase formation due to the suppression of β-Sn nucleation: β-Sn dendrites

The most striking microstructural anomaly of undercooled Sn-base alloys is the large volume fraction of β-Sn dendrites. Inspection of a typical SAC micrograph, such as Fig. 4, suggests the presence of hundreds of small, individual β-Sn dendrites within a SAC solder joint. Surprisingly, however, it turns out that relatively few of these dendrites are crystallographically distinct. As was first shown by Telang et al. [34, 35], and later investigated further by LaLonde et al. [23], Lehman et al. [24] and Henderson et al. [25], what appear to be myriad, small β-Sn dendrites within a solder joint are in fact secondary and tertiary arms of relatively few crystallographically distinct dendrites.

Figure 5a, b, taken from [23], shows a polarized light microscopy (PLM) image (a) and an electron backscatter diffraction (EBSD) orientation imaging map (b) of the same solder ball. These two techniques

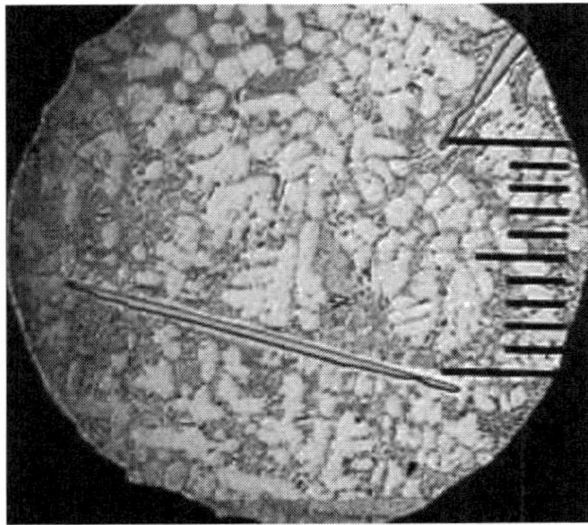

Fig. 4 Optical micrograph of a Sn–3.4wt.%Ag–0.9wt.%Cu solder solidified at a rate of 0.25 K/s, showing the formation of large, primary Ag_3Sn plates. (Scale marker equals 50 microns.) Note that the plate in the lower section of the solder ball nearly subtends the ball

are able to show contrast based on crystallographic orientation of β-Sn. It is readily seen that the images are comparable, and that only five regions of independent crystallographic orientation are visible in this 900 micron diameter solder ball, even though optical microscopy appears to show individual β-Sn dendrite cells that are nominally 40 microns in size.

A detailed analysis by LaLonde et al. using both EBSD and PLM showed that on average, one SAC ball grid array (BGA) solder joint, comprising a 900 micron diameter solder ball such as that shown in Fig. 5a, b, would be expected to contain at most 8 individual β-Sn dendrites, independent of cooling rate over the range studied (0.35–3.0 K/s). The work of Henderson et al. [25] yielded similar conclusions based on PLM analysis. Additionally, LaLonde et al. measured the misorientation between neighboring β-Sn dendrites for more than 200 adjacent dendrites, and found that their relative orientations were not random, an observation consistent with the earlier work of Telang et al. [34, 35]. A histogram of misorientation angle between neighboring dendrites is given in Fig. 6. The histogram shows that about 60% of the relative misorientations of adjacent dendrites lie within just three angle ranges: 5–15°, 20–25° and 55–65°. These results indicated that even among the eight "independent" dendrites present in a single SAC solder joint, many of these dendrites are related by special crystallographic orientations, suggesting that they may spring from a common origin during solidification. Several mechanisms have been reported in the literature for the bifurcation of dendrites during solidification, including twinning, kinking of primary dendrite arms due to void coalescence and collapse into dislocation loops, and simple bending of primary dendrite arms owing to impingement [36, 37].

The paucity of independent Sn dendrites in SAC solder joints is again consistent with inhibition of β-Sn nucleation during solidification. Fundamental studies by Rosenberg and Winegard [32] of the growth velocity of elemental β-Sn dendrites from pure liquid Sn have shown that the growth is extremely rapid, and has an exponential relationship with liquid undercooling. The data of Rosenberg and Winegard have been fitted by Christian [38] to the relationship: $v \sim 0.7(\Delta T^-)^{1.8}$, where v is the growth velocity in cm/s and ΔT^- is the undercooling in Kelvins. In SAC alloys, the β-Sn growth rate is undoubtedly lower than what would be predicted by the preceding equation, owing to the presence of Ag and Cu in the liquid phase. Nevertheless, as discussed by LaLonde et al., given the small dimensions of a typical solder joint (less than 1 millimeter), at undercoolings in the vicinity of 25 K,

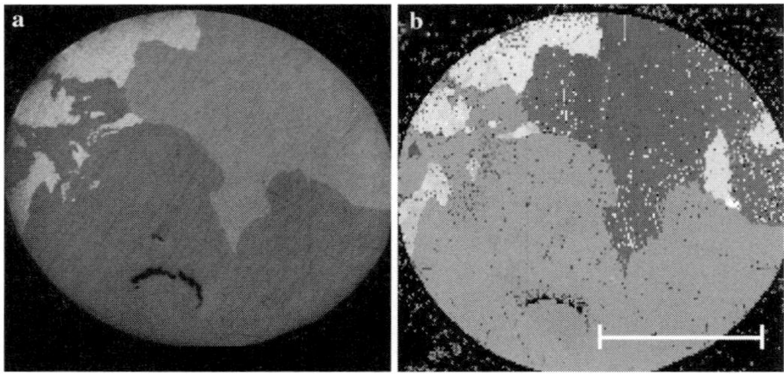

Fig. 5 **(a)** Polarized Light Microscopy (PLM) image of a Sn–3.4wt.%Ag–0.9wt.%Cu solder joint, cooled at K/s, and showing contrast based on crystallographic orientation of β-Sn dendrites. **(b)** Electron Backscattered Diffraction (EBSD) image of the same solder ball, again showing contrast based on crystallographic orientation of β-Sn dendrites. (Scale marker equals 50 microns)

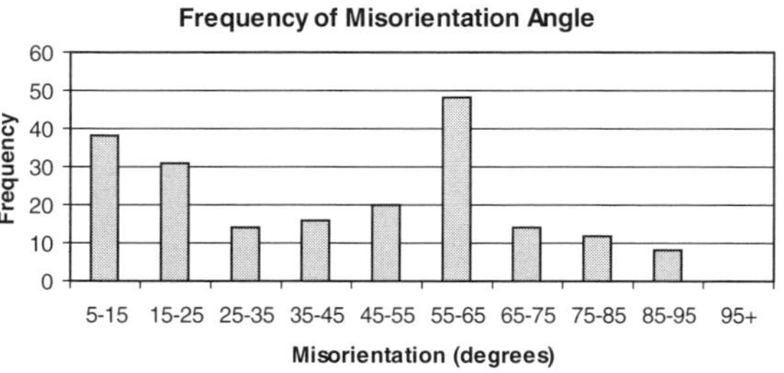

Fig. 6 Histogram of misorientation angle between adjacent β-Sn dendrites, based on EBSD data from more than 40 solder balls and 200 adjacencies. Note the prevalence of misorientations in the ranges of 55–65, 5–15 and 15–25 degrees

these data for the growth of β-Sn from pure liquid Sn suggest that a dendrite could probably subtend even a SAC solder joint in times on the order of 0.5 milliseconds. This implies that once β-Sn nucleation commences, solidification is so rapid that there is almost no time left for additional β-Sn nuclei to form. Moreover, release of the latent heat of fusion during β-Sn dendrite growth increases the temperature of the remaining liquid, which reduces the thermodynamic driving force for further β-Sn nucleation once a nucleation event is initiated. Finally the rapidity of solidification would also likely contribute to the generation of crystallographic "mistakes", such as twins, that would contribute to dendrite bifurcation, consistent with the preponderance of special misorientation angles between adjacent β-Sn dendrites.

Because β-Sn dendrites comprise the highest volume fraction of all microconstituents in a typical SAC solder joint, one would expect them to have a substantial effect upon the properties of the joint. Moreover, a solder joint consisting primarily of a few unique β-Sn dendrites should be more monocrystalline (anisotropic) than polycrystalline (isotropic) in character. For example, the thermal fatigue resistance of such a joint might depend strongly on the crystallographic orientation of the joint with respect to the applied stresses, and disadvantageously oriented joints would be more

prone to early fatigue failure than would other joints. Specific issues related to thermomechanical fatigue mechanisms for monocrystalline-like β-Sn solders have been discussed extensively by Telang and Bieler [34] and Henderson et al. [25].

4 Strategies for minimizing microstructural phenomena caused by the suppression of β-Sn nucleation

Nucleation theory [14, 15] suggests that the rate of nucleation depends upon both thermodynamic and kinetic factors. It stands to reason, therefore, that approaches to dealing with nucleation-related phenomena might focus on either thermodynamics or kinetics. Here, three strategies are presented for ameliorating the microstructural phenomena related to suppressed β-Sn nucleation: one based on thermodynamics, and two based on kinetics. In the thermodynamic approach, one may use the appropriate metastable phase diagram and the known degree of liquid undercooling to modify the solder composition. This strategy will minimize the temperature range over which primary or primary-like intermetallic phase formation occurs, in turn minimizing the final volume fraction of primary microconstituents. However, this first approach cannot prevent the eventual development of a high volume fraction of β-Sn dendrites. In the second strategy, one may attempt to use a β-Sn inoculant, whereby the undercooling prior to β-Sn nucleation is greatly reduced. In principle, this strategy should render the microstructure to be nearly the same as that which would be expected under equilibrium solidification conditions. In the third approach, one may increase the cooling rate of the liquid. This will have the effect of refining (reducing the phase dimensions of) the microstructure by limiting growth times, but may not eliminate nonequilibrium microconstituents. Each of these strategies is discussed below.

4.1 Modification of the solder composition based on prevailing metastable phase equilibria

As discussed in Sect. 3.2, the simultaneous presence of large, primary or primary-like Ag_3Sn plates and Cu_6Sn_5 needles in SAC solders is attributable to β-Sn nucleation difficulties, and the liquid composition following metastable portions of the Ag_3Sn liquidus surface and $Ag_3Sn + Cu_6Sn_5$ monovariant liquidus valley until eventual β-Sn nucleation at high undercoolings. If one cannot enhance β-Sn nucleation, one may still avoid formation of these primary intermetallic particles by using the metastable Sn–Ag–Cu phase diagram shown in Fig. 3a as a guide for alloy selection, rather than the equilibrium phase diagram depicted in Fig. 1a. Such an approach was first suggested by Henderson et al. [21], whose main concern was elimination of Ag_3Sn plate formation.

Henderson et al. showed that primary Ag_3Sn formation may be avoided by substantially reducing the Ag content of SAC alloys. It would stand to reason that reducing the Ag content of SAC solders would reduce the amount of Ag_3Sn formed, since Ag_3Sn is the only Ag-bearing phase present in near-eutectic SAC alloys. However, Henderson et al. were able to quantify the required reduction in Ag content using the following reasoning. It is known that SAC solders undercool about 20 K below the eutectic temperature prior to β-Sn nucleation. The amount of undercooling prior to soldification is relatively insensitive to the solder composition. Additionally, as may be seen in Figs. 1a and 3a, both the stable and metastable portions of the Ag_3Sn liquidus surface decrease sharply in temperature with decreasing Ag composition. Therefore, if one were to choose a SAC alloy composition for which the metastable Ag_3Sn liquidus surface was at least 20 K below the equilibrium eutectic temperature, primary Ag_3Sn would be thermodynamically precluded from forming prior to β-Sn nucleation, because the alloy

would lie within the metastable, single phase liquid region of the phase diagram. In other words, Henderson et al. recognized the existence of a metastable thermodynamic constraint on the system; for any phase transformation, even a metastable phase transformation, to occur spontaneously, there must still be a thermodynamic driving force, i.e., a lowering of the Gibbs energy of the system.

Using the thermodynamic model of Moon et al. [8] to extrapolate the position of the Ag_3Sn liquidus surface to metastable regions (similar to Fig. 3a), Henderson et al. estimated that if the Ag_3Sn content were reduced to below 2.7 wt.%, the Ag_3Sn liquidus surface would reside at least 20 K below the eutectic temperature. Their subsequent experiments using low Ag content SAC alloys proved that by restricting the Ag content to this level, Ag_3Sn plate formation could be avoided, regardless of cooling rate, because thermodynamics would forbid primary Ag_3Sn formation.

In principle, this strategy could also be utilized to preclude formation of primary or primary-like Cu_6Sn_5. However, no such investigations have been reported in the literature. Additionally, as is confirmed via inspection of the micrographs in Henderson et al.'s paper, this strategy cannot be employed to prevent primary β-Sn dendrite formation, since the β-Sn remains highly constitutionally supercooled, as discussed previously. Figure 7, from Buckmaster et al. [22], depicts a slowly cooled (0.25 K/s) Sn–2.5wt.%Ag–0.9wt.%Cu alloy, which in agreement with Henderson et al. has no Ag_3Sn plates but is rife with β-Sn dendrites.

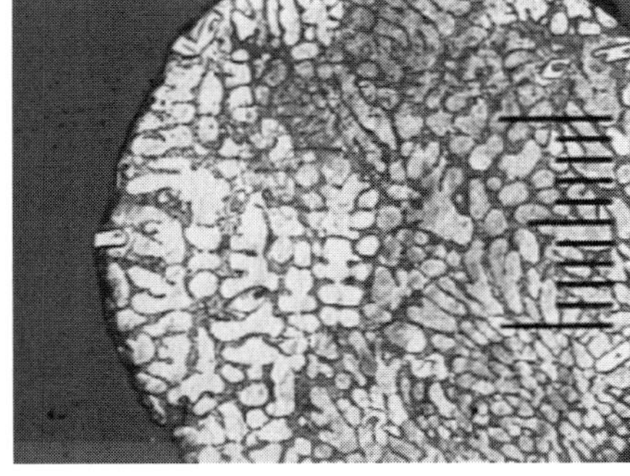

Fig. 7 Optical micrograph of a Sn–2.5wt.%Ag–0.9wt.%Cu solder solidified at a rate of 0.25 K/s, showing that Ag_3Sn plates do not form from low Ag content alloys, even at low cooling rates. (Scale marker equals 50 microns.) It is noteworthy, however, that reducing the Ag content of the alloy does not prevent formation of Cu_6Sn_5 needles and β-Sn dendrites

4.2 Addition of a β-Sn inoculant to promote solidification at small liquid undercoolings

A second strategy for rendering the microstructure of near-eutectic SAC alloys more like the expected equilibrium structure would be to add a β-Sn inoculant to the solder. An inoculant is a material that, even in small quantities, promotes rapid heterogeneous nucleation of solid phases from the liquid at low to modest undercoolings. In practice, inoculants generally take the form of insoluble, solid particles whose melting temperatures far exceed those of the liquid alloy in which they promote nucleation. Inoculants are widely used in the metal casting industry to refine the grain or dendrite size of alloys. (For example, titanium is routinely used as an inoculant in Al-base alloys [39]).

Clearly, employment of a β-Sn inoculant is a straightforward solution to the problems described in the present paper, since they all stem from β-Sn nucleation difficulties. However, inoculants are found via trial and error, and their efficacies may be sensitive to processing conditions. To date, only one study has been reported in the literature pertaining to the development β-Sn inoculants. Ohno and Motegi [7] studied the effects of minor alloying additions on the undercooling of high purity Sn prior to its solidification. In their experiments, 0.1 wt.% of several elements was added individually to filtered, 99.999% pure Sn (the impurities present were not specified) and the undercooling of the melt was measured as a function of superheating. Ohno and Motegi found that the elements Al, Sb and Zn were effective in reducing the undercooling of high purity Sn from $\Delta T^- \sim 18$ K to 3–5 K, 3–7 K and 3–7 K, respectively, depending on the amount of superheating employed prior to solidification. They also observed that Sb and Zn became ineffective as inoculants upon superheating beyond 500 K and 640 K, respectively, above the melting temperature of Sn. This "fading" of inoculating capability may be attributable to preferential evaporation of Sb and Zn at higher temperatures. The inoculating effect of Al additions, on the other hand, was insensitive to the degree of superheating over the entire superheating range investigated (up to $\Delta T^+ = 800$ K).

Since the β-Sn phase in Sn–base solders is essentially elemental Sn, it is reasonable to assume that Al, Sb and Zn might also be effective inoculants when added to alloys such as the SAC alloy, barring their strong chemical interactions with other alloying additions present. Indirect confirmation of this hypothesis could be found in the lead-free solder literature. For example, McCormack and coworkers [40, 41] modified the Sn–3.5Ag binary eutectic solder with 0.5–2.0 wt.% Zn additions, and noted substantial refinement of the microstructure; in particular, they observed a decrease in the incidence of β-Sn dendrites and an increase in the fraction of eutectic microconstituent. These are precisely the microstructural changes one would expect if β-Sn nucleation were facilitated so as to occur at small undercoolings. However, their investigations did not incorporate studies of undercooling; rather, only changes in melting temperature due to the Zn additions were noted.

Buckmaster et al. [22] and Kang and et al. [42] were the first to attempt to correlate changes in undercooling in Zn-modified SAC alloys with concomitant changes in microstructure, using metallography and thermal analysis. Figure 8 depicts differential scanning calorimetry (DSC) data from Buckmaster et al., which show the effects of an 0.1 wt.% Zn addition on the undercooling and melting temperatures of several near-eutectic SAC solder alloys, as well as on pure Sn. The figure shows that these small Zn additions are effective in reducing undercooling prior to solidification over a wide range of SAC alloy compositions. The undercooling prior to solidification was decreased from 25–30 K to 5–10 K. Additionally, it is noteworthy that the cooling rates were relatively low in these DSC experiments (about 0.25 K/s).

Figures 9a, b, again taken from Buckmaster et al., show the microstructure of a Sn–3.4wt.%Ag0.9wt.%Cu alloy cooled at 0.25 K/s (a) and the microstructure of the same alloy inoculated with 0.1 wt.% Zn and cooled at the same rate (b). Upon inspection of these micrographs, one may see that while Ag_3Sn plates are present in the ternary eutectic solder, they have not formed in the Zn-modified solder. This absence of Ag_3Sn plate formation in Zn-bearing solders was verified by Buckmaster et al. based upon the metallographic observation of dozens of Zn-modified solder joints solidified over a range of cooling rates.

Further comparison of Fig. 9a, b shows that the volume fraction of ternary eutectic microconstituent has been increased substantially upon addition of Zn to the solder; in fact, the ternary eutectic has gone from being a minority microconstituent in Fig. 9a to the majority microconstituent in Fig. 9b. It must be emphasized, however, in comparing these figures that even in the presence of Zn inoculant, β-Sn dendrites still persist, although the Zn addition has dramatically reduces their volume fraction. This may be attributable to the relatively high growth rate of β-Sn from the melt, even at these reduced (5–10 K) undercoolings. (Indeed, the data of Rosenberg and Winegard [32] suggest that even at an undercooling of 5 K, the growth velocity of a β-Sn dendrite would be more than 10 cm/s

Fig. 8 Melting temperatures (T_m) and solidification temperatures (T_s) of SAC alloys of varying Ag content, with and without 0.1wt.%Zn alloying additions. Note that Zn additions have a negligible effect on T_m and a substantial effect upon T_s

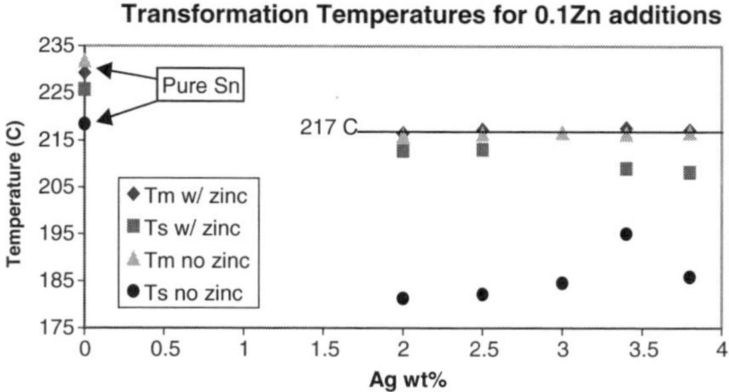

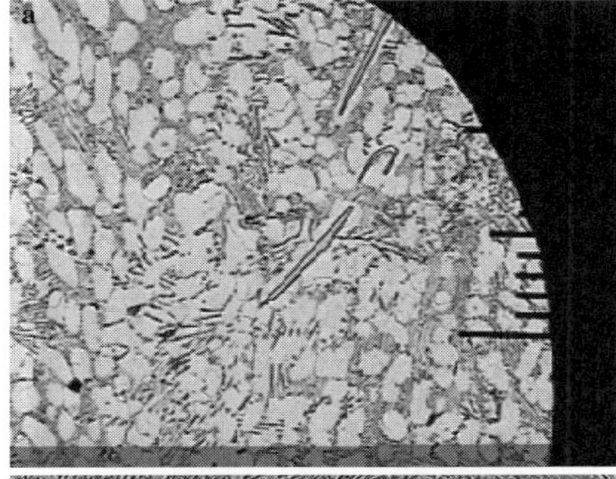

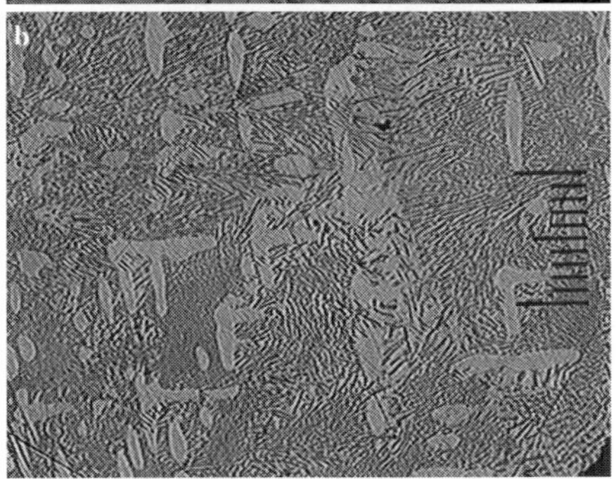

Fig. 9 (a) Optical micrograph of a Sn–3.4wt.%Ag–0.9wt.%Cu solder ball, cooled at 0.25 K/s. (Scale marker equals 100 microns) Note the high volume fraction of β-Sn dendrites and the presence of Ag₃Sn plates. **(b)** Optical micrograph of a Sn–3.4wt.%Ag–0.9wt.%Cu+0.1wt.%Zn solder ball, cooled at 0.25 K/s. (Scale marker equals 100 microns.) In this Zn-modified solder, fewer β-Sn dendrites are present, and no Ag₃Sn plates are observed

from pure Sn. Therefore, if growth velocities of β-Sn dendrites in SAC solders are of similar magnitude, a β-Sn dendrite could still traverse a ~1 mm diameter solder joint in about 0.01 s.)

Additionally, the average β-Sn dendrite size appears to be larger in the Zn-modified alloys than it is in the unmodified alloys. This observation was also made by Kang et al. [42]. However, as was discussed in Sect. 3.2.2, one should not rely upon metallography to ascertain β-Sn dendrite size in these solders; instead, one should utilize a technique such as EBSD that identifies β-Sn dendrites based on crystallographic orientation.

At present, it is unclear why Zn is an effective inoculant for β-Sn. Given that Zn promotes heterogeneous nucleation equally well in both pure Sn and SAC alloys, its inoculating effect is not attributable to interactions between Zn and Ag or Cu. It is possible that Zn, given its strong affinity for oxygen, forms ZnO in the presence of liquid Sn or Sn-base alloys. ZnO has a high melting temperature, and is likely to be insoluble in Sn-base liquids at temperatures near the melting point of Sn. Small, solid ZnO particles might act as heterogeneous nucleation sites for β-Sn, owing to a low liquid/ZnO interfacial energy. Whatever the physical basis for its inoculating effect, Zn addition shows great promise for promoting formation of the eutectic microconstituent in SAC solders.

A related point of interest is the distribution of Zn within the SAC solder subsequent to solidification. To address this issue, Buckmaster et al. performed X-ray mapping during scanning electron microscopy (SEM) analysis of Zn-modified solders to pinpoint the distribution of Ag, Cu and Zn within Zn-modified Sn–3.4wt.%Ag–0.9wt.%Cu solders. A typical result is shown in Fig. 10, which depicts a Cu₆Sn₅ particle embedded in the β-Sn matrix. X-ray mapping shows appreciable accumulation of Zn within the Cu₆Sn₅ phase, indicating a preference for Cu–Zn interaction. No accumulation of Zn was observed within the Ag₃Sn phase (not shown) or the β-Sn phase. Preferential interaction between Cu and Zn is of great practical importance, given that Cu is among the most common solder pad metals. Concerns related to Zn-(Cu pad) interactions will be addressed at a later point in the paper.

Fig. 10 Scanning electron micrograph and X-ray mapping of the Cu_6Sn_5 phase in a Zn-modified solder ball. (Scale marker equals 10 microns) This analysis shows that Zn is preferentially partitioned into the Cu_6Sn_5 phase. No accumulation of Zn was observed in β-Sn or Ag_3Sn

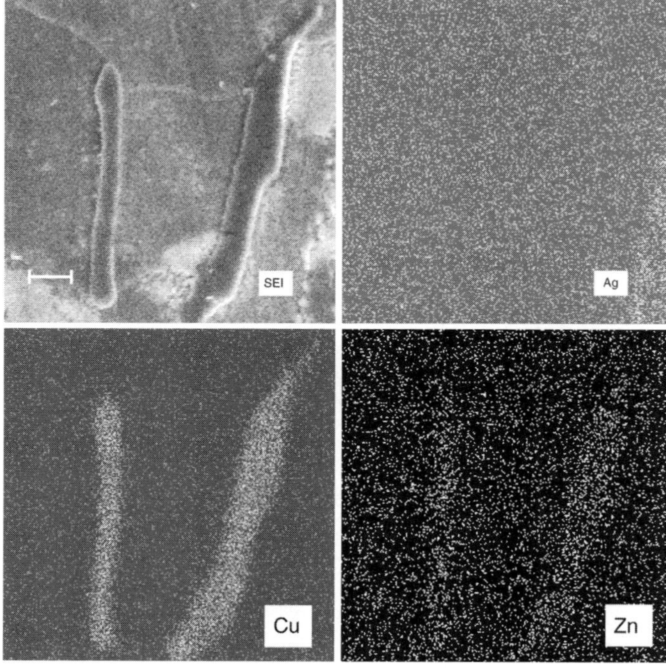

4.3 Increasing the cooling rate to limit growth time for primary phases

The final strategy for limiting nonequilibrium microstructural development due to suppressed β-Sn nucleation is to increase the cooling rate of solders during the solidification process. This approach has been investigated by several researchers [9, 21, 33]. A higher cooling rate might have two effects on the early stages of solidification (e.g., those stages involving the formation of primary and monovariant microconsituents). First, a high cooling rate will limit, or possibly preclude, the generation of stable nuclei during primary and monovariant solidification. Second, for those stable nuclei that do form, a high cooling rate will limit the time available for phase growth. Together, these effects should reduce the volume fraction of primary and monovariant phases present at the onset of β-Sn nucleation; moreover, continued rapid cooling during the β-Sn phase nucleation and growth should affect the β-Sn phase in a similar manner. Ultimately, one would anticipate a refinement of the microstructure under conditions of sufficiently fast cooling.

This structural refinement effect has been observed by Henderson et al. [21] and Kim et al. [33] in regard to primary Ag_3Sn plate formation. These investigators found that at sufficiently high cooling rates, Ag_3Sn plate formation could be prevented, even from SAC alloys of relatively high Ag content as shown in Fig. 11. The most plausible explanation for their observations is likely a growth limitation rather than a nucleation limitation; in other words, a high cooling rate does not allow sufficient time for primary Ag_3Sn particles to develop a large, plate-like morphology [21].

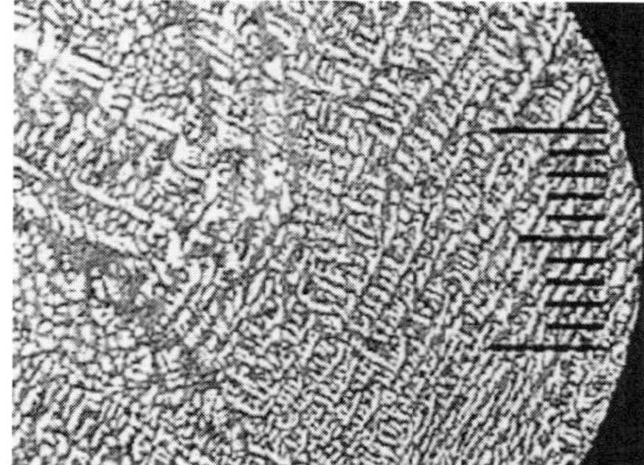

Fig. 11 Optical micrograph of a Sn–3.4wt.%Ag–0.9wt.%Cu solder ball, cooled at 4.0 K/s. (Scale marker equals 200 microns) At high cooling rates, Ag_3Sn plates do not form; however, both β-Sn dendrites and Cu_6Sn_5 needles are still observed

Based upon metallography, these researchers also noted a general refinement in the scale of all microconstituents, including β-Sn dendrites, for cooling rates in excess of about 1.5 K/min. However, the work of LaLonde et al. [23] showed that although β-Sn dendrites appear to be smaller in rapidly cooled solder joints, based on metallography, EBSD analysis proved that fast cooling rates merely reduce the average primary and secondary dendrite arm dimensions, and do not increase the number of crystallographically distinct β-Sn dendrites present. This was proven true for cooling rates of up to 3 K/s which encompasses the rates studied by the researchers cited above.

While this strategy may be effective for eliminating or reducing primary intermetallic phase formation, it does have several disadvantages when compared with the other strategies for minimizing the development of nonequilibrium microconstituents. First, solidification at increased cooling rates does not prevent β-Sn dendrite formation, nor does it refine their true size. Therefore, all potential problems related to β-Sn dendrites outlined in Sect. 3.2.2 remain. Second, in many instances it may prove impossible in practice to increase the cooling rate during microelectronics assembly to a level sufficiently high to affect the solder microstructure. Typical cooling rates observed during assembly processes range from 0.25–4 K/s [21], and much of this range is too low to prevent Ag_3Sn plate formation, for example, based on the work cited above. Finally, high cooling rates increase the likelihood of thermal stress-related defects and failures, owing to substantial differences in coefficients of thermal expansion between board and card components. Overall, then, this strategy is probably of limited utility for microelectronics applications.

5 Suggested future work

Of the three strategies described in Sect. 4.0 for minimizing the development of nonequilibrium microconstituents in Sn-base solders, the most promising would seem to be utilization of β-Sn inoculants. This strategy directly addresses the root cause of nonequilibrium microstructural evolution, namely, suppressed β-Sn nucleation. Therefore, in principle one could modify Sn-base solders to include inoculants and thus obtain equilibrium microstructures under a variety of processing conditions. In practice, however, many concerns need to be addressed simultaneously when reformulating solder compositions. In this section, such concerns are outlined as they relate to inoculant

development. They include inoculant identification, optimization of inoculant content, potential chemical interactions between inoculants and metal solder pads during reflow (melting and resolidification of the solder) and use, and the effects of inoculants on the microstructure of β-Sn dendrites. It is hoped by the author that these discussions may serve as a basis for future research pertaining to inoculant development for Sn-base solders.

5.1 Inoculant identification

To date, only Zn has been studied as an inoculant for SAC solders [22, 42]. Ideally, however, several possible inoculants would be identified, a few of which would likely prove superior to the others when one takes into account additional considerations for alloy formulation, such as those outlined in the remainder of the present paper. As has already been mentioned, inoculants are found via trial and error. Nevertheless, data that have already been reported in the literature suggest directly or indirectly several alloying additions, including Al, Ce, Co, Fe, La, Sb, and Zn, that are likely to serve as β-Sn inoculants.

Ohno and Motegi [7] proved that Al and Sb act as effective inoculants for β-Sn in pure Sn melts. Clearly, both of these elements should therefore also be studied as potential β-Sn inoculants for SAC solders. The remaining evidence for alloying additions that serve as β-Sn inoculants is less direct, and is based on several reports of the effects of minor (generally less than 0.5 wt.%) alloying additions on the microstructures of SAC alloys. These reports include Co(0.15–0.45 wt.%) [5, 43], Fe(0.2 wt.%) [43], and Ce + La(0.025–1.0 wt.%) [44] additions. In each case, minor alloying additions have been shown to provide substantial microstructural refinement, and/or to eliminate or greatly reduce nonequilibrium primary phase formation. Given that these microstructural effects are the same as would be caused by inoculation, it is possible that these alloying additions are in fact effective β-Sn inoculants. However, studies of solder undercooling with and without alloying additions, which would provide strong evidence of an inoculation effect, were not generally carried out in these investigations. The sole exception is the work by Anderson et al. [5], who used DSC to show that 0.45 wt.% Co additions to Sn–3.6wt.%Ag–1.0wt.% Cu solder reduced the undercooling prior to solidification by about 20 K when the solder was in the form of fine, emulsified droplets of unspecified size.

It is also noteworthy that in Pound and La Mer's [11] nucleation studies of pure Sn, it was found that the addition of Fe_2O_3 powder to an array of micron scale

Sn droplets reduced the maximum observed undercooling prior to solidification by more than a factor of two. That observation might suggest indirectly that Fe, to the extent that the Fe_2O_3 was attacked and reduced by the liquid Sn, acts as an effective inoculant for β-Sn as well.

Overall, careful thermal analysis studies should be undertaken on all SAC-related alloys containing minor alloying additions, comparing melting temperatures upon heating with solidification temperatures upon cooling, to see whether undercooling has been reduced over unmodified alloys.

5.2 Optimization of inoculant content

Once an inoculant is identified, the next question to be addressed is the establishment of the appropriate inoculant content. Buckmaster et al. [22] used DSC to study the undercooling of Sn–3.4wt.%Ag–0.9wt.%Cu solder as a function of Zn content, in order to ascertain the minimum Zn content required for effective β-Sn inoculation. The results of their investigation are shown in Fig. 12. This figure indicates that once Zn additions reach a level of about 0.1 wt.%, further Zn additions do not have an appreciable effect upon the level of undercooling prior to solidification. Therefore, one would consider ~0.1wt.% Zn to be the optimal inoculant content for this alloy.

However, the undercooling data in Fig. 12 are for chemically isolated solders. Different results might be obtained in actual solder joints, in which metal pad materials undergo chemical reactions with the solder. This was in fact shown to be the case by Kang et al. [42], who investigated the microstructures of Zn-modified Sn–3.8wt.%Ag–0.9wt.%Cu solders solidified in contact with Cu pads. They found that subsequent to one reflow at a cooling rate of 0.2 K/s, Ag_3Sn plate formation occurred in a solder containing 0.1 wt.% Zn, along with a high volume fraction of β-Sn dendrites. However, under identical reflow conditions, a solder comprising 0.7 wt.% Zn contained no Ag_3Sn plates and had a high volume fraction of eutectic microconstituent, in addition to lesser amounts of β-Sn dendrites.

Kang et al. rationalized these observations based on the concept of preferential reaction between Zn and the Cu pads during reflow. The preferential partitioning of Zn within the Cu_6Sn_5 phase of SAC solder has already been shown in Fig. 10, suggesting a strong chemical affinity between these elements. Moreover, the preference of Cu reaction with Zn over Sn has been noted previously in the literature in studies of metal pad reactions with Sn–Zn solders, which showed Cu–Zn intermetallic formation during reflow (see, for example, [45]). Given the small Zn content of the 0.1 wt.% Zn-bearing solder, Kang et al. postulated that preferential Cu–Zn reaction depleted the solder of Zn to a concentration that was no longer able to inoculate β-Sn nucleation. Alternatively, while preferential Cu–Zn reaction also occurred in the solder joint containing 0.7 wt.% Zn, subsequent to intermetallic compound (IMC) formation the solder presumably still contained sufficient Zn to provide the inoculation effect.

During microelectronics assembly, multiple reflows of solder joints are often required. Every reflow brings with it the probability of further IMC formation between the solder and metal pads. Therefore, in an industrial setting an inoculated solder may have to contain alloying additions far in excess of what would be required to provide β-Sn inoculation during a single reflow, if the inoculant reacts preferentially with metal pad material. Therefore, once potential β-Sn inoculants are identified, systematic tests should be conducted in which SAC alloys containing various amounts of inoculant are reflowed numerous times when in contact with all common metal pads (Cu, Ni, Au, Pd, etc). Such a test matrix may elucidate required inoculant contents for various numbers of solder reflows.

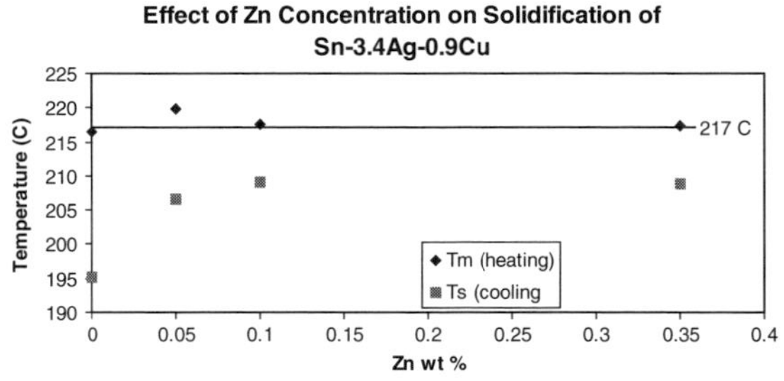

Fig. 12 Melting temperature (T_m) and solidification temperature (T_s) of Sn–3.4wt.%Ag–0.9wt.%Cu as a function of Zn alloying addition. Zn additions beyond 0.1 wt.% appear to have little additional effect upon the undercooling prior to β-Sn nucleation

5.3 Intermetallic phase formation during reflow and aging

Preferential reaction between inoculants and metal pad materials may have a substantial effect upon IMC formation, both during reflow and in the course of subsequent aging of solder joints, when compared with IMC formation in unmodified solders. These effects fall into three related categories. First, such interaction may alter the diffusion path (i.e., the sequence of phases formed as a function of time and temperature in a diffusion couple subject to certain boundary conditions; see, for example, [46]) of the joint. Formation of different phases, or the same phase with a different composition, may substantially alter the properties of a solder joint.

Second, chemical interaction between inoculant additions and pad metals may change the rate-limiting step of IMC formation, which is typically tied to the value of the chemical diffusivity of each phase [46]. Alteration of the rate limiting step may be good or bad, depending upon whether it enhances or inhibits IMC growth. In a qualitative sense, if inoculant additions decrease the chemical diffusivities of IMC's below their values in unmodified solders, the rate of IMC formation will decrease consequently the lifetime of a solder joint will increase, all other factors being equal.

For example, Lin and Hsu [47] modified eutectic Sn–9wt.%Zn alloy with an 0.45 wt.% Al addition. When they reflowed this solder on a Cu pad, they found that the Al reacted preferentially with the Cu to form a ternary IMC, $Al_{4.2}Cu_{3.2}Zn_{0.7}$. During subsequent 150°C aging experiments, they observed the additional formation of Cu–Zn IMC's. Thus, a minor Al addition changed both the diffusion path and most likely the rate limiting step of IMC formation.

Finally, interaction between inoculant additions and metal pads may change the morphologies of IMC layers. A classic example of such a morphological change is the switch between a planar reaction layer interface and a nonplanar reaction layer interface [48]. Drastic changes in IMC layer morphology have been observed after minor alloying additions to SAC solders. Anderson et al. [43] observed that the replacement of 0.3 wt% Cu with 0.3 wt.% Co in Sn–3.7wt.%Ag–0.9wt.%Cu solder led to a change in Cu_6Sn_5 IM formation from a scallop-like morphology to a fine, coral-like morphology. Chuang and Lin [49] observed that minor (0.07 wt.%) Ni additions to Sn–3.5wt.% Ag0.5wt.%Cu solder increased the Cu IMC growth rate substantially during reflow over that of the base SAC solder, and also changed its morphology to be more coral- or worm-like. In both studies, no new phases were found to form; however, the Cu_6Sn_5 IMC phases exhibited solubility of the minor alloying elements.

The results described above highlight how minor changes in solder alloy composition can have a dramatic effect upon chemical reactions between the solder and metal pad materials. Investigators seeking to establish workable inoculant additions for SAC solders must be cognizant of this important issue.

5.4 Crystallographic analysis of β-Sn dendrites in inoculated SAC solders

Finally, it is of interest to determine whether, even if the microstructure of a eutectic SAC solder can be fabricated as pure eutectic microconstituent, the β-Sn matrix is highly polycrystalline and refined, or if it is more monocrystalline in character, as has been found in highly undercooled SAC alloys containing a large volume fraction of β-Sn dendrites. As was mentioned in Sect. 3.2.2, this subject is of utmost importance for interpreting and predicting the mechanical properties of SAC solder joints. Unpublished, prelimary data by LaLonde et al. [23] involving EBSD data on Zn- and Fe-modified SAC solders suggested that even in nominally 100% eutectic β-Sn + Ag_3Sn + Cu_6Sn_5 microstructures, there are still relatively few crystallographically independent β-Sn dendrites. However, EBSD analysis is of such solders was extremely difficult, since the dimensions of the β-Sn regions were found to be on the order of the minimum specimen area from which diffraction data could be collected with the equipment LaLonde et al. had available. It may be that extensive transmission electron microscopy (TEM) analysis will prove necessary to resolve this issue.

References

1. C.J. Evans, Metallurgia **51**, 11 (1984)
2. C.M. Miller, I.E. Anderson, J.F. Smith, J. Electron. Mater. **23**, 595 (1994)
3. J. Bath, C. Handwerker and E. Bradley, Circuits Assembly (May 2000) 31.
4. D. Suraski, K. Seeling, IEEE Trans. Electron. Packag. Manuf. **24**, 244 (2001)
5. I.E. Anderson, J.C. Foley, B.A. Cook, J.L. Harringa, R.L.Terpstra, O. Unal, J. Electron. Mater. **30**, 1050 (2001)
6. T.B. Massalski, *Binary Alloy Phase Diagrams*, 2nd edn. (ASM International, Materials Park OH, 1990)
7. A. Ohno, T. Motegi, J. Japan Inst. Metals **37**, 777 (1973)
8. K.-W. Moon, W.J. Boettinger, U.R. Kattner, F.S. Biancaniello, C.A. Handwerker, J. Electron. Mater. **29**, 1122 (2000)
9. S.K. Kang, W.K. Choi, D.-Y. Shih, D.W. Henderson, T. Gosselin, A. Sarkhel, C. Goldsmith, K.J. Puttlitz, Proc. 53rd ECTC Conf. (IEEE, Piscataway, NJ, 2003), p. 64

10. B. Vonnegut, J. Colloid. Sci. **3**, 563 (1948)
11. G.M. Pound, V.K. LA Mer, J. Amer. Chem. Soc. **74**, 2323(1952)
12. M. Volmer, A. Weber, Z. Phys. Chem. **119**, 227 (1926)
13. R. Becker, W. Döring, Ann. Phys. **24**, 719 (1935)
14. D. Turnbull, J.C. Fisher, J. Chem. Phys. **17**, 71 (1949)
15. J.H. Holloman, D. Turnbull, Progress Metal Phys. **4**, 333 (1953)
16. J.H. Perepezko, Mater. Sci. Eng. **65**, 125 (1984)
17. K.F. Kelton, in *Solid State Physics*, vol. 45, ed. by F. Seitz, D. Turnbull (1991), p. 75.
18. O. Kubaschewski, C.B. Alcock, Metallurgical Thermochemistry, 5th edn. (Pergamon Press, Oxford UK 1979)
19. D. Turnbull, J. Appl. Phys. **21**, 1022 (1950)
20. D. Lewis, S. Allen, M. Notis, A. Scotch, J. Electron. Mater. **31**, 161 (2002)
21. D.W. Henderson, T. Gosselin, A. Sarkhel, S.K. Kang, W.-K. Choi, D.-Y. Shi, C. Goldsmith, K.J. Puttlitz, J. Mater. Res. **17**, 2775 (2002)
22. K.L. Buckmaster, J.J. Dziedzic, M.A. Masters, B.D. Poquette, G.W. Tormoen, D. Swenson, D.W. Henderson, T. Gosselin, S.K. Kang, D.Y. Shih and K.J. Puttlitz, Presented at the TMS 2003 Fall Meeting, Chicago, IL, November 2003.
23. A. LaLonde, D. Emelander, J. Jeannette, C. Larson, W. Rietz, D. Swenson, D.W. Henderson, J. Electron. Mater. **33**, 1545 (2004)
24. L.P. Lehman, S.N. Athavale, T.Z. Fullem, A.C. Giamis, R.K. Kinyanjui, M. Lowenstein, K. Mather, R. Patel, D. Rae, J. Wang, Y. Xing, L. Zavalij, P. Borgesen, E.J. Cotts, J. Electron. Mater. **33**, 1429 (2004)
25. D.W. Henderson, J.J. Woods, T.A. Gosselin, J. Bartelo, D.E. King, T.M. Korhonen, M.A. Korhonen, L.P. Lehman, E.J. Cotts, S.K. Kang, P. Lauro, D.-Y. Shih, C. Goldsmith, K.J. Puttlitz, J. Mater. Res. **19**, 1608 (2004)
26. M.E. Loomans, M.E. Fine, Metall. Mater. Trans. A **31A**, 155 (2000)
27. F.R. Mollard, M.C. Flemings, TMS AIME **239**, 1526 (1967)
28. D.G. Mccartney, J.D. Hunt R.M. Jordan, Metall. Trans. 11A (1980)
29. T. Himemiya, T. Umeda, Mater. Trans. JIM **40**, 665 (1999)
30. J.D. Hunt, K.A. Jackson, Trans. TMS AIME **236**, 843 (1966)
31. W.A. Tiller, K.A. Jackson, J.W. Rutter, B. Chalmers, Acta Metall. **1**, 428 (1953)
32. A. Rosenberg, W.G. Winegard, Acta Metall. **2**, 342 (1954)
33. K.S. Kim, S.H. Huh, K. Suganuma, Mater. Sci. Engin. A **A333**, 106 (2002)
34. Telang A.U., Bieler T.R., Choi S., Subramanian K.N. (2002) J. Mater. Res. 17: 2294
35. A.U. Telang, T.R. Bieler, J.P. Lucas, K.N. Subramanian, L.P. Lehman Y. Xing, E.J. Cotts, J. Electron. Mater. **33**, 1412 (2004)
36. E.O. Hall, Twinning and Diffusionless Transformations in Metals. (Buttersworth Scientific Publications, London, UK, 1954)
37. E. Teghtsoonian, B. Chalmers, Can. J. Phys. **30**, 388 (1952)
38. J.W. Christian, The Theory of Transformations in Metals and Alloys. (Pergamon Press, Oxford UK, 1965)
39. F.A. Crossley, L.F. Mondolfo, Trans. TMS AIME **191**, 1143 (1951)
40. M. McCormack, S. Jin, G.W. Kammlott, H.S. Chen, Appl. Phys. Lett. **63**, 15 (1993)
41. M. McCormack, S. Jin, J. Electron. Mater. **23**, 635 (1994)
42. S.K. Kang, D.-Y. Shih, D. Leonard, D.W. Henderson, T. Gosselin, S.-I. Cho, J. Yu, W.K. Choi, J. Metals **56**, 34 (2004)
43. I.E. Anderson, B.A. Cook, J. Harringa, R.L. Terpstra, J. Electron. Mater. **31**, 1166 (2002)
44. Z.G. Chen, Y.W. Shi, Z.D. Xia, Y.F. Yan, J. Electron. Mater. **31**, 1122 (2002)
45. P. Harris, Surface Mount Tech. (U.K.), 11 (1999) 46.
46. J.S. Kirkaldy, D.J. Young, Diffusion in the Condensed State. (Inst. Of Metals, London UK, 1985)
47. K.-L. Lin, H.-M. Hsu, J. Electron. Mater. **30**, 1068 (2001)
48. C. Wagner, J. Electrochem. Soc. **103**, 571 (1956)
49. C.-M. Chuang, K.-L. Lin, J. Electron. Mater. **32**, 1426 (2003)

Development of Sn–Ag–Cu and Sn–Ag–Cu–X alloys for Pb-free electronic solder applications

Iver E. Anderson

Published online: 27 September 2006
© Springer Science+Business Media, LLC 2006

Abstract The global electronic assembly community is striving to accommodate the replacement of Pb-containing solders, primarily Sn–Pb alloys, with Pb-free solders due to environmental regulations and market pressures. Of the Pb-free choices, a family of solder alloys based on the Sn–Ag–Cu (SAC) ternary eutectic ($T_{\mathrm{eut.}}$ = 217°C) composition have emerged with the most potential for broad use across the industry, but the preferred (typically near-eutectic) composition is still in debate. This review will attempt to clarify the characteristic microstructures and mechanical properties of the current candidates and recommend alloy choices, a maximum operating temperature limit, and directions for future work. Also included in this review will be an exploration of several SAC + X candidates, i.e., 4th element modifications of SAC solder alloys, that are intended to control solder alloy undercooling and solidification product phases and to improve the resistance of SAC solder joints to high temperature thermal aging effects. Again, preliminary alloy recommendations will be offered, along with suggestions for future work.

1 Introduction

The replacement of Sn–Pb solders for assembly of electronic systems is being driven by impending environmental regulations [1–3] and global market pressures to utilize Pb-free solders. During this major transition involving substitution for a near-universal joining material system, eutectic or near-eutectic Sn–Pb solder, there is also the opportunity to make a major improvement in joint reliability for challenging operating environments, i.e., high temperatures and stress levels, as well as impact loading situations. To help realize this opportunity, investigations into a promising alloy "family" of eutectic and near-eutectic Sn–Ag–Cu (SAC) solders [4, 5] have increased at many different laboratories, worldwide [6–9]. Since the Sn–Ag–Cu ternary eutectic temperature of 217°C is significantly higher than 183°C for Sn–37Pb (wt.%), it should be said that Pb-free SAC solder is not a "drop-in" replacement for Sn–Pb in the highly refined electronics assembly sequence, e.g., for surface mount technology (SMT) [2, 9]. While the enhanced reflow temperatures of 235–255°C for SAC near-eutectic solder, compared to approximately 220°C for Sn–Pb eutectic, are within the capabilities of current reflow ovens, some of the component packages and circuit board materials, primarily polymeric, need to be upgraded to withstand these higher processing temperatures. In addition, the transition must be made to an alternative Pb-free component lead coating to replace Sn–Pb. Also, a Pb-free high/low temperature solder hierarchy pair, to replace Sn–95Pb/Sn–37Pb, needs to be developed that includes (presumably) SAC solder and another Pb-free alloy to permit effective multi-chip module assembly [2, 9]. In spite of these complications, the electronics industry has seized the challenge and is proceeding forward rapidly to develop the assembly techniques and to generate the reliability data for Sn–Ag–Cu as a preferred Pb-free solder in many electronic assembly applications [2, 9].

I. E. Anderson (✉)
Ames Laboratory (USDOE), Iowa State University,
222 Metals Development Building, Ames, IA 50011, USA
e-mail: andersoni@ameslab.gov

Compared to Sn–Pb solders that have been limited typically to low stress joints and reduced temperature service because of the soft Pb phase that is prone to coarsening and ductile creep failure [10], the high Sn content and strong intermetallic phases of a well designed Sn–Ag–Cu alloy solder can promote enhanced joint strength and creep resistance [11], and can permit an increased operating temperature envelope for advanced electronic systems and devices. Results of SAC alloy development have demonstrated increased shear strength at ambient and elevated temperatures, e.g., 150°C, resistance to isothermal fatigue [12], and resistance to thermal aging during temperature excursions up to about 150°C, the current test standards for under-the-hood automotive electronics [13]. In terms of mechanical properties, shear strength [14], rather than joint tensile strength, will be the focus of the property characterization results that are included in this review due to the importance of shear failures in electronic systems with mismatched coefficients of thermal expansion [2, 15]. Also, early tensile failure studies showed that essentially all solder joints with Sn-based solders would exhibit the same type of localized parting at fairly low stress of the solder matrix and the Cu_6Sn_5 intermetallic layer, with little possibility of discriminating between solder alloy effects [10]. Although important for reliability testing of specific electronic assembly applications [16], thermal–mechanical fatigue analysis results also will not be included in this review since many of the novel SAC alloying concepts, with some exceptions [16], have not been tested for TMF, but typically have been tested in shear. In the expanding world of portable electronics and miniaturization of electronic devices, the ability of circuitry to remain undamaged after a drop impact [9, 17–21] is becoming another key objective for SAC solder joint microstructure design. Modified Izod impact test results will be reviewed because such impact testing is simple to practice and very useful for quantitative ranking in alloy design and thermal aging studies [17, 18], compared to board level drop testing [20].

Superior levels of these solder joint mechanical properties can be accomplished by microstructural control approaches, starting with the as-solidified solder joint [8, 22–28], i.e., tailoring of the as-solidified intermetallic interface with the substrate and controlling of solidification nucleation for the solder matrix. While substrate/solder interface tailoring is practiced by alloy additions to the SAC solder or metallization (coating) of the substrate, the issue of nucleation control for the solder matrix has been addressed by both SAC alloy variations and by fourth element additions. It should be noted that the solidification microstructures of Sn–Pb eutectic and near-eutectic solders are not nearly as sensitive to cooling rate and composition variations as the near-eutectic Sn–Ag–Cu solders [9]. This sensitivity is due primarily to the characteristic of Sn and Sn-enriched alloys for high undercooling prior to solidification [9]. It should be noted that a significant portion of the fundamental research on microstructural control of SAC solder joints has been done for the general case of bonding to Cu conductors and this review will be limited to the Cu substrate case. The common industrial case of solder bonding to one of several types of multi-layer metallization surfaces and the consequences of multiple reflow cycles has also been studied in detail, but has been covered in other reviews [29–34].

Another focus of this review will be on microstructural control for the thermally aged solder matrix/intermetallic interface region of SAC solder joints, in particular. Motivation for these recent studies was provided [20, 35, 36] by reported problems with pore development/coalescence and brittle fracture along the intermetallic (Cu_3Sn) interface with a Cu substrate. Recent studies [9, 35, 36] on accelerated aging of SAC alloy joints report that the suppression of voids, and especially suppression of void coalescence, in the intermetallic interfacial region is promoted by fourth element additions, eliminating an apparent embrittlement precursor in coarsened SAC solder joints. Interestingly, some of the same fourth element additions have been useful for both nucleation control and for resistance to aging effects.

2 Ternary Pb-free solder joint microstructures and mechanical properties

A closely related set of Sn–Ag–Cu near-eutectic alloys have risen through the ranks of many experimental studies to stand as the most likely candidates for widespread replacement of Sn–Pb solders [9, 26, 37]. As an improvement over the previous Sn–Ag eutectic solder in Pb-free electronic assembly applications, the Sn–Ag–Cu solders offer a reduced melting temperature (about 4°C lower) and additional tolerance for variations in cooling rate after reflow [9, 26, 37]. Prior calorimetric studies [26] revealed the very similar melting behavior of several closely related Sn–Ag–Cu alloys, consistent with the phase diagram studies [6] on this system. Compared to other common choices, the Cu alloy addition to Sn–Ag is abundant and low cost, e.g., lower cost than In, is an aid to wetting and does not increase drossing, unlike Zn [29], is compatible

with common no-clean paste fluxes [38], and is not a by-product of Pb mining, e.g., unlike Bi and Sb [29].

As an indication of the Sn–3.5Ag (wt.%) baseline solder alloy joint characteristics, Fig. 1a shows a low magnification optical micrograph of an etched cross-section [39]. Sn dendrites appear to extend directly across the full 70 μm width of this joint with a secondary dendrite arm spacing (SDAS) of about 5 μm. Optical microscopy seems to be better suited than SEM for revealing a long range dendritic solidification morphology in these joints, clarifying the distinctions between cells and dendrites [39]. The typical Cu_6Sn_5 intermetallic phase at the interface of the Cu substrate and the solder matrix exhibits some faceting and a thickness of approximately 2 μm, as revealed in Fig. 1b, consistent with previous results [26].

2.1 Solidification microstructures and shear strength for near-eutectic SAC solders

In comparison to Sn–Pb solders, a successful Sn–Ag–Cu alloy solder should exhibit enhanced shear strength at ambient temperature. Perhaps more importantly, the Sn–Ag–Cu solder should display improved creep strength and resistance to thermal–mechanical fatigue during temperature excursions up to about, e.g., 150°C, the current maximum operating temperature limit of under-the-hood automotive electronics [13]. This can be accomplished by microstructural strengthening approaches for the solder matrix. Essentially, the matrix strengthening strategy is to avoid or refine Sn dendrites, to minimize or refine any pro-eutectic intermetallic phases, and to increase the fraction of fine eutectic in the as-solidified joint [40]. To implement the matrix strengthening approach, four Sn–Ag–Cu solder alloys were selected for a previous study [39] which fall in a composition region that includes and closely surrounds the calculated ternary eutectic composition [6] of Sn–3.7Ag–0.9Cu (wt.%), as shown in Fig. 2. Two

alloy choices, Sn–3.0Ag–0.5Cu (outlined by a triangle in Fig. 2) and Sn–3.9Ag–0.6Cu (outlined by a circle in Fig. 2), were chosen to correspond to alloys being studied actively by JEIDA and NEMI [37], respectively. Another alloy, Sn–3.6Ag–1.0Cu (outlined by a hexagon in Fig. 2), was included to provide a correspondence to other previous studies [26]. The melting behavior of three of the four alloys (excluding Sn–3.9Ag–0.6Cu) was also studied by differential thermal analysis [26] and was determined to be closely related, with a clear eutectic melting onset at 217°C and a slightly extended liquidus melting signal, consistent with near-eutectic alloys.

It should be noted that the solidification microstructures of other ternary Sn–Ag–Cu alloys certainly have been studied and some of these will be cited in this review. However, this set of solder alloy joints were prepared under identical conditions and were intended as a representative survey of the different relevant regions of the ternary phase diagram that are likely to be used for solder applications. Although a hand soldering technique [39] was used, the temperature ramp, peak temperature (255°C), cooling conditions (1–3°C/s), and Cu substrates were intended to simulate typical reflow parameters for common circuit assembly practice, to provide a comparison of the microstructure results with practical technological value.

The optical micrographs [39] of Fig. 3 show the significant effects on the as-solidified joint microstructures of fairly minor variations in Ag and Cu content in the near-eutectic Sn–Ag–Cu solder alloys used to make the joints. The rather coarse Sn dendrites for Sn–3.0Ag–0.5Cu in Fig. 3a, with a secondary dendrite arm spacing (SDAS) of about 6–10 μm, can be compared to the extremely fine Sn dendrites for Sn–3.9Ag–0.6Cu in Fig. 3b, with an SDAS of about 2 μm. Also, the Sn dendrite pattern associated with Sn–3.7Ag–0.9Cu in Fig. 3c has a very similar spacing to the Sn dendrites of Fig. 2a, associated with Sn–3.5Ag. In contrast, the

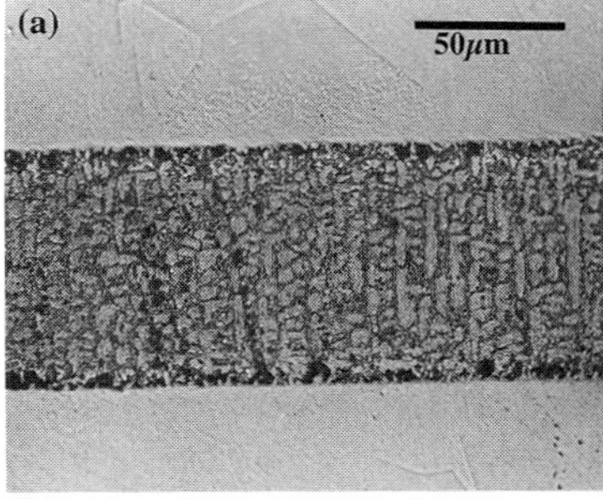
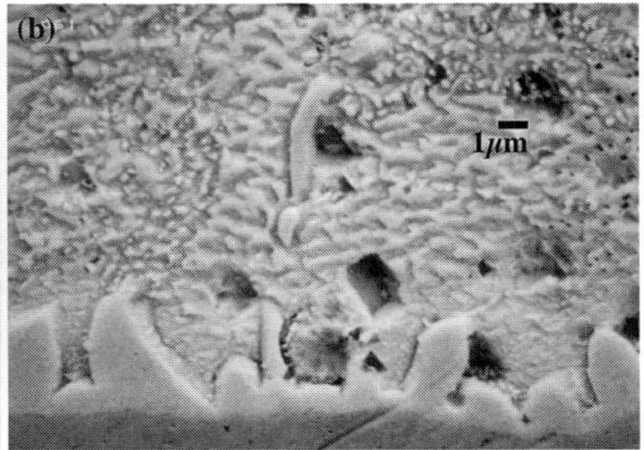

Fig. 1 Cross-section microstructure of as-solidified solder joints made from Sn–3.5Ag (wt.%), including, (**a**) optical micrograph of joint with Cu substrate on top and bottom, (**b**) SEM micrograph (backscattered electron imaging-BEI) with Cu substrate on bottom [39]

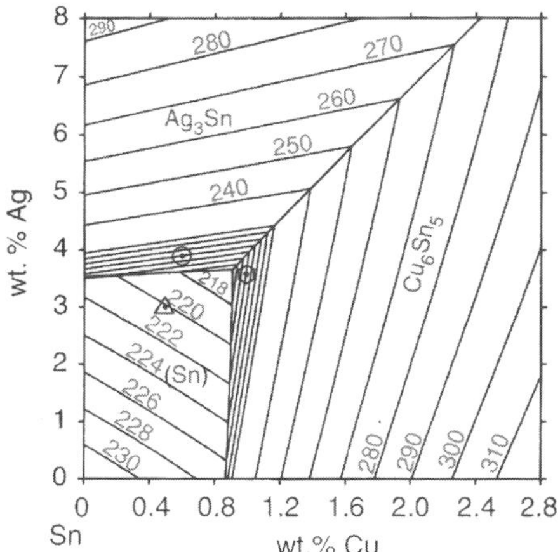

Fig. 2 The composition region that closely surrounds the calculated ternary eutectic composition of Sn–3.7Ag–0.9Cu (wt.%), as seen from the calculated liquidus surface [6, 39]

solidification morphology associated with Sn–3.6Ag–1.0Cu in Fig. 3d does not appear dendritic, certainly not with any extended Sn dendrites. Moreover, the large Ag$_3$Sn "needle" phases in Fig. 3b, as confirmed by energy dispersive spectroscopy (EDS) in the SEM, that project from the upper intermetallic layer into the solder matrix, distinguished this Sn–3.9Ag–0.6Cu microstructure from the others in Fig. 3.

The higher magnification and atomic number contrast of the SEM-BEI micrographs [39] in Fig. 4 provide additional information on the as-solidified joint microstructures shown in Fig. 3. For example,

from Fig. 4d it seems that, rather than Sn, Cu$_6$Sn$_5$ is the apparent primary (pro-eutectic) phase in the solidification of Sn–3.6Ag–1.0Cu solder, as confirmed by EDS in the SEM, followed by solidification of an extremely fine ternary eutectic. Also, the coarser Sn dendrite patterns of Fig. 4a (Sn–3.0Ag–0.5Cu) and Fig. 4c (Sn–3.7Ag–0.9Cu) allow for wider regions of ternary eutectic, compared to Fig. 4b (Sn–3.9Ag–0.6Cu). An interesting difference in the intermetallic at the Cu/solder interface is also apparent, where the partially faceted Cu$_6$Sn$_5$ fingers of Fig. 4b (Sn–3.9Ag–0.6Cu) and 4c (Sn–3.7Ag–0.9Cu) contrast with the finely spaced rounded stubs of Fig. 4d (Sn–3.6Ag–1.0Cu) and the widely spaced projections of Fig. 4a (Sn–3.0Ag–0.5Cu).

Although the ternary eutectic alloy, Sn–3.7Ag–0.9Cu, calculated from experimental and thermodynamic data [6] would be expected to have an equilibrium structure that consists of mixture of three phases: β-Sn, Ag$_3$Sn, and Cu$_6$Sn$_5$, with volume fractions consistent with the phase diagram [6], the actual joint microstructure in Figs. 3c and 4c is more complex. It appears that the joint made from Sn–3.7Ag–0.9Cu exhibits Sn primary dendrites with a fine uniform ternary eutectic structure in the interdendritic regions. If the calculated ternary eutectic composition is correct, the appearance of primary Sn dendrites in this alloy is evidence for some significant undercooling of the solder joint [6], where the Sn phase probably solidified at a temperature beneath the coupled eutectic growth region [41]. Some studies on undercooling of bulk samples of similar solder alloys, along with pure Sn and Pb, demonstrated that a Sn–3.8Ag–0.7Cu near-eutectic alloy can under-

Fig. 3 Optical micrographs of as-solidified solder joints made from (**a**) Sn–3.0Ag–0.5Cu, (**b**) Sn–3.9Ag–0.6Cu, (**c**) Sn–3.7Ag–0.9Cu, and (**d**) Sn–3.6Ag–1.0Cu [39]

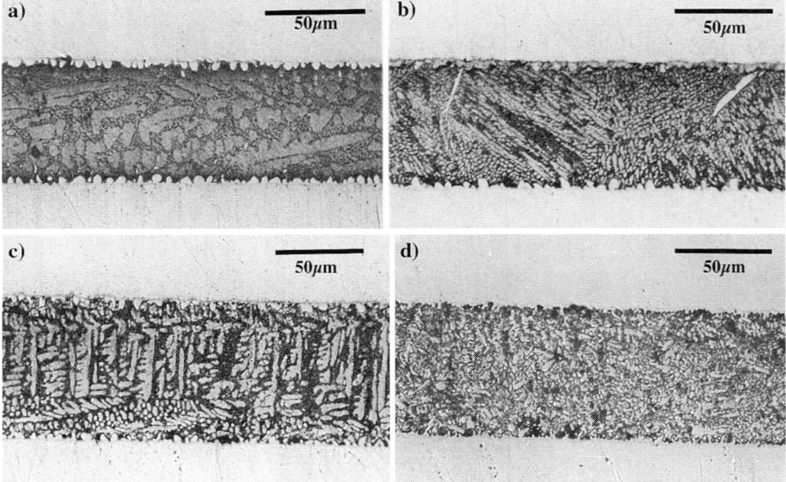

Fig. 4 SEM micrographs (BEI) of as-solidified solder joints made from (**a**) Sn–3.0Ag–0.5Cu, (**b**) Sn–3.9Ag–0.6Cu, (**c**) Sn–3.7Ag–0.9Cu, and (**d**) Sn–3.6Ag–1.0Cu [39]

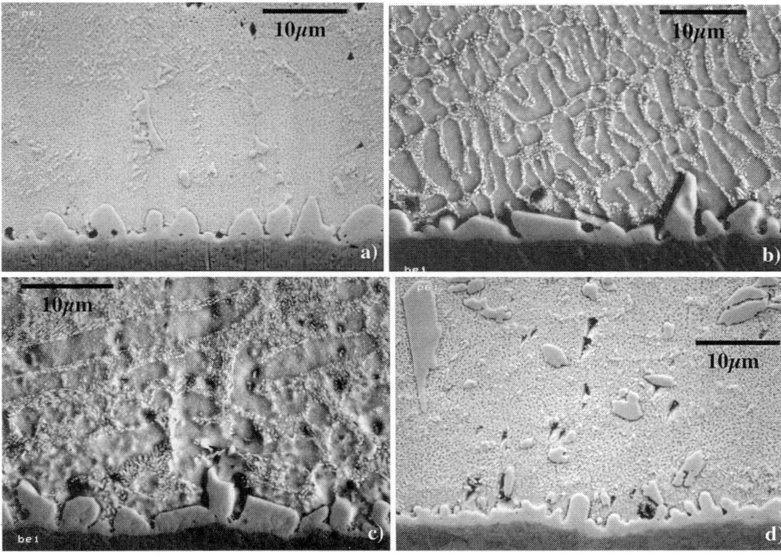

cool nearly 30°C before nucleation of solidification and other similar SAC alloys had undercoolings from 18°C to 35°C [42]. Moreover in the Sn–Ag–Cu (SAC) alloy system, both intermetallic phases in the eutectic, Ag_3Sn and Cu_6Sn_5, are known to be faceted [41], which would be expected to skew the coupled eutectic growth region, promoting primary Sn phase nucleation and growth for a eutectic alloy even at modest undercoolings [6].

The tendency for high undercooling and Sn dendrite formation of the ternary eutectic SAC alloy can be contrasted with the observed solidification behavior for the Sn–Pb solders, which do not undercool appreciably, only 3–4°C, and exhibit a fairly uniform coupled (irregular) eutectic structure over a wide range of cooling rates [9]. Generally, in Sn–Pb alloys an increased cooling rate during solidification results in a refined eutectic microstructure with enhanced strength, in a continuous trend. For eutectic and most near-eutectic SAC solder alloys, instead of coupled eutectic solidification, melt undercooling and the nucleation and growth of Sn dendrites is a less controlled phenomenon. However, it is generally true that rapid cooling of SAC alloys (>5–10°C/s) typically will result in a highly refined microstructure with fine dendrites and interdendritic ternary eutectic that has significantly enhanced strength [9]. On the other hand, slow cooling of SAC alloys (<1–3°C/s) will result in Sn dendrites, probably with a larger dendrite arm spacing, and interdendritic eutectic, but formation of massive

pro-eutectic intermetallic phases may precede Sn nucleation, as will be discussed below. Thus, slow-cooled SAC solder joints may be more ductile or more brittle, depending on whether massive intermetallic phases are able to nucleate and grow.

To illustrate the composition dependence of these SAC solidification characteristics, three near-eutectic SAC alloy were used to make solder joints that were all cooled at the same rate, 1–3°C/s from the same reflow peak temperature, 255°C. Melt undercooling and Sn dendrite nucleation and growth appear to have affected the solidification of two of the three near-eutectic alloys shown above, Sn–3.0Ag–0.5Cu and Sn–3.9Ag–0.6Cu in different ways, but not to the extent that it changed the equilibrium primary (pro-eutectic) phases expected from the calculated phase diagram in Fig. 2 [39]. In other words, the Sn dendrites in Fig. 3a are the predicted primary phase in Sn–3.0Ag–0.5Cu and the coarse dendrite arm spacing implies that they grew at a reduced velocity, consistent with a reduced undercooling [39]. In addition, the Ag_3Sn primary phase plates in Fig. 3b also are the predicted primary (pro-eutectic) phase for Sn–3.9Ag–0.6Cu, but the very fine Sn dendrites in the majority of the solder matrix are consistent with rapid growth from a highly undercooled melt. Although the undercooling could not be estimated qualitatively from the microstructure for the Sn–3.6Ag–1.0Cu alloy in Figs. 3d and 4d, the lack of Sn dendrites and the formation of primary (pro-eutectic) Cu_6Sn_5 phase particles, apparently randomly dispersed

 Springer

in a ternary eutectic matrix, is expected from the phase diagram [39].

Based on this representative set of microstructure observations, it is useful to compare the criteria described above for ideal microstructural strengthening of the solder matrix to the results for the as-solidified solder joints. To restate, the matrix strengthening strategy is to avoid or refine Sn dendrites, to avoid or refine any pro-eutectic intermetallic phases, and to increase the fraction of fine eutectic in the as-solidified joint. Of the four choices above, including the calculated eutectic composition, Sn–3.7Ag–0.9Cu, the most promising appears to be Sn–3.6Ag–1.0Cu which has a joint microstructure that is dominated by a highly refined ternary eutectic and is decorated by small (10 μm or less) pro-eutectic Cu_6Sn_5 phase regions. In terms of microstructural strengthening, this type of as-solidified microstructure would be expected to have an enhanced shear strength that is dominated by the fine eutectic and would tend to deform uniformly if stressed beyond the yield point. In contrast, a joint formed from Sn–3.0Ag–0.5Cu solder with coarse Sn dendrites and interdendritic ternary eutectic would be expected to have a reduced shear strength that is dominated by the low yield point of the nearly pure Sn phase and would deform in a ductile manner within the dendritic regions.

A summary [39] of the ambient temperature shear strength results for all of the alloys that are illustrated above are shown in Fig. 5, along with the results from two modified SAC alloys containing either Co or Fe additions that will be discussed below. An indication of the repeatability of the maximum shear strength values is given by the narrow range (± 2.5–5 MPa) of the standard deviation of the measurements, where at least seven specimens were tested for each type of solder joint. The weakest of the solder joints was made from the Sn–3.0Ag–0.5Cu, while the strongest SAC solder joints were made from Sn–3.6Ag–1.0Cu, consistent with the predictions of the microstructural analysis from above. The shear strength of the baseline Sn–3.5Ag solder joints fell in the middle of the range of values. This apparent strengthening may be due to the sensitivity of the Sn–3.5Ag solder microstructure to reflow and solidification conditions and the Cu dissolution phenomenon, as discussed below [26, 43–45]. The difference between the maximum shear strength and the apparent yield strength is an indication of the reasonable ductility exhibited by these samples, although the yield strength values had an increased standard deviation. It is also interesting to note that the joints made with Sn–3.6Ag–1.0Cu also exhibited the highest yield strength with the lowest scatter of the data, suggesting a very consistent microstructure.

In addition to selection of a SAC solder alloy based on as-solidified joint strength, where increased strength is preferred within reasonable limits [16], another criteria should be the type of joint failure mechanism, where uniform yielding and ductile failure are preferred. It is important to note that the solidification of relatively large plates of Ag_3Sn in a Sn–Ag–Cu solder microstructure was shown to be particularly detrimental to the fracture behavior and thermal-fatigue life of solder joints, promoting brittle failure along interfaces between Ag_3Sn and β-Sn at reduced stress and shorter joint life [42]. Motivated by these observations, an extensive effort has been devoted to improved SAC solder alloy design to eliminate the occurrence of Ag_3Sn pro-eutectic nucleation [9, 42], leading to a patent [46]. This effort quantified the

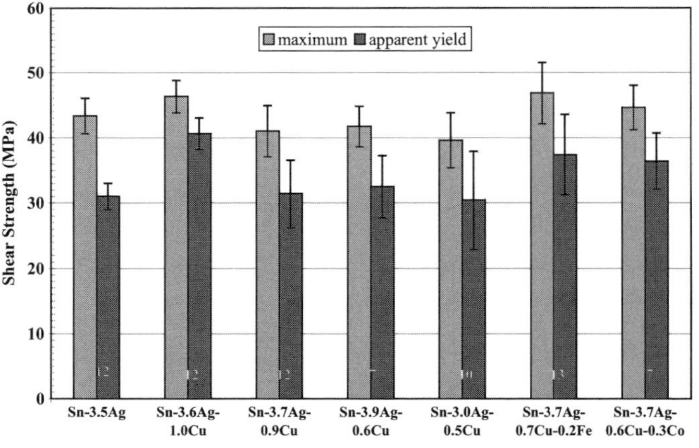

Fig. 5 Summary of asymmetric four point bend (AFPB) shear strength tests at ambient temperature on as-solidified solder joints made from the alloys listed along the x-axis. The number of repeat samples for each alloy is indicated in light contrast on each set, along with the standard deviation of the values at the top of each bar [39]

effect of independent variations of Ag and Cu content on the formation of Ag$_3$Sn and used 3.8Ag and 0.9Cu as maximum levels of each element [42]. In one part of the study, the composition dependent solidification of independent solder alloy balls was characterized and the same solder balls were also solidified between Cu substrates to form joints. From the results, some general recommendations were to utilize a Ag content of 2.0–3.0 wt.% and a Cu content of about 0.7 wt.%, which would suppress Ag$_3$Sn formation and still maintain a pasty range (difference between liquidus and solidus temperatures) less than 3.5°C, i.e., near-eutectic behavior [42]. For example, one SAC alloy mentioned specifically was Sn–2.7Ag–0.7Cu as representing an upper limit (for Ag content) for possible formation of Ag$_3$Sn plates, even at extremely slow-cooled conditions (0.02°C/s) [42]. From the position of this alloy on the liquidus surface of the calculated phase diagram (see Fig. 2), the solidified microstructure of a joint prepared from Sn–2.7Ag–0.7Cu in the same manner as the ones shown in Figs. 2–4 should be similar to Figs. 3a and 4a that were made from Sn–3.0Ag–0.5Cu.

2.2 Preliminary recommendation for SAC solders

Thus, to avoid the formation of Ag$_3$Sn plates according to the findings of the previous extensive study [9, 42], one must choose a solder joint of reduced strength, perhaps Sn–2.7Ag–0.7Cu. Alternatively, to avoid nucleation of massive Ag$_3$Sn plates and to maintain enhanced solder joint strength, a SAC composition of Sn–3.6Ag–1.0Cu could be picked [39], which exhibits pro-eutectic nucleation of a well distributed Cu$_6$Sn$_5$ phase, followed by ternary eutectic solidification, effectively precluding nucleation of massive Ag$_3$Sn plates and long range Sn dendrites. However, it is recommended that future work with this type of SAC alloy should focus on a small composition region with slightly lower Ag and Cu, perhaps Sn–3.5Ag–0.95Cu. The advantage could be a similar joint solidification path, a reduced amount of pro-eutectic Cu$_6$Sn$_5$, and a significantly smaller pasty range. According to the calculated liquidus surface in Fig. 2, Sn–3.6Ag–1.0Cu has a pasty range of about 5°C, compared to about 3°C for Sn–3.5Ag–0.95Cu, assuming a solidus temperature that is equivalent to the ternary eutectic of 217°C. If this apparent mechanism is used properly, it should be possible to generate a very high fraction of ternary eutectic structure from a very slightly off-eutectic

composition by accepting the inclusion of a small fraction of well-distributed Cu$_6$Sn$_5$ pro-eutectic phase.

It should be noted also that the effect of Cu dissolution from the substrates on the solder matrix alloy composition is not included in this discussion. Thus, the possible precision for adjustment of Cu content in some SAC alloy solders is open to question, i.e., precise control of the final joint composition. A previous estimate of the Cu dissolution effect was an increase of roughly 0.5% Cu to the solder joint above the initial alloy composition, depending on the reflow conditions [47]. Some recent studies that involve the Cu pick-up phenomenon in Sn-enriched solder alloys have been reported [36, 45], but the difficulty of probing this effect is well recognized, especially by micro-analytical techniques [48]. One data point for Cu solubility in Sn is the maximum level (2.2 wt.%) that can be retained by rapid quenching, followed by cryogenic X-ray diffraction of the structure [49]. Vianco reported [36] that a maximum local concentration enhancement of about 2% was measured by a wavelength dispersive spectroscopy (WDS) line scan method with an electron microprobe using a highly focused beam method, after soldering of Cu substrates with Sn–3.9Ag–0.6Cu by reflowing at a peak temperature of 260°C. This local maximum Cu content was reported to decrease continuously from the inner edge of the typical Cu$_6$Sn$_5$ interfacial layer toward the center of the joint, proceeding from about 2% to a baseline (nominal) value of about 0.6% over a distance of about 10 μm [36]. Also, the existence of interior Cu$_6$Sn$_5$ phases (within the joint matrix) was also detected in the Cu and Sn line scans, as peaks and valleys, respectively, by Vianco [36] and others [35, 40]. In a microstructural sense, to measure the true pick-up of Cu into the joint matrix region one must include excess Cu dissolved in Sn and the interior content of Cu$_6$Sn$_5$, both as independent phases and as part of the ternary eutectic structure. A broad beam WDS method, most common in analysis of geological specimens [50, 51], has recently been applied to this difficult problem [45]. One encouraging result from this study is that, although low Cu content SAC alloys, e.g., Sn–3.0Ag–0.5Cu, may pick up 0.5–0.7Cu, a solder joint made with Sn–3.7Ag–0.9Cu did not have any significant change in Cu content [45]. Thus, the prospect of closely controlling the final composition (Cu content) of a joint made with Sn–3.5Ag–0.95Cu appears more likely than that of a joint made with a SAC alloy containing only 0.5–0.6Cu.

2.3 Thermally aged microstructures and shear strength for near-eutectic SAC solders

The demand for some electronic assemblies to remain reliable for extremely long times in harsh environments, particularly at high temperatures, is an accelerating trend, e.g., the desired positioning of automotive control systems immediately adjacent to the internal combustion engines that they control [16]. In other words, in these situations it is highly desirable for a Pb-free solder to exhibit increased strength and sufficient ductility, compared to Sn–Pb eutectic solder, at temperatures in the neighborhood of 150°C, as mentioned above [13, 16]. To provide a preliminary test of high temperature performance of a partial set of the SAC solder joints that are included in the discussion above, shear testing of such solder joint samples was conducted at 150°C [39]. A comparison of the maximum shear strength values at ambient temperature to the shear strength values at 150°C is shown in Fig. 6. The results of this elevated temperature test indicated a drop of at least half of the shear strength of the ambient temperature measurements to a range between about 15 MPa and 20 MPa [39]. Due to the increased standard deviation of the measurements, no significant differences can be observed between the elevated temperature shear strengths of the solder joints made from any of the alloys tested. Actually, the collapse of all of the 150°C shear strength values to a single range of 15–20 MPa suggests that the high temperature mechanical properties probably are controlled by the decreased strength of the Sn phase in the solder matrix [39], related to Sn grain growth and Sn inter-granular de-cohesion [52]. Any more detailed analysis of these results is hindered by the dynamic nature of microstructural coarsening that must have occurred during

these high temperature tests. Thus, it was decided to conduct future shear strength tests at ambient temperature after fixed periods of high temperature aging and to correlate these results to observations of the aged solder joint microstructures.

As mentioned above, ambient temperature shear strength provided a mechanical property test for ranking of solder joints at two different stages of accelerated (isothermal) aging at 150°C (approximately 0.86 T_m). A summary of the ambient temperature shear test results for the 100 h and 1,000 h aged solder joints of all seven alloys that were shown in Fig. 7, compared to the as-soldered shear strength [35, 53]. In both the as-soldered and 100 h aged conditions, the three weakest solder joints were made from Sn–3.9Ag–0.6Cu, Sn–3.7Ag–0.9Cu, and Sn–3.0Ag–0.5Cu, while the three strongest aged joints were made from Sn–3.7Ag–0.6Cu–0.3Co, Sn–3.6Ag–1.0Cu, and Sn–3.7Ag–0.7Cu–0.2Fe, with the joints made from Sn–3.5Ag in the mid-range of shear strength [35, 53]. After 1,000 h of aging the weakest group still retained the same ranking, but one of the strong alloys, Sn–3.7Ag–0.6Cu–0.3Co, moved to the mid-range and the Sn–3.5Ag results moved into the strong group. If all of the alloy results in Fig. 7 are averaged, the maximum shear strength of these Pb-free solder joints dropped about 20% from the as-soldered condition to the 100 h aged condition, and dropped only another 16% from the 100 h aged condition to the 1,000 h aged condition [35]. Thus, after 1,000 h of aging at 150°C, the average maximum shear strength of the solder joints made from all the alloys in this study retained about two thirds of the as-soldered maximum shear strength. In contrast to the previous 150°C shear testing [39], careful control of the thermally aged microstructures of these joint samples and the use of shear testing at

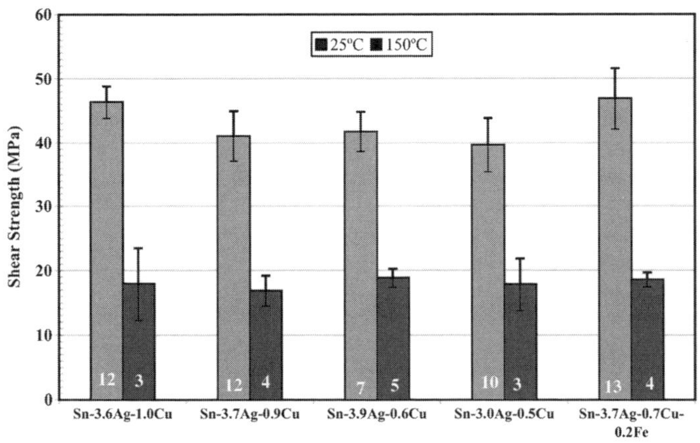

Fig. 6 Summary of asymmetric four point bend tests at ambient and elevated (150°C) temperature on solder joints made from the alloys listed along the *x*-axis. The number of repeat samples for each alloy is indicated in light contrast on each bar, along with the standard deviation of the values at the top of each bar [39]

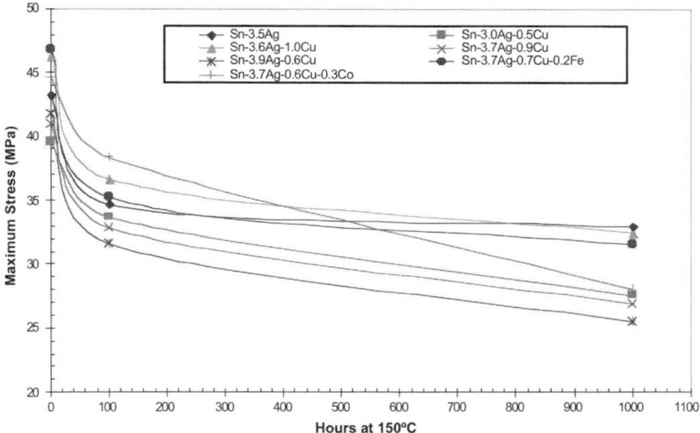

Fig. 7 Summary of maximum shear strength from AFPB tests at ambient temperature on as-soldered and aged solder joints made from the indicated alloys [35, 53]

room temperature provided an improved opportunity [35] to explore any measurable differences between this fairly representative set of SAC solder alloys, given the distinctions already noted in their as-solidified microstructures and properties [39, 53].

To more fully understand the results for maximum shear strength after aging, the load versus elongation data was analyzed and correlated with the corresponding solder joint microstructures and macroscopic joint failure characteristics [35]. Further distinctions between the joint sample microstructures were derived from analysis in both the as-aged and post-shear test conditions [35]. The strongest solder joints after 100 h aging, e.g., made from Sn–3.6Ag–1.0Cu, exhibited considerable ductility and plastic yielding before the maximum shear strength level and considerable plastic yielding after the maximum shear strength was reached (see Fig. 8a). The companion post-shear microstructure in Fig. 8b shows that a significant amount of homogeneous plastic flow had occurred in the solder matrix, suggested by the apparent bending and alignment of the elongated intermetallic phases with the plastic flow direction of the solder matrix [35].

On the other hand, the localized ductile shear behavior of the "weak" class of joint samples, represented by the results for Sn–3.7Ag–0.9Cu in Fig. 7, was typical of the less uniform ductile failure mechanism displayed by many of the 100 h aged specimens, both weak and strong [35]. The load versus shear elongation data in Fig. 9a, shows only a small amount of elastic deformation before plastic yielding at about 22 MPa and maximum shear strength at about 32 MPa. As reported by several investigators of shear failures in aged and as-solidified SAC solder joints [11, 54, 55], this non-uniform ductile shear displacement of the

solder joint matrix appears to be localized to a portion of the continuous Sn phase that is immediately adjacent to the Cu_6Sn_5 intermetallic interface (see Fig. 9b). There also appears to be some minor amount of intermetallic cracking across projections of the Cu_6Sn_5 and along the interfaces between the Cu_6Sn_5 and the Cu_3Sn layer, which is well known to form during initial aging at such high temperatures [36]. Thus, the failure of both strong and weak solder joints does not seem to involve fracturing of the Cu_3Sn layer, although linking of voids during crack propagation may contribute to weakening of the Cu_6Sn_5/Cu_3Sn interface [35].

In the initial analysis of solder joints after 1,000 h of aging at 150°C there was noted an additional loss of shear strength, averaged for all of the solder alloys studied, of approximately 16% beyond the initial drop after 100 h, as given in Fig. 7. Over the range of alloys studied and within the uncertainty of the intermetallic compound (IMC) layer measurement, neither the thickness of the Cu_3Sn layer nor the total thickness of both intermetallic layers, $Cu_3Sn + Cu_6Sn_5$, had any apparent relationship to the ranking of maximum shear strength after 1,000 h of aging [35], as reported in some studies [36]. However, in contrast to the as-solidified and 100 h aged results, two different types of mechanical deformation behavior were observed for each alloy in the 1,000 h aged results (e.g., Fig. 10a) in most cases, leading to a greater standard deviation of the strength data [35]. The strong joint exhibited almost the same maximum shear strength, ductility, and plastic yielding behavior as the 100 h aged sample in Fig. 8a. Also, the post-shear microstructure in Fig. 10b and c revealed that a significant amount of homogeneous plastic flow had occurred in the solder matrix, similar to Fig. 8b, along the entire length of the joint [35].

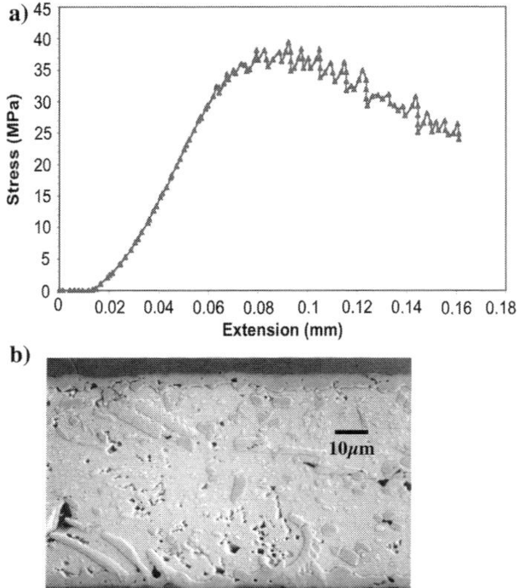

Fig. 8 Shear test (AFPB) results for a solder joint sample made from Sn–3.6Ag–1.0Cu that had been aged for 100 h at 150°C, showing: (**a**) the stress versus strain data, and (**b**) a post-AFPB microstructure of a central region of the joint in cross-section with the Cu substrates on the top and bottom [35]

On the other hand, the weaker example of the Sn–3.6Ag–1.0Cu joints (in Fig. 10a) exhibits a higher yield point than the joint made from Sn–3.7Ag–0.9Cu (see Fig. 9a), but has an immediate drop in load bearing capability, i.e., the maximum shear strength equals the yield strength [35]. This mechanical behavior is consistent with the obvious failure (see Fig. 10d and e) of a long region of the Cu_3Sn/Cu interface, shortly after the crack enters through the ductile matrix on both ends of the joint (see Fig. 10e). The micrograph in Fig. 10d, taken from the middle of the joint, shows no apparent signs of joint matrix flow or intermetallic cracking in addition to the obvious interface debonding at the bottom of the micrograph [35]. This extreme debonding example is actually fairly isolated and ductile failure of the SAC alloy joints is the dominant observation. In other words, localized ductile yielding was observed after aging for 1,000 h in greater than 65% (on average) of solder joint samples made from Sn–3.5Ag and all of the SAC alloys [35]. Figure 11 reveals that although the Sn–3.6Ag–1.0Cu solder joints had the most observations of partial interface debonding (half of the joint samples), the strength level of the joints that failed in a ductile manner was high enough to push it into the strong joint class of the 1,000 h aged samples (see Fig. 7). Figure 11 also indicates that all of the solder joints made from the Co- and Fe-modified SAC alloys exhibited [40] a fully ductile shear failure mode, as will be explained in the following section.

Fig. 9 Summary of the shear test (AFPB) results for a solder joint sample made from Sn–3.7Ag–0.9Cu that had been aged for 100 h at 150°C, showing: (**a**) the stress versus strain data, (**b**) a central region of the joint in cross-section that reveals the localized shear failure mode [35]

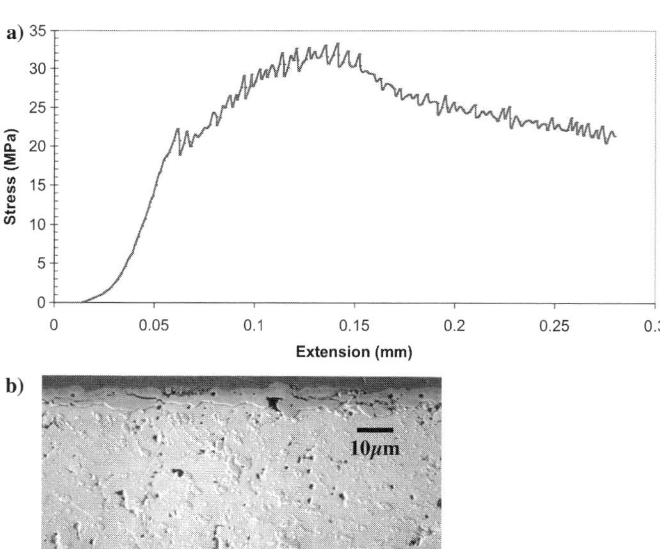

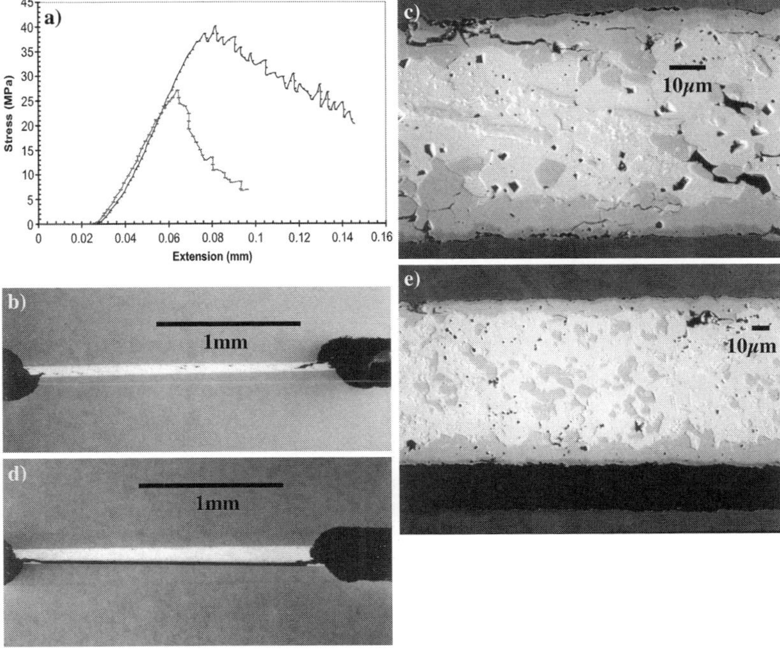

Fig. 10 A summary of the shear failure observations for two solder joints soldered from Sn–3.6Ag–1.0Cu, showing: (**a**) the shear stress versus elongation data for both joints, (**b**) a cross-section SEM micrograph of the strong joint, (**c**) an SEM macrograph of the full width of the strong joint, (**d**) a cross-section SEM micrograph of the weak joint, and (**e**) an SEM macrograph of the full width of the weak joint [35]

Since the first reports of the phenomenon of embrittlement in solder joints made from SAC solder alloys after prolonged high temperature aging in 2003 [56], these thermal aging effects have been the subject of some conflicting reports. Vianco et al. [36] reported on a thorough, well controlled study of Sn–3.9Ag–0.6Cu solder solidified on a Cu substrate, that was subject to thermal aging at several temperatures (70, 100, 135, 170, and 205°C) for a series of prolonged times (1–400 days) to study IMC formation and growth. Earlier studies [36] on thermal aging by the authors had included Sn–3.5Ag and Sn–0.5Ag–4.0Cu solders and pure Sn, also solidified on a Cu substrate. First, the Sn–3.9Ag–0.6Cu solder joint required an aging temperature of at least 135°C to develop a measurable layer of Cu_3Sn between the Cu substrate and the Cu_6Sn_5 IMC layer [36]. However, no void formation at the Cu/Cu_3Sn interface, a precursor observation to embrittlement [35], was detected in these samples even for aging times out to 350 days [36]. The onset of void formation at the Cu/Cu_3Sn interface was observed at 170°C after an aging time in excess of

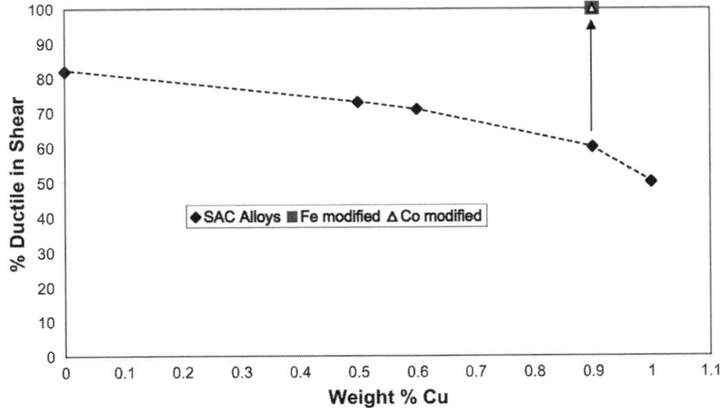

Fig. 11 Summary of the apparent failure mechanisms in the shear test results of the solder joints that were aged for 1,000 h at 150°C, as described in Fig. 7 [40]. Note that the arrow points to the data for the Fe- and Co-modified SAC solder joints

150 days (3,600 h) and was attributed to Kirkendall porosity [57], where longer times and higher aging temperatures promoted void growth and coalescence [36]. Although thermal aging at 150°C was not studied, the results for 170°C would have been expected to exhibit pore formation at a shorter aging time, presumably less than 1,000 h, to be completely consistent with the results in Fig. 11. However, the initial void formation is difficult to detect by microstructural observation alone, since metallographic preparation methods can produce anomalous porosity [35]. Also, shear testing was helpful in detecting the outcome of void formation and, especially, void coalescence, since the shear failure process tends to expose such joint weaknesses [35]. These reasons again may explain why a previous study [36] also failed to detect void formation at the Cu/Cu$_3$Sn interface in similar samples made from Sn–3.5Ag solder in variance with the results summarized in Fig. 11, which found that 20% of the Sn–3.5Ag joint samples exhibited noticeable embrittlement [35].

For completeness, another study of Sn–Ag–Cu solder joints with Cu also studied embrittlement induced by thermal aging at several temperatures (100, 125, 150, and 175°C) for 3–80 days [20], but did not specify the solder alloy or the reflow conditions of the solder ball samples. Interestingly, this study used a type of high strain rate (drop) tensile loading of components on a board and rapid tensile (pull) and shear testing of individual component (ball) joints to detect the onset of the embrittlement, along with microstructure analysis to observe void development associated with the IMC layers [20]. The authors reported a lack of correlation of the mechanical tests of individual ball joints with the board level failures due to the diversity of loading modes of the joints. However, they did find a consistent association between cross-section microstructure observation of advanced void coalescence at the Cu$_3$Sn/Cu interface and the percentage of ductile (bulk solder) failure on fracture surface of solder ball pull failures [20]. The void formation was attributed to the Kirkendall effect [57]. The associated data showed that aging conditions of at least 125°C and 20 days (480 h) were needed to promote void coalescence into continuous regions of interface separation and an obvious drop of joint ductility. This thermal aging threshold is lower than even the aging temperature (135°C) for detection of significant Cu$_3$Sn IMC layer formation by Vianco et al. [36], adding some uncertainty to the results, but the observation of Cu$_3$Sn/Cu void formation and coalescence is consistent with both the results in Fig. 11 and the microstructure observations of Vianco et al. [36].

2.4 Impact strength of SAC solder joints

In another study of embrittlement of solder joints on thermal aging [17, 18], Sn–9Zn and Sn–8Zn–3Bi solder balls were reflowed three times onto two different substrates, Cu and Au/Ni(P), and subsequently subject to thermal aging at 150°C for up to 1,000 h. While several aspects of this study were significantly different than the other SAC solder studies described above, the use of a well-instrumented (pendulum type) impact tester is an important advance in the quantitative testing of solder joints. The impact specimen was an individual solder ball sample that was joined to a substrate on the bottom only and the miniature "hammer" at the end of the pendulum struck the solder ball below the mid-point of the sphere, in a pseudo-Izod type of specimen geometry [17, 18]. Quantitative energy absorption values were obtained for each impact test and the substrate fracture surface of each specimen was analyzed to categorize the fracture mode, i.e., within the bulk (ductile), at the IMC interface (brittle), or mixed, and the results were strongly affected by void formation and IMC layer formation and growth after thermal aging [17, 18]. While the actual results are not important for this report, the use of an instrumented impact tester on an Izod-type specimen to investigate the resistance of a solder joint to impact loading, simulating an accidental dropping incident, is a very worthwhile experimental approach to study this type of reliability problem.

It should be noted that the miniature pendulum-type impact tester discussed above is a custom laboratory instrument [58] and is not broadly available. However, commercial impact test instruments, normally used to test small polymer specimens in an Izod configuration (e.g., Tinius-Olsen model 92-T), are available of a suitable load range to study small solder joint specimens of the type that were used for the asymmetric four point bend (AFPB) shear tests in Fig. 11 [59]. The rectangular specimens (40 mm × 3 mm × 4 mm) were notched only at the solder joint on the leading edge (facing the hammer impact point) and tested at room temperature, although an environmental chamber can be fitted to the instrument to permit testing at sub-ambient and elevated temperatures, similar to a recent Charpy impact study of bulk solder specimens [21]. Some very recent results of such Izod impact testing (see Fig. 12) were published on as-soldered SAC solder and SAC + X joints (based on Sn–3.7Ag–0.9Cu) with Cu [59], but more testing is on-going with thermally aged joint specimens of these SAC joints and some new SAC + X solder joint samples that will be discussed below.

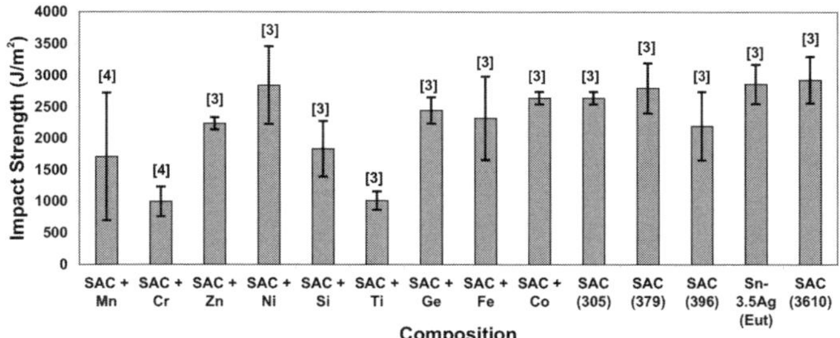

Fig. 12 Summary of the Izod impact strength results from the as-soldered joints in this study, where the standard deviation is superimposed on the bar for each alloy and the number of repeat samples is noted above each bar. The SAC (305) notation, for example, is for Sn–3.0Ag–0.5Cu [59]

2.5 Summary recommendation for SAC solders

In light of the expanding evidence of a reliability issue after extensive aging at high temperatures for solder joints made with SAC alloys, a modified recommendation should be provided beyond the SAC alloy composition, Sn–3.5Ag–0.95Cu, that was provided above. This composition seems preferred based on control of the joint solidification path, i.e., to avoid nucleation of massive Ag₃Sn plates and Sn dendrites and to promote formation of a maximum volume fraction of coupled ternary eutectic microstructure (with an enhanced shear strength) over a range of typical cooling rates. In addition to these highly desirable characteristics for widespread use, some limitations need to be placed on the upper operating temperature of such SAC solder joints. From analysis of all of the accumulated results on thermal aging, it is reasonable to recommend that SAC solder joints (with Cu substrates) of all common compositions should be used for applications with typical operating temperatures less than about 135°C. Actually, it is likely that a higher upper use threshold temperature may be more closely defined between 135°C and 150°C with more study. In fact, it is well known that substrate metallization layers, e.g., Ni or Ni(P) [29] can extend the upper use temperature and it was recently discovered that minor alloying of SAC solders, producing SAC + X solders, also can eliminate the embrittlement problem observed at 150°C out to at least 1,000 h of exposure, as described in the section below.

3 Quaternary Pb-free solder joint microstructures and mechanical properties

In the section above, a thorough review was presented of the use of composition modifications within the Sn–Ag–Cu ternary system to control the solidification path and product phases, especially in solder joints that join

Cu substrates. It was also stated that the least desirable solidification product phase seems to be pro-eutectic Ag₃Sn in the form of massive plates or "blades," since it has been identified as a severe limitation on the plastic deformation properties of SAC solder joints [8, 9]. Two basic SAC alloy design strategies to avoid this problem were presented involving either solute-poor (especially low Ag) or slightly hyper-eutectic (Cu-rich) ternary solder alloys. A second problem with SAC solder joints also was identified in the section above that was related to partial embrittlement of joints after extensive high temperature aging. Recognition of this problem lead to the additional recommendation that an upper limit on SAC solder joint temperature exposure be set at approximately 135–150°C, particularly for exposure times on the order of 1,000 h. Therefore, the logic can be seen clearly for the development of a third alloy design strategy involving minor alloy additions to a baseline SAC solder to permit both avoidance of Ag₃Sn pro-eutectic and high reliability use at increased temperatures, perhaps higher than 150°C for times exceeding 1,000 h. This section will present a brief review of the so-called "SAC + X" alloy design strategies and the effects of alloy additions on solidification (as-soldered) microstructures and properties, as well as the alloy design efforts targeted at the thermal aging problem.

3.1 Solidification microstructures for SAC + X solders

High undercooling was recognized as a common occurrence and a major influence on the solidification microstructure of Sn–Ag–Cu alloys in bulk melts during early work that explored the existence of a ternary eutectic reaction [4, 7, 29] and in subsequent phase diagram analysis work [6]. At the calculated ternary eutectic composition of Sn–3.7Ag–0.9Cu, the difficulty of nucleation and growth of a coupled ternary eutectic structure has prevented its observation in as-solidified

ingots and solder joints [6]. In fact, the most common as-solidified structure in near-eutectic solder ingots and solder joints (see Fig. 3a–c) is a divorced eutectic microstructure with Sn dendrites as the primary phase and an interdendritic ternary eutectic that freezes subsequently [6]. The high undercooling needed for nucleation of the β-Sn phase in common near-eutectic SAC alloys with reduced Cu ($\leq$0.9Cu) also presents the opportunity for nucleation of Ag_3Sn pro-eutectic plates at a reduced undercooling, producing an undesirable solder joint microstructure, as described above [8].

Thus, one approach to modification of SAC alloys with minor 4th element additions was to minimize the undercooling needed for nucleation of β-Sn phase. As a basis for this recent work, it was noted that low level additions (0.1 wt.%) of some elements, e.g., Zn, Al, Sb, and Bi, had been observed previously to catalyze β-Sn nucleation at significantly reduced undercooling in otherwise pure bulk Sn melts [9, 60]. Of these, SAC + Zn has been pursued aggressively [8] and a Zn addition appeared to promote solidification of coarse Sn dendrites at undercoolings of only 3–4°C in bulk alloy samples, compared to approximately 30°C for the unmodified Sn–3.8Ag–0.7Cu solder alloy. Moreover, when two levels of the Zn addition, 0.1 and 0.7Zn (wt.%), were added to the same SAC solder alloy, the complete elimination of Ag_3Sn plates was achieved for the higher Zn level when the solder was applied to a Cu substrate, even at a cooling rate as slow as 0.02°C/s [8]. Another type of minor 4th element addition, a rare earth (RE) like La or Ce, also seems to catalyze β-Sn (undercooling was not reported), but its primary function [61] was intended to be formation of RE-Sn intermetallic compounds at the Cu interface for improvement of creep properties [61].

Alternatively, some success at nucleation control of a SAC + X solder alloy was achieved by using a Hume-Rothery approach to design a series of possible 4th element additions to a SAC solder alloy, initially Sn–3.6Ag–1.0Cu, that was based on their ability to match closely ($<5\%$ mismatch) with the atomic size of Cu [26, 62]. Copper was selected as the object of substitutional alloying because a Cu_6Sn_5 intermetallic compound typically is formed first during SAC solder joint solidification on a Cu substrate, usually as a substrate interface layer. The original intent of the substitutional element in SAC + X was both to modify the as-solidified morphology of the initial Cu_6Sn_5 interfacial layer and to limit growth of the eventual $Cu_6Sn_5 + Cu_3Sn$ interfacial layers under thermal aging conditions [25, 26]. Of the original "X" elements tested, two levels of Co (0.15 and 0.45 wt.%) were added to Sn–3.6Ag–1.0Cu and were found to produce an unusual "corral-

like" intermetallic (apparently Cu_6Sn_5) interface when applied to a Cu substrate, as shown in Fig. 13 [26]. Other X additions to SAC + X solders that produce a similar effect on the intermetallic interface when applied to a Cu substrate include Fe, Ni, and Au [23, 25, 27, 28]. Solder joint solidification with SAC + Co also resulted in formation of massive pro-eutectic Cu_6Sn_5 phases within a complex ternary eutectic matrix. Elemental mapping by electron microprobe indicated that Co was segregated to the tips of some massive pro-eutectic Cu_6Sn_5 phase particles within the solder joint matrix, consistent with a nucleation catalysis effect [22, 26] for Cu_6Sn_5. Subsequently, the undercooling ability of the Sn–3.6Ag–1.0Cu–0.45Co alloy was compared to Sn–3.6Ag–1.0Cu by a fine droplet emulsion technique [26, 63] and was found to decrease undercooling by about 15%, from 142°C to 122°C, confirming the ability of Co to reduce undercooling. It should be noted that much larger undercoolings typically are accessible in solder alloys by the droplet emulsion technique, which may be desirable for increased discrimination of alloy catalyst activity [63].

3.2 Preliminary recommendations for SAC + X solders

Generally, the understanding of minor alloy additions to control nucleation and the solidification pathway of SAC solder alloys is still incomplete, including questions regarding the solidification catalysis mechanism and the minimum required concentration of an active addition. For β-Sn catalysts like Zn, the early formation of Cu–Zn [42] or Cu–Zn- -Sn compounds within the solder melt or at a Cu substrate interface may

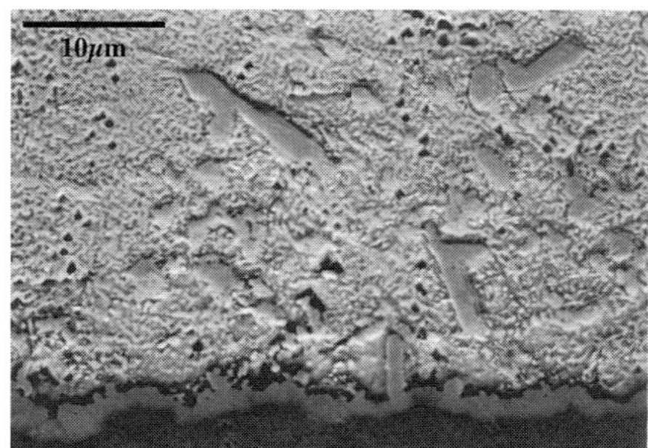

Fig. 13 SEM micrograph, using backscattered electron imaging, showing Cu substrate (bottom), intermetallic Cu/solder interface, and solder matrix microstructures at moderate magnification of joint cooled at 1–3°C/s and made from Sn–3.6Ag–1.0Cu–0.45Co [26]

provide preferred sites for Sn nucleation and dendritic growth. After initial Sn dendrite growth starts, the interdendritic regions are filled with a ternary eutectic microstructure. The addition of Si, Ti, Cr, Mn, Ni, and Ge to a SAC solder also appear to promote the formation of β-Sn in solder joints with Cu [35]. On the other hand, Co appears to catalyze nucleation of Cu_6Sn_5, presumably by the early formation of Co–Sn or Co–Cu–Sn compounds that become a "seed" crystal for Cu_6Sn_5 [26]. After the initial Cu_6Sn_5 (pro-eutectic) particles form, the regions between Cu_6Sn_5 particles are filled with a ternary eutectic microstructure. For completeness, it should be noted that the initial work on Co additions was performed with alloys containing a sufficient excess of solute, i.e., Sn–3.6Ag–1.0Cu–0.15Co (at minimum), to promote formation of massive Cu_6Sn_5 phase particles without a catalysis effect. However, more recent work [35] focused on an alloy closely correlated to the calculated ternary eutectic, Sn–3.7Ag–0.6Cu–0.3Co. Without a Co addition the base alloy (Sn–3.7Ag–0.9Cu) joint microstructure exhibits Sn dendrites and interdendritic ternary eutectic, as given in Figs. 3c and 4c. However, the joint microstructure of the near-eutectic SAC + Co solder joint is characterized by small pro-eutectic Cu_6Sn_5 phases and a complex ternary eutectic phase assembly, as given in Fig. 14a and b [39]. Interestingly, the intermetallic interface in Fig. 14a is also very similar to that shown for a solder joint made from a Sn–3.6Ag–

1.0Cu–0.45Co solder alloy under equivalent solidification conditions in Fig. 13. These microstructural effects indeed are consistent with the catalytic action of the Co addition, which seems to control the solder joint solidification pathway at this reduced (solute-lean) SAC solder composition. Although such interesting observations have been made, much work remains to determine the limits of both types of catalytic effects and to understand their mechanisms, permitting the maximum benefits for initial solder joint performance and extended reliability. These experiments should include other choices for a fourth element addition (aiming at minimizing the required concentration), extensive calorimetry experiments (to quantify the undercooling effect), and detailed microstructural analysis, including TEM of selected SAC + X solder joints (to probe the nucleant identity and mechanism).

3.3 Thermally aged microstructures and shear strength for SAC + X solders

As thoroughly described in the previous section, mechanical (shear) testing of a series of SAC solder joints and post-test microstructural analysis indicated that after aging at 150°C for 1,000 h, the typical ductile yielding behavior of SAC solder joints was accompanied by observations of a finite probability of some degree of embrittlement in samples made from Sn–3.5Ag and all of the SAC alloys. Microstructural analysis of as-aged (before shear testing) joint samples revealed that Cu_3Sn/Cu interface void coalescence appears to be a precursor condition for the joint embrittlement mechanism. However, the shear test results in Fig. 11 also indicated that all of the solder joints made from the Co- and Fe-modified SAC alloys exhibited a fully ductile shear failure mode. Figure 15 shows an example [35] of the fully ductile shear failure mode for a solder joint of reduced shear strength made with the SAC + Co alloy after aging at 150°C for 1,000 h, which may be compared to the example of SAC joint embrittlement in Fig. 10. This section will summarize the alloy design strategy and beneficial capabilities of these minor alloy additions (and others) to SAC solders for thermal aging resistance.

To probe the mechanism that resulted in resistance to thermal aging embrittlement, initial analysis [35] by line scans with EDS in the SEM indicated that Co and Fe preferentially segregated into the Cu_3Sn and Cu_6Sn_5 intermetallic layers between the solder matrix and the Cu substrate. The preferential segregation was confirmed later by WDS with measurements of increased fidelity [40]. Thus, substitutional alloying of

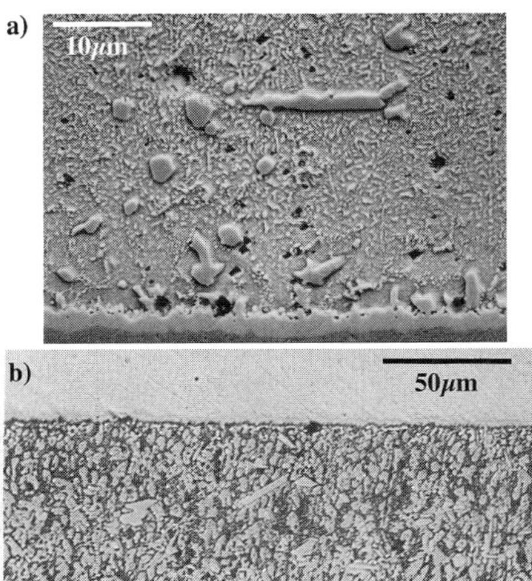

Fig. 14 Microstructure of as-solidified solder joints made from Sn–3.7Ag–0.6Cu–0.3Co, shown as (**a**) SEM micrograph (BEI), and (**b**) optical micrograph [39]

Fig. 15 A summary of the shear failure observations for two typical solder joints made from Sn–3.7Ag–0.6Cu–0.3Co, showing: (**a**) the shear stress versus elongation data for both strong and weak joints, (**b**) a cross-section SEM micrograph of a weak (but ductile) joint, (**c**) an SEM macrograph of the full width (nearly) of the weak joint [35]

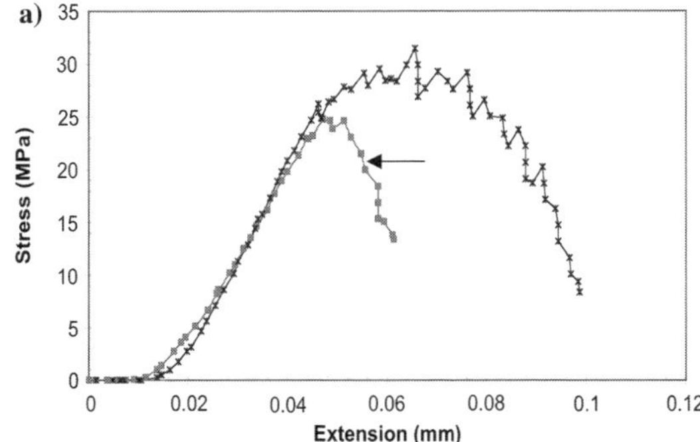

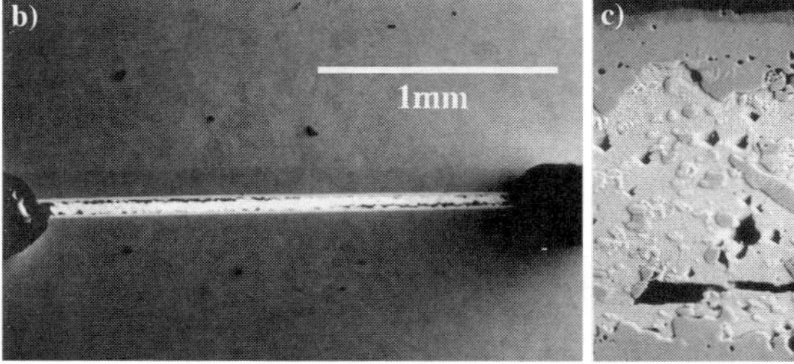

Co and Fe for Cu appeared to take place in these intermetallic layers, either during joint solidification or after aging. It should be noted that the apparent substitution of these elemental additions for Cu in the Cu-based intermetallic layers was intended by the alloy design approach, which was based on the Hume-Rothery criteria for extensive solid solubility, i.e., the atomic (metallic) radii mismatch of Fe and Co compared to Cu [24] are − 0.3% and − 1.9%, respectively, i.e., far less than 15% mismatch that is the criteria maximum [62]. More specifically, the alloy design objective was to use the enhanced lattice strain that results from substitutional alloying into the intermetallics to reduce the (vacancy diffusion) rate for inter-diffusion of Sn and Cu that is needed for coarsening of the joint microstructure and for void formation and coalescence. The void formation has been presumed to be due to a Kirkendall mechanism [57], where the diffusion flux of Cu is thought to be greater than that of Sn to cause the Cu/Cu$_3$Sn interface to be the preferred location of the voids [9]. Consistent with this objective, further analysis of the aged microstructures of these joints revealed that both Co and Fe additions reduced the growth rate of the Cu$_3$Sn layer, although growth of the combined (Cu$_3$Sn + Cu$_6$Sn$_5$) intermetallic layer seemed to be increased, as given in Fig. 16 [35]. Clearly, more study of the diffusion fluxes that cause these apparently contradictory observations is required. However, both of these alloying additions do

appear to minimize the formation and coalescence of voids at the Cu$_3$Sn/Cu interface after thermal aging, preventing the interfacial weakening that is the precursor to joint embrittlement. These initial observations also motivated an extension of the study to look for other additions that can suppress the embrittlement precursor condition and to seek more understanding of the Cu and Sn diffusion flux changes.

In one case, further systematic expansion of the list of possible SAC solder additives that could suppress thermal aging effects utilized a Darken-Gurry ellipse [40] as a further refinement of the Hume-Rothery criteria by adding close electronegativity agreement for predicting considerable substitutional solid solubility for Cu in the Cu$_3$Sn and Cu$_6$Sn$_5$ phases. Thus, a new set of fourth element additions to SAC solder alloys, i.e., SAC + X (X = Si, Ti, Cr, Mn, Ni, Ge, and Zn), were selected on the basis of their predicted ability [40] to substitute for Cu in the intermetallic phase layers that form on a Cu substrate after thermal aging, as a direct extension of the original reason for selection of Co and Fe [24]. A second research effort [64] that sought to address the thermal aging issue in SAC solder joints also identified Ni as a good candidate to add to a SAC solder. However, this effort [64] suggested two criteria; (1) "relatively large" solubility in the Cu$_3$Sn intermetallic phase, as demonstrated by an existing Cu–Sn–X phase diagram, and (2) the ability to increase the liquidus temperature of the Cu$_3$Sn phase with alloying, requiring

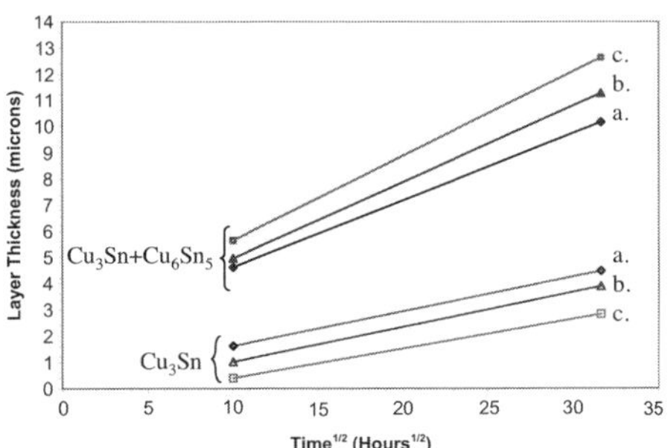

Fig. 16 Comparison of the Cu$_3$Sn and total (Cu$_3$Sn + Cu$_6$Sn$_5$) intermetallic interface thickness measurements as a function of isothermal aging time at 150°C for solder joints made from Sn–3.7Ag–0.9Cu (**a**), Sn–3.7Ag–0.7Cu–0.2Fe (**b**), and Sn–3.7Ag–0.6Cu–0.3Co (**c**) [35]

also an existing Cu–Sn–X phase diagram. To begin a broader test of the embrittlement suppression hypothesis, the former research effort provided microstructural measurements which indicated that all of the fourth element additions, except Ge, did reduce the measured thickness of the Cu$_3$Sn layer after aging for 1,000 h at 150°C compared to the base SAC alloy, Sn–3.7Ag–0.9Cu [40]. It should be noted that the previous report did not include data for the growth from thermal aging of the Cu$_3$Sn layer in SAC + Zn solder joints. A follow-up study [59] reported that these samples actually exhibited no continuous Cu$_3$Sn layer in the microstructure after 100, 1,000, or 2,000 h of aging at 150°C. The second study [64] also showed microstructural evidence that Ni "doping" (<1 wt.%) to a different base SAC alloy, Sn–1.0Ag–0.5Cu, suppressed growth of the Cu$_3$Sn layer on aging for 120 h at 150°C.

Further investigation by WDS in the former study [40] was conducted of the segregation tendency of X in some SAC + X joints (X = Ni, Mn, and Ge) that had been aged for 1,000 h at 150°C to compare to the segregation behavior in aged joints [35] made from the SAC + Co and SAC + Fe solder alloys. The results of WDS analysis of a SAC + Ni solder joint after extensive thermal aging is shown in Fig. 17 [40]. A later extension of this study also investigated the segregation tendency after thermal aging of Zn in SAC + Zn joints, using the same SAC base composition [59], as given in Fig. 18. As Figs. 17 and 18 demonstrate, both the Ni and Zn additions to the base SAC solder alloy exhibited strong segregation to the Cu/Cu$_3$Sn interface and into the Cu$_3$Sn and Cu$_6$Sn$_5$ intermetallic layers, similar to the segregation behavior of Co and Fe after the same thermal aging treatment [35]. Figures 17b and 18b also illustrate the difference in the growth behavior

of the Cu$_3$Sn intermetallic phase along the Cu substrate interface after aging for 1,000 h at 150°C. The Cu$_3$Sn phase grows as an obvious layer in the SAC + Ni case (Fig. 17b), but it grows as a series of disconnected projections from the Cu interface in the SAC + Zn case into the Cu$_6$Sn$_5$ phase layer. Also, the follow-up study reported that 2,000 h of aging at 150°C was not enough to promote growth of a continuous layer of Cu$_3$Sn in the SAC + Zn case [59].

A summary and complete discussion has been given elsewhere [35] of the ambient temperature shear test results for solder joints made with SAC + X solders (X = Fe, Co, Si, Ti, Cr, Mn, Ni, Ge, and Zn), showing the effect of thermal aging, compared to a representative set of SAC solder joints. The average shear strength results generally fall within the range of the data reported in Fig. 7 that includes two of the SAC + X results. Extracted from the SAC + X data, the shear test results for SAC + Ni and SAC + Zn solder joints provided an interesting comparison, where the stress versus elongation curves of joint samples made from the two alloys showed a difference in consistent ductility after the start of plastic deformation. In other words, the SAC + Ni joints had a few samples (one in particular) with less ductility than the rest [40], while the SAC + Zn joints all displayed extended plastic flow at a high stress level [59]. The difference in shear failure behavior between these two cases was illustrated clearly by cross-section micrographs of post-shear test samples, with the least ductile (and weakest) SAC + Ni joint given in Fig. 19 and the least ductile SAC + Zn joint given in Fig. 20 [59]. Obviously, the SAC + Ni joint failed in a fairly brittle manner, while the SAC + Zn joint exhibited fully (localized) ductile shear [59].

Taking the observations in total, some general conclusions may be offered regarding the shear strength behavior and microstructural coarsening and segregation effects after extended isothermal aging at 150°C of solder joint samples (with Cu) made from SAC + X alloys, where X = Co, Fe, Si, Ti, Cr, Mn, Ni, Ge, and Zn. All SAC + X solder joints in the as-soldered, 100 h aged, and 1,000 h aged condition, with the exception of one SAC + Ni sample, experienced shear failure in a ductile manner by either uniform shear of the solder matrix (in the highest strength solders) or by a more localized shear of the solder matrix adjacent to the Cu$_6$Sn$_5$ interfacial layer, consistent with many other observations on SAC solder joints [59]. After 1,000 h of aging, some weakening of the Cu$_3$Sn/Cu interface seemed to occur in only one solder joint made with SAC + Ni that lead to partial debonding during shear testing [59]. Without this isolated observation, only

Fig. 17 Typical line scan results for WDS measurements of solder joint made with SAC + Ni alloy that had been aged for 1,000 h at 150°C, showing (**a**) a collection of individual element concentration profiles, and (**b**) an SEM micrograph with a horizontal white line that indicates the transverse path of the elemental scans. Note that the calibration of the Ni composition (right scale on the *y*-axis) has not been adjusted to zero, affecting the absolute values of the concentration [40]

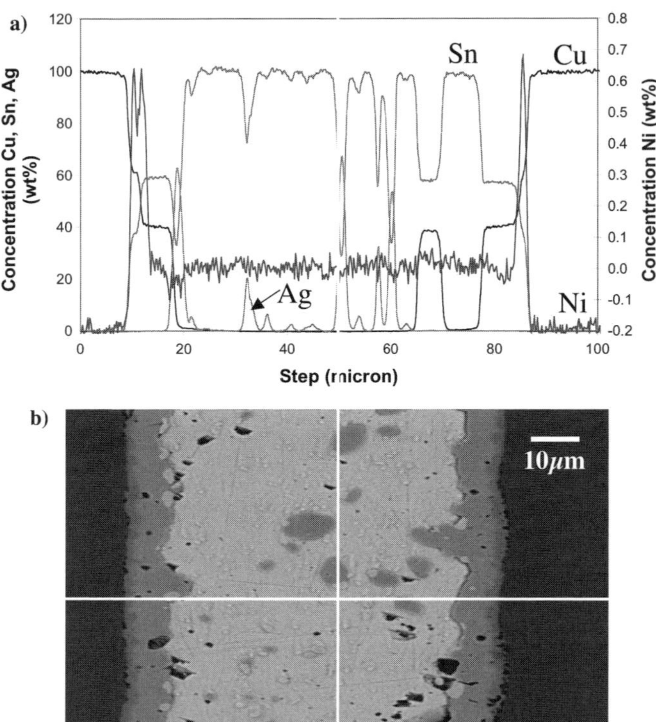

ductile failure was observed in all solder joints made from the SAC + X alloys after aging for 1,000 h. Analysis of aged joint samples by WDS suggests that substitutional alloying of the X addition into the intermetallic layers takes place between the solder matrix and the Cu substrate, with SAC + Ge as the only counter example [35]. This apparent substitutional alloying of the X element for Cu seems to depress the diffusion rate of Cu (most likely) and to minimize the formation and coalescence of voids at the Cu_3Sn/Cu interface, preventing interfacial weakening. Thus, the strategy of modifying a strong (high Cu content) SAC solder alloy with a substitutional alloy addition for Cu seems to be effective for producing a solder joint with Cu that retains both strength and ductility for extremely long periods of isothermal aging (to at least 1,000 h) at high temperatures (to at least 150°C).

3.4 Impact strength of thermally aged SAC + X solder joints

As with the SAC solder joint samples discussed above, the need to add impact testing to the charac-
terization suite of new SAC + X solder alloys has been demonstrated in some recent work [17–19] on the effects of thermal aging on Pb-free solder joints. In other words, to exhibit high reliability a Pb-free solder joint should retain high impact strength and a ductile failure mode after extensive use at high temperatures. As described above, preliminary Izod impact testing was performed [59] on joints samples made from a reduced set of SAC + X alloys with X = Si, Ti, Cr, Mn, Ni, Zn, and Ge and it provided some interesting observations (see Fig. 21). Of these samples, the most impressive were the results after aging for SAC + Ni and SAC + Zn, which maintained impact strength values above 1,500 J/m^2. In fact, the SAC + Zn joints lost only 3% of their as-soldered impact strength after 1,000 h of aging at 150°C. The impact strength retention of aged SAC + Ni joints was not nearly as impressive, with a loss of about 40% of the as-soldered value (see Fig. 21). Broader conclusions will be possible when additional results are available from an increased set of SAC and SAC + X solder alloys.

The Izod sample bar configuration [24] used for the results in Figs. 12 and 21 places the solder joint

Fig. 18 Typical line scan results for WDS measurements of solder joint made with SAC + Zn alloy that had been aged for 1,000 h at 150°C, showing (**a**) a collection of individual element concentration profiles, and (**b**) an SEM micrograph with a horizontal white line that indicates the transverse path of the elemental scans [59]

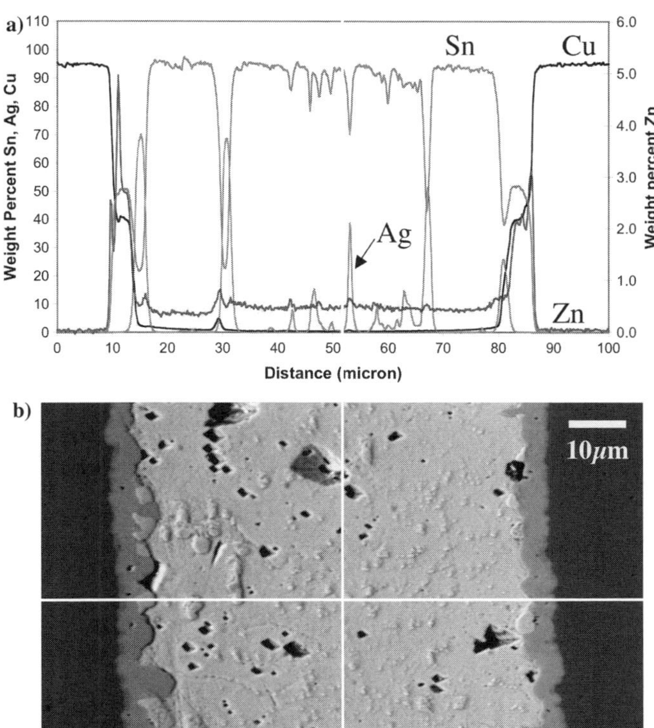

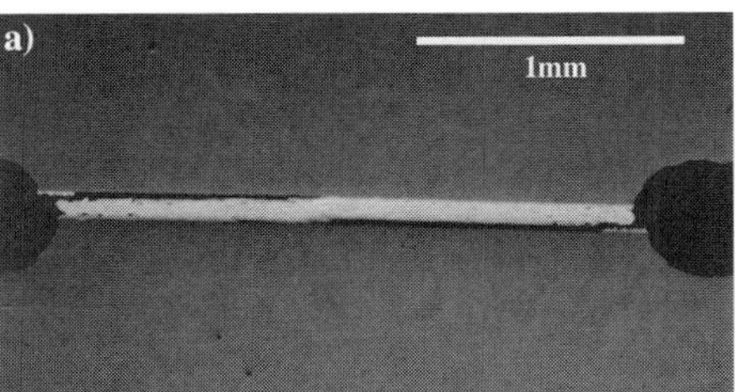

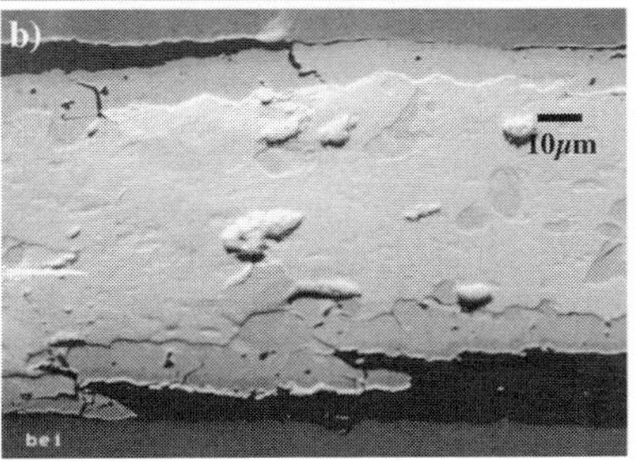

Fig. 19 SEM micrographs of post-AFPB solder joint made with SAC + Ni that displayed reduced ductility, where (**a**) shows the full joint width and (**b**) is a high magnification of the central joint region [59]

line well below the impact point of the pendulum striker, which provides a combined tensile and shear (mixed mode) loading along the joint at an extreme strain rate. A machined notch, aligned on the joint centerline, is also placed on the striker side of the bar, as recommended for the typical Izod impact sample [65] to localize the fracture initiation point. Compared to impact testing of single BGA joints [17, 18], the current Izod sample configuration is more representative of bulk solder joint properties in realistic impact situations and can be practiced on joints that have experienced a wide range of reflow temperatures and cooling rates, simulating either reflow oven practice (as in this study) or the slow cooling of BGA reflow. However, the previous study [17, 18] also accompanied the impact strength measurements with a microstructural assessment of the fracture surface type, i.e., either ductile, brittle, or mixed, which provides a more complete characterization than the current study. Such an assessment is recommended for further Izod impact studies of solder joints.

 Springer

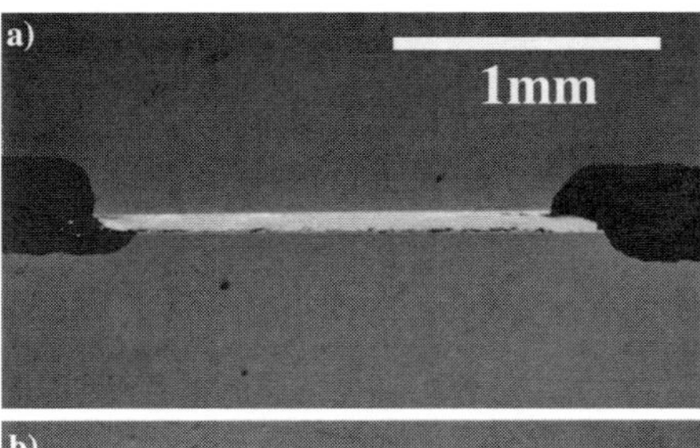

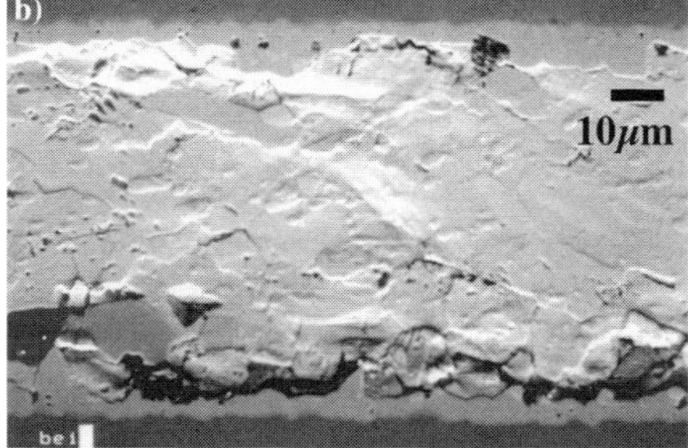

Fig. 20 SEM micrographs of post-AFPB solder joint made with SAC + Zn that displayed minimum ductility, where (**a**) shows the full joint width and (**b**) is a high magnification of the central joint region [59]

3.5 Summary recommendations for SAC + X solders

Although interesting observations have been made on the effects of minor 4th element additions, much work remains to determine the limits of the types of solidification catalyst effects and to understand their mechanisms, permitting the maximum benefits for initial solder joint performance and extended reliability. One goal of this work could be to promote the initial solidification of a high fraction of ternary eutectic microstructure, i.e., minimizing extensive Sn dendritic growth, which may enhance isotropic joint strength and reliability. Of the choices studied, the Co addition and, perhaps, Zn, seem to satisfy this objective. The challenges ahead include discovery of the minimum addition needed to catalyze the desired effect and determination of whether the catalysis effect can persist through multiple reflow cycles. For long-term joint reliability, modifying a strong (high Cu content) SAC solder alloy with a substitutional alloy addition for Cu seems effective for producing a solder joint that retains both strength and ductility after extreme (at least 1,000 h) aging at high temperatures (up to 150°C). Of the choices tested, Co, Fe, and Zn substitutions for Cu seem most attractive from the standpoint of the microstructure (IMC segregation effects and suppression of Cu_3Sn growth and Cu/Cu_3Sn void formation/coalescence) and mechanical properties (retained shear strength and ductility). Although Ni has the same strong interfacial segregation behavior of the others, the discovery of a brittle failure case and the reduced impact strength on thermal aging seem to take it out of prime consideration. Both Fe and Zn have minimal cost, but if reduced amounts of Co prove effective, it may also have an acceptable cost. At this point only anecdotal evidence can be reported of the good (sufficient) solderability (wetting, oxidation, and melting range) of Co, Fe, and Zn, but quantitative studies are needed promptly to facilitate extensive application testing. It is clear that Zn requires a reduced superheat for alloying into a SAC + X solder. Also, Zn retained a remarkable percentage of impact strength after thermal aging, but aged impact results for SAC + Co and SAC + Fe are still in-progress. Confirmation of these preferences by further testing and analysis should be completed to establish the final ranking of the SAC + X choices. Especially if low level additions are proven effective, an expanded testing program that includes SAC + Co, SAC + Fe, and SAC + Zn seems

Fig. 21 Comparison of the as-soldered impact strength results with the available results from the 1,000 h aged joints in the study, where the standard deviation is superimposed on the bar for each alloy and the repeat samples are noted above each bar [59]

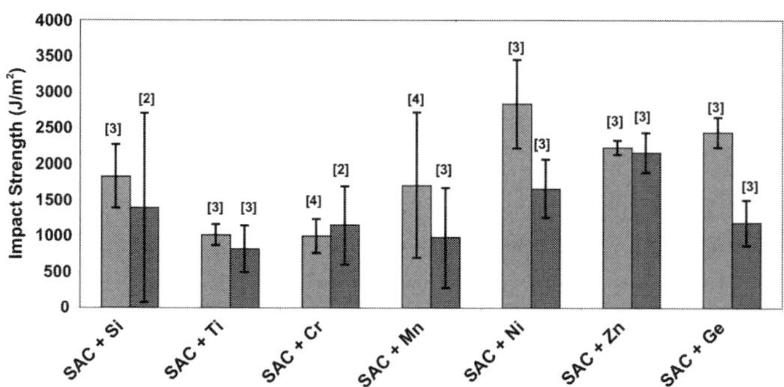

appropriate to evaluate the performance of these solders in many types of electronic assembly applications, including BGA assemblies.

Acknowledgements Collaborators in this work including Joel Harringa, Bruce Cook, Bob Terpstra, Jack Smith, Tammy Bloomer, Ozer Unal, Chad Miller, and Fred Yost must be recognized for their valuable contributions to these studies. The author is grateful to the Materials Preparation Center of the Ames Laboratory for experimental assistance in solder alloy and wire preparation and in solder joint thermal aging and mechanical testing. Specifically, appreciation is due to Hal Sailsbury for metallographic preparation of the solder joint samples and to Fran Laabs and Alfred Kracher for taking the SEM micrographs and WDS measurements. Bill Wing of the Ames Lab machine shop is also thanked for his careful electro-discharge machining of the joint specimens. Support for the fundamental aspects of this work is gratefully acknowledged from USDOE-BES, Materials Science Division under Contract No. W-7405-Eng-82. Support for other applied aspects of this work is gratefully acknowledged from DOE-EM, under TTP No. CH133001. For recent studies, support is much appreciated from the Iowa State University Research Foundation, under Grant No. 405-25-02. The Advanced Industrial Materials Program of the Office of Industrial Technologies of DOE-EE is also acknowledged for sponsoring a graduate student fellowship during an earlier stage of this study.

References

1. Refer to website at http://www.rohs-news.com/ for information on Restriction on use of certain Hazardous Substances (RoHS) regulations
2. K.N. Tu, A.M. Gusak, M. Li, J. Appl. Phys. **93**(3), 1335–1353 (2003)
3. Refer to websites at http://www.lead-free.org/legislation/, http://www.nemi.org/PbFreepublic/, and http://www.jeida.org.jp/english/information/pbfree/roadmap.html/ for information on relevant organization activities
4. C.M. Miller, I.E. Anderson, J.F. Smith, J. Electron. Mater. **23**, 595 (1994)
5. I.E. Anderson, F.G. Yost, J.F. Smith, C.M. Miller, R.L. Terpstra, U.S. Patent No. 5,527,628, June 18, 1996
6. K.-W. Moon, W.J. Boettinger, U.R. Kattner, F.S. Biancaniello, C.A. Handwerker, J. Electron. Mater. **29**, 1122 (2000)
7. M.E. Loomans, M.E. Fine, Metal. Mater. Trans. A **31A** (4), 1155–1162 (2000)
8. S.K. Kang, D.-Y. Shih, D. Leonard, D.W. Henderson, T. Gosselin, S-I. Cho, J. Yu, W.K. Choi, JOM **56**(6), 34–38 (2004)
9. S.K. Kang, P.A. Lauro, D.-Y. Shih, D.W. Henderson, K.J. Puttlitz, IBM J. Res. Dev. **49**(4/5), 607–620 (2005)
10. D.R. Frear, J.W. Jang, J.K. Lin, C. Zhang, JOM **53**(6), 28–32 (2001)
11. S. Choi, J.G. Lee, K.N. Subramanian, J.P. Lucas, T.R. Bieler, J. Electron. Mater. **31**, 292 (2002)
12. C. Andersson, Z. Lai, J. Liu, H. Jiang, Y. Yu, Mater. Sci. Eng. A **394**, 20–27 (2005)
13. Dr. T. PAN, Ford Motor Company, Dearborn, Michigan, private conversation, February 12, 2001
14. O. Unal, I.E. Anderson, J.L. Harringa, R.L. Terpstra, B.A. Cook, J.C. Foley, J. Electron. Mater. **30**(9), 1206–1213 (2001)
15. W.L. Winterbottom, JOM **45** 20 (1993)
16. F.W. Gayle, G. Becka, J. Badgett, G. Whitten, T.-Y. Pan, A. Grusd, B. Bauer, R. Lathrop, J. Slattery, I.E. Anderson, J. Foley, A. Gickler, D. Napp, J. Mather, C. Olson, JOM **53**(6), 17–21 (2001)
17. M. Date, T. Shoji, M. Fujiyoshi, K. Sato, K.N. Tu, Scripta Mater. **51**, 641–645 (2004)
18. M. Date, K.N. Tu, J. Mater. Res. **19**(10), 2887–2896 (2004)
19. B. Wang, S. Yi, J. Mat. Sci. Lett. **21**, 697–698 (2002)
20. T.-C. Chiu, K. Zeng, R. Stierman, D. Edwards, K. Ano, in *Proceedings of the 54th Electronic Components and Technology Conference* (IEEE, 2004), pp. 1256–1262
21. P. Ratchev, T. Loccufier, B. Vandevelde, B. Verlinden, S. Teliszewski, D. Werkhoven, B. Allaert, in *Proceedings of EMPC 2005, Brugge, Belgium* (IMAPS, 2005)
22. I.E. Anderson, B.A. Cook, J. Harringa, R.L. Terpstra, J.C. Foley, O. Unal, Mater. Trans. **43**, 1827 (2002)
23. J.-Y. Park, R. Kabade, C.-U. Kim, T. Carper, S. Dunford, V. Puligandla, J. Electron. Mater. **32**, 1474 (2003)
24. I.E. Anderson, T.E. Bloomer, J.C. Foley, R.L. Terpstra, in *Proceedings of IPC Works '99 IPC*, Northbrook, IL, paper No. S-03-5 (1999)
25. I.E. Anderson, R.L. Terpstra, U.S. Patent No. 6,231,691 B1, May 15, 2001
26. I.E. Anderson, J.C. Foley, B.A. Cook, J.L. Harringa, R.L. Terpstra, O. Unal, J. Electron. Mater. **30**, 1050 (2001)
27. J.Y. Tsai, Y.C. Hu, C.M. Tsai, C.R. Kao, J. Electron. Mater. **32**, 1203–1208 (2003)
28. C.-M. Chuang, K.-L. Lin, J. Electron. Mater. **32**, 1426–1431 (2003)
29. K. Zeng, K.N. Tu, Mater. Sci. Eng. R **38**, 55–105 (2002)
30. S.K. Kang, D.Y. Shih, K. Fogel, P. Lauro, M.-Y. Yim, G.G. Advocate Jr., M. Griffin, C. Goldsmith, D.W. Henderson, T.A. Sosselin, K.E. King, J.J. Konrad, A. Sarkhel, K.J. Puttlitz, IEEE Trans. Electron. Packag. Manufac. **25**(3), 155–161 (2002)
31. L.C. Shiau, C.E. Ho, C.R. Kao, Solder. Surf. Mount Technol. **14**(3), 25–29 (2002)
32. C.-W. Hwang, K.-S. Kim, K. Suganuma, J. Electron. Mater. **32**, 1249–1256 (2003)
33. K.Y. Lee, M. Li, J. Electron. Mater. **32**, 906–912 (2003)
34. S.K. Kang, R.S. Rai, S. Purushothaman, J. Electron. Mater. **25**, 1113–1120 (1996)
35. I.E. Anderson, J.L. Harringa, J. Electron. Mater. **33**, 1485–1496 (2004)
36. P.T. Vianco, J.A. Rejent, P.F. Hlava, J. Electron. Mater. **33**, 991 (2004)
37. J. Bath, C. Handweker, E. Bradley, "Research update: lead-free solder alternatives," Circuits Assembly (2000), pp. 31–40
38. C.A. Drewien, F.G. Yost, S.J. Sackinger, J. Kern, M.W. Weiser, "Progress Report: High Temperature Solder Alloys for Underhood Applications," Sandia Report, SAND95-0196.UC-704, February 1995
39. I.E. Anderson, B.A. Cook, J. Harringa, R.L. Terpstra, J. Electron. Mater. **31**, 1166 (2002)
40. I.E. Anderson, J.L. Harringa, J. Electron. Mater. **35**(1), 1–13 (2006)
41. R.J. Schaefer, D.J. Lewis, Metal. Mater. Trans. A **36A**, 2775–2783 (2005)
42. S.K. Kang, W.K. Choi, D.-Y. Shih, D.W. Henderson, T. Gosselin, A. Sarkhel, C. Goldsmith, K.J. Puttlitz, JOM **55**(6), 61–65 (2003)
43. M. McCormack, S. Jin, JOM 36–40 (1993)
44. C.M. Liu, C.E. Ho, W.T. Chen, C.R. Kao, J. Electron. Mater. **30**(9), 1152–1156 (2001)

45. A. Garg, I.E. Anderson, J.L. Harringa, A. Kracher, D. Swenson, "Dissolution of Copper from Substrate Surfaces into Lead-Free Solder Joints," poster presentation at 2006 TMS Annual Meeting, San Antonio, Texas (March 2006)
46. D.W. Henderson, T. Gosselin, S.K. Kang, W.K. Choi, D.Y. Shih, C. Goldsmith, K. Puttlitz, U.S. Patent 6,805,974, October 19, 2004
47. I.E. Anderson, K. Kirkland, W. Willenberg, Surf. Mount Technol. **14**(11), 78–81 (2000)
48. W.J. Boettinger, C.E. Johnson, L.A. Bendersky, K.-W. Moon, M.E. Williams, G.R. Stafford, Acta Mater. **53**, 5033–5050 (2005)
49. R.H. Kane, B.C. Giessen, J.J. Grant, Acta Metal. **14**, 605–609 (1966)
50. X. Llovet, G. Galan, Am. Mineral. **88**, 121–130 (2003)
51. S.J.B. Reed, *Electron Microprobe Analysis* (Cambridge University Press, Cambridge, 1993)
52. A.U. Telang, T.R. Bieler, JOM **57**(6), 44–49 (2005)
53. I.E. Anderson, B.A. Cook, J.L. Harringa, R.L. Terpstra, JOM **54**, 26 (2002)
54. S.K. Kang, W.K. Choi, M.J. Yim, D.Y. Shih, J. Electron. Mater. **31**, 1292 (2002)
55. H.-T. Lee, M.-H. Chen, H.-M. Jao, T.-L. Liao, Mat. Eng. A **358**, 134 (2003)
56. I.E. Anderson, B.A. Cook, J.L. Harringa, R.L. Terpstra, in *Proceedings of the International Brazing and Soldering Conference*, ed. by F.M. Hosking (American Welding Society), CD-ROM, 3.7 (2003)
57. R.E. Reed-Hill, in *Physical Metallurgy Principles* (D. Van Nostrand Company, New York, NY, 1973), pp. 386–397
58. T. Morita, R. Kajiwara, K. Yamamoto, K. Sato, M. Date, T. Shoji, I. Ueno, S. Okabe, in *Proceedings of the 16th JIEP Annual Meeting* (JIEP, Tokyo, Japan, 2000), p. 107
59. I.E. Anderson, J.L. Harringa, in *Brazing and Soldering, Proceedings of the 3rd International Brazing and Soldering Conference*, eds. by J.J. Stephens, K.S. Weil (ASM International, Materials Park, Ohio, 2006), pp. 18–25
60. A. Ohno, T. Motegi, Nippon Kinzoku Gakkaishi **37**(7), 777–780 (1973)
61. B. Li, Y. Shi, Y. Lei, F. Guo, Z. Xia, B. Zong, J. Electron. Mater. **34**(3), 217–224 (2005)
62. W. Hume-Rothery, R.E. Smallman, C.W. Haworth, "The Structure of Metals and Alloys," Institute of Metals (1969), p. 124
63. J.H. Perepezko, B.A. Mueller, K. Ohsaka, Undercooled Alloy Phases, eds. by E.W. Collings, C.C. Koch (Metallurgical Society, Inc., Warrendale, PA, 1986), p. 289
64. L. Garner, S. Sane, D. Suh, T. Byrne, A. Dani, T. Martin, M. Mello, M. Patel, R. Williams, Intel Technol. J. **9**(4), 297–308 (2005)
65. ASTM Standard, D-256, ASTM International, West Conshohocken, Pennsylvania, pp. 1–20 (2002)

Rare-earth additions to lead-free electronic solders

C. M. L. Wu · Y. W. Wong

Published online: 12 September 2006

Abstract The research in lead(Pb)-free solder alloy has been a popular topic in recent years, and has led to commercially available Pb-free alloys. Further research in certain properties to improve aspects such as manufacturability and long term reliability in many Pb-free alloys are currently undertaken. It was found by researchers that popular Pb-free solders such as Sn–Ag, Sn–Cu, Sn–Zn and Sn–Ag–Cu had improved their properties by doping with trace amounts of rare earth (RE) elements. The improvements include better wettability, creep strength and tensile strength. In particular, the increase in creep rupture time in Sn–Ag–Cu–RE was 7 times, when the RE elements were primarily Ce and La. Apart from these studies, other studies have also shown that the addition of RE elements to existing Pb-free could make it solderable to substrates such as semiconductors and optical materials. This paper summarizes the effect of RE elements on the microstructure, mechanical properties and wetting behavior of certain Pb-free solder alloys. It also demonstrates that the addition of RE elements would improve the reliability of the interconnections in electronic packaging. For example, when Pb-free-RE alloys were used as solder balls in a ball grid array (BGA) package, the intermetallic compound layer thickness and the amount of interfacial reaction were reduced.

1 Introduction

In recent years, many countries had introduced environmental related regulations such as the restriction of certain hazardous substances (RoHS) and the waste electrical and electronic equipment (WEEE). Lead (Pb) is one of the chemical elements in this category. The anticipation to the imminent introduction of legislations such as RoHS and WEEE had accelerated the research and development activities in developing Pb-free solders. As a result, a number of Pb-free alloys are commercially available. Modern microelectronics products demand for high reliability, and electronic solder materials such as Pb-free alloys need to be reliable over long term use. Although many of the commercially available Pb-free solders possess good attributes such as high strength, there is still room for improvement in certain properties to improve manufacturability and long term reliability. For example, improvements in wettability and creep properties in many Pb-free alloys are desired.

The use of electronic solders is still one of the most popular methods for the interconnection and packaging of virtually all electronic devices and circuits. Pb-bearing solders and especially the eutectic or near-eutectic SnPb alloys have been used extensively in the assembly of modern electronic circuits [1]. Due to the mandatory requirements for Pb-free imports introduced by countries such as Japan and those in the European Union, many Pb-free solders were developed to replace Pb-containing ones. Recently, the International Printed Circuit (IPC) association of the US reviewed the availability of Pb-free solder alloys and indicated two choices for electronics manufacturers [2]. These two alloys are the Japanese adopted

C. M. L. Wu (✉) · Y. W. Wong
Department of Physics and Materials Science, City
University of Hong Kong, Hong Kong SAR, P. R. China
e-mail: Lawrence.Wu@cityu.edu.hk

tin0–3.0 wt%silver–0.5 wt%copper (Sn–3.0Ag–0.5Cu) alloy and the North American Electronics Manufacturing Initiative (NEMI) Sn–3.9Ag–0.6Cu alloy. The Japanese and the US manufacturing associations both believe that their particular choice is the best candidate for replacing Pb-containing solders. Both of these alloys as well as many others available commercially have undergone significant research and development. For example, the basic physical and mechanical properties have been studied, and long-term reliability tests have been carried out. While these alloys have many merits, there is still scope to replace them with better alloys.

It is also clear that, apart from the development of proper alloy compositions for the new solder systems, suitable fluxes and assembly process for Pb-free solders is also needed [3].

It has been known that the addition of a small amount of rare earth (RE) elements in metals can greatly enhance their properties. They have played an important role in the development of magnetic materials, high temperature superconductors, hydrogen storage alloys as well as in ceramics and glasses [4]. Due to the unique properties of RE elements, some researchers have tried to improve the properties of solders by adding RE elements. In this paper, we review the recent developments in Pb-free solder alloys doped with RE elements. The development includes the effect on microstructure, tensile properties, creep behavior and wetting performance. The mechanism of the refinement of the microstructure is explained. To indicate their potential of application in electronic packaging, a review of the changes at the solder/substrate interface is given. Other practical considerations such as using these solder materials as ball grid array (BGA) solder balls, and for the direct soldering onto substrates which are regarded as difficult to be soldered are also reviewed.

2 Melting behavior

When the addition of RE elements are mainly Ce and La, the melting temperatures are 220.9 and 220.8°C for SnAg–0.25RE and SnAg–0.5RE, respectively, as compared with 221.4°C for eutectic SnAg [5]. It was also shown in Ref. [6] that the addition of 0.5–2 wt% of Lu did not significantly affect the melting point of the eutectic SnAg alloy. It was found that the onset transformation temperature for the exothermic descent of the curve, which represents the onset of melting, does not change markedly. It is understood that by adding the Lu element the alloy composition has

moved away from the eutectic value, as shown by the effect of melting point peak broadening from the DSC results [6].

For SnZn–0.05RE and SnZn–0.1RE, their melting temperatures are 198.7 and 199.0°C, compared with 198.9°C for SnZn without the addition of RE elements.

It was found in Ref. [7] that the addition of 0.25 and 0.5 wt% of RE elements to SnCu eutectic alloy did not change the melting temperature. It stayed at the value of 227°C.

The melting behavior of Sn–3.8Ag–0.7-xRE alloys was studied in Ref. [8]. As shown in Table 1, the liquidus was increased by less than 2°C when the amount of RE elements is 0.1 wt%. When the RE content is increased to 1 wt%, the liquidus was increased by 5°C.

3 Microstructural change of Pb-free alloys after adding RE elements

Most of the Pb-free alloys developed were Sn-rich, and alloyed with elements such as Ag, Cu, Zn, Bi, In and Sb. The more popular binary eutectic Pb-free alloys are SnAg, SnZn and SnCu [3, 9]. As for ternary alloys, those in the SnAgCu system are preferred [10, 11]. It is quite interesting to note that these solders are eutectic alloys, or near eutectic ones, and have similar microstructures. Coarse β-Sn phase is formed at a cooling rate of 15–20Ks^{-1} [5, 12].

From the melting temperature point-of-view, the eutectic Sn–9Zn alloy is one of the best alternatives to PbSn with a melting temperature of 199°C, as compared with 183°C of PbSn. Its microstructure consists of two phases: a body centered tetragonal Sn matrix phase and a secondary phase of hexagonal Zn containing less than 1 wt% Sn in solid solution. The solidified microstructure exhibits large grains with a fine uniform two-phase eutectic colony [3, 12, 13]. The microstructure of Sn–9Zn and Sn–9Zn–0.05RE are shown in Fig. 1a, b respectively. Figure 1a shows the microstructure of Sn–9Zn under scanning electron microscope (SEM) examination. The eutectic colonies of Sn–9Zn consist of fine needlelike Zn-rich phase and dark-colored β-Sn matrix phase. Figure 1b shows that when 0.05% RE elements were added to Sn–Zn, the

Table 1 Melting Behaviors of Sn–3.8Ag–0.7-xRE alloys [8]

Amount of RE elements, x (wt%)	0	0.025	0.05	0.1	0.25	0.5	1.0
Solidus (°C)	217.7	217.9	218.4	218.5	218.8	218.9	220.1
Liquidus (°C)	219.8	220.5	221.3	221.8	223.2	223.6	225.0

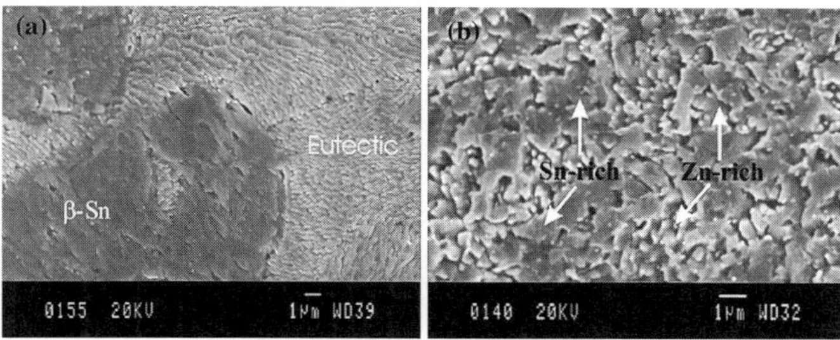

Fig. 1 Secondary electron micrographs of (a) Sn–9Zn; (b) Sn–9Zn–0.05RE

coarse β-Sn grains disappeared and the shape of the Sn-rich phase was rodlike. It is also noted that the microstructure had become very uniform [12].

The eutectic SnAg alloy has a melting temperature of 221°C. The microstructure consists of primary β-Sn grains and an Ag_3Sn intermetallic compound (IMC) in the form of thin platelets, as shown in Fig. 2a [3, 5, 9]. With the addition of RE elements, the β-Sn grain size decreases, but has little effects on the Sn–Ag inter-metallics. In faster-cooled solder alloys, fine Sn–RE IMCs can be found in the eutectic colonies. When the amount of RE element addition is high enough and with a low cooling rate during solidification, the small RE-bearing IMC particles would agglomerate together to form an island shape, as shown in Fig. 2b. In the same figure the rod-like Ag_3Sn phases can also be seen to form uniformly in the β-Sn matrix. In [5, 11, 12, 14] the microstructure of the samples are obtained at a cooling rate of 15–20 Ks^{-1}, which is close to that during conventional soldering. From the phase diagrams of Ce–Sn and La–Sn, the RE-bearing phase was a mixture of $CeSn_3$ and $LaSn_3$[15, 16]. It was found that the RE-bearing phase was quite hard. The average micro-hardness of the phase in a slow-cooled alloy has a large value of 48 HV, compared with the average micro-hardness of the matrix of 13.5 HV [5].

The microstructure of SnCu alloy as shown in Fig. 3a is similar to that of SnAg with a more elongated Cu_6Sn_5IMC [14]. The additional of RE elements again decrease the β-Sn grain size, as shown in Fig. 3b for Sn–0.7Cu–0.25RE.

The SnAgCu Pb-free solders are quite important because they are generally recognized as the first choice for a Pb-free solder. The microstructure of chill cast Sn–3.5Ag–0.7Cu, as shown in Fig. 4a, mainly consisted of phases of β-Sn, Cu_6Sn_5 and Ag_3Sn. Under this cooling rate, the β-Sn phases were distributed in the ternary eutectic region. The grain size of the β-Sn phases was about 10–20 µm. In some regions, small Ag–Sn IMCs coexisted with the β-Sn phases. When RE was added, the β-Sn grain size decreased to 5–10µm, as shown in Fig. 4b [11]. The RE-bearing phase could not be detected due to the small amount of RE addition and its fine dispersion.

4 Adsorption effect of RE

It was mentioned above that with the addition of trace RE elements, the more uniform microstructures of SnAg, SnCu, SnZn, SnBiAg, SnAgCu alloys have been obtained [5, 8, 11, 12, 17–20]. The reason is due to the unique properties of RE elements. As active elements, RE will accumulate at the interface of IMC particles. It is well known that the adsorption phenomenon plays an important role during solidification processes of alloys and would greatly affect their microstructures [21].

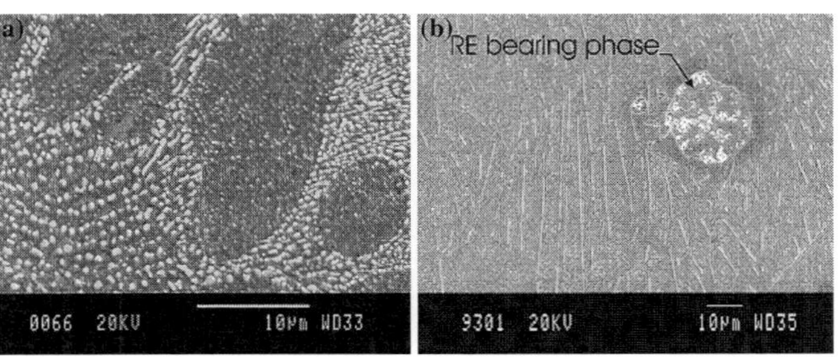

Fig. 2 Secondary electron micrographs of (a) Sn–3.5Ag; (b) Sn–3.5Ag–0.5RE by slow cooling

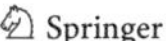

Fig. 3 Secondary electron micrographs of (**a**) Sn–0.7Cu; (**b**) Sn–0.7Cu–0.25RE

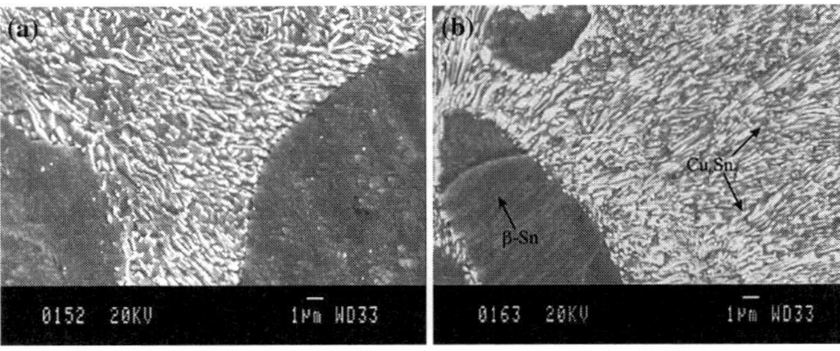

The adsorption of surface-active elements at various planes is different. The plane with maximum surface tension grows the fastest, while the amount of surface-active material adsorbed is maximized. The amount of surface-active material adsorbed at plane L is [22] :

$$\Gamma^L = -\frac{C}{RT}\frac{d\gamma^L}{dC} \tag{1}$$

where Γ^L is the adsorption of surface-active material at plane L, C is the concentration of surface-active material, R is the universal gas constant, T is the absolute temperature, and γ^L is the surface tension of plane L.

When the plane L adsorbs a layer of active material, the surface tension can be deduced from the integral of equation (1) :

$$\gamma_C^L = \gamma_0^L - RT \int_0^c \frac{\Gamma^L}{C}dC \tag{2}$$

where γ_C^L is the surface tension of plane L with adsorption of active material, γ_0^L is the surface tension of the initial plane L without adsorption. Thus the surface free energy of the whole crystal is :

$$\sum_L \gamma_C^L A_L = \sum_L \left(\gamma_0^L - RT \int_0^C \frac{\Gamma^L}{C}dC \right) A_L \tag{3}$$

where A_L is the area of plane L. Therefore, in order to minimize the surface free energy,

$$\sum_l A_L \int \frac{\Gamma^L}{C}dC \to \text{max.} \tag{4}$$

The relation indicates that the plane with the largest Γ^L value plays the most important role in minimizing the free energy of the entire interface. The nature of metamorphosis is from the adsorption of active elements on certain planes. This adsorption will not only decrease the difference of surface energy of the crystal, but also prevent the latter from further growth on these planes. In [5, 14], the addition of trace RE elements decrease the size of Cu_6Sn_5 and Ag_3Sn IMC particles and at the same time the β-Sn grains also become finer. This uniform microstructure was then be able to enhance the hardness, tensile and creep properties.

5 Improvement of mechanical properties of solder alloys and solder joints

With adequate design, Pb-free solders are capable of withstanding normal operating stress and strain, and creep deformation. The ultimate tensile strength (UTS) is the maximum engineering stress which a material can withstand in tension. It was shown in

Fig. 4 Secondary electron micrographs of (**a**) Sn–3.5Ag–0.7Cu; (**b**) Sn–3.5Ag–0.7Cu–RE

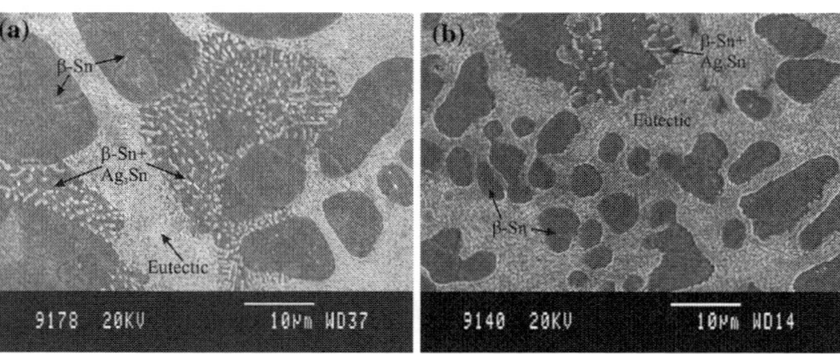

Section 4 that the microstructure was generally refined when RE elements were added. With this microstructural change, it is foreseen that the tensile properties will be improved.

The results of tensile tests on SnCu, SnCu–0.5RE, SnAg, SnAg–0.5RE, SnZn and SnZn–0.5RE Pb-free solder alloys are shown as Fig. 5 [5, 12, 14]. In these cases the RE elements added were mainly Ce and La. In Ref. [23], the UTS of SnAgCu–0.25RE was found to be 64 MPa. In general, the tensile strengths are improved by RE additions. However, if the amount of RE elements added is increased, the elongation to failure of the solder alloys becomes lower for SnAg, SnCu and SnZn. This was not the case for Sn–3.5Ag–0.7Cu–0.1RE and Sn–3.5Ag–0.7Cu–0.25RE, as their elongations were higher than that of Sn–3.5Ag–0.7Cu [23].

Figure 6 shows how a typical creep curve is modified by adding RE elements, e.g. to SnCu [14]. The creep curve is substantially improved in both creep time and creep strain to failure. It may also be noticed that the steady-state creep rate is substantially decreased. It can be seen in Fig. 6 that the creep rupture time for 303 K (30°C) at 16 MPa is about the same as that for 393K (120°C) at 8 MPa.

For a low melting point alloy, its creep behavior is important because it is used at high homologous temperature. Creep data are generally analyzed by relating the steady-state strain rate to the stress through a power law relation. The stress exponent n and the activation energy (Q_c) for creep can be calculated from the Dorn equation [24]:

$$\dot{\varepsilon} = \frac{AGb}{RT}\left(\frac{\sigma}{G}\right)^n \exp\left(\frac{-Q_c}{RT}\right) \tag{5}$$

where Q_c represents the activation energy for creep, n is the power-law stress exponent, G is the temperature-

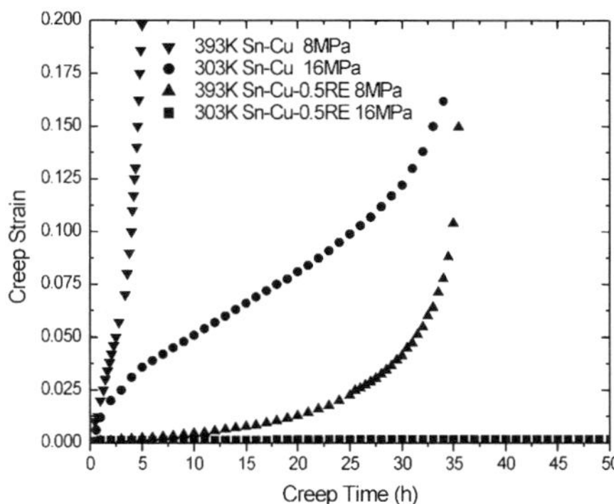

Fig. 6 Creep curves of Sn–Cu and Sn–Cu–0.5RE alloys

dependent shear modulus, b is the Burgers vector, A is a material-dependent constant, R is the universal gas constant and T is the temperature. The values of n and Q_c are representative of the dominant creep mechanism.

For SnAg–RE, the Ag_3Sn intermetallic particles play two different roles. They may strengthen the alloy matrix and prevent the formation of large dislocation pile-ups at grain boundaries. On the other hand, the higher the number of particles in a given matrix, the more matrix/intermetallics interfaces it contains, leading to a higher likelihood of microcrack nucleation that can speed up the failure process. In [25], the eutectic SnAg alloy was investigated for its creep behavior at three temperatures ranging from 303 to 393 K, under the tensile stress range of $\sigma/E = 10^{-4}$–10^{-3}. One of the major results confirmed that the steady-state creep rate of SnAg was controlled by the dislocation-pipe diffusion in the Sn matrix.

The data of steady state creep under different temperatures and stresses of SnAg, SnAg–0.25RE are presented in Fig. 7 [5]. With the addition of the RE elements of mainly Ce and La, the creep strain decreased substantially. From Fig. 7, the stress exponents, n, at 393, 348 and 303 K (120, 75 and 30°C) are found to be 9.9, 11.3 and 12.3 respectively for SnAg. As for SnAg–0.25RE, the n values are 9.1, 12.0, and 12.1 respectively. Large values of stress exponent have been observed in dispersion strengthened alloys [24, 26] and the same also applies to the cases in [5] at 303 and 348 K (30 and 75°C). A large n value obtained from a low test temperature, e.g. 303 K (30°C), reflects that the dispersion precipitation strengthening is more dominant in this temperature range than at high temperatures. It is well known that precipitation strengthening can be sustained below $0.7T_m$, which is about 348 K (75°C) for SnAg. At 393 K (120°C, i.e.

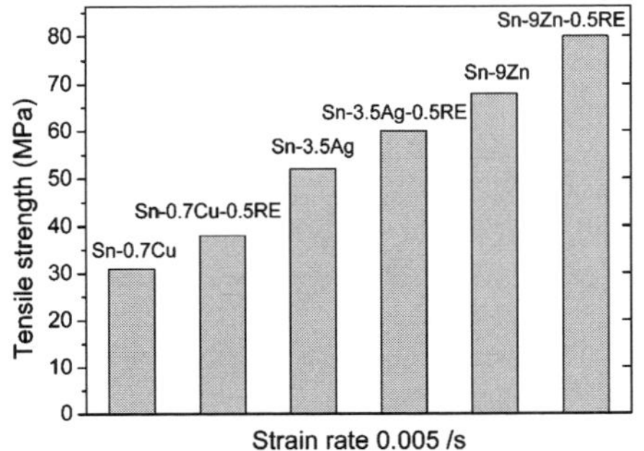

Fig. 5 Tensile strength of various Pb-free solders with and without addition of RE elements

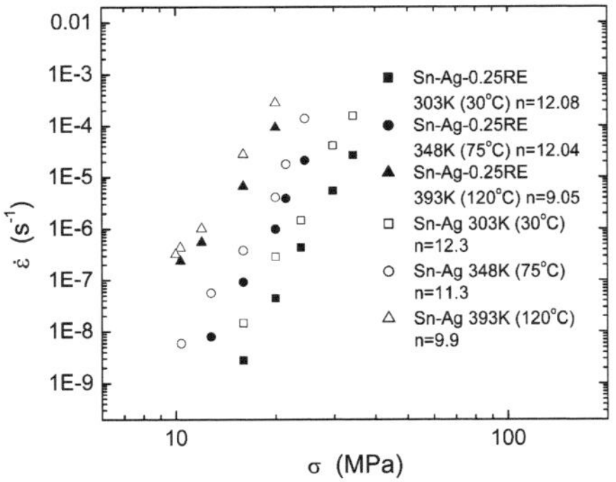

Fig. 7 Comparison of steady-state creep strain with creep time for Sn–Ag and Sn–Ag–0.25RE

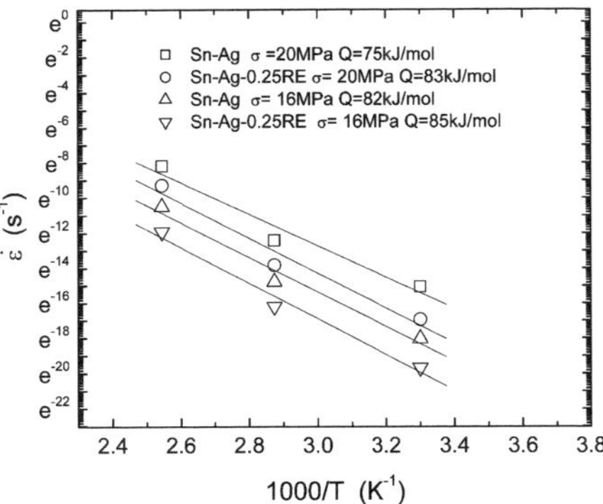

Fig. 8 Effect of temperature on steady-state creep rate of Sn–Ag and Sn–Ag–0.25RE

$0.8T_m$), when precipitation strengthening is not guaranteed, the value of stress exponents for both SnAg and SnAg–0.25RE are found to be significantly lower than those at 303 and 348 K (30 and 75°C). This shows that the results in [5, 24, 26] are in good agreement for dispersion strengthened alloys. Also, at a high temperature such as 393 K (120°C), the improvement of creep resistance at low stress is not significant when compared with those at higher applied stresses.

The creep rate, $\overset{\bullet}{\gamma}$, that relates to the microstructure in alloys with precipitates can be expressed as [27] :

$$\overset{\bullet}{\gamma} = \frac{b\rho\lambda^2}{\Lambda\Delta t} \tag{6}$$

where ρ is the dislocation density, Λ is the spacing of obstacles along an arrested dislocation under stress, Δt is the time required to by-pass an obstacle, and λ^2 is the average area per particle on the slip plane and is approximately the average area slipped per activated event. With the addition of RE elements, there is no doubt that the number of IMC particles would increase. As discussed in Section 3, with 0.25% RE addition for SnAg, a significant refinement of particles is seen, leading to a sharp decrease of λ. As the creep rate is proportional to λ^2, the addition of RE elements, which refines the particle size and spacing, has then provided an alloy with high creep resistance. At the same time, it is essential to minimize particle coarsening during prolonged exposure at elevated temperatures and to ensure that the finely dispersed IMC particles are still stable at higher temperature, so as to control the creep rate [5].

Figure 8 shows the effect of temperature on the steady-state creep rate at 16 and 20 MPa. The activation energies were calculated from these data. The results indicate that at both stress levels, the activation energy increases with RE addition. There is a strong indication that the improvement of creep behavior of eutectic SnAg with 0.25% RE doping is due to an increase in activation energy for creep.

It is well known that RE elements are active and they like to agglomerate at the grain boundaries and at the interfaces of phases. The energy of surface or grain boundary will decrease after the addition of RE. Thus the RE elements would play an important role on preventing the movement of dislocations and the nucleation of microcracks. Consequently the fine dispersive particles and the active properties of RE improves the creep behavior of the SnAg alloy [5].

The Sn–3.8Ag–0.7Cu alloy was doped with 0.1% RE elements in [8]. As a result, the creep rupture life of the solder joint at room temperature increased by about 7 times. In [28], solder joints made up from Sn–3.8Ag–0.7Cu–xRE solders were creep-tested. It was found that Sn–3.8Ag–0.7Cu–0.1RE had the lowest creep rupture time. For this solder at low stress, the true activation energy indicates that creep deformation was dominated by the rate of lattice self-diffusion, with stress exponent of 8.2. At high stress, this solder had stress exponent of 14.6, and the creep behavior was dominated by the rate of dislocation-pipe diffusion.

6 Wetting enhancement

Researches have shown that the wettability of Pb-free alloys on Cu is generally inferior to that of SnPb [29–31]. Nevertheless, a successful Pb-free alloy should have adequate wettability during hand soldering, infrared reflow, convection reflow and wave soldering. In other research work, the wetting behaviors of Sn–37Pb solder and Pb-free solders with or without RE addition were reported by using the wetting balance method [5, 8, 11, 14, 18, 23, 32,33]. The results of these wetting tests are compared in Table 2. In general, for a given type of solder using a particular flux, the wettability is improved as the soldering temperature is increased. The SnPb, SnAg, SnAg–0.25RE, SnAg–0.5RE, SnCu, SnCu–0.5RE, SnAgCu, SnAgCu–0.05RE, SnAgCu–0.1RE and SnAgCu–0.25RE, in which the RE elements are mainly Ce and La, can all be soldered successfully using the rosin mildly activated (RMA) flux [5, 11, 14, 23, 32]. The SnZn solder, with or without the addition of RE elements of mainly Ce and La, can only be soldered using the rosin activated (RA) flux [18]. The SnAg, SnCu and SnAgCu alloys would perform adequately when soldering onto Cu. The SnZn alloy has the worst wettability with Cu. The addition of RE elements helps to reduce the wetting angle. It can

be seen that the amount of RE elements needs to be optimized for the best wetting performance. For example, the alloys with their corresponding best wetting performance are SnAg–0.25RE (250°C), SnCu–0.5RE (260°C), SnAgCu–0.1RE (250°C) and SnZn–0.05RE (260°C) [5, 11, 18, 32].

The wettability of SnAg, SnCu, SnZn and SnAgCu with and without the addition of RE elements is compared in Table 2 [5, 11, 18, 32]. It can be seen that the SnAg, SnAg–0.25RE and SnAg–0.5RE alloys have roughly the same wetting time. The effect of adding an appropriate amount of RE elements of mainly Ce and La into the alloy is clearly demonstrated by noting that the wetting force achieved by SnAg–0.25RE is very close to that of SnPb. However, an excessive amount of RE addition has the effect of lowering the wetting force. This effect is confirmed by the wetting angle

Table 2 Results of wetting balance test from the literature

Solder Alloy	Substrate	Soldering temperature (°C)	Flux[†]	Contact Angle (°)	Wetting time (s)	Wetting force (mN)	Reference
Sn–37Pb	Cu	215	RMA	22			[29]
	Cu	215	WS	12			[29]
	Cu	215	NC	31			[29]
	Cu	220	RMA	12			[30]
	Cu	235	RMA	35	0.7	5.9	[5]
	Cu	260	RMA	17			[31]
Sn–3.5Ag	Cu	245	RMA	41			[29]
	Cu	245	WS	63			[29]
	Cu	245	NC	39			[29]
	Cu	250	RMA	47	0.7	5.0	[5]
	Cu	254	RMA	27			[30]
Sn–3.5Ag–0.25RE	Cu	250	RMA	43	0.7	5.7	[5]
Sn–3.5Ag–0.5RE	Cu	250	RMA	54	0.75	4.3	[5]
Sn–0.7Cu	Cu	260	RMA	49	1.05	4.7	[14]
Sn–0.7Cu–0.5RE	Cu	260	RMA	42	0.7	5.9	[14]
Sn–3.5Ag–0.7Cu	Cu	250	RMA	48	0.7	5.0	[32]
Sn–3.5Ag–0.7Cu	Matte Sn on Cu	260	RMA	very small			[33]
Sn–3.5Ag–0.7Cu	Bright Sn on Cu	260	RMA	very small			[33]
Sn–3.5Ag–0.7Cu	SnBi on Cu	260	RMA	very small			[33]
Sn–3.5Ag–0.7Cu–0.05RE	Cu	250	RMA	49	0.7	5.1	[32]
Sn–3.5Ag–0.7Cu–0.1RE	Cu	250	RMA	40	0.75	6.0	[32]
Sn–3.5Ag–0.7Cu–0.1RE	Matte Sn on Cu	260	RMA	very small			[33]
Sn–3.5Ag–0.7Cu–0.1RE	Bright Sn on Cu	260	RMA	very small			[33]
Sn–3.5Ag–0.7Cu–0.1RE	SnBi on Cu	260	RMA	very small			[33]
Sn–3.5Ag–0.7Cu–0.25RE	Cu	250	RMA	48	1.05	5.05	[32]
Sn–9Zn	Cu	245	RA	83	0.6	0.45	[18]
	Cu	260	RA	77			[18]
	Cu	290	RA	72			[18]
Sn–9Zn–0.05RE	Cu	245	RA	68	0.45	2.3	[18]
	Cu	260	RA	59			[18]
	Cu	290	RA	56			[18]
Sn–9Zn–0.1RE	Cu	245	RA	70	0.65	2.05	[18]
	Cu	260	RA	65			[18]
	Cu	290	RA	56			[18]

[†] RMA: rosin mildly activated; RA: rosin activated; WS: water soluble; NC: no-clean.

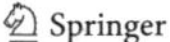

results, such that SnAg–0.25RE has the lowest wetting angle among SnAg, SnAg–0.25RE and SnAg–0.5RE, as shown in Table 2.

The wettability of the SnCu alloy is improved with the addition of 0.5 wt% of RE elements of mainly Ce and La. In fact, the wetting time of the SnCu–0.5RE alloy is smaller than that of SnPb, and its wetting force is nearly the same as that of SnPb [32]. As for Sn–Zn, the addition of 0.1 wt% RE elements has provided good wetting time and force. The choice of flux between RA, RMA, R and volatile organic compounds-free (VOC-free) is very important for this alloy, as indicated in Figs. 9, 10, such that only the RA flux is successful in providing wetting to the samples [18].

The SnAgCu alloy also has the same wetting improvement with 0.1 wt% of RE elements addition. As shown in Table 2, its wetting performance is then comparable to that of SnPb [11].

This enhancement obtained by adding a proper amount of RE can be explained through the effects of RE on the interfacial tensions. According to the Ce–Cu phase diagram [34], below the soldering temperature of 300°C, interaction between Ce and Cu is very low. It is therefore impossible to reduce the interfacial tension between the solder and the Cu substrate by the interaction of RE and Cu. On the other hand, as the same type of flux and substrate are used in each test, the interfacial tension between substrate and flux is a constant. As RE is an active element, it is easier for it to accumulate at the solder/flux interface in the molten state, and subsequently, the interfacial surface energy is decreased. Therefore, the interfacial tension between solder and flux, and thereby the contact angle can be reduced. However, an excessive amount of RE addition into the solder appears to lower the wetting angle due to the increase in solder viscosity. This is because RE elements tend to be oxidized easily during soldering, and an excessive amount of these oxides

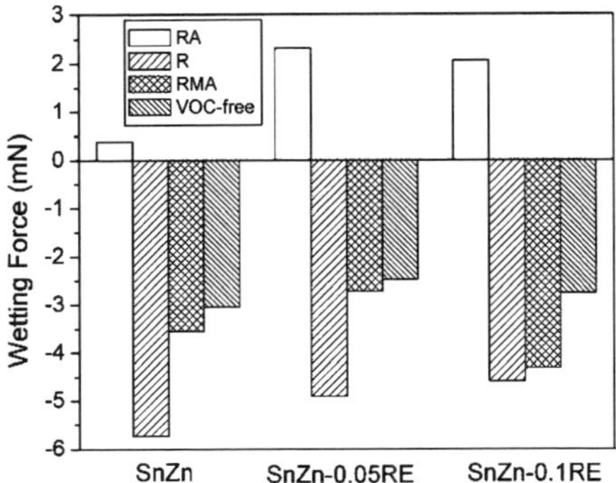

Fig. 10 The wetting forces of Sn–Zn–RE solders with using four different types of fluxes

increases the surface tension of the solder bath and inhibits the wetting behavior.

In protecting and enhancing the wettability of the Cu traces on the printed circuit board (PCB), some material is deposited onto the Cu traces as final surface finishing. The currently available methods of PCB surface finishing are hot air solder leveling (HASL), organic solder protectants (OSP), immersion finishing with Ag or Sn, and electroless Ni immersion Au (ENIG) [35]. There also exist finishes with electroplated Sn (matte and bright) and Sn–x wt%Bi layers. It is expected that the addition of RE elements in Pb-free solder, or even including it in the surface finishes of PCB will enhance wetting and reliability of electronic assemblies. A study along this direction was carried out with SnAgCu and SnAgCu–0.1RE alloys soldered onto Cu. The Cu substrate contained surface finishing of either matte Sn, bright Sn or Sn–3 wt%Bi (SnBi) electroplated layer [33]. The results of this wetting test can be found in Table 2. It was found that the wetting angle between the substrate and solder was very small, i.e. of the order of a few degrees.

7 Interfacial considerations

As metals are normally used on electronic circuits to carry electrical signals and currents, it is necessary to understand the interfacial reaction of solder alloy with these metals. In many forms of interconnections such as the surface mount technology (SMT) or BGA, the electronic solder becomes the major component of a solder joint. The solder alloy is required to provide mechanical support and heat dissipation, and it is required to have high reliability. The interaction between solder and substrate metal during soldering has been studied during the last few decades. The major

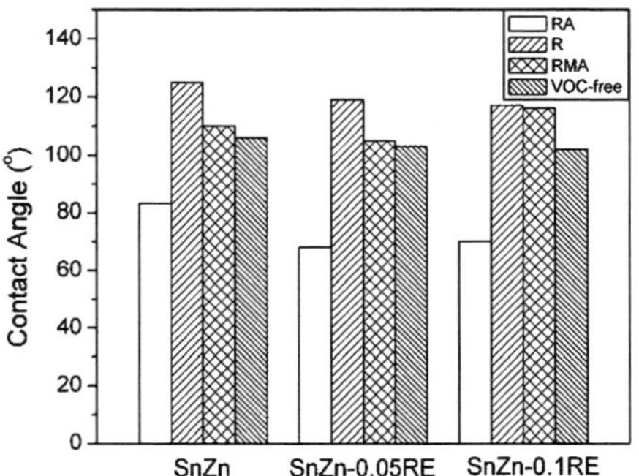

Fig. 9 The contact angles of Sn–Zn–RE solders with using four different types of flux

issue in soldering is the formation and growth of IMC particles between the solder and substrate [3]. Recent investigations have reported that the growth of the IMC layer (IML) would degrade solder joint reliability, thermal fatigue life, tensile strength and fracture toughness of solder joints [36, 37]. As Sn is the main component in Pb-free solders, their Sn content tends to react with the substrate material to form Cu_6Sn_5 and Cu_3Sn IMLs, especially during thermal aging. In this sense the solder interface needs to be investigated more systematically from the soldering stage to the in-service stage.

The studies of the IML in SnAg–RE, SnCu–RE and SnAgCu–RE solders were carried out in Refs. [38] and [11]. Cu_6Sn_5 and Cu_3Sn IMLs were confirmed to present in SnAg–RE and SnCu–RE solders on Cu [38]. The cross sections of as-reflowed SnAgCu and SnAgCu–0.25RE solders are shown in Fig. 11a, b, respectively [11]. For these two solders, the Cu–Sn intermetallic layer in the shape of scallops was formed between the copper substrate and solder. This Cu–Sn intermetallic was identified to be Cu_6Sn_5 by energy dispersive x-ray (EDX) analysis. The Ag_3Sn IMC was also detected in the solder matrix.

One way to consider the long-term reliability of solder alloys is by long-term thermal aging test. For example, test coupons of SnAgCu–RE solders after soldering were thermally aged at 170°C for 100, 200, 500, and 1,000 h [11]. The samples aged were also cross-sectioned to investigate the IML thickness. After 100 h of aging, a layer of Cu_3Sn was found between the Cu_6Sn_5 layer and the Cu substrate. Many Kirkendall voids were found between the Cu_3Sn IMC layer and the Cu substrate. The formation of these voids was due to the different diffusion properties of Sn and Cu during the aging process. Also, the scallop-shaped Cu_6Sn_5 layer was changed into a layer with fairly constant thickness during aging. Upon further aging (for 200, 500, and 1,000 h), both the Cu_6Sn_5 and Cu_3Sn intermetallic layers increased their thicknesses with the aging time. It was found that the coarsened Ag_3Sn IMC was embedded into the Cu_6Sn_5 layer. This observation is consistent with that of previous studies [39, 40]. Fig. 11c, d are the cross-sections of aged 1,000 h SnAgCu and SnAgCu–0.25RE solders, respectively. Since the coarsened Cu_6Sn_5 IML was brittle, cracks were easily initiated during mechanical grinding.

The as-soldered total IML thickness of the SnAg–RE and SnCu–RE solders is nearly the same as their counterparts without the addition of RE elements. Then, by aging the samples, the total thickness of the IML were increased. Figure 12 shows the variation of total IML thickness with aging time for SnAg–xRE solders. The same for SnCu–RE is shown in Fig. 13. It is seen clearly that that the total IML thickness increased as the aging time increased. Figure 14 shows the variation of the total IML thickness with aging time for SnAgCu–xRE. The same trends as those for SnAg–RE and SnCu–RE were obtained. Further, it can be seen that throughout the aging times considered, ie. up to 1,000 h, the IML growth was inhibited as a result of the addition of RE elements into SnAg, SnCu and SnAg–RE solders. The linear relationship observed from the aging curves suggested that a thermal

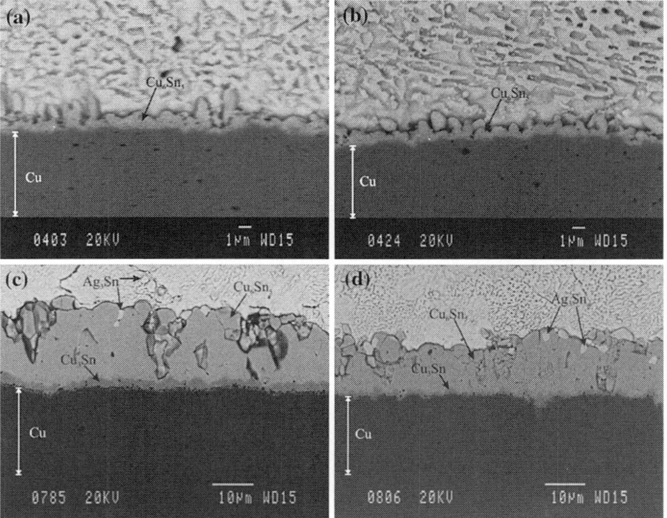

Fig. 11 BSE cross sections of as-soldered (**a**) Sn–3.5Ag–0.7Cu; (**b**) Sn–3.5Ag–0.7Cu–0.25RE; and of aged 1,000 h (**c**) Sn–3.5Ag–0.7Cu; and (**d**) Sn–3.5Ag–0.7Cu–0.25RE

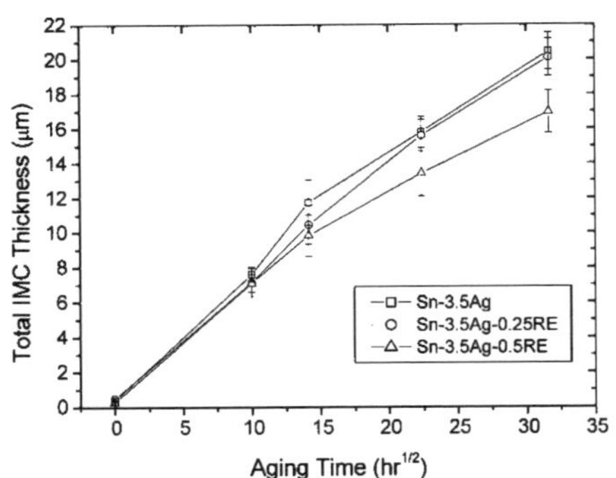

Fig. 12 Variation of Sn–3.5Ag–xRE total IMC thickness with aging time

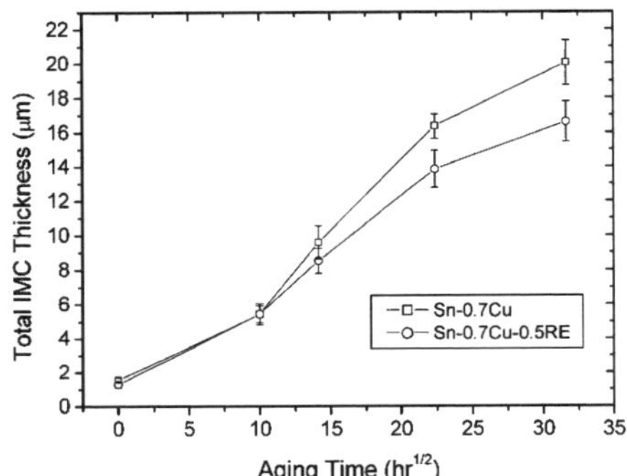

Fig. 13 Variation of Sn–0.7Cu–xRE total IMC thickness with aging time

activated diffusion process had taken place for the IML growth in these Pb–free–RE systems.

The reduction of the IML thickness of Pb-free solders with RE addition is attributable to the fact that

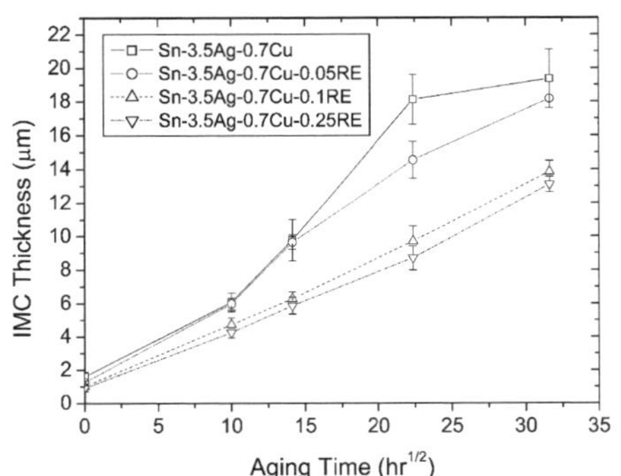

Fig. 14 Variation of Sn–3.5Ag–0.7Cu–xRE total IMC thickness with aging time

RE reacts with Sn at the interface between solder and the Cu_6Sn_5 IML. These Sn–RE compounds are very fine in size and can only be observed and accumulated under a slow cooling solidification. Tu et al. [41, 42] pointed out that Cu is the dominant species in forming Cu_6Sn_5 IML by marker displacement. This means that the Cu–Sn IMC is formed by the transport of Cu atoms across the Cu_6Sn_5 IMC. The reaction between Sn and Cu takes place at the Cu_6Sn_5–Sn interface. Since the activity of Sn at the interface of Cu_6Sn_5-solder is lowered by Sn–RE compound formation, lowering the activity of Sn will subsequently depress the growth of the Cu_6Sn_5 IML upon aging.

8 Electronic packaging considerations

In this section, the areas of potential applications of Pb-free alloys with the addition of RE elements are outlined.

In the area of microelectronics, interconnections are applied on semiconductors (e.g. Si, GaAs, SiC, and doped diamond), diffusion barriers (e.g. TiN and TaN), dielectrics (e.g. SiO_2, Ta_2O_5 and Si_3N_4), electrical conductors (e.g. Al, W, Cu and $CoSi_2$) and heat sinks (e.g. Al, AlN, diamond). The reliability requirements of the joints provided by the interconnection material cannot be over-emphasized. For telecommunication applications, materials need to be interconnected on an assembly substrate with sub-micrometer accuracy and stability in alignment to ensure maximum signal transmission [43]. It would be convenient if Pb-free solders could be developed to provide the above interconnections. The surfaces of these materials contain nitrides, carbides, oxides, etc., and they are known to be very difficult to bond with solders. Although certain electronic solders such as SnPb, SnAg, SnAu and SnBi have been considered for these interconnections, the results have not been satisfactory as their direct bonding to oxides is poor. One approach is to incorporate RE elements in solders to promote bonding because RE metals have a stronger affinity for oxygen than most metals and tend to reduce metal oxides via formation of RE oxides. For example, it has been proposed in [43, 44] that solders doped with RE elements are useful for bonding microelectronic device materials. As the researchers found that RE materials can promote chemical reactions at the bond interface and provide a very strong bond during soldering, RE materials such as Lu, Ce and Er were used for their studies. Two Pb-free solders Au-20 wt% Sn eutectic solder and eutectic SnAg solder were selected to dope with 2–2.5 wt% of RE elements of Lu, Er and Ce.

They synthesized these solders by employing a magnetic remote maneuvering device to plunge solid RE into molten base solder under $\sim10^{-5}$ Torr vacuum. The RE-bearing solders produce powerful bonding to various electronic materials and optical materials. The shear stress for fracture was typically in excess of 6.9 MPa [6]. It was mentioned in [6] that solders containing other RE elements such as Er and Ce can also produce bond strengths comparable to those obtained by the Lu-containing solders.

To demonstrate the suitability of the Pb-free-RE allows for general electronic packaging applications, SnAg–RE, SnCu–RE and SnZn–RE solders have been formed as BGA solder balls and soldered onto Au/Ni/ Cu under bump metallization (UBM) for studies [45–47]. The solder balls had diameter of ~760 μm. The pad metallization was an electroplated Ni/Au layer underneath Cu trace with a Au and Ni thickness of about 1 and 6 μm, respectively. The interfacial microstructure was investigated for the as-soldered samples. The variation of ball shear strength with aging time was also investigated.

At the interface of as-reflowed SnAg solder ball and the UBM, a whole Au layer is completely dissolved to the solder, and Ni–Sn IML is formed [45]. This IML is identified as Ni_3Sn_4 layer, as shown in Fig. 15a. After addition RE elements, Ni_3Sn_4 IML is also detected at the interface, as shown in Fig. 15b. Both $AuSn_4$ and Ag_3Sn IMCs in SnAg and SnAg–0.5RE solders are dispersed in the same way what are observed in the respective bulk solder ball. No Sn–RE phase is detected at the interface and in the bulk SnAg–0.5RE

solder ball after bonding. Upon aging, $(Au,Ni)Sn_4$ phases are found near the interface and Ag_3Sn IMCs increase their size in both solders. Fig. 15c, d illustrate the aged 500 h SnAg and SnAg–0.5RE solders, respectively. Insular Sn–RE phases consist of Sn, Ce and La were detected in SnAg–0.5RE solder, close to the Ni_3Sn_4 IML interface in the aged 500 h sample, as shown in Fig. 15d. After 1,000 h aging, the microstructures of SnAg and SnAg–0.5RE solders are similar to the aged 500 h respective ones. Because of the dissolution of the whole Au layer, Ni reacts with the molten solder directly to form Ni_3Sn_4 layer at the interface. With further aging, Ni atoms from interface diffuse into the $AuSn_4$ IMCs to form $(Au,Ni)Sn_4$ compounds. Since $(Au,Ni)Sn_4$ compounds were observed in both Sn–3.5Ag and Sn–3.5Ag–0.5RE solders, it means that the RE elements do not have any effect on $(Au,Ni)Sn_4$ compounds.

At the interface of as-reflowed SnCu and SnCu–0.5RE solders with the UBM, the same Cu–Ni–Sn ternary layer is formed, as shown in Fig. 16a, b, respectively [46]. According to the atomic composition of the Cu–Ni–Sn IML, it is regarded as the $(Cu,Ni)_6Sn_5$ layer [48–50], which has a structure based on Cu_6Sn_5 and verified by x-ray diffraction analysis [49]. The $(Cu,Ni)_6Sn_5$ layer thicknesses are about 1.5–2.0 μm in both solders. Cu_6Sn_5 and $AuSn_4$ IMCs are distributed near the interface. They have similar microstructure to the corresponding bulk ones. With long time aging, both coarsened $AuSn_4$ and Cu_6Sn_5 IMCs are seen and located near the interface. The $(Cu,Ni)_6Sn_5$ layer in Sn–0.7Cu solder increases its thickness to 2–2.5 μm

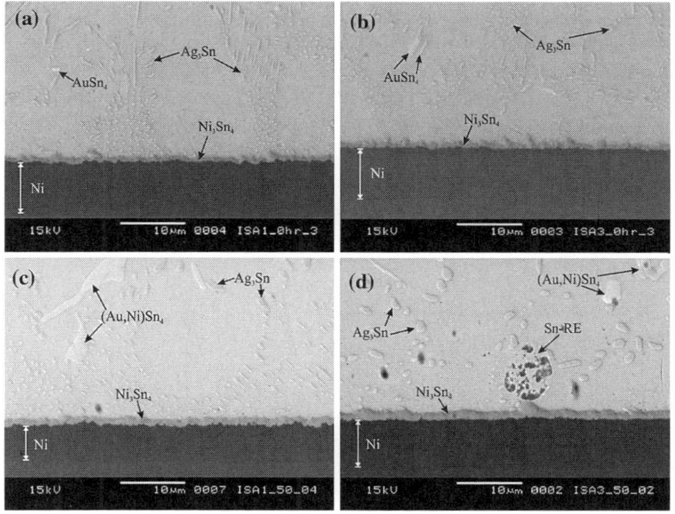

Fig. 15 BSE micrographs of interfacial microstructure (**a**) as-reflowed Sn–3.5Ag; (**b**) as-reflowed Sn–3.5Ag–0.5RE; (**c**) aged 500 h Sn–3.5Ag; and (**d**) aged 500 h Sn–3.5Ag–0.5RE

Fig. 16 BSE micrographs of interfacial microstructure (**a**) as-reflowed Sn–0.7Cu; (**b**) as-reflowed Sn–0.7Cu–0.5RE; (**c**) aged 1,000 h Sn–0.7Cu; and (**d**) aged 1,000 h Sn–0.7Cu–0.5RE

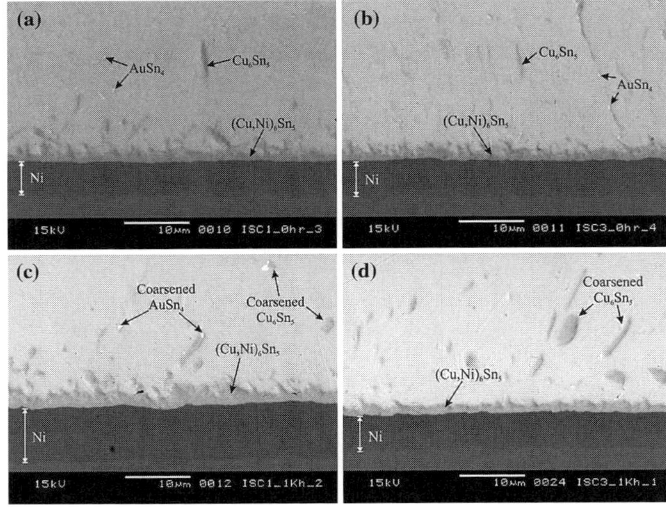

after 1,000 h aging, as shown in Fig. 16c. However, Sn–0.7Cu–0.5RE solder has a lower IML thickness than that of Sn–0.7Cu solder even under the same aging time, as shown in Fig. 16d.

At the interface of the as-reflowed SnZn solder ball on UBM, a smooth Ni–Zn–Sn IML was found with thickness of about 1 μm, as shown in Fig. 17a [47]. AuSn$_4$ IMCs were observed and were found to be similar to those in the bulk microstructure. Furthermore, circular Au–Zn precipitates of 3–4 μm in size were found near the interface. At the interface of the SnZn–0.5RE solder after reflow, the AuSn$_4$ IMCs and Au–Zn phases were not found. Instead, a spalled 1.5–2 μm thick Au–Zn layer above the Ni layer is seen, as shown in Fig. 17b. This spalled layer is thicker than the original Au layer just after reflow. It is interesting to note that the weight percentage of the spalled layer is similar to that of the Au–Zn phase found near the interface in the Sn–9Zn solder. The gap seen between the Au–Zn layer and Ni layer is found to be nearly pure Sn.

During thermal aging of the SnZn solder ball on UBM, the Ni–Zn–Sn IML begins to change, such that the top part of the IML becomes the NiZn$_3$ IMC, which appears darker than the Ni–Zn–Sn IML under SEM examination. The thickness of the NiZn$_3$ IMC increases with aging time. Also, the overall thickness of the Ni–Zn–Sn/ NiZn$_3$ IML increases from 1 to 1.5 μm after 500 h aging, as shown in Fig. 18a. This shows that the Ni layer is a good diffusion barrier layer for the Sn–9Zn alloy. Elongated AuSn$_4$ IMCs and coarsened Au–Zn phase are also observed close to the interface. The interfacial microstructure of the solder aged for 1,000 h is similar to that aged for 500 h. Upon aging of the Sn–9Zn–0.5RE sample, the spalled Au–Zn layer still exists and did not dissolve in the bulk solder. Since the Au–Zn layer is spalled from the Ni layer, the Sn–9Zn–0.5RE solder reacted with the Ni layer directly. This implies that the fine (Zn) phases are able to diffuse through the spalled Au–Zn layer to react with the Ni layer during thermal aging. As a result, a Zn-rich NiZn$_3$ layer forms on top of the Ni layer during the

Fig. 17 BSE micrographs of interfacial microstructure of as-reflowed (**a**) Sn–9Zn with line scan; and (**b**) Sn–9Zn–0.5RE

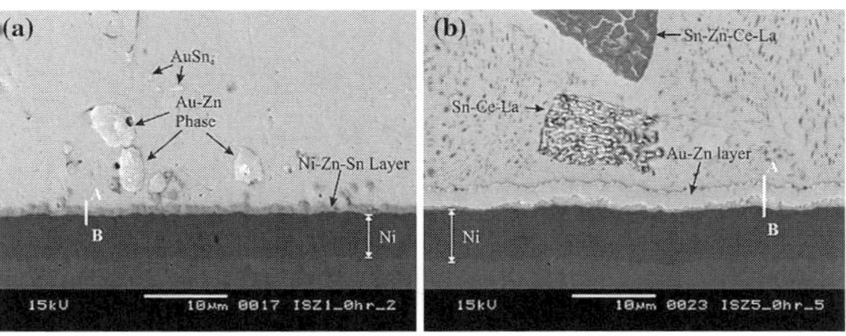

Fig. 18 BSE micrographs of (**a**) Sn–9Zn at the interface after 500 h of aging; (**b**) Sn–9Zn–0.5RE at the interface after 500 h of aging; (**c**) Sn–9Zn–0.5RE at the interface after 1,000 h of aging

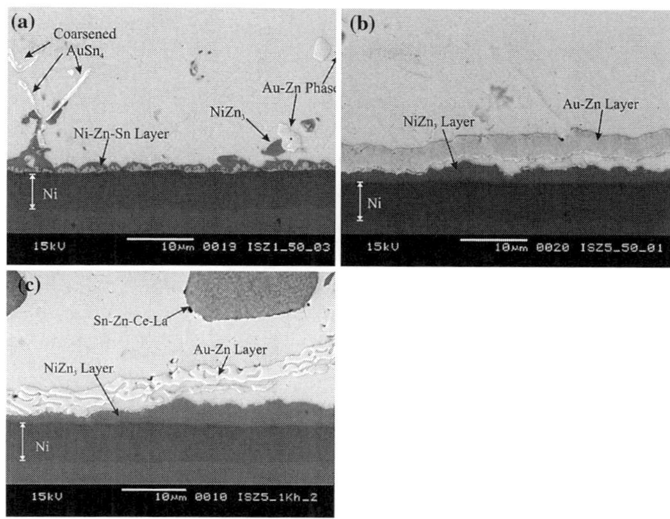

aging process. The remaining Ni layer thickness is about 5.5 µm after aging for 500 h, as shown in Fig. 18b. This means that about the top 0.5 µm of the Ni layer reacts with Zn to form the $NiZn_3$ layer of about 2 µm thick. After 1,000 h of aging, the spalled Au–Zn layer starts to dissolve in the bulk solder, as revealed in Fig. 18c. At this moment the averaged $NiZn_3$ layer thickness is more than 3 µm.

SnAg–RE, SnCu–RE and SnZn–RE solder ball packages were heat treated in a furnace at 150°C up to 1,000 h. The heat-treated packages were tested with a ball shear tester to determine the shear strength. Fig. 19a–c show the variation of the ball shear force with aging time for SnAg–RE, SnCu–RE, and SnZn–RE, respectively [45–47]. Both SnAg–RE and SnCu–RE solders have higher shear strengths than solders without RE addition for all aging times. This is attributable to the refinement of the microstructure of SnAg–RE and SnCu–RE. On the other hand, the SnZn solder alloy has higher shear strength than the SnZn–RE solder alloy for all aging times. As mentioned before, there is no IML formation between the spalled Au–Zn layer and the Ni layer in SnZn–RE after bonding. Instead, a nearly pure Sn layer exists between the Au–Zn layer and the substrate in the aged SnZn–RE solder alloy. However, for the SnZn solder, a NiSnZn ternary layer forms at the interface. This IML is believed to play an important role in providing a higher shear strength for SnZn than that of SnZn–RE.

It is worth noting that apart from the efforts above, the microstructure of SnAgCu–RE solder joint was studied in [51]. The research results indicate that when the RE content is 0.1 wt%, the needle-like phase is dominant in the eutectic like structure. However, when the RE content is 0.05 wt% or 0.25 wt%, the lamellar structure is the majority. It is predicted that these microstructural changes will affect the creep properties of SnAgCu–RE.

9 Concluding remarks and remaining issues

There have been a few advantages found with Pb-free alloys doped with RE elements. These include the improvement in tensile strength, creep strength and wettability. The long-term reliability was predicted to improve as interfacial studies reflected a more controlled growth of IMC layer. Further, these RE-doped alloys were seen to possess good interfacial behavior when used in BGA balls, as well as solder for microelectronic materials which were regarded as difficult to be soldered. With their excellent creep resistance and improvement in wettability on difficult substrates, they are seen to be viable for optoelectronic packaging applications.

It is seen that Pb-free solder alloys doped with RE elements are now ready for reliability studies to prove their feasibility as working solders.

Just like other Pb-free solders, Pb-free-RE solders need to undergo vigorous long-term reliability tests before their introduction for field use. These include the normally adopted thermal cycling and thermal aging tests. As electronic components are getting smaller and smaller, and at the same time the number

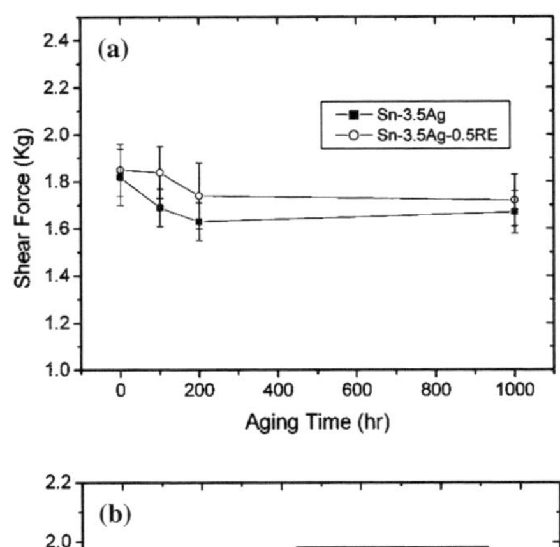

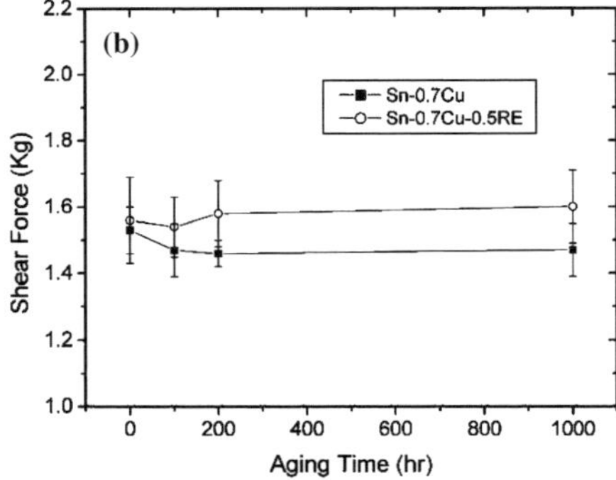

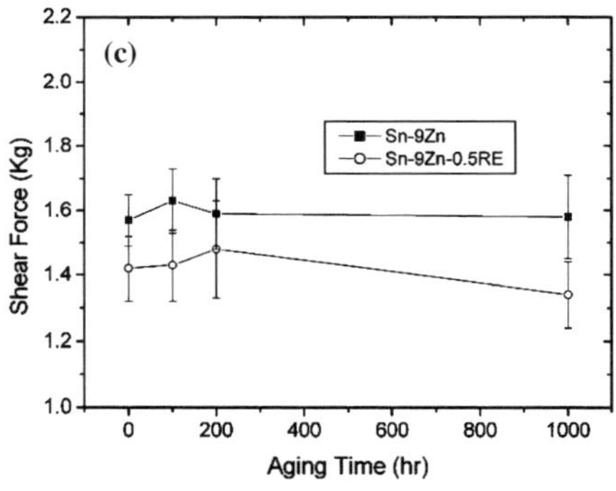

Fig. 19 Variation of ball shear strength with aging time in system of (**a**) Sn–Ag–RE; (**b**) Sn–Cu–RE; and (**c**) Sn–Zn–RE

of inputs and outputs per unit area is increased, the current density of many modern electronic interconnections becomes high. Therefore, the electromigration effects due to high current density need to be investigated. These tests are currently available, and they can be adopted for tests on Pb-free-RE alloys.

In addition to long-term reliability, the more immediate issues, i.e. their behavior when soldered onto certain common surface finishing materials on substrate metallizations for Pb-free soldering needs to be studied.

At the same time, the potential problems that would arise during production are required to be studied. For example, the RE elements are known to be easily oxidized. This effect on common soldering processes such as SMT needs to be thoroughly understood.

Optoelectronic packaging activities have recently grown substantially and there is a strong demand for cheaper solder alloy than the currently used one, e.g. SnAu, for optoelectronic packaging applications. The development of Pb-free-RE solders for successful direct bonding on optoelectronic materials would be seen to be an important activity.

Acknowledgements The work described in this report was fully supported by a grant from the Research Grants Council of the Hong Kong Special Administrative Region, P. R. China [Project No. CityU 1026/04P].

References

1. K.N. Tu, K. Zeng, Mater. Sci. Eng. R **34**(1), 1 (2001)
2. International Printed Circuit Association Solder Products Value Council, "White Paper: IPC-SPVC-WP-006 Round Robin Testing and Analysis, Lead Free Alloys: Tin, Silver, Copper", 12 August 03
3. M. Abtew, G. Selvaduray, Mater. Sci. Eng. R **27**(5/6), 95 (2000)
4. Z.G. Yu, M.B. Chen, in *Rare earth elements and their applications* (Metallurgical Industry Press, China, 1995), pp. 138
5. C.M.L. Wu, D.Q. Yu, C.M.T. Law, L. Wang, J. Mater. Res. **31**(9), 3146 (2002)
6. H. Mavoori, A.G. Ramirez, S. Jin, Appl. Phys. Lett. **78**(19), 2976 (2001)
7. C.M.T. Law, PhD Thesis on "Reliability and Interfacial Reaction of Lead-free Solder Alloys Doped with Rare Earth Elements", Department of Physics and Materials Science, City University of Hong Kong (2004)
8. Z.G. Chen, Y.W. Shi, Z.D. Xia, Y.F. Yan, J. Electron. Mater. **32**(4), 235 (2003)
9. J. Glazer, J. Electron. Mater. **23**(8), 693 (1994)
10. K. Suganuma, Curr. opin. solid mater. sci. **5**, 55 (2001)
11. C.M.T. Law, C.M.L. Wu, D.Q. Yu, L. Wang, J.K.L. Lai, J. Electron. Mater. **35**(1), 89 (2006)
12. C.M.L. Wu, D.Q. Yu, C.M.T. Law, L. Wang, J. Electron. Mater. **31**(9), 921 (2002)
13. K. Suganuma, K. Niiara, J. Mater. Res. **13**(10), 2859 (1998)
14. C.M.L. Wu, D.Q. Yu, C.M.T. Law, L. Wang, J. Electron. Mater. **31**(9), 928 (2002)
15. H. Baker et al., (ed.), Alloy phase diagrams, ASM Handbook 3, Materials Park, OH, 1990, pp. 137
16. H. Baker et al., (ed.), Alloy phase diagrams, ASM Handbook 3, Materials Park, OH, 1990, pp. 275
17. X. Ma, F. Yoshida, Mater. Lett. **56**(4), 441 (2002)
18. C.M.L. Wu, D.Q. Yu, C.M.T. Law, L. Wang, J. Electron. Mater. **32**(29), 63 (2003)
19. Z.D. Xia, Z.G. Chen, Y.W. Shi, N. Mu, N. Sun, J. Electron. Mater. **31**(6), 564 (2002)

20. Z.G. Chen, Y.W. Shi, Z.D. Xia, Y.F. Yan, J. Electron. Mater. **31**(10), 1122 (2002)
21. E. Gebhardt, G. Petzow, **50**, 597 (1959)
22. Q.J. Zhai, S.K. Guan, Q.Y. Shang, *Alloy Thermo-Mechanism: Theory and Application* (Metallurgy Industry Press, Beijing, 1999)
23. D.Q. Yu, J. Zhao, L. Wang, J. Alloys Compd. **376**, 170 (2004)
24. R.J. Mccabe, M.E. Fine, JOM **52**(6), 33 (2000)
25. M.L. Huang, L. Wang, C.M.L. Wu, J. Mater. Res. **17**(11), 2897 (2002)
26. V.I. Igoshev, J.I. Kleiman, D. Shanguan, C. Lock, S. Wong, J. Electron. Mater. **27**(12), 1367 (1998)
27. W.C. Oliver, W.D. Nix, Acta Metall. **30**(7), 1335 (1982)
28. Z.G. Chen Y.W. Shi, Z.D. Xia, J. Electron. Mater. **33**(9), 964 (2004)
29. P.T. Vianco, A.C. Claghorn, Solder. Surf. Mt. Technol. **8**(3), 12 (1996)
30. C.C. Tu, M.E. Natishan, Solder. Surf. Mt. Technol. **12**(2), 10 (2000)
31. P.T. Vianco, D.R. Frear, JOM, **23**(7), 14 (1993)
32. C.M.L. Wu, D.Q. Yu, C.M.T. Law, L. Wang, *Materials Science and Engineering Reports*, R44/1 pp. 1, April 2004
33. H.C.B Woo, in MSc Thesis of "Solderability & Microstructure of Lead-free Solder in Leadframe Packaging", Department of Physics and Materials Science, City University of Hong Kong (2005)
34. P. Nash, A. Nash, H. Baker et al., (ed.), Alloy phase diagrams, ASM Handbook 3, Materials Park, OH, 1990, Section 2, p. 32
35. IPC, IPC Roadmap on Lead-free soldering, 3rd draft, (2003)
36. D.R. Frear, JOM **48**(5), 49 (1996)
37. J.K. Shang, D. Yao, J. Electron. Packag. **118**(3), 170 (1996)
38. C.M.T. Law, C.M.L. Wu, D.Q. Yu, K.Y. Lee, M. Li, in *Proceedings of the Materials Science & Technology* 2003 "Solderability and Growth of Intermetallic Compounds upon Aging of Sn–Ag–RE and Sn–Cu–RE Lead-free Alloys" Conf., Chicago, 9–12 Nov (2003)
39. D.R. Flanders, E.G. Jacobs, R.F. Pinizzotto, J. Electron. Mater. **26**, 883 (1997)
40. S. Choi, T.R. Bieler, J.P. Lucas, K.N. Subramanian, J. Electron. Mater. **28**, 1209 (1999)
41. K.N. Tu, R.D. Thompson, Acta Metall. **30**, 947 (1982)
42. K.N. Tu, Phys. Rev. B **49**, 2030 (1994)
43. H. Mavoori, A.G. Ramirez, S. Jin, J. Electron. Mater. **31**(11), 1160 (2002)
44. A.G. Ramirez, H. Mavoori, S. Jin, Appl. Phys. Lett. **80**(3), 398 (2002)
45. C.M.T. Law, C.M.L. Wu, in *Proceedings of the 6th IEEE CPMT Conference on High Density Microsystem Design and Packaging and Component Failure Analysis*, 30 June–3 July 2004, Shanghai, pp. 60–65
46. C.M.L. Wu, C.M.T. Law, in *Proceedings of the "Microstructure and Shear Strength of Aged Sn–Cu–RE BGA Solder Bumps", Conf. Materials Science & Technology*, Chicago, 9–12 Nov, (2003)
47. C.M.T. Law, C.M.L. Wu, D.Q. Yu, M. Li, D.Z. Chi, IEEE Trans. Adv. Pack. **28**(2), 252 (2005)
48. A. Zribi, A. Clark, L. Zavalij, P. Borgesen, E.J. Cotts, J. Electron. Mater. **30**, 1157 (2001)
49. W.T. Chen, C.E. Ho, C.R. Kao, J. Mater. Res. **17**, 263 (2002)
50. C.E. Ho, R.Y. Tsai, Y.L. Lin, C.R. Kao, J. Electron. Mater. **31**, 584 (2002)
51. B. Li, Y. Shi, Y. Lei, F. Guo, Z. Xia, B. Zong, J. Electron. Mater. **34**, 217 (2005)

Compression stress–strain and creep properties of the 52In–48Sn and 97In–3Ag low-temperature Pb-free solders

Paul T. Vianco · Jerome A. Rejent · Arlo F. Fossum ·
Michael K. Neilsen

Published online: 23 September 2006
© Springer Science+Business Media, LLC 2006

Abstract Lead (Pb)-free, low melting temperature solders are required for step-soldering processes used to assemble micro-electrical mechanical system (MEMS) and optoelectronic (OE) devices. Stress–strain and creep studies, which provide solder mechanical properties for unified creep-plasticity (UCP) predictive models, were performed on the Pb-free 97In–3Ag (wt.%) and 58In–42Sn solders and counterpart Pb-bearing 80In–15Pb–5Ag and 70In–15Sn–9.6Pb–5.4Cd alloys. Stress–strain tests were performed at 4.4×10^{-5} s^{-1} and 8.8×10^{-4} s^{-1}. Stress–strain and creep tests were performed at –25, 25, 75, and 100°C or 125°C. The samples were evaluated in the as-fabricated and post-annealed conditions. The In–Ag solder had yield stress values of 0.5–8.5 MPa. The values of ΔH for steady-state creep were 99 ± 14 kJ/mol and 46 ± 11 kJ/mol, indicating that bulk diffusion controlled creep in the as-fabricated samples (former) and fast-diffusion controlled creep in the annealed samples (latter). The In–Sn yield stresses were 1.0–22 MPa and were not dependent on an annealed condition. The steady-state creep ΔH values were 55 ± 11 kJ/mol and 48 ± 13 kJ/mol for the as-fabricated and annealed samples, respectively, indicating the fast-diffusion controlled creep for the two conditions. The UCP constitutive models were derived for the In–Ag solder in the as-fabricated and annealed conditions.

1 Introduction

Currently, the electronics industry is being tasked with developing lead (Pb)-free solders as alternatives to the traditional tin–lead (Sn–Pb) alloys for second-level (printed wiring assembly) interconnections. However, there are also applications that require lower temperature joining processes, which will also require Pb-free solders. For example, some high-valued, high-reliability military and space electronics are assembled with low-melting temperature solders simply to minimize damage to heat-sensitive components. On the other hand, there are complex assemblies that require low-temperature alloys as part of a multiple step-soldering process. In addition, circuitry constructed of precious metal conductors (e.g., hybrid microcircuits) use the low-temperature solders to avoid the rapid dissolution of conductor traces and pads that would occur with molten (Sn-based) solders that require higher process temperatures.

At present, there is a developing need for Pb-free, low-temperature soldering methods for advanced optoelectronic (OE), micro-electrical mechanical systems (MEMS), and micro-optical electrical mechanical system (MOEMS) devices. These products are highly complex, often having several devices in a single package, and as such, require multiple attachment steps. Also, a number of these components are

---

Sandia is a multiprogram laboratory operated by Sandia Corporation, a Lockheed Martin Company, for the US Dept. of Energy's National Nuclear Security Administration under contract DE-AC04-94AL85000.

P. T. Vianco (✉) · J. A. Rejent · A. F. Fossum ·
M. K. Neilsen
Sandia National Laboratories, PO Box 5800, Albuquerque, NM 87185, USA
e-mail: ptvianc@sandia.gov

heat-sensitive, thereby requiring the use of low-temperature soldering processes.

Stringent package environments and device alignment requirements can be accommodated by the use of low-temperature, Pb-free solders in OE, MEMS, and MOEMS components. Low-temperature solders can reduce the temperature excursions required to assemble such packages. Also, they readily deform by creep, thereby reducing the residual stresses that can build up because of thermal expansion mismatches between joined structures. These residual stresses can give rise to the catastrophic failure of brittle components (e.g., GaAs laser diodes) immediately after assembly or generate misalignment between components later on, which degrades product performance in the field. Lastly, because solder joints are often essential for thermal management in OE packages, their integrity is critical towards the long-term reliability of several components (e.g., laser diodes).

Two low-melting temperature solders have been used rather extensively for a number of applications, including high-reliability printed circuit assemblies. Those materials, together with their respective solidus temperatures (T_s) and liquidus temperatures (T_l), are the 80In–15Pb–5Ag (wt.%) alloy, $T_s = 142°C$, $T_l = 149°C$; and the 70In–15Sn–9.6Pb–5.4Cd solder, $T_s = T_l = 125°C$. The compositions were abbreviated as In–Pb–Ag and In–Sn–Pb–Cd, respectively. These solders are particularly suited for Au-based conductors because the primary component, In, has a lower dissolution rate for Au than do Sn-based solders. Also, the In–Pb–Ag and In–Sn–Pb–Cd alloys exhibit excellent solderability and have provided the required service reliability.

Unfortunately, Cd and Pb have been targeted for elimination from electronic products. Therefore, Pb- and Cd-free alternative solders were sought for the In–Pb–Ag and In–Sn–Pb–Cd alloys. Based upon the need for similar liquidus and solidus temperatures, the two Pb-free solders, 97In–3Ag ($T_s = T_l = 143°C$) and 52In–48Sn ($T_s = T_l = 118°C$), were identified to replace the In–Pb–Ag and In–Sn–Pb–Cd alloys, respectively. Both alloys, which were abbreviated as In–Ag and In–Sn, respectively, are eutectic compositions [1].

The microstructures and primary phases of the In–Ag and In–Sn solders appear in Figs. 1 and 2, respectively. All samples were cooled at 10°C/min. The optical micrographs were digitally modified to enhance the second phase particles due to the difficulty with preparing In-based alloys for metallographic examination. In Fig. 1, the In–Ag alloy has a fine dispersion of very small $AgIn_2$ particles (light) within an essentially 100% In matrix (dark). The In–Sn alloy solidified

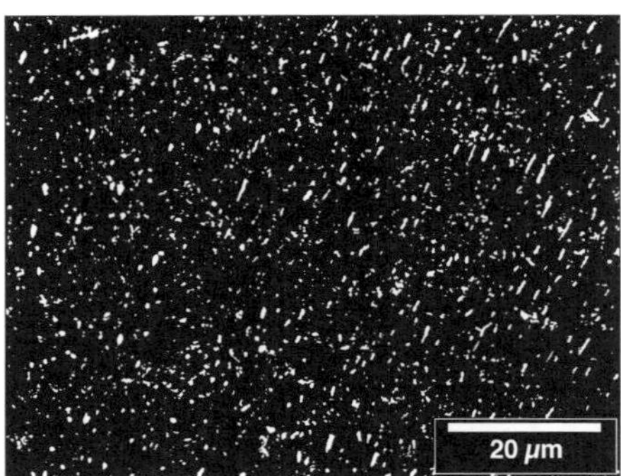

Fig. 1 Optical micrograph showing the In–Ag solder microstructure after solidification at 10°C/min. The image was digitally modified to enhance the light, Ag_3Sn particle phase against the darker, In-rich matrix phase

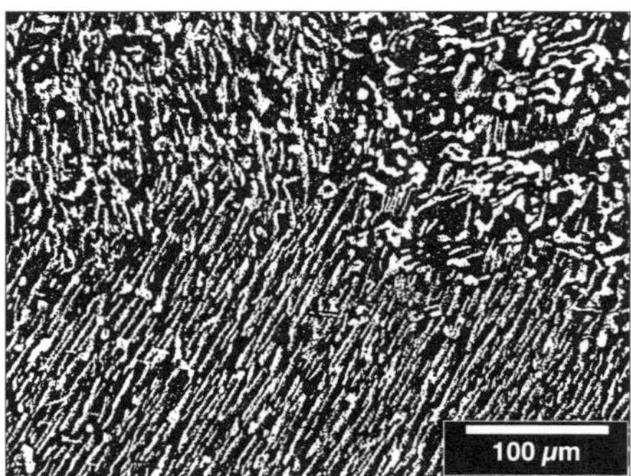

Fig. 2 Optical micrograph showing In–Sn solder microstructure after solidification at 10°C/min. The image was digitally modified to enhance the light, Sn-rich lamellae phase against the darker, In-rich matrix phase

into a more traditional lamellar microstructure. Each of the In-rich β phase (dark) and Sn-rich γ phase (light) had relatively high solubility limits for the other element—approximately 30 wt.% Sn and 23 wt.% In, respectively.

A second requirement of the candidate Pb-free solders was that they exhibit thermal mechanical fatigue (TMF) properties that were comparable to, or better than, those of the Pb- and Cd-bearing materials. Given the wide variety of potential applications of these solders in OE, MEMS, and MOEMS technologies, it became apparent that predicting the long-term reliability of the solder joints would require the use of computational models. Only through such models could the variety of solder joint geometries, substrate materials, and service environments be addressed in a timely and cost-effective manner. The basis of a TMF computational model is the unified creep-plasticity

(UCP) constitutive model that combines creep and plastic strain rates into a single inelastic strain rate. Experiments must be performed at a variety of temperatures and strain rates to obtain the material parameters for such models.

Several studies have investigated the mechanical properties of the low-temperature, Pb-free solders. Mei and Morris examined the creep behavior of 52In–48Sn alloy at temperatures of 20, 65, and 90°C, using a nine-solder joint, double-lap shear specimen [2]. The applied shear stresses ranged from 0.34 MPa to 6.9 MPa. The authors fitted the steady-state shear strain rate values, $d\gamma/dt$, to the power law (stress) expression $A\tau^{n}\exp(-\Delta H/RT)$ where A is a constant; τ is the applied shear stress; n is the power law exponent; ΔH is the apparent activation energy; R is the universal gas constant; and T is temperature. The values of A, n, and ΔH were 1.94×10^{4} s^{-1}, 3.2, and 96 kJ/mol, respectively. The relatively high ΔH value suggested that bulk diffusion was the active mechanism during creep.

Goldstein and Morris examined the shear stress–strain and creep behaviors of the 52In48Sn eutectic alloy using the double-lap shear (single joint) specimen [3]. The creep tests were performed at temperatures of 0–75°C. The shear stress–strain tests were performed at strain rates of 2×10^{-4} s^{-1} to 8×10^{-4} s^{-1}; the creep tests used stresses from 1 MPa to 12 MPa. Shear yield stresses ranged from 7.8 MPa to 11.4 MPa between the respective strain rate limits, when tests were performed at 40°C. The stress–strain curves of these particular tests exhibited strain-softening plastic deformation. The creep parameters were determined according to the power law expression. The value of n was 3.3, which was very similar to that determined in [2]. The ΔH was 70 kJ/mol, which was less than the value obtained in that prior study, suggesting that a fast-diffusion process (e.g., grain boundary diffusion) contributed to creep deformation.

Hwang and Vargas examined the stress–strain and creep properties of the 60In–40Sn off-eutectic composition ($T_{s} = 122$°C, $T_{l} = 118$°C) using bulk tensile test samples [4]. The authors measured the room temperature 0.2% yield strength, ultimate tensile strength, and uniform elongation to be 4.6 MPa, 7.7 MPa, and 5.5%, respectively. (The authors did not note which of the two cited strain rates was used to collect that data: 8.8×10^{-4} s^{-1} or 4.2×10^{-3} s^{-1}.)

The TMF performance of the In–Sn solder was documented with those of three other low-temperature solders in a study by Seyyed [5]. The test methodology was the thermal cycling of surface mount printed wiring assemblies. The strength of the gull-wing In–Sn solder joints exhibited a 75% decrease between 0 and 6,000 cycles (–10°C to 70°C; 1 min dwell; 20°C/min ramps). Similar behaviors were observed with the other low-temperature solders. The J-lead solder joint exhibited very little degradation for the In–Sn as well as the other solders. The strengths of the In–Sn solder joints were generally one-half the strengths of the baseline Sn–Pb solder joints.

It was concluded that there was insufficient stress–strain and creep data on the In–Sn or In–Ag alloys in the literature to develop the required UCP constitutive models. Therefore, a series of constant strain-rate, uniaxial stress–strain and creep compression experiments were performed on these Pb-free alloys. Parallel studies were also performed on the Pb-bearing counterparts, In–Pb–Ag and In–Sn–Pb–Cd, respectively. The yield stress and static elastic modulus were calculated from the true-stress, true-strain curves. The yield stress was also used to determine the applied stresses for the creep tests.

The quantitative metric of the creep tests was the *minimum creep rate*, $d\varepsilon/dt_{min}$, which characterizes the secondary or steady-state creep stage. Regardless of whether the exhaustion theory or the strain hardening theory governs primary creep, it has been proposed that similar deformation kinetics likely govern the two regimes [6–8]. Thus, the rate kinetics, which were based upon $d\varepsilon/dt_{min}$, could be used to predict both primary and secondary creep through the UCP constitutive model.

The UCP equation cannot address the tertiary creep stage. This stage, which is marked by an accelerating creep rate with time, results from the generation of cracks that quickly lead to creep rupture. Nevertheless, this study examined the strain–time curves for the onset of tertiary creep.

The physical metallurgy of the solders is an important facet of mechanical deformation. Microstructural features such as grain boundaries and dislocation structures determine the strain response of the material to the applied stress. Also, an understanding of the individual microstructures provides the means to more fully understand the stress–strain and creep behaviors of the solder. However, a detailed analysis of the physical metallurgy of the candidate solders was not performed in the current study. Solder microstructures were examined when there was the need to confirm tertiary creep behavior.

2 Experimental procedures

The solder compositions that were evaluated in this study were: 52In–48Sn (abbreviated as In–Sn),

70In–15Sn–9.6Pb–5.4Cd (In–Sn–Pb–Cd), 97In–3Ag (In–Ag), and 80In–15Pb–5Ag (In–Pb–Ag). All compositions were given in weight percent. Liquidus and solidus temperatures were cited in the Introduction section. The alloys were tested in the as-fabricated (i.e., as-cast) condition as well as after one of the three annealing treatments. The annealing temperatures were 52°C for the In–Sn and In–Sn–Pb–Cd solders, and 67°C for the In–Ag and In–Pb–Ag alloys. The annealing times were 8, 16, and 24 h. The temperature values were calculated, based upon approximately 82% of the solidus temperatures (K) from the two alloy groups. Imposing an annealing treatment prior to testing allowed for an evaluation of the effects that microstructural stabilization have on the mechanical properties.

The compression test method was used to collect stress–strain and creep data. The nominal dimensions of the cylindrical samples were 10 mm diameter and 19 mm length. The stress–strain tests were performed at one of two (engineering) strain rates: 4.4×10^{-5} s^{-1} or 8.8×10^{-4} s^{-1}. These relatively low values were selected to measure the deformation properties at strain rates that were commensurate with typical TMF conditions. The test temperatures were listed in Table 1 for each of the solder compositions. Duplicate samples were tested per each time, temperature, and annealing condition. Further details of the experimental procedures have been presented in previous publications [9, 10].

The stress–strain tests were analyzed to obtain the following properties: (1) yield stress, (2) static elastic modulus, and (3) yield strain. The yield strain data will not be reported here. Engineering stresses and strains were used to calculate the yield stress because they were suitable approximations to true stresses and true strains at these low strain values [11]. The yield stress was determined, using the 0.2% offset method [12]. Error terms were based upon scatter of the load–displacement curve at the intersection between the plot and the 0.2% intercept line.

The static elastic modulus was calculated from the linear portion of the stress–strain curve. A linear least-squares fit provided the slope (Δload/Δdisplacement). That slope value was multiplied by l_0/A_0 to determine the static elastic modulus, where l_0 and A_0 were the initial (gage) length and cross sectional area, respectively.

Lastly, the load–displacement plots were converted to true stress (σ) and true strain (ε) curves. The true stress and true strain values were calculated from the load–displacement data using Eqs. 1 and 2, respectively, shown below [11]:

$$\sigma = (L/A_0)(\Delta l/l_0 + 1) = (L/A_0)(l/l_0) \tag{1}$$

$$\varepsilon = \ln(l/l_0) \tag{2}$$

Compression creep tests were performed, using the same test temperatures as listed in Table 1. The samples were tested in the as-fabricated condition or after annealing for 16 h at one of the two temperatures noted above. The initial applied stresses were identified from the yield stresses measured at the slower strain rate, 4.4×10^{-5} s^{-1}, per each of the test temperatures. Specifically, those stress values were 20%, 40%, 60%, and 80% of the mean yield stress measured between samples in both the as-fabricated and 16 h annealing condition. The stresses are listed in Table 2 as a function of test temperature.

The creep experiments were allowed to progress through the secondary or steady-state creep stage to determine the minimum creep rate, $d\varepsilon/dt_{min}$. At very low stress for which, primary creep occurred over a longer time period, the test was concluded after approximately 3300 min (2.3 days). In such cases, the minimum creep rate was calculated from data taken at the end of the test. The parameters $d\varepsilon/dt_{min}$ as well as the applied stress, σ, and temperature, T, were expressed by the sinh law creep equation (3) below:

$$d\varepsilon/dt_{min} = f_0 \sinh^p(\alpha\sigma) \exp(-\Delta H/RT) \tag{3}$$

where f_0 is a constant (s^{-1}); p is the sinh term exponent (a constant also); α is the stress coefficient (MPa^{-1}); ΔH is the apparent activation energy (J/mol); R is the universal gas constant (8.314 J/mol-K) ; and T is the temperature (K). The applied stress was calculated from cross section of the deformed sample, using Eq. 1. The stress coefficient, α, is temperature dependent. However, its dependence was sufficiently weak such that it was considered a constant. (The product, $\alpha\sigma$, is often referred to as an *effective stress*.) The sinh law was selected because of its success for representing secondary creep deformation in metals and alloys, including solders, by avoiding the power law breakdown phenomenon encountered when attempting to use the simpler power law representation over a substantial stress range [13, 14].

Table 1 Compression test temperatures used for each solder alloy composition

Solder	Test temperature (°C)
In–Sn	−25, 25, 75, 100
In–Sn–Pb–Cd	−25, 25, 75, 100
In–Ag	−25, 25, 75, 125
In–Pb–Ag	−25, 25, 75, 125

Table 2 Compression creep stresses used for each temperature and solder alloy composition

Solder	Annealing conditions	Nominal stress (MPa)	Test temperature (°C)
In–Sn	52,°C, 16 h	3.3, 6.7, 10, 13	–25
		3.7, 5.6, 7.5	25
		0.86, 1.3, 1.7	75
		0.47, 0.71, 0.94	100
In–Sn–Pb–Cd	52°C, 16 h	5.4, 7.2, 11	–25
		2.7, 4.0, 5.4	25
		0.50, 0.70, 1.0	75
		0.27, 0.41, 0.55	100
In–Ag	67°C, 16 h	1.2, 2.3, 3.5, 4.6	–25
		0.75, 1.5, 2.3, 3.0	25
		0.84, 1.3, 1.7	75
		0.16, 0.33, 0.49, 0.65	125
In–Pb–Ag	67°C, 16 h	3.9, 7.8, 12, 16	–25
		1.9, 3.7, 5.6, 7.4	25
		0.50, 1.0, 1.5, 2.0	75
		0.13, 0.27, 0.40, 0.54	125

The values of f_0, Q/R, and p were calculated, using a multivariable, linear regression analysis that was performed on the logarithm of equation (3). The parameters $1/T$ and $\ln[\sinh(\alpha\sigma)]$ were the independent variables and $\ln(d\varepsilon/dt_{\min})$ was the dependent variable. The regression analysis was performed for different values of α. The value of α was selected, which maximized the square of the correlation coefficient, R^2, at the 95% confidence interval.

3 Results and discussion

3.1 In–Ag and In–Pb–Ag solders—compression stress–strain data

Shown in Fig. 3 are the true-stress, true-strain curves for the In–Ag solder. The plots represent the as-fabricated condition and the annealing treatment of 16 h at 67°C. The stress–strain curves, which corresponded to the other annealing times, were omitted for clarity as their trends were similar to those shown here. Both strain rates, 4.4×10^{-5} s^{-1} (a, b) and 8.8×10^{-4} s^{-1} (c, d) were represented in the plots. Under *all* conditions, work hardening was most significant at –25°C and then diminished with increased test temperature. At the slower strain rate of 4.4×10^{-5} s^{-1}, the as-fabricated (Fig. 3a) and annealed (Fig. 3b) conditions resulted in nearly identical stress–strain curves at each test temperature. At the faster strain rate of 4.4×10^{-5} s^{-1}, the extent of work hardening was slightly greater, as expected. Nevertheless, the as-fabricated samples (Fig. 3c) and annealed samples (Fig. 3d) exhibited similar degrees of work hardening.

Two general trends were observed in Fig. 3. First, the degree of work hardening decreased with increasing test temperature. This behavior resulted from dynamic recovery and recrystallization processes resulting from tests being performed at a high homologous temperature, T_h, where $T_h = T_{\text{test}}$ (K)$/T_{\text{solidus}}$ (K). The value of T_h was 0.60–0.96 for test temperatures in the range of –25 to 125°C. The second trend was the negligible effect of annealing treatment on the stress–strain curves. This behavior indicated that the In–Ag microstructure and responsible deformation mechanisms were not affected by the relatively "impressive" annealing conditions.

The yield stress was plotted as a function of annealing time in Fig. 4 for the In–Ag specimens tested at a strain rate of 4.4×10^{-5} s^{-1}. The open symbols were the mean values; the corresponding solid symbols were the minimum and maximum values. This plot illustrated three points: First of all, as expected, the yield stress decreased with increasing test temperature. Second, there was a relatively small scatter between the two yield stress measurements made per test condition. The same two trends were observed in the yield stress measured at the faster strain rate of 8.8×10^{-4} s^{-1}. Third, the effect of the annealing treatment time, which was to decrease the yield stress, was significant only at the 25 and –25°C test temperatures. At 25°C, the decrease of yield stress was observed only after the 24 h annealing treatment. At –25°C, both 16 and 24 h heat treatments caused a drop of yield stress. The 8 h heat treatment produced a slightly higher yield stress than the as-fabricated condition. However, the difference remained within experimental error. At the fast strain rate, annealing treatment produced a noticeable decrease of yield stress at 25 and –25°C, as well, but only following the 24 h period.

The entire set of *mean* yield stress values obtained for the In–Ag solder were plotted as a function of test temperature in Fig. 5. Overall, the yield stress values were in the range of 0.5–8.5 MPa and the plot shows

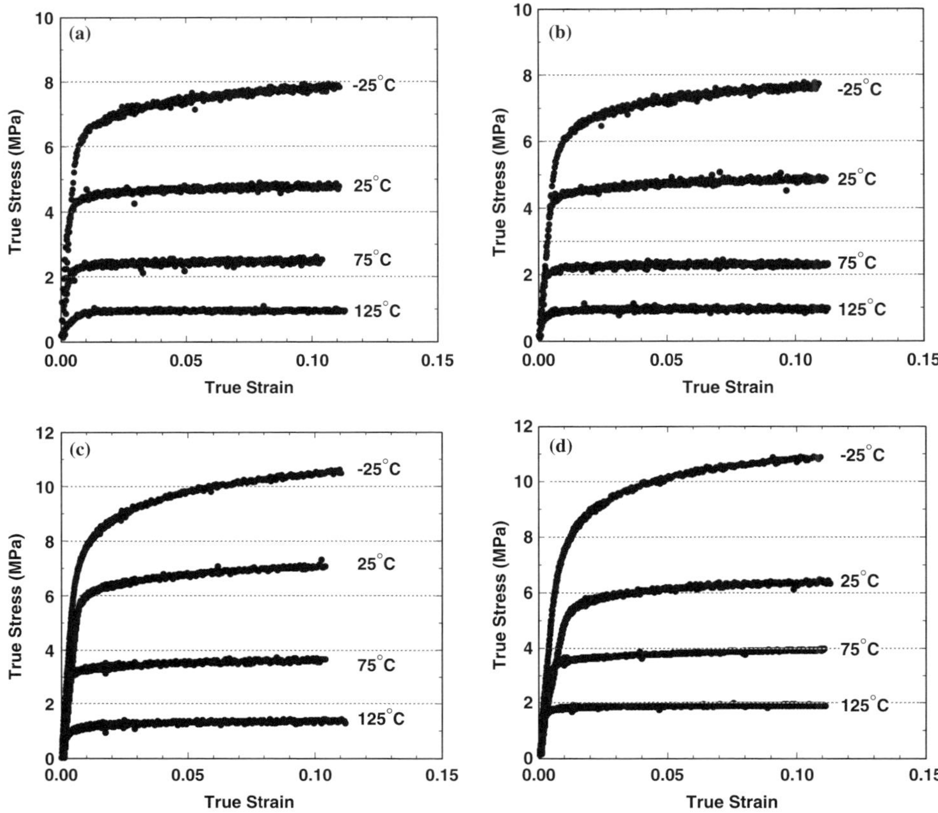

Fig. 3 Stress–strain curves for the In–Ag solder for the following combinations of sample condition and strain rate: (**a**) as-fabricated, 4.4×10^{-5} s^{-1}; (**b**) annealed (67°C, 16 h), 4.4×10^{-5} s^{-1}; (**c**) as-fabricated, 8.8×10^{-4} s^{-1}; and (**d**) annealed (67°C, 16 h), 8.8×10^{-4} s^{-1}

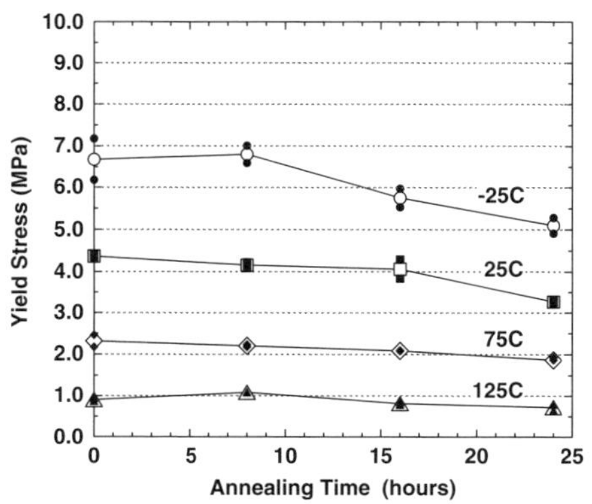

Fig. 4 Yield stress as a function of annealing time (67°C) for the In–Ag solder tested at a strain rate of 4.4×10^{-5} s^{-1}. The open symbols are the mean values; the corresponding solid symbols are the minimum and maximum values

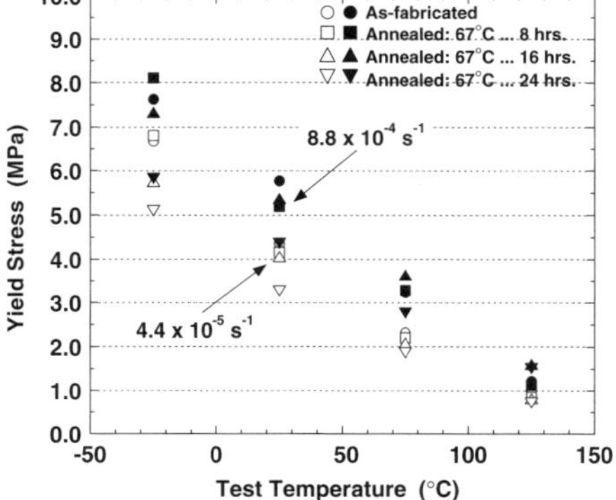

Fig. 5 Yield stress as a function of test temperature and annealing times (67°C) for the In–Ag solder. The open and closed symbols indicated the 4.4×10^{-5} s^{-1} and the 8.8×10^{-4} s^{-1} strain rates, respectively

the magnitude of the strain rate effect. Although the slower strain rate produced a lower yield stress, that difference became less significant with increased test temperature. The trends observed in Fig. 4, regarding the effects of annealing time on the yield stress at the slow strain rate, were also observed at the fast strain rate. In particular, there was the non-monotonic dependence of yield stress on annealing time at –25°C as well as the overall increasing insensitivity to annealing treatment with higher test temperatures.

In summary, the In–Ag solder exhibited the expected trends of decreasing yield stress with slower strain rates and increasing test temperature. Only the 16 and 24 h annealing times (67°C) caused noticeable changes of yield stress; moreover, those changes were limited to decreased values at only the 25 and –25°C test temperatures. This trend indicated that yield stress was most sensitive to the starting microstructure, as established by the annealing condition, at the lower test temperatures (–25 and 25°C). The strain rate effect became less distinct with increased testing temperature.

Understanding the above trends provided insight into the performance of the In–Ag solder in actual OE, MEMS, and MOEMS application, albeit to a limited degree because only the constant strain rate deformation has thus far been considered. In the foreseeable future, OE and MEMS devices will be exposed to service condition primarily in the low-temperature regime, in the present case, –25 to 25°C. As noted above, the stress–strain curves and, in particular, the yield stresses were most sensitive to the initial microstructure, as established by the annealing conditions in these experiments (or the process cooling rate in actual assemblies). The variability of the yield stress must be considered when constructing the UCP constitutive model, likely adding to the latter's complexity. The alternative approach is to define an "average" parameter and accept the possible loss of prediction fidelity.

The constant strain rate, uniaxial compression experiments were also evaluated for the counterpart Pb-bearing solder, In–Pb–Ag. At a strain rate of 4.4×10^{-5} s^{-1} and the as-fabricated condition, the In–Pb–Ag stress–strain curves exhibited work hardening only at –25°C. Work hardening was very minimal at 25°C. At 75°C or 125°C, the stress fluctuated with the progression of deformation. The stress varied by 0.3 MPa, peak-to-peak, over a strain "period" of 0.05 at 75°C, as shown in Fig. 6. The fluctuations diminished in magnitude at 125°C. At the faster strain rate of 8.8×10^{-4} s^{-1}, work hardening was recorded at –25 and 25°C and a monotonic strain-softening occurred at 75 and 125°C. The overall complex

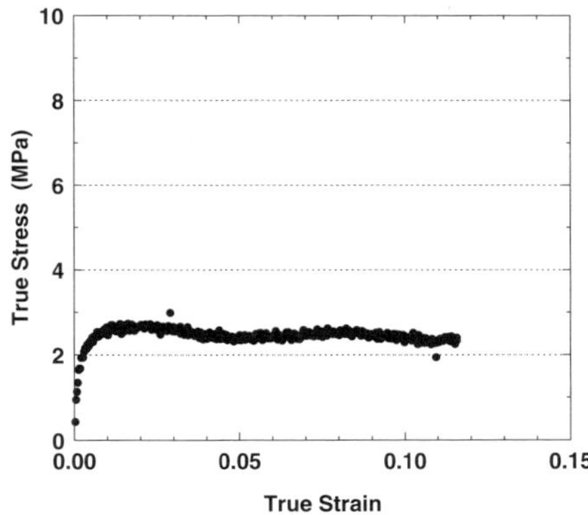

Fig. 6 Stress–strain curve for the as-cast In–Pb–Ag that was obtained at 75°C and a strain rate of 4.4×10^{-5} s^{-1} showing fluctuations of stress as a function of strain

stress–strain behavior of the In–Pb–Ag solder suggested that it had a less stable microstructure during deformation than did the In–Ag solder. Like the In–Ag solder, the annealing treatments had a minimal effect on the stress–strain curves.

The yield stress data were plotted in Fig. 7 for the In–Pb–Ag solder. The yield stress was in the range of 1.0–23 MPa, which was nearly twice that of the In–Ag solder (Fig. 5), indicating a greater temperature sensitivity. The majority of that difference between the two solders occurred at –25 and 25°C for either strain rate. The two solders exhibited comparable yield strengths at 75 and 125°C. At the faster strain rate of 8.8×10^{-4} s^{-1}, the In–Pb–Ag solder was stronger than the In–Ag solder at all test temperatures.

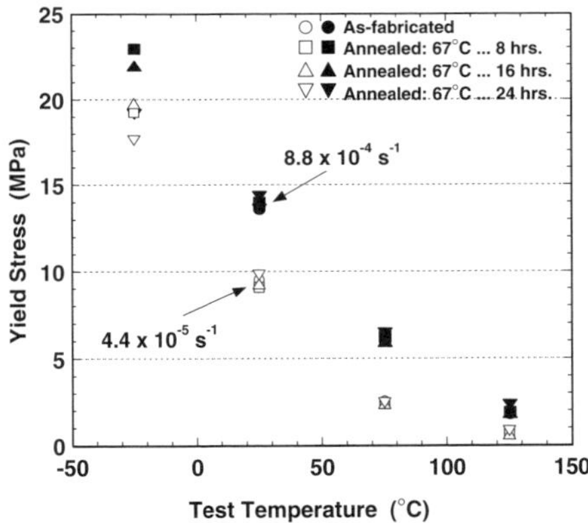

Fig. 7 Yield stress as a function of test temperature and annealing times (67°C) for the In–Pb–Ag solder. The open and closed symbols indicated the 4.4×10^{-5} s^{-1} and the 8.8×10^{-4} s^{-1} strain rates, respectively

The yield stress values of the In–Pb–Ag solder were even less sensitive to the annealing treatments than were those measured for the In–Ag solder. The 24 h heat treatment caused a significant strength decrease, but only at –25°C.

In summary, the yield stress of the In–Pb–Ag solder exhibited many of the same general trends as were noted for the In–Ag solder as a function of strain rate, test temperature, and annealing treatment. The differences rested primarily in the magnitudes of the effects. The In–Pb–Ag solder was generally stronger than the In–Ag solder, especially at the –25 and 25°C test temperatures. Because that difference diminished significantly at 75 and 125°C, the In–Pb–Ag solder yield stress had a greater sensitivity to test temperature than the In–Ag solder. On the other hand, the In–Pb–Ag solder exhibited a lesser sensitivity to the annealing treatment. If one were making a prediction of the ability of the In–Ag versus In–Pb–Ag to relieve residual stresses at low temperatures ($<75°C$), based solely on a lower yield stress, it would appear that the In–Ag solder would be the better selection.

The static elastic modulus was determined from the initial, linear segment of the compression stress–strain curves. The linearity of that early portion of each curve was confirmed by the R^2 values of a linear regression analysis, which were rarely below 0.95. The static elastic moduli were plotted as a function of test temperature in Fig. 8 for the In–Ag solder. The strain rate was 4.4×10^{-5} s^{-1}. The samples were tested in the as-fabricated condition and following the annealing treatment for 24 h. The large symbols are the mean values; the smaller symbols are the individual measurements. The purpose of this figure was to demonstrate the scatter observed in these measurements. The scatter was greatest with the as-fabricated specimens and least with those annealed for 24 h. However, the scatter did not monotonically decrease for 8 and 16 h. The largest decrease in scatter occurred between the 16 h and 24 h time periods. Very similar trends were observed for samples tested at the faster strain rate of 8.8×10^{-4} s^{-1}. Therefore, variability of the static elastic modulus values indicated that the 24 h heat treatment was required to adequately stabilize the In–Ag microstructure.

It was observed in Fig. 8 that the static elastic modulus values were nearly an order of magnitude lower than those expected from solders or other metals and alloys [15]. The same behavior was observed in a study of the Sn–Ag–Cu solder [9]. In that study, an exhaustive assessment was made to determine whether the testing procedures were the source of the reduced moduli. Factors such as load train compliance as well as sample alignment and possible calculation errors were ruled out[1]. Therefore, it was concluded that the In–Ag solder exhibited either anelastic or significant inelastic deformation that accompanied the purely linear elastic response of the sample, at the initial, low stress portion of the stress–strain curve.

Unfortunately, existing UCP models assume that at low stress levels, the material response is linear elastic. These models do not account for any anelastic or significant inelastic strain at low stress levels. This model deficiency could have a significant negative impact on the accuracy of model predictions, especially in the present circumstance of the material being used at a high homologous temperature.

Next, the static modulus values of the In–Ag solder were examined as a function of test temperature and annealing time (67°C). The data were compiled in Fig. 9a for the slower strain rate of 4.4×10^{-5} s^{-1}. In the as-fabricated condition, the samples exhibited a maximum in the static elastic modulus at 25°C. This same behavior was observed for the stress–strain data of the Sn-Ag-Cu solder, using similar test parameters [9]. For those samples annealed for 8 and 16 h, a similar maximum was observed; but, it shifted to 75 and 125°C, respectively, albeit the latter case remained somewhat tentative in the absence of data beyond 125°C. A monotonically decreasing modulus was observed for those samples that were annealed for 24 h,

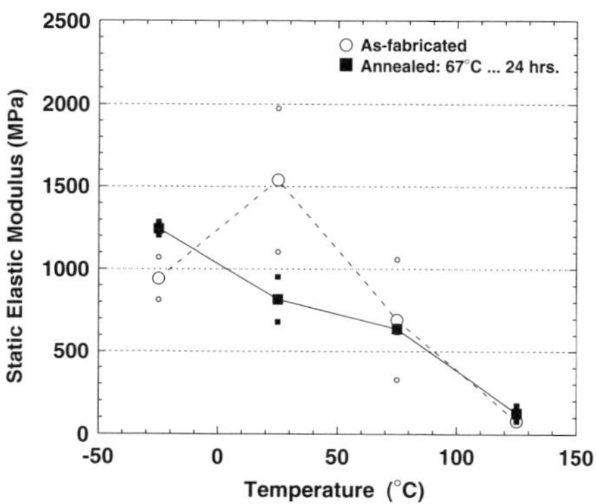

Fig. 8 Static elastic modulus as a function of test temperature for the In–Ag solder. The strain rate was 4.4×10^{-5} s^{-1}. The two sample conditions, as-fabricated and annealed at 24 h (67°C) were represented. The large symbols are the mean values; the smaller symbols are the two individual maximum and minimum points

---

[1] Tests were preformed on Sn–Ag–Cu samples at considerably faster strain rates of up to 10^{-2} s^{-1} [16]. The static elastic moduli values had similar magnitudes.

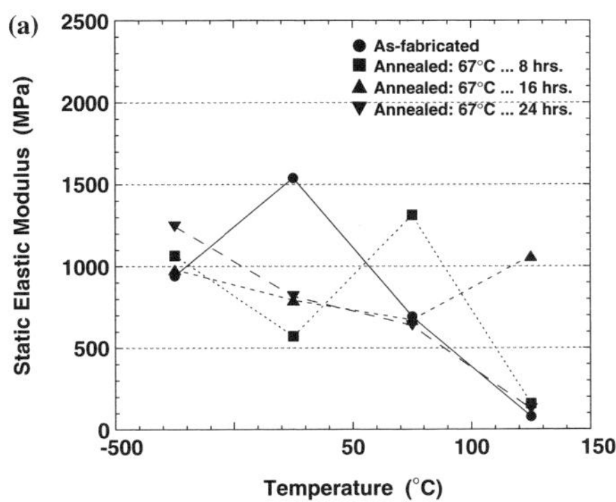

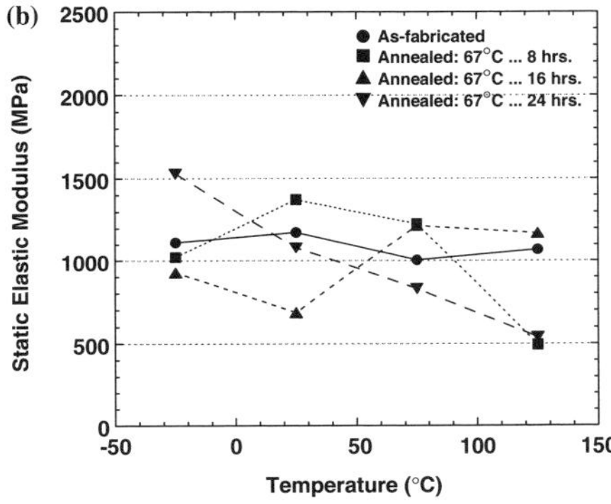

Fig. 9 Static elastic modulus (mean) as a function of test temperature for the In–Ag solder. The samples were in the as-fabricated condition as well as after having been annealed at 67°C for 8, 16, and 24 h. The two plots showed the moduli for the two strain rates (**a**) 4.4×10^{-5} s^{-1} and (**b**) 8.8×10^{-4} s^{-1}

The presence of an anelastic or inelastic deformation concurrent with the linear elastic deformation was further substantiated by the different behaviors observed between Fig. 9a and b. Although the strain rates differed by nearly an order of magnitude, such a difference would not be expected to affect either the magnitude, or the temperature dependence, of strictly the linear elastic modulus to the degree observed in the plots of Fig. 9.

In summary, the static elastic modulus was evaluated for the In–Ag solder. It was concluded that the unexpectedly low values (by nearly an order-of-magnitude) indicated that anelastic or inelastic deformation must have accompanied the expected linear elastic deformation in the stress–strain curve. The scatter of the modulus values, as well as their test temperature dependence, indicated that the 24 h annealing treatment (67°C) was required to provide some stability to the In–Ag microstructure, resulting in a monotonic decrease of modulus with increasing test temperature. Finally, the variations in the static elastic modulus would complicate UCP model development. "Average" static modulus values could be used in the constitutive equation to reduce model complexity, but with a potential loss of prediction fidelity.

The static elastic modulus values were similarly investigated for the In–Pb–Ag solder. Shown in Fig. 10 are the moduli measured at the slow strain rate of 4.4×10^{-5} s^{-1}. The samples were tested in the as-fabricated condition as well as after having been annealed for the longest interval of 24 h (67°C). When compared to similar data for the In–Ag solder (Fig. 8), the

implying that the maximum had been eliminated or moved to a temperature beyond the current test regime.

An analysis was also made of the In–Ag static elastic modulus values measured at the faster strain rate of 8.8×10^{-4} s^{-1}. The mean static elastic moduli were plotted as a function of test temperature and heat treatment in Fig. 9b. The values representing the as-fabricated condition fluctuated very little versus temperature when compared to the slower strain rate (Fig. 9a). The 8 h annealing treatment resulted in a maximum modulus between 25 and 75°C. The 16 h annealing treatment resulted in a general decrease of modulus with increasing test temperature, but with an apparent maximum at 75°C. A monotonically decreasing static elastic modulus was observed versus test temperature for samples annealed for 24 h prior to testing. However, the decrease was not as sharp as that observed at the slower strain rate.

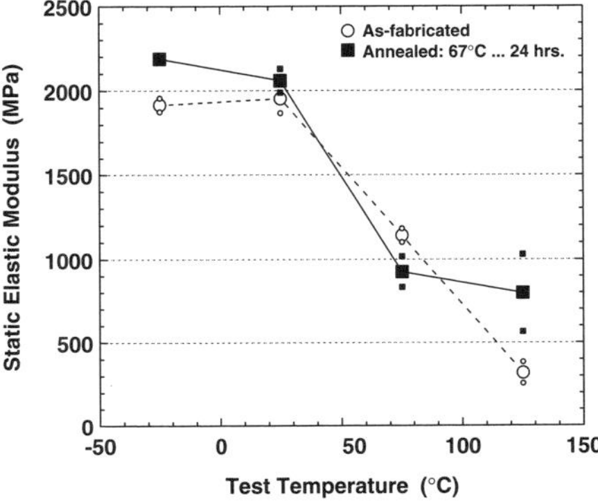

Fig. 10 Static elastic modulus as a function of test temperature for the In–Pb–Ag solder. The strain rate was 4.4×10^{-5} s^{-1}. The two sample conditions, as-fabricated and annealed at 24 h (67°C) were represented. The large symbols are the mean values; the smaller symbols are the two individual maximum and minimum points

In–Pb–Ag values exhibited slightly less scatter, particularly for the as-fabricated condition. At the intermediate heat treatments and faster strain rate, the In–Pb–Ag and In–Ag solders exhibited similar data scatter.

A comparison of Figs. 8 and 10 revealed several observations. First of all, the modulus values were higher for the In–Pb–Ag solder at each of the temperatures and both the as-fabricated and annealed conditions. The stiffness difference was particularly significant at the two lowest test temperatures of –25 and 25°C The higher modulus of the In–Pb–Ag solder placed this alloy at a disadvantage vis-à-vis the In–Ag solder in terms of its ability to relieve residual stresses caused by thermal expansion mismatch between materials in the joint, particularly at the low temperatures. This detriment is especially important in the low-temperature regime where creep mechanisms are ineffective at relieving residual stresses.

Although the In–Pb–Ag solder showed a general decrease of static elastic moduli with increased temperature, the values appeared to be segregated into two regimes: The high values at –25 and 25°C stepped down to the low values at 75 and 125°C. This behavior indicated that there was a transition in anelastic or inelastic deformation that occurred simultaneously with the linear elastic behavior. It was unlikely that there was a change to the linear elastic deformation because there were no phase changes reported to occur in this material over this temperature regime.

It was observed in Fig. 10 that static elastic modulus values of the In–Pb–Ag solder exhibited very little difference between the as-fabricated condition and the longest annealing time of 24 h (67°C). The absence of a consistent dependence of the static elastic modulus on annealing treatment was observed at both strain rates, as can be observed in Fig. 11a (4.4×10^{-5} s^{-1}) and 11b (8.8×10^{-4} s^{-1}). The static elastic moduli values were plotted as a function of test temperature. In Fig. 11a, the annealing times of 8 and 24 h indicated the step in the modulus values between 25°C and 75°C. The exception was the 16 h annealing treatment and, specifically, 25°C datum. This step behavior appeared to have shifted to the 75°C temperature at the faster strain rate (Fig. 11b). The different trends observed between the two strain rates—Fig. 11a versus Fig. 11b—were likely caused by the anelastic or inelastic deformation that combined with the linear elastic deformation to form the initial linear segment of the stress–strain curve.

In summary, the static elastic modulus was evaluated for the In–Pb–Ag solder counterpart to the Pb-free, In–Ag alloy. The modulus values were

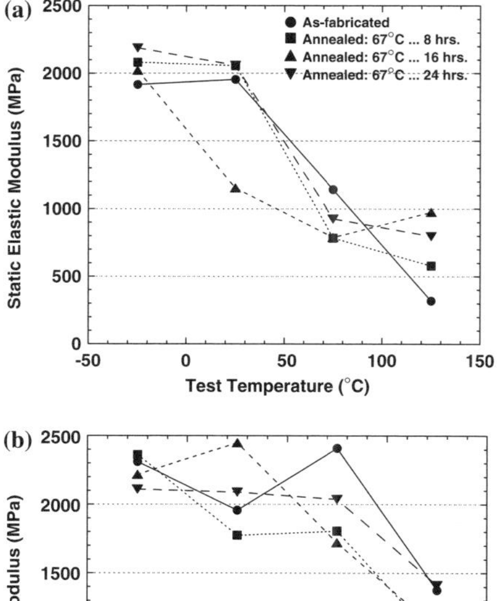

Fig. 11 Static elastic modulus (mean) as a function of test temperature for the In–Pb–Ag solder. The samples were in the as-fabricated condition as well as after having been annealed at 67°C for 8, 16, and 24 h. The two plots showed the moduli for the two strain rates (**a**) 4.4×10^{-5} s^{-1} and (**b**) 8.8×10^{-4} s^{-1}

generally higher than those of the In–Ag solder, particularly at the lower temperatures. Unlike the In–Ag solder, the In–Pb–Ag alloy exhibited a step-wise decrease of the modulus values as a function of test temperature between 25°C and 75°C for the slower strain rate. At the faster strain rate, the stepped decrease of static elastic modulus occurred at 75°C. With only a single exception, this behavior was independent of annealing condition.

3.2 In–Sn and In–Sn–Pb–Cd solders—compression stress–strain data

The yield stress and static elastic modulus properties were examined for the Pb-free, lower melting temperature In–Sn solder and the counterpart Pb-(and Cd-) containing In–Sn–Pb–Cd alloy. Shown in Fig. 12a, b are true-stress, true-strain curves representing the case of as-fabricated samples tested at –25°C for the strain

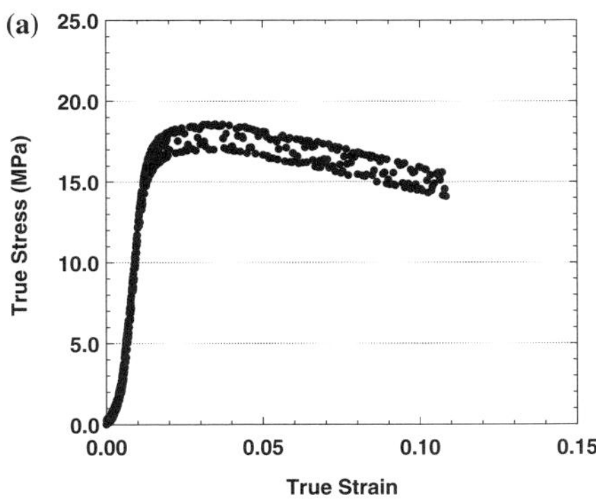

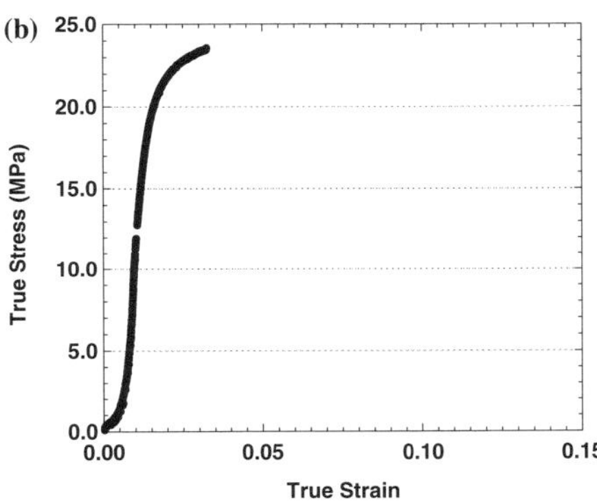

Fig. 12 Stress–strain curve of the In–Sn solder, tested in the as-fabricated condition at –25°C and strain rates of (**a**) 4.4×10^{-5} s^{-1} and (**b**) 8.8×10^{-4} s^{-1}

rates 4.4×10^{-5} s^{-1} and 8.8×10^{-4} s^{-1}, respectively. Three phenomenon were apparent in Fig. 12a. First, there was the slow ramp-up in the stress with increased strain. Sample misalignment was ruled out. Secondly, the plot in Fig. 12a showed fluctuations in the stress value as a function of deformation. Those fluctuations were not caused by jitter in the test frame nor by chatter in the electrical signal. Rather, this behavior was caused by a ratcheting effect of the deformation within the microstructure of the solder. Third, post-yield stress deformation exhibited a substantial strain softening effect. The strain-softening effect was noted in the shear stress–strain tests reported by Goldstein and Morris [3]. There were no indications that crack development was responsible for the decreasing stress. These three phenomena were also observed in samples exposed to one of the three annealing conditions prior to testing at –25°C. There were no consistent trends to suggest that these behaviors had a significant dependence on the annealing time at 52°C.

The stress–strain curve in Fig. 12b showed the effect of the faster strain rate of 8.8×10^{-4} s^{-1} on the deformation of the as-fabricated, In–Sn solder at –25°C. The ratcheting effect was eliminated from the curve, albeit, this particular plot did not show the deformation after yield stress. The extent of the initial stress ramp-up behavior was slightly reduced at the faster strain rate and it exhibited no observable dependence on the annealing treatment. Also, as was the case at the slower strain rate, the stress ramp-up effect was unique to the tests performed at –25°C. Tests that were continued past the yield stress exhibited strain softening.

The ratcheting and slow stress ramp-up phenomena were considered with respect to microstructural mechanisms. The ratcheting behavior has typically been associated with the start-stop movement of dislocations. The "stop" segment would also be a source of work hardening in the material. At the relatively high homologous temperatures at which these tests were performed, thermal activation should allow dislocations to jump past barriers, thereby lessening the likelihood for a ratcheting phenomenon. Moreover, dynamic recovery/recrystallization process would become apparent through strain-softening effects. This discrimination between low and high homologous temperature behaviors did not apply to the In–Sn solder at –25°C since *both* phenomena—ratcheting and strain softening—were observed in the samples.

The second phenomenon, the slow stress ramp-up at the beginning of deformation, suggests that there was a significant number of low-activation energy "defect sites" that were associated only with anelastic or inelastic deformation at –25°C. As the strain increased, those sites were used up so that the higher activation energy sites then contributed to the anelastic or inelastic deformation prior to yielding, increasing the slope of the plot. (Of course, the anelastic or inelastic deformation was occurring simultaneously with the linear elastic deformation.) At the faster strain rate (and higher test temperatures where the phenomenon was not observed), these low-activation sites would be rapidly used up, resulting in a loss of the slow, stress ramp-up effect. This mechanism would not be applicable to linear elastic deformation, since the latter does not occur by thermal activation, per se. Therefore, the stress ramp-up phenomenon provided further evidence that an anelastic or inelastic deformation mode contributed to the initial, linear deformation of the material before yielding took place.

The stress–strain curves were also investigated for the higher test temperatures. In all of those cases, the annealing treatments had very little impact on the

deformation performance. Also, the stress ramp-up behavior, which was observed at –25°C, did not reappear at the higher test temperatures. Shown in Fig. 13 are the stress strain curves obtained from as-fabricated specimens that were tested at (a) 4.4×10^{-5} s^{-1} and (b) 8.8×10^{-4} s^{-1}. Significant strain softening was observed after the yield stress in both cases. In fact, the extent of strain softening appeared to be greater at the faster strain rate, indicating that dynamic recovery/recrystallization was overwhelming work hardening in the solder. Similar behaviors were observed at 75 and 100°C; the latter case being represented by the stress–strain curves in Fig. 14. The extent of strain softening was greater at the faster strain rate. As noted above, the annealing treatments did not significantly change the stress–strain behavior of the In–Sn solder.

The stress–strain curves in Figs. 12–14 and associated discussion indicated that the In–Sn solder experienced significant strain softening during stress–strain

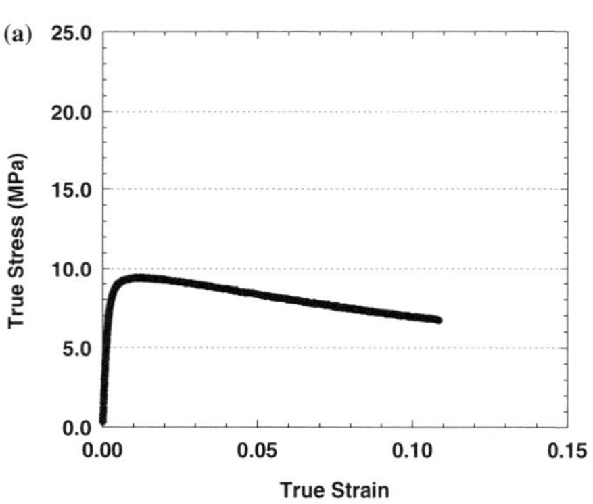

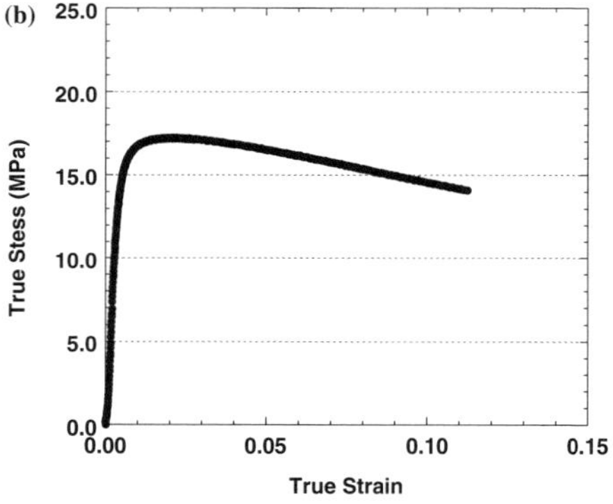

Fig. 13 Stress–strain curve of the In–Sn solder, tested in the as-fabricated condition at 25°C and strain rates of (**a**) 4.4×10^{-5} s^{-1} and (**b**) 8.8×10^{-4} s^{-1}

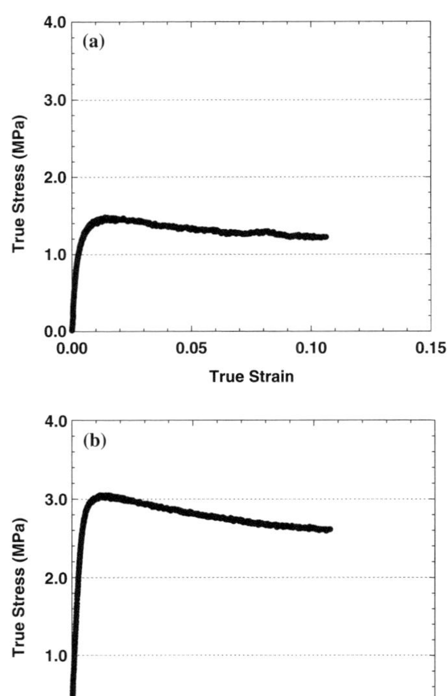

Fig. 14 Stress–strain curve of the In–Sn solder, tested in the as-fabricated condition at 100°C and strain rates of (**a**) 4.4×10^{-5} s^{-1} and (**b**) 8.8×10^{-4} s^{-1}

testing. Strain softening can be taken into account in the development of a constitutive model representing the deformation map of this material. From the applications point-of-view, strain softening provides a mechanism to relieve residual stresses effectively in the solder joint in the event that the solder surpasses its yield stress in a particular application. It is not conclusive that, because of strain softening, the In–Sn solder will have an inherently shorter cyclic lifetime. Such a conclusion must await a more extensive program of fatigue testing for this material.

The yield stress behavior was documented for the In–Sn solder. The data scatter was very similar to that observed for the In–Ag solder in Fig. 4 and, thus, will not be discussed further here. The mean yield stress as a function of test temperature was plotted in Fig. 15. Both strain rates and all annealing times (52°C) were represented in the graph. As expected, the yield stress decreased with increased test temperature. Also, the slower strain rate resulted in lower yield stresses than

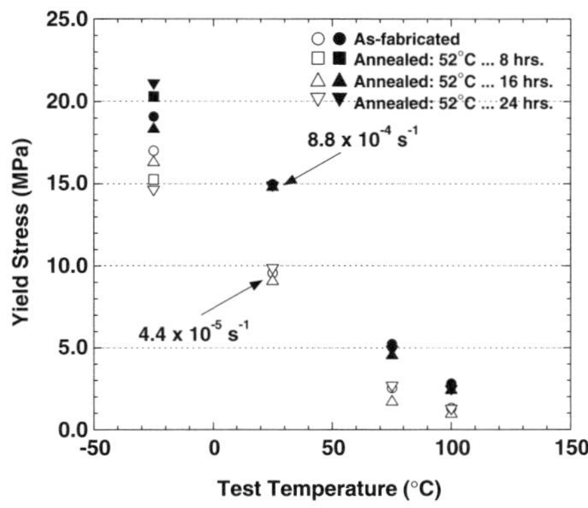

Fig. 15 Yield stress as a function of test temperature and annealing times (52°C) for the In–Sn solder. The open and closed symbols indicated the 4.4×10^{-5} s^{-1} and the 8.8×10^{-4} s^{-1} strain rates, respectively

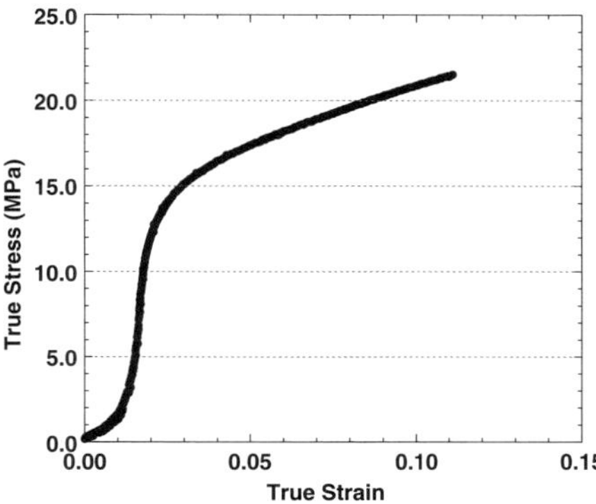

Fig. 16 Stress–strain curve of an as-fabricated In–Sn–Pb–Cd solder specimen tested at –25°C and at a strain rate of 8.8×10^{-4} s^{-1}

the faster strain rate; that difference diminished with increasing test temperature. The annealing time appeared to have a significant effect on the yield stress only for tests performed at –25°C. And, even in that sole case, there was no consistent trend between annealing time and yield stress value. In summary, the yield stress of the In–Sn solder, like the stress–strain deformation overall, was insensitive to the starting microstructure of the solder as established by the annealing conditions. The one possible exception, albeit quite weak, was observed for tests performed at –25°C.

A comparison was made between the yield stress values of the In–Sn solder (Fig. 15) and those of the higher melting temperature, Pb-free In–Ag solder (Fig. 5). At test temperatures of 75 and 100°C (In–Sn) or 125°C (In–Ag), the two alloys exhibited very similar yield stresses. However, in stress–strain tests at 25 and –25°C, the In–Sn solder, in spite of its lower melting temperature, had a higher yield strength than did the In–Ag solder at both strain rates. This comparison showed that the melting temperature was a poor indicator of the stress–strain behavior.

The stress–strain curves were examined for the In–Sn–Pb–Cd alloy. Although the topic of this paper is Pb-free solders, it was interesting to note the stress–strain curves from this particular alloy versus test temperature, strain rate, and annealing time. Shown in Fig. 16 is the stress–strain curve of the as-fabricated sample tested at the fast strain rate of 8.8×10^{-4} s^{-1} and test temperature of –25°C. The plot began with a stress ramp-up as was similarly observed for the In–Sn solder tested under these conditions. After the yield stress, the material exhibited work hardening, which was contrary to the strain softening experienced by the

In–Sn solder. A plot that was nearly identical to that in Fig. 16, was observed for the In–Sn–Pb–Cd solder when tested at the slower strain rate, except for a slightly reduced rate of work hardening. At both strain rates, the annealing treatment did not significantly affect the stress–strain response.

The stress–strain behavior of the In–Sn–Pb–Cd solder changed when the test temperature was raised to 25°C. At both strain rates, the slow stress ramp-up prior to linear deformation was lost. This observation was illustrated in Fig. 17, which shows the stress–strain curve of the as-fabricated sample tested at 4.4×10^{-5} s^{-1}. It was also noted in Fig. 17 that the post-yield stress deformation began with a slight degree of strain softening followed with a small amount of work hardening. At the faster strain rate, the entire plastic deformation

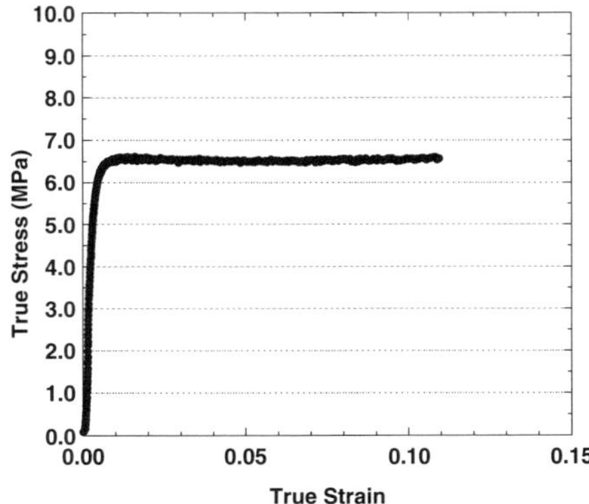

Fig. 17 Stress–strain curve of the as-fabricated In–Sn–Pb–Cd solder tested at 25°C and a strain rate of 4.4×10^{-5} s^{-1}

regime exhibited work hardening, albeit, to a lesser degree than was observed in Fig. 16. At both strain rates, the annealing times did not significantly affect the stress–strain behavior of the In–Sn–Pb–Cd solder.

The stress–strain tests performed at 75°C resulted in entirely different deformation behaviors. At the 4.4×10^{-5} s^{-1} rate, the as-fabricated specimens exhibited fluctuations in the stress values as shown in Fig. 18a. The peak-to-peak difference was 0.1 MPa and the strain period was approximately 0.02. The magnitude of the fluctuations diminished with increased annealing time, resulting in a stress–strain curve very similar to that in Fig. 17. The stress–strain curve of the as-fabricated sample tested at the faster strain rate of 8.8×10^{-4} s^{-1} was shown in Fig. 18b. A peak in the stress was followed by strain softening, which in turn, was followed by a strain hardening behavior. The annealing treatment had no noticeable effect on the stress–strain plot at the faster strain rate; all curves were similar to Fig. 18b.

Lastly, the stress–strain curves were investigated for the In–Sn–Pb–Cd solder exposed to tests at 100°C. Plots representing the two strain rates and the as-fabricated condition were provided in Fig. 19. Both curves had generally the same shape except that the sequence of a stress maximum and strain softening followed by work hardening was clearly more distinct at the faster strain rate (Fig. 19b). Very slight fluctuations appeared in the curves generated by the annealed samples tested at 4.4×10^{-5} s^{-1}. At the faster strain rate, the annealing treatments caused a decrease in the stress maximum after yielding. These were the only effects manifested by the annealing treatments on the In–Sn–Pb–Cd stress–strain behavior.

In summary, the stress–strain behavior was examined for the In–Sn–Pb–Cd solder. A particular emphasis was placed on this Pb-and Cd-bearing alloy to illustrate the different deformation behaviors that can occur with these more complex, low-temperature metal alloys. It was fortuitous that such variable

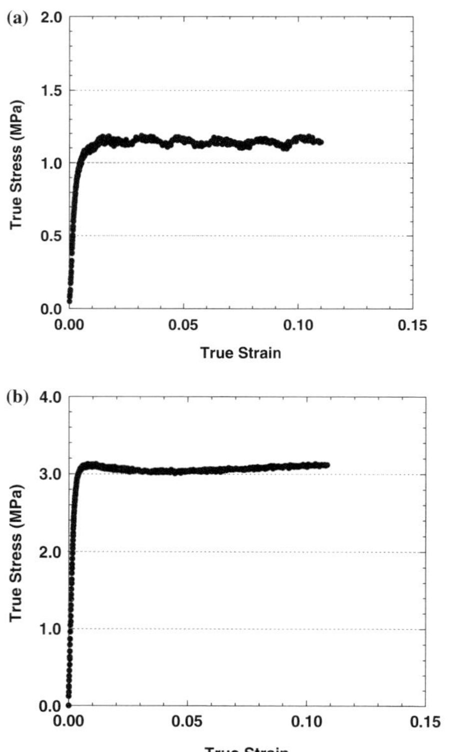

Fig. 18 Stress–strain curve of the In–Sn–Pb–Cd solder, tested in the as-fabricated condition at 75°C and strain rates of (a) 4.4×10^{-5} s^{-1} and (b) 8.8×10^{-4} s^{-1}

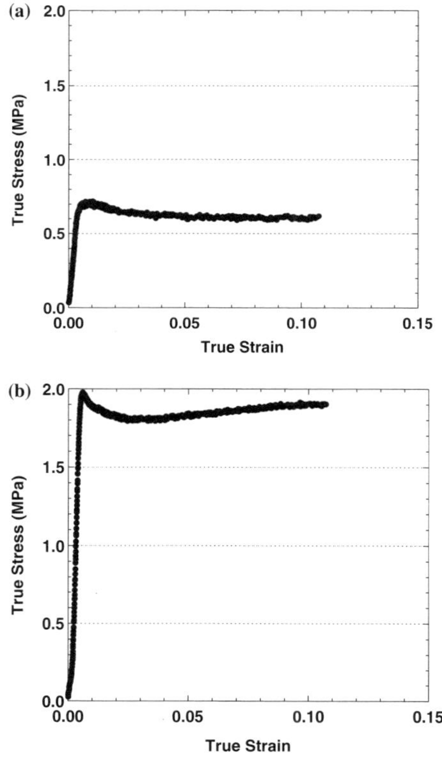

Fig. 19 Stress–strain curve of the In–Sn–Pb–Cd solder, tested in the as-fabricated condition at 100°C and strain rates of (a) 4.4×10^{-5} s^{-1} and (b) 8.8×10^{-4} s^{-1}

behavior happened to a lesser degree with the Pb-free, In–Sn solder, implying that the latter alloy had a more stable microstructure. That stability will provide for a more predictable response to an applied stress, which facilitates the development of a higher fidelity computational model. For either solder, it was interesting to note that the annealing treatments had, in general, very little effect on the stress–strain curve. This observation showed that the individual solder microstructures that were generated by the casting process, were relatively stable.

The yield stress values as a function of test temperature and annealing treatment were compiled for the In–Sn–Pb–Cd solder in Fig. 20. The trends were unique vis-à-vis the other solders discussed, thus far. The yield stresses were nearly identical at –25°C for *both* strain rates and *all* annealing times (at 52°C). Then, the yield stresses were more distinguished by strain rate with increasing test temperatures. However, that difference was greatest at 25°C; it then diminished with the 75 and 100°C test temperatures. Also, at the test temperatures of 25, 75, and 100°C, the yield stress remained insensitive to annealing time.

The yield stress data for the In–Sn–Pb–Cd solder were compared to the In–Sn solder (Fig. 15). The aforementioned strain rate sensitivity of the In–Sn–Pb–Cd solder, particularly at –25°C, was contrary to the behavior of the In–Sn solder. The values of the In–Sn–Pb–Cd solder were less than those of the In–Sn solder, more so at the two lowest test temperatures of –25 and 25°C. On the other hand, both solders were similar with respect to the absence of a sensitivity to annealing treatment at test temperatures of 25, 75, and 100°C.

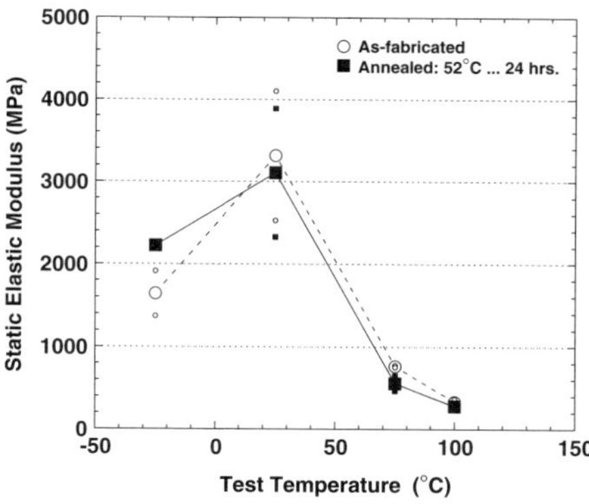

Fig. 21 Static elastic modulus as a function of test temperature for the In–Sn solder. The strain rate was 4.4×10^{-5} s^{-1}. The two sample conditions, as-fabricated and annealed at 24 h (52°C) were represented. The large symbols are the mean values; the smaller symbols are the two individual maximum and minimum points

The static elastic modulus was also investigated for the In–Sn and In–Sn–Pb–Cd solders. Shown in Fig. 21 are the modulus values as a function of test temperature for the In–Sn solder, representing the as-fabricated and the 24 h annealing time (52°C). Figure 21 exemplified the scatter observed for the individual values. The variability was greatest at 25°C; it was slightly less at –25°C and became negligible at 75 and 100°C. The scatter was largely independent of the sample annealing conditions. At the faster strain rate, the maximum variability was observed at –25°C. The scatter decreased with increasing temperature. Lastly, differences of static elastic modulus between the two strain rates decreased with higher test temperature. Although difficult to explain from a microstructural standpoint, it is clear that these trends reflect the contribution of the anelastic or inelastic deformation process. A more consistent behavior would have been reflected in this data if only linear elastic deformation were the controlling phenomenon. This point is further evidenced in the following discussion.

The In–Sn static elastic modulus was plotted as a function of test temperature in Fig. 22a and b for the 4.4×10^{-5} s^{-1} and 8.8×10^{-4} s^{-1} strain rates, respectively. All of the annealing conditions were represented in the graph. In Fig. 22a, the modulus appeared to be sensitive to the annealing time only at –25 and 25°C, but not in a consistent manner. All values were nearly the same at the two highest test temperatures. Overall, the modulus values did not change monotonically with test temperature. In fact, the plot indicated two regimes; the high modulus values for –25 and 25°C

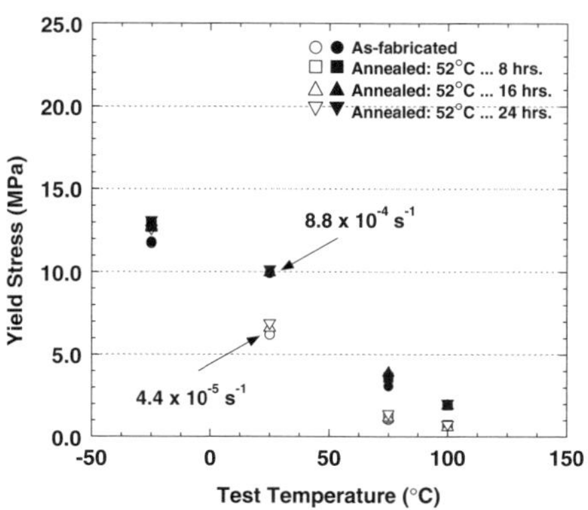

Fig. 20 Yield stress as a function of test temperature and annealing time (52°C) for the In–Sn–Pb–Cd solder. The open and closed symbols indicated the 4.4×10^{-5} s^{-1} and the 8.8×10^{-4} s^{-1} strain rates, respectively

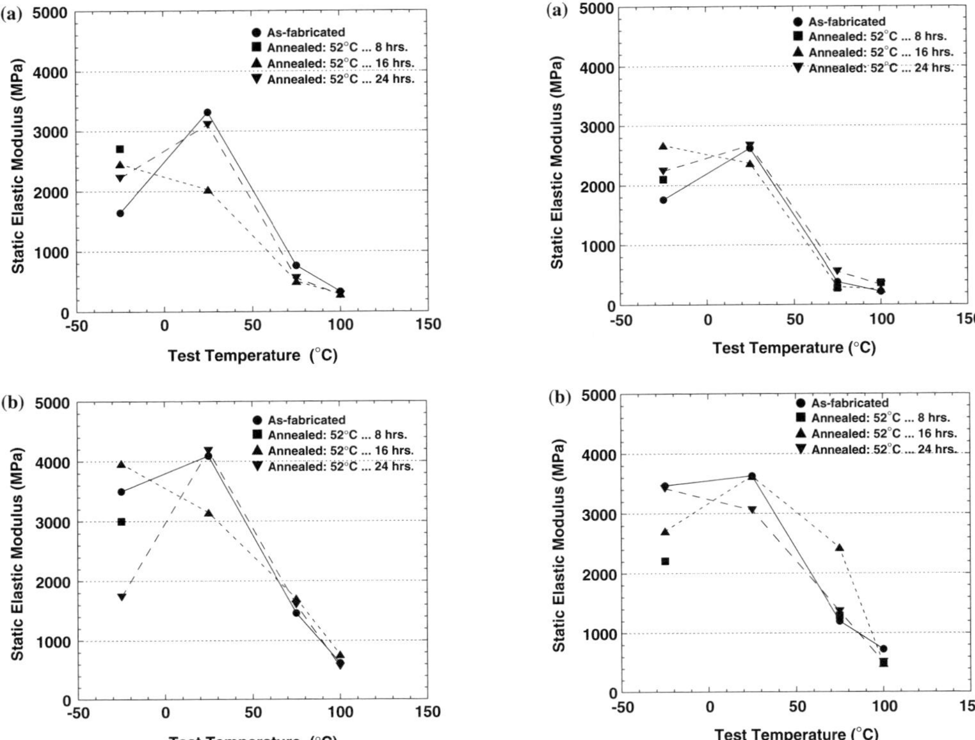

Fig. 22 Static elastic modulus (mean) as a function of test temperature and annealing condition for the In–Sn solder. The two plots showed the modulus for the two strain rates **(a)** 4.4×10^{-5} s^{-1} and **(b)** 8.8×10^{-4} s^{-1}

Fig. 23 Static elastic modulus (mean) as a function of test temperature and annealing condition for the In–Sn–Pb–Cd solder. The two plots showed the modulus values for the two strain rates; **(a)** 4.4×10^{-5} s^{-1} and **(b)** 8.8×10^{-4} s^{-1}

and low modulus values for 75 and 100°C. The same trends were observed at the faster strain rate (Fig. 22b). In the latter case, the modulus values were slightly higher at each of the test temperatures.

The static elastic modulus values were also examined for the In–Sn–Pb–Cd solder. The scatter ranges were similar to those observed for the In–Sn solder except for less spread at 25°C. Shown in Figs. 23a and b are the modulus values of the In–Sn–Pb–Cd solder for the two strain rates. The trends and magnitudes observed in Fig. 23a and b were nearly identical to those observed for the In–Sn solder in the corresponding Fig. 22a and b, respectively.

In summary, the yield stress and static elastic modulus values were compiled for the In–Sn and In–Sn–Pb–Cd solders. Both solders exhibited a decreasing yield stress with increased temperature and slower strain rate. The In–Sn solder had higher yield stress values than the In–Sn–Pb–Cd solder, particularly at the two lowest temperatures. In both cases, the annealing

treatment did not have either a significant or consistent effect on the yield stress. The static elastic modulus values were lower by an order of magnitude than expected, due to the anelastic or inelastic deformation mode that was hypothesized to have accompanied linear elastic deformation. The modulus dependence on test temperature exhibited two regimes, a low-temperature regime of higher values at –25 and 25°C and a high-temperature regime of lower values at 75 and 100°C. The step-wise temperature dependence was more distinct at the slower strain rate. To within experimental error, the modulus values were similar between the two solders.

It was possible to hypothesize some approximate effects of the solder deformation properties on interconnection reliability and the development of computational models for the In–Sn and In–Sn–Pb–Cd solders. (The effects are only approximate in the sense that creep deformation, which is discussed below, has an important role, as well.) The static modulus values

were generally comparable between the two solders at each temperature so that both alloys have similar capacities to relieve residual stresses by linear anelastic/elastic deformation. Assuming that the stress levels are sufficiently high so that inelastic deformation occurs in the solders, the higher yield stress of the In–Sn alloy would cause it to be less capable of relieving those stresses than the Pb- and Cd-bearing counterpart alloy, particularly at the lower temperatures.

Finally, it was observed that the yield stresses and static elastic moduli were relatively insensitive to the annealing treatments. This trend will improve reliability predictions because the material properties are less likely to depend upon the processing parameters used to make electrical interconnections.

3.3 In–Ag and In–Pb–Ag solders—compression creep data

Compression creep tests were performed on the Pb-free, In–Ag solder and its Pb-bearing counterpart, the In–Pb–Ag alloy. The strain-time curves of the In–Ag solder have been summarized in Fig. 24a–d. Only the creep curves representing the as-fabricated condition were presented because the generalized strain-time behavior was similar for those samples exposed to the 67°C, 16 h annealing treatment before creep testing. In a number of cases, in particular, the reduced stresses at the lowest temperatures, the positive creep strain was negligible. Therefore, the associated strain-time curves were simply omitted from the composite plots described below vis-à-vis the targeted stresses listed in Table 2.

The purpose of discussing the curves in Fig. 24 was to appreciate the long-term deformation behavior of these solders. Recall that creep will comprise a large part of the deformation that occurs during the service life (and accelerated aging) of the solder joints. In addition, the characteristics of the strain–time curves, and the stability of those curves with respect to the starting microstructure (annealing treatment), determine the complexity that is required of the UCP

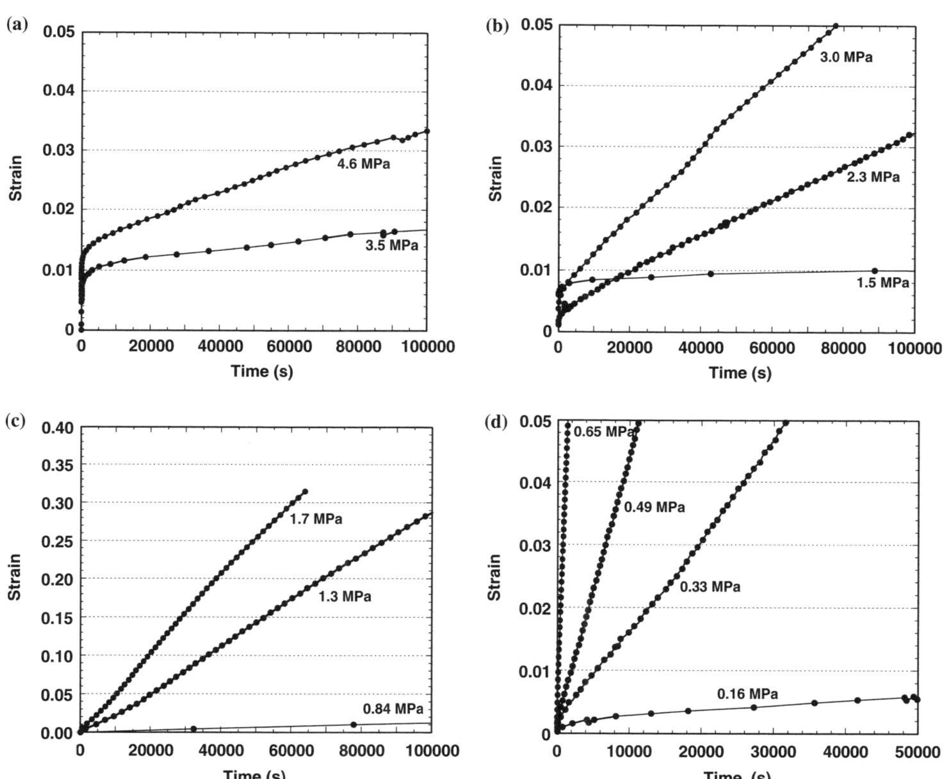

Fig. 24 Creep curves of the In–Ag solder tested in the as-fabricated, representing the following temperature conditions: (**a**) –25°C; (**b**) 25°C; and (**c**) 75°C, and (**d**) 125°C

constitutive model to accurately predict the creep component ("C" in UCP) of the overall deformation behavior.

It should be recalled that the applied stresses were not selected arbitrarily. Rather, the stress values were selected, based upon approximately 20%, 40%, 60%, and 80% of the yield stress. Therefore, when the material yield stress is used as an upper-limit design criterion for mechanical strength, an interconnection would be expected to experience stresses at levels that are commensurate with those percentages in engineering applications.

The In–Ag creep curves representing the temperatures of –25 and 25°C were shown in Fig. 24a and b. The strain–time plots showed both primary and secondary or steady-state creep stages. The primary creep stage became a lesser proportion of the curves when the test temperature was raised from –25 to 25°C. The creep curves representing tests performed at 75 and 125°C were shown in Fig. 24c and d and indicated that the deformation was largely steady-state creep. There was very little contribution by primary creep except for the 0.16 MPa test performed at 125°C. At 75°C (Fig. 24c), the creep curves had a sigmoidal shape. Sigmoidal creep indicates that there was initially an insufficient quantity of defects present in the material to support deformation, resulting in a delayed start of deformation [7]. Therefore, it was necessary to first generate the defects that could then respond to the applied stress, resulting in the subsequent increase in the strain rate.

Lastly, the creep curves in Fig. 24 did not show an obvious tertiary stage. Tertiary creep, which is indicated by an accelerating strain rate with time, occurs when large-scale damage, i.e. cracks, have been generated in the microstructure. Although the current methodology is based on compression testing, critically resolved shear stresses (CRSSs) can prevail in the specimen microstructure that are capable of creating damage that culminates into crack development. Similarly, diffusion-based mechanisms, primarily based on the movement of vacancies, can form discontinuities that develop into cracks, irrespective of the "sign" of the applied stress. Therefore, the absence of a tertiary creep stage in the deformation curves in Fig. 24 indicated that large-scale damage was not generated in the samples.

The minimum strain rate observed during steady-state creep, $d\varepsilon/dt_{min}$ (s^{-1}) was evaluated as a function of temperature, T (K), and stress, σ (MPa), per equation (3). The resulting expressions for the as-fabricated and post-annealed conditions are shown as Eqs. 4 and 5 below:

$$d\varepsilon/dt_{min} = 2.8 \times 10^8 \sinh^{3.2 \pm 0.5}(1.075\sigma)\exp(-99 \pm 14/RT) \tag{4}$$

$$d\varepsilon/dt_{min} = 2.8 \sinh^{1.1 \pm 0.4}(1.500\sigma)\exp(-46 \pm 11/RT) \tag{5}$$

The R^2 values for Eqs. 1 and 2 were 0.90 and 0.69, respectively. The error terms represented the 95% confidence interval. The reduced R^2 value for the annealed samples was somewhat unexpected. It was anticipated that the annealing treatment would stabilize the solder microstructure to produce a more consistent creep behavior. It is evident that the aging treatment had the opposite effect in this regard. All parameters except for the stress coefficient, α, were reduced by the aging treatment; the opposite trend was observed in the case of α.

Some insight was sought into the possible physical mechanisms active during creep from the sinh law parameters. The sinh law exponent (p) values were commensurate with those observed for the Sn–Pb eutectic solder [14]. Unfortunately, only relative comparisons can be drawn between the values of p since the latter has not been correlated to particular creep mechanisms or microstructural features. A similar situation prevailed with respect to f_0 and α.

On the other hand, the apparent activation energy, ΔH, can provide some insight into the rate-controlling mechanism of creep. It has been observed by other investigators that, when lattice or bulk diffusion is the controlling mechanism, ΔH will have values that are typically 90–110 kJ/mol for metals and alloys [17, 18]. When the controlling mechanism is fast or short-circuit diffusion, such as the movement of vacancies or interstitials along grain and/or interface boundaries, the value of ΔH is typically 0.4–0.6 of the bulk diffusion value, or about 40–60 kJ/mol [19–21]. According to these benchmarks, lattice diffusion controlled creep of the as-fabricated In–Ag samples. On the other hand, the annealing treatment caused the creep rate-controlling mechanism to shift to short-circuit diffusion. Two microstructural changes could have likely resulted from the annealing treatment. (1) The annealing treatment eliminated point and line defects such as vacancies and dislocations, respectively, through recovery. Because these defects supported lattice diffusion, the creep deformation had to be carried by grain and phase boundary processes. (2) The second hypothesis was that the annealing treatment simply increased the number of grain and phase boundaries through the early stages of recrystallization in which newly created grains have not yet begun to consume the older grains.

The poor correlation accompanying Eq. 5 can often be traced to a temperature dependence of the rate kinetics. In order to test this hypothesis, the data set was divided into two regimes, a low-temperature regime comprised of the −25 and 25°C results and the high-temperature regime that included the 75 and 125°C data. The resulting kinetics equations were described by Eqs. 6a, b, respectively:

$$[-25, 25\,°\text{C}] \quad d\varepsilon/dt_{min}$$
$$= 0.19 \sinh^{0.9 \pm 0.3}(1.500\sigma)\exp(-37 \pm 12/RT) \quad (6a)$$

$$[75, 125\,°\text{C}] \quad d\varepsilon/dt_{min}$$
$$= 3.6 \times 10^9 \sinh^{1.8 \pm 0.6}(1.500\sigma)\exp(-113 \pm 23/RT) \quad (6b)$$

For simplicity, the α value was allowed to remain unchanged, since the prior analyses used to obtain Eq. 5 showed a minimal degree of sensitivity to it. The higher R^2 values were 0.81 and 0.83, respectively, indicating that the steady-state creep rate kinetics were dependent on temperature. The data were also evaluated by grouping the 25°C results together with the 75 and 125°C data. A low R^2 value of 0.55 was obtained, indicating that the 25°C data was better grouped with the −25°C results and that the mechanism change occurred between 25°C and 75°C.

A comparison of Eqs. 6a and b indicated that the sinh term exponent increased from the low-temperature to the high temperature regime; however, the difference was not statistically significantly. On the other hand, the high-temperature regime exhibited a higher apparent activation energy value, indicating that bulk diffusion controlled the creep deformation. Fast-diffusion remained the rate-controlling mechanism for creep in the low-temperature regime. Similar trends of ΔH have been observed in the creep behavior of other engineering alloys [10].

The effect of the annealing treatment on steady-state creep was illustrated by the plot in Fig. 25 that shows the minimum strain rate calculated by Eqs. 4 and 5 over a 0–4 MPa stress range and concurrent test temperatures. The annealing treatment caused the strain rate to become less sensitive to temperature because the curves were closer together in the vertical direction. The strain rate was also less sensitive to the applied stress after the annealing treatment as indicated by the reduced slope of the curves. Beyond these summarizations, the fact that, for the same test temperature, the curves corresponding to the as-fabricated and annealed conditions crossed one-another indicated that no hard-and-fast rules could be formulated for

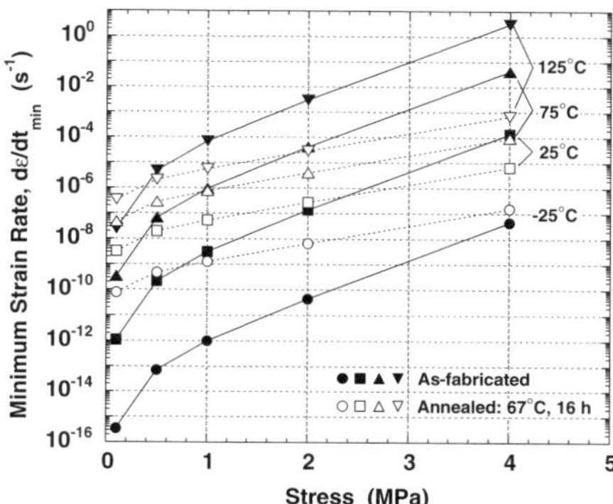

Fig. 25 Plot of the sinh law equations of steady-state creep strain rate for the In–Ag solder in the as-fabricated condition and after annealing at 67°C for 16 h

predicting, in general, the effect of annealing treatment on the steady-state creep rate.

The strain-time creep curves are shown in Fig. 26 for the In–Pb–Ag solder in the as-fabricated condition. As was the case with the In–Ag solder, there was very little primary creep exhibited for the applied stresses and temperatures. Steady-state creep dominated the deformation from the onset of testing for all, but the −25°C experiments in which a slightly greater proportion of primary creep was observed. Also, the creep curves did not exhibit a tertiary stage that was indicative of large-scale damage.

The creep behavior of the In–Pb–Ag solder was compared to that of the In–Ag solder by examining Figs. 24 and 26, respectively. The In–Ag solder experienced a greater degree of creep strain for the stress range of approximately 3–8 MPa and comparable test temperatures. The greater creep strain implied that the In–Ag solder could more readily relieve residual stresses in engineering structures than could its Pb-bearing counterpart.

The strain rate kinetics of the In–Pb–Ag solder were represented by the sinh law expression shown in Eq. 7 for the as-fabricated condition:

$$d\varepsilon/dt_{min} = 3.3 \times 10^9 \sinh^{2.9 \pm 0.3}(0.050\sigma)\exp(-74 \pm 5/RT) \quad (7)$$

The R^2 value of the regression analysis was 0.97 showing a very good correlation. For the sake of brevity, this analysis was not discussed for the In–Pb–Ag solder following the annealing treatment. A comparison was made between Eq. 7 and 4, the latter representing the In–Ag solder in the as-fabricated condition. The sinh term exponents were similar;

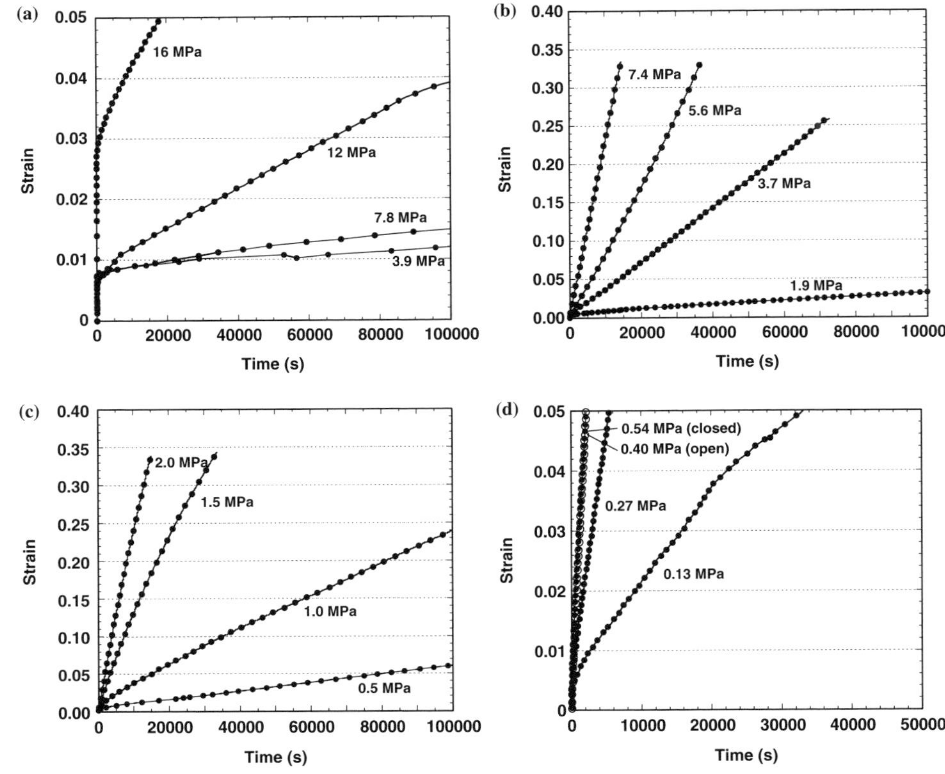

Fig. 26 Creep curves of the In–Pb–Ag solder tested in the as-fabricated, representing the following temperature conditions: (**a**) –25°C; (**b**) 25°C; and (**c**) 75°C, and (**d**) 125°C

however, the lower apparent activation energy of Eq. 6 indicated that short-circuit or fast-diffusion mechanisms governed In–Pb–Ag creep in the as-fabricated condition. Also, the stress coefficient, α, equal to 0.05 for the In–Pb–Ag alloy was nearly two orders of magnitude less than 1.075 for the In–Ag solder.

A direct comparison was made of the sinh law strain rate Eqs. 4 and 7 representing the In–Ag and In–Pb–Ag solders, respectively, in the as-fabricated condition. The corresponding plot is shown in Fig. 27. Because the curves were closer together, the steady-state strain rate of the In–Pb–Ag solder was less sensitive to test temperature than was the In–Ag solder. Similarly, the steady-state strain rate was less sensitive to stress as was evidenced by the reduced slope at stresses greater than 0.5 MPa. Overall, a comparison of creep behaviors between the two solders depends upon temperature and applied stress. At stresses below 2 MPa, the In–Ag solder exhibited a lower strain rate than the In–Pb–Ag alloy. The relative ranking went through a

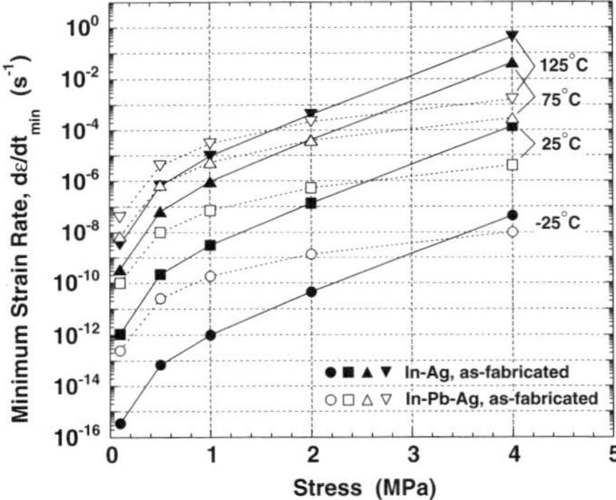

Fig. 27 Plot of the sinh law equations of steady-state creep strain rate comparing the In–Ag and In–Pb–Ag solders in the as-fabricated condition

transition in the stress range of 2–4 MPa so that, at stresses greater than 4 MPa, the In–Ag solder clearly had a faster strain rate than its Pb-bearing counterpart.

Therefore, in terms of steady-state creep, which appeared to predominate the creep curves of these two alloys, the relative effectiveness of reducing residual stresses in complex OE devices through creep deformation will be a function of the residual stress level and the temperature conditions.

In summary, the creep behavior was investigated for the In–Ag and counterpart In–Pb–Ag solders. The strain time curves of the In–Ag solder showed largely the steady-state stage. Only a limited degree of primary creep was observed and the tertiary stage was absent from the curves of both alloys. Similar strain-time trends were also observed in the samples following the annealing treatment. In the case of the In–Ag solder, the decrease of ΔH after the annealing treatment changed the creep mechanism from bulk diffusion to a short-circuit diffusion mechanism. The creep curves of the In–Pb–Ag solder exhibited the same qualitative trends. It was clear that the degree of creep strain exhibited by the In–Ag solder with respect to the In–Pb–Ag solder depended upon the applied stress and

temperature conditions. It was safe to conclude that, above 4 MPa, the In–Ag solder exhibited a greater propensity for creep deformation than did its Pb-free counterpart.

3.4 In–Sn and In–Sn–Pb–Cd solders—compression creep data

The creep behaviors were examined for the two lower melting temperature solders, In–Sn and In–Sn–Pb–Cd. The strain-time plots of the In–Sn solder were shown in Fig. 28. Creep deformation at –25°C (Fig. 28a) exhibited very little primary creep; the deformation was almost entirely steady-state creep. The plots representing the 10 and 13 MPa stresses showed a slight up-turn that suggested the beginning of tertiary creep. The appearance of tertiary creep, and whether it was caused by the accumulation of large-scale damage or was the result of more subtle changes within the In–Sn microstructure—e.g., defect density and/or defect velocity—will be explored below.

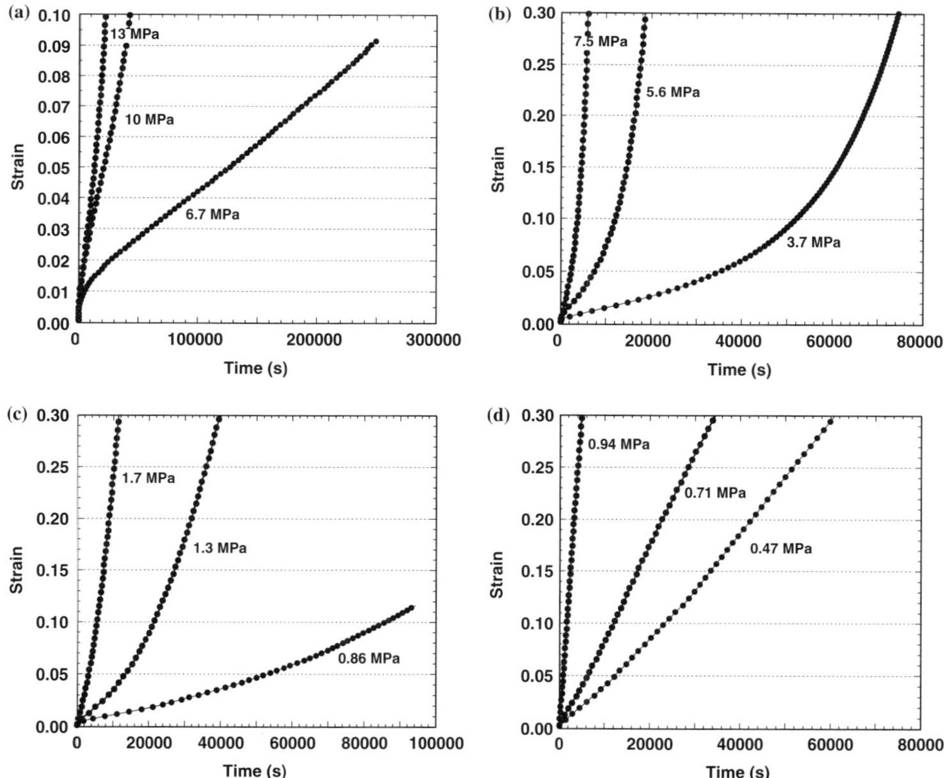

Fig. 28 Creep curves of the In–Sn solder tested in the as-fabricated, representing the following temperature conditions: (**a**) –25°C; (**b**) 25°C; and (**c**) 75°C, and (**d**) 100°C

The creep curves obtained from tests at 25°C were shown in Fig. 28b. Primary creep was nearly absent from the plots; rather, there was a short duration of steady-state creep followed by an extensive tertiary stage. The creep curves obtained at 75 and 100°C are shown in Fig. 28c and d, respectively. At 75°C, the shapes of the creep curves were very similar to those obtained at 25°C. There was an absence of significant primary stage, a short steady-state creep regime, and a significant tertiary creep stage. Unexpectedly, the tests performed at 100°C (Fig. 28d) showed a *reduced* extent of the tertiary stage. In fact, the tertiary behavior appeared to diminish with increased stress. Because primary creep was largely absent, these curves were primarily the steady-state stage.

An extensive microstructure analysis of the solders was not planned at the time of this study. However, the prevalence of the tertiary creep behavior in the In–Sn solder warranted an investigation of the In–Sn microstructure in order to determine whether large-scale damage was responsible for this tertiary behavior. Shown in Fig. 29 are low and high magnification opti-

cal micrographs of the as-fabricated In–Sn specimen that was creep tested at –25°C and 13 MPa. Individual cells, which are distinguished by different lamellae size and/or orientations, were separated by coarsened grain boundaries of both the In-rich β phase (dark) and Sn-rich γ phase (light). The coarsened boundaries were denoted by the white arrows in Fig. 29a. The micrograph in Fig. 29a, as well as the higher magnification image in Fig. 29b, clearly showed crack development along those coarsened cell boundaries. There was no evidence of phase boundary sliding or cracking within the cells, themselves. Similar coarsened boundary cracking was observed in the samples tested at 25 and 75°C, the latter case being shown in Fig. 30a. Therefore, at each of the three temperatures of –25, 25, and 75°C, microstructural damage in the form of cracking along coarsened, cell boundaries was responsible for the tertiary creep behavior observed in Fig. 28a–c.

Recall that, in Fig. 28d, which represented the creep tests performed at 100°C, tertiary stage was significantly reduced or absent altogether. Shown in Fig. 30b is an optical micrograph of the post-creep tested

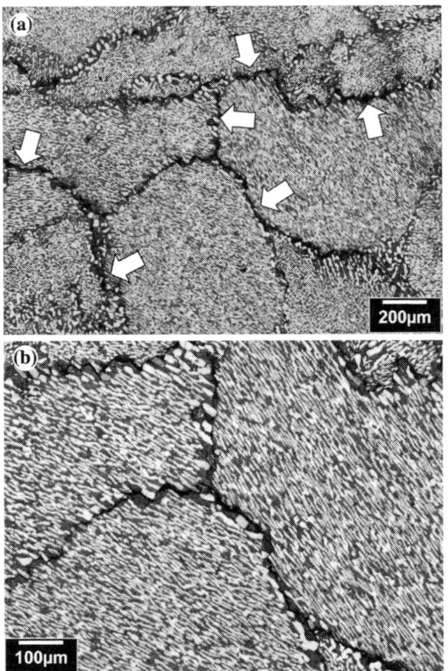

Fig. 29 Optical micrographs showing the In–Sn solder microstructure after creep testing at –25°C and 13 MPa, using two magnifications. The sample was tested in the as-fabricated condition

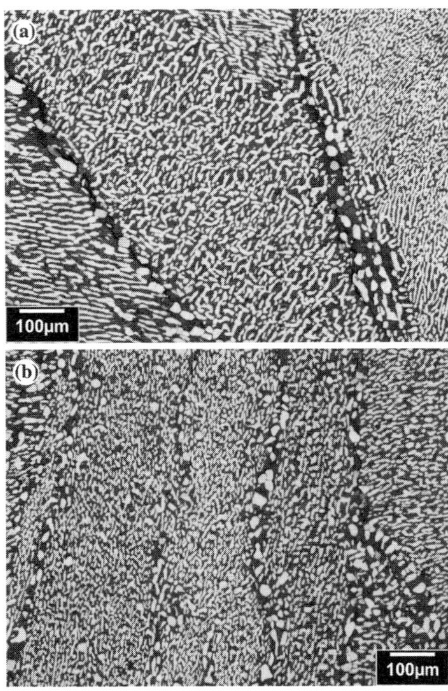

Fig. 30 Optical micrographs showing the In–Sn solder microstructure after creep testing under the following conditions: (**a**) 75°C, 1.7 MPa and (**b**) 100°C, 0.94 MPa. In both cases, the specimens were tested in the as-fabricated condition

sample exposed to a stress of 0.94 MPa at 100°C (see Fig. 28d). The coarsened boundaries were still present, although, they were less distinguishable from the generally more coarsened intracellular microstructure. Also, the extent of cracking in those coarsened boundaries was significantly reduced as compared to other temperatures. It was hypothesized that the coarser cell (interior) microstructure had taken up a greater share of the creep deformation, thereby relieving the boundaries of the need to generate cracks in response to the applied stresses.

The creep strain observed with the In–Sn solder (Fig. 28) was compared to that of the In–Ag solder (Fig. 24). The In–Sn solder exhibited a greater degree of creep strain at the equivalent test temperature and stress. This result was not surprising, given the In–Sn alloy's lower melting temperature.

The creep rate kinetics were also examined for the In–Sn solder. Equations 8 and 9 below express the minimum or steady-state creep rate as a function of applied stress and temperature for the as-fabricated and annealed (52°C, 16 h), respectively:

$$d\varepsilon/dt_{\min} = 5313 \sinh^{1.7\pm0.4}(0.325\sigma) \exp(-55 \pm 11/RT) \tag{8}$$

$$d\varepsilon/dt_{\min} = 687 \sinh^{1.5\pm0.6}(0.295\sigma) \exp(-48 \pm 13/RT) \tag{9}$$

The R^2 values were 0.83 and 0.73, respectively. It was interesting to note that the annealing treatment did not significantly change the sinh law exponent or the apparent activation energy. The value of α decreased only slightly following the annealing treatment; there was a greater decrease in the pre-exponential coefficient, f_0.

As was done in the case of the In–Ag solder, the low R^2 value of the annealed samples warranted further analysis to address the possibility that the apparent activation energy was dependent on temperature. The α value was kept the same. The results, which were very interesting, were presented by the three cases in Eq. 10a–c:

$$[-25, 25 \deg C] \quad d\varepsilon/dt_{\min}$$
$$= 27 \sinh^{1.5\pm0.6}(0.295\sigma) \exp(-41 \pm 13/RT) \tag{10a}$$

$$[75, 125 \deg C] \quad d\varepsilon/dt_{\min}$$
$$= 1.2 \times 10^{22} \sinh^{4.3\pm1.7}(0.295\sigma) \exp(-168 \pm 73/RT) \tag{10b}$$

$$[25\text{--}125 \deg C] \quad d\varepsilon/dt_{\min}$$
$$= 2.2 \times 10^{22} \sinh^{3.2\pm1.0}(0.295\sigma) \exp(-113 \pm 35/RT) \tag{10c}$$

The R^2 values were 0.81, 0.74, and 0.73, respectively. A higher correlation was obtained when the two lowest test temperature data sets (–25 and 25°C) were paired together. This was the same case for the In–Ag solder. The kinetics parameters were very similar between Eqs. 9 and 10a. On the other hand, very little improvement was observed in the R^2 value of the Eq. 10b when the two higher test temperatures were grouped together (75 and 100°C). By means of the apparent activation energy, Eq. 10a and b indicated that lattice diffusion controlled creep at the higher test temperatures and fast-diffusion controlled creep at the lower temperatures. However, from the viewpoint of developing a UCP constitutive model, Eq. 9 would provide a suitable representation of the creep behavior of the annealed In–Sn alloy.

The creep kinetics described above were compared to those obtained by Mei and Morris [2]. Although the latter authors tested the In–Sn solder in shear and described steady-state creep by a power law expression, it was expected that the kinetics would be similar. Recall that the cited tests were performed at 20, 65, and 90°C. The stress exponent and apparent activation energy values were 3.2 and 96 kJ/mol, respectively. These values were each considerably higher than those recorded in Eqs. 8 or 9. However, when compared to Eq. 10c for which, the –25°C data were eliminated, the kinetics parameters were similar to those in [2]. This comparison further substantiated the difference in creep mechanisms between the low and high-temperature regimes.

Equations 7 and 8 were plotted as a function of stress and temperature in Fig. 31. As expected, there was very little difference in the predicted strain rate between the as-fabricated and annealed conditions. These trends were in sharp contrast to the In–Ag solder for which, the annealing treatment caused a significant change in the steady-state creep kinetics.

The values of ΔH in Eqs. 7 and 8 were commensurate with grain or phase boundary (fast) diffusion during steady-state creep. The optical micrographs in Figs. 29 and 30 showed the extensive network of phase boundaries that prevailed in the two-phase microstructure of each cell. Those intra-cell diffusion processes culminated in the defects that accumulated at the cell boundaries, resulting in the formation of cracks that were the source of the pronounced tertiary creep in the In–Sn samples.

 Springer

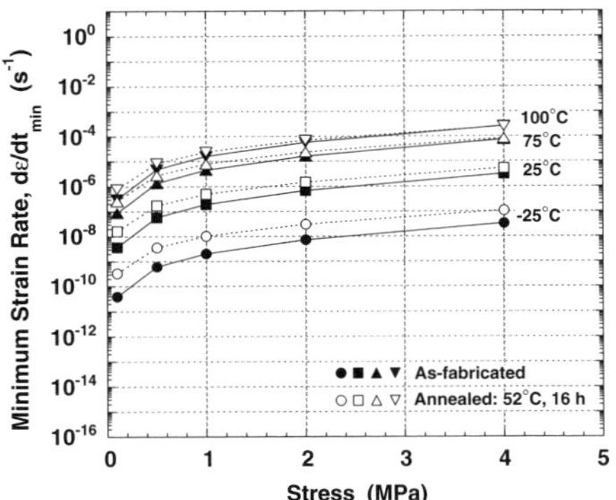

Fig. 31 Plot of the sinh law equations of steady-state creep strain rate for the In–Sn solder in the as-fabricated condition and after annealing at 52°C for 16 h

The strain-time curves for the In–Sn–Pb–Cd solder appear in Fig. 32 for each of the four test temperatures. At –25°C (Fig. 32a), there was significant primary creep preceding steady-state creep at each of the applied stresses. Although primary creep was also present in the creep curves generated at 25°C (Fig. 32b), the extent of primary creep diminished and the plots showed largely steady-state creep with increasing stress level. This same trend was observed at the test temperatures of 75°C (Fig. 32c) and 100°C (Fig. 32d). At all test temperatures and stresses, tertiary creep was not observed in the strain-time curves.

A comparison was made between the strain-time curves of the In–Sn solder (Fig. 28) and In–Sn–Pb–Cd alloy (Fig. 32) for comparable stresses and temperatures. At –25°C, the In–Sn solder shows slightly more creep strain than did the In–Sn–Pb–Cd alloy. However, at the three higher temperatures of 25, 75, and 100°C, the In–Sn solder exhibited noticeably less creep strain, in spite of the accelerated strain rate associated with the tertiary creep stage. The difference was particularly noticeable at 100°C where the tertiary stage was largely absent from the creep response of the In–Sn solder.

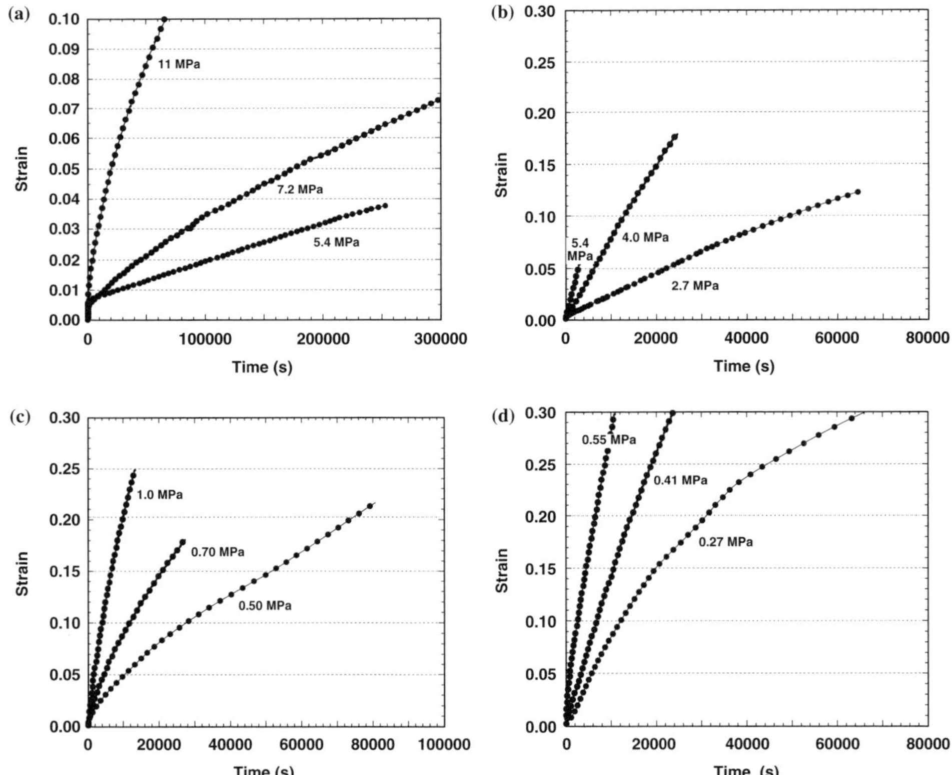

Fig. 32 Creep curves of the In–Sn–Pb–Cd solder tested in the as-fabricated, representing the following temperature conditions: (**a**) –25°C; (**b**) 25°C; and (**c**) 75°C, and (**d**) 100°C

Therefore, with the exception of tests performed at –25°C, the In–Sn solder appeared to be more creep resistant than its Pb-bearing counterpart.

The secondary creep rate kinetics of the In–Sn–Pb–Cd solder were represented by Eq. 9 below:

$$d\varepsilon/dt_{min} = 7.2 \times 10^{6} \sinh^{2.1\pm0.3}(0.182\sigma)\exp(-66\pm7/RT) \tag{11}$$

The R^2 value for the regression analysis was 0.97 demonstrating an excellent correlation between the independent and dependent variables. The sinh law exponent was slightly higher than the values calculated for the In–Sn solder in Eqs. 8 and 9. Also, the apparent activation energy in Eq. 11 was still commensurate with a fast-diffusion mechanism.

A direct comparison was made of the steady-state creep rates between the as-fabricated conditions of the In–Sn solder [Eq. 8] and In–Sn–Pb–Cd alloy [equation (11)]. The corresponding plot was shown in Fig. 33. The In–Sn–Pb–Cd solder showed a slightly faster strain rate than the In–Sn solder. That difference increased with both test temperature as well as applied stress. These results corroborated the earlier analysis of creep strains, that the In–Sn solder was more creep resistant than its Pb-bearing counterpart.

In summary, the compression creep behaviors were evaluated for the low-melting temperature In–Sn and In–Sn–Pb–Cd solders. The In–Sn solder creep curves exhibited extensive tertiary stage behavior, which correlated with the development of cracking along coarsened cell boundaries. The annealing treatment had very little effect on either the strain-time behavior or the steady-state creep rate kinetics of the In–Sn solder. With the exception of creep at –25°C, the In–Sn solder was more creep resistant than was the In–Sn–Pb–Cd solder. The apparent activation energy indicated that the rate kinetics of creep for both solders was controlled by a fast diffusion mechanism.

4 Applications note

The creep and plastic deformation properties were evaluated for the two Pb-free solders, In–Ag and In–Sn. These two alloys represented melting temperature regimes that could be used to develop a step-soldering process for the assembly of OE, MEMS, and MOEMS devices. Several *qualitative* comparisons were made with respect to relative abilities of the Pb-free solders to deform when compared to the Pb-bearing counterpart alloy—e.g., for relieving residual stresses. In the case of the In–Ag and In–Pb–Ag solders, a comparison of creep and plastic deformation properties indicated that, in general, the In–Ag solder would deform to a greater degree than the In–Pb–Ag solder. On the other hand, the In–Sn solder had the lesser capacity to deform under an applied load versus the In–Sn–Pb–Cd material.

Nevertheless, it has been recognized that, in most applications, both plastic and creep deformation can occur to varying degrees. Secondly, it is necessary to have a *quantitative* prediction of the ensuing deformation for engineering design and reliability analyses as would be provided by the single UCP constitutive model. Therefore, a long-range goal is to develop a UCP constitutive model for each of the Pb-free solders that would form the basis of a TMF predictive (numerical) model. Of course, the UCP constitutive equation would describe the deformation leading up to crack initiation, only. At this stage, the equation would not be capable of describing damage in the form of crack propagation.

The associated constitutive equations were developed for the In–Ag solder, representing both the as-fabricated and annealed conditions. This as-fabricated case will be highlighted below. The one-dimensional equation for the inelastic strain rate in the UCP model is:

$$d\varepsilon/dt_{11} = f_0\{\sinh^p[(|\sigma_{11} - B_{11}|/\alpha D)]\}(\sigma_{11} - B_{11})\exp(-H/RT) \tag{12}$$

where the subscript (11) denotes the uniaxial compression direction; B_{11} is the directional hardening which, at this point in the study, was taken to be a constant; and D represents the internal state variable of isotropic hardening/recovery, which is described by

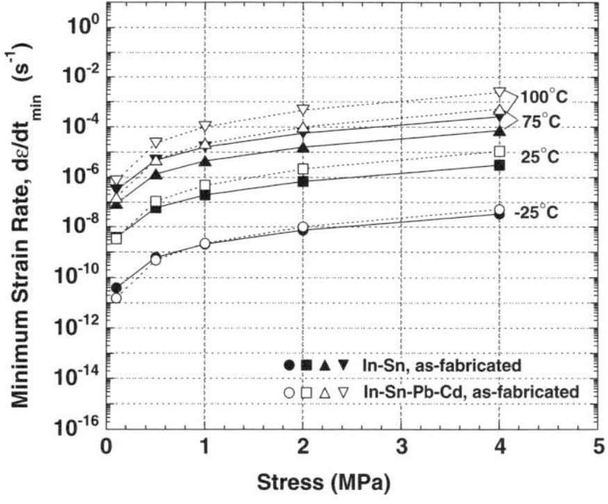

Fig. 33 Plot of the sinh law equations of steady-state creep strain rate comparing the In–Sn and In–Sn–Pb–Cd solders in the as-fabricated condition

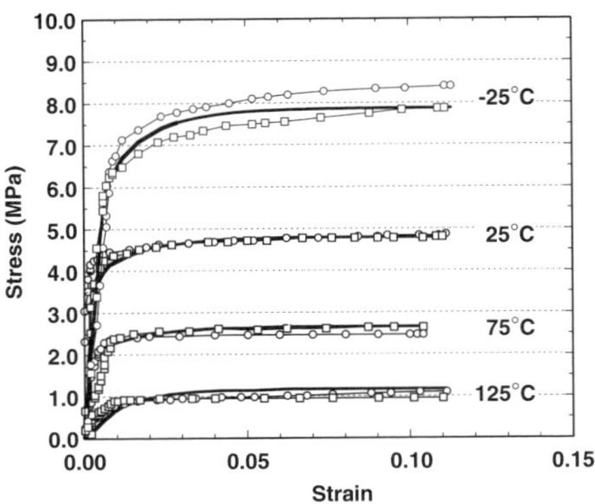

Fig. 34 Graph comparing the stress–strain behavior of the as-fabricated In–Ag solder at 4.4×10^{-5} s^{-1} (symbols) to the predictions calculated by the UCP constitutive model (solid lines)

$$dD/dt = \{[A_1|de/dt_{11}|]/[(D - D_0)_3^A]\} - A_2(D - D_0)^2 \tag{13}$$

where A_1, A_2, A_3, and D_0 are material constants. These material constants were initially constructed from the steady-state creep equation and then adjusted to calibrate the model prediction with both creep and plastic deformation experimental data. Thus, in Eq. 12, the values of f_0, α, p, and H were the same as those determined in Eq. 4, except that the value of α used in Eq. 12 was approximated to 1.0.

The prediction capability of Eq. 12, when populated with the suitable parameters, was assessed using the stress-stain behavior of the In–Ag solder. Shown in Fig. 34 is a plot that shows the two experimental stress–strain curves (symbols) for the as-fabricated In–Ag solder per each test temperature. The strain rate was 4.4×10^{-5} s^{-1}. The model prediction was represented by the solid lines. The UCP model adequately represented the plastic deformation. Similar correlations were observed at the faster strain rate as well as for both strain rates after the In–Ag solder in the annealed condition. Therefore, Eqs. 12 and 13 would be suitable for the UCP constitutive model in a finite element computational model for predicting the deformation behavior of In–Ag in a solder joint.

5 Summary

1. Lead (Pb)-free, low melting temperature solders are required for the step-soldering process used to assemble complex micro-electrical mechanical system (MEMS), OE, and micro-optical, electrical, mechanical system (MOEMS) devices. Stringent alignment protocols as well as the long-term stability of component placement necessitate the use of computational models to predict solder deformation following the assembly process as well as in response to anticipated service conditions.

2. The development of suitable computational models for such solder joints, begins with constructing a UCP constitutive model that describes both plastic and creep deformation properties of the solder alloy.

3. The stress–strain and creep properties were measured for the low-temperature, Pb-free 97In-3Ag (wt.%) and 58In-42Sn solders using compression testing techniques. The companion, Pb-bearing solders, 80In–15Pb–5Ag and 70In–15Sn–9.6Pb–5.4Cd, respectively, were similarly tested for comparison purposes.

4. Compression stress–strain tests were performed at strain rates of 4.4×10^{-5} s^{-1} and 8.8×10^{-4} s^{-1} and temperatures of –25, 25, 75, and 100°C or 125°C. Compression creep tests were performed at the same four temperatures. The minimum strain rate at the steady-state creep stage was represented by the equation: f_0 sinh^p $(\alpha\sigma)$ exp($-\Delta H/RT$). Samples were evaluated in the as-fabricated (cast) condition and after thermal annealing.

5. The stress–strain plots of the In–Ag solder exhibited reduced strain hardening with increased test temperature. The yield stress values were in the range of 0.5 MPa to 8.5 MPa and decreased with both test temperature and strain rate. The yield stresses were sensitive to the annealing treatments only at –25 and 25°C. The static elastic modulus generally decreased with test temperature, although not in a monotonic manner.

6. The values of ΔH for steady-state creep were 99 ± 14 kJ/mol and 46 ± 11 kJ/mol, indicating that bulk diffusion controlled creep in the as-fabricated samples and fast-diffusion controlled creep of the annealed samples, respectively.

7. The stress–strain plots of the In–Sn solder showed strain softening that resulted from dynamic recovery/recrystallization process. The yield stress ranged from 1.0 MPa to 22 MPa and decreased with both test temperature and strain rate. The annealing treatments did not significantly nor consistently change the yield stress. Two temperature regimes distinguished the static elastic modulus values.

8. The values of ΔH for steady-state creep were 55 ± 11 kJ/mol and 48 ± 13 kJ/mol for the as-fabricated and annealed conditions, respectively, indicating that fast diffusion, was the controlling mechanism in both cases.

9. UCP constitutive models were derived for the In–Ag solder representing the as-fabricated and 67°C, 16 h annealing conditions. The respective UCP equations represented very well the stress–strain behavior of the solder at both strain rates.

Acknowledgments The authors wish to thank A. Kilgo with assistance in metallographic sample preparation, M. Dvorack for his thorough review of the manuscript, and the US Army, Redstone Arsenal, AL for their financial support of this study.

References

1. T. Massalski (ed.), *Binary Alloy Phase Diagrams* (ASM, International, Materials Park, OH; 1986); vol. 1, p. 34 and vol. 2, p. 1401.
2. Z. Mei, J. Morris, J. Electron. Mater. **21**, 401 (1992)
3. J. Goldstein, J. Morris, J. Electron. Mater. **23**, 477 (1994)
4. J. Hwang, R. Vargas, Solder Surface Mount Technol. **5**, 38 (1990)
5. J. Seyyed, Solder Surface Mount Technol. **13**, 26 (1993)
6. F. Garofalo, *Fundamentals of Creep and Creep Rupture in Metals* (MacMillian, New York, NY, 1965), p. 156
7. A. Krausz, H. Eyring, *Deformation Kinetics* (McGraw-Hill, New York, NY), p. 190
8. F. Garofalo, op. cit., p. 25
9. P. Vianco, J. Rejent, A. Kilgo, J. Electron. Mater. **32**, 142 (2003)
10. P. Vianco, J. Rejent, A. Kilgo, J. Electron. Mater. **33**, 1389 (2004)
11. G. Dieter, *Mechanical Metallurgy* (McGraw-Hill, New York, NY, 1976), p. 75
12. Standard Test Methods for Young's Modulus, Tangent Modulus, and Chord Modulus, *ASTM E111–82* (American Society for Testing and Materials, West Conshohocken, 1995)
13. O. Sherby, P. Burke, Progress Mater. Sci. **13**, 340 (1968)
14. J. Stephens, D. Frear, Metall. Mater. Trans. A **30A**, 1301 (1999)
15. *Solder Alloy Data–Mechanical Properties of Solders and Soldered Joints* Report 656 (Inter. Tin Research Institute, Uxbridge, Middlesex, UK, 1986), pp. 12, 38
16. D. Shangguan (ed.), *Lead-Free Solder Interconnect Reliability* (ASM, Inter., Materials Park, 2005), p. 67
17. J. Askin, *Tracer Diffusion Data* (Plenum, New York, 1970)
18. *Smithells Metal Reference Book*, 6th edn. (Butterworth and Co. New York, 1983) pp. 13–18
19. P. Shewmon, *Diffusion in Solids*, 2nd edn. (TMS, Warrendale, 1989), p. 197
20. P. Shewmon, *Transformations in Metals* (McGraw-Hill, New York, NY, 1969), p. 61
21. J. Christian, *The Theory of Transformations in Metals and Alloys: Part I–Equilibrium and General* (Pergamon, Oxford, UK, 1975), p. 411

Sn–Zn low temperature solder

Katsuaki Suganuma · Kuen-Soo Kim

Published online: 1 September 2006
© Springer Science+Business Media, LLC 2006

Abstract Low temperature soldering is one of the key technologies before the accomplishment of total lead-free conversion in electronics industries. While Sn-Zn eutectic alloy has excellent properties as low temperature solder, it has some drawbacks. Damage by heat exposure and corrosion in humidity are two of the main concerns. Zn has an important role in chemical properties. The material physical properties, wetting, chemical stabilities and various reliabilities have been well understood on this alloy system through the numerous past works. The understanding of both materials and processing aspects enables one to manufacture sound electronic products without any serious problems. The basic properties and the current understandings on the limit of the application of this solder are reviewed in this paper.

1 Introduction

Sn–Ag–Cu ternary alloys are known to possess good solderability and mechanical property, and have been widely used as reliable lead-free solder. Even though these alloys can be applied to a wide variety of applications, high melting temperature of Sn–Ag–Cu ternary alloys still limits the adoption of these alloys to certain applications. It is required to establish reliable low temperature soldering techniques, especially for

temperature-sensitive components, optoelectronics modules, step soldering process and thin printed wiring boards (PWBs) [1]. In addition, low processing temperature is rather desirable even for many "heat-resistant" components currently fabricated with Sn–Ag–Cu because excess heating during reflow treatment always induces some damage to electronic devices and PWBs, which may influence the long–term lives of assemblies. This also gives a reason for the adoption of the low melting temperature alloys, i.e., Sn–Zn, Sn–Ag–Bi, and Sn–Ag–In for reflow soldering. In fact, the percentage of the production of these alloys against the total lead–free solders exceeds 10% in Japanese electronic industries [1]. Among these alloys, Sn–Zn has a significant benefit on cost as well as excellent mechanical properties.

There have already been several publication reporting on the various properties of the Sn–Zn alloys [2–8]. The effort on the development of the suitable flux for the Sn–Zn system achieved the practical adoption of this alloy in the market. Since 1999, Sn–Zn solder paste has been used in the actual production of commercial products [9]. Figure 1 shows a typical example of the PWB soldered with Sn–8Zn–3Bi alloy, where the excellent wetting feature of this alloy is clearly observed.

Besides the fascinating aspects of Sn–Zn alloy as lead–free solder, it has some drawbacks such as poor oxidation resistance in humid/high temperature condition and poor compatibility with Cu substrate at elevated temperature. The recent research and development revealed the mechanisms of such degradation or incompatible natures of these alloys. Although, individual companies such as solder manufactures, component suppliers and electronics equipment

K. Suganuma (✉) · K.-S. Kim
Institute of Scientific & Industrial Research, Osaka University, Mihogaoka 8-1, Ibaraki, Osaka 567-0047, Japan
e-mail: suganuma@sanken.osaka-u.ac.jp

Fig. 1 Commercial notebook computer PWB soldered with Sn–8Zn–3Bi

manufactures as they are doing, can overcome those weak points systematically, certain kinds of standardized testing tools, reliable database and process guidelines will greatly help them to promote and accomplish total lead–free conversion. In addition, solving some of the critical issues requires the basic understanding of the phenomena with the aid of scientific experts. Thus, it is worthy to carry out a collaborative research under a national or worldwide support to promote low temperature soldering. In the current paper, the recent understanding on Sn–Zn alloy system is briefly summarized.

2 Solidification and microstructure

Sn–Zn alloy is one of the typical eutectic alloys. The eutectic composition is Sn–8.8 wt.%Zn (hereafter, the unit "wt.%" is omitted) and both elements hardly dissolve in each other. Bi is usually added to Sn–Zn binary alloy in order to improve wetting property as well as to lower melting temperature. Figure 2 shows

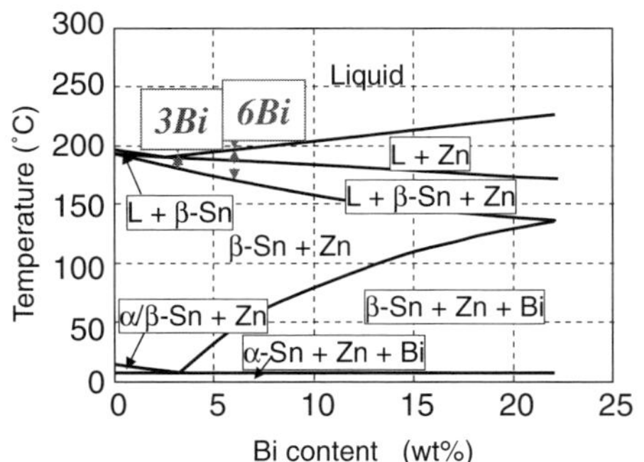

Fig. 2 Sn rich corner of Sn–8Zn–xBi ternary calculated phase diagram (Pandat/Adamis)

the Sn rich corner of the calculated phase diagram of Sn–8Zn–xBi ternary alloy. Because the addition of Bi enhances brittleness of Sn–Zn, the maximum content of Bi should be kept below 6 wt.%. The composition of the commercial Sn–Zn paste used currently is Sn–8Zn–3Bi. By the addition of Bi, the pasty range becomes wider and primary Zn phase will appear during solidification. Table 1 summarizes some basic properties of Sn–Zn–Bi alloys. As is expected from the phase diagram, the microstructure of the alloy changes as shown in Fig. 3. Zn phase disperses as platelet or fibrous precipitates in the eutectic phase and large primary Zn platelets appears with increasing Bi content. The precipitation of primary Zn degrades the mechanical properties of Sn–Zn alloy because of the brittleness of Zn phase. This solidification microstructure is greatly influenced by cooling rate. Slow cooling enhances the formation of primary Zn resulting in degradation of mechanical property.

The undercooling of this alloy is not so large as compared with Sn–Ag–Cu or Sn–Cu alloys. Figure 4 shows the undercooling temperature of Sn–Ag–xBi and Sn–Zn–xBi alloys as a function of Bi content [7]. Though the Bi addition lowers undercooling temperature for both alloys, the Sn–Zn binary alloy has much lower undercooling than Sn–Ag binary alloy.

3 Mechanical properties

Tensile properties of Sn–Zn alloys are excellent. Figure 5 compares typical tensile properties of Sn–Zn alloys as a function of Bi content up to 8 wt.% Zn. As increasing Bi content, 0.2% proof stress and ultimate tensile strength increases from 50 MPa at 0 wt.% Zn to 90 MPa at 8 wt.% Zn while elongation decreases monotonously.

Strain rate dependence of proof stress is also realized in Sn–Zn alloys. Figure 6 shows the strain rate dependence of 0.2% proof stress [7]. The well known power law, $\dot{\varepsilon} = A\sigma^n \exp\left[-\frac{Q}{RT}\right]$, where $\dot{\varepsilon}$ is strain rate, A is constant, σ is 0.2% stress, Q is activation energy, R is gas constant, and T is temperature, can be applied to

Table 1 Liquidus/solidus temperatures determined by DSC and calculated density of selected alloys [7]

Alloys	Liquidus (°C)	Solidus (°C)	Density g/cm³
Sn–9Zn	199	199	7.28
Sn–8Zn–1Bi	199	192	7.30
Sn–8Zn–3Bi	197	187	7.34
Sn–8Zn–6Bi	194	178	7.40
Sn–37Pb	183	183	8.64

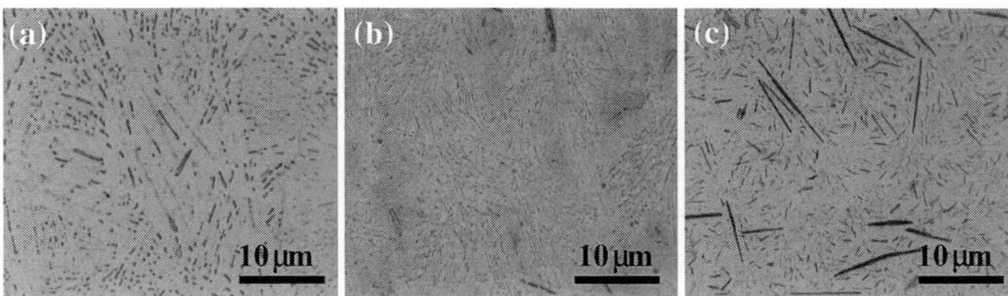

Fig. 3 Optical micrographs of Sn–Zn–Bi alloys [7] (**a**) Sn–9Zn, (**b**) Sn–8Zn–3Bi and (**c**) Sn–8Zn–6Bi

these alloys. n values are listed in Table 2 and compared with those of other alloys. n values of Sn–Zn slightly increases with increasing Bi content, which are in a similar range of Sn–Ag–Cu.

It is well known that creep resistance and fatigue resistance of Sn–Zn alloys are also excellent [2]. For instance, the activation energy for Sn–9Zn eutectic alloy as well as Sn–3.5Ag is close to 100 kJ/mole, which is equivalent to that for Sn–Pb alloys, indicating matrix creep mechanisms controlled by conventional disloca-

tion climb [3]. In fact, the excellent thermal and mechanical fatigue resistance of chip scale package (CSP) mounting can be noted in Fig. 7 [1].

4 Interface reaction and microstructure

Because of the high activity of Zn, Sn–Zn alloys behave differently in interface formation with electrodes. Zn reacts firstly at the interface. For instances, on Cu substrate, γ–Cu_5Zn_8 becomes the thick primary layer and, beneath this compound, thin β'–CuZn is

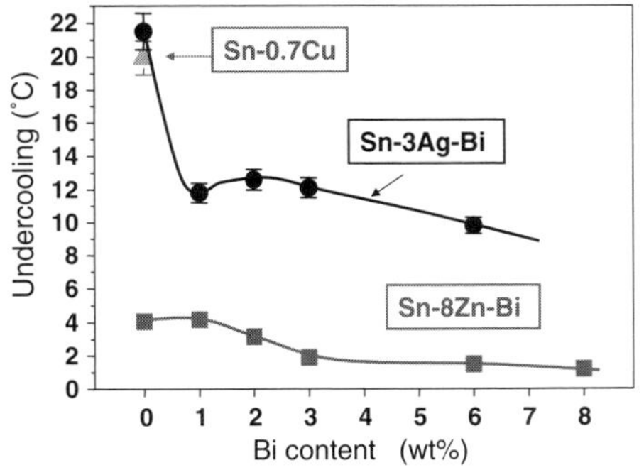

Fig. 4 Undercooling of Sn–3Ag–Bi and Sn–8Zn–Bi alloys as a function of Bi content [7]

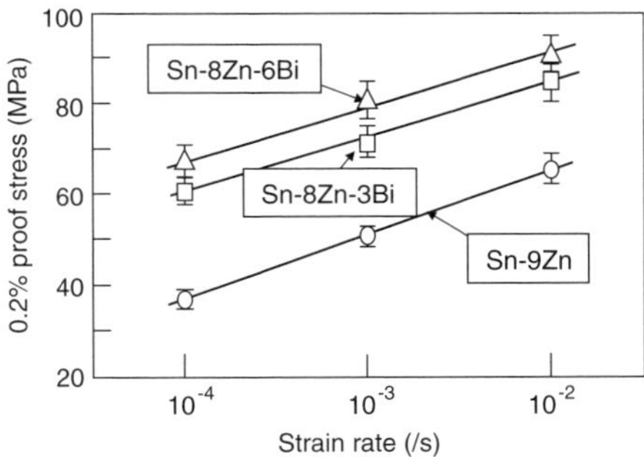

Fig. 6 Strain rate dependence of 0.2% proof stress of Sn–Zn–Bi alloys [7]

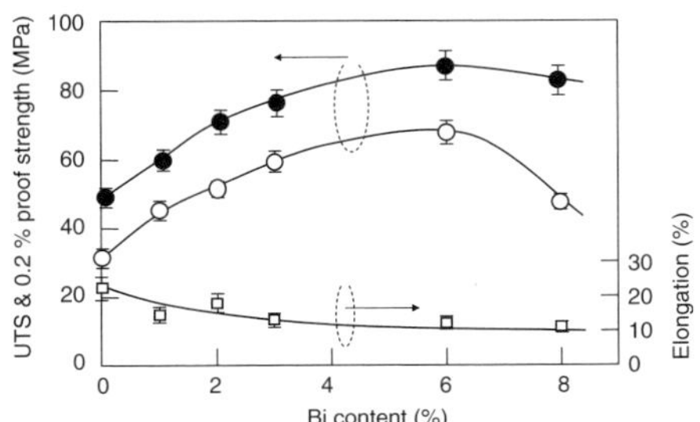

Fig. 5 Ultimate tensile strength (●), 0.2% proof stress (○) and elongation (□) of Sn–8Zn–Bi alloy as a function of Bi content [7]

Table 2 Typical n values for various Sn solders obtained by tensile test

Alloys	n values	References
Sn–9Zn	8.2	[7]
	8.1	[3]
Sn–8Zn–3Bi	14.4	[7]
Sn–8Zn–6Bi	15.0	[7]
Sn–3Ag	12	[3]
Sn–3Ag–0.5Cu	12.5	[7]
Sn–3.5Ag–0.7Cu	9.1	[7]
Sn–3.9Ag–0.6Cu	12.5	[7]
Sn–0.7Cu	17.4	[7]
Sn–0.7Cu–0.5Ag	11.5	[7]
Sn–38Pb	~20	–

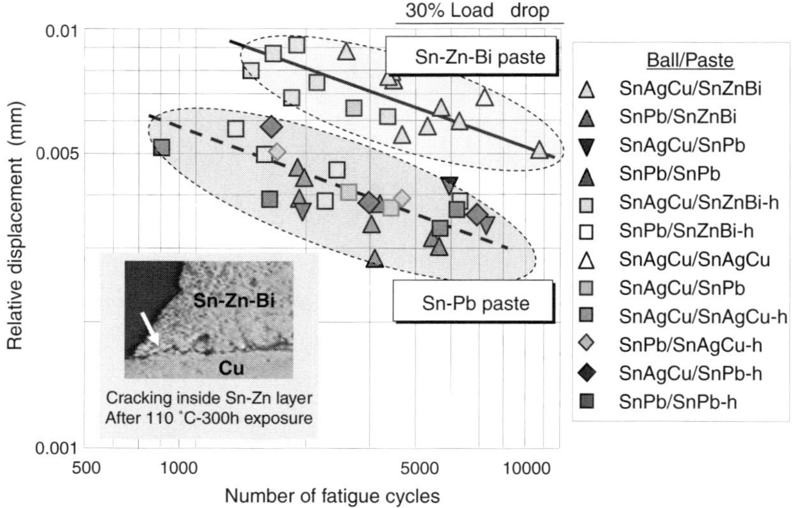

Fig. 7 Mechanical fatigue life of various CSP joints after 125°C exposure for 300 h [1]

formed adjacent to Cu [4]. The Cu–Zn intermetallic compounds are not stable at temperatures as high as 150°C. At high temperatures, Zn continuously diffuses toward interfaces or fillet surfaces because of its high activity and high diffusivity in Sn matrix. At the same time, Cu diffuses from the substrate to the Sn–Zn solder, easily. The diffusion constant of Cu in Sn matrix at 125°C is reported as 1.1×10^{-11} (m²/s) [10], while that of Zn is as 1.7×10^{-15} (m²/s) [11]. During early stages of high temperature annealing, Zn platelet dispersions formed in Sn matrix turn to be Cu_5Zn_8 while the interface Cu–Zn intermetallic compound layer becomes slightly thicker as shown in Fig. 8 [6]. When all the Zn in the solder is consumed, Sn diffusion toward Cu substrate through Cu–Zn intermetallic interface layer becomes dominant resulting in Sn–Cu intermetallic compound formation at the interface

between Cu substrate and Cu–Zn intermetallic layer. Kirkendall voids are formed along the interface between the Cu–Zn interface layer and the solder. This reaction sequence is schematically illustrated in Fig. 9 [5, 6]. Because of large void formation with intermetallic growth at the interface, this reaction degrades the solder joints. Thus, one needs to know this interface feature and some suitable treatment is required for high temperature applications of this solder [5].

On thin Au/Ni–P plating, Zn reacts to form AuZn compound at the interface after reflow treatment while most of lead–free and leaded alloys react to form Ni–Sn compounds after Au dissolution into solders [9]. When the Au layer becomes thick, the spalling of the intermetallic layer occurs as shown in Fig. 10a. The spalling does not influence the interface strength for Sn–Zn alloys. The intermetallic compound is $AuZn_3$ in this reaction. Thus, the interface reaction is not uniform even in the same solder fillet. This is caused primarily by the limited volume of a solder drop on the limited substrate plating thickness. Figure 10 also shows the difference in interface intermetallic compounds depending on the position inside a solder drop.

5 Corrosion by humidity exposure

Due to its high activity, Zn is easily oxidized even in Sn matrix when humidity in the environment is high enough at elevated temperature. Zn diffuses toward the surface of the solder fillet or the interface by humidity/temperature exposure. Figure 11 shows the

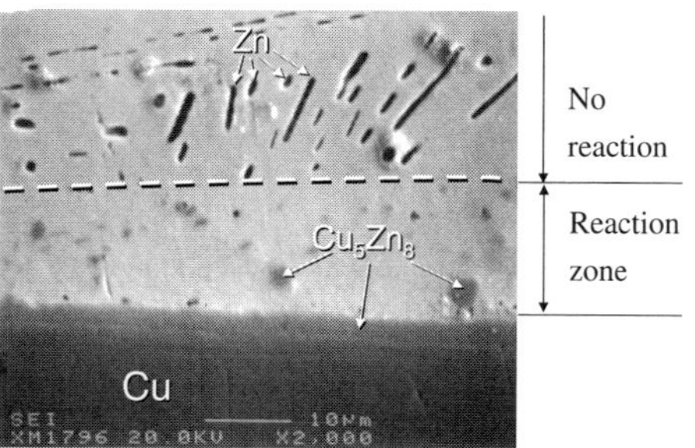

Fig. 8 Interface microstructure of Sn–8Zn–3Bi/Cu joint after 135°C—50 h exposure [6]

Fig. 9 Schematic illustration of heat exposure degradation of Sn–Zn/Cu interface [5, 6]

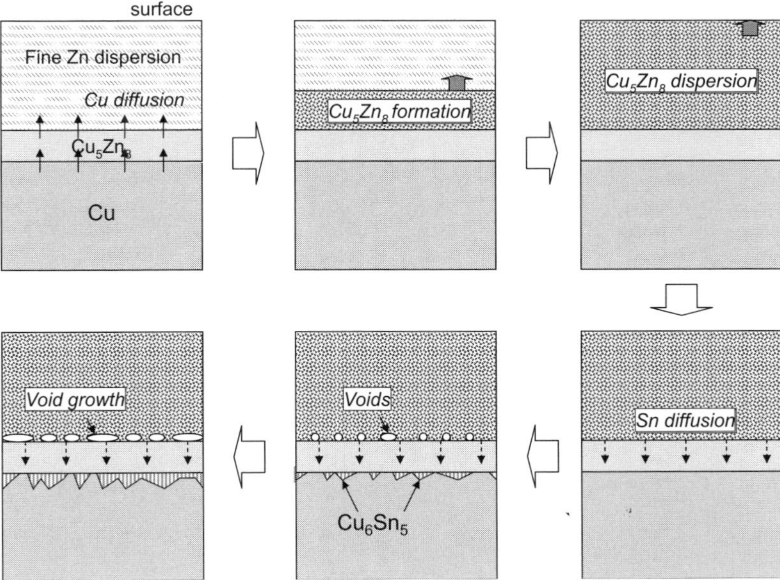

typical example. Though the mechanism of this oxidation is not clear yet, it is easy to imagine that Galvanic corrosion can occur at the Zn/Sn interface or at the Cu_5Zn_8/Sn interface.

The exposure of Sn–Zn–Bi to humidity/heat conditions of 85°C/85%RH for up to 1,000 h promotes Zn and O segregation in the grain boundary while 60°C/90%RH does not [9]. The addition of Pb or of Bi to Sn–Zn alloys accelerates the change from eutectic Zn phase into ZnO phase. Sn whiskers are sometimes formed on the surface of Sn–Zn alloys during the heat/humidity exposure. The formation of ZnO phases along grain boundaries near the free surface of a solder fillet generates compressive stress in/on a Sn grain, and, as a result, Sn whiskers grow from the surface of Sn grains to release this compressive stress. Such

Fig. 10 Peculiar wetting behavior of Sn–9Zn on thick Au (0.25 μm)/Ni plating and interface microstructures (SEM). (**a**) Sn–Zn drop appearance on Au/Ni plating and interfaces of (**b**) central region, (**c**) middle region and (**d**) outer region

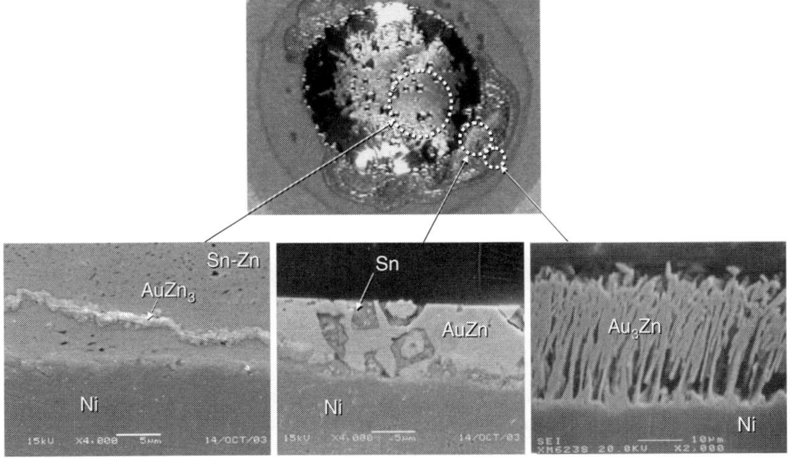

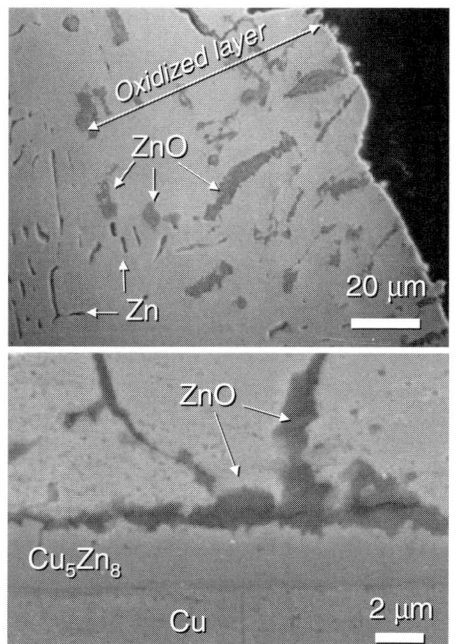

Fig. 11 ZnO formation due to humidity exposure at 85 C/ 85%RH for 1,000 h [12]

oxidation-driven whisker formation has been also observed not only for Sn–Zn alloy but also for Sn–Pb alloys and Sn–Ag–In alloy [13]. At 85°C/85%RH

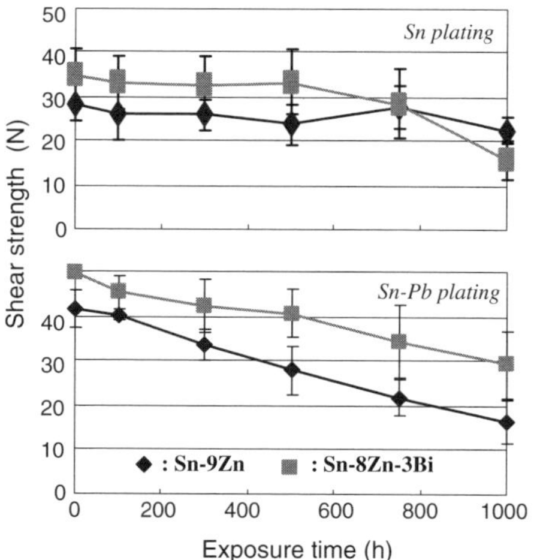

Fig. 12 Humidity at 85C/85%RH and Pb contamination influence on shear strength of chip component on PWB [12]

exposure, there is serious degradation of the joint strength for Sn–Zn alloys due to the formation of ZnO oxide at interface. Figure 12 shows the shear strength change of surface mounted chip components by humidity/heat exposure. It is apparent that Bi and Pb enhance this degradation.

6 Summary and future

Sn–Zn solders posses several fascinating features such as low cost as well as low reflow temperature of 220°C. Due to the drastic improvement of flux technology for Sn–Zn system, Sn–Zn–Bi solder pastes have been widely applied in many products, i.e., notebook computers, desktop computers, printers, TV tuners, electronics dictionaries, etc. For the practical uses, unfortunately, one needs to think of several specific phenomena caused by the high activity of Zn. The formation of Kirkendall voids at the interface and the corrosion in humid environment are two of the important aspects. The condition, in which such serious degradation occurs, it will be not practical. Even though it is true, some accelerated tests to decide the solder can be adopted or not for a given life time for a product is still needed. The scientific information is still lacking to understand these unsolved issues. Moreover, one needs the prevention methods for those possible degradations in the market.

In order to provide such useful information for Sn–Zn as well as for the other low temperature solders; so that one can utilize these solders with confidence, a national project on the development and standardization of low temperature soldering organized by JEITA is now under the way. The main goal of the current project is, of course, to provide useful information, process knowledge, and tools for various reliability evaluations to the database. The basic idea of this project lies in the scientific understanding that will lead to the practical solutions for industries.

From the other standpoint, the trend in electronic packaging is also in seeking lowering process temperature besides the lead-free soldering movement. Even lower than 150°C can be required for high-density thin silicon dies, thin foil applications, and flat panel displays in near future. For such value-added applications, the R&D effort on low temperature soldering can provide the first step to explore the future technology.

Acknowledgements This work was carried out under the both support of the 21COE program of the Japan Ministry of Education, Culture, Sports, Science and Technology and of METI (Ministry of Economy, Trade and Industry) Low Temperature Soldering Project organized by JEITA.

References

1. K. Suganuma, T. Kiga, M. Takeuchi, Q. Yu, K. Tanabe, K. Toi, H. Tanaka, Y. Kato, K. Sasaki, K. Takahashi, M. Tadauchi, T. Tsukui, T. Suga, T. Makimoto, Current Technology of Low Temperature Lead-Free Soldering and JIEP Project, *International Conference on Lead Free Electronics "Towards Implementation of the RoHS Directive"*, IPC/SolderTech, Brussels, June 10–11, 2003, pp. 97–104
2. M. McCormack, S. Jin, J. Electron. Mater. **23**, 635–640 (1997)
3. H. Mavoori, J. Chin, S. Vaynman, B. Moran, L. Keer, M. Fine, J. Electron. Mater. **26**, 783–790 (1997)
4. K. Suganuma, K. Niihara, T. Shoutoku, Y. Nakamura, J. Mater. Res. **13**, 2859–2865 (1998)
5. K. Suganuma, T. Murata, H. Noguchi, Y. Toyoda, J. Mater. Res. **15**, 884–891 (2000)
6. K.S. Kim, Y.S. Kim, K. Suganuma, H. Nakajima, J. Jpn. Inst. Electron. Packag. **5**(7), 666–671 (2002)
7. Y.S. Kim, K.S. Kim, C.W. Hwang, K. Suganuma, J. Alloys Compd. **352**(1–2), 237–245 (2003)
8. M. Suzuki, Y. Matsuoka, E. Kono, H. Sasaki, M. Igarashi, K. Onodera, Mate 2000, 325 (2000)
9. C.W. Hwang, K.S. Kim, K. Suganuma, J. Electron. Mater. **32**(11), 1249–56 (2003)
10. B.F. Dyson, T.R. Anthony, D. Turnbull, J. Appl. Phys. **38**, 3408–3409 (1967)
11. F.H. Huang, H.B. Huntington, Phys. Rev. **9B**, 1479–1488 (1974)
12. Kim K.S., Matsuura T., Suganuma K. (2006). J. Electron Mater. **35**(1), 41–47
13. K. Suganuma, T. Imanishi, K.S. Kim, M.Ueshima, *Properties of low temperature solder Sn–Ag–In–Bi, presented at Materials Science & Technology 2005 Conference and Exhibition (MS&T'05)*, TMS, Pittsburgh, USA, Sep. 25–28, 2005

Composite lead-free electronic solders

Fu Guo

Published online: 22 September 2006
© Springer Science+Business Media, LLC 2006

Abstract Composite approaches have been developed in lead-free solder research in an effort to improve the service temperature capabilities and thermal stability of the solder joints. This article overviews the background for composite lead-free solder development, the roles of reinforcements and their requirements, composite solder processing techniques, as well as current status of composite solder research. Examples of several representative lead-free composite solders produced with various methods and reinforcement types are presented. The effects of reinforcement addition on processing and mechanical properties are analyzed. Difficulties and problems that exist in composite solder research are proposed and tentative solutions are attempted with examples of newly emerging novel lead-free composite solders.

1 Introduction

Several challenges are faced in the development of lead-free solders since they are not just drop-in substitutes for traditionally used leaded solders. These challenges may be related to the solder melt temperature, processing temperature, wettability, mechanical and thermo-mechanical fatigue (TMF) behaviors, etc. Knowledge base on leaded solders gained by experience over a long period time is not directly applicable to lead-free solders. As a result, a database for modeling for reliability predictions of lead-free solders is not currently available [1]. Most of the lead-free solder developments for electronic applications are aimed at arriving at suitable alloy compositions [2–5]. Sn-Ag alloy system, with or without small alloy additions such as Cu, is believed to have significant potential [2–6]. The binary Sn-Ag eutectic temperature is 221°C, and the ternary Sn-Ag-Cu eutectic temperature is 217°C, both being reasonably higher than the Sn-Pb binary eutectic temperature of 183°C. Although the processing parameters have to be modified to accommodate this increase in eutectic temperature, use of such solders also provides higher service temperature capability to the solder joints. Results of several typical studies on such solders are recently reported in the published literature [2–6].

1.1 Harsh service conditions

Solders in general operate at high homologous temperature ranges. During turning on and off operations, the electrical circuitry heats up or cools down, experiencing low cycle TMF due to stresses that develop as a consequence of CTE mismatch between the solder/substrate/components. Mechanical vibration of other entities to which the electronic components are mechanically attached, such as automotive engines, can create higher frequency vibrational fatigue conditions. When an automobile/tank hits a pot-hole/major obstruction, or landing of an airplane, it can impose impact loading on the solder joint. Although fine-grained microstructure may be beneficial for

F. Guo (✉)
College of Materials Science and Engineering, Beijing University of Technology, Beijing 100022, The People's Republic of China
e-mail: guofu@bjut.edu.cn

mechanical fatigue considerations, it may not be ideal for creep resistance since creep deformation at the service temperature (high homologous temperature for the solder alloys) will be by grain boundary sliding. In addition to these opposing requirements, the highly in-homogenous as-joined solder joint microstructure coarsens during service. This aging process causes growth of solder/substrate interface intermetallic layer and coarsening of microstructural constituents within the solder joint. Such evolving microstructure continuously alters the mechanical properties of the solder joint resulting in significant hurdles in reliability prediction modeling. The presence of fine, stable, and system-compatible dispersoids located at the grain boundaries can retard coarsening, enhance mechanical fatigue behavior, and reduce creep rate by decreasing grain boundary sliding tendency. Dispersoids present at a grain boundary represent obstacles which resist sliding between grains which share the boundary.

None of the typical lead-free solder alloys can be utilized at high service temperatures (~270°C) except 80Au-Sn solder, which would have a limited range of applications, since it is expensive [7]. How to increase the service temperatures which existing lead-free solders can endure, or arrive at new solder compositions for such applications, is one of the serious problems that had remained largely unaddressed. However, the National Center for Manufacturing Systems (NCMS) launched a lead-free solder program to explore their use in applications with service temperatures up to about 150°C [8]. For example, in automotive under-the-hood applications, significant performance advantages could be derived if electronic components/circuit boards were mounted on the engine manifold. This would significantly reduce the amount of wiring, and minimize several complications in the electrical circuitry. Similar conditions also exist in aerospace and defense applications. Although the processing parameters must be modified to accommodate the temperature increase, these solders may or may not provide solder joints with a higher service temperature capability depending on TMF characteristics. The results of several mechanical and thermal studies of solders whose structures remain stabilized during elevated temperature applications are reported in the literature [9–13].

Solder joints experience even more severe conditions in applications that combine both high current densities [14, 15] and high temperatures as in applications such as automotive alternators or rectifiers. At present there is no substitute, which is both suitable and economical, for the high-lead/low-tin solders used for these applications [16]. Typical solders based on Sn-Au alloys that can withstand these service conditions are cost prohibitive in large-scale automotive type manufacturing situations. Given the need to satisfy these design requirements, it is likely this issue will be addressed in the near future. This aspect of lead-free solders has not received much attention because solder joints in consumer and hand-held communication electronics typically are not required to withstand the harsh environmental conditions encountered in automotive, defense, and aerospace applications.

1.2 Needs to develop composite solders targeting at the improvement of the service temperature capability of lead-free solders

Solders with intentionally incorporated reinforcements are termed composite solders. Composite approach was developed mainly to improve the service performance including service temperature capability. In other words, the basic purpose of this methodology is to engineer and stabilize a fine-grained microstructure, and homogenize solder joint deformation, so as to improve the mechanical properties of the solder joint, especially creep and thermo-mechanical fatigue resistance. An important additional feature is that the reinforcements do not alter the melting point of the solder matrix, but may effectively increase the service temperature of the base solder materials by improving the creep or thermo-mechanical fatigue properties of the solder matrix.

The methodology that will be employed to enhance the service performance of the solder joint should not affect the well laid out current metallurgical processes in the electronic manufacturing. It is essential that the incorporation of dispersoids does not significantly alter the solderability, solder/substrate wetting characteristics, melting temperature, etc.

Although one would prefer a strong solder joint for creep resistance considerations, it may not be ideal in electronic applications. If the solder in the joint is not able to dissipate the stresses that develop, failure of the electronic components will result. Therefore, a tradeoff is required between a strong solder joint to provide creep resistance but also one which is sufficiently compliant to dissipate stresses which could damage a device, that is, a solder joint must be capable of dissipating induced stresses, while still maintaining electrical functionality and mechanical integrity. Although these two requirements appear to be mutually exclusive, both of them can be satisfied by appropriate microstructural engineering of solders with suitable dispersoids.

2 Composite solder fabrication

2.1 Selection of compatible reinforcements—important considerations

Reinforcements added to a solder matrix should satisfy certain conditions to enhance the service performance of the solder. Reinforcing phases should bond to the solder matrix, and the bonding can be weak or strong depending on the requirements for the specific application. Reinforcements must have minimal or no solubility in molten solder under normal reflow temperatures to maintain stability of the reinforcements during reflow. Reinforcements and the solder matrix densities should be closely matched to facilitate uniform distribution of the reinforcing phase throughout the matrix. If the densities are not matched, settling or floating of the reinforcements will cause the reinforcements to segregate within the solder, resulting in a non-uniform distribution [17]. Good wetting between the reinforcements and the solder matrix is also essential, since the interfacial strength could significantly alter the over-all mechanical behavior of the composite solder. The size of reinforcing phases should be optimal in order to stabilize the microstructure. It has been determined that reinforcement particles of size ~1μm or smaller tend to stabilize a very fine-grained microstructure [18, 19]. Reinforcement particles should not be prone to significant coarsening during reflow and service. Due to thermo-dynamic considerations, fine particles tend to coalesce to reduce the high interfacial energy associated with small particles. Finally, reinforcements should neither significantly alter the processing temperature nor alter solderability (i.e., wetting characteristics of a solder with a chip carrier pad).

2.2 Fabrication techniques

Usually, reinforcement particles used in the composite solder can be grouped into two categories. One kind of reinforcement addition involves the intermetallic particles. These intermetallic reinforcements can be incorporated to the solder matrix either by adding preformed intermetallic particles (like Cu_6Sn_5, Cu_3Sn, or Ni_3Sn_4 [18, 20–27]) or by converting from elemental particles (like Cu, Ni, or Ag [20–24]) by their reaction with Sn during fabrication or the subsequent aging and reflow process. Another kind of reinforcement addition involves those having a low solubility and diffusivity in Sn, or even non-reactive with Sn. Examples of such reinforcements could be Fe particles [28] or oxide particles like Al_2O_3 or TiO_2 [19]. Choices of such reinforcements serve several purposes. Proper choice of foreign reinforcement additions could desirably introduce uniformly distributed intermetallic hard particles or non-coarsening particles. Well-dispersed reinforcements can serve as obstacles to grain growth, crack growth, and dislocation motion, so as to strengthen the solder against creep and fatigue deformation [18].

Two possible ways to introduce the desired reinforcements to the solder matrix are the in-situ method and the mechanical mixing method. In-situ method refers to the technique by which reinforcing phases, like Cu_6Sn_5 or Ni_3Sn_4 intermetallic particles, are readily formed upon processing the bulk solder itself [29–31]. Details of the procedures used to achieve such in-situ composite solders are not available in the current published literature, since the patent application for the process by the inventors is still pending. The mechanical mixing method is more related to extrinsically adding reinforcement particles into the solder matrix, usually in the form of solder paste. The composite solder paste is usually prepared by mechanically blending the mixture for certain length of time to achieve uniform distribution of the reinforcements. After such significant mixing this mixture is melted to produce the composite solder. Traditionally, the reinforcement is usually added to the molten solder to produce the composite [20]. These reinforcements could be stable intermetallics or metallic particles that can form intermetallics during processing [17, 18, 21, 24, 26, 32, 33]. The main aim of these approaches is to introduce compatible intermetallic reinforcements in the solder matrix. It is essential that these reinforcements do not significantly alter their size, shape and volume fraction during service.

3 Current status of composite solder research and development

3.1 Early Sn-Pb based composite solder development

Very early efforts involved the research activities carried out to improve the comprehensive properties of lead-bearing solders using composite approach [18–28]. Microstructural analysis as well as mechanical testing of such composite solders have been reported. Certain composite solders did show improved mechanical properties sought by electronic/automobile industries.

Marshall et al. were among the very first a few groups in early 1990s that carried out studies in microcharacterization of composite solders [20–24].

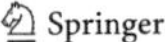

Their composite solders were primarily prepared by mixing Cu_6Sn_5 (10, 20, 30 wt.%), Cu_3Sn (10,20, 30 wt.%), Cu (7.6 wt.%), Ag (4 wt.%), or Ni (4 wt.%) particles with the eutectic Sn-37Pb solder paste. The microstructure features of these bulk composite solder specimens showed that Cu-Sn, Ag-Sn, and Ni-Sn intermetallics were developed in the composite solders around Cu, Ag, and Ni particles respectively. Cu_6Sn_5 layer formed around Cu_3Sn particles in the Cu_3Sn reinforced composite solder, while no more new intermetallic formed in the Cu_6Sn_5 particle reinforced composite solder. The microstructural analysis showed good bonding of the particulate reinforcements to the solder matrix suggesting that the resulting composite solders might exhibit enhanced strength. Intermetallic formation at the solder/copper interface was studied for the above composite solder samples aged at 140°C for 0–16 days, as reported by Pinizzotto et al. [34]. Similar studies were carried out by Wu et al. with aging temperatures of 110°C–160°C for 0–64 days [35]. Intermetallic formation near the Cu substrate was found to be greatly affected by these particle additions. Sn sink theory, i.e., the particles act as Sn sinks which remove Sn from the solder and decrease the amount of Sn for reaction at the interface, were proposed for the effects of Cu-containing particles and Ag particles on the kinetics of intermetallic formation.

Dispersion strengthened in-situ composite solders of Sn-Pb-Ni and Sn-Pb-Cu alloys containing 0.1–1.0 μm dispersoids/reinforcements were produced by induction melting and inert gas atomization, by Sastry et al. [25]. It was found that, upon reflow of the solder specimens, the fine spherical dispersoids in rapidly-solidified Sn-Pb-Cu alloys coarsen to > 1 μm platelets, however, the dispersoids in Sn-Pb-Ni alloys remain spherical and be stable with a size of < 1 μm. The difference in stability of dispersoids in Cu- and Ni-containing solders was explained on the basis of the difference in solubilities and diffusivities of Cu and Ni in Sn-Pb matrix. These composite solders showed an increase of 25–180% in yield stress and 20–80% in the modulus values compared to eutectic Sn-37Pb solder.

Another type of dispersion strengthened composite solder was formulated by Betrabet et al. by adding 2.2 wt.% of Ni_3Sn_4 intermetallic particles into the Sn-40Pb solder matrix [18]. Mechanical alloying, a solid-state high-energy milling process developed for superalloy manufacture, provided the means to process such dispersion strengthened solders. The presence of Ni_3Sn_4 dispersoids resulted in a smaller grain size in the as-cast microstructure and after aging at 100°C for 29 h. Their subsequent study of Cu_9NiSn_3 intermetallic particles reinforced Sn-40Pb composite solder showed

an increase of the strain to failure in shear by 40% while the ultimate shear strength essentially remained unchanged [26]. They claimed this as an indication of improved fatigue resistance because it was believed that fine, uniformly dispersed phases would stabilize microstructures by pinning grain boundary dislocations and by restricting grain boundary motion.

Mavoori and Jin prepared their composite solders by mixing 3 vol.% of 10 nm sized Al_2O_3 powders or 3 vol.% of 5 nm sized TiO_2 powders with 35 μm sized eutectic Sn-37 Pb solder powder [19]. Nano-sized, non-reacting, non-coarsening oxide particles formed uniform coatings of solder after repeated plastic deformation for rearrangement of the particles. Three orders decrease in the steady-state creep rate was achieved by this approach. Such composite solder was found to be much more creep resistant than their control sample, the eutectic Sn-80Au solder. This has great significance in replacing the conventional high melting point (278°C) Sn-80Au solder for its creep resistant applications such as optical or optoelectronic packaging.

Clough et al. reported that, with properly controlled porosity, Cu_6Sn_5 particle reinforced eutectic Sn-37Pb solders exhibited twice the yield strength without significant ductility loss [27]. It was also shown in their study that the creep rate of the composite solder was nearly an order of magnitude less than that of unreinforced solders. The boundary layer fracture behavior was studied using single shear lap specimens using the same composite solder. The specimens failed as shear fracture ran in from opposite edges about 10 μm inside of the interfaces. These boundary layer fractures were characterized and a fracture model was developed. Composite strengthening was shown to significantly improve the ductility, creep life and properties associated with improved reliability and creep-fatigue life [36].

The effects of phase additions on the microstructure, wettability and other mechanical properties of the composite solders have also been reported in other studies [32, 37]. In general, composite solders tend to render improved properties. All the reported investigations were basically exploratory in nature, and the extent of improvement must be weighed against environmental and economic factors before widespread adoption can be realized. However, studies on lead-free Sn-Ag based composite solders have received attention only recently.

3.2 Classic lead-free composite solders

Several lead-free solders such as binary Sn-Ag, Sn-Bi, ternary Sb-Ag-Cu, etc. are being considered as potential alternatives for lead-bearing solders. Composite approaches used in lead-bearing solders can also be applied to lead-free solders. Although studies on composite solders utilizing lead-free solders are somewhat limited compared to those for solder alloys themselves, development of lead-free composite solders is essential based on the reliability considerations. The following sections present the effects of reinforcement addition in lead-free solders on the microstructure evolution, processing, and mechanical properties of the matrix solder materials.

3.2.1 Microstructure modifications

Subramanian et al. have reported the microstructural evolution in the Cu_6Sn_5 particle reinforced eutectic Sn-3.5Ag based composite solders made by in-situ method [29]. Figures 1 and 2 provide microstructures of eutectic Sn-Ag solder joints containing Cu_6Sn_5, Ni_3Sn_4, and Fe-Sn reinforcement particles in the as-joined condition and after long-term aging [38, 39]. Intermetallic particles of Ni_3Sn_4 and Fe-Sn incorporated into the solder matrix by the in-situ method were found to be relatively stable even after 1,400 h of aging at 180°C as compared to Cu_6Sn_5 intermetallic particles, an attribute desirable for a composite approach. The Cu_6Sn_5 intermetallic reinforcements were observed to coarsen under long-term conditions, 5,000 h at temperatures above 120°C. They did not coarsen during short-term aging consisting of a few hundred hours as shown in Fig. 3. This may indicate different mechanism of coarsening at low and high temperatures.

The effect of the Cu_6Sn_5, Ni_3Sn_4, and $FeSn_2$ particles on the solder/substrate interfacial intermetallic compound (IMC) layer growth in the eutectic Sn-Ag composite solder joints was investigated by carrying out aging studies up to a few thousand hours at several temperatures [38, 39]. Figure 4a shows an example of the total growth of interfacial IMC layers (Cu_6Sn_5 and Cu_3Sn) during aging at 150°C up to 5,000 h. The solder joints exhibited only Cu_6Sn_5 intermetallic layer at initial condition, and the Cu_6Sn_5 and Cu_3Sn intermetallic layers are clearly visible after aging as shown in Fig. 2. The interfacial IMC layers normally exhibited rapid growth at the initial stage of aging, but the growth rate decreased as the layer thickness increased. The reinforcement particles were observed to affect the formation of interfacial intermetallic layers during reflow, resulting in a thinner initial interfacial layer compared with the same solder without reinforcements. Figure 4b shows IMC layer growth rate constant of the Cu_6Sn_5 intermetallic layer in the eutectic Sn-Ag and Cu_6Sn_5 particulate reinforced Sn-Ag solder joints during aging at 150°C up to 5,000 h at five temperatures. The Cu_6Sn_5 particulate reinforced Sn-Ag solder joints showed higher IMC layer growth rate constants at temperatures above 120°C, while the IMC layer growth rate constants were lower below 120°C. A similar behavior was observed for Ni_3Sn_4 and $FeSn_2$ particulate reinforced Sn-Ag solder joints. This was a temperature dependent effect in that the reinforcement

Fig. 1 Scanning electron microscope secondary electron emission photographs of the initial microstructure of Sn-Ag eutectic solder joints containing 20 vol.% (**a**) Cu_6Sn_5, (**b**) $FeSn_2$, and (**c**) Ni_3Sn_4 reinforcement particles (After S. Choi et al., Refs. [38, 39], 1999/2000)

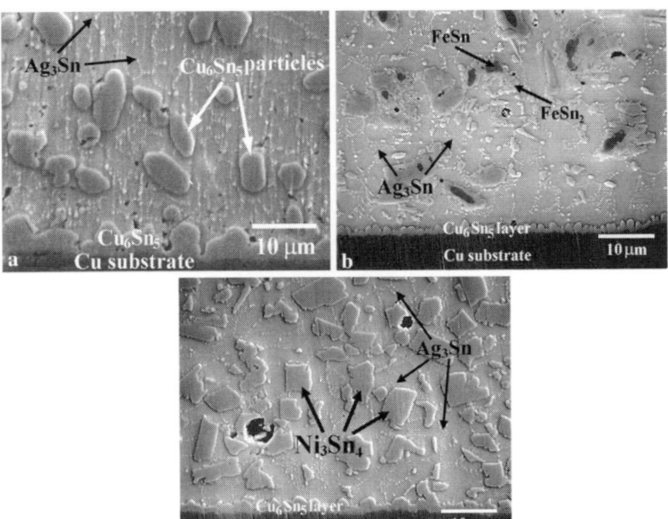

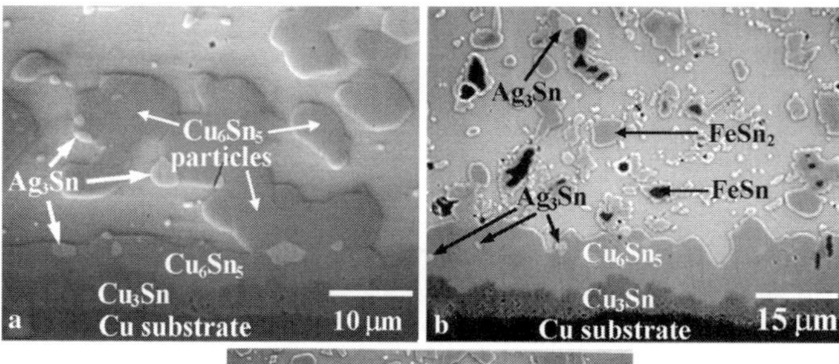

Fig. 2 Scanning electron microscope secondary electron emission photographs of the microstructure of in-situ composite solder joints after aging (**a**) Cu_6Sn_5 composite solder at 150°C for 5,000 h, (**b**) $FeSn_2$ composite solder at 180°C for 1,400 h, and (**c**) Ni_3Sn_4 composite solder at 180°C for 1,400 h (After S. Choi et al., Refs. [38, 39], 1999/2000)

particles were effective in retarding the interfacial intermetallic layer growth at temperatures below 120°C, but not above this temperature, as shown in Fig. 4b.

The stability of mechanically-incorporated metallic reinforcement particles in the Sn-Ag composite solders was examined after multiple reflows and during aging. The composite solders were prepared by mechanically blending nominally 15 vol.% of micron-sized Cu, Ag, or Ni particles with eutectic Sn-Ag solder to form a mixture with a uniform distribution of reinforcement particles [40, 41]. Figures 5 through Fig. 7 illustrate the microstructure of Cu, Ag, and Ni particles in as-fabricated solder joints, after aging at 150°C for 500 h, and after three reflows at 280°C, respectively.

The intermetallic layers formed around the particles due to the reaction of Sn in the solder matrix with the metallic particle during the first reflow can be seen in Fig. 5. The Cu-Sn intermetallic layers consist of Cu_3Sn and Cu_6Sn_5 layers formed on the surface of Cu particles, and Ag_3Sn intermetallic layer formed on the surface of Ag particles. In contrast, a ternary Cu-Ni-Sn intermetallic layer formed around the Ni particles. This indicates that Cu diffused to Ni particles from the Cu substrate during reflow, and participated in the formation of the ternary Cu-Ni-Sn intermetallic layer. It was also observed that binary Cu-Sn and ternary Cu-Ni-Sn intermetallic layers were much thicker on Cu or Ni particles than the Ag_3Sn intermetallic layer on Ag particles. Similar differences in the thickness of the

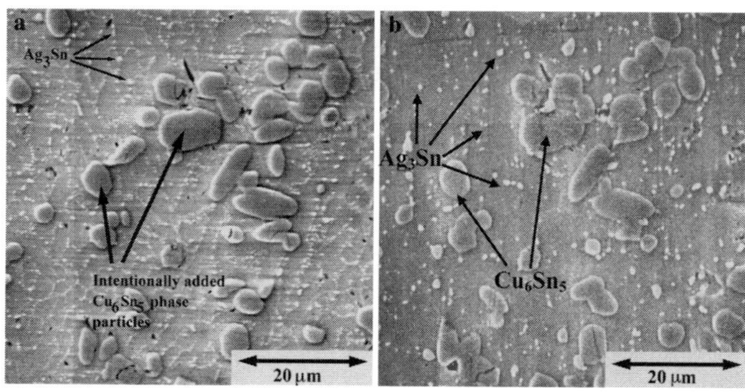

Fig. 3 Scanning electron microscope secondary electron images depicting microstructural stability of Cu_6Sn_5 particles during short-term aging of a composite Sn-Ag solder (**a**) before aging, (**b**) after aging at 150°C for 100 h. (After A.W. Gibson et al., Ref. [3], 1997)

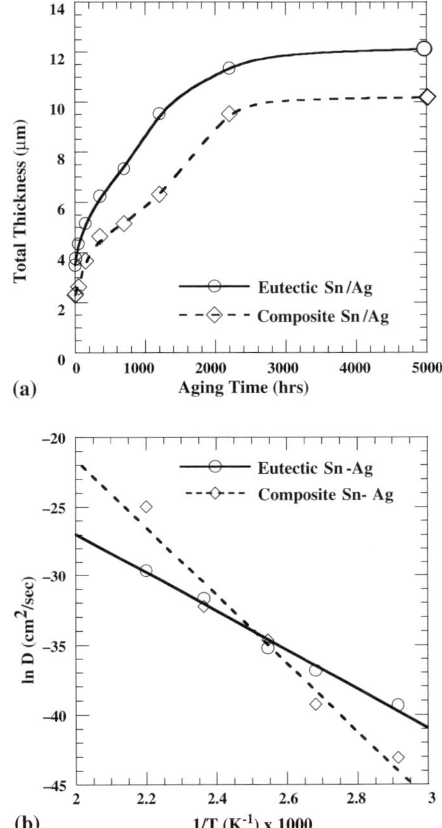

Fig. 4 (**a**) Graph which depicts the effect of aging time at 150°C on the total thickness of interfacial IMC layers (Cu_6Sn_5 + Cu_3Sn) for eutectic Sn-Ag and composite Sn-Ag solder joints which contain Cu_6Sn_5 intermetallic particles introduced by the in-situ method (After S. Choi et al., Ref. [38], 1999). (**b**) Graph which depicts the IMC layer growth rate constants at five different temperatures for the Cu_6Sn_5 intermetallic layer in the eutectic Sn-Ag solder joints and Cu_6Sn_5 particulate reinforced Sn-Ag solder joints during aging at 150°C up to 5,000 h (After S. Choi et al., Ref. [38], 1999)

intermetallic layers around particles was also noted after isothermal aging at 150°C, or reflow at 280°C as shown in Fig. 6 and Fig. 7.

As noted in Fig. 6, the growth of intermetallic layers around particles during aging was in the following order: Ni particles > Cu particles > Ag particles. In the study conducted by Guo et al. [40], Cu particles were largely consumed by Sn in the solder matrix after aging at 150°C for 1,000 h, while there were no significant changes in the case of Ag particles under similar conditions. In contrast, Ni particles were completely

converted to ternary Cu-Ni-Sn intermetallics after aging for 500 h [41]. These studies attribute the difference in growth kinetics of intermetallic layers that surrounded Ni, Cu, and Ag particles to their difference in diffusion coefficients. The diffusion of Ni, Cu, and Ag in Sn is in the following order: Ni > Cu > Ag under the same conditions. This large difference in diffusion capabilities may account for the observed difference in intermetallic layer growth around reinforcement particles. Since Ag particles consume less amount of Sn, it is expected that change in composition of solder matrix would be least in Ag reinforced composite solders compared to composite solders containing Cu and Ni reinforcements. Ag particles also exhibit higher resistance to coarsening than Cu and Ni reinforcement particles due to its lower diffusion coefficient in a Sn matrix.

On the other hand, Fig. 7 illustrates that the growth of intermetallic layers around particles during multiple reflow was: Cu particles > Ni particles > Ag particles. The intermetallic layers present around metal reinforcement particles appear to control the growth of the intermetallic layers under such conditions.

3.2.2 Processing properties

The methodology employed to enhance the service performance of a solder joint should be compatible with the standard processes and assembly procedures practiced by the microelectronics industry. It is also essential that the incorporation of dispersoids does not significantly alter solderability, melting temperature, etc.

The wettability of in-situ composite solders was observed not to degrade up to 20% volume fraction of Cu_6Sn_5, the limit investigated [29]. The contact angle is a measure of the wettability of solder to the substrate. Several micrographs of the spherical caps of the solders produced by reflow of disk shaped solder preforms on Cu substrate were analyzed with PhotoShop® to arrive at average contact angles [29]. The measured contact angles for composite and non-composite solders on a Cu substrate are given in Table 1. The eutectic and composite eutectic solders exhibited similar contact angles, indicating no substantial difference in wettability.

The comparison of average wetting angles of mechanically-incorporated Cu, Ni, Ag particle-reinforced composite solders measured after four reflows are given in Table 2 along with the data for the eutectic Sn-Ag solder, and in-situ Cu_6Sn_5 reinforced composite solders [41, 42]. The wettability of composite solder with 6 vol.% mechanically-incorporated Cu particles is comparable to 15 vol.% Ag-particle reinforced composite solder produced by a similar method, and

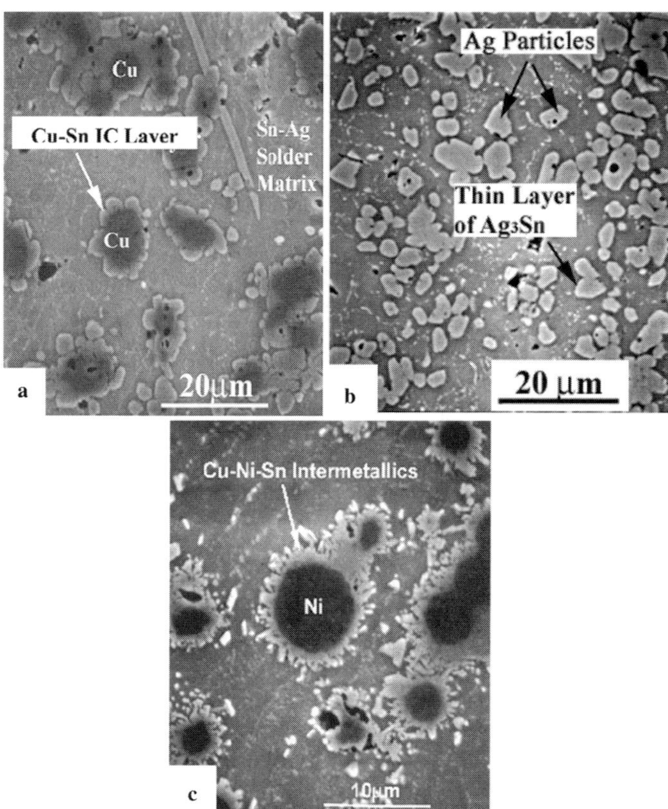

Fig. 5 Scanning electron microscope secondary electron images depicting the as-fabricated microstructure of Sn-Ag eutectic solder joints with **(a)** Cu reinforcement, **(b)** Ag reinforcement, and **(c)** Ni reinforcement, which were mechanically added. In (a) a petal-like layer of CuSn IMC with a light appearance surrounding the dark appearing Cu particles added. In (b) the added Ag particles have a gray appearance and are surrounded by a thin layer of light appearing Ag_3Sn. The Ni particles added in (c) appear dark and are surrounded by light appearing Cu-Ni-Sn intermetallics (After F. Guo et al., Refs. [40, 41], 2001)

20 vol.% of in-situ Cu_6Sn_5-reinforced composite solder. This was attributed to the fact that the volume fraction of Cu-Sn intermetallics was nearly doubled after four reflows, resulting in an increased solder viscosity and wetting angle during reflow. On the other hand, the wetting angle of the 15 vol.% Ni-reinforced composite solder (12.5°) was comparable to eutectic Sn-Ag solder (10.5°), but significantly better than 15 vol.% Ag (21.1°) and 15 vol.% Cu (47.3°) reinforced solders. The total effective volume fraction of Ni reinforcements was observed to remain nearly unchanged during multiple reflow processes.

3.2.3 Mechanical properties

As stated previously, certain composite solders have exhibited enhanced strength and other desired properties sought by the electronics industry. Tensile, creep, and stress relaxation behavior of lead-free composite solder joints have been extensively investigated. Creep properties are one of the most important properties frequently reported in the literature in the development of composite solders.

3.2.3.1 Tensile and steady-state creep properties McCormack et al. have developed composite solders by adding 2.5 wt.% of ~2 µm magnetic Fe powders into pure Sn and eutectic Sn-Bi solder [28]. The idea of adding Fe powders lies in the fact that Fe has low solubility and diffusivity in Sn-based solder and thus is impervious to coarsening. A fine, uniform dispersion of particles was obtained by imposing a magnetic field during the solidification process. The composite solder made by adding 2.5 wt.% of Fe powders to pure Sn exhibited ~60–100% higher ultimate tensile strength than the dispersion-free solder materials. More importantly, its creep resistance at 100°C showed an increase by a factor of 20. Fe particles reinforced eutectic Sn-Bi composite solder exhibited 10% higher tensile strength and five times improvement in creep resistance under similar conditions.

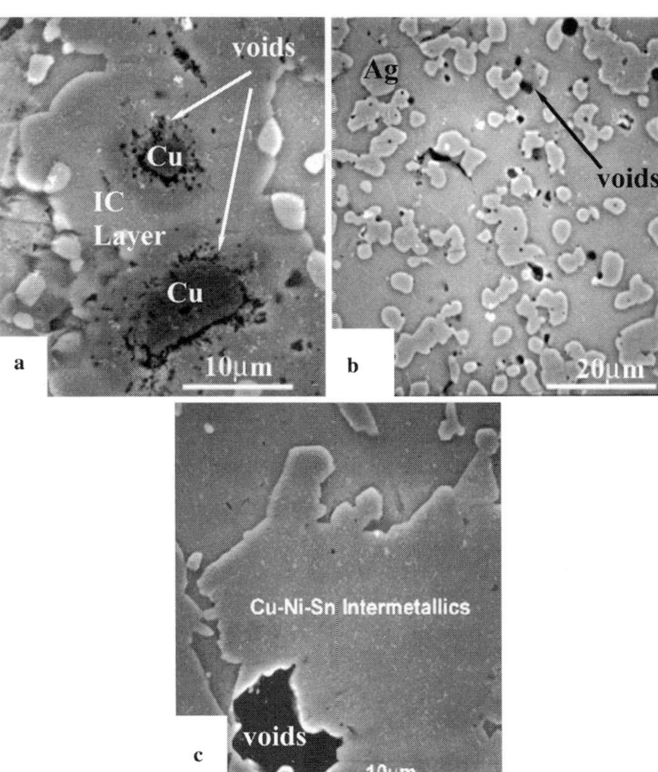

Fig. 6 Scanning electron microscope secondary electron images depicting the thermally-aged microstructure of Sn-Ag eutectic solder joints with (**a**) Cu reinforcement, (**b**) Ag reinforcement, and (**c**) Ni reinforcement, which were mechanically added. The solder joints were aged at 150°C for 500 h (After F. Guo et al., Refs. [40, 41], 2001)

Results of creep testing performed on Sn-Ag composite solder joints containing 20 vol.% Cu_6Sn_5 particles introduced by the in-situ method in as-fabricated and pre-aged conditions for 100 h at various temperatures, have been reported by Subramanian et al., and Choi et al. [29, 43]. Figure 8 illustrates creep behavior of the composite solder joints in as-fabricated and pre-aged condition along with data of eutectic Sn-Ag solder joints from Darveaux et al., and McDougall et al. The 20 vol.% in-situ Cu_6Sn_5 particle reinforced composite solder has about two to three orders of magnitude better creep resistance as compared to non-composite solder at room temperature and lower strain-rates, representing conditions that will exist during lower temperature extreme in a thermo-mechanical cycle. Although aging reduces the creep resistance of these solders, the composite solder possesses better creep resistance as compared to the non-composite solder even under aged conditions. At higher temperatures and higher strain rates region, the composite solder approaches the non-composite solder in creep behavior. At higher temperatures, the ability

of dislocations to climb over particles is apparently fast enough to render the particles as ineffective dislocation barriers. At lower temperatures, the time needed for a dislocation to climb around a particle (for a given stress) is much longer, so the creep resistance is significantly improved.

Choi et al. have performed constant-load creep tests on composite solder joints containing 20 vol.% in-situ Cu_6Sn_5, Ni_3Sn_4, and $FeSn_2$ reinforcement particles [44]. Figure 9 compares the normalized creep behavior of eutectic Sn-Ag and each composite solder joint, earlier creep data obtained from cyclic creep testing, and the data of Darveaux and Banerji [45]. For the eutectic Sn-Ag solder joints, the activation energy for creep was determined to be 60 kJ/mol, which is about half the activation energy for tin self-diffusion. This suggests that grain boundary or dislocation pipe diffusion may control deformation of Sn-Ag eutectic solder joints. A value of 120 kJ/mol was found in composite solder joints containing Cu_6Sn_5 or Ni_3Sn_4 reinforcements, but only 38.5 kJ/mol for composite solder joints with $FeSn_2$ reinforcements. Figure 9

 Springer

Fig. 7 Scanning electron microscope secondary electron images depicting the multiple reflowed microstructure of Sn-Ag eutectic solder joints with **(a)** Cu reinforcement, **(b)** Ag reinforcement, and **(c)** Ni reinforcement, which were mechanically added. The solder joints were reflowed three times at 280°C with a total time of about 30 s above the melting point of eutectic Sn-Ag (MP = 221°C) (After F. Guo et al., Refs. [40, 41], 2001)

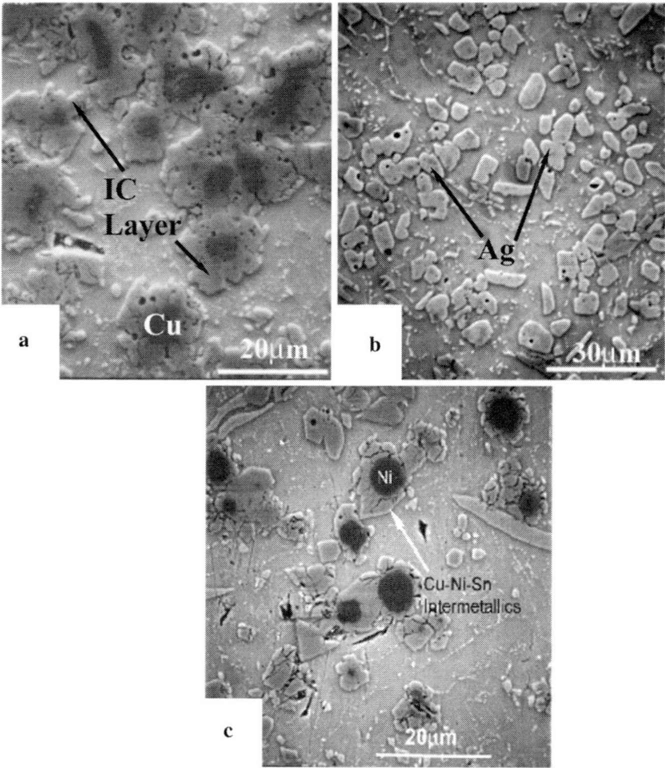

Table 1 Contact angle measurements of Eutectic Sn-Ag Solder and 20 vol.% in-situ Cu$_6$Sn$_5$ Reinforced Sn-Ag Composite Solders on a copper substrate (After K.N. Subramanian, Ref. [29], 1999)

Contact angle measurement	Specimen	Actual contact angle
Composite Sn/Ag solder 5-disk stack disk radius = 1500 μm disk thickness = 165 μm	# 1	16.62, 17.15
	# 2	21.17, 19.55
	Average	18.63
Eutectic Sn/Ag solder 5-disk stack disk radius, = 1500 μm disk thickness = 190 μm	# 1	18.52, 19.35
	# 2	19.67, 19.62
	Average	19.29

Table 2 Average wetting angles of metallic particle reinforced composite solders, along with Eutectic Sn-Ag Solder, and Sn-Ag In-Situ composite solders after four reflows on a copper substrate (After F. Guo et al., Refs. [41, 42], 2001/2000)

Solders	Methods M-Mech. I- in-situ	Average wetting angle after four reflows on Cu substrate
Eutectic Sn-Ag Solder paste	N/A	10.5°
Cu particle reinforced composite solder	M	47.3° (15 vol.%) 18.0° (6 vol.%)
Ag particle reinforced composite solder (15 vol.% Ag)	M	21.1°
Ni particle reinforced composite solder (15 vol.% Ag)	M	12.5°
In-situ Cu$_6$Sn$_5$ Reinforced composite solder (20 vol.% Cu$_6$Sn$_5$)	I	17.6°

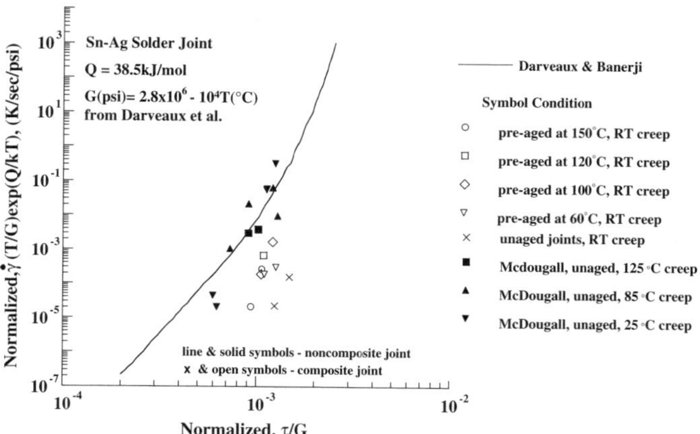

Fig. 8 Plot of normalized shear strain rate vs. normalized shear stress data showing creep behavior of Sn-Ag composite solder joints in the as-fabricated, and in pre-aged condition, along with data from Darveaux et al., and McDougall et al. (After K.N. Subramanian et al., Refs. [29], 1999; and S.L. Choi et al., Ref. [43], 1997)

shows that Sn-Ag composite solder joints typically exhibit the same properties as Sn-Ag eutectic solder joints at elevated temperatures, but not at lower temperatures. Cu_6Sn_5 and Ni_3Sn_4 composite solder joints appear to be weaker at room temperature, while $FeSn_2$ composite solder joints are effectively stronger than the eutectic solder joints at equivalent normalized stresses.

Data from Choi et al.'s prior work [43] on the same in-situ Cu_6Sn_5 composite solder joints showed substantially better creep resistance even under pre-aged conditions as shown in Fig. 9a. These data points were obtained by unloading the specimen to obtain micrographs of a particular scratch placed on a solder joint to quantify the shear displacement. Recovery that can occur during these periods of interrupted unloading apparently reduce the mobile dislocation density available for deformation on reloading the sample.

A summary of the creep properties of composite solders reinforced with Cu, Ag, and Ni metallic elements introduced by mechanical mixing methods is given in Table 3, as are eutectic Sn-Ag and in-situ Cu_6Sn_5 composite solders for comparison [42, 46, 47]. Solder joints made with Cu-particle reinforced composite solders are noted to exhibit a lower steady-state creep rate at all testing temperatures compared with Ag-particle reinforced composite solder joints. The creep behavior of the Ag-particle reinforced composite solder joints is comparable to eutectic Sn-3.5Ag solder. Ni-particle reinforced composite solder joints are about five times more creep resistant than composite solder joints reinforced with Cu, and about 30 times better than eutectic Sn-Ag and composite solder reinforced with Ag. Even at 105°C, the Ni-reinforced composite solder joints were stronger than Cu and Ag-reinforced composite solder joints.

3.2.3.2 The strain at the onset of tertiary creep The strain at the onset of tertiary creep is another important parameter investigated besides the steady-state creep behavior. As indicated in Table 3, eutectic Sn-Ag solder joints exhibited a significantly higher strain for the onset of tertiary creep than all the composite solder joints tested. Among the composite solder joints, the tertiary creep of in-situ Cu_6Sn_5 composite solder joints occurred at strain value comparable to eutectic Sn-Ag solder joints. The relatively higher strain value required for the onset of tertiary creep observed in in-situ Cu_6Sn_5-reinforced composite solder joints is attributed to the weak interfacial bonding between Cu_6Sn_5 particles and the solder matrix identified from nano-indentation tests [48]. It was concluded that the weak interface of uniformly distributed Cu_6Sn_5 particles in the solder matrix provided multiple initiation sites for deformation throughout an entire solder joint. This condition promoted homogeneous deformation, leading to the observed higher ductility of in-situ composite solders as compared to composite solder joints prepared by mechanical mixing.

3.2.3.3 The SGL parameter Use of reinforcement particles did provide a viable means to improve the creep properties and service temperature capabilities of solders. However, neither the steady-state creep rate, nor the strain at the onset of tertiary creep, can be used as an independent criterion to evaluate the creep properties of a solder joint, nor can they be used to predict service life. Guo et al. recently proposed a so-called "SGL parameter" for reliability predictions of composite solder joint based on their creep properties [49]. The SGL parameter can use the local

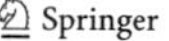

Fig. 9 Graphical depiction of normalized steady-state creep behavior of eutectic Sn-Ag and composite solder joints. Data is presented for **(a)** the eutectic and Cu6Sn5 composite solders, **(b)** the eutectic and Ni3Sn4 composite solders, and **(c)** the eutectic and FeSn2 composite solder joint. Note that **(a)** contains data from Choi for solder joints in the pre-aged condition (After S. Choi et al., Ref. [44], 2001; and R. Darveaux et al., Ref. [45], 1992)

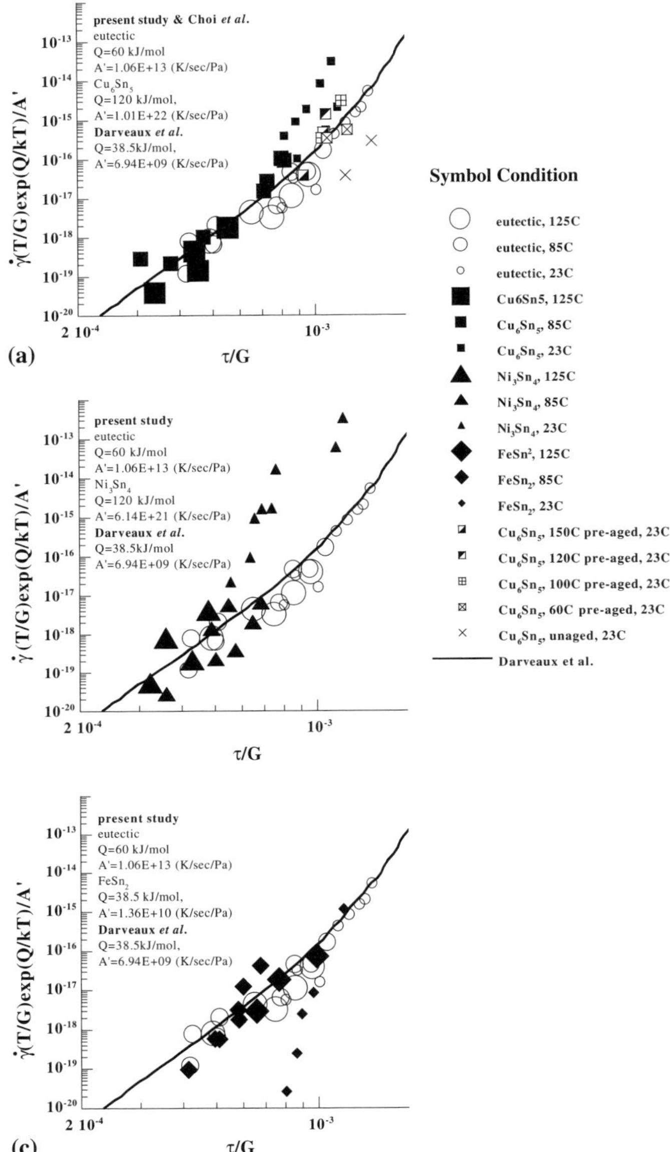

conditions, in addition to the global conditions, for reliability predictions. The aim of creating the SGL parameter is to find a single criterion so that creep properties of a solder joint (not bulk solder) can be assessed accordingly. Both steady-state creep rate and strain for the onset of tertiary creep, as well as their relation to the applied stress are taken into account when this parameter is established. The definition of the SGL parameter is

$$P_{\mathrm{SGL}} = \frac{\varepsilon_{III}}{\dot{\varepsilon}} \times \sigma^{m-n},$$

where ε_{III} is the strain for the onset of tertiary creep (note: the strain at failure for solder joint was not

Table 3 Summary of creep properties of Eutectic Sn-Ag and composite solders prepared by In-Situ and mechanical mixing methods (After F. Guo et al., Refs. [42, 46, 47], 2000/2001)

Solders	Steady-state creep rate (S^{-1}) (at 17 MPa nominal stress)			Average strain for the onset of tertiary creep (at 17MPa)	Methods M-Mech. I-in-situ
	25°C	65°C	105°C		
Sn-3.5Ag solder paste	2.62×10^{-5}	1.50×10^{-4}	1.90×10^{-3}	0.54	–
Cu particle reinforced composite solder (15 vol.% Cu)	4.07×10^{-6}	2.03×10^{-5}	3.95×10^{-4}	0.08	M
Ag particle reinforced composite solder (15 vol.% Ag)	1.91×10^{-5}	8.37×10^{-5}	1.80×10^{-3}	0.19	M
Ni particle reinforced composite solder (15 vol.% Ag)	8.02×10^{-7}	1.08×10^{-5}	1.54×10^{-4}	0.11	M
In-situ Cu_6Sn_5 reinforced composite solder (20 vol.% Cu_6Sn_5)	7.6×10^{-6} (25 MPa)	85°C 9.8×10^{-5} (13 MPa)	125°C 5.8×10^{-4} (12 MPa)	0.45	I

used), $\dot{\varepsilon}$ is the steady-state creep rate, σ is applied stress, m is the stress exponent in Norton's equation, and $-n$ is the stress exponent relating the time at the onset of tertiary creep and the applied stress. This parameter can be comprehensively used to determine which solder material has better creep property in the joint configuration. Therefore, the single SGL parameter proposed, provides a means for lifetime predictions of solder joints under realistic service condition, provided the failure is dominated by creep.

4 Problems and solutions—newly emerged lead-free composite solders

Several novel types of composite solders have emerged based on various difficulties and problems encountered in conventional composite solder research. Different processing methods and choices of reinforcements have been selected in an effort to tackle the problems.

4.1 Nano-sized particle reinforced composite solders

As the development of nano technology has become active lately, various nano-sized particles were chosen as reinforcements in producing composite lead-free solders. The nano-sized particles were added in the hope that they could possibly be more effective, as compared to the micron-sized reinforcements, in the way they key the grain boundaries if they are located at the Sn-Sn grain boundaries so as to serve as obstacles to dislocation motion as well as the grain boundary diffusion during creep.

Inspired by earlier studies adding stiffer materials to conventional solder alloys to enhance their performance, Zhong et al. recently reported the synthesis of high strength lead-free composite solder materials using nano Al_2O_3 as reinforcement [50]. Powder metallurgy was used to synthesize the Sn-based (91.4Sn-4In-4.1Ag-0.4Cu) composite solders containing different volume fractions of nano-sized alumina particulates. The emphasis was placed on the effects of increasing presence of nano Al_2O_3 reinforcement and variation in extrusion temperature on the microstructure, physical, and mechanical properties of the solder alloy matrix. The composite solders exhibited lower density values and CTE values compared with the unreinforced solder matrix. The presence of nano Al_2O_3 particulates assisted in increasing both the 0.2%YS and UTS of the solder matrix all extrusion temperatures employed.

In another recent research, nano-sized Ag reinforcement particles were incorporated into a promising lead-free solder, Sn-0.7Cu, by mechanical means in an effort to improve the comprehensive property of the Sn-0.7Cu solder [51]. This investigation was aimed at developing a series of composite solders with improved mechanical, especially creep, properties, using a cheap base materials and very small amount of reinforcement additions. Ag particles, though relatively expensive, were recognized as good reinforcements because of their non-coarsening nature in the solder due to their low inter-diffusion characteristics with Sn at service temperatures [40]. Under such considerations, the low cost advantage of the Sn-0.7Cu solder and its processing properties are expected to retain in the composite solders. Experimental results indicated that the composite solders and their joints showed better wettability, mechanical properties, as well as longer creep-rupture lives than Sn-0.7Cu solder. The composite solder with 1 vol.% Ag reinforcement addition exhibited the best comprehensive properties as compared to the composite solders with other reinforcement volume

fractions. Significant enhancement of the creep-rupture lives were found in the the the 1 vol.% Ag reinforced Sn-0.7Cu based composite solder joints under different stress and temperature combinations as compared to the Sn-0.7Cu solder joint.

4.2 Lead-free solders reinforced with shape memory alloys

In response to damage accumulation existed in the above traditional particle reinforced composite solder joints, Dutta et al. reported their recent research effort in developing "smart" or "adaptive" solder alloys reinforced by NiTi shape-memory alloy (SMA) particles [52]. Such efforts were reported earlier by Silvain et al. of incorporating NiTi SMA particles into SnPbAg matrix, in which they have shown that the concept of incorporating SMA particles into solder alloys can improve thermo-mechanical characteristics of the solders [53, 54]. The purpose of such composite approach was to fabricate NiTi particulate-containing solder alloys with the objective of exploiting the superelastic properties of austenitic NiTi (i.e., its ability to sustain large recoverable strains) to reduce the stresses in the solder matrix immediately adjacent to the reinforcements. Significant efforts were focused on processing issues related to incorporating NiTi reinforcements in the solder, which has been very difficult because of the poor wetting between NiTi and solder. Copper coated NiTi SMA particles were incorporated into Sn-3.8Ag-0.7Cu lead-free solder paste in order to improve the mechanical performances of its solder joint, reported by Fouassier et al. [55], at the same time to improve bonding between the reinforcement and the solder matrix. It has been noted that the wetting characteristics deteriorate with time and temperature exposure to the solder melt because of dissolution of the coating [54]. This suggests that the coating approach may not be suitable for microelectronic applications, where multiple reflows are common. Dutta et al. successfully fabricated a composite solder paste from which adaptive solders with a uniform distribution of about 5 vol.% of NiTi particulates may be produced [52]. The phase transformations occurring in NiTi were observed to reduce the inelastic-strain range to which a NiTi solder joint is subjected by ~25%, without enhancing the resultant stress range (i.e., without making the joint stiffer). The TMF experiments, in conjunction with DSC studies, clearly established that the M → A transformation occurring in the NiTi fiber during the heating segment imposes a shear strain opposite in sense to the applied shear stress on the solder joint. Similar inelastic strain range reduction

was also reported for SMA wire reinforced Sn-3.5Ag composite solder [56].

4.3 Development of nano-structured lead-free solder materials

In the typical composite solder fabrication process when intermetallic particles or metallic particles are incorporated as reinforcements, size of the IMC reinforcements realized by both these methods tends to be several microns. In addition, IMC reinforcements tend to coarsen during service and deteriorate their effectiveness. Service performance improvements achieved by IMC reinforcements tend to be varied depending on the type of the IMC reinforcement, method used to incorporate, such reinforcements, and their coarsening kinetics during service. As a consequence IMC particulate reinforced composite solders have not been implemented in the actual electronic solder interconnects.

In another composite approach when inert particulates are introduced to the solder matrix, one problem with this methodology is the agglomeration of the reinforcements. During the reflow process the particulates become large clusters with pores. Attempts such as mechanical working by rolling to break up and disperse these agglomerated particulates have been tried with associated problems of interface cracking between the reinforcements and the matrix [19]. Another serious problem associated with incorporation of such inert reinforcements is lack of any chemical bonding between the reinforcement and the solder matrix, which makes them not very effective to enhance the service performance. These extra manufacturing steps increase the cost of the solder. Also these methods are applicable for making solder preforms only, and thus limiting its viability. As a consequence such composite solders with inert reinforcements have not been implemented in practice.

In microelectronic applications, as the current density increases, the migration of ions between electrodes results in void formation which cause failure of the solder joint [57, 58] Similar issues in computer industry have been successfully tackled by incorporation of copper atoms in the grain boundaries of aluminum lines [59–61]. However, there is no known solution for the same problem in the case of lead-free solders. Therefore, a need for solders with sub-micron reinforcements becomes a necessity. These sub-micron size reinforcements when present at the grain boundaries can minimize grain boundary sliding, the predominant mode of TMF damage that occurs during the high

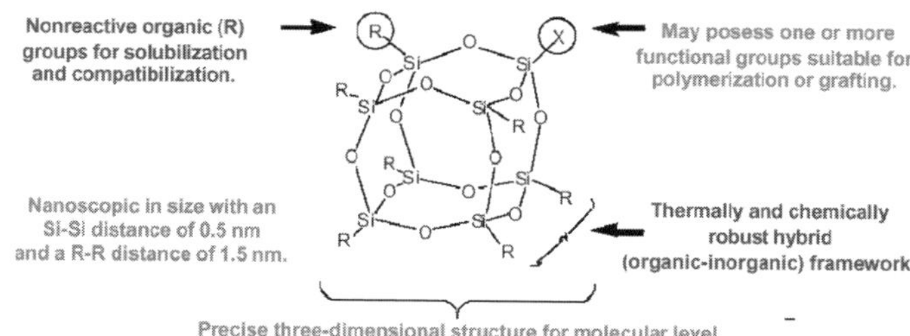

Fig. 10 Anatomy of POSS trisilanol for use in modification of solders. (After A. Lee et al., Ref. [62], 2005)

temperature dwell in a TMF cycle, by keying the grain boundaries.

In order to provide possible solutions to the above stated problems associated with composite solder development, Lee et al. has incorporated surface active, inert reinforcements in the solder matrix [62]. Such an approach will facilitate the initial bonding of reinforcement with the matrix during reflow, and leave the inert particulates from reacting with matrix any further. The nano-structured materials technology of polyhedral oligomeric silsesquioxanes (POSS®), with appropriate organic groups, can produce suitable means to promote bonding between nano- reinforcements and the metallic matrix. Figure 10 shows a representative 3-D structure of such molecular level reinforcement. In the microstructural analysis of the POSS® reinforced Sn-3.5Ag composite solder, shown in Fig. 11, the sub-micron-size (about 50 nm) POSS particulates were visibly well dispersed and were present in higher concentration along the grain boundaries. The results of mechanical and thermo-mechanical study on the POSS®-solder composites with various POSS® additions validated the concept of using surface active, inert nano-structured chemical rigid cages for pinning the grain boundary of solder alloys, which lead to enhanced mechanical performance at elevated temperatures and improved service reliability. An example of shear strength improvement in the POSS® reinforced Sn-3.5Ag solder joints is given in Table 4.

5 Summary

Use of dispersoids is a viable means to improve the properties and service temperature capabilities of solders to be utilized in microelectronic packages and assembles. Composite approaches can provide improvements without significantly affecting current solder-joint fabrication practices. Dispersoids must be compatible with the solder matrix material and remain relatively stable when solder joints are in service. Reinforcement materials can be introduced to the solder by either in-situ methods or by converting mechanically-mixed elemental metallic particles into stable intermetallic compounds. The conversion into intermetallics occurs due to a chemical reaction when the solder is molten during solder joint formation or during reflow. The presence of dispersoids aids in stabilizing solder-joint microstructures by retarding the aging process. All dispersoids tend to improve solder creep strength by several orders of magnitude. An ideal dispersoid should enhance solder ductility without significantly strengthening it. Since the microstructure of solder joints typically is highly inhomogeneous, the deformation within a joint tends to be highly localized. Dispersoids such as the in-situ Cu_6Sn_5 particles [29] create weakly-bonded heterogeneities which promote the initiation of deformation at many locations within a solder joint, resulting in deformation which is much more homogeneous than similar non-reinforced solders. In other words, reinforcements aid in improving the overall ductility of a solder joint. These features render solder joints more compliant by increasing their capacity to accommodate

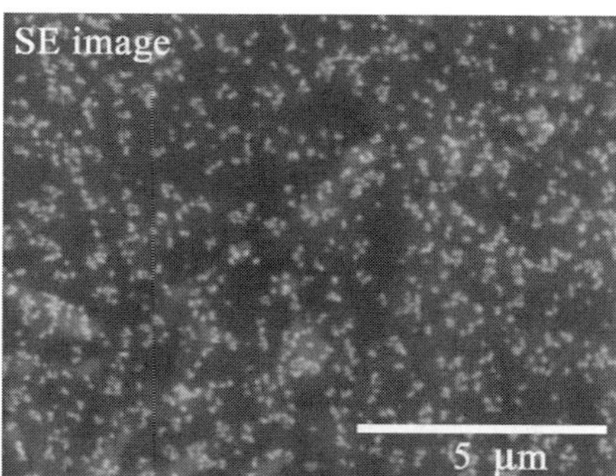

Fig. 11 The SEM image reveals the dimensions of POSS trisilanols and their distribution in eutectic Sn-Ag matrix. (After A. Lee et al., Ref. [62], 2005)

Table 4 Shear strength of Nano-Composite solder joints tested at 25°C with simple shear strain rate of 10^{-2} s^{-1}(After A. Lee et al., Ref. [62], 2005)

Types of POSS used	Shear strength (MPa)
No POSS (eutectic Sn-Ag)	42 ± 4
2 wt.% Cyclohexenyl-triol	58 ± 5
3 wt.% Cyclohexenyl-triol	60 ± 2
2 wt.% Phenyl-triol	60 ± 3
3 wt.% Phenyl-triol	61 ± 4
3 wt.% Ethyl-triol	60 ± 2
2 wt.% Cyclohexyl-diol	64 ± 3
3 wt.% Cyclohexyl-diol	62 ± 4

stresses by relaxation while delaying the onset of tertiary creep. These factors can result in a significant enhancement of TMF resistance of solder joints. Reinforcements introduced by in-situ methods appear to offer the best opportunity for achieving this goal.

When the current rush to identify suitable lead-free solder candidates to replace Pb-based solders succeeds, attention will be focused on improving their service temperature capabilities. It is anticipated that activities related to lead-free composite solders will increase dramatically since they show promise in providing properties difficult to achieve but necessary for high temperature applications. Although there is sufficient evidence at present to indicate the potential of these composite solders, they have not received sufficient attention. The technology to produce lead-free solders with dispersoids, either by in-situ or mechanical means already exists. However, implementation of these solders to produce joints in large-scale manufacturing settings still needs to be developed. Also, compatible no-clean solder flux to be used with these composite solders needs to be identified. These solders conveniently lend themselves to be utilized in either a preform or paste format. Atomization techniques must be developed to convert in-situ composite solders into a powder form to make paste. However, in the mechanical mixing method, the particles can be added to the metal powder during paste making process. These hurdles are manageable, and lead-free composite solders are likely to gain initial importance for solder joints that experience harsh environments, and then perhaps move globally to provide enhanced fatigue resistance in high-reliability applications.

Acknowledgements Several of the results presented in this article are based on the research activity primarily at Michigan State University and Beijing University of Technology by the authors and their colleagues. The author wishes to thank Composite Materials and Structures Center at Michigan State University, National Science Foundation (NSF-DMR-0081796), Technical New Star Program by Beijing Science and Technology Commission (2004B03), and New Century Excellent Talents Program by China Ministry of Education, for financially supporting this effort.

References

1. NEMI, NIST, NSF, TMS, *Report on the Workshop on Modeling and Data Needs for Lead-Free Solders, TMS 2001 Annual Meeting*. New Orleans Convention Center, New Orleans, LA, Feb. 15, 2001
2. McCormack M., Jin S., JOM **45**, 36 (1993)
3. A.W. Gibson, S.L. Choi, K.N. Subramanian, T.R. Bieler, Issues Regarding Microstructural Coarsening due to Aging of Eutectic Tin-Silver Solder, *Design & Reliability of Solders and Solder Interconnections—Proceedings of Minerals, Metals & Materials Society (TMS) Symposium.* (Orlando, FL, 1997), pp. 97–103
4. W.L. Winterbottom, Journal of The Minerals, Metals & Materials Society **45**, 20 (1993)
5. C. Melton, Journal of The Minerals, Metals & Materials Society **45**, 33 (1993)
6. C.M. Miller, I.E. Anderson, J.F. Smith, J. Electron. Mater. **23**(7), 595 (1994)
7. P.T. Vianco, Solder Materials. *AWS Soldering Handbook*, 3rd edn. Chapter 2, (American Welding Society, Miami, 1999)
8. F.W. Gayle, et al., JOM **53**(6), 17–21 (2001)
9. S. Choi, K.N. Subramanian, J.P. Lucas, T.R. Bieler, J. Electron. Mater. **29**(10), 1249–1257 (2000)
10. Y. Kariya, Y. Hirata, M. Otsuka, J. Electr. Mater. **28**(11), 1263–1269 (1999)
11. J.P. Lucas, F. Guo, J. McDougall, T.R. Bieler, K.N. Subramanian, J.K. Park, J. Electr. Mater. **28**(11), 1270–1275 (1999)
12. D. Shangguan, Sold. Surf. Mount Technol. **11**(3), 27–32 (1999)
13. G. Selvaduray, M. Abtew, Mater. Sci. Eng. **R27**, 95–141 (2000)
14. C.Y. Liu, C. Chen, C.N. Liao, K.N. Tu, Applied Physics Letters **75**(1), 58 (1999)
15. C.Y. Liu, C. Chen, K.N. Tu, Journal of Applied Physics **88**(10), 5703 (2000)
16. B.P. Richards, C.L. Levoguer, C.P. Hunt, K. Nimmo, S. Peters, P. Cusack, *Lead-Free Soldering*, DTI (Department of Trade and Industry), (UK, 1999), p. 21
17. J.A. Wasynczuk, G.K. Lucey, Shear Creep of Cu_6Sn_5 Eutectic Composites, *Proceedings of the Technical Program, National Electronic Packaging and Production Conference (NEPCON WEST)*, vol. 3, (Des Plains, IL, 1992), pp. 1245–1255
18. H.S. Betrabet, S.M. McGee, J.K. McKinlay, Scripta Metall. Mater. **25**, 2323–2328 (1991)
19. H. Mavoori, S. Jin, J. Electron. Mater. **27**(11), 1216–1222 (1998)
20. J.L. Marshall, J. Calderon, Sold. Surf. Mount Technol. **9**(2), 22 (1997)
21. J.L. Marshall, J. Sees, J. Calderon, Microcharacterization of Composite Solders, *Proceedings of Technical Program—Nepcon West Conference*, Anahein, CA, 1992, Cahners Exhibition Group, Des Plains, IL, **3**, 1992, pp. 1278–1283
22. J.L. Marshall, J. Calderon, J. Sees, G. Lucey, J.S. Hwang, IEEE Trans. Comp. Hyb. Manufact. Technol. **14**(4), 698 (1991)
23. J.L. Marshall, J. Calderon, Sold. Surf. Mount Technol. **9**(3), 6 (1997)

24. J.L. Marshall, J. Calderon, Sold. Surf. Mount Technol. **9**(3), 11–14 (1997)
25. S.M.L. Sastry, T.C. Peng, R.J. Lederich, K.L. Jerina, C.G. Kuo, Microstructures and Mechanical Properties of In-situ Composite Solders, *Proceedings of Technical Program—Nepcon West Conference*, Anahein, CA, 1992, Cahners Exhibition Group, Des Plains, IL, **3**, 1992, p. 1266
26. H.S. Betrabet, S. McGee, Towards Increased Fatigue Resistance in Sn-Pb Solders by Dispersion Strengthening, *Proceedings of Technical Program—Nepcon West Conference*, Anahein, CA, 1992, Cahners Exhibition Group, Des Plains, IL, **3**, 1992, pp. 1276–1277
27. R.B. Clough, R. Petel, J.S. Hwang and G. Lucey, Preparation and Properties of Reflowed Paste and Bulk Composite Solder, *Proceedings of Technical Program—Nepcon West Conference*, Anahein, CA, 1992, Cahners Exhibition Group, Des Plains, IL, **3**, 1992, p. 1256
28. M. McCormack, S. Jin, G.W. Kammlott, (1994) IEEE Trans. Comp. Packag. Manufact. Technol.—Part A **17**(3), 452
29. K.N. Subramanian, T.R. Bieler, J.P. Lucas, J. Electron. Mater. **28**(11), 1176–1183 (1999)
30. J.H. Lee, D.J. Park, J.N. Heo, Y.H. Lee, D.H. Shin, Y.S. Kim, Scripta Materialia **42**(8), 827–831 (2000)
31. S.Y. Hwang, J.W. Lee, Z.H. Lee, J. Electron. Mater. **31**(11), 1304–1308 (2002)
32. C.G. Guo, S.M.L. Sastry, K.L. Jerina, Tensile and Creep Properties of In-Situ Composite Solders, *1st Int'l Con'f on Microstructural and Mechanical Properties of Aging Materials—Proceedings of Minerals, Metals and Materials Society (TMS) Symposium*, (Chicago, IL, 1993), pp. 409–415
33. C.G. Guo, S.M.L. Sastry, K.L. Jerina, Fatigue Defoirmation of In-Situ Composite Solders, *1st Int'l Con'f on Microstructural and Mechanical Properties of Aging Materials—Proceedings of Minerals, Metals and Materials Society (TMS) Symposium*, (Chicago, IL, 1993), pp. 417–423
34. R.F Pinizzotto, Y. Wu, E.G. Jacobs, L.A. Foster, Microstructural Development in Composite Solders Caused by Long Term, High Temperature Annealing, *Proceedings of Technical Program—Nepcon West Conference, Anahein*, CA, 1992, Cahners Exhibition Group, Des Plains, IL, **3**, 1992, p. 1284
35. Y. Wu, J.A. Sees, C. Pouraghabragher, L.A. Foster, J.L. Marshall, E.G. Jacobs, R.F. Pinizzotto, J. Electron. Mater. 22(7):769 (1993)
36. R.B. Clough, A.J. Shapiro, A.J. Bayba, G.K. Lucey Jr., *Boundary Layer Fracture of Copper Composite Solder Interfaces, Advances in Electronic Packaging*, ASME, New York, NY, EEP 4–2, 1993, p. 1031
37. C.G. Kuo, S.M.L. Sastry K.L. Jerina, *Fatigue Deformation of In-situ Composite Solders, Microstructures and Mechanical Properties of Aging Material*, ed. by P.K. Liaw, R. Viswanathan, K.L. Murty, E.P. Simonen, D. Frear (eds), TMS, Warrendale, PA, p. 417
38. S. Choi, T.R. Bieler, J.P. Lucas, K.N. Subramanian J. Electron. Mater. **28**(11), 1209–1215 (1999)
39. S. Choi, J.P. Lucas, K.N. Subramanian, T.R. Bieler, J. Mater. Sci.-Mater. El. **11**(6), 497–502 (2000)
40. F. Guo, S. Choi, J.P. Lucas, K.N. Subramanian, Sold. Surf. Mount Technol. **13**(1), 7–18 (2001)
41. F. Guo, J. Lee, S. Choi, J.P. Lucas, T.R. Bieler, K.N. Subramanian, J. Electron. Mater. **30**(9), 1073–1082 (2001)
42. F. Guo, S. Choi, J.P. Lucas, K.N. Subramanian, J. Electr. Mater. **29**(10), 1241–1248 (2000)
43. S.L. Choi, A.W. Gibson, J.L. McDougall, T.R. Bieler, K.N. Subramanian, Mechanical Properties of Sn-Ag Composite Solder Joints Containing Copper-Based Intermetallics, *Design & Reliability of Solders and Solder Interconnections—Proceedings of Minerals, Metals & Materials Society (TMS) Symposium*, (Orlando, FL, 1997), pp. 241–245
44. S. Choi, J.G. Lee, F. Guo, T.R. Bieler, K.N. Subramanian, J.P. Lucas, J. Miner. Metal. Mat. Soc. **53**(6), 22–26 (2001)
45. R. Darveaux, K. Banerji, IEEE Transactions on Components, Hybrids, and Manufacturing Technology **15**(6), 1013–1024 (1992)
46. F. Guo, J.P. Lucas, K.N. Subramanian, J. Mater. Sci.-Mater. El. **12**(1), 27–35 (2001)
47. F. Guo, J. Lee, S. Choi, J.P. Lucas, K.N. Subramanian, T.R. Bieler, J. Electron. Mater. **30**(9), 1222–1227 (2001)
48. J.P. Lucas, A.W. Gibson, K.N. Subramanian, T.R. Bieler, Nano-indentation Characterization of Microphases in Sn-3.5Ag Eutectic Solder Joints, *Proc. Mater. Res. Soc. Conf.*, vol. 522, (Pittsburgh, PA, 1998), pp. 339–345
49. F. Guo, J.G. Lee, K.N. Subramanian, Sold. Surf. Mount Technol. **15**(1), 39–42 (2003)
50. X.L. Zhong, M. Gupta, Adv. Eng. Mater. **7**(11), 1049–1054 (2005)
51. F. Tai, F. Guo, Z.D. Xia, Y.P. Lei, Y.F. Yan, J.P. Liu, Y.W. Shi, J. Electron. Mater. **34**(11), 1357–1362 (2005)
52. I. Dutta, B.S. Majumdar, D. Pan, W.S. Horton, W. Wright, Z.X. Wang, J. Electron. Mater. **33**(4), 258–270 (2004)
53. J.F. Silvain, J. Chazelas, M. Lahaye, S. Trombert, Mater. Sci. Eng. A **273–275**, 818–823 (1999)
54. S. Trombert, J. Chazelas, M. Lahaye, J.F. Silvain Compos. Interfaces **5**(5), 479 (1998)
55. O. Fouassier, J. Chazelas, J.F. Silvain, Composites: Part A **33**, 1391–1395 (2002)
56. Z.X. Wang, I. Dutta, B.S. Majumdar, Scripta Materialia **54**(4), 627–632 (2006)
57. C.M. Chen, S.W. Chen, J. Appl. Phys. **90**(3), 1208–1214 (2001)
58. S.W. Chen, C.M. Chen, JOM **55**(2), 62–67 (2003)
59. C.Y. Liu, C. Chen, C.N. Liao, K.N. Tu, Appl. Phys. Lett. **71**(5), 58–60 (1999)
60. Y.J. Park, V.K. Andleigh, C.V. Thompson, J. Appl. Phys. **85**(7), 3546–3555 (1999)
61. Z.B. Zhang, J.S. Huang, M. Twiford, E. Martin, N. Layadi, A. Salah, B. Bhowmik, D. Vitkavage, S. Lytle, E.C.C. Yeh, K.N. Tu, J. Electrochem. Soc. **149**(5), G324–G329 (2002)
62. A. Lee, K.N. Subramanian, J. Electron. Mater. **34**(11), 1399–1407 (2005)

Processing and material issues related to lead-free soldering

Laura J. Turbini

Published online: 12 September 2006

Abstract The European requirement for lead-free electronics has resulted in higher soldering temperature and some material and process changes. Traditional tin–lead solder melts at 183°C, where as the most common lead-free alternatives have a much higher melting temperature—tin–copper (227°C), tin–silver (221°C) and tin–silver–copper (217°C). These have challenged the ingenuity of the materials and process engineers. This chapter will explore some of the issues that have come up in this transition, and which these engineers have understood and addressed. As we enter the lead-free era, we see changes as printed wiring board (PWB) substrates which were designed for lower soldering temperatures are being replaced by newer materials. Factors such as glass transition temperature (T_g), decomposition temperature (T_d) and coefficient of thermal expansion must be considered. Many electronic components are made for lower peak temperatures than those required by the new solders. Solder flux chemistries are changing to meet the needs of the new metal systems, and cleaning of flux residues is becoming more of a challenge. Finally, there is a potential reliability problem—an increased potential for the growth of conductive anodic filament (CAF), an electrochemical failure mechanism that occurs in the use environment.

1 Introduction

The age of lead-free soldering for electronics is upon us. Legislators in the European Union have demanded this, with few exceptions. While much research has been done on the reliability of the new solder systems, little has been said about the other materials which are affected. This chapter will review the questions and issues that must be understood relative to printed wiring substrate materials, soldering fluxes and pastes, the soldering process itself, the cleaning materials, and the changes that must take place because of the higher soldering temperatures required for most lead-free alternatives.

2 Alloy selection

The solder alloys most commonly used for electronics assembly contain tin (Sn) and lead (Pb) with a standard eutectic composition being Sn63Pb37. Compared to the tin–lead solder alloy, most lead-free alloys melt at much higher temperatures, while only a few melt at lower temperatures. Some solder replacement candidates are eutectic alloys melting at a single temperature, while others are non-eutectic alloys that melt over a temperature range. Table 1 lists the melting point or range for Sn/Pb eutectic compared with the other major alloys selected by most companies as the replacement.

3 Thermal processing requirements

Most lead-free alloys under investigation melt at temperatures that are 30–40°C higher than that of eutectic Sn/Pb solder. Several issues become important when

L. J. Turbini (✉)
University of Toronto, Toronto, Ontario, Canada
e-mail: turbini@ect.utoronto.ca

Table 1 Melting temperature of SnPb and Pb-free solder alloys

Alloy	Melting point (°C)
Sn–Pb37	183
Sn–Ag3.8–Cu0.7	217
Sn–Ag3.0–Cu0.5	217–220
Sn96.2Ag2.5Cu0.8Sb0.5	216
Sn–Cu0.7	227
Sn–Cu0.7 + Ni	227
Sn–Ag3.5	221
Sn–Zn9.0	198.5
Sn–Bi58	138

these higher soldering temperatures are used. These include the effect of these higher temperatures on components, soldering flux or paste chemistry, cleaning, and substrate material properties.

3.1 Printed wiring boards

Traditionally, FR-4 epoxy/glass boards have been used as the workhorse in the industry. Difunctional brominated epoxy-glass resins have a glass transition (T_g) temperature of 125–135°C. The poly functional epoxy has a T_g of 140–150°C, and the high temperature, one-component epoxy system has a T_g of ~ 180°C. Newer board materials [1] for lead-free soldering include FR-4 with phenolic (T_g 180), modified FR-4 (T_g 190–220), polyimide (T_g 250), polyphenylene oxide (T_g 180), polyphenylene ether (T_g 180) and others. The higher soldering temperatures required by lead-free processes may necessitate the use of a higher T_g laminate in most applications. These other substrate materials are available, but cost more and have other electrical and mechanical properties which provide renewed challenges.

Another property of the laminate that must also be considered is the decomposition temperature (T_d), a measure of the actual chemical physical degradation of the substrate system. Measured using a thermogravametric analysizer (TGA) T_d is defined as the temperature at which 5% of the mass of the sample is lost to decomposition [2].

The soldering process creates a difference in the expansion of the materials due to differences in their coefficient of thermal expansion (CTE). This is the fractional increase in length per unit length, over the temperature excursion range required for lead-free soldering. It is usually expressed as ppm/°C or 10^{-6}/K. The higher temperatures exacerbate those differences.

An electronic assembly contains a number of different components. The laminate is usually a composite of polymer and e-glass. This creates a material which is constrained in the X-Y direction (CTE 15–18), and thus expands in the Z direction upon heating (CTE-45–60). The base metallization on the board is copper (CTE-17), and this is also used in plating holes and vias. The silicon chip with a CTE of 2.6 may be packaged in an hermetic ceramic package (CTE 4–8) or a non-hermetic plastic package using epoxy molding compound (CTE 14–20). Alloy 42 which is used to connect the device electrically to the board has a CTE of 43. If the device is packaged as a BGA or micro-BGA, the CTE of the alloy can range between 20 and 30. Often an Underfill is used to mitigate the CTE mismatch between the chip, the package and the board. The underfill CTE is designed to match that of the solder. Table 2 lists the CTE of a few of the many materials that become part of the electronic assembly. It is clear that there are a number of thermal stresses that take place during the manufacturing process and that the higher soldering temperatures for lead-free soldering exacerbate these stresses. An increased scrap rate can be caused by board warpage, delamination, and material degradation.

Higher process temperatures will result in an increased scrap rate of FR-4 epoxy-glass printed wiring boards due to board warpage, delamination and material degradation (charring). In addition, boards processed at higher temperatures are prone to conductive anodic filament failure (CAF). In many cases new chip carrier materials will be required.

Table 2 Coefficient of thermal expansion values for a number of materials used in electronic assemblies

Material	Coefficient of thermal expansion ppm/°C	
Copper [3]	17	
Alloy 42 [4]	4.3	
Lead [3]	29	
Sn [5]	23.5	
Sn63Pb37 [6]	21.6–28.9	
Sn98.8Cu0.7Sb0.5 [6]	17.4–22.1	
Sn95.5Ag3.8Cu0.7 [6]	17.6–18.8	
Sn96.2Ag2.5Cu0.8Sb0.5 [6]	26.9	
Sn96.5Ag3.5 [7]	20.2–22.9	
Sn99.3Cu0.7 (Nihon Superior, private communication)	26.5	
Alumina [3]	7	
Silicon [3]	2.6	
Epoxy molding compound [3]	14–20	
Ceramic [8]	4–8	
Underfill	20–29	
Laminates [9]	X, Y axis	Z axis
Epoxy/e-glass	15–18	45–60
Polyimide/e-glass	15–18	45–60

Engelmaier [10] has proposed a Soldering Temperature Impact Index (STII) to take into account the relative effect of these factors.

$$\text{STII} = (T_g + T_d)/2 - \%\ \text{Z-axis expansion}\ (\Delta T)$$

where ΔT represents the temperature excursion during the soldering process (50–260°C). Engelmaier proposes a minimum index of 215 for lead-free applications. Table 3 gives an example of eight commercially available substrate materials. It can be seen that the materials A, B, C and E fall below this value and would not be suitable for lead-free applications.

Material A and B have significantly different index values because of the improved T_d for material B. Material C has a low T_g but a high T_d, while Material E has a high T_g but poor T_d. In both cases, this leads to a low STII. Material D's low Z-axis expansion and F's higher T_d make these the highest rated materials. Kelley [6] notes that materials that have high T_d values survive more thermal processing cycles.

Plated through holes, and vias become a concern as well [11]. New laminate materials make processing more difficult. In addition, it is more difficult to plate the small vias used in today's electronics. One material supplier [12] has defined the difficulty factor in plating vias as:

$$\text{Difficulty Factor} = L^2/D$$

where L is the board thickness and D is the diameter of the hole.

Another factor is the thickness of the copper in the hole or via. A minimum of 25 microns is essential to insure that the via or barrel does not completely dissolve during the soldering process since the lead-free alloys are high Sn materials and Cu dissolves in this rapidly. Also, the thicker the multilayer board, the greater the stress on the plated holes during the soldering cycle, especially in the z-direction.

Table 3 Comparison of the thermal property values of eight commercially available laminate materials, with the calculated value of their soldering temperature impact index (STII)

Material	T_g°C	T_d °C	% Z-axis Expansion	STII
A	140	320	4.4	186
B	142	350	4.3	203
C	150	345	3.4	211
D	170	345	2.7	231
E	172	310	3.4	208
F	175	350	3.2	231
G	180	350	3.2	226
H	180	350	3.5	222

3.2 Surface finish

There are several laminate surface finishes available for lead-free applications. These include organic solubility preservatives (OSP), immersion silver, electrolytic nickel/gold, electroless nickel/immersion gold (ENIG), and immersion tin. The surface finish chosen should be compatible with the soldering flux.

3.3 Flux chemistry

Soldering [13] is defined as the process of joining metallic surfaces with solder without the melting of the basis metal. In order for this joining to take place, the metal surfaces must be clean of contamination and oxidation. This cleaning action is performed by the flux [2] a chemically active compound which, when heated, removes minor surface oxidation, minimizes oxidation of the basis metal, and promotes the formation of an intermetallic layer between solder and basis metal.

Solder fluxes and pastes have gone through significant evolution since the early 1980s. Before then, soldering fluxes were traditionally rosin-based and they conformed to military specifications and nomenclature: R—rosin, RMA—rosin mildly active, RA—rosin active and RSA—rosin super activated. The activation levels were determined by an extract resistivity test among others. There were also water soluble fluxes used for some applications. In the past, most fluxes contained 25 to 30% solids. Today, new flux formulations use weak organic acids and have much lower solids content (1.5–5 %). These low residue fluxes are often not cleaned—thus the term No Clean Flux. In North America about 70% of the fluxes are not cleaned, 25% are water soluble and 5% are rosin based for military applications [14].

The IPC J-STD-004 [15] for soldering fluxes defines a series of test that are to be used to characterize fluxes. These tests are designed to evaluate the corrosive characteristics of the flux and the flux residues. Fluxes are then defined by their main constituent under one of four categories: R0—Rosin, RE—Resin, OR—Organic, and IN—inorganic. They are further categorized as: L—low flux or flux residue activity, M—moderate flux or flux residue activity and H—high flux or flux residue activity with zero or one being added to identify whether halide has been added to the flux. Thus, an ROL1 flux is a rosin, low-activity flux which contains some halide ($<0.5\%$).

The role of the flux is to remove oxides and other contaminants on the metal surfaces to be soldered. The flux contains several ingredients:

- Activators—react with and remove the metal oxides.
- Vehicle—coats the surface to be soldered, dissolves the metal salts produced when the activator reacts with the oxides, and provides a covering for the cleaned metal surface to prevent further oxidation until soldering takes place.
- Solvent—dissolves the activators and vehicle and deposits them uniformly on the board and component surfaces.
- Special additives—rheological agents and other special ingredients are added to fluxes used in solder pastes, paste flux, and cored wire flux.

The flux becomes active as it is heated. In traditional flux chemistry for Sn/Pb solder, the assembly is preheated to around 100–125°C to remove the solvent and begin to activate the chemicals used to remove the metal oxide. After this plateau, the temperature is increased above the melting point of the solder (183°C) to 240°C for sufficient time to reflow the solder paste, and then the assembly is cooled, solidifying the solder and creating a metallurgical bond between the board metallization and the components. For lead-free soldering the preheat plateau temperature is higher – 150–200°C – and the peak temperature is 245–260°C. This requires solvents that evaporate at a higher temperature, and activators that become chemically active at a higher temperature. In addition, new activators are needed to address the new metallurgy on board surfaces, and new lead-free solders containing Ag, Cu and much higher levels of Sn.

3.4 Cleaning

For some applications, the removal of the solder flux residues is essential, e.g. before conformal coating, or for reasons of reliability. In the 1970s chlorinated solvents such as perchloroethylene, trichloroethylene and methyl chloroform were used to remove flux residues. When they became suspect as potential carcinogens, the chlorofluorocarbon (CFC) based solvents became prominent, and surfactant based water cleaning processes were used. The elimination of CFCs in 1994 let to the prominent use of low solids/no clean fluxes. Water soluble fluxes were cleaned with water, and surfactants and semi-aqueous solvents were used to remove rosin or resin flux residues. The challenge for cleaning in the lead-free era comes from the higher soldering temperatures that create residues that are more difficult to remove. New cleaning agents are being developed to address these issues.

3.5 Components

In manufacturing complex, printed wiring assemblies (PWAs), the thermal process chosen must take into account the thermal mass of the assembly, the component density, the solder flux/paste characteristics, and the maximum temperature limitation of the components. Most of the components assembled to the board in the soldering process, have a maximum temperature of 240°C for traditional Sn/Pb soldering. Some electronic components, such as electrolytic capacitors and plastic-encapsulated components are not rated to experience the high temperatures required to process with lead-free solders. The resulting heat-induced degradation can result in early field failures. Also, the higher temperatures required for lead-free solders are not compatible with many optoelectronic components. The increased heat can cause a variety of conditions with these components, among them: electrical variances, changes in silver-epoxy die attach properties, delamination between plastic and lead-frame parts, deformation of plastic encapsulants and plastic lenses, damage to lens coatings and changes in the light transmission properties.

IPC/JEDEC [16] has developed a recommended reflow profile for nonhermetic packaged semiconductor components. These are based on the temperature taken at the top-side of the packaged device. These recommendations are based on package volume excluding external leads, or solder balls in the case of ball grid arrays (BGAs), and non-integrated heat sinks. Table 4 lists the recommended range for Sn/Pb and Pb-free assemblies.

The maximum recommended temperature for the package depends on the package thickness and volume. Table 5 lists the recommended reflow temperatures for Sn/Pb processing while Table 6 lists the recommendations for Pb-free processing.

Table 4 Reflow profiles recommended for Sn/Pb and Pb-free Assemblies based on temperatures taken on the package body

Profile feature	Sn/Pb assemblies	Pb-free assemblies
Average ramp-up rate	3°C/s max	3°C/s max
Preheat		
- Temperature min	100°C	150°C
- Temperature max	150°C	200°C
- Time	60–120 s	60–180 s
Time maintained above melting temperature	60–150 s	60–150 s
Time within 5°C of peak temperature (T_p)	10–30 s	20–40 s
Ramp-down rate	6°C/s max	6°C/s max
Time 25°C to T_p	6 min max	8 min max

Table 5 Recommended maximum package reflow temperatures for Sn/Pb process

Package thickness	Volume < 350 mm^3	Volume ≥ 350 mm^3
<2.5 mm	$240 + 0/ - 5$°C	$225 + 0/ - 5$°C
≥ 2.5 mm	$225 + 0/ - 5$°C	$225 + 0/ - 5$°C

Table 6 Recommended maximum package reflow temperatures for Pb-free process

Package thickness	Volume <350 mm^3	Volume 350–2000 mm^3	Volume >2000 mm^3
<1.6 mm	$260 + 0$°C	$260 + 0$°C	$260 + 0$°C
1.6–2.5 mm	$260 + 0$°C	$250 + 0$°C	$245 + 0$°C
≥ 2.5 mm	$250 + 0$°C	$245 + 0$°C	$245 + 0$°C

4 Conductive anodic filament formation

One failure mode in printed wiring boards that is enhanced by the higher temperatures needed for lead-free soldering is conductive anodic filament (CAF) formation [17]. This failure mode was first reported in 1976 by researchers at Bell Labs [18]. It involves the electrochemical growth of a copper-containing filament subsurface along the polymer-glass interface of a PWB, from anode to cathode. A model developed by the Bell Labs researchers [19] in the late 1970s [20] details the mechanism by which CAF formation and growth occurs. The first step is a physical degradation of the glass/epoxy bond. Moisture absorption then occurs under high humidity conditions. This creates an aqueous medium along the separated glass/epoxy interface that provides an electrochemical pathway and facilitates the transport of corrosion products. A close up of this phenomenon for a real assembly is shown in Fig. 1.

Despite the projected lifetime reduction due to CAF, field failures were not identified in the 1980s. More recently, however, field failures of critical equipment have been reported [21]. Factors that affect this failure mode are substrate choice, conductor configuration, voltage gradient, and storage and use environment. Certain soldering fluxes [22] and HASL fluids, high humidity either in the storage or the use environment, and high voltage gradient enhance this failure mechanism. A recent study indicates that the higher reflow temperatures needed for lead-free soldering will result in significantly higher incidents of CAF in the future [17].

The objective of the study was to evaluate a series of water-soluble fluxes for their propensity to enhance CAF and to determine the effect of reflow temperature on the number of CAF observed. Specifically, it looked

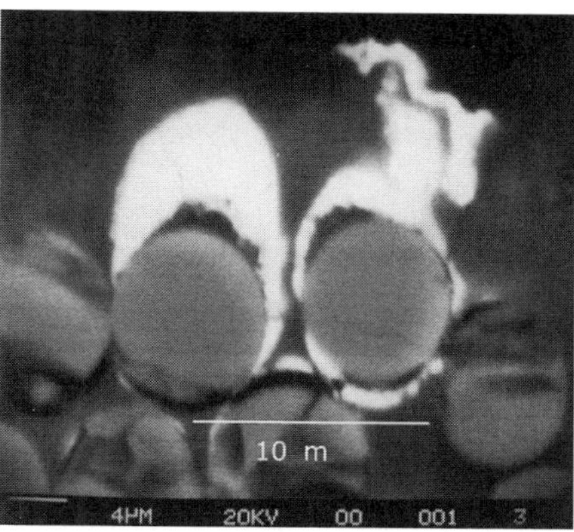

Fig. 1 Cross section of a PWB showing CAF growing along the epoxy/glass interface

at 201°C as the peak temperature experienced by a PWB during wave soldering with Sn/Pb solder versus 241°C peak expected with lead-free wave soldering. The fluxes in this study contained 20 w% of one of the following vehicles: polyethylene glycol [PEG], polypropylene glycol [PPG], polyethylene propylene glycol MW1800 [PEPG 18] and polyethylene propylene glycol MW2600 [PEPG26], glycerine [GLY], octyl phenol ethoxylate [OPE] and a modified linear aliphatic polyether [LAP] dissolved in isopropyl alcohol (IPA). Flux formulations containing 20 w% of the different flux vehicles were also tested with 2 w% HBr or HCl activators, to see what effects the presence of the halide had on CAF formation.

The test boards were IPC-B-24 boards (Fig. 2) containing four comb patterns per board. Two boards for each flux were processed and cleaned. The coupons were placed in a temperature humidity chamber at 85°C and 85% RH, and surface insulation resistance (SIR) measurements were taken for all the boards at 24-hour intervals, over a 28-day period. The SIR testing was done using a bias voltage and a test voltage of 100 V and the same polarity. At the end of 28 days, each board was examined under an optical microscope using back-lighting and the number of CAF counted. Figure 3 shows how the CAF appears as dark shadows originating at the anode when viewed with back lighting.

Table 7 shows the average SIR levels at the end of the 28-day test for boards reflowed at 201°C and at 241°C. Most of the electrical readings were the same for both reflow temperatures. Exceptions to that include PEG/HCl and PEG/HBr which had acceptably high SIR readings (high 10^8) for the 241°C reflow

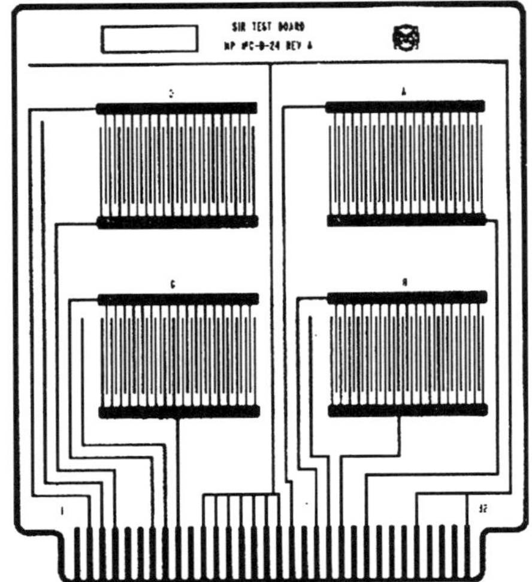

Fig. 2 IPC-B-24 Test Board

conditions but failed electrically ($<10^6$) at the 201°C reflow temperature. Additionally, glycerine (GLY) gave slightly lower SIR readings (high 10^9 vs. $>10^{10}$) under the higher temperature reflow conditions. Table 6 also shows the total number of CAF observed on two boards for each flux chemistry under each of the reflow conditions.

The following observations were made:

- PEG: CAF only forms when no halide activator was present. Also, the numbers of CAF at the *lower* reflow temperature were almost twice as many as at the *higher* reflow temperature. For all PEG fluxes the SIR levels were below the value of the limiting resistor in the circuit, i.e. 10^6 indicating that they

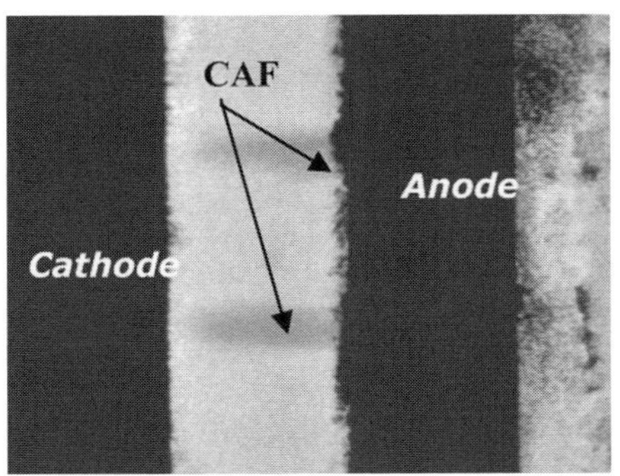

Fig. 3 Using back lighting CAF appears as dark shadows coming from the copper anode to the cathode. The spacing between the anode and cathode is 0.5 mm

have failed the SIR electrical test, except the halide formulations at higher reflow.

- PPG: CAF was almost non-existent at the lower reflow temperature. But many hundreds were observed for all three flux formulations at the higher reflow temperature.

- PEPG 18: There were 13–400 × as many CAF caused by the higher reflow temperature. At the higher reflow temperature the halide-free formulation had the largest number of CAF. At the lower temperature the pattern was: Cl⁻ activated > Br⁻ activated. None were observed for the halide free flux.

- PEPG 26: At the higher temperature the number of CAF followed the pattern: Cl⁻ activated > halide-free > Br⁻ activated flux. At the lower temperature the number of CAF followed a different pattern: Cl⁻ activated > Br⁻ activated. None were observed for the halide-free flux. Also, the total number of CAF observed at both temperatures were significantly less than those noted for the lower molecular weight PEPG 18 flux formulations.

- GLY: CAF is predominantly associated at the higher reflow temperature with Cl⁻ activated > Br⁻ activated > halide-free. At the lower reflow temperature, only the Br⁻ activated gave a few CAF.

- OPE: At the higher reflow temperature, Br⁻ activated flux >> halide-free > Cl⁻ activated. At the higher reflow temperature the number of CAF was 4–300 × as many as at the lower reflow temperature. And, at the lower reflow temperature the Cl⁻ activated flux performed the worst.

- LAP: At the higher reflow temperature Br⁻ activated > Cl⁻ activated, whereas at the lower reflow temperature only the Cl⁻ activated flux showed CAF and this was less than 10 × as many as for the higher temperature.

It is clear from the above data that the interactions of the flux and processing temperature with the test boards is complex and needs further study. Diffusion of polyglycols into the PWB substrate occurs during soldering. Since the diffusion process follows Arrhenius behavior, the length of time the board is above the glass transition temperature will have an effect on the amount of polyglycol absorbed into the epoxy and that will, in turn, affect its electrical properties. Diffusion will also depend upon the specific chemistry of the flux vehicle and its interaction with the substrate. Brous [23] linked the level of polyglycol in a board to surface insulation resistance (SIR) measurements. Jachim re-

Table 7 Comparison of SIR levels and number of CAF associated with two different reflow temperatures

Flux	SIR (Ω) 201°C reflow	SIR (Ω) 241°C reflow	#CAF at 201°C reflow	#CAF at 241°C reflow
Polyethylene glycol-600(PEG)	$<10^6$	$<10^6$	90	55
PEG/HCl	$<10^6$	High 10^8	None	None
PEG/HBr	$<10^6$	High 10^8	None	None
Polypropylene glycol 1200 (PPG)	$>10^{10}$	$>10^{10}$	None	455
PPG/HCl	$>10^{10}$	$>10^{10}$	None	379
PPG/HBr	$>10^{10}$	$>10^{10}$	1	423
Polyethylene propylene glycol 1800 (PEPG 18)	High 10^9	High 10^9	1	406
PEPG 18/HCl	High 10^9	High 10^9	10	135
PEPG 18/HBr	10^{10}	High 10^9	9	279
Polyethylene propylene glycol 2600 (PEPG 26)	High 10^9	High 10^9	None	91
PEPG 26/HCl	High 10^9	High 10^9	6	218
PEPG 26/HBr	10^{10}	High 10^9	None	51
Glycerine (GLY)	$>10^{10}$	High 10^9	None	56
GLY/HCl	$>10^{10}$	High 10^9	None	583
GLY/HBr	$>10^{10}$	High 10^9	3	104
Ocyl phenol ethoxylate (OPE)	Low 10^9	Low 10^9	None	83
OPE/HCl	Low 10^9	Low 10^9	14	62
OPE/HBr	$>10^{10}$	High 10^9	2	599
Linear aliphatic polyether (LAP)	Low 10^9	Not Tested	None	Not Tested
LAP/HCl	Low 10^9	Low 10^9	15	203
LAP/HBr	Low 10^9	Low 10^9	None	272

ported on water-soluble flux-treated test coupons that were prepared using two different thermal profiles. Those which experienced the higher thermal profile exhibited a SIR level that was an order of magnitude lower than those processed under less aggressive thermal conditions. It is clear that the higher the soldering temperature, the greater the polyglycol absorption. Similarly, for each thermal excursion that occurs, the bonding between the epoxy and glass fibers is weakened due to different coefficient of thermal expansion characteristics of these two materials.

One way of quantifying the effect of the reflow temperature on CAF is to examine the thermal strain (ε) associated with the difference in coefficient of thermal expansion (ΔCTE) between the adjacent materials. Table 8 details that comparison for copper versus FR-4 substrate and e-glass versus epoxy where:

$$\varepsilon = \Delta\text{CTE}\,\Delta T$$

It is clear from this table that the higher reflow temperature creates a severe strain on the epoxy/glass

Table 8 Thermal strain ($\times 10^{-6}$) assuming an initial temperature of 25 °C

Material	ΔCTE ($\times 10^{-6}$/K)	ε at 201°C reflow	ε at 241°C reflow
Cu/FR-4	2	352	432
e-glass/epoxy	15	2640	3240

interface, weakening the bond and in general, enhancing the rate of CAF formation. This explains the much higher level of CAF observed for the higher reflow temperature.

Higher board processing temperatures result in increased numbers of CAF for most of the fluxes tested. The 241°C peak temperature represents the wave soldering peak temperature for a typical lead-free solder alloy. Reflow temperatures for solder pastes will be even higher.

5 Summary

The move to lead-free electronics involves a number of material and process issues which are being addressed. These issues are driven by the higher soldering temperatures required for most lead-free solders. Traditional tin–lead solder melts at 183°C, where as the most common lead-free alternatives have a much higher melting temperature—tin–copper (227°C), tin–silver (221°C) and tin–silver–copper (217°C). Other materials are also affected. These include the printed wiring board substrate, the components, the flux and cleaning chemistries, among others.

A failure mode in PWBs that is enhanced by the higher soldering temperatures, conductive anodic filament formation has been described and discussed. This failure is due to electrochemical migration in the use environment. The enhancement related to various flux

chemistries has been described. While the CTE mismatch between epoxy and glass place stress on the board, further work is in progress to understand the complicated flux interactions.

References

1. B. McGrath, *The effects of Lead-free on PCB Fabrication*, (Printed Circuit Design and Manufacture, February, 2005), pp. 44–47
2. E. Kelley, in *Proceedings of Nordic Conference on Printed Circuit Board Quality*, Lillestróm, Norway, 21–22, September 2004
3. Louis T. Mazione, *Plastic Packaging of Microelectronic Devices*. (Van Nostrand Reinhold, New York, 1990), p. 63
4. C.J. Smithells, *Metals Reference Book*, vol. 3, 4th edn. (Butterworths, London, 1967), p. 687
5. *Electronics Materials Handbook*, vol. 1, Packaging, (ASM International, Materials Park, OH, 1989), p. 639
6. C.A. Handwerker, F.W. Gayle, E.E. de Kluizenar, K. Sugauma, in *Handbook of Lead-Free Solder Technology for Microelectronic Assemblies*, ed. by K.J. Puttlitz, K.A. Stalter (Marcel Dekker, New York, 2004), p. 698
7. http://www.metallurgy.nist.gov/solder/clech/Sn-Ag_Other.htm#Coefficients
8. *Electronics Materials Handbook*, vol. 1, Packaging, (ASM International, Materials Park, OH, 1989), p. 468
9. V.T. Brzozowski, R.N. Horton, in *Electronic Packaging and Interconnect Handbook*, ed. by Charles A. Harper (McGraw Hill, New York, 1991), p. 8.44
10. W. Engelmaier (2005), IPC Technet Archives. http://www.listserv.ipc.org/archives/technet.html. Cited 13 Dec 2005
11. J. Fellman, in *Proceedings of International Conference on Lead-free Soldering*, Toronto, Ontario,16–18, May 2006
12. G.L. Fisher, W. Sonnenberg, R. Bernards, Printed Circuit Fabri. **12**(4) (1989)
13. *Terms and Definitions for Interconnecting and Packaging Electronic Circuits*, (ANSI/IPC-T-50) published by the IPC, 3000 Lakeside Drive, 309 S, Bannockburn, IL 60015
14. Peter Biocca, in *Proceedings of SMTA International*, Chicago, Illinois, 30 September 2001
15. *Requirements for Soldering Fluxes*, published by IPC, 3000 Lakeside Drive, 309 S, Bannockburn, IL 60015
16. *Moisture/Reflow Sensitivity Classification for Nonhermetic Solid State Surface Mount Devices*, IPC/JEDEC J-STD-020C, published jointly by IPC, 3000 Lakeside Drive, 309 S, Bannockburn, IL 60015
17. L.J. Turbini, W.R. Bent, W.J. Ready, J. Surf. M. Technol. **13**, 4 (2000)
18. P.J. Boddy, R.H. Delaney, J.N. Lahti, E.F. Landry, in *14th Annual Proceedings of Reliability Physics*, 108 (1976)
19. D.J. Lando, J.P. Mitchell, T.L. Welsher, in *17th Annual Proceedings of Reliability Physics*, 51 (1979)
20. J.N. Lahti, R.H. Delaney, J.N. Hines, in *17th Annual Proceedings of Reliability Physics*, 39 (1979)
21. W.J. Ready, L.J. Turbini, L.J., S.R. Stock, B.A. Smith, in *34th Annual Proceedings of Reliability Physics*, 267 (1996)
22. J.A. Jachim, G.F. Freeman, L.J. Turbini, IEEE T. Compon. Pack. B, **20**(4), 443 (1997)
23. J. Brous, in *Proceedings of NEPCON West 1992*, 386 (1992)

Interfacial reaction issues for lead-free electronic solders

C. E. Ho · S. C. Yang · C. R. Kao

Published online: 22 September 2006
© Springer Science+Business Media, LLC 2006

Abstract The interfacial reactions between Sn-based solders and two common substrate materials, Cu and Ni, are the focuses of this paper. The reactions between Sn-based solders and Cu have been studied for several decades, and currently there are still many un-resolved issues. The reactions between Sn-based solders and Ni are equally challenging. Recent studies further pointed out that Cu and Ni interacted strongly when they were both present in the same solder joint. While this cross-interaction introduces complications, it offers opportunities for designing better solder joints. In this study, the Ni effect on the reactions between solders and Cu is discussed first. The presence of Ni can in fact reduce the growth rate of Cu_3Sn. Excessive Cu_3Sn growth can lead to the formation of Kirkendall voids, which is a leading factor responsible for poor drop test performance. The Cu effect on the reactions between solders and Ni is then covered in detail. The knowledge gained from the Cu and Ni effects is applied to explain the recently discovered intermetallic massive spalling, a process that can severely weaken a solder joint. It is pointed out that the massive spalling was caused by the

shifting of the equilibrium phase as more and more Cu was extracted out of the solder by the growing intermetallic. Lastly, the problems and opportunities brought on by the cross-interaction of Cu and Ni across a solder joint is presented.

1 Introduction

Soldering has been the key assembly and interconnection technology for electronic products since the dawn of the electronic age, and will remain so in the foreseeable future. Solder joints have long been recognized as the weak links in electronic products, and the reliability of each individual joint can control the overall lifespan of an electronic product. Soldering by definition involves the chemical reaction(s) between the solder and the two surfaces to be joined together [1], and consequently the importance of understanding the chemical reactions between solders and bonding surfaces cannot be overemphasized.

Figure 1 is a schematic illustration showing the key elements of the solder joints in an up-to-date electronic package. Here, the chip is connected to a ball grid array (BGA) substrate through an array of solder joints (flip-chip joints) with a diameter of around 90 μm. The BGA substrate is then connected to the printed circuit board (PCB) through another array of solder joints (BGA joints) that are one order of magnitude larger in diameter. This one order of magnitude difference in joint diameter actually translates into a 1,000 times difference in solder volume. Since the heat transfer during soldering is not a weak function of distance (and thus volume), the thermal histories of the flip-chip joints and the BGA joints could be very

C. E. Ho
Department of Chemical Engineering & Materials Science, Michigan State University, East Lansing, MI 48824, USA
e-mail: ceho1975@hotmail.com

S. C. Yang
Department of Chemical & Materials Engineering, National Central University, 320 Jhongli City, Taiwan
e-mail: tree12281228@yahoo.com.tw

C. R. Kao (✉)
Department of Materials Science & Engineering, National Taiwan University, 106 Taipei, Taiwan
e-mail: crkao@ntu.edu.tw

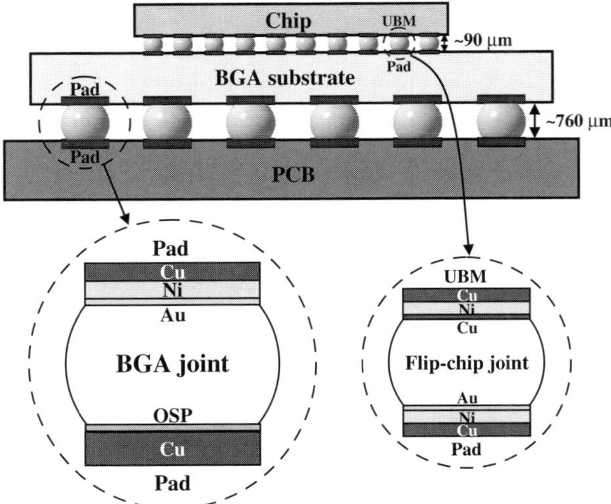

Fig. 1 Schematic drawing showing two types of solder joints, flip-chip joints and BGA joints, used in microelectronic packaging

different. This difference in thermal history could in turn result in different extents of the chemical reaction, and possibly different solidification microstructures if both types of joints become molten during soldering. This effect is just one of many reasons why electronic soldering is becoming an ever more challenging technology.

Those regions on the surface of a chip that are to be in direct contact with the solders are the so-called under bump metallurgy (UBM) regions. Copper is the most popular choice for the surface layer of the UBM, mainly due to its good wetting property with solders [2, 3]. In Fig. 1, Cu is selected for illustration as the surface layer of the UBM, and the Ni layer beneath serves as the diffusion barrier layer. During assembly or normal service of the device, the Cu layer will be consumed completely, exposing the Ni layer to the solder. Those regions on the BGA and PCB substrates that are to be in contact with solders are known as the soldering pads, and the base metal of the soldering pads is always Cu with a few rare exceptions. The Cu base metal on the soldering pads of the BGA and PCB substrates has to be coated with a surface finish (or final finish) to preserve the wetting property during the storage period before assembly. Popular surface finishes include Au/Ni, immersion Sn, immersion Ag, and organic surface preservative (OSP). In the case of Au/ Ni, the Au layer will be dissolved into the solder rapidly during soldering [4–15], exposing the Ni layer below. In the cases of the other three surface finishes, Sn, Ag, or OSP will be dissolved into the solder (immersion Sn and immersion Ag) or be displaced from the interface (OSP) during soldering, leaving the

Cu layer exposed to the solder. In Fig. 1, Au/Ni is used as the surface finish for the BGA substrate, and OSP for the PCB substrate. Summarizing all the popular material choices for the UBM and the surface finishes, one can expect that the most common material sequences across a post-assembly solder joints are, Cu/ solder/Cu, Cu/solder/Ni, and Ni/solder/Ni. The Ni layer here can be deposited by electroplating, electroless plating, or sputtering. In the electroless process, a few atomic percents of P are often co-deposited with Ni, and the Ni layer is customarily labeled as Ni(P). In the sputtering process, V is often added to the sputtering target to ease off the Ni magnetic effect to enhance the deposition rate, resulting in the co-deposition of one or two atomic percents of V into the Ni layer. This type of Ni layer is customarily labeled as Ni(V), or simply NiV. The presence of P or V in the Ni layer do complicate the chemical interactions between Ni and solders [16–38], but these reactions still share enough common features to allow a discussion of them together in this paper.

Several good review articles for electronic solders have been published in the past few years [3, 11, 29, 39–44]. The intention here is not to duplicate their efforts. The focus of this paper is to review and to present original data on three key issues related to the interfacial reactions between common lead-free solders and Ni, Cu, or Ni + Cu. The first issue is about the effect of Ni on the interfacial reactions between solders and Cu (Sect. 2). Recently, the formation of Kirkendall voids during the solders/Cu reactions had been observed. These Kirkendall voids have been attributed to being the root cause for joint fracture during the drop test.

The addition of Ni into solders has a strong effect on the Kirkendall voids in the solders/Cu reactions. The second issue is related to the effect of the Cu concentration in solders on the reactions between solders and Ni (Sect. 3). The problem caused by the limited supply of Cu, known as the solder volume effect, will also be discussed. The third issue is about the cross-interaction between Cu and Ni in Cu/solder/Ni sandwich joints (Sect. 4).

2 Effects of Ni on the reactions between Sn-based solders and Cu

Among all the binary systems, the Cu–Sn is the most important one in many respects. Historically, Cu–Sn alloys have been used as bronze for 6,000 years, and soldering based on the Cu–Sn reaction for Jewellery making has a history of over 4,000 years. Today, Cu–Sn reaction is relied on for electronic interconnections that involve electronic products with a total commercial value second to none. It is no wonder that the Cu–Sn reaction is probably the most extensively studied one in binary metallic systems. Nevertheless, even for such an important system with such a long history, there are still many un-resolved problems. The brittle fracture induced by the Kirkendall voids is one such problem that recently is receiving significant attention.

2.1 Kirkendall voids formation in solders/Cu reactions

It is well known that at temperatures greater than 50~60°C the reactions between Cu and Sn-based solders (pure Sn, eutectic PbSn, and SnAgCu, with the exception of SnZn solders) will produce two reaction products, Cu_6Sn_5 and Cu_3Sn. At temperatures lower than 50~ 60°C, only Cu_6Sn_5 is detected at the interface [29]. The Kirkendall voids tend to form with Cu_3Sn, and the formation of Cu_6Sn_5 alone does not induce the formation of such voids [29, 45–51]. Zeng et al. [50] reported that a large number of Kirkendall voids formed when electroplated Cu was reacted with eutectic PbSn solder at 100–150°C. As shown in Fig. 2(a), these voids located not only at the Cu/Cu_3Sn interface but also within the Cu_3Sn layer. The formation of the Kirkendall voids is not limited to the eutectic PbSn solder. The Kirkendall voids were also reported to form when electroplated Cu was reacted with SnAgCu solder [46], as shown in Fig. 2(b), or pure Sn [29]. Without question, the rapid diffusion of Cu out of the Cu_3Sn layer was the major contributing factor for the formation of these voids. It had been pointed

out that both Cu and Sn were mobile within Cu_3Sn, although the Cu flux was somewhat greater than that of the Sn flux (three times greater at 200°C) [52].

The Kirkendall voids accompanying the Cu_3Sn growth raise serious reliability concerns because excessive void formation increases the potential for brittle interfacial fracture [45–47, 50]. The voids in Fig. 2(b) had reached such a high population that they almost aggregated into continuous regions. Such a reliability threat is especially serious at high temperature because Cu_3Sn only grows at temperatures greater than 50~60°C. In brief, there is a need for ways to reduce the Cu_3Sn growth so that excessive Kirkendall void formation can be avoided.

2.2 Ni addition to solders

It had been shown that Ni addition to Sn3.5Ag (3.5 wt.% Ag, balance Sn) in amounts as minute as 0.1 wt.% was able to substantially hinder the Cu_3Sn growth during soldering [53] as well as during the sequential solid-state aging [54–57]. As shown in Fig. 3,

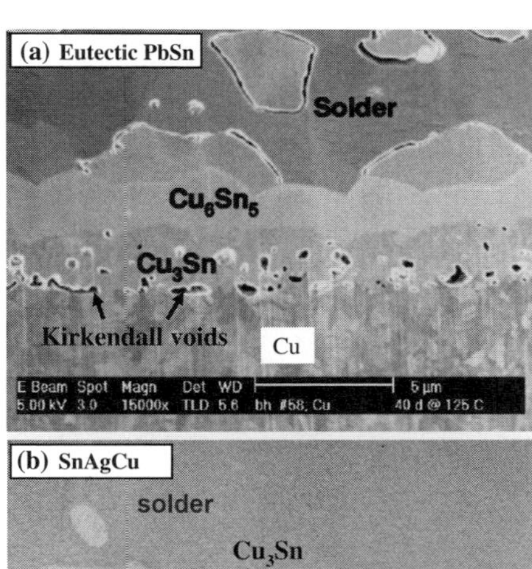

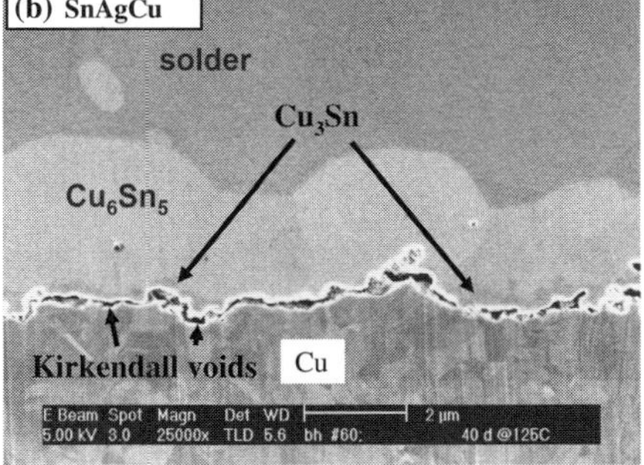

Fig. 2 (a) Secondary electron image showing the Kirkendall voids located within the Cu_3Sn layer in the reaction between eutectic PbSn and Cu at 125°C for 40 days [50]. (b) Secondary electron image showing the continuously aggregated Kirkendall voids in the reaction between SnAgCu and Cu at 125°C for 40 days [46]

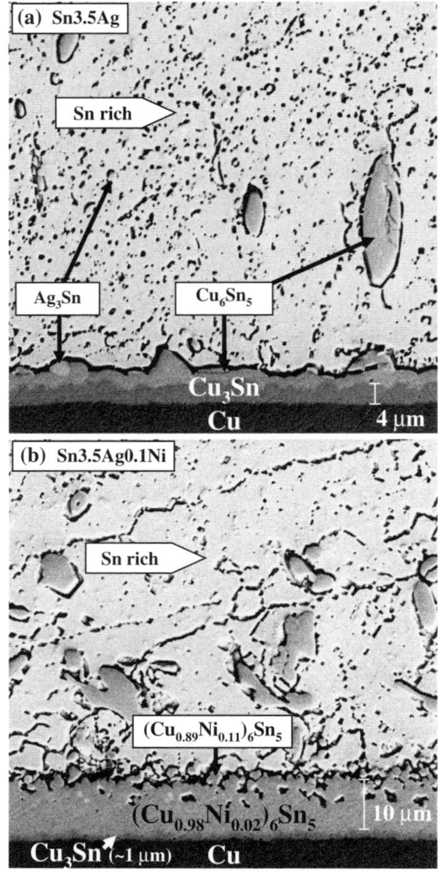

Fig. 3 Backscatter electron micrographs for the Sn3.5Ag/Cu interface (**a**) and the Sn3.5Ag0.1Ni/Cu interface (**b**) that had been aged at 150°C for 1,000 h. [54]

the Cu_3Sn thickness remained very thin even after aging at 150°C for 1,000 h [54]. As the growth of Cu_3Sn is the culprit for the formation of the Kirkendall voids, the addition of Ni can therefore reduce the amount of the Kirkendall voids. In fact, recently it had been shown that the interfacial strength from drop test indeed increase for solders joints with just a small amount of Ni addition (< 1 wt.%) [56].

Figure 3 also shows that Ni addition increased the amount of Cu_6Sn_5 at the interface compared to that without Ni addition. In addition, the Cu_6Sn_5 phase now possessed a small amount of Ni, and became $(Cu_{1-x}Ni_x)_6Sn_5$. A close examination of Fig. 3(b) reveals that there were two distinct $(Cu_{1-x}Ni_x)_6Sn_5$ regions [54]. The inner region was a dense layer of $(Cu_{0.98}Ni_{0.02})_6Sn_5$, and the outer region was actually composed of an aggregate of $(Cu_{0.89}Ni_{0.11})_6Sn_5$ particles. The compositions noted were carefully determined by using the electron probe microanalysis (EPMA). The space between these $(Cu_{0.89}Ni_{0.11})_6Sn_5$ particles was occupied by the solder. It was proposed that the inner region formed during the solid-state aging period, and the outer region was the remains of the $(Cu_{1-x}Ni_x)_6Sn_5$ formed during the reflow period. This proposition was supported by the micrographs for the as-reflow samples shown in Fig. 4 [54]. It can be seen that Ni addition induced a thicker layer of $(Cu_{1-x}Ni_x)_6Sn_5$ during the reflow stage. During the solid-state aging, the region of the aggregate near the Cu side was converted into the dense $(Cu_{1-x}Ni_x)_6Sn_5$ layer, and what was left of the aggregate showed up in Fig. 3(b). The outer $(Cu_{1-x}Ni_x)_6Sn_5$ region in Fig. 4(b) apparently served as a short cut for Sn diffusion, that enhanced the Cu_6Sn_5 growth but inhibited the growth of Cu_3Sn. Consequently, the un-desirable growth of Cu_3Sn was hindered. Another possibility for the shrinkage of Cu_3Sn could be attributed to the substitution of Ni into the Cu sublattice of Cu–Sn compound(s), to reduce the interdiffusion coefficient [56].

The observation that Ni addition could induce a larger amount of $(Cu_{1-x}Ni_x)_6Sn_5$ during reflow deserves more attention. It is proposed that this phenomenon is closely related to the Cu metastable solubility and the Cu equilibrium solubility in the solder. Here, these two types of solubility will be briefly discussed first. During the reaction of Cu with molten solders, it is often observed that in some regions bare Cu was in direct contact with the molten solders without being covered

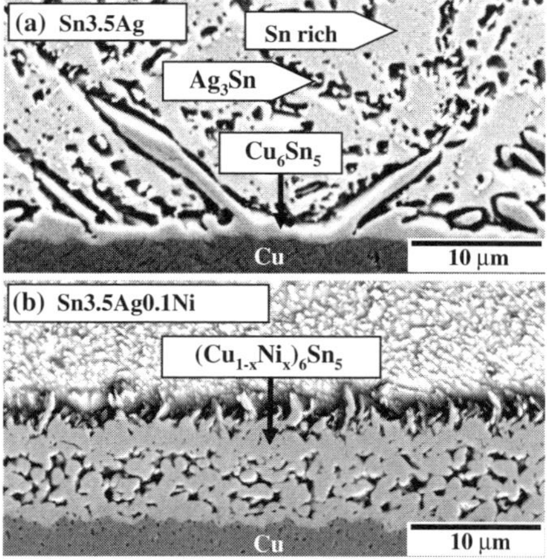

Fig. 4 Backscatter electron micrographs for the Sn3.5Ag/Cu interface (**a**) and the Sn3.5Ag0.1Ni/Cu interface (**b**) after reflow soldering at 240°C for 2 min. [54]

by Cu_6Sn_5 or Cu_3Sn. Under such circumstance, Cu is in a local, metastable equilibrium with solders, and a metastable solubility of Cu in solders will eventually be established. According to thermodynamic arguments, this metastable solubility is always higher than the stable solubility of Cu in solders at the same temperature. As had been pointed out by Laurial et al. [29], it is the metastable solubility that determines the dissolution rate of Cu into molten solders because during dissolution bare Cu is in direct contact with the solder. The metastable solubility cannot be determined experimentally, and has to be calculated by approaches such as the CALPHAD method [29]. As for the equilibrium solubility, some experimental results are available. Shown in Fig. 5 is the Sn-rich corner of the Cu–Ni–Sn isotherm at 240°C. This isotherm is drawn based on the previously published isotherms [19, 29, 41] and the solubility data from the Cu–Sn [58, 65] and Ni–Sn [59] binary systems. This isotherm shows that the addition of Ni into molten Sn can substantially decrease the Cu solubility in Sn. While the equilibrium solubility of Cu in Sn is sensitive to the presence of Ni, the metastable solubility of Cu in Sn must be less sensitive to minor Ni additions from the arguments of the common tangent construction of the Gibbs free energies of the phases involved. Under the assumption that the Cu metastable solubility in Sn is indeed insensitive to the presence of Ni, one can conclude that the Cu dissolution rate into Sn is insensitive to the presence of Ni. In other words, adding Ni to Sn will not decrease the Cu dissolution flux into molten solders. Because the Cu flux entering Sn does not decrease with Ni addition and because the equilibrium Cu solubility

does decrease substantially with the Ni addition, the amount of $(Cu_{1-x}Ni_x)_6Sn_5$ at the interface has to increase as had been observed in Fig. 4.

The way Ni was introduced into the reacting system did not seem to matter as far as hindering the Cu_3Sn growth was concerned. It had been reported that the growth of Cu_3Sn was hindered even if Ni-alloyed Cu substrates were used. These substrates included Cu alloyed with 6–9 wt.% Ni [60] and Cu alloyed with 15 at.% Ni [29]. This fact seems to support the view that dissolution plays a very important role during liquid/solid reaction because Ni in Cu–Ni alloys must dissolve into Sn first to function as an additive of the solders.

3 Effects of Cu on the reactions between Sn-based solders and Ni

According to the binary phase diagram [59], there are three stable intermetallic compounds, Ni_3Sn, Ni_3Sn_2, and Ni_3Sn_4, in the Ni–Sn system. However, when Ni reacted with Sn at temperatures relevant to soldering, only Ni_3Sn_4 was observed [29, 61, 62]. The other two compounds were observed only at higher temperatures, outside the usual operating range of soldering [29, 61, 62]. The reaction rate of Ni with solder is roughly one order of magnitude lower than that of Cu, and this is one of the reasons for Ni being a popular diffusion barrier material [3].

By using the microstructure features of the grains, it had been proposed that Ni was the dominant diffusing species in Ni_3Sn and Ni_3Sn_2, and Sn was the dominant diffusing species in Ni_3Sn_4 [63]. A marker experiment had been performed to verify if Sn was indeed the dominant diffusing species in Ni_3Sn_4 [64]. In that study, a layer of 2 μm Ti film was deposited on a Ni substrate. The Ti layer was then patterned into 10 μm wide thin strips to serve as the markers. As shown in Fig. 6(a), the spacing between neighboring Ti strips was 40 μm. The sample was then dipped into pure molten Sn at 250°C for 10 min to produce the diffusion couple shown in Fig. 6(b). The resulting diffusion couple after aging at 225°C for 1,000 h is shown in Fig. 6(c). From the location of the Ti markers after aging, one can conclude that Sn is indeed the dominant diffusing species in Ni_3Sn_4.

Although the reaction between Ni and pure Sn only produces Ni_3Sn_4 at temperatures relevant to soldering, the reaction between Ni and Sn alloys with a small amount of Cu is very complicated (see Sect. 3.1). Unfortunately, the common lead-free solders recommended by various national or international organizations all

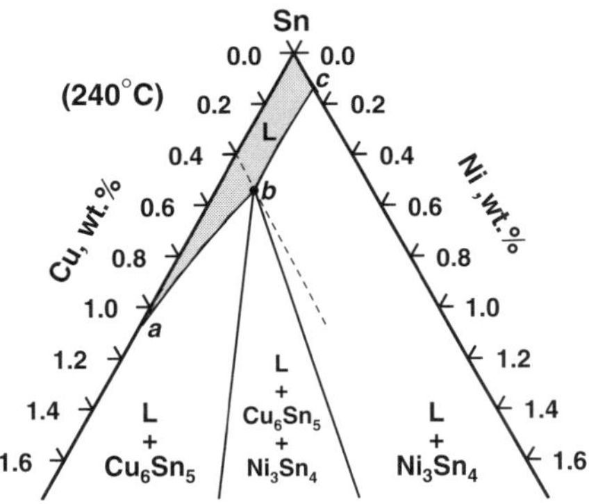

Fig. 5 Schematic drawing showing the Sn-corner of the Cu–Ni–Sn isotherm at 240°C. This isotherm is drawn based on the previously published isotherms [19, 29, 41] and the solubility data from the Cu–Sn [58, 65] and Ni–Sn [59] binary systems

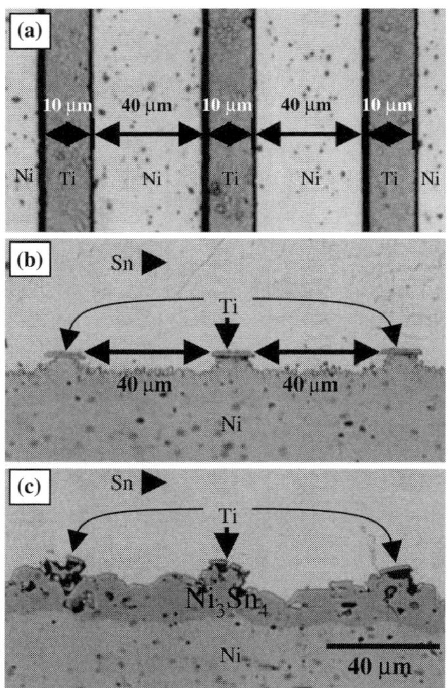

Fig. 6 (a) Optical micrograph showing the plane-view of a Ni substrate that had been deposited with Ti stripes as markers. (b) Cross-sectional view of the Ni/Sn diffusion couple before aging. (c) After aging at 225°C for 1,000 h. [64]

have Cu as a minor constituent (Table 1), and the allowance for the concentration variation for each element is typically $\pm$ 0.2 wt.%, a customary adoption of the practice from the eutectic PbSn solder. As will be shown below, this $\pm$ 0.2 wt.% difference can change the compound formed at the interface after reflow.

3.1 Uncomplicated Cu concentration effect

During the reflow of SnCu or SnAgCu solders over those soldering pads that have the Ni, Ni(P), and Ni(V) underlayer, the reaction product(s) at the interface are

very sensitive to the Cu concentration in the solders. After the first report [73] of this Cu sensitivity a few years ago, over forty papers had been published on this subject. As summarized in Table 2, the results of these studies as a whole are quite consistent, even though different reflow conditions and different Ni substrates, including Ni, Ni(P), Ni(V), or Au/Ni, had been used. As shown in Table 2, when the Cu concentration was low (<0.3 wt.%), only $(Ni_{1-y}Cu_y)_3Sn_4$ formed at the interface. When the Cu concentration increased to 0.4–0.5 wt.%, both $(Ni_{1-y}Cu_y)_3Sn_4$ and $(Cu_{1-x}Ni_x)_6Sn_5$ formed. When the Cu concentration increased above 0.5 wt.%, only $(Cu_{1-x}Ni_x)_6Sn_5$ formed. In a few specific cases [28, 86, 89–91, 96, 104, 105], where the reported results seemed to differ from the trend shown in Table 2 at the first sight, the inconsistency in fact could be attributed to the so-called "solder volume effect," which is to be discussed in the next section.

To exclude other factors that may obscure the discussion, the reaction between a large amount of solder and a piece of thick, high purity Ni substrate will be examined first. Since the amount of solder was large, one can then assume that the solder composition remained constant during the reaction even though Cu was being extracting out of the solder to be incorporated into the reaction product(s). In addition, since the Ni substrate was thick, the Ni substrate was never totally consumed. In some aspects, the system approached the limit of classical infinite/infinite diffusion couple. As shown in Fig. 7(a), the reaction product at the interface was a continuous layer of $(Ni_{1-y}Cu_y)_3Sn_4$ when the Cu concentration was 0.2 wt.%. When the Cu concentration increased to 0.4 wt.%, discontinuous $(Cu_{1-x}Ni_x)_6Sn_5$ particles began to form over the $(Ni_{1-y}Cu_y)_3Sn_4$ continuous layer, Fig. 7(b). When the Cu concentration increased to 0.5 wt.%, both $(Cu_{1-x}Ni_x)_6Sn_5$ and $(Ni_{1-y}Cu_y)_3Sn_4$ were continuous, Fig. 7(c). When the Cu concentration reached 0.6 wt.%, only a continuous $(Cu_{1-x}Ni_x)_6Sn_5$ layer was present, Fig. 7(d). The crystal structures of the products had been verified by the X-ray diffraction (XRD) patterns shown in Fig. 8 [73, 106]. To obtain such

Table 1 Lead-free solders recommended by various national or international organizations

SnAgCu composition (wt.%)	Recommending organizations
Sn-(3.5 $\pm$ 0.3)Ag-(0.9 $\pm$ 0.2)Cu	NIST (ternary eutectic) [65]
Sn-(3.9 $\pm$ 0.2)Ag-(0.6 $\pm$ 0.2)Cu	NEMI (N. America) [66]
Sn-(3.4~ 4.1)Ag-(0.45~ 0.9)Cu	Soldertec-ITRI (UK) [67]
Sn-3.8Ag-0.7Cu	IDEALS (EU) [68]
Sn-3.0Ag-0.5Cu	JEITA (Japan) [69]
Sn-4.0Ag-0.5Cu	–
[a]Sn-2.5Ag-0.8Cu-0.5Sb	AIM, CASTIN alloy [70]
[a]Sn-3.5Ag-0.5Cu-1.0Zn	NCMS [71]
[b]Sn-3.0Ag-0.6Cu-0.019Ce	China RoHS Standard Committee [72]

[a] SnAgCu with additive(s)

[b] Under consideration

Table 2 Summary of the reported reaction products between SnAgCu solders and various Ni-bearing soldering pads after reflow

Cu (wt.%)	Ag (wt.%)	Sn (wt.%)	Base metal	Intermetallic(s)	References
0.0	3.5–3.9	Balance	Ni and Ni(P)	Ni_3Sn_4	[26, 74–77]
			Au/Ni and Au/Ni(P)	Ni_3Sn_4	[20, 22, 23, 25, 28, 31, 78–85]
0.1	0	Balance	Ni	Ni_3Sn_4	[86]
0.2	0–3.9	Balance	Ni and Ni(P)	$(Ni,Cu)_3Sn_4$	[32, 73, 75, 87, 88]
0.3	0–3.0	Balance	Ni	$(Ni,Cu)_3Sn_4$	[86, 89]
0.4	0–3.9	Balance	Ni	$(Ni,Cu)_3Sn_4/(Cu,Ni)_6Sn_5$	[75, 87–91]
			Au/Ni(P)	$(Ni,Cu)_3Sn_4/(Cu,Ni)_6Sn_5$	[33]
0.5	1.0–4.0	Balance	Ni and Ni(P)	$(Cu,Ni)_6Sn_5$	[89–91]
			Au/Ni and Au/Ni(P)	$(Cu,Ni)_6Sn_5$	[21, 25, 34, 82, 83, 85, 92]
			Ni and Ni(P)	$(Ni,Cu)_3Sn_4/(Cu,Ni)_6Sn_5$	[32, 75, 88]
			Au/Ni and Au/Ni(P)	$(Ni,Cu)_3Sn_4/(Cu,Ni)_6Sn_5$	[21, 28, 81, 93]
0.6	0–3.9	Balance	Ni	$(Cu,Ni)_6Sn_5$	[73, 75, 88–91]
0.7	0–3.8	Balance	Ni and Ni(P)	$(Cu,Ni)_6Sn_5$	[27, 32, 75, 86, 87, 94, 95]
			Au/Ni and Au/Ni(P)	$(Cu,Ni)_6Sn_5$	[19, 28, 30, 96–101]
0.75	3.5	Balance	Au/Ni and Au/Ni(P)	$(Cu,Ni)_6Sn_5$	[102]
			Ni and Ni(P)	$(Cu,Ni)_6Sn_5$	[22, 23, 81, 102]
0.8	0–3.9	Balance	Au/Ni and Au/Ni(P)	$(Cu,Ni)_6Sn_5$	[75, 88]
			Ni	$(Cu,Ni)_6Sn_5$	[25]
0.9	0	Balance	Ni	$(Cu,Ni)_6Sn_5$	[86]
1.0	3.5–3.9	Balance	Ni and Ni(P)	$(Cu,Ni)_6Sn_5$	[32, 75, 87]
			Au/Ni	$(Cu,Ni)_6Sn_5$	[103]
1.5	0	Balance	Ni	$(Cu,Ni)_6Sn_5$	[86]
1.7	4.7	Balance	Ni	$(Cu,Ni)_6Sn_5$	[103]
3.0	0–3.9	Balance	Ni	$(Cu,Ni)_6Sn_5$	[75, 94]
			Au/Ni	$(Cu,Ni)_6Sn_5$	[100]

Only those studies that used reflow conditions similar to the industry practices are included in this table

patterns, the solders had to be etched away first by using a proper etching solution, and the remaining reaction products as well as the Ni substrate were then subjected to the measurements. The amount of the discontinuous $(Cu_{1-x}Ni_x)_6Sn_5$ particles in Fig. 7(b) was too small to allow for a positive identification through XRD, but the identity of these particles was established by using the transmission electron microscopy (TEM) shown in Fig. 9 [106]. The discontinuous $(Cu_{1-x}Ni_x)_6Sn_5$ particles in Fig. 7(b) and the continuous $(Cu_{1-x}Ni_x)_6Sn_5$ layer in Fig. 7(d) might have different formation mechanisms. Those $(Cu_{1-x}Ni_x)_6Sn_5$ grains in Fig. 7(d) formed through the direct reaction of Ni and the solder, and accordingly these grains had a preferred growth direction of 0001 as shown in Fig. 10 [106]. This preferred growth direction had also been observed during the reaction between Ni and the eutectic PbSn solder with 0.5 wt.% Cu dopant [107]. On the other hand, those discontinuous $(Cu_{1-x}Ni_x)_6Sn_5$ particles in Fig. 7(b) did not show any preferred orientation.

To understand this strong Cu concentration dependency, one needs to have the relevant phase diagram information for the Sn–Ag–Cu–Ni system. Although Ag is an important constituent controlling the solidification microstructure of the solder itself [44, 93, 108–112], it has been shown that Ag is inert as far as the interfacial reaction is concerned [73, 75, 87]. Accordingly, the Sn–Cu–Ni ternary phase diagram is sufficient for the

present purpose. The Sn–Cu–Ni isotherm had been measured by two independent groups [113, 114], and the results are reasonably consistent. The 240°C Sn–Cu–Ni isotherm basing on these two studies is shown in Fig. 11. The Sn-rich corner of this isotherm is shown in Fig. 5. There is some evidence for the existence of a ternary compound (Ni26Cu29Sn45, atomic percent) [115]. If this compound is indeed stable, then the isotherm in Fig. 11 is only a metastable isotherm [29, 41]. Nevertheless, as far as soldering is concerned, the isotherm shown in Fig. 11 is still adequate, as results from most soldering reaction experiments were observed to follow the phase relationships shown in Fig. 11.

As shown in Fig. 5, the molten Sn phase (L) has a phase boundary a–b–c, which is composed of two segments a–b and b–c. Along a–b, L is in equilibrium with the Cu_6Sn_5 phase, and L is in equilibrium with Ni_3Sn_4 along b–c. The point b represents that L is in equilibrium with both Cu_6Sn_5 and Ni_3Sn_4. As indicated by the dash line passing through b, b has a Cu concentration of about 0.4 wt.%. In other words, when the Cu concentration in the solder is 0.4 wt.%, this solder is in equilibrium with both Cu_6Sn_5 and Ni_3Sn_4. This explains why both these two phases form as shown in Fig. 7(b). When the Cu concentration is less than 0.4 wt.%, the solder in this range is in equilibrium with only Ni_3Sn_4, and consequently only this phase forms as shown in Fig. 7(a). When the Cu concentration is

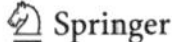

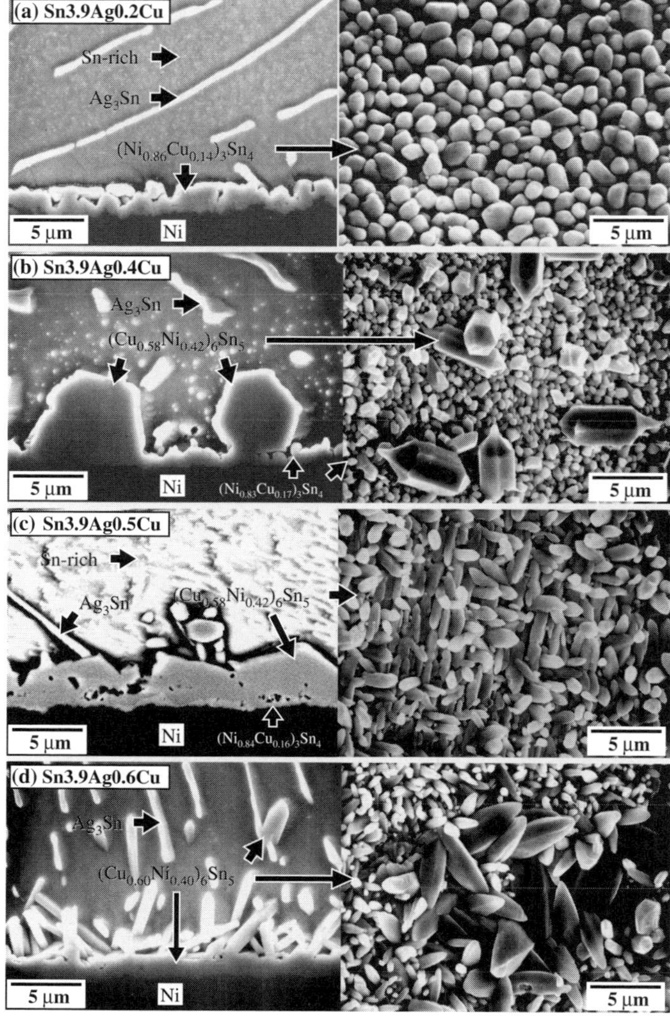

Fig. 7 Micrographs showing the Sn3.9AgxCu/Ni interfaces after 250°C soldering for 10 min. The left column is cross-sectional view, and the right column is top view. The reaction couples here approached the classical infinite/infinite diffusion couples. [73, 75]

higher than 0.4 wt.%, L is in equilibrium with only Cu_6Sn_5, and only this phase can be immediately next to L, as shown in Fig. 7(c) and (d).

3.2 Solder volume effect during reaction with molten solders (reflow)

During reflow and the subsequent aging, Cu atoms are incorporated into the intermetallic(s) [i.e., $(Cu_{1-x}Ni_x)_6Sn_5$ and $(Ni_{1-y}Cu_y)_3Sn_4$] at the interface. As the intermetallic(s) grow, more Cu atoms are being consumed. The results presented in the previous section were for the condition that the amount of the solder was relatively large so that the supply of Cu was also large. The average Cu concentration in the solder could consequently remain nearly constant. The thermodynamic condition at the interface was thus static, and the formation of one compound or another was more or less dictated by the thermodynamics. For a real solder joint in array–array packages, the supply of Cu is actually very limited because the solder volume is quite small in the first place, and secondly the Cu concentration in SnAgCu solder is always less than a few atomic percents (see Table 2). The Cu concentration can decrease noticeably as the intermetallic(s) grow [89–91]. As the Cu concentration changes, the type of the equilibrium intermetallic at the interface might also change. Now the condition at the

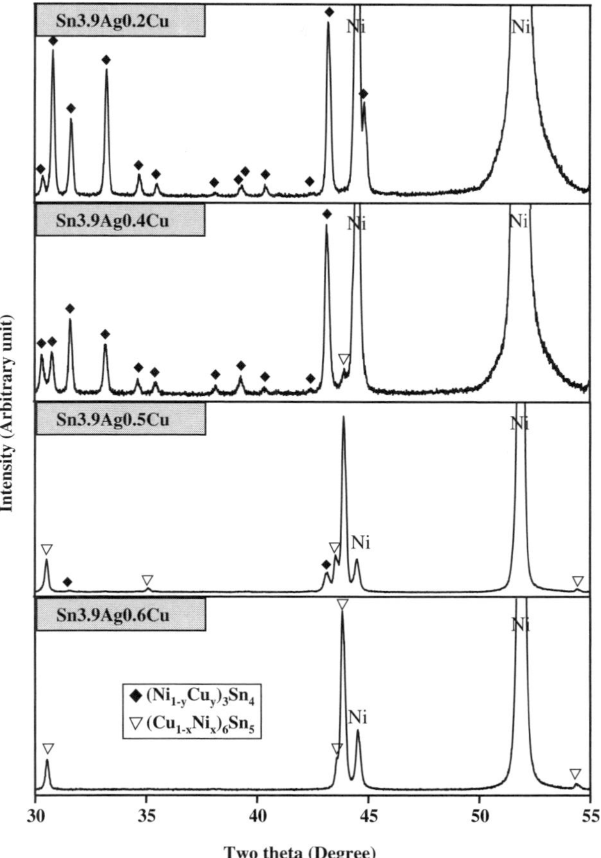

Fig. 8 XRD patterns for the reaction products in Fig. 7(a)–(d). The Ni signals originated from the Ni layer beneath the intermetallic compounds [73, 106]

interface becomes dynamic. The volume effect must be considered as the volume of the solder determines the total available Cu. As the size of the joints shrinks, the supply of Cu becomes more limited, and the decrease in Cu concentration becomes more critical. Next, the combined effects of the limited Cu supply (solder volume effect) and the Cu concentration (Cu concentration effect) will be presented.

In the reaction between SnAgCu solder balls and Ni substrate (Fig. 12), the original amount of Cu in the solder ball before reflow equals the remaining Cu in the solder plus the Cu which is incorporated into the intermetallic(s). Neglecting the Cu atoms in those intermetallic particles located inside the solder, one can obtain the following equation [89–91]:

$$W_{Cu} - W_{Cu}^o \cong -40 \frac{d_{pad}^2}{d_{joint}^3} T_{IMC}(wt.\%) \tag{1}$$

where W_{Cu}^o and W_{Cu} represent the Cu concentration (in wt.%) in SnAgCu before reflow and the remaining Cu concentration after reflow, respectively. The symbols d_{joint} and d_{pad} represent the diameters (in μm) of the solder ball and the pad's opening of the substrate, respectively. The symbol T_{IMC} represents the thickness (in μm) of intermetallic compound (IMC) at the interface. The compounds $(Cu_{1-x}Ni_x)_6Sn_5$ and $(Ni_{1-y}Cu_y)_3Sn_4$ are the only two Cu-bearing intermetallics that can be present at the interface, and their compositions had been determined by EPMA to be around $(Cu_{0.60}Ni_{0.40})_6Sn_5$ and $(Ni_{0.80}Cu_{0.20})_3Sn_4$ for those solders with Cu within 0.4–0.6 wt.% [73, 75, 87, 89–91]. Here, those Cu in $(Ni_{1-y}Cu_y)_3Sn_4$ could be ignored because the thickness of this compound was thin and its Cu concentration was low (~6 wt.%). The drop of the Cu concentration according to Eq. 1 for several different ball/pad

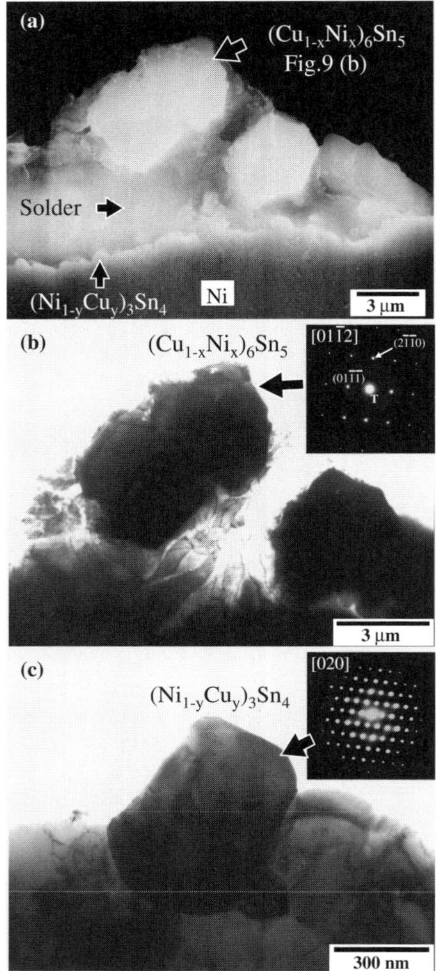

Fig. 9 (**a**) Scanning electron image showing the reaction zone of Sn3.9Ag0.4Cu/Ni after 250°C soldering for 10 min. (**b**) TEM bright field image of (**a**) and selected area diffraction pattern of (Cu$_{1-x}$Ni$_x$)$_6$Sn$_5$. (**c**) TEM bright field image of (**a**) and selected area diffraction pattern of (Ni$_{1-y}$Cu$_y$)$_3$Sn$_4$. [106]

combinations ranging from the BGA to flip-chip dimensions is plotted in Fig. 13. For example, if 2 µm (Cu$_{1-x}$Ni$_x$)$_6$Sn$_5$, which is the thickness commonly seen in real solder joints, forms in the 100 µm/80 µm combination, the Cu concentration will drop by as large as 0.51 wt.%. Under such a condition, the residue Cu concentration in solder will be less than 0.3 wt.% for most SnAgCu solder compositions listed in Table 1. If the Cu concentration indeed become less than 0.3 wt.%, then the Cu$_6$Sn$_5$/molten solder interface is no longer thermodynamically stable

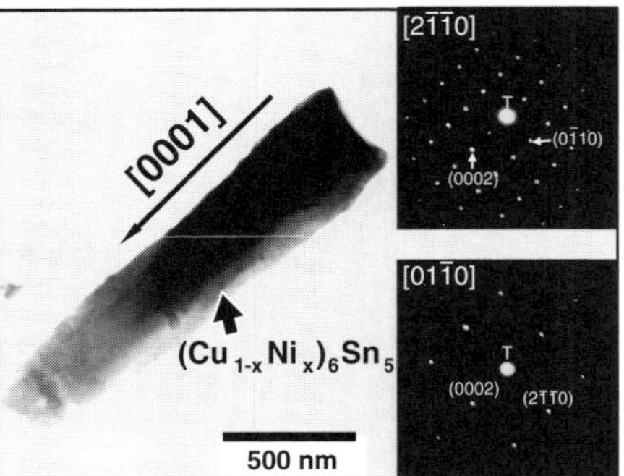

Fig. 10 TEM bright field image and selected area diffraction patterns for a (Cu$_{1-x}$Ni$_x$)$_6$Sn$_5$ needle shown in Fig. 7(d) [106]

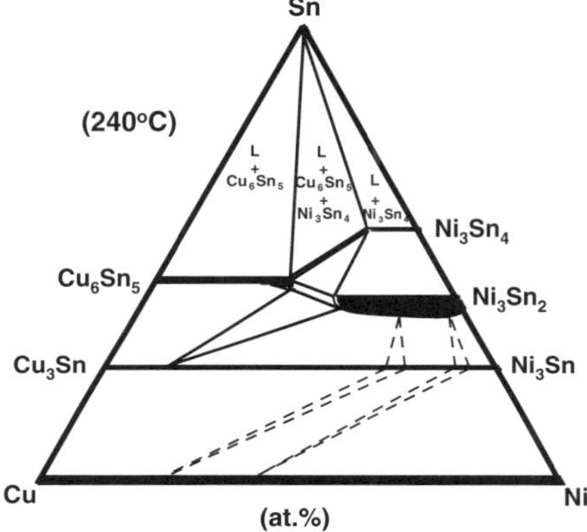

Fig. 11 Cu–Ni–Sn isotherm at 240°C. This isotherm is re-drawn basing on the literature data [113, 114]

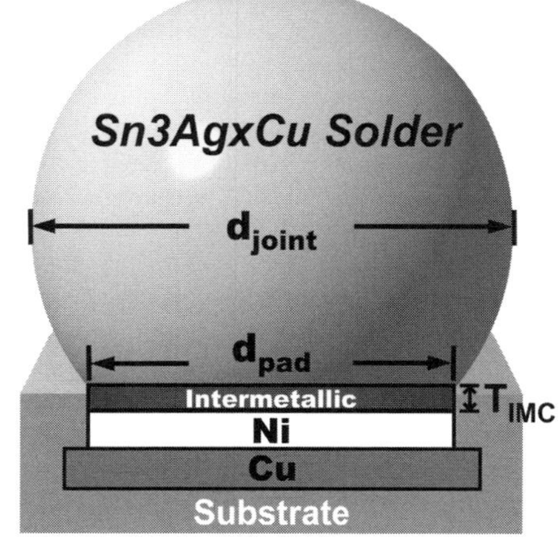

Fig. 12 Schematic diagram showing a Sn3AgxCu solder ball soldered onto a Ni/Cu pad

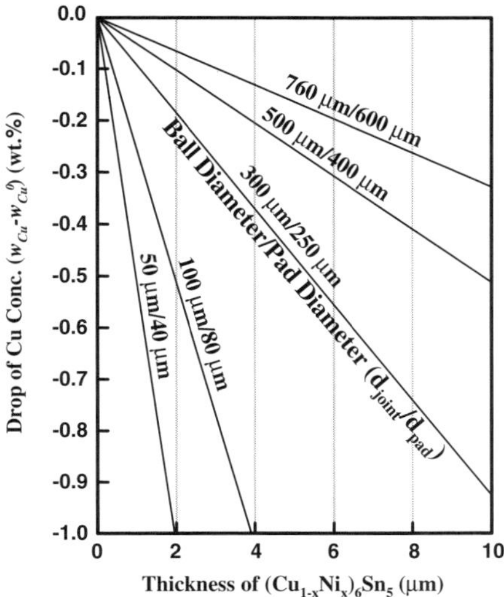

Fig. 13 Drop of Cu concentration according to Eq. 1 for different d_{joint}/d_{pad} combinations

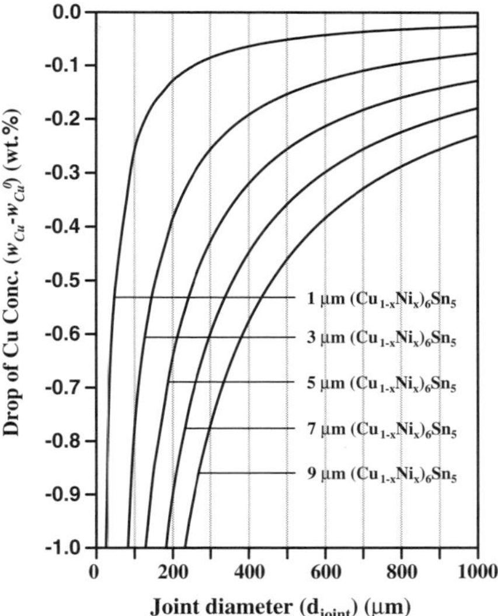

Fig. 14 Drop of Cu concentration according to Eq. 2 for different d_{joint}

according to Fig. 5. There will be a huge driving force for Ni_3Sn_4 [or $(Ni_{1-y}Cu_y)_3Sn_4$] to form.

For BGA or flip-chip joints, the design rule usually calls for the condition that $d_{joint} \approx 1.2 d_{pad}$, therefore, the following equation can be obtained from Eq. 1:

$$W_{Cu} - W_{Cu}^o \approx -28 \frac{T_{IMC}}{d_{joint}} \propto \frac{1}{d_{joint}} \quad (2)$$

This joint size dependence can be clearly seen in Fig. 14, where the drop of the Cu concentration is plotted against the joint size. In short, the Cu concentration drop rapidly increases as the joint becomes smaller.

Additional experimental evidence of the solder volume effect is presented in Fig. 15 [89–91]. The experimental setup in Fig. 15 was used with d_{pad} being kept constant at 375 μm, and d_{joint} being varied from 760 to 500, and to 300 μm. In other words, three different d_{joint}/d_{pad} values were used. The solder used here was Sn3Ag0.6Cu, and the samples had been reflowed with a typical profile (235°C peak temperature, 90 s molten solder duration). Using Eq. 1 and the $(Cu_{1-x}Ni_x)_6Sn_5$ thicknesses measured from the three cases shown in Fig. 15 (1.2, 1.5, and 2.2 μm, respectively), one can calculate the Cu concentration drops to be 0.02, 0.05, and 0.48 wt.% for 760, 500, and 300 μm d_{joint}, respectively. As a result, the remaining Cu concentrations of the solders after reflow were 0.58, 0.55, and 0.12 wt.% for d_{joint} = 760, 500, and 300 μm, respectively. As expected, the intermetallic compound for the first two cases shown in Fig. 15(a) and (b) was still

$(Cu_{1-x}Ni_x)_6Sn_5$. However, for the 300 μm case, the Cu concentration was so low that a new layer of $(Ni_{1-y}Cu_y)_3Sn_4$ had nucleated beneath the $(Cu_{1-x}Ni_x)_6Sn_5$ layer that formed in the early stage of the reaction when the Cu concentration was still high. In Fig. 15(c), a series of voids can be seen, separating these two intermetallic compounds. As will be shown in the next section, an advanced development, when the reflow time is increased longer (see Fig. 18) or the solder joint is shrunk even smaller (see Fig. 17), of this effect is the total separation (spalling) of the upper $(Cu_{1-x}Ni_x)_6Sn_5$ layer from the interface. That is, the consequence of the shifting equilibrium phase leads to the massive spalling of $(Cu_{1-x}Ni_x)_6Sn_5$.

In summary, in the reaction between Ni substrate and the SnAgCu solder with a small volume, the supply of Cu becomes a problem. Under such conditions, the Cu concentration is no longer constant during the reaction. In fact, it is possible that the interface experiences several phase equilibrium conditions during the course of the reaction. This solder volume effect is more severe for smaller solder joints and for joints with a thicker $(Cu_{1-x}Ni_x)_6Sn_5$ at the interface (i.e., joints with a longer reaction time).

3.3 Massive spalling of intermetallic compound from the interface

Spalling refers to the detachment of a compound from the interface into one of the reacting phases. One

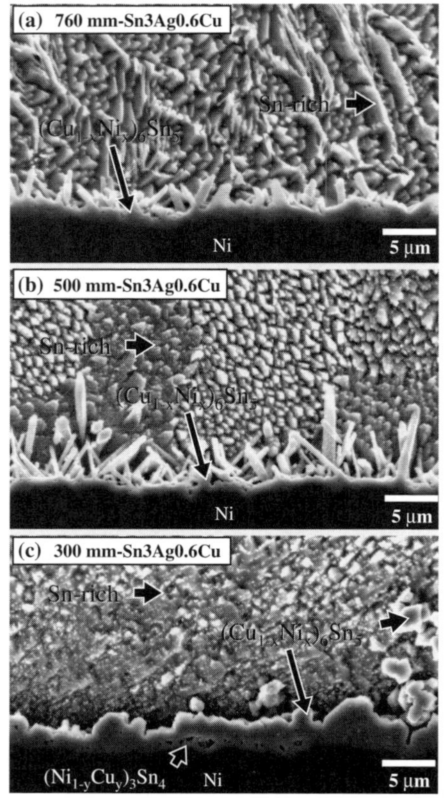

Fig. 15 Micrographs showing the interface after a typical reflow. The value of d_{pad} was kept constant at 375 μm, and d_{joint} was 760 μm (**a**), 500 μm (**b**), and 300 μm (**c**). The solder was 96.4Sn3Ag0.6Cu (wt.%). [89–91]

classical example, shown in Fig. 16, is the spalling of Cu_6Sn_5 during the reaction of eutectic PbSn solder with Au/Cu/Cu–Cr thin film [3, 116]. Spalling here was due to the exhaustion of the Cu film, and the poor wetting between Cu_6Sn_5 and the remaining Cr (or SiO_2) caused the compound to detach itself from the interface. The "massive spalling' introduced in this section is not caused by the exhaustion of a reaction layer. Instead, it is caused by the consumption of one element inside the reacting phase, as had been discussed in the previous section. The shifting of the interfacial equilibrium causes the product phase to separate itself from the interface [89–91].

The massive spalling has a higher tendency to occur in smaller joints because these joints will experience a lager Cu concentration drop. Fig. 17 shows an example of massive spalling in its early stage. The reflow con-

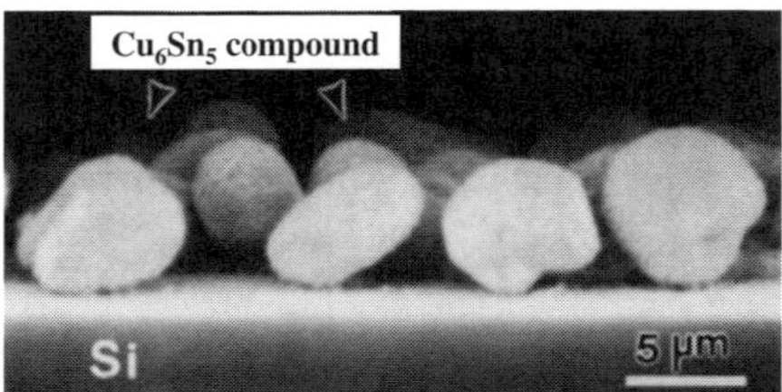

Fig. 16 SEM image showing the spalling of Cu_6Sn_5 during the reaction between eutectic SnPb and Au/Cu/Cu–Cr substrate [3, 116]

ditions were the same as those in Fig. 15, but d_{pad} and d_{joint} were 175 and 200 μm, respectively. As can be seen in Fig. 17, a new $(Ni_{1-y}Cu_y)_3Sn_4$ layer had formed over the Ni substrate, pushing the original $(Cu_{1-x}Ni_x)_6Sn_5$ layer away from the interface, and a gap had appeared between these two layers. It should be stressed again that even though the reflow conditions were the same, such spalling did not occur for the larger joints in Fig. 15(a) and (b). Longer reflow time also favors the spalling, as shown in Fig. 18. Here, the sample had the same configuration as that of Fig. 15(c), but the reflow time was increased from 90 s to 20 min. Longer reflow time had increased the separation between $(Ni_{1-y}Cu_y)_3Sn_4$ and $(Cu_{1-x}Ni_x)_6Sn_5$ from a series of voids in Fig. 15(c) to a gap in Fig. 18.

A very revealing example of the massive spalling is shown in Fig. 19, where a d_{joint}/d_{pad} = 760 μm/600 μm joint had been reflowed at 235°C for 5 min. In this

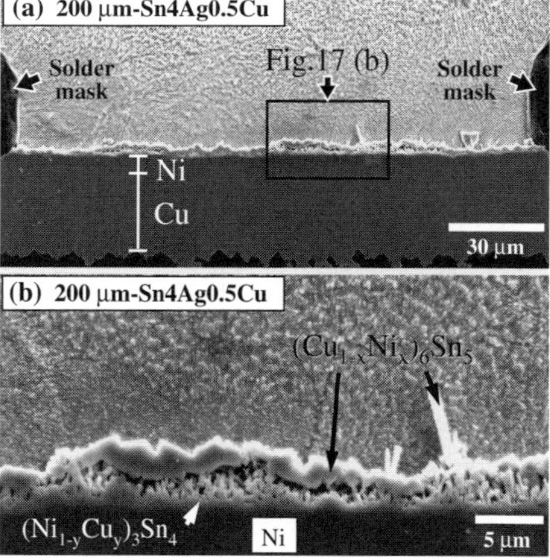

Fig. 17 (**a**) Massing spalling of $(Cu_{1-x}Ni_x)_6Sn_5$ after a typical reflow (235°C peak temperature, 90 s molten solder duration) in a solder joint with d_{joint}/d_{pad} = 200 μm/175 μm. (**b**) Zoom-in view of (**a**). [91]

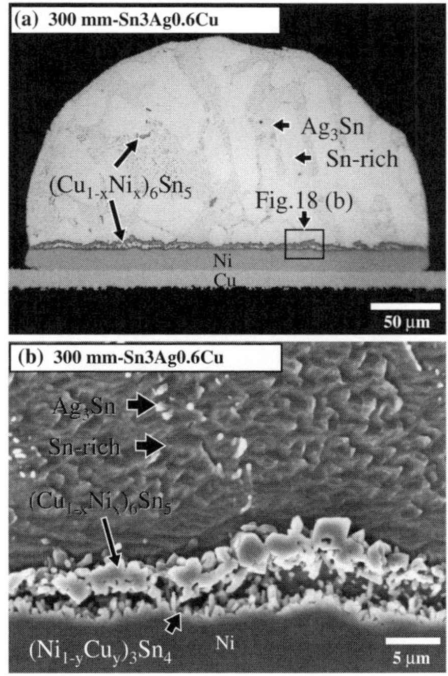

Fig. 18 (**a**) Massing spalling of $(Cu_{1-x}Ni_x)_6Sn_5$ after reflow at 235°C for 20 min in a solder joint with d_{joint}/d_{pad} = 300 m/ 375 μm. (**b**) Zoom-in view of (**a**). [89–91]

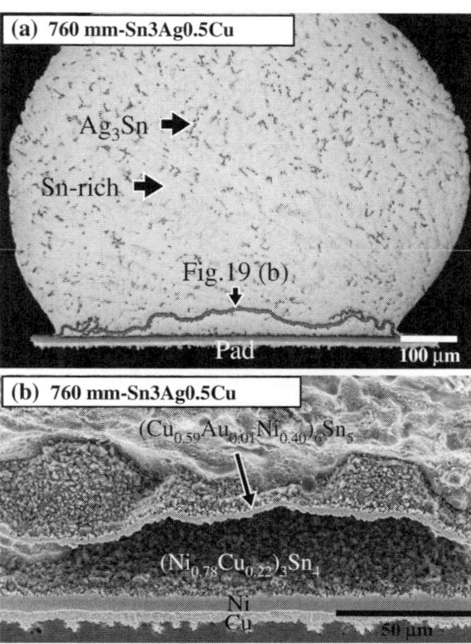

Fig. 19 (**a**) Cross-section view of a solder joint that had been reflowed at 235°C for 5 min. A 760 μm Sn3Ag0.5Cu sphere was attached to a 600 μm pad with the Au(1.2 μm)/Ni surface finish. (**b**) Zoom-in view of (**a**) after etching away the solder. This micrograph had been tilted 45°

particular sample, a 1.2(± 0.1) μm Au layer had been coated over the Ni layer of the substrate before reflow. The images in Fig. 19(a) and (b) were obtained from the same specimen, but the solder in Fig. 19(b) had been etched away. As can be seen here, the detached Au-bearing $(Cu_{1-x}Ni_x)_6Sn_5$ layer was almost continuously. The extensive spalling shown in this pair of micrographs demonstrates the justification of naming the phenomenon massive spalling. It should be pointed out that our recent data suggests that Au does have the effect of stimulate early massive spalling. More studies are needed to clarify this effect.

There are two approaches to inhibit the massive spalling during reflow. The first is to use solders with a higher Cu concentration. The second method is to provide an infinite Cu source, such as a thick Cu layer on either side of a solder joint [89–91]. Our latest results also supported this view.

3.4 Solder volume effect during reaction with solid solders (aging)

The solder volume effect also plays an important role during the solid-state aging. Fig. 20 shows the solder joints with the same reflow conditions (235°C peak temperature, 90 s molten solder duration), the same aging conditions (160°C for 1,000 h), and the same d_{pad} value (375 μm), but with different d_{joint} values and different Cu compositions in solder. When the joint was relatively large (760 μm), the only intermetallic compound at the interface was $(Cu_{1-x}Ni_x)_6Sn_5$ for all the solder compositions, as shown in Fig. 20(a)–(c). This is due to the fact that the supply of Cu was relatively abundant for the larger solder joints.

For the solder joints with the medium diameter (500 μm), the supply of Cu became somewhat limited. There was still only the $(Cu_{1-x}Ni_x)_6Sn_5$ layer at the interface if the joint's Cu concentration was high enough (0.5 and 0.7 wt.%), as shown in Fig. 20(d) and (e). But when the applied Cu concentration became very low (0.3 wt.%), as shown in Fig. 20(f), a layer of $(Ni_{1-y}Cu_y)_3Sn_4$ formed beneath $(Cu_{1-x}Ni_x)_6Sn_5$. Given more time, the remaining $(Cu_{1-x}Ni_x)_6Sn_5$ will also disappear, leaving $(Ni_{1-y}Cu_y)_3Sn_4$ as the only layer. This is because the Cu atoms in $(Cu_{1-x}Ni_x)_6Sn_5$ are extracted out to form more $(Ni_{1-y}Cu_y)_3Sn_4$.

When the solder joints became even smaller (300 μm), the supply of Cu became very limited. For

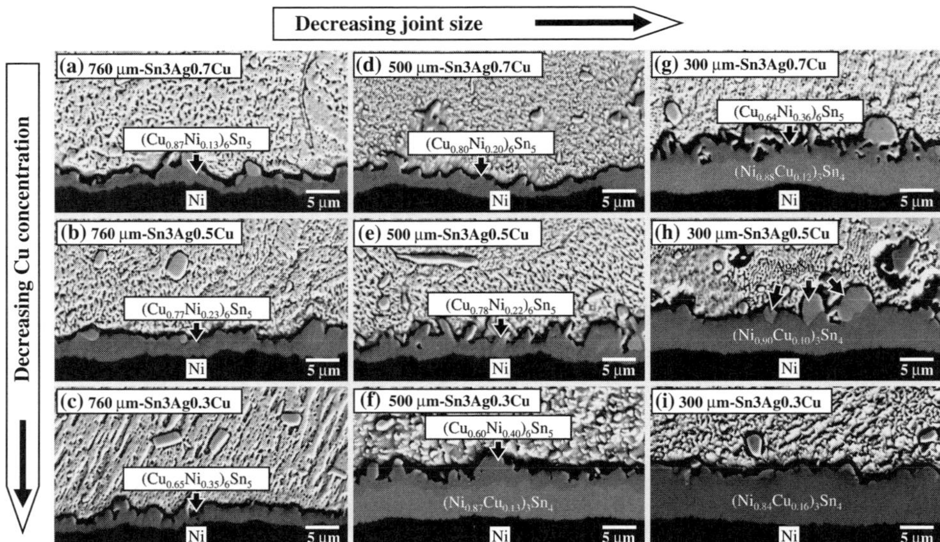

Fig. 20 Micrographs showing solder joints with the same reflow conditions (235°C peak temperature, 90 s molten solder duration), the same aging conditions (160°C for 1,000 h), and the same d_{pad} value (375 µm), but with different d_{joint} values and different Cu compositions. Pictures in the first, second, and third row have Cu concentrations equal 0.7, 0.5, and 0.3 wt.%, respectively. Pictures in the first, second, and third column have d_{joint} equal 760, 500, and 300 µm, respectively

the joint with the highest Cu concentration (0.7 wt.%), shown in Fig. 20(g), the amount of Cu was still enough to sustain a $(Cu_{1-x}Ni_x)_6Sn_5$ layer over the $(Ni_{1-y}Cu_y)_3Sn_4$, as shown in Fig. 20(g). But when the Cu concentration became lower (0.5 and 0.3 wt.% Cu), shown in Fig. 20(h) and (i), the amount of Cu was not enough to sustain the $(Cu_{1-x}Ni_x)_6Sn_5$ layer, and only a $(Ni_{1-y}Cu_y)_3Sn_4$ layer was present. It should be noted that in these two cases $(Cu_{1-x}Ni_x)_6Sn_5$ did appear over the $(Ni_{1-y}Cu_y)_3Sn_4$ layer once during certain time frame, but $(Cu_{1-x}Ni_x)_6Sn_5$ eventually disappeared and be converted into $(Ni_{1-y}Cu_y)_3Sn_4$. It can be anticipated that the $(Cu_{1-x}Ni_x)_6Sn_5$ layer in Fig. 20(g) will also disappear if this sample is subjected to longer aging.

From our recent study, the formation of both $(Ni_{1-y}Cu_y)_3Sn_4$ and $(Cu_{1-x}Ni_x)_6Sn_5$ layers at the interface after aging has a negative impact on the strength of the joints. The joints tend to fail along the interface of these two compounds. This is because both these two compounds are quite brittle, and the interface between two brittle intermetallics unfortunately tends to be weak, especially for two compounds originate from two different binary systems. To avoid the formation of two compounds simultaneously, both the Cu concentration effect and the solder volume effect have to be taken into account at the same time.

4 Cross-interaction between Cu and Ni across a solder joint

In Sects. 2 and 3, the effect of Ni on the reaction between Cu and solder and the effect of Cu on the reaction between Ni and solder are discussed, respectively. In this section, the interaction between Cu and Ni across a solder joint will be considered. The importance of such cross-interaction had been reported in several publications available in the literature [53, 80, 117–120].

4.1 Cross-interaction during reaction with molten solders (reflow)

To investigate the cross-interaction during the reflow stage more carefully, the solder joints that were assembled by two different paths are compared. As illustrated in Fig. 21, in Path I, the solder (Sn3.5Ag) was attached to the Cu substrate in the first reflow, and then the Ni substrate was attached to the other side of the solder joint in the second reflow. In path II, the solder was attached to the Ni substrate first (the first reflow), and in the second reflow Cu substrate was attached to the other side of the solder joint.

Shown in Fig. 22 are the solder/Cu and Ni/solder interfaces for Path I (left column) and path II (right

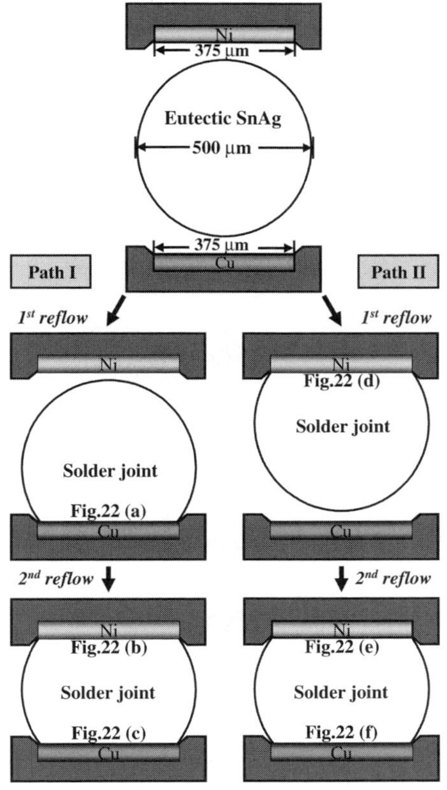

Fig. 21 Illustration showing the two sequences that the solder joints were assembled

column) after the first reflow, and the second reflow. For Path I, the solder/Cu interface after the first reflow, Fig. 22(a), is typical of the reaction between Cu and Sn-based solders: formation of Cu_6Sn_5 scallops along the interface. A typical reflow actually dissolves a large quantity of Cu from the substrate. Our own data shows that as high as 7.5($\pm$0.3) μm of Cu can be dissolved into the Sn3.5Ag during a typical reflow (90 s reflow time with a peak temperature of 235°C of a 760 μm BGA joint, making the Cu concentration in solder become as high as ~1.2 wt.%. These dissolved Cu atoms precipitated out as long Cu_6Sn_5 rods inside the solder when the solder solidified, as shown in Fig. 23. During the second reflow of Path I, the dissolved Cu atoms had an effect on the Ni/solder reaction. As shown in Fig. 22(b), the reaction product under such conditions was $(Cu_{0.89}Ni_{0.11})_6Sn_5$. This microstructure shows marked difference with that of Fig. 22(d), which shows fresh Sn3.5Ag soldered over the Ni substrate. The

second reflow also influenced the solder/Cu interface. As shown in Fig. 22(c), a small amount of Ni (~2 at.%) can be detected in Cu_6Sn_5. The only Ni source was from the Ni substrate. This result clearly shows that during the second reflow Ni had the capacity to have itself dissolved at the Ni/solder interface, migrate across the solder joint, and incorporate itself into Cu_6Sn_5. Nevertheless, the amount of Ni that was able to reach the solder/Cu interface, it was apparently not large enough to cause the effect described in Sect. 2.2. In short, during the assembly of a Cu/solder/Ni solder joint through Path I, the two interfaces can cross-interact during the reflow stage.

For Path II after the first reflow, a thin layer of Ni_3Sn_4 formed at the Ni/solder interface, as shown in Fig. 22(d). A small amount of Ni also dissolved into the molten solder. During the second reflow, the dissolved Ni atoms had an effect on the solder/Cu reaction. As shown in Fig. 22(f), the reaction product at the solder/Cu interface had two distinct regions with their own compositions, $(Cu_{0.81}Ni_{0.19})_6Sn_5$ and $(Cu_{0.93}Ni_{0.07})_6Sn_5$. This microstructure is the same as that shown in Fig. 4(b). Apparently, the reflow condition was sufficient to dissolve enough amount of Ni into solder to produce the Ni effect as discussed in Sect. 2.2. Quite surprisingly, the dissolved Cu during its fist reflow (i.e., the second reflow in path II) was also high enough to induce the Cu concentration effect. As shown in Fig. 22(e), the outer intermetallic at the Ni/solder interface had changed from Ni_3Sn_4 to $(Cu_{1-x}Ni_x)_6Sn_5$. This fast transformation from one compound to another in time as short as 90 s does not seem reasonable. A careful cutting by using the focused ion beam (FIB) technique was able to show that this transformation was indeed fast, but was not complete. As shown in Fig. 24, there was a very thin layer of $(Ni_{1-y}Cu_y)_3Sn_4$ beneath the $(Cu_{1-x}Ni_x)_6Sn_5$ layer. The FIB technique was also applied to other interface of Fig. 22, but only one type of intermetallic was observed. As shown in Fig. 25, the crystal structures of the Ni_3Sn_4 compound and the Cu_6Sn_5-based compound in Fig. 22(d) and (e) had also been positively identified by the glancing angle XRD measurements.

Comparing Path I and II, one can conclude that even the sequence of assembly has an effect on the intermetallic microstructures.

4.2 Cross-interaction during reaction with solid solders (aging)

Figure 26 shows the microstructures of the interfaces in Fig. 22 after aging at 160°C for 1,000 h. The Ni/solder interface and the solder/Cu interface that had

 Springer

Fig. 22 Micrographs showing the solder/Cu and Ni/solder interfaces for Path I (left column) and path II (right column) illustrated in Fig. 21. The micrographs in (**a**) and (**d**) had been tilted by 30°

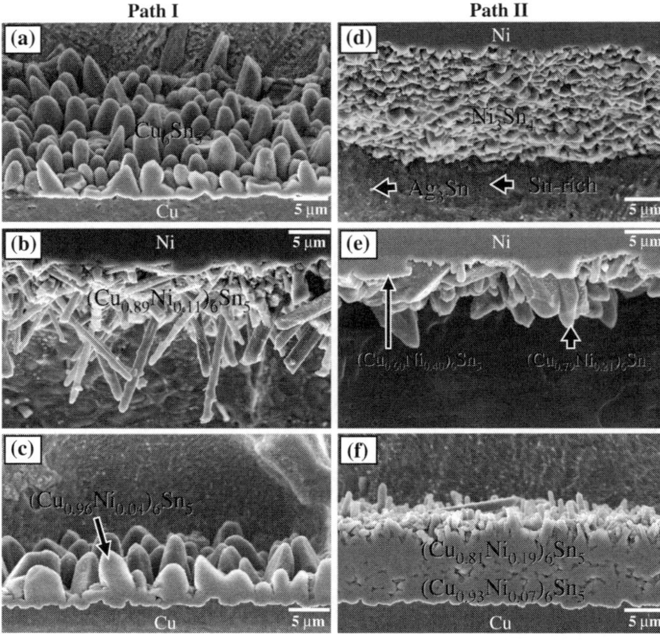

been soldered only once and consequently had no cross-interaction during aging were shown in the left column for comparison. In the middle column and the right column, the corresponding interfaces for the Path I and Path II after aging are shown, respectively. It can be seen that Cu_3Sn in Fig. 26(d) and (f) are thinner than that in Fig. 26(b). This is due to the Ni effect presented in Sect. 2.2. Moreover, Cu_3Sn in Fig. 26(f) is also thinner than that in Fig. 26(d). This is reasonable because in Path II more Ni was dissolved because Ni in Path II went through one more reflow than in Path I. In addition, no Kirkendall void was observed in Fig. 26(d) and (f) even by the means of FIB sample preparation. It should also be pointed that the cross-interaction accelerated the consumption rate of the Cu substrate and reduced the consumption rate of the Ni substrate [120]. The Ni consumption rate decreased because the growth of $(Cu_{1-x}Ni_x)_6Sn_5$ consumed a less amount of Ni than did the growth of $(Ni_{1-y}Cu_y)_3Sn_4$ [81, 107, 117, 120].

5 Summary

The addition of Ni to Sn, SnCu, SnAg, and SnAgCu in amounts as minute as 0.1 wt.% is able to substantially hinder the Cu_3Sn growth in the reaction between these solders and the Cu substrate. The growth of Cu_3Sn often accompanies the formation of Kirkendall voids, which has been linked to the weakening of the solders joints. Accordingly, Ni has been proposed as a useful alloying additive to these solders.

Likewise, Cu has a strong effect on the reactions between these solders and Ni-based substrates. When the amount of solder is large so that the supply of Cu is not an issue, the interfacial reaction is dictated by the Cu–Ni–Sn phase equilibrium, and the types of the reaction products formed depend on the Cu concentration. Under such a condition, when the Cu concentration is low (≤0.3 wt.%), only $(Ni_{1-y}Cu_y)_3Sn_4$ forms at the interface. When the Cu concentration increases to 0.4–0.5 wt.%, both $(Ni_{1-y}Cu_y)_3Sn_4$ and $(Cu_{1-x}Ni_x)_6Sn_5$ form. When the Cu concentration increases above 0.5 wt.%, $(Cu_{1-x}Ni_x)_6Sn_5$ forms. However, in BGA or flip-chip solder joints, where the solder joints are small and the Cu supply becomes limited, the solder volume effect becomes important because the Cu concentration is no longer constant during the reaction. An equation, Eq. 1, for the Cu concentration drop has been derived as a function of the joint size and the intermetallic compound thickness. As the Cu concentration decreases, the equilibrium phase at the interface may change, and the local thermodynamic equilibrium at the interface is no longer static. The

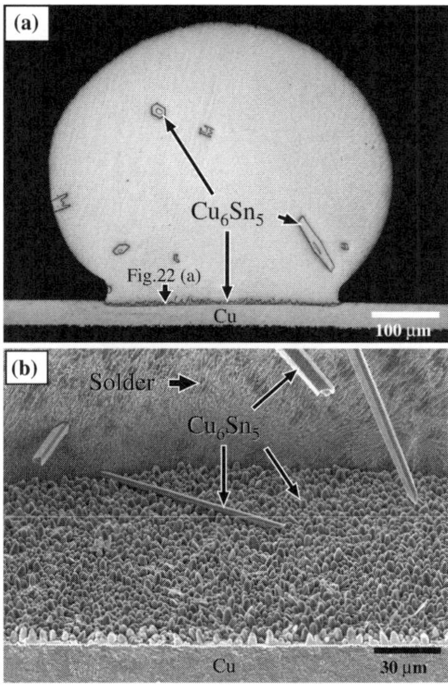

Fig. 23 (**a**) Micrograph showing the solder/Cu interface after the first reflow. The Cu atoms in those Cu_6Sn_5 needles inside the solder came from the dissolved Cu atoms when the solder was molten. (**b**) Long Cu_6Sn_5 needles described in (**a**). Part of the solder had been etched away

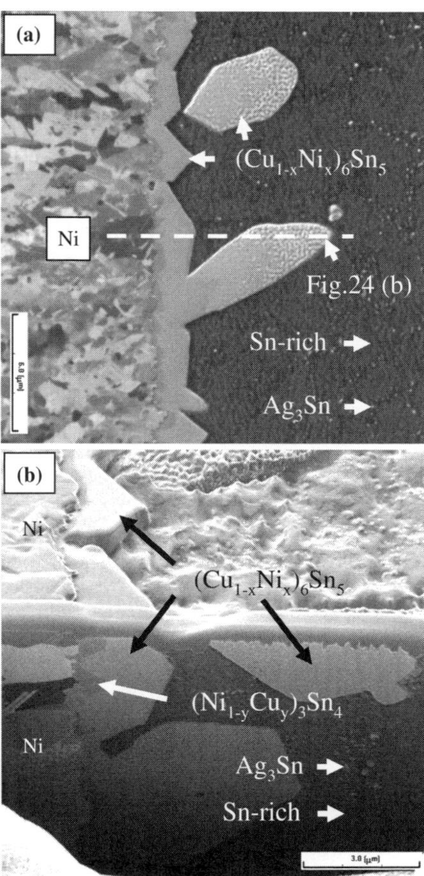

Fig. 24 (**a**) Plain view micrograph showing the Ni/solder interface in Fig. 22(e) after FIB polishing. (**b**) Cross-section view micrograph from the dash line in (**a**)

shifting of the equilibrium phase at the interface could cause the massive spalling of the intermetallic from the interface.

Lastly, the effect of cross-interaction of the Cu and Ni substrate across a solder joint was highlighted. This cross-interaction can occur during the reflow stage. Consequently, the sequence of assembly has an effect on the microstructure of the solder joints.

Acknowledgements This work was supported by the National Science Council of R.O.C. through grants NSC-94-2216-E-008-001 and NSC-94-2214-E-008-005. The author (CRK) would like to thanks his students for their contributions to the study of soldering reactions (C. M. Liu, J. Y. Tsai, C. W. Chang, W. C. Luo, W. T. Chen, and L. C. Shiau).

References

1. A. Rahn (ed.), in *The Basics of Soldering* (John Wiely & Sons, New York, 1993)
2. J.H. Lau (ed.), in *Flip Chip Technology* (McGraw Hill, New York, 1996)
3. K.N. Tu, K. Zeng, Mater. Sci. Eng. **R34**, 1 (2001)
4. W.G. Bader, Weld. J. Res. Suppl. **28**, 551s (1969)
5. C.J. Thwaites, Electroplat. Met. Finish. **26**, 10 (1973)
6. W.A. Mulholland, D.L. Willyard, Weld. J. Res. Suppl. **54**, 466s (1974)
7. R. Duckett, M.L. Ackroyd, Electroplat. Met. Finish. **29**, 13 (1976)
8. H. Heinzel, K.E. Saeger, Gold Bull. **9**, 7 (1976)
9. D.M. Jacobson, G. Jumpston, Gold Bull. **22**, 9 (1989)
10. P.A. Kramer, J. Glazer, J.W. Morris, Jr., Metall. Mater. Trans. **25A**, 1249 (1994)
11. J. Glazer, Inter. Mater. Rev. **40**, 65 (1995)
12. F.G. Yost, Gold Bull. **10**, 94 (1977)
13. C.E. Ho, Y.M. Chen, C.R. Kao, J. Electron. Mater. **28**, 1231 (1999)
14. C.E. Ho, S.Y. Tsai, C.R. Kao, IEEE Trans. Adv. Packag. **24**, 493 (2001)
15. Z. Huang, P.P. Conway, C. Liu, R.C. Thomson, J. Electron. Mater. **33**, 1227 (2004)
16. J.W. Jang, P.G. Kim, K.N. Tu, D.R. Frear, P. Thomson, J. Appl. Phys. **85**, 8456 (1999)

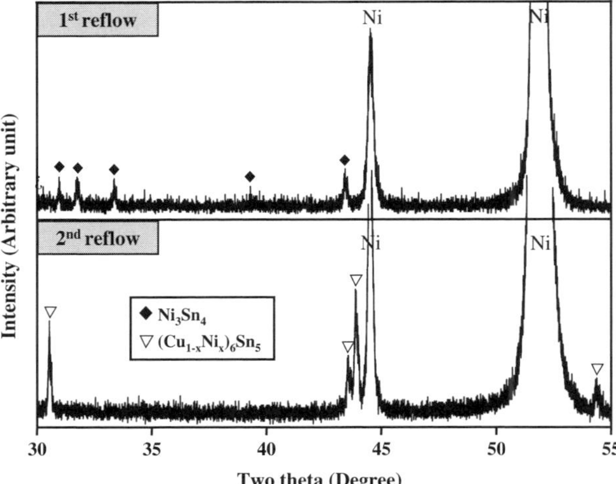

Fig. 25 Glancing angle XRD pattern of the interface in Fig. 21(d) and (e). The crystal structures of the Ni_3Sn_4 compound [Fig. 21(d)] and the $(Cu_{1-x}Ni_x)_6Sn_5$ compound [Fig. 21(e)] were positively identified

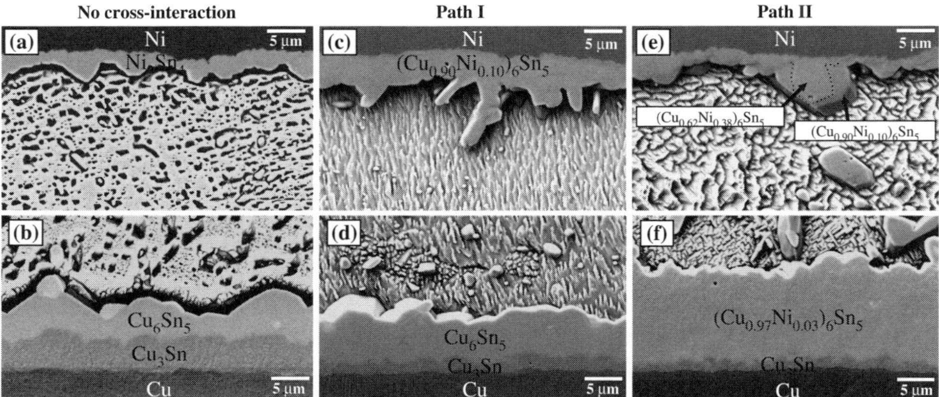

Fig. 26 Micrographs showing the microstructures of the interfaces in Fig. 22 after aging at 160°C for 1,000 h. Each micrograph here corresponds to the after-aging microstructure of the interface that has the same label in Fig. 22

17. P. Liu, Z. Xu, J.K. Shang, Metall. Mater. Trans. **31A**, 2857 (2000)
18. H. Matsuki, H. Ibuka, H. Saka, Sci. Technol. Adv. Mater. **3**, 261 (2002)
19. K. Zeng, V. Vuorinen, J.K. Kivilahti, IEEE Trans. Electron. Packag. Manufact. **25**, 162 (2002)
20. S.K. Kang, D.Y. Shih, K. Fogel, P. Lauro, M.J. Yim, G.G. Advocate, Jr., M. Griffin, C. Goldsmith, D.W. Henderson, T.A. Gosselin, D.E. King, J.J. Konrad, A. Sarkhel, K.J. Puttlitz, IEEE Trans. Electron. Packag. Manufact. **25**, 155 (2002)
21. Y.D. Jeon, S. Nieland, A. Ostmann, H. Reichl, K.W. Paik, J. Electron. Mater. **32**, 548 (2003)
22. N. Torazawa, S. Arai, Y. Takase, K. Sasaki, H. Saka, Mater. Trans. **44**, 1438 (2003)
23. T. Hiramori, M. Ito, M. Yoshikawa, A. Hirose, K.F. Kobayashi, Mater. Trans. **44**, 2375 (2003)
24. C.W. Hwang, K. Suganuma, M. Kiso, S. Hashimoto, J. Mater. Res. **18**, 2540 (2003)
25. C.W. Hwang, K. Suganuma, M. Kiso, S. Hashimoto, J. Electron. Mater. **33**, 1200 (2004)
26. M. He, Z. Chen, G. Qi, Acta Mater. **52**, 2047 (2004)
27. J.W. Yoon, S.W. Kim, S.B. Jung, J. Alloys Compd. **385**, 192 (2004)
28. Y.D. Jeon, K.W. Paik, A. Ostmann, H. Reichl, J. Electron. Mater. **34**, 80 (2005)
29. T. Laurila, V. Vuorinen, J.K. Kivilahti, Mater. Sci. Eng. **R49**, 1 (2005)
30. T.T. Mattila, J.K. Kivilahti, J. Electron. Mater. **34**, 969 (2005)
31. A. Kumar, M. He, Z. Chen, Surf. Coat. Tech. **198**, 283 (2005)
32. S.T. Kao, J.G. Duh, J. Electron. Mater. **34**, 1129 (2005)
33. J.W. Yoon, S.B. Jung, J. Alloys Compd. **396**, 122 (2005)

34. D.G. Kim, J.W. Kim, S.B. Jung, Mater. Sci. Eng. B **121**, 204 (2005)
35. C.Y. Liu, K.N. Tu, T.T. Sheng, C.H. Tung, D.R. Frear, P. Elenius, J. Appl. Phys. **87**, 750 (2000)
36. P.S. Teo, Y.W. Huang, C.H. Tung, M.R. Marks, T.B. Lim, in Proc. of 2000 IEEE Electron. Comp. Tech. Conf. (ECTC), p. 33
37. F. Zhang, M. Li, C.C. Chum, K.N. Tu, J. Mater. Res. **17**, 2757 (2002)
38. F. Zhang, M. Li, C.C. Chum, Z.C. Shao, J. Electron. Mater. **32**, 123 (2003)
39. M. Abtew, G. Selvaduray, Mater. Sci. Eng. **R27**, 95 (2000)
40. K. Suganuma, Curr. Opin. Solid State Mater. Sci. **5**, 55 (2001)
41. K. Zeng, K.N. Tu, Mater. Sci. Eng. **R38**, 55 (2002)
42. K.N. Tu, A.M. Gusak, M. Li, J. Appl. Phys. **93**, 1335 (2003)
43. C.M.L. Wu, D.Q. Yu, C.M.T. Law, L. Wang, Mater. Sci. Eng. **R44**, 1 (2004)
44. S.K. Kang, P.A. Lauro, D.-Y. Shih, D.W. Henderson, K.J. Puttlitz, IBM J. Res. & Dev. **49**, 607 (2005)
45. S. Ahat, M. Sheng, L. Luo, J. Electron. Mater. **30**, 1317 (2001)
46. T.-C. Chiu, K. Zeng, R. Stierman, D. Edwards, K. Ano, in Proceedings of 2004 IEEE Electron. Comp. Tech. Conf. (ECTC), p. 1256
47. M. Date, T. Shoji, M. Fujiyoshi, K. Sato, K. N. Tu, Proceedings of 2004 IEEE Electron. Comp. Tech. Conf. (ECTC), p. 668
48. P.T. Vianco, J.A. Rejent, P.F. Hlava, J. Electron. Mater. **33**, 991 (2004)
49. P. Borgesen, D. W. Henderson, Report of Universal Instruments (http://www.uci.com), (2004)
50. K. Zeng, R. Stierman, T.-C. Chiu, D. Edwards, K. Ano, K.N. Tu, J. Appl. Phys. **97**, 024508 (2005)
51. Z. Mei, M. Ahmad, M. Hu, G. Ramakrishna, in Proceedings of 2005 IEEE Electron. Comp. Tech. Conf. (ECTC), p. 415
52. M. Oh, Doctor Dissertation, Lehigh University, (1994)
53. S.W. Chen, S.H. Wu, S.W. Lee, J. Electron. Mater. **32**, 1188 (2003)
54. J.Y. Tsai, Y.C. Hu, C.M. Tsai, C.R. Kao, J. Electron. Mater. **32**, 1203 (2003)
55. C.M. Chung, P.C. Shih, K.L. Lin, J. Electron. Mater. **33**, 1 (2004)
56. L. Garner, S. Sane, D. Suh, T. Byrne, A. Dani, T. Martin, M. Mello, M. Patel, R. Williams, Intel Technol. J. **9**, 297 (2005)
57. I.E. Anderson, J.L. Harringa, J. Electron. Mater. **35**, 94 (2006)
58. T.B. Massalski (ed.), in *Binary Alloy Phase Diagrams* (ASM International, Metal Park, OH, 1990) p. 1481
59. P. Nash, A. Nash, Bull. Alloy Phase Diag. **6**, 350 (1985)
60. E.K. Ohriner, Weld. J. Res. Suppl. **7**, 191 (1987)
61. S. Bader, W. Gust, H. Hieber, Acta. Metall. Mater. **43**, 329 (1995)
62. D. Gur, M. Bamberger, Acta Mater. **46**, 4917 (1998)
63. J.A. van Beek, S.A. Stolk, F.J. J. van Loo, Z. Metallkde **73**, 441 (1982)
64. C.M. Liu, M.S. Thesis, National Central University, Taiwan (2000)
65. K.-W. Moon, W.J. Boettinger, U.R. Kattner, F.S. Biancaniello, C.A. Handwerker, J. Electron. Mater. **29**, 1122 (2000)
66. NEMI (National Electronics Manufacturing Initiative)- *Workshop on Modeling and Data Needs for Lead-Free solders*, (New Orleans, LA, February 15th 2001)
67. Soldertec-ITRI, *Lead-free alloys-the way forward*, October 1999 (http://www.lead-free.org)
68. IDEALS (International Dental Ethics and Law Society), Improved design life and environmentally aware manufacturing of electronics assemblies by lead-free soldering, Brite-Euram contract BRPR-CT96-0140, project number BE95-1994 (1994–1998)
69. JEITA (Japan Electronics and Information Technology Industries Association), Lead-Free Roadmap 2002, v 2.1, (2002)
70. K.F. Seeling, D.G. Lockard, United States Patent, Patent No. 5352407, (Oct 1994)
71. IPC Roadmap, Assembly of Lead-Free Electronics, 4th draft, IPC, (Northbrook, IL, June 2000)
72. L.S. Bai, Taiwan Printed Circuit Association (TPCA) Magazine **31**, 21 (2006)
73. C.E. Ho, Y.L. Lin, C.R. Kao, Chem. Mater. **14**, 949 (2002)
74. J.W. Jang, D.R. Frear, T.Y. Lee, K.N. Tu, J. Appl. Phys. **88**, 6359 (2000)
75. C.E. Ho, R.Y. Tsai, Y.L. Lin, C.R. Kao, J. Electron. Mater. **31**, 584 (2002)
76. S.M. Hong, C.S. Kang, J.P. Jung, IEEE Trans. Adv. Packag. **27**, 90 (2004)
77. G.Y. Jang, J.G. Duh, J. Electron. Mater. **34**, 68 (2005)
78. C.M. Liu, C.E. Ho, W.T. Chen, C.R. Kao, J. Electron. Mater. **30**, 1152 (2001)
79. M. Li, K.Y. Lee, D.R. Olsen, W.T. Chen, B.T. C. Tan, S. Mhaisalkar, IEEE Trans. Electron. Packag. **25**, 185 (2002)
80. S.K. Kang, W.K. Choi, M.J. Yim, D.Y. Shih, J. Electron. Mater. **31**, 1292 (2002)
81. L.C. Shiau, C.E. Ho, C.R. Kao, Solder. Surf. Mount Tech. **14/3**, 25 (2002)
82. M.O. Alam, Y.C. Chan, K.N. Tu, Chem. Master. **15**, 4340 (2003)
83. K.Y. Lee, M. Li, J. Electron. Mater. **32**, 906 (2003)
84. C.B. Lee, J.W. Yoon, S.J. Suh, S.B. Jung, C.W. Yang, C.C. Shur, Y.E. Shin, J. Mater. Sci.: Mater. Electron. **14**, 487 (2003)
85. A. Sharif, M.N. Islam, Y.C. Chan, Mater. Sci. Eng. B **113**, 184 (2004)
86. D.Q. Yu, C.M. L. Wu, D.P. He, N. Zhao, L. Wang, J.K.L. Lai, J. Mater. Res. **20**, 2205 (2005)
87. W.T. Chen, C.E. Ho, C.R. Kao, J. Mater. Res. **17**, 263 (2002)
88. W.C. Luo, C.E. Ho, J.Y. Tsai, Y.L. Lin, C.R. Kao, Mater. Sci. Eng. A **396**, 385 (2005)
89. C.E. Ho, W.C. Luo, S.C. Yang, C.R. Kao, in Proceedings of IMAPS Taiwan 2005 International Technical Symposium (Taipei, June 2005), p. 98
90. C.E. Ho, Y.W. Lin, S.C. Yang, C.R. Kao, in Proceedings of the 10th International Symposium on Advanced Packaging Materials: Processes, Properties and Interface, IEEE/CPMT (Irvine, March 2005), p. 39
91. C.E. Ho, Y.W. Lin, S.C. Yang, C.R. Kao, D. S. Jiang, J. Electron. Mater. **35**, 1017 (2006)
92. M.O. Alam, Y.C. Chan, K.N. Tu, J.K. Kivilahti, Chem. Mater. **17**, 2223 (2005)
93. K.S. Kim, S.H. Huh, K. Suganuma, J. Alloys Compd. **352**, 226 (2003)
94. J.S. Ha, T.S. Oh, K.N. Tu, J. Mater. Res. **18**, 2109 (2003)
95. C.H. Wang, S.W. Chen, Acta Mater. **54**, 247 (2006)
96. S.K. Kang, W.K. Choi, D.Y. Shih, P. Lauro, D.W. Henderson, T. Gosselin, D.N. Leonard, in Proceedings of 2002 IEEE Electron. Comp. Tech. Conf. (ECTC), p. 146
97. Y. Zheng, C. Hillman, P. McCluskey, in Proceedings of 2002 IEEE Electron. Comp. Tech. Conf. (ECTC), p. 1226
98. M.D. Cheng, S.Y. Chang, S.F. Yen, T.H. Chuang, J. Electron. Mater. **33**, 171 (2004)

99. J.H. L. Pang, T.H. Low, B.S. Xiong, X. Luhua, C.C. Neo, Thin Sol. Films, **462–463**, 370 (2004)

100. J.W. Yoon, S.W. Kim, J.M. Koo, D.G. Kim, S.B. Jung, J. Electron. Mater. **33**, 1190 (2004)

101. J.W. Yoon, S.W. Kim, S.B. Jung, J. Alloys Compd. **391**, 82 (2005)

102. C.B. Lee, S.B. Jung, Y.E. Shin, C.C. Shur, Mater. Trans. **43**, 1858 (2002)

103. A. Zribi, A. Clark, L. Zavalij, P. Borgesen, E.J. Cotts, J. Electron. Mater. **30**, 1157 (2001)

104. G. Ghosh, Acta Mater. **49**, 2609 (2001)

105. G. Ghosh, J. Electron. Mater. **33**, 229 (2004)

106. C.E. Ho, Doctor Dissertation, National Central University, Taiwan, (2002)

107. C.E. Ho, L.C. Shiau, C.R. Kao, J. Electron. Mater. **31**, 1264 (2002)

108. L.P. Lehman, S.N. Athavale, T.Z. Fullem, A.C. Giamis, R.K. Kinyanjui, M. Lowenstein, K. Mather, R. Patel, D. Rae, J.Wang, Y. Xing, L. Zavalij, P. Borgesen, E.J. Cotts, J. Electron. Mater. **33**, 1429 (2004)

109. D.W. Henderson, T. Gosselin, A. Sarkhel, S.K. Kang, W.K. Choi, D.Y. Shih, C. Goldsmith, K. Puttlitz, J. Mater. Res. **17**, 2775 (2002)

110. S.K. Kang, W.K. Choi, D.Y. Shih, D.W. Henderson, T. Gosselin, A. Sarkhel, C. Goldsmith, K. Puttlitz, in Proceedings of 2003 IEEE Electron. Comp. Tech. Conf. (ECTC), p. 64

111. S.K. Kang, W.K. Choi, D.Y. Shih, D.W. Henderson, T. Gosselin, A. Sarkhel, C. Goldsmith, K. Puttlitz, J. Minerals Metals Mater. Soc. **55**, 61 (2003)

112. S. Terashima, Y. Kariya, T. Hosoi, M. Tanaka, J. Electron. Mater. **32**, 1527 (2003)

113. C.H. Lin, S.W. Chen, C.H. Wang, J. Electron. Mater. **31**, 907 (2002)

114. C.Y. Li, J.G. Duh, J. Mater. Res. **20**, 3118 (2005)

115. P. Oberndorff, Doctoral Dissertation, Technical University of Eindhoven, (2001)

116. G.Z. Pan, A.A. Liu, H.K. Kim, K.N. Tu, P.A. Totta, Appl. Phys. Lett. **71**, 2946 (1997)

117. C.R. Kao, C.E. Ho, L.C. Shiau, Solder Point with Low Speed of Consuming Nickel, R.O.C. patent, patent No. 181410, (2003)

118. S.J. Wang, C.Y. Liu, J. Electron. Mater. **32**, 1303 (2003)

119. T.L. Shao, T.S. Chen, Y.M. Huang, C. Chen, J. Mater. Res. **19**, 3654 (2004)

120. C.M. Tsai, W.C. Luo, C.W. Chang, Y.C. Shieh, C.R. Kao, J. Electron. Mater. **33**, 1424 (2004)

Microstructure-based modeling of deformation in Sn-rich (Pb-free) solder alloys

N. Chawla · R. S. Sidhu

Published online: 12 September 2006
© Springer Science+Business Media, LLC 2006

Abstract The mechanical properties of Sn-rich solder alloys are directly related to their heterogeneous microstructure. Thus, numerical modeling of the properties of these alloys is most effective when the microstructure is explicitly incorporated into the model. In this review, we provide several examples where 2D and 3D microstructures have been used to model the material behavior using finite element modeling. These included (a) 3D visualization of the solder microstructure, (b) 3D microstructure-based modeling of tensile behavior, (c) 2D modeling of the effect of intermetallic volume fraction and morphology on shear behavior of solder joints, and (d) prediction of crack growth in solder joints. In all these cases, the experimentally observed behavior matches very well with the microstructure-based models.

1 Introduction

Increasing concerns over the environmental and health hazards of Pb–Sn solders, used in electronic packaging, have prompted the need for Pb-free solder alternatives [1–3]. The design and development of Pb-free solders for electronic packaging requires a thorough understanding, and careful control, of microstructure and its effect on properties [2–7]. This is particularly challenging given the multiphase and heterogeneous nature of most Sn-rich alloys. Traditionally, the mechanical behavior of materials has been quantified by large and costly sets of experiments. Experiments are and will always be an important necessary part of research and development. Nevertheless, modeling of the behavior of materials can be used as a versatile, efficient, and low cost tool for developing an understanding of material behavior. The robustness and accuracy of the model used can and should, of course, be verified by experimental results.

A review of the literature (presented in the next section) shows that current analytical and numerical techniques simplify the heterogeneous microstructure of multiphase solder materials. These simplifications make modeling and analysis more efficient and straightforward, but fail to accurately predict the effective properties and local damage behavior which are inherently dependent on microstructure. It follows that accurate prediction of macroscopic deformation behavior and modeling of localized damage mechanisms can only be accomplished by capturing the microstructure of the material as a basis for the model.

This paper reviews a novel methodology that addresses the critical link between microstructure and deformation behavior, by using two-dimensional (2D) and three-dimensional (3D) virtual microstructures as the basis for a robust model to simulate damage caused by deformation. Such an approach provides a unique and compelling means for visualization, modeling of damage, and prediction of the macroscopic stress–strain behavior, using the microstructure of the material as the basis for the model.

N. Chawla (✉) · R. S. Sidhu
School of Materials, Fulton School of Engineering, Arizona State University, Tempe, AZ 85287-8706, USA
e-mail: nchawla@asu.edu

2 Analytical and conventional numerical modeling

Modeling and predicting the macroscopic deformation and local damage mechanisms of Sn-rich solders are very complex problems. Analytical and empirical models can provide an effective means of predicting the effective macroscopic properties of solders (e.g., Young's modulus). There is a considerable amount of variability in the reported Young's moduli of Sn-rich solders [8–10]. Chawla et al. [11], showed that the variability in modulus measurements was significantly influence by porosity in the solder. They measured the modulus by two techniques [11]: (a) loading-unloading measurements in tension, and (b) non-destructive resonant ultrasound spectroscopy. The latter technique was used since the linear elastic portion of the tensile stress–strain curve, in most Sn-rich alloys, is quite limited. Figure 1 shows the experimentally determined Young's modulus of a Sn-3.5wt%Ag solder. Both techniques yielded similar values for the Young's modulus. As expected, a gradual decrease in modulus was observed with increasing porosity. The analytical model by Ramakrishna and Arunachalam [12] (R-A model), was used to predict the Young's modulus, E, with a given fraction of porosity, p:

$$E = E_O \left[\frac{(1 - p)^2}{1 + \kappa_E p} \right] \tag{1}$$

where E_o is the Young's modulus of the fully dense solder (taken by extrapolating the experimental data to zero porosity ~50 MPa), and κ_E is a constant in terms of the Poisson's ratio of the fully-dense material, v_o:

$$\kappa_E = 2 - 3v_O \tag{2}$$

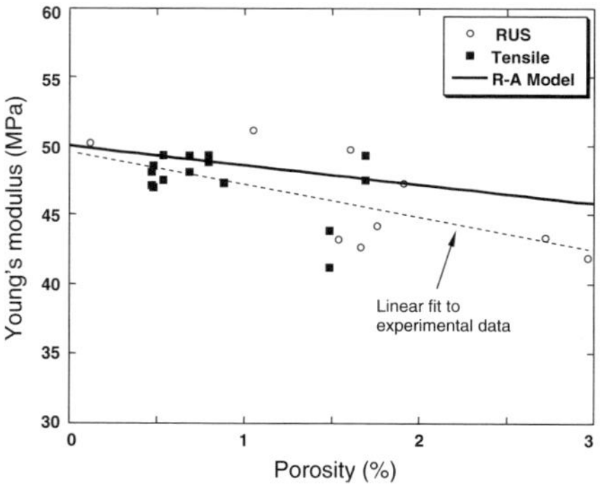

Fig. 1 Comparison of predictions from analytical model of Ramakrishnan and Arunachlam to the experimental measurements of Young's modulus for Sn-3.5wt%Ag solder. A reasonable correlation between experiment and prediction is observed [11]

The results predicted by the R-A model are in reasonable agreement with the experimental data at porosity levels less than 1%, but show somewhat larger deviations from the experiments at porosity levels greater than 1%. This example helps to illustrate the point that, while analytical solutions provide some insight into macroscopic solder behavior, they cannot predict the experimental behavior precisely due to the simplifying assumptions of the microstructure; i.e., an assumption of uniformly distributed regular shaped spherical pores. Furthermore, an understanding of the local stress/strain state cannot be obtained through analytical solutions.

In multiphase materials, numerical modeling is often more effective than analytical modeling since these materials are not readily amenable to closed-form theoretical analyses. Another advantage of numerical modeling is that deformation and damage characteristics, particularly on a local scale, can be revealed. Several studies have utilized the finite element method (FEM) to study complex states of stress at the solder/substrate interface for flip chip and Ball Grid Array (BGA) packages [13–28]. Kim et al. [14] correlated the complex stress state in a solder bump and showed that the highest region of plastic strain was near the surface of the Ni–P bump. The plastic deformation was particularly noticeable in solder near the solder/metallization interface, parallel to the chip surface. The modeled behavior also corroborated experimental observations of fractured specimens. Non-linear FEM analysis, using an elastic–viscoplastic constitutive model, has also been used to study the effect of ball shear speed on the shear forces of flip chip solder bumps [17].

Finite element modeling has also been utilized extensively to characterize damage arising from temperature cycling conditions [20–28]. Zhai et al. [20] used finite element analysis to study the effects of ramp rate and dwell times on solder joint fatigue life. Chen et al. [28] used both linear and non-linear numerical methods to model deformation behavior of the package. They showed that since the underfill and solder bump material properties exhibit large non-linearity at higher temperatures (~125°C), using a linear elastic assumption was not very accurate. They highlighted this effect by comparing the differences in interfacial stress and strain between the linear and non-linear model, and highlighted the importance of elastic-plastic analysis for these ductile materials.

All of the examples cited above focused on the solder joint as a homogeneous material system, without explicitly considering the microstructure of the solder or the nature of the intermetallic formed at the solder/metallization interface. Thus, while some insight into the general nature of damage accumulation was obtained, the models were unable to accurately predict the effective properties and local damage characteristics that are inherently dependent on the complex solder microstructure. In order to accurately predict macroscopic deformation behavior and understand localized damage mechanisms, a means of capturing the microstructure of the material as a basis for the model is essential. In the next few sections, we review some methodologies for: (a) 3D microstructure visualization of Sn-rich solders and (b) 2D and 3D microstructure-based finite element modeling of deformation in Sn-rich solders.

3 Microstructure visualization and microstructure-based modeling

3.1 3D Microstructure visualization of Sn-rich solder alloys

Most eutectic Sn-rich alloys consist of a discontinuous intermetallic phase embedded in a Sn-rich matrix. In order to fully understand the physical, electrical, and mechanical properties of the solder, it is necessary to accurately characterize the intermetallic size, distribution, morphology and orientation within the Sn-rich matrix. Ochoa et al. [6, 7], for example, determined that cooling rate has a significant effect on intermetallic size and morphology. At relatively fast cooling rates (24°C/s), a fine distribution of spherical Ag_3Sn particles was observed. At slower cooling rates (0.08°C/s), however, the intermetallic had a needle-like morphology. Traditional methods of visualizing the microstructure, such as two-dimensional (2D) images, are not fully representative of the three-dimensional (3D) microstructure of the material. The size and aspect ratio of spherical microstructural features can be characterized adequately by 2D circles [29]. In the case of needles, however, a 2D section perpendicular to the major axis of the needle can significantly underestimate the size and aspect ratio of the needle.

Clearly, a technique for 3D visualization is required for accurate characterization of non-spherical features, such as Ag_3Sn needles in a Sn matrix. Serial sectioning of 2D microstructural images with computer-aided reconstruction and visualization has increasingly been used as a technique for 3D visualization of microstructures [29–42]. Yamaguchi et al. [39] utilized a 3D reconstruction technique to quantify pathogenic yeast

cells and reveal variations in their morphology before and after freezing. In steels, Yokomizo et al. [35] studied the 3D distribution, morphology, and nucleation sites of intragranular ferrite and inclusions formed in a Fe-0.1C-1.5Mn steel alloy. The technique has also been utilized in composites, where in Chawla and collegues [30] used serial sectioning to visualize and model the microstructure of SiC particles in an Al matrix.

Sidhu and Chawla [31] used serial sectioning to reconstruct and visualize the 3D microstructure of Sn-3.5wt%Ag solder. The specimens were heated to 240°C (approximately 20°C above the eutectic point of the solder) and cooled at a rate of 0.08°C/s. This cooling rate produced a nearly eutectic microstructure, consisting of Ag_3Sn needle-like intermetallics in a Sn-rich matrix [7]. A serial sectioning method was employed to acquire 2D images of the microstructure as a basis for reconstructing the 3D microstructure of the solder. The basic steps of the serial sectioning process and modeling consisted of sample preparation, fiducial marking by indentation, serial polishing, imaging and image segmentation, image stacking and visualization, Fig. 2. In general, selecting the region of interest is relatively subjective. In their study, the goal was to encompass a range of intermetallic colonies that formed at different orientations. Thus, the size of the microstructural region of interest was taken as approximately 180 μm × 170 μm, and a depth of approximately 60 μm. Fiducial marks were placed on the sample to define the region of interest and to measure the material loss during serial sectioning. Fiducial marks were placed on the sample to define the region of interest, to align the indentations during reconstruction, and to measure the material loss during serial sectioning. This method of quantifying the material removal has been proven to be quite effective, since the cross-sections of the indentations are nearly square, making it relatively simple to measure the length of the diagonals [30, 32, 34–36, 38]. The approximate depth (h) of the indentation was determined by the following equation:

$$h = \frac{D}{2\tan(\varphi/2)} \tag{3}$$

where D is the average of the indentation diagonals (D_1 & D_2) on a 2D projection and (φ) is the angle between the two diagonals. A removal rate of about 0.5 μm per cycle was sufficient to adequately describe the morphology of the Ag_3Sn particles. Details of the indentation and polishing routine can be obtained elsewhere [31].

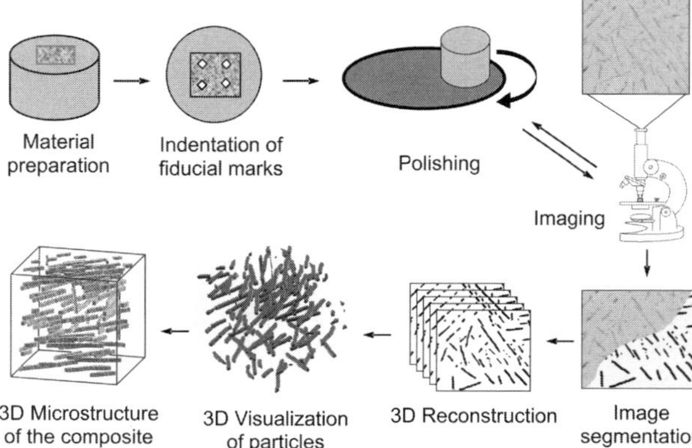

Fig. 2 Flow chart of serial sectioning process for Sn-rich solders

Optical micrographs were taken after each polishing cycle and analyzed using conventional image analysis software. The segmented serial sections were stacked, and using a 3D reconstruction software called MIMICS (Ann Arbor, MI), reconstruction was conducted and visualization of the 3D virtual microstructure achieved. The software enabled graphical analysis and viewing of the 3D volumetric data by contouring individual serial sections, meshing contours between these sections, and generating a high-resolution 3D virtual microstructure. Figure 3(a) shows a multi-needle reconstruction produced by MIMICS, where several of the needles are aligned along given colonies. The reconstructed virtual microstructure reproduces the actual size and morphology of Ag_3Sn very well. Table 1 shows the analysis of the Ag_3Sn volume fraction for the 3D virtual microstructure versus 2D slices of the 3D microstructure, and that calculated from the equilibrium diagram. The results show that the measurements from the 3D virtual microstructure are closest to that predicted by the phase diagram, while the 2D images either underestimate or overestimate the volume fraction. Overestimation of the volume fraction takes place when the 2D plane is taken parallel to the Ag_3Sn orientation. If the 2D image slice is normal to the needle's long axis, the fraction of Ag_3Sn is underestimated.

The size and aspect ratios of the particles were quantified using computer-aided design software, allowing measurement of the length of the major axis and minor axis (i.e., diameter) for individual particles. These results were compared to 2D measurements of size and aspect ratios of the Ag_3Sn particles, shown in Table 2. The distribution of aspect ratio measurements, from 2D images and the 3D model, is shown in Fig. 4. The 2D sections exhibited a broader distribution and higher distribution of needles with aspect ratios between 1 and 4. This can be attributed to measurements of "sliced needles" represented in a 2D plane. The higher variability in the 2D measurements can also be observed in the larger error in measurements, shown in Table 2. The 3D model shows a narrower distribution, with very few lower aspect ratio needles, and thus, appears to yield a more accurate representation of the Ag_3Sn particles perpendicular and parallel to the polishing plane, when compared to the 2D analysis.

3.2 3D Microstructure-based modeling of tensile behavior

While visualization of the 3D microstructure of the material is important, prediction of the properties and local damage characteristics of the material is equally important. Recently, Chawla and co-workers [43–45] have shown that 3D microstructures constructed from serial sectioning can be incorporated into commercial finite element method (FEM) codes to model the deformation behavior of composites. By comparing the 3D microstructure-based modeling approach with conventional unit cell modeling it was shown that the virtual microstructure provided a superior quantitative understanding of localized damage phenomena, as well as excellent correlation to macroscopic behavior, measured experimentally.

The 3D virtual microstructure described in section 3.1 was used as a basis for the finite element model. The particles were imported into a dedicated meshing program (HyperMesh®, Troy, MI) to mesh the particles and "create" the matrix surrounding the Ag_3Sn particles. The 3D meshed model was exported into a finite element analysis program (ABAQUS/CAE, version 6.4-1, HKS, Inc., Pawtucket, RI). Ten noded,

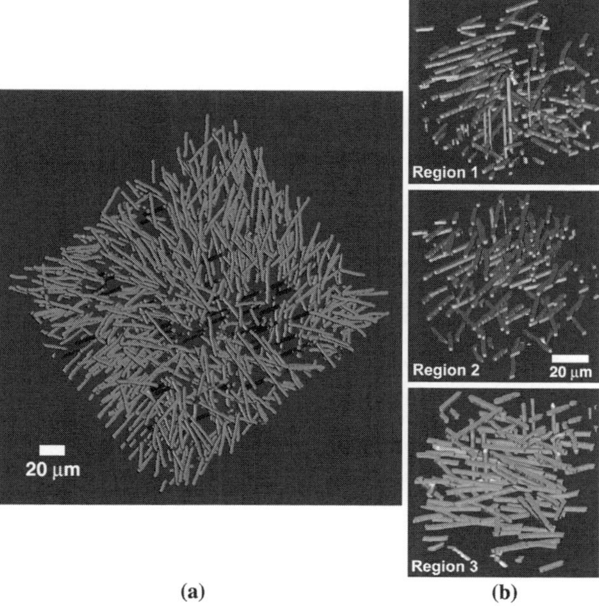

(a)   **(b)**

Fig. 3 3D microstructure reconstruction from serial sections: **(a)** 3D model of the microstructure (180 µm × 170 µm × 60 µm),

(b) Three random regions of interest used for FEM modeling (60 µm × 60 µm × 60 µm)

Table 1 Predicted volume fraction from the 3D virtual microstructure and 2D sections in comparison to experiment

	Volume % of Ag₃Sn Particles
Experimental	4.7
3D Virtual Microstructure	4.7
2D – View 1	5.2 ± 2.2
2D – View 2	3.3 ± 2.4
2D – View 3	3.9 ± 1.6

modified quadratic tetrahedral elements were used to conform to the irregular shape of the particles. A typical number of elements in the 3D model was about 95,000 elements. The Ag₃Sn needles were modeled as linear elastic and Sn-rich matrix modeled as an elastic-plastic material, using the experimentally determined stress–strain behavior of each phase [46].

Three models (60 µm × 60 µm × 60 µm models with approximately 100 particles each) were built from three different regions of the microstructure of the material (Fig. 3b), to quantify the extent of microstructural variability on both local and macroscopic stress–strain behavior. These models were compared with conventional 3D unit cell model where the Ag₃Sn intermetallic was modeled as a single sphere. The local stress state of the Ag₃Sn needles for the three virtual microstructures is inherently different, and is intimately linked to the relative orientation of the particles (Fig. 5a). In general, the needles aligned parallel to the loading axis carried a higher load. The distribution of von Mises stress in the Ag₃Sn needles and equivalent plastic strain in Sn matrix is shown in Fig. 5(b, c) (the histograms are obtained from stress/plastic strain in the elements of the models). The microstructure that exhibited greater alignment of needles along the loading axis, i.e., Region 2, had a higher fraction of elements with higher stress. The distribution of the equivalent plastic strain in the elements in the matrix,

Table 2 Summary of the quantitative microstructure characterization

	# of particle measured	Mean diameter (µm)	Mean length (µm)	Mean aspect ratio
2D section	250	1.4 ± 0.3	6.5 ± 6.9	4.9 ± 4.7
3D parallel	90	1.9 ± 0.2	15.6 ± 4.0	8.0 ± 2.0
3D perpendicular	15	2.1 ± 0.2	12.5 ± 2.2	6.1 ± 1.1

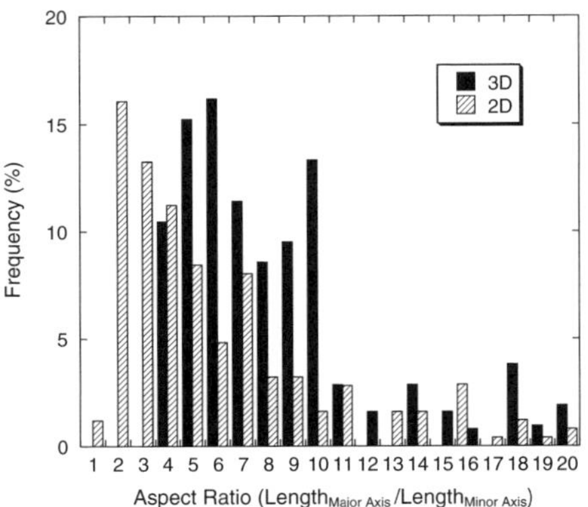

Fig. 4 2D and 3D characterization of the Ag$_3$Sn intermetallic aspect ratio for the Sn-3.5Ag solders. The 2D measurements contain low aspect ratio measurements, due to visualization of "sliced" Ag$_3$Sn needles

for Region 2, also shows a correspondingly lower fraction of strain than the other two regions. The local plastic strain in the matrix (Fig. 6) shows that deformation is greater in Sn-rich regions. Sn matrix regions between agglomerates of Ag$_3$Sn were more constrained, which resulting in a lower plastic strain in these regions.

The advantage of using a 3D microstructure-based model is shown by a comparison of the overall stress–strain curve of the 3D virtual microstructure simulation

with experiment, as well as unit cell simulations, shown in Fig. 7. The microstructure-based model predicts the experimental behavior quite well, while the spherical unit cell model predicts lower strength than the experiment. More importantly, the localized plasticity that results from the needle-like Ag$_3$Sn intermetallics, can only be captured in the microstructure-based model. A similar analysis conducted by Chawla et al. [44, 45] on particle reinforced metal matrix composites showed that the 3D virtual microstructure provided the most accurate representation of the experimental behavior. The predicted Young's modulus of the solder, once again, shows that the microstructure-based model prediction is closest to the experiment [7], yielding a modulus of 51.5 GPa (Table 3).

3.3 Modeling of intermetallic morphology and thickness on deformation behavior and crack growth in Sn-rich solder joints

An important component of solder joint deformation is the intermetallic layer that forms between the Sn-rich solder and the Cu substrate. This is particularly important in Sn-rich solders, relative to Pb–Sn solders, because of the larger amount of Sn, which has a higher affinity for reaction with Cu. In recent years, a fundamental understanding of the nature and growth mechanisms of interfacial products formed between Sn-rich alloys and Cu has been developed [47–55]. The

Fig. 5 (**a**) von Mises stress distribution in three different virtual microstructure regions (each 60 μm × 60 μm × 60 μm), highlighting the complex stress state due to randomly distributed needle-like Ag$_3$Sn intermetallics. Particles aligned parallel to the loading axis exhibited a higher stress due to more load transfer, (**b**) histogram of the von Mises stress distribution, (**c**) histogram of the equivalent plastic strain (PEEQ) distribution in the matrix of the three microstructural regions

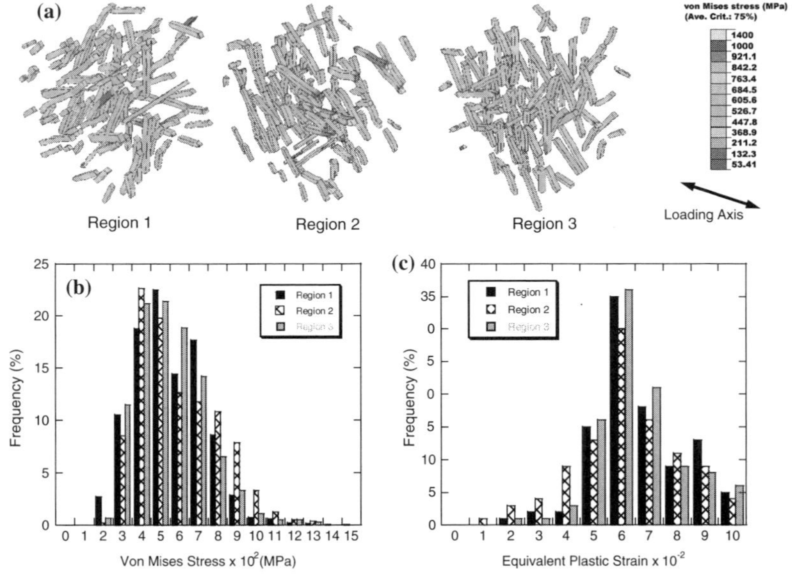

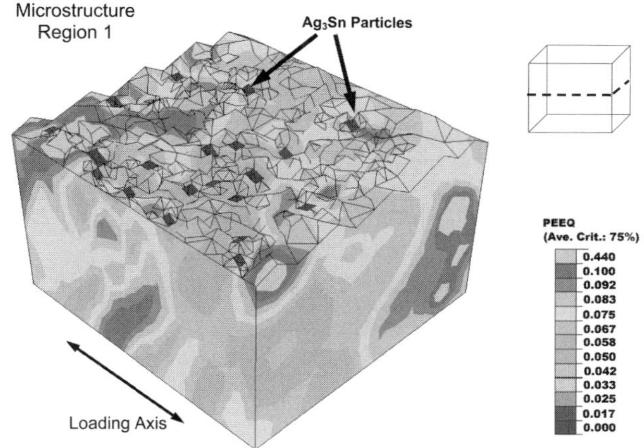

Fig. 6 Evolution of equivalent plastic strain in the 3D virtual microstructure (region 1, 60 μm × 60 μm × 30 μm) showing the plasticity within the Sn-rich matrix. Regions where the particles are more closely spaced, exhibit lower plasticity due to the higher degree of constraint on the matrix

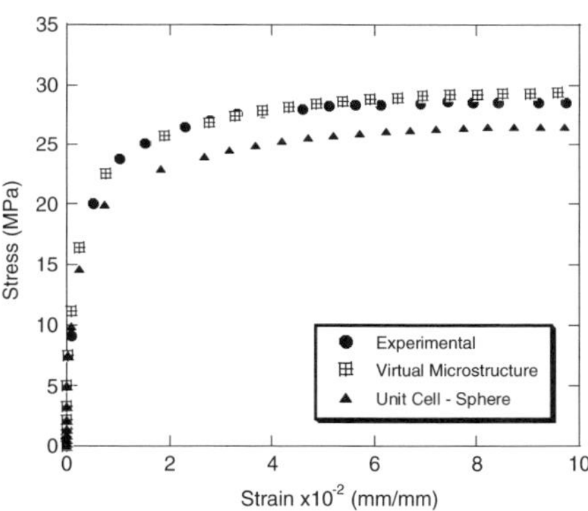

Fig. 7 Comparison of stress–strain predictions from unit cell and 3D microstructure models with experiment. The 3D virtual microstructure is most accurate in predicting the experimentally observed behavior

reaction between Sn and Cu generally results in the formation of two types of intermetallics: Cu_6Sn_5 (η) and Cu_3Sn (ε) [50, 56, 57]. The Cu_6Sn_5 phase is typically formed above the melting point of the solder, Fig. 8, while the Cu_3Sn phase is generally produced during low temperature thermal aging (100–175°C) by diffusion between Cu and Cu_6Sn_5 [48, 49]. The thickness of the intermetallic layer increases with an

Table 3 Young's modulus predicted by the finite element models compared to experiment

Type of model	Young's modulus (GPa)
Experiment [7]	51.3
3D Microstructure	51.5
Unit Cell – Sphere	48.6

increase in thermal aging time and/or an increase in aging temperature [58–61]. A similar dependence of intermetallic thickness has been observed with increasing reflow time [47, 50–55]. Increasing the reflow temperature, reflow time, and/or the number of reflow cycles results in an increase in the intermetallic thickness [47, 50–55]. The morphology and thickness of the intermetallics are also influenced by the cooling rate of the joint [49]. Deng et al. [49] showed that at faster cooling rates, a thinner and more planar intermetallic morphology was obtained, while slower cooling rates yielded a thicker and more nodular intermetallic morphology. Cooling rate also has a significant influence on the solder microstructure since a faster cooling rate produced a finer microstructure, while slower cooling rate resulted in a coarser solder microstructure [6, 7, 49].

Although a thin layer of intermetallic between solder and Cu substrate is beneficial to wettability and bonding between solder and substrate [59, 62], thicker intermetallics may have a negative effect on the toughness of solder joints [2, 63, 64]. Several authors have investigated the effect of the intermetallic compounds on the shear behavior of solder/Cu joints [58, 59, 65, 66]. Tu et al. [66] studied the shear fatigue behavior of 63Sn-37Pb solder/Cu joints after aging at 150°C for various times. They concluded that the low fracture toughness of the brittle intermetallic layer, which grew thicker with time, decreased the lifetime of the joint. Similarly, Lee et al. [58] showed a decrease in the shear strength of solder joints (Sn-4.11Ag-1.86Sb) with increased aging time and temperature. They concluded that the degradation in mechanical properties was a result of increased roughness of the solder/intermetallic interface during intermetallic growth.

 Springer

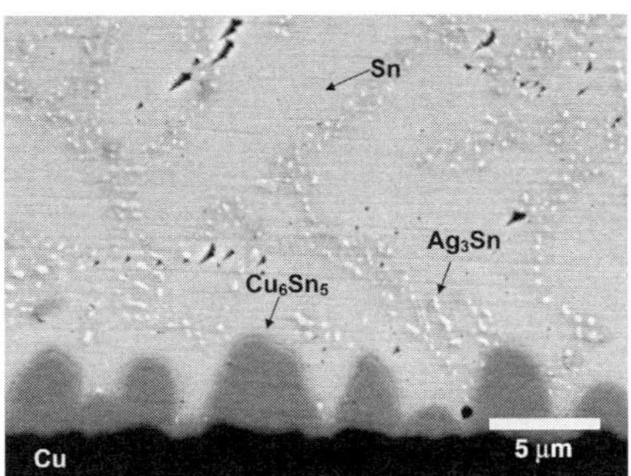

Fig. 8 Microstructure of Sn-3.5Ag/Cu joints, showing nodular Cu₆Sn₅ layer formed at the Sn-rich/Cu interface

It is clear that the mechanical behavior of the joint is dependent on the characteristics of both the intermetallic and the solder. Deng et al. [67] used finite element analysis to model the effects of intermetallic thickness, morphology, and solder yield strength. They used a two dimensional (2D) model, shown in Fig. 9, that consisted of the two Cu bars, and a Sn-3.5Ag solder interlayer. The size of the model conformed to a typical lap shear sample ($b = c = 6.35$ mm, and $t = 250$ µm). The load was applied through a displacement on right end of the copper substrate, while the constraint was applied to the left end of the other substrate. The displacement in the direction marked '2' is equal to zero, which is an accurate representation of the experimental behavior in the lap shear test. Shear strain was taken as the displacement on the right end of the copper substrate divided by solder thickness. Shear stress was measured by the load on the right end of the copper substrate divided by solder/Cu interface area. Quadrilateral elements were used to mesh the solder region and a triangular mesh was used for the Cu bar. A very fine mesh was employed at the solder/Cu interface to ensure accuracy of the numerical simulation. The mesh size was refined to minimize stress singularities and to achieve convergence. Mesh refinement was conducted until the overall shear stress–strain curve output from the model was unchanged. The Young's moduli for Cu and solder were taken as 114 GPa [46] and 50 GPa [7] respectively. Cu was assumed to be linear elastic in the simulation. To include work hardening of the solder, experiments from tensile tests of bulk Sn–Ag solder [7] were used for the constitutive behavior of the solder.

The effect of intermetallic thickness and intermetallic morphology was examined, while keeping the solder properties constant. Figure 9(b, c) shows the intermetallic morphologies employed in this investi-gation: Nodular and planar. The volume fraction of intermetallic with respect to the total solder joint was varied from 0 to 0.2 by tailoring the ratio of intermetallic thickness t_i to the total solder joint thickness, t_j. In this study, Cu₆Sn₅ was modeled as linear elastic, and the intermetallic was assumed to be homogeneous (the presence of Cu₃Sn was neglected since the elastic plastic properties of Cu₆Sn₅ and Cu₃Sn are similar [68]). The intermetallic layers and Cu substrates were defined as pure elastic materials, since no appreciable plastic deformation occurs in either Cu or the intermetallic layer [46]. The solder in these simulations was assumed to behave as elastic–plastic material as determined from experiments. The boundary conditions used in the FEM are shown in Fig. 9a.

Figure 10 shows the effect of intermetallic thickness and morphology on the lap shear behavior. The influence of both intermetallic thickness and morphology is quite significant. For both planar and nodular intermetallics, the larger intermetallic thickness leads to higher shear stress at the same shear strain. The nodular morphology of intermetallic results in a higher shear strength than planar intermetallics at the same intermetallic thickness. It should be noted that the experimental data shows that an increase in intermetallic thickness decreased the shear strength to some degree. This difference between FEA and experiment is due to the fact that FEA analysis presented here does not consider fracture, and is based on defect-free solder joints. In reality, the density of processing-induced defects (pores, voids, etc.) may increase with the reflow time and may have contributed to crack formation, decreasing the shear strength significantly.

Figure 11 shows the equivalent plastic strain (PEEQ) in solder joints at 15% nominal shear strain. For the solder joints with negligible intermetallic, a maximum in plastic deformation occurs at the corner of solder joint, which is the most likely site for crack nucleation. For both the solder joints with thin and thick planar intermetallics, the plastic deformation is uniform along the solder/Cu interface. The solder joints with nodular intermetallics, however, exhibited a more complex plastic strain distribution. It is interesting to note that plastic deformation in the solder in "intermetallic valleys" is absent. Thus, the plastic deformation in the solder region occurs only above the nodular peaks of the intermetallic, so the nodular intermetallics reduce the "actual" volume of solder that undergoes plastic deformation region. Thus, an increase in nodularity will increase the work hardening rate in the solder and contribute to early strain localization and crack formation. Figure 12 shows the significant effect of intermetallic morphology on the stress

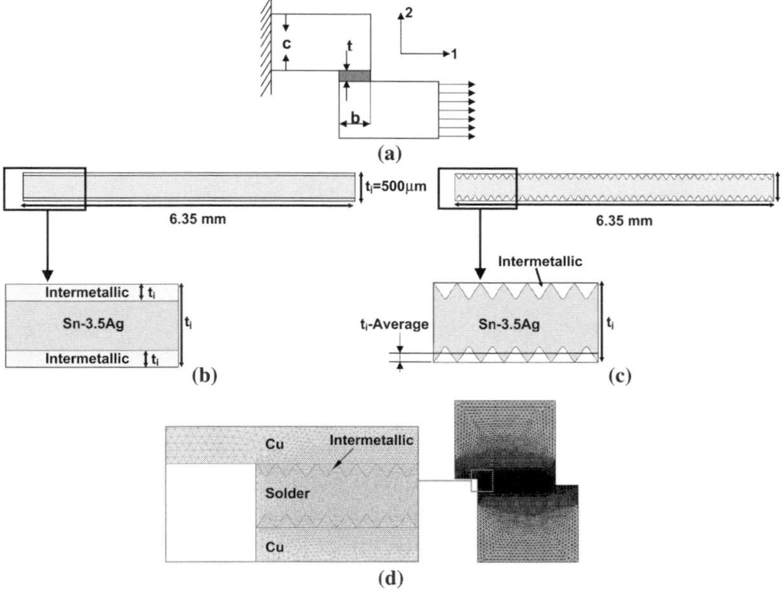

Fig. 9 (**a**) Lap shear configuration, morphology of intermetallics at solder/Cu joints, (**b**) planar and (**c**) nodular intermetallics, and (**d**) mesh detail used in finite element simulations

concentration within the solder joints. Compared with the planar intermetallic, the nodular intermetallic had a significantly higher stress concentration at the nodule valleys, which likely contributed to the onset of crack initiation between intermetallics observed in experiments.

The microstructure of Ag_3Sn particle in the Sn matrix can also be incorporated in the model, Fig. 13. Here a representative micrograph of the solder joint is used as a basis for the model. The microstructure is segmented and duplicated several times, to obtain a larger, more realistic model. This model is then meshed and the analysis conducted. Notice that while the stress contours were relatively homogeneous, when the solder is modeled as a homogeneous solid, the Ag_3Sn particles significantly "disrupt" the contours, Fig. 13(c). This can be rationalized by the fact that

Ag_3Sn particles are much stiffer than pure Sn (~90 GPa for Ag_3Sn [46] versus 50 GPa for Sn [11, 46]). This can be seen in part (d) of Fig. 13, which shows a 3D plot of the stresses in Ag_3Sn, which are much higher than that in the Sn matrix.

3.4 Modeling crack growth in Sn-rich solder joints

Crack growth in Sn-rich solder alloys can also be modeled numerically. Crack propagation in two-dimensional structures can be modeled by finite element analysis using a specialized package called FRANC2D/L (Manhattan, KS) [69, 70]. In FRANC2D/L, crack growth can be achieved by a nodal release and re-meshing method, Fig. 14. The re-meshing method is better suited for Sn-rich alloy systems since the crack path is not known priori. Linear elastic fracture

Fig. 10 FEA shear stress–shear strain curves for the effect of (**a**) planar intermetallic and (**b**) nodular intermetallic with different thickness when the work hardening behavior of solder is included in the simulation. The volume fraction of intermetallics is relative to the solder volume

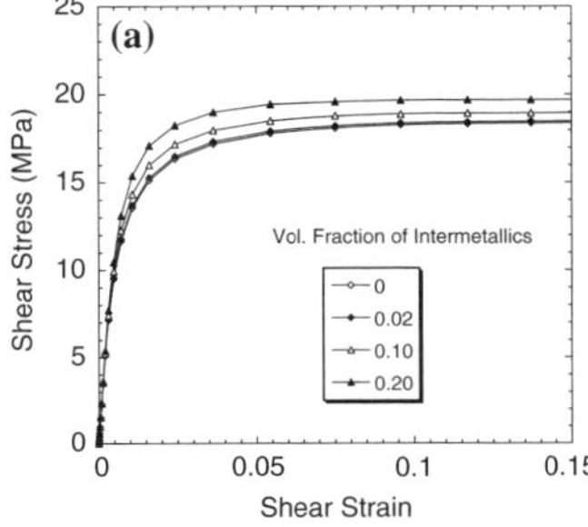

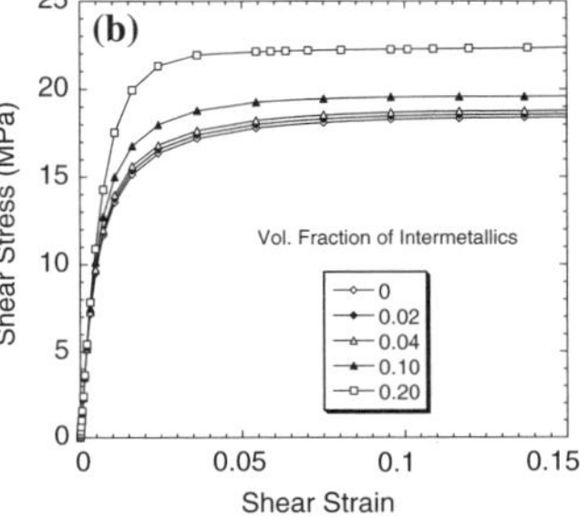

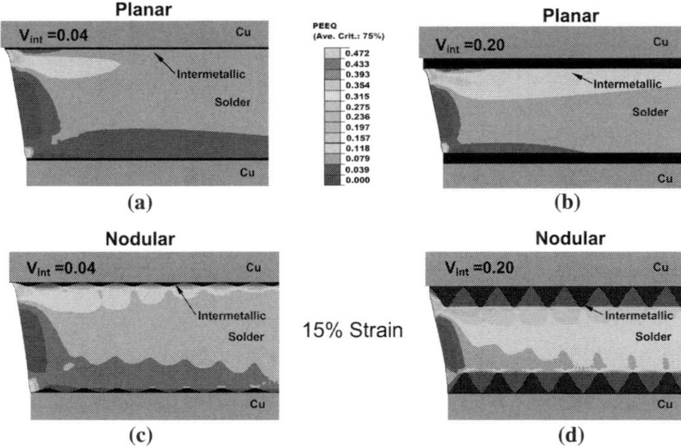

Fig. 11 Equivalent plastic strain distribution in Sn-3.5Ag solder/Cu joint, with different intermetallic morphology and volume fraction, (**a**) planar, *4%*, (**b**) planar, *20%*, (**c**) nodular, *4%*, and (**d**) nodular, *20%*. The volume fraction of intermetallic is relative to the solder region. The contours of plastic deformation apply to the solder only (the deformation of intermetallic and Cu are assumed to be elastic)

mechanics (LEFM) principles were used to propagate the crack and obtain the stress state in the microstructure. Two parameters are required in LEFM for the prediction of crack growth: (a) the stress intensity factor (SIF) that governs the fracture process and (b) the direction of crack propagation. FRANC2D/L is capable of computing the stress intensity factors by three methods—Displacement correlation method, Modified crack closure integral method, and the J-integral method. Modified crack closure integral method [71] was used in this study as the primary method for computation of stress intensity factors. This is an indirect method in which the SIF values (K) are computed after computing the strain energy release rates (ς), Fig. 15(a). The strain energy release rate in Mode I, ς_I, then, is given by:

$$\varsigma_I = \frac{K_I^2}{E} \beta \tag{4}$$

where $\beta = 1$ for plane stress and $\beta = 1 - v^2$ for plane strain. Numerically, ς_I can be evaluated by the following equation:

$$\varsigma_I = \lim_{\Delta c \to 0} \frac{1}{2\Delta c} \bar{F}_c \cdot (v_c - v_d) \tag{5}$$

where F_c is the force at the nodes, Δc is the crack growth increment, and v_c and v_d are the node displacements at the crack tip. The principle behind this method is that when a crack extends by a small amount Δc, the energy absorbed in creating the new crack surface is equal to the work required to close the crack to its original length. The advantage of this method is that both mode I and II stress intensity values can be computed from a single analysis. Crack propagation direction was

computed using the maximum circumferential tensile stress theory ($\sigma_{\theta \, max}$) [72]. This theory states that the crack will grow from the tip in a radial direction along which σ_θ is a maximum and $\tau_{r\theta}$ is zero, Fig. 15(b). This criterion is suitable for tensile dominated fracture, i.e., Mode I situation, and can handle localized mixed mode stress intensity factors. The equations for the stress components at the crack tip are given as follows:

$$\sigma_r = \frac{1}{(2r)^{1/2}} \cos\frac{\theta}{2} \left[k_1 \left(1 + \sin^2\frac{\theta}{2} \right) + \frac{3}{2} k_2 \sin\theta - 2k_2 \tan\frac{\theta}{2} \right] \tag{6}$$

$$\sigma_\theta = \frac{1}{(2r)^{1/2}} \cos\frac{\theta}{2} \left[k_1 \cos^2\frac{\theta}{2} - \frac{3}{2} k_2 \sin\theta \right] \tag{7}$$

$$\tau_{r\theta} = \frac{1}{2(2r)^{1/2}} \cos\frac{\theta}{2} [k_1 \sin\theta + k_2 (3\cos\theta - 1)] \tag{8}$$

The methodology used for modeling microstructure-based crack growth in Sn-rich alloys was as follows. The first step is to obtain a micrograph of high resolution to delineate reinforcement particle morphology. Optical micrographs were imported into a mesh generating software. After the geometry was discretized into a grid of finite elements, it was converted to be readable by FRANC2D/L. Material properties were defined, boundary conditions were applied and finite element analysis was performed. A crack was initiated at the tip of a notch, which was a stress concentration point. Equilibrium was established and the new stress state was obtained. The crack was propagated based on fracture mechanics principles and continued until the final desired crack

Fig. 12 von Mises stress distribution in Sn-3.5Ag solder/Cu joint, with 20% intermetallic of varying intermetallic morphology, (**a**) planar and (**b**) nodular. The volume fraction of intermetallic is relative to the solder region. Note that stress concentrations in the nodule are observed, which likely serve as cites for crack initiation

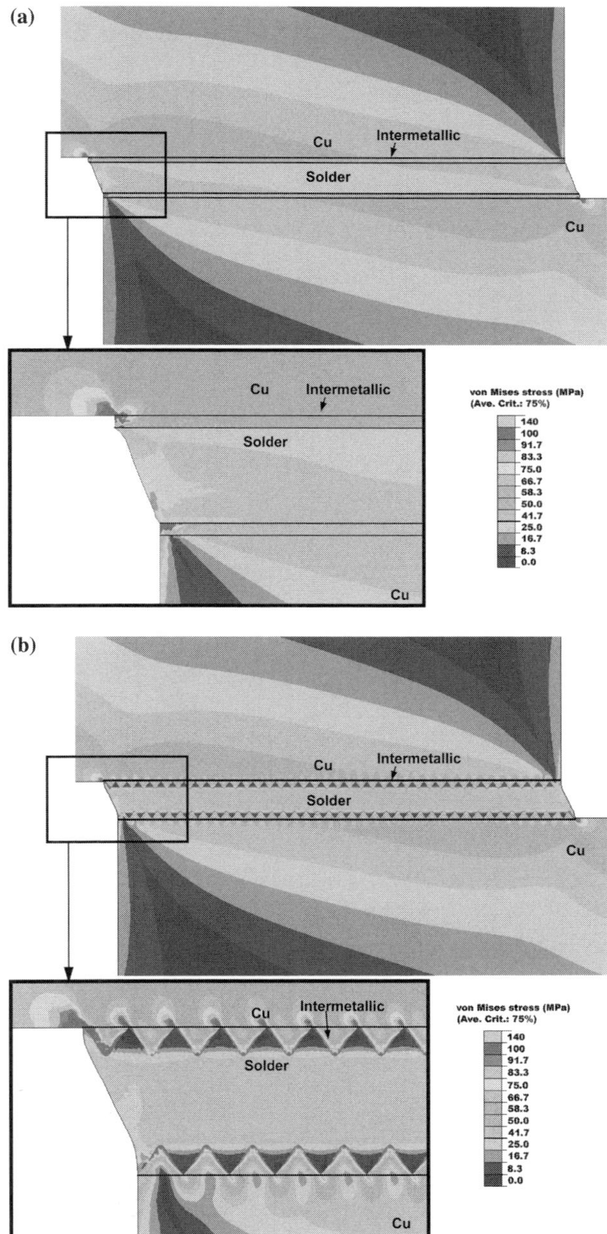

length was obtained. A similar approach was used when modeling crack growth in particle reinforced metal matrix composites [73].

Figures 16 and 17 show predicted crack paths in a solder/Cu joint, with a Cu_6Sn_5 intermetallic formed between solder and Cu. When a single crack is initiated at the tip of the joint (the point of highest stress concentration in the lap shear geometry, as shown in Fig. 12), the crack propagates parallel to the solder/Cu interface. This correlates very well with the

Fig. 13 Microstructure based finite element modeling (2D): (**a**) scanning electron micrograph, (**b**) meshed model microstructure, (**c**) equivalent plastic strain in lap shear (**d**) von Mises stress distribution in lap shear

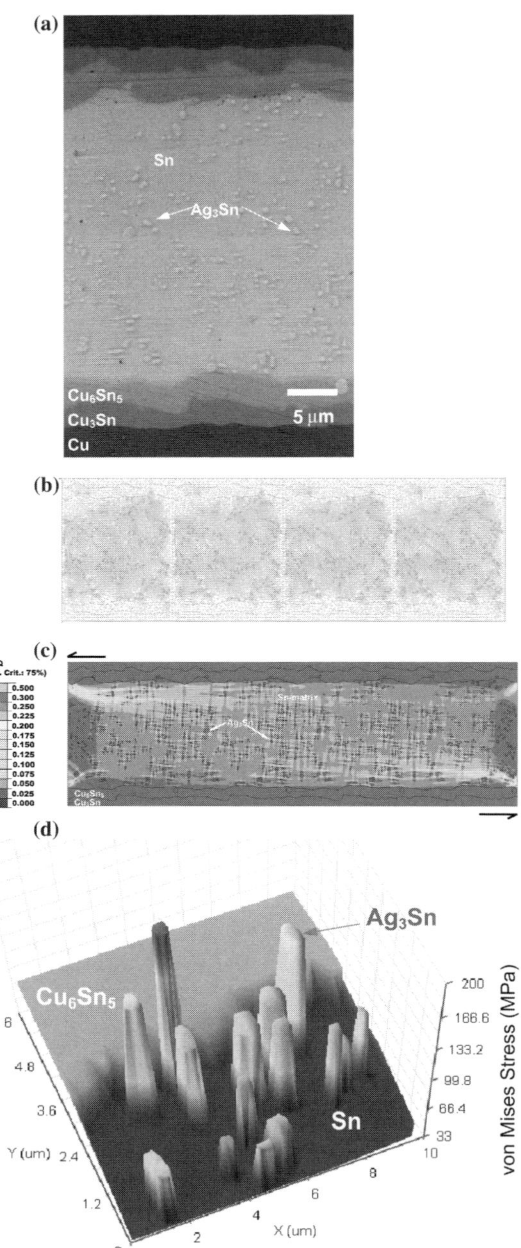

experimentally observed behavior by Kerr and Chawla [5], Figs. 16(b) and 17(b). In some cases, experimental observations show that the crack propagates at an angle. This can be rationalized by a model where the crack tip initiates at both tips of the joint, on opposite sides. Here the cracks propagate at an angle and meet in the center of the joint, as shown experimentally.

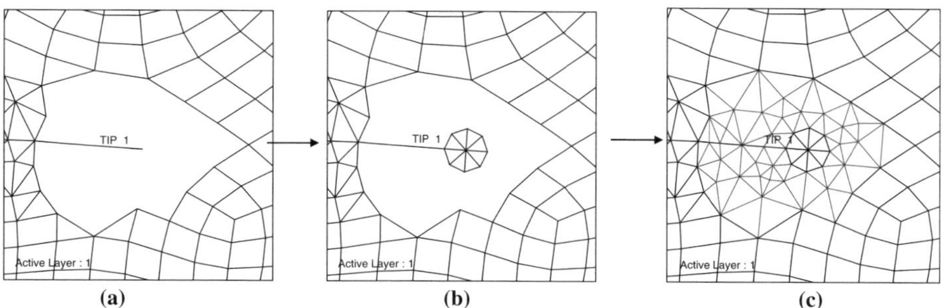

Fig. 14 Crack growth using the re-meshing methodology. (**a**) elements in the vicinity of the crack tip are deleted (**b**) crack-tip is extended and rosette is formed (**c**) trial mesh is inserted

Fig. 15 (**a**) Finite element nodes near the crack tip using the Modified Crack Closure Integral method [71], (**b**) stress components near the crack tip in cylindrical coordinates [72]

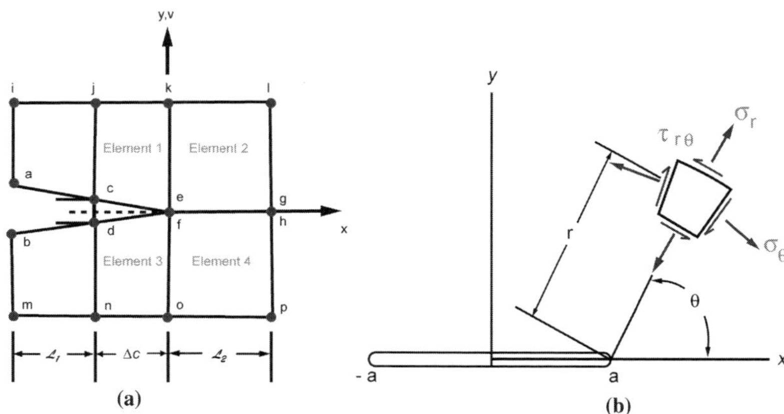

Fig. 16 (**a**) Crack profile predicted by FEM for single crack initiation, (**b**) experimental single crack initiation results in crack propagating close to the solder/intermetallic interface and around the Ag_3Sn particles

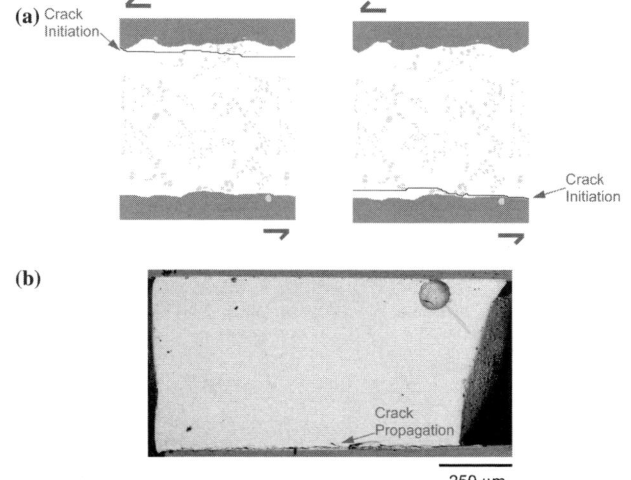

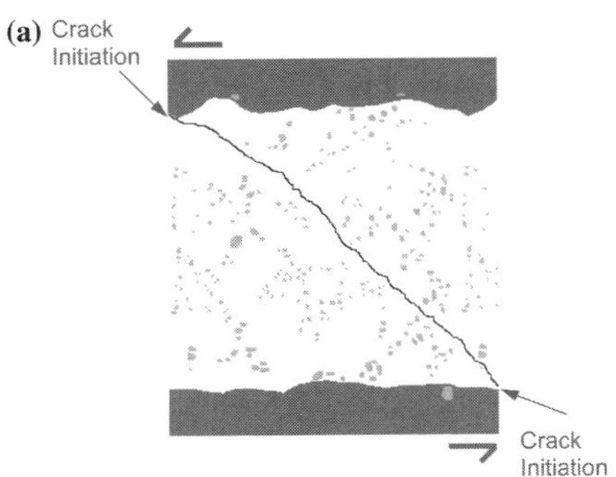

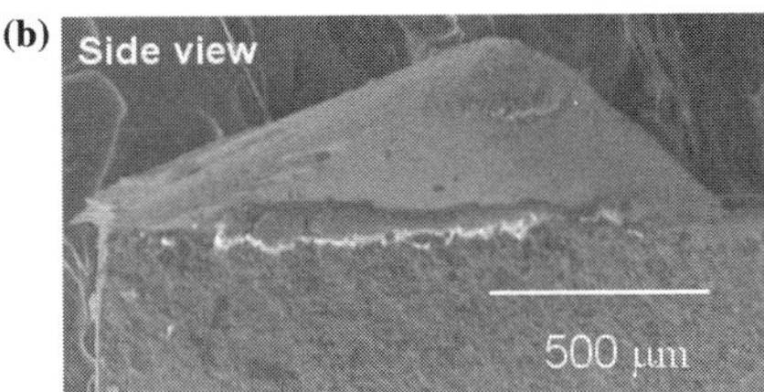

Fig. 17 (**a**) Prediction of crack propagation when cracks initiation occurs at both ends, (**b**) Experimental fracture surfaces corroborating the FEM results

4 Summary

The mechanical properties of Sn-rich solder alloys are directly related to their heterogeneous microstructure. Thus, numerical modeling of the properties of these alloys is most effective when the microstructure is explicitly incorporated into the model. This review, provides several examples where 2D and 3D microstructures have been used to model the material behavior using finite element modeling. These included (a) 3D visualization of the solder microstructure, (b) 3D microstructure-based modeling of tensile behavior, (c) 2D modeling of the effect of intermetallic volume fraction and morphology on shear behavior of solder joints, and (d) prediction of crack growth in solder joints. In all these cases, the experimentally observed behavior matched very well with the microstructure-based models.

Acknowledgements The authors gratefully acknowledge financial support for support of this research from the National Science Foundation under contract #DMR-0092530 (Drs. H. Chopra, S. Ankem, B. Macdonald and K.L. Murty, program managers).

References

1. S. Kang, A.K. Sarkhel, J. Electron. Mater. **23**, 701 (1994)
2. D.R. Frear, P.T. Vianco, Metall. Mater. Trans. A **25**, 1509 (1994)
3. J. Glazer, Inter. Mater. Rev. **40**, 65 (1995)
4. W.J. Plumbridge, C.R. Gagg, Proc. Inst. Mech. Engrs. L, J. Mater.: Des Appl. **214**, 153 (2000)
5. M. Kerr, N. Chawla, Acta Mater. **52**, 4527 (2004)
6. F. Ochoa, J.J. Williams, N. Chawla, J. Electron. Mater. **32**, 1414 (2003)
7. F. Ochoa, J.J. Williams, N. Chawla, JOM **55**, 56 (2003)
8. R.J. McCabe, M.E. Fine, Scripta Mater. **39**, 189 (1998)
9. H. Rhee, J.P. Lucas, K.N. Subramanian, J. Mater. Sci. **13**, 477 (2002)
10. C. Basaran, J. Jiang, Mech. Mater. **34**, 349 (2002)
11. N. Chawla, F. Ochoa, S. Scaritt, M. Koopman, K.K. Chawla, V.V. Ganesh, X. Deng, J. Mater. Sci.: Mater. Electron. **15**, 385 (2004)
12. N. Ramakrishna, V.S. Arunachalam, J. Am. Ceram. Soc. **76**, 2745 (1993)
13. W.M. Sherry, J.S. Erich, M.K. Bartschat, F.B. Prinz, IEEE Trans. Comp., Hybrids Manuf. Tech. **8**, 417 (1985)
14. D.G. Kim, H.S. Jang, J.W. Kim, S.B. Jung, J. Mater. Sci.: Mater. Elec. **16**, 603 (2005)
15. S. Ling, A. Dasgupta, Trans. ASME **118**, 72 (1996)
16. D.G. Kim, J.W. Kim, S.B. Jung, Microelec. Eng. **82**, 575 (2005)
17. J.W. Kim, D.G. Kim, S.B. Jung, Microelec. Rel. **46**, 535 (2006)
18. J.W. Kim, S.B. Jung, Microelec. Eng. **82**, 554 (2005)
19. H. Ye, C. Basaran, D.C. Hopkins, Inter. J. Solid. Struct. **41**, 4959 (2004)
20. C.J. Zhai, Sidharth, R. Blish II, IEEE Trans. Device Mater. Rel. **3**, 207 (2003)
21. M.P. Rodriquez, N.Y.A. Shammas, A.T. Plumpton, D. Newcombe, D.E. Crees, Microelec. Rel. **40**, 455 (2000)
22. J.H. Lau, IEEE Trans. Comp., Pack., Manuf. Tech. B **19**, 728 (1996)
23. V. Sarihan, IEEE Trans. Comp., Pack., Manuf. Tech. B **17**, 626 (1994)
24. C.G. Schmidt, J.W. Simons, C.H. Kanazawa, D.C. Elrich, IEEE Trans. Comp., Pack., Manuf. Tech. A **18**, 611 (1995)
25. B.Z. Hong, J. Elec. Mater. **26**, 814 (1997)
26. B.Z. Hong, J. Elec. Mater. **28**, 1071 (1999)
27. B.Z. Hong, L.G. Burrell, IEEE Trans. Comp., Pack., Manuf. Tech. A **18**, 585 (1995)
28. S.C. Chen, Y.C. Lin, C.H. Cheng, J. Mater. Proc. Tech. **171**, 125 (2006)
29. E.E. Underwood, in *Quantitative Microscopy*, ed. by R.T Dehoof, F.N. Rhines (McGraw-Hill, New York, 1968), p. 149
30. B. Wunsch, X. Deng, N. Chawla, in *Computational Methods in Materials Characterisation*, ed. by A.A. Mammoli, C.A. Brebbia (WIT Press, Boston, 2004), pp. 175–184
31. R.S. Sidhu, N. Chawla, Mater. Charact. **52**, 225 (2004)
32. M. Li, S. Ghosh, T.N. Rouns, H. Weiland, O. Richmond, W. Hunt, Mater. Charact. **41**, 81 (1998)
33. M. Li, S. Ghosh, O. Richmond, H. Weiland, T.N. Rouns, Mater. Sci. Eng. A **265**, 153 (1999)
34. M.V. Kral, M.A. Mangan, G. Spanos, R.O. Rosenberg, Mater. Charact. **45**, 17 (2000)
35. T. Yokomizo, M. Enomoto, O. Umezawa, G. Spanos, R.O. Rosenberg, Mater. Sci. Eng. A **344**, 261 (2003)
36. C.Y. Hung, G. Spanos, R.O. Rosenberg, M.V. Kral, Acta Mater. **50**, 3781 (2002)
37. A.C. Lund, P.W. Voorhees, Acta Mater. **50**, 2582 (2002)
38. K.M. Wu, M. Enomoto, Scripta Mater. **46**, 569 (2002)
39. M. Yamaguchi, S.K. Biswas, Y. Suzuki, H. Furukawa, K. Takeo, FEMS Microbio. Lett. **219**, 17 (2003)
40. A. Tewari, A.M. Gokhale, Mater. Charact. **46**, 329 (2001)
41. M.V. Kral, G. Spanos, Acta Mater. **47**, 711 (1999)

42. J. Alkemper, P.W. Voorhees, Acta Mater. **49**, 897 (2001)
43. N. Chawla, K.K. Chawla, J. Mater. Sci. **41**, 913–925 (2006)
44. N. Chawla, V.V. Ganesh, B. Wunsch, Scripta Mater. **51**, 161 (2004)
45. N. Chawla, R.S. Sidhu, V.V. Ganesh, Acta Mater. **54**, 1541 (2006)
46. X. Deng, N. Chawla, K.K. Chawla, M. Koopman, Acta Mater. **52**, 4291 (2004)
47. W. Yang, L.E. Felton, R.W. Messler, J. Electron. Mater. **24**, 1465 (1995)
48. K.N. Tu, R.D. Thompson, Acta Metall. **30**, 947 (1982)
49. X. Deng, G. Piotrowski, J.J. Williams, N. Chawla, J. Electron. Mater. **32**, 1403 (2003)
50. K.H. Prakash, T. Sritharan, Acta Mater. **49**, 2481 (2001)
51. W.K. Choi, H.M. Lee, J. Electron. Mater. **29**, 1207 (2000)
52. F. Guo, S. Choi, J.P. Lucas, K.N. Subramanian, J. Electron. Mater. **29**, 1241 (2000)
53. D. Ma, W.D. Wang, S.K. Lahiri, J. Appl. Phys. **91**, 3312 (2002)
54. C.R. Kao, Mater. Sci. Eng. A **238**, 196 (1997)
55. S. Chada, R.A. Fournelle, W. Laub, D. Shangguan, J. Electron. Mater. **29**, 1214 (2000)
56. Z. Mei, A.J. Sunwoo, J.W. Morris Jr, Metall. Trans. A **23**, 857 (1992)
57. W.K. Choi, H.M. Lee, J. Electron. Mater **29**, 1207 (2000)
58. H. Lee, M. Chen, H. Jao, T. Liao, Mater. Sci. Eng. A **358**, 134 (2003)
59. Y.C. Chan, A.C.K. So, J.K.L. Lai, Mater. Sci. Eng. B **55**, 5 (1998)
60. H.L.J. Pang, K.H. Tan, X.W. Shi, Z.P. Wang, Mater. Sci. Eng. A **307**, 42 (2001)
61. H.W. Miao, J.G. Duh, Mater. Chem. Phys. **71**, 255 (2001)
62. P. Protsenko, A. Terlain, V. Traskine, N. Eustathopoulos, Scripta Mater. **45**, 1439 (2001)
63. D.R. Frear, JOM **48**, 49 (1996)
64. R.E. Pratt, E.I. Stromswold, D.J. Quesnel, J. Electron. Mater. **23**, 375 (1994)
65. C.K. Alex, Y.C. Chan, IEEE Trans. CPMT-B **19**, 661 (1996)
66. P.L. Tu, Y.C. Chan, J.K.L. Lai, IEEE Trans. CPMT-B **20**, 87 (1997)
67. X. Deng, R.S. Sidhu, P. Johnson, N. Chawla, Metall. Mater. Trans A **36**, 55 (2005)
68. X. Deng, M. Koopman, N. Chawla, K.K. Chawla, Mater. Sci. Eng. **364**, 241 (2004)
69. M.A. James, D. Swenson, FRANC2D/L: A Crack Propagation Simulator for Plane Layered Structures, available from http://www.mne.ksu.edu/~franc2d/.
70. V.V. Ganesh, N. Chawla, Mater. Sci. Eng. A **391**, 342 (2005)
71. E.F. Rybicki, M.F. Kanninen, Eng. Frac. Mech. **9**, 931 (1977)
72. F. Erdogan, G.C. Sih, J. Basic Eng. (1963) 519
73. A. Ayyar, N. Chawla, Comp. Sci. Tech. **66**, 1980 (2006)

Deformation behavior of tin and some tin alloys

Fuqian Yang · J. C. M. Li

Published online: 13 October 2006
© Springer Science+Business Media, LLC 2006

Abstract Plastic deformation, creep and deformation twinning of β-tin and some tin alloys related to Pb-free solder applications are reviewed. The results are summarized and evaluated among conflicting findings and conclusions. The studies are helpful for the search of the best Pb-free solder with reliability and long service life. The areas which need more information are pointed out.

1 Allotropic forms, bct unit cell, slip systems, elastic constants and self-diffusion

1.1 Allotropic forms

Tin can exist in 3 allotropic forms, α the gray diamond cubic tin, β the white tetragonal tin and γ the rhombic tin. As mentioned by Tyte [1] the transition between the first two is at 18°C and that between the latter two is at 202.8°C. However, the Wikipedia (http://www.en.wikipedia.org/wiki/Tin) listed the first transition at 13.2°C. The transition from white tin to gray tin upon cooling is called "tin pest" or "tin disease" which can be prevented by adding impurities such as Sb or Bi. The effect of cyclic deformation on the $\beta \rightarrow \alpha$ transformation was investi-

gated by Löhberg and Moustafa [2] with decreased incubation time and increased growth rate and transformed area. For solder applications, the bct (body centered tetragonal) β tin is the most important.

1.2 The bct unit cell

The bct unit cell is shown in Fig. 1 taken from Chu and Li [3]. The dimensions are $a = 0.58194$ nm and $c = 0.31753$ nm ($c/a = 0.54564$). In addition to the 8 atoms at the corners and one at the center, there are 4 more atoms on the four faces with locations such as [0, 1/2, 3/4] and [1/2, 0, 1/4]. So there are 4 atoms per unit cell. Each atom has 4 nearest neighbors at 0.30160 nm, 2 next nearest neighbors at $c = 0.31753$ nm and 8 third nearest neighbors at 0.44106 nm. The lattice is a distorted diamond cubic lattice (see [4]). When $c/a = \sqrt{2}$ it is equivalent to the diamond cubic lattice.

1.3 Thermal expansion

Subramanian and Lee [5] quoted from Barrett [6] the coefficient of linear thermal expansion as 15.4×10^{-6} in the a direction and 30.5×10^{-6} in the c direction. They believe that this difference in thermal expansion will cause internal stress to develop in solders from thermal excursions.

1.4 Slip systems

As shown in Fig. 1 there are a total of 6 possible common families of slip systems [7] and [8]:

(a) $(110)[\bar{1}11]$ (b) $(110)[001]$ (c) $(100)[010]$
(d) $(100)[001]$ (e) $(101)[\bar{1}01]$ (f) $(121)[\bar{1}01]$

F. Yang
Department of Chemical and Materials Engineering,
University of Kentucky, Lexington, KY 40506, USA

J. C. M. Li (✉)
Department of Mechanical Engineering, University
of Rochester, Rochester, NY 14627, USA
e-mail: li@me.rochester.edu

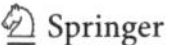

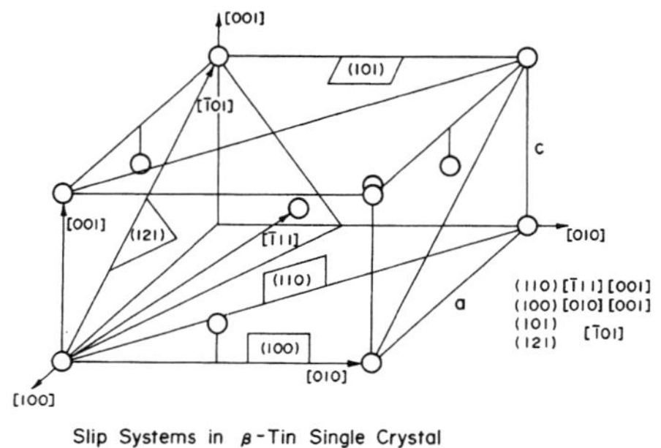

Fig. 1 The unit cell of bct tin and some common slip systems, Chu and Li [3]

However from the slip line observations around an indenter, Chu and Li [3] showed that only the first 3 are possible. Tyte [1] mentioned some earlier work of Mark and Polanyi [9] who found that type (b) was the most frequent but all others were possible. Obinata and Schmid [10] repeated their experiments and found types (b) and (d) are most frequent but types (e) and (f) are also possible. Weertman and Breen [11] observed type (c) in tensile creep of [110] single crystals. Kirichenko and Soldatov [12] did tensile creep at below 78 K for single crystals oriented near [110] and found types (a), (c), (e) and (f) slip systems operating. Ojima and Hirokawa [13] found the slip system of type (c) operating in a tensile test of single crystals. They found that these dislocations can form polygonized walls at 15°C. Nagasaka [14] showed in a tensile test of single crystals slip systems for types (a) and (c). From X-ray topography, Fiedler and Lang [15] found dislocations in bct tin with only [001] and 0.5[111] types consistent with the calculations of Fiedler and Vagera [16] that these two types have the lowest energy based on Stroh's [17] anisotropic elasticity results. More recently Duzgun and Aytas [18] summaried previous results including the work of Honda [19–21] with 6 more possibilities:

(g) $(100)[011]$ (h) $(101)[11\bar{1}]$ (i) $(101)[010]$
(j) $(001)[100]$ (k) $(001)[110]$ (l) $(121)[1\bar{1}1]$

However, their own observation included only (c), (f) and (g) types. Nagasaka [14] also observed the (g) types. For the size of the Burgers vector, $b[111] = 0.4411$ nm, $b[001] = 0.3157$ nm, $b[100] = 0.5819$ nm and $b[101] = 0.66293$ nm. Hence it is likely only the first 3 families of slip systems and the type (g) are operating at low temperatures and the rest at high temperatures and/or high stresses.

1.5 Elastic constants

The Young modulus of polycrystalline tin is given by Rotherham et al. [22]:

$$E(\text{MPa}) = 76087 - 109T(\text{K}) \tag{1}$$

More recently Fraizier [23] used laser ultrasonics to determine the adiabatic moduli of tin at different temperatures and presented the following results in GPa between 305 K and 501 K:

$$B = 70.5 - 15.6T/T_\text{m} \quad E = 71.0 - 32.4T/T_\text{m}$$
$$G = 26.5 - 12.6T/T_\text{m} \tag{2}$$

where T_m is the absolute melting temperature, 505.08 K, B is the bulk modulus, E is Young's modulus and G is shear modulus. They also presented the following densities for solid and liquid tin in kg/m^3 between 100 K and 500 K:

$$\rho_\text{S} = \rho_0/[1 + \alpha_\text{th}(T - T_0)]^3$$
$$\rho_\text{L} = 6980 - 0.61(T - T_\text{m}) \tag{3}$$

where T_0 is 293 K and ρ_0 is 7,282 kg/m^3. The density of liquid tin is from Takamishi and Roderick [24].

Mathew et al. [25] calculated the 2nd and 3rd order elastic constants for bct tin from deformation theory and compared with experimental results. His results for the 2nd order elastic constants are given below in GPa:

$c_{11} = 73.5(73.5, 83.7)$ $c_{12} = 44.2(23.4, 48.7)$
$c_{13} = 40.7(28, 28.1)$ $c_{33} = 103.0(87, 96.7)$
$c_{44} = 38.3(22, 17.5)$ $c_{66} = 42.8(22.65.7.41)$

The experimental results are in the parentheses in which the first one is from Mason and Bommel [26] and the second one is from Bridgeman [27]. The 3rd order elastic constants in GPa are:

$c_{111} = -410.0(-410)$ $c_{112} = -246.9(-553)$
$c_{113} = -202.3(-467)$ $c_{123} = -100.2(128)$
$c_{133} = -259.4(-156)$ $c_{144} = -76.0(-162)$
$c_{155} = -208.1(-177)$ $c_{166} = -242.3(-191)$
$c_{333} = -929.5(-1427)$ $c_{344} = -200.4(-212)$
$c_{366} = -72.2(-78)$ $c_{456} = -72.2(-52)$

The experimental values in the parentheses are from Swartz et al. [28]. Based on finite strain elasticity the pressure derivatives of the 2nd order elastic constants are:

$$dC'_{11}/dp = 3.32(7.49, 6.48) \quad dC'_{12}/dp = 3.46(6.71, 5.61)$$
$$dC'_{13}/dp = 3.04(--, 2.96) \quad dC'_{33}/dp = 5.68(9.87, 8.87)$$
$$dC'_{44}/dp = 1.68(3.22, 2.22) \quad dC'_{66}/dp = 2.30(1.61, 1.61)$$

The values in the parentheses are from Swartz et al. [28] measured experimentally and from Rao and Padmaja [29] calculated using the same equations. Since c_{33} is greater than c_{11} the bindings between atoms in the ab plane are somewhat weaker than along the c direction. Furthermore, since c_{333} has a higher magnitude than others, it shows a greater anharmonicity along the c axis.

1.6 Self-diffusion in β-Sn

Self-diffusivities in β-Sn were measured first by Fensham [30] using a radioactive isotope Sn^{113} which had a half life of about 100 days. During preparation of the isotope by neutron irradiation, another radioactive species Sn^{125} was formed also which decayed in 9 min to become Sb^{125} by emitting a β ray. This impurity Sb^{125} must be carefully removed by precipitation. Single crystals of pure tin (99.998%) were grown, cut into samples and annealed at 150°C for a few hours. One face of the crystal was electro-polished before the application of the isotope layer. Their results showed the self-diffusivity in the c direction 2–3 times higher than that perpendicular to the c direction. This is inconsistent with the fact that binding in the c direction is stronger than that perpendicular to the c direction. The activation energies are 25 kJ/mol in the c direction and 44 kJ/mol perpendicular to the c direction. These energies did not agree with the activation energies for high temperature creep.

In view of these discrepancies, Meakin and Klokholm [31] repeated these experiments. Contrary to earlier findings, they found the self-diffusivity in the c direction only about 50% of that perpendicular to the c direction. The activation energies were 107 J/mol in the c direction and 98 J/mol perpendicular to the c direction. These are more consistent with atomic bindings and creep data. However in the same year, Chomka and Andruszkiewicz [32] reported diffusivities of Sn, Zn and Co in single and polycrystals of β-tin. While they did not differentiate between c and a directions, they found self-diffusivity of Sn in polycrystals slightly larger (about 50%) than that in single crystals. But the activation energy was 45 kJ/mol, more in line with Fensham's data perpendicular to the c direction. So the discrepancy remained.

In 1964, Coston and Nachtrieb [33] determined the activation volume for self-diffusion of Sn by measuring the self-diffusivity under high pressure. They grew large single crystals of pure tin (99.999%). The quality was checked with a modified Laue back-reflection technique with a beam large enough to cover a 1 cm^2 area. Only crystals showing nearly perfect spots were used. They found that the self-diffusivity in the c direction is about 45% of that perpendicular to the c direction. This ratio appeared independent of pressure up to 10 kbars for about 30 measurements. The activation energies were 107 kJ/mol in the c direction and 105 kJ/mol perpendicular to the c direction. These findings were consistent with those of Meakin and Klokholm [31].

However in 1967, Pawlicki [34] studied self-diffusion in polycrystals of tin by measuring continuously the radioactivity on the other side of the specimen without sectioning. The sample was a thin disk of 2 cm diameter and 0.1 mm thick. The starting grain size was 0.5 mm but annealing for 4 h at 200°C made them grow to about 1 mm. Thus the grains were much larger than the thickness of the specimen implying only one grain through the thickness. The activation energy was found to be 32 kJ/mol.

In an attempt to understand the discrepancies among the various measurements, all the published data are plotted in Fig. 2. While Pawlicki [34] claimed that his activation energy agreed with earlier results of Fensham [30] and Chomka and Andruszkiewicz [32] by ignoring the work of Meakin and Klokholm [31] and that of Coston and Nachtrieb [33], his diffusivities were an order of magnitude higher than earlier results. It is likely that Pawlicki was measuring the self-diffusivity in the grain boundaries which traversed the thickness of his specimen. On the other hand, the results of Meakin and Klokholm [31] agreed very well with those of Coston and Nachtrieb [33] both in the c direction and perpendicular to the c direction. Both sets of data covered a large temperature range with a single activation energy. The crystals were nearly perfect in both cases, especially those of Coston and Nachtrieb [33] who took special caution to make sure nearly perfect crystals by X-ray inspection of a large area.

The other sets of data did not agree with each other except those of Chomka and Andruszkeiwicz [32] for single crystals whose high temperature data seemed to approach those of Meakin and Klokholm [31] or those of Coston and Nachtrieb [33]. The high temperature data of Fensham [30] perpendicular to the c direction behaved similarly. Such behavior is typical of two parallel processes with different activation energies taking place simultaneously (see [35]). One of them is the lattice self-diffusion and the other is most likely the self-diffusion along individual dislocations or along the

 Springer

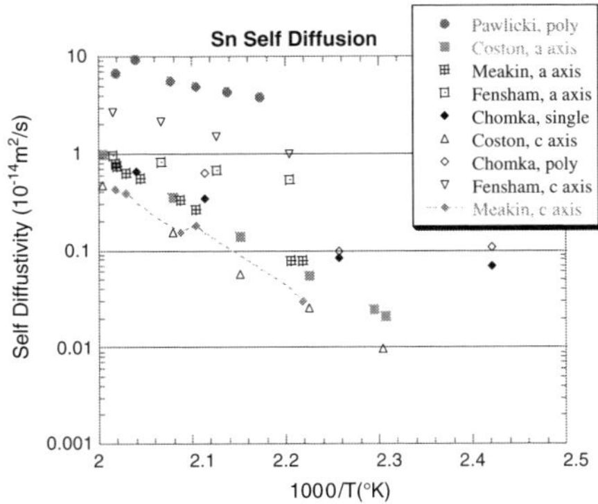

Fig. 2 Self-diffusion of β-Sn [30–34]

subboundary dislocations. The transition temperature will depend on the density of these short-cut paths. For the case of Chomka and Andruszkeiwicz [32] and the case of Fensham [30] perpendicular to the c direction, the density of fast paths was low so that the transition temperature was about 500 K. However, for the case of Pawlicki [34] and the case of Fensham [30] along the c direction, the density of fast paths was high so that the transition temperature was even higher than 500 K. We suggest that the diffusion experiments for single crystals be repeated with different levels of dislocation densities.

Our conclusion is that the best set of lattice self-diffusion data is those of Coston and Nachtrieb [33] who showed that the self-diffusivities along the c direction were about 45% slower than those perpendicular to the c direction for all temperatures and pressures. The activation volume is about a third of atomic volume, 5.3 ± 0.3 cc/g-atom, independent of temperature. This activation volume agrees with that for creep as measured by DeVries et al. [36] (see Section 3.6 later). The activation energy is about 106 kJ/mol for both directions which agrees with that obtained for high temperature (>150°C) creep.

2 Low temperature deformation

Natsik et al. [37] studied tensile creep of single crystal β-Sn in the temperature range of 0.45–4.2 K for the slip system of type (c) (see Section 1.4 above) and concluded that below about 1 K, the flow stress is athermal at about 10 MPa. They thought that the nucleation of kink pairs occurs by a quantum tunneling effect through the Peierls barrier. The deformation was in a normal state (rather than superconducting) by using an

above-critical longitudinal magnetic field. By examining the logarithmic creep behavior, the mechanism above 1 K appeared different from that below 1 K and could be the thermal nucleation of double kinks. The estimated energy barrier was 0.005 eV or 480 J/mol.

The effect of superconducting transition on the creep behavior was reported by Soldatov et al. [38]. Briefly when a normal conducting single crystal β-Sn was creeping under stress at 1.6 K, a transition to superconducting by removing the magnetic field caused a sudden increase in creep strain of about 0.5% which decreased with increasing prior strain. They claimed to have 3 stages and discussed some more in a subsequent publication [39]. Nothing happened in the reverse transition from superconducting to the normal state.

Earlier Kirichenko and Soldatov [12] made tensile creep measurements at 1.5–78 K for single crystals of pure (99.9995) tin near [110] orientation with possible slip systems of types (c) and (f). They showed logarithmic creep at each temperature and different amounts of prior strain at the same temperature. The temperature dependence of flow stress had a sudden drop at 65 K. Later Kirichedo et al. [40] showed solid solution hardening with Cd up to 0.53 at.% with the sudden drop disappeared completely at this concentration. More recently Diulin et al. [41] studied deformation of Sn single crystals with Cd, In and Zn impurities between 1.6 K and 150 K with the slip system of type (c) operating. The extra hump at about 60 K in the yield stress–temperature relation found in pure tin still existed for alloys of low concentration even though the yield stress for alloys was lower (solid solution softening). However, the hump disappeared when the concentration of impurity was sufficiently high, 0.53 at.% for all three alloys. The stress dependence of activation strain volume had a peak at about 6 K for pure tin. This peak shifted to higher temperature for low concentration alloys and disappeared for alloys of sufficiently high concentration, 0.53 at.% for all three alloys.

Fujiwara and Hirokawa [42] deformed single crystal of $2 \times 5 \times 75$ mm³ of 99.999%Sn in tension at 288 K with the 75 mm direction parallel to [110] and the 5×75 face parallel to (001). The (001) surface was etched after deformation to count the dislocations. Then a layer of the surface (10–20 μm) was removed by chemical polishing and etched again to reveal the dislocations inside the crystal. They found that the density of dislocations was high near the surface than in the interior by a factor or 10 or more. The ratio was less if the deformation took place in a chemical polishing solution which removed the surface at 250 nm/min. The dislocation density on the surface was higher for higher applied stress. The surface hardening layer was about 60–80 μm. This implies that the dislocations are initiated from the surface sources and any indentation test especially nano indenters will sense only the surface layer with high dislocation densities.

Fujiwara [43] made indentations on (001), (554) and (110) surfaces of tin (99.999%) by conical, Vickers and a wedge shaped indenters. For the conical indenter, they observed possible active slip systems of types (b) and (d) on (001) and (554) surfaces and of types (a), (c) and (g) on (554) and (110) surfaces. For the Vickers indenter on the (110) surface, the slip systems did not seem to change with the orientation of the indenter. For a wedge shaped indenter on the (110) surface, the slip systems just under the indenter seemed to be of types (c) and (g) and those near the pileup surface type (a). The slip systems changed when the indenter sank further into the surface.

Nagasaka [14] showed tensile stress-strain curves at 1.7×10^{-5}/s for 3 different orientations of single crystal tin between 77 (liquid nitrogen) and 435 K. In one crystal, the slip systems were of type (g) at 77 K but changed to type (c) at 200, 295 and 435 K. For another crystal oriented at [100] the slip system was of type (g) at 77 K but changed to type (a) at 160, 295 and 435 K. The slip system of type (a) appeared to have the strongest temperature dependence. The critical resolved shear stress changed from 11 MPa at 160 K to 50 KPa at 435 K. Later Nagasaka [44] studied the strain rate $(4 \times 10^{-5} - 4 \times 10^{-3}$/s) and temperature (200–500 K) dependences of yield stress for a single crystal of [100] orientation. The slip system of type (a) was found operating. He reported two yield points below 300 K, the first appeared to arise from a sudden multiplication of dislocations and the second the dissociation of the cell structure. The strain rate $(\dot{\gamma})$ and temperature (T) effects gave the following stress exponents n and activation energy Q:

Temperature (K)	First yield point		Second yield point	
	n	Q, kJ/mol	n	Q, kJ/mol
200–250	4.2	48	–	–
250–480	3.8	48	3.9	48
480–500	3.8	128	3.9	128

Note that Q was obtained by plotting $\log(\dot{\gamma}T)$ versus $1/T$.

Ekinci et al. [45] showed stress–strain curves in a tensile test at 293 K for single crystals of 3 different orientations [001], [110] and [$1\bar{1}0$] in a range of strain rates between 1×10^{-4}/s to 4.5×10^{-3}/s. From the strain rate dependence of yield stress a stress exponent of 4.5 can be obtained for the [001] orientation. The slip lines were observed but no slip systems determined.

3 Creep deformation

3.1 Uniaxial creep

Tyte [1] conducted the constant-load creep test of tin wire in the stress range of 0.73–12.83 MPa and the temperature range of 292–480 K. A constant rate of extension was observed in the small strain range at each stress and temperature. A plot of the log rate of extension versus load showed two straight line portions with different temperature dependences. His empirical equations relating creep rate with stress and temperature did not yield any quantities with physical meaning since at the time our understanding of creep was still very limited. Using Tyte's data Breen and Weertman [46] found that the activation energy was around 31 kJ/mol at low temperatures (below about 150°C).

Breen and Weertman [46] reported the constant-stress creep test of polycrystalline tin in the stress range of 4.34–9.61 MPa and the temperature range of 294.1–497.5 K. The composition of the specimen was 99.9%Sn, 0.01%Cu, 0.01%Pb, 0.01%Bi, 0.01%Fe. Two groups of specimens were used with different grain sizes, 53.0 grains/mm^2 and 715 grains/mm^2. Little transient creep was observed. The creep rate of the fine grain tin was much higher than that of the coarse grain tin. They demonstrated the difference between the constant-load creep and constant-stress creep. Clearly, there was no steady state creep for the constant-load creep due to the change of cross-sectional area. From the creep rate at different temperatures and tensile stresses, they were able to determine the temperature dependence of the steady-state creep. Two activation energies were obtained of 46 kJ/mol at low temperatures and 109 kJ/mol in the temperature range of 363–433 K. The stress exponent was 4.6 at low temperatures.

Using the constant-stress creep under tensile loading, Weertman and Breen [11] determined the activation energy of the creep deformation for a series of tin single crystals oriented with the specimen axis close to a [110] direction in the temperature range of 304–492 K and the stress range of 0.6–6 MPa. Specimens oriented with the specimen axis near a [110] direction deformed by extensive double slip at all temperatures on the two type (c) slip systems. Two activation energies were found. One was 92.1 kJ/mol for temperature above 403 K, and the other 46.1 kJ/mol for temperature below 403 K. The stress exponent varied from

3.6 to 5.1. Later, Weertman [47] studied the compressional creep of tin single crystals and obtained similar results. There were two activation energies, one was 102.6 kJ/mol for temperature above 393 K, and the other 50.2 kJ/mol for temperature below 393 K.

Frenkel et al. [48] evaluated the creep deformation of tin under constant load in the temperature range of 298–478 K. They obtained the activation energy of 87.9 kJ/mol, which was close to the activation energy, 80.4 kJ/mol, of the grain boundary shearing given by Puttick and King [49].

Using the technique of temperature changes during a creep test, Wiseman et al. [50] determined the activation energy of the creep deformation of tin in the forms of single crystals and polycrystals. The testing temperatures were in the range of 351–394.7 K for single crystals and 320–325 K for polycrystals. Both single crystals and polycrystalline tin had the same activation energy of 96.2 kJ/mol. They suggested that the creep deformation was controlled by the grain-boundary shearing arising from crystallographic mechanisms of deformation in the vicinity of the grain boundaries, since the activation energy was closer to the activation energy, 80.4 kJ/mol, of the grain boundary shearing in tin.

Bonar and Craig [51], using the cycling of temperature and the constant-stress creep, studied the creep behavior of tin in the temperature range of 300–350 K. Spectrographically pure tin (98.998%) and zone-refined tin (99.999%+) were used in the tests. The zone-refined tin had higher creep rate than the spectrographically pure tin. Both had the same activation energy of 37.7 kJ/mol for the creep deformation. They concluded that the activation was not measurably affected by minor impurities, as pointed out by Sherby and Dorn [52].

To determine whether cross-slip or other non-diffusional controlled processes are the rate controlling mechanisms in the creep deformation of tin, Suh et al. [53, 54] used X-ray diffraction to study the misorientation of subgrains during the creep of tin single crystals. The creep tests were performed on pure tin single crystals with the [100] direction as the tensile axis in the temperature range of 303–473 K and the constant stresses of 1 MPa and 1.57 MPa. Two activation energies were observed with the transition temperature of 423 K. Above the transition temperature the activation energy of the creep deformation was 98–118 kJ/mol, which was very close to the activation energy of self-diffusion in tin. Below the transition temperature, the activation energy was 40–52 kJ/mol, which was about one half of that of the self-diffusion. They found that the average subgrain misorientation angles for specimens crept at 303 K and 373 K increased continuously with the creep strain. No abrupt change of angle was observed at the transition temperature. They concluded that a cross-slip controlled mechanism can not control the creep deformation below the transition temperature and dislocation climb likely controls the creep of tin both above and below the transition temperature. They proposed that dislocation pipe diffusion might involve in the control of the creep deformation. One of the problems involving the dislocation pipe diffusion is that there have no direct experimental results on the measurement of diffusivity for the dislocation pipe diffusion.

Adeva et al. [55] used the technique of constant strain rate of 0.1/min to do tensile test of polycrystalline tin in the temperature range of 20–190°C. The wire sample had a diameter of 3 mm and a gage length of 2 cm. The sample annealed and deformed at 20°C had a grain size of about 0.3 mm but that annealed and deformed at 190°C had a grain size of 0.7 mm. The fracture was of chisel point type at 20°C and of shear type with cracking along a shear band at 190°C. The low temperature deformation had higher ductility with double elongation to fracture. The authors thought it was due to more grains in the cross section for the low temperature specimen. In situ tensile deformation of a thin film specimen (40 μm thick, 3 mm wide at a strain rate of 0.02/min) in the SEM showed that the prevailing mechanism is slip. Grain growth was evident at 190°C but dynamic recrystallization was not important in controlling the deformation. They used the change of strain rate to determine the stress exponent which was about 6 in the stress range of 1.8–18 MPa. An activation energy of 35 kJ/mol was obtained in the temperature range. They suggested that the slip process was controlled by dislocation pipe diffusion.

McCabe and Fine [56] studied the creep behavior of tin (99.86%) using the constant stress creep and stepped stress tests. The tensile stress was in the range of 2–78 MPa, and the temperature in the range of 123–175°C. Two regions of different stress exponents and activation energies were observed. They obtained a stress exponent of 8.6 and an activation energy 73 kJ/mol at intermediate stresses while they found a stress exponent of 6 and an activation energy 36 kJ/mol at the high stresses. Their results differed from most reported in the literature probably due to the impure tin they used.

Mathew et al. [57] recently determined the characteristics of the creep deformation in tin in the temperature range of 23–200°C and the stress range of 1–30 MPa. The constant-load creep was used in the tests. When a steady state was reached, the stress was increased to reach another steady state. A power-law break down was observed in the high stress region at room temperature. By using the Garofalo [58] empirical relation, they were able to combine the low and high stress data into a single power law relation of a hyperbolic sine function of stress. They obtained a stress exponent of 7.6 and an activation energy of 60.3 kJ/mol. Their results were compatible with those of McCabe and Fine [56] at intermediate stresses. It is unclear what purity of tin was used in their work.

3.2 Impression creep

Chu and Li [3] studied the impression creep of β-tin single crystals in three orientations: [001], [100] and [110] in the temperature range of 313–481 K and the punching stress range of 7.53–47.7 MPa. They observed two parallel processes in controlling the impression creep of tin. The activation energy for all three orientations at high temperatures was 104.7–108.9 kJ/mol, and showed no obvious effect of stress. The value was comparable to that for lattice diffusion. For the deformation process at low temperatures, the activation energy was stress dependent, which was 34.3 kJ/mol for the [001] orientation at 16–20 MPa, 41.9 kJ/mol for [100] at 12–16 MPa, and 39.4 kJ/mol for [110] at 16–20 MPa. For all orientations and all temperatures, the stress dependence of the impression velocity followed a power law with stress exponents between 3.6 and 5.0. Slip lines developed around the impression were more numerous at low temperatures than at high temperatures; and pencil slip was observed at low temperatures. They concluded that the deformation process at low temperatures was dislocation slip and at high temperatures dislocation climb.

Park et al. [59] compared the impression creep and compression creep of polycrystalline β-tin in the temperature range of 323–423 K. The punching stress was in the range of 10–50 MPa, and the compressive stress in the range of 3–20 MPa. The tin sample had a grain size of 20–30 μm. For impression test, the tungsten carbide punch had a diameter of 100 μm. For compression test the sample had 3 mm length and 3 mm diameter. They found stress exponent 5 and activation energy 43 kJ/mol for the impression test and stress exponent 5 and activation energy 41 kJ/mol for the compression test. These results are consistent with the those given by Chu and Li [3] for the impression creep of β-tin single crystals below 150°C. They concluded that the deformation mechanism was dislocation core diffusion. But the core diffusion is a climb process and Chu and Li [3] observed extensive slip lines around the impression below 150°C.

Long et al. [60] did finite element modeling of impression creep and found the plastic zone was centered at about 0.5ϕ below the punch where ϕ is the punch diameter. The size of the plastic zone was about 3ϕ. These findings are consistent with the observation of etch pits in LiF reported earlier by Yu and Li [61]. The simulated stress exponent, 5.87–5.95 and activation energy, 56–60 kJ/mol for impression creep were very close to the input values, 6 (stress exponent) and 61 kJ/mol for the deformation of each element. Similar findings were reported earlier by Yu and Li [62] and others, see a review by Li [63].

3.3 Indentation creep

Mayo and Nix [64] used a micro-indentation technique to determining the strain rate sensitivity of tin at room temperature. They defined the indentation stress as the load divided by the projected contact area and the indentation strain rate to be the indentation velocity divided by its corresponding depth. They used constant indentation loading rate in evaluating the stress dependence of the strain rate. For indentations made in the middle of large Sn grains the stress exponent was 11.4, which was much higher than the results reported in conventional creep tests. They suggested that single grain indentation tests were not equivalent to the tension creep of single crystal, since several slip systems must operate during indentation. For the indentations of small grained tin, the stress exponent was 6.3. In view of the fact that impression creep requires some time to reach steady state, indentation creep has the risk of measuring transient rates all the time.

Differing from the approach used by Mayo and Nix [64] in indentation test, Raman and Berrich [65] performed constant load indentation test of polycrystalline Sn at room temperature. The variation of the indentation depth with time was monitored, from which the indentation strain rate (depth rate divided by depth) was calculated. The stress exponent obtained was in the range of 6.7–8.1. They also performed the load change test and found larger changes of strain rate compared to the individual tests. The results showed an absence of the effect of grain size on the stress exponent. They concluded that grain boundary sliding was unimportant while dislocation creep was operative in the nanoindentation creep of tin without characterizing the temperature dependence of the indentation creep. As mentioned earlier, the lack of steady state is a problem with such tests.

Fujiwara and Otsuka [66–67] studied the indentation creep of β-tin single crystal in the temperatures range of 303–483 K and the indentation loads between 0.2 N

and 0.49 N. The stress exponents obtained were 6.5, 6.1–5.2 and 4.5 for the temperature ranges of 303–403, 414–416 and 418–483 K, respectively. The activation energies were 49 kJ/mol in the temperature range of 303–403 K and 110 kJ/mol in the high temperature range of 418–483 K. They suggested that the pipe diffusion was controlling in the low temperature region and lattice diffusion dominated the indentation creep in the high-temperature region. But core diffusion is a climb process inconsistent with the observation of slip lines. Here again since indentation creep shows no steady state, it could be in the transient stage all the time. The results must be compared with a tensile creep test or impression creep test both of which show steady states.

3.4 Shear testing and grain boundary sliding

It was demonstrated by Kê [68–71] that, in polycrystals of Al, the internal friction is low at low temperatures, but increases rapidly to a maximum with increasing temperature, then falls to low values again with further increase in temperature. The peak temperature depends on the frequency. However in single crystals the internal friction steadily increases with temperature at low values without any peak. Kê concluded that the peak of the internal friction in polycrystals is due to the grain boundary sliding controlled by the viscous flow of the grain boundary layer. Using the same internal friction technique, Rotherham et al. [22] examined polycrystalline pure tin crystals (99.992%) in the temperature range of 283–423 K. They obtained an average activation energy of 80 kJ/mol. To extrapolate to the viscosity of liquid tin at its melting point (232°C), 0.002 Pa.s, the thickness of the grain boundaries must be assumed to be 0.6 nm using Kê's equation or 0.4 nm using the modified equation used by Rotherham et al. [22]. Hence grain boundary fluidity is a possible mechanism for creep at high temperatures.

Puttick and King [49] investigated the grain boundary sliding in bicrystals of tin in the temperature range of 453–498 K and the shear stress range of 0.049–0.206 MPa for straight boundaries subjected to a pure shear stress. They observed three types of relative motion over the boundaries. For the motion of type A, the displacement occurred relatively rapidly at first and then slowed down continuously to become very small after 30–40 h. The initial rate of the displacement was of the order of 10 μm/h, and the initial portion of the displacement versus time appeared to be linear. At lower temperatures this initial steady state rate might persist for some hours; at higher temperatures the

linear portion was not so clearly defined. For the motion of type B, there was a short period during which the displacement was a linear function of time. For the motion of type C, the displacement started off similar to that of the type A, after the initial rate decreased for a certain time, the rate increased again, only to decrease and then increase once more before finally attaining a low value. At each temperature, they observed that the displacement rate was proportional to the stress (the stress exponent was one) and obtained the activation energy of 80.4 kJ/mol consistent with internal friction studies.

Mohamed et al. [72] used the double shear technique to study the creep deformation of pure tin (99.999%) in the temperature range of 402 ~495 K and the stress range of 0.034–0.69 MPa. They observed a transition of the stress dependence from the stress exponent of 6.6 at high stresses to 1 at low stresses. The shear stress for the transition was 0.22 MPa at 495 K. The activation energy was determined to be 93.6 kJ/mol in the high stress region. The creep rate was two orders of magnitude above the Nabarro–Herring creep. They attributed the creep at low stress levels to the Harper–Dorn creep which could have a dislocation mechanism.

3.5 Torsional creep

Zama et al. [73] performed the torsional creep-test of tin. The torsional test eliminated the variation of the cross-sectional area during the test and simplified the maintenance of a uniform temperature along the length of the specimen. A constant torque was applied to the specimen to create a constant angular shear rate. At temperatures below ~488 K the steady-state creep rate at constant torques gave the activation energy of 94.2 kJ/mol and the stress exponent of 4.5–5.0. The activation energy was consistent with those reported by using tensile creep tests and the shear creep tests. They suggested that, at high temperatures, grain boundary slip could contribute to the overall deformation. The problem with torsional test is that the applied stress is not as uniform as in a tension or compression test.

3.6 Bending creep under pressure

DeVries et al. [36] did bending creep of tin under high pressure (up to 7 kbar) to obtain the activation volume. The beam sample had 2.4 cm long, 0.275 cm wide and 0.25–0.35 cm thick with a grain size of about 0.1 mm. The loading of about 1.5 kg was at the center. The pressure medium was kerosene or Dow Corning 200 fluid. The activation volume obtained was about a third of atomic volume, $8.5 \pm 2\ \text{Å}^3$ or 5.1 ± 1.2 cc/g-atom in the temperature range between 0 °C and 27°C. This agrees with the activation volume for self diffusion determined by Coston and Nachtrieb [33], 5.3 ± 0.3 cc/g-atom, for both a and c directions in a single crystal, see Section 1.6 before.

4 Creep deformation of Sn–Ag based alloys

Due to the increasing environmental and health concerns, many countries have regulated the use of traditional tin-lead (Sn–Pb) based solders in microelectronic packaging and assembly [74, 75]. The tin-silver (Sn–Ag) based alloys have become a possible replacement of Sn–Pb based solders. In service, the temperature of interconnection in microelectronic devices and systems ranges between 293 K and 373 K, corresponding to high homologous temperatures of the Sn–Ag based alloys (0.6–$0.8\ T_\text{m}$). In this temperature range, the creep deformation is important for structural integrity of microelectronic interconnections.

4.1 Uniaxial testing

Plumbridge and Gagg [76] performed tensile tests of Sn3.5 wt.%Ag, Sn0.5 wt.%Cu and Sn37 wt.%Pb in the temperature range of 263–348 K and the strain rate range of 10^{-3}~10^{-1} s^{-1}. The Pb–Sn alloy exhibited a cup-and-cone type of fracture, the Cu alloy failed in shear and the Ag alloy produced a chisel type of profile. They found that the strain rate hardening exponent ($1/n$) saturated to a constant value at strains above 1%. The stress exponents were 10, 8 and 9 at -10, 22 and 75°C respectively. These numbers were 25, 20 and 5 for the Pb alloy and 8, 13 and 14 for the Cu alloy. The work hardening exponents were small, varying between 0.005 and 0.43 depending on both the temperature and strain rate.

Amagai et al. [77] did tensile tests on Sn3.5Ag0.75-Cu and Sn2Ag0.5Cu alloys. The specimen had 1 cm diameter and 5 cm gage length with a total length of 14 cm. The temperature range was -25°C to 125°C and elongation rate was 1, 10 and 100 mm/min. They used a hyperbolic sine function raised to a power of about 6 to describe the stress dependence of strain rate. The activation energy obtained was about 70 kJ/mol for both alloys.

Huang et al. [78] studied the constant-load tensile creep of Sn3.5Ag alloy at three temperatures of 30, 75 and 120°C in the stress range of 8–40 MPa. The microstructure of the as solidified eutectic Sn–Ag solder consisted of dendritic β-Sn and spherical particles (0.5 µm) of Ag$_3$Sn finely dispersed in the matrix of β-Sn. The Ag$_3$Sn dispersed regions were separated from β-Sn regions giving the appearance of a network structure with the β-Sn elongated "grains" (10–20 µm) surrounded by the dispersed regions (1–2 µm thick). The weight fraction of Ag$_3$Sn is about 4.8% and the volume fraction, 4%. At 23°C and 0.3/min strain rate, the UTS was 52 MPa and the 0.2% yield stress was 42 MPa, the elongation to fracture was 48% and the reduction in area was 83%. There appeared to be a power-law breakdown at high stresses – 23.5 MPa at 120°C, 27.3 MPa at 30°C and 24.5 MPa at 75°C. The stress exponent at low stresses increased with decreasing temperature and was 12.3 at 30°C, 11.0 at 75°C, and 10.1 at 120°C. The activation energy was 83 kJ/mol and 75 kJ/mol at constant stresses of 15.9 MPa and 20 MPa, respectively. They attributed the higher values of the stress exponent to the strengthening effect of the second phase, Ag$_3$Sn, as suggested by McCabe and Fine [56] in the study of the creep behavior of Sb-solution-strengthened tin and SbSn-precipitate-strengthened tin. Assuming that the creep rate was controlled by detachment of dislocations from particles, they expressed the steady-state creep rate as (Cadek [79]):

$$\dot{\varepsilon} = \frac{\alpha G b}{kT}\left(\frac{\sigma - \sigma_\text{th}}{G}\right)^n \exp\left(-\frac{Q}{RT}\right) \tag{4}$$

where σ_th is the threshold stress which was found to be about 10 MPa at 30°C, 6 MPa at 75°C and 4.1 MPa at 120°C. These stresses did not scale with the shear modulus. The stress exponent n was found to be 7. They used the following equation to calculate the "true" creep activation energy, Q_t,

$$Q_\text{t} = Q - nR\frac{T^2}{E}\frac{\text{d}E}{\text{d}T} \tag{5}$$

where E is Young's modulus of the Sn3.5Ag alloy; and they obtained "true" creep activation energy of 79 kJ/mol at 30°C, 86 kJ/mol at 75°C, and 97 kJ/mol at 120°C. They concluded that the steady-state creep was controlled by the dislocation-pipe diffusion.

Yu et al. [80] evaluated the constant-load tensile creep rupture behavior of Sn3.5Ag based ternary alloys at temperature of 373 K. From the stress effect of minimum creep rate (n) and the stress effect of rupture

time (m), their results for the various alloys are shown in the following table:

Alloy	$B(\text{MPa}^{-n}/10^{10}\text{s})$	n	$C(10^8\text{s}/\text{MPa}^m)$	m	ϕ
Sn–3.5Ag	3.24	4.5	8.9	4.9	6.6
Sn3.5Ag0.5Cu	0.08	5.9	0.154	5.9	7.4
Sn3.5Ag0.75Cu	0.126	7.6	121	5.3	9.1
Sn3.5Ag1.0Cu	1.00	4.2	81.7	3.9	5.0
Sn3.5Ag1.5Cu	6.27	3.9	6.26	3.5	6.8
Sn3.5Ag2.5Bi	9.33	3.8	9.9	4.5	7.9
Sn3.5Ag4.8Bi	10.7	4	10.9	4	6.9
Sn3.5Ag7.5Bi	33.9	3.4	3.0	3.4	5.7
Sn3.5Ag10Bi	661	2	0.522	3	2.4

The minimum creep rate and rupture time were expresssed by:

$$\dot{\varepsilon}_{\min} = B\sigma^n \quad t_f = C\sigma^{-m} \tag{6}$$

The constant load creep curve was fitted by the following equation:

$$\frac{\varepsilon}{\varepsilon_f} = 1 - \left(1 - \frac{t}{t_f}\right)^{1-\frac{n}{1+\phi}} \tag{7}$$

The 0.75% Cu alloy showed the lowest creep rate and longest rupture time while the 10% Bi alloy showed the highest creep rate and shortest rupture time. The values of n and m are similar. The large n for the two Cu alloys could be related to the finely dispersed Cu_6Sn_5 particles. The rupture strain for the Cu alloys was about 0.5 ± 0.15.

Kim et al. [81] used the constant-strain rate tensile test to study the effects of cooling speed on the tensile properties of Sn–Ag–Cu alloys at room temperature and at three strain rates of 10^{-4}, 10^{-3} and 10^{-2}/s. The compositions of the alloys were Sn3.0Ag0.5Cu, Sn2.5Ag0.7Cu, and Sn3.9Ag0.6Cu. Three cooling rates were used: 0.012, 0.43 and 8.3°C/s after casting. The strain rate exponents were 12 for Sn3.0Ag0.5Cu and 12 for Sn3.9Ag0.6Cu. For Sn3.5Ag0.7Cu, the stress exponent was 9 and 17 for the cooling rates of 8.3°C/s and 0.012°C/s, respectively.

Vianco et al. [82–83] studied the compression creep behavior of Sn3.9Ag0.6Cu in the as cast condition as well as after aging. The sample was a cylinder of 1 cm diameter and 1.9 cm long under a stress of 2–45 MPa and at a temperature between – 25°C and 160°C. The stress dependence of creep rate could be described by a hyperbolic sine power law or just a power law. In each case there appeared to have two processes involved. For the hyperbolic sine power law, the exponent was 4.4 ± 0.7 and the activation energy was 25 ± 7 kJ/mol for the low temperature process (<75°C) and 5.2 ± 0.8 and 95 ± 14 kJ/mol for the high temperature process (>75°C). A threshold stress was not needed and grain boundary sliding or decohesion was not observed. However, while creep at low temperatures (<75°C) did not change the microstructure, creep at 75°C produced a "boundary" of coarsened Ag_3Sn particles with depletion zones to either side of the boundary in which the small Ag_3Sn particles disappeared. Creep at higher temperatures produced further coarsening of Ag_3Sn particles. A comparison of the microstructure after annealing without creep indicated that creep enhanced the coarsening process. For specimens aged at 125°C for 24 h, the hyperbolic sine exponent was 5.3 ± 0.7 and the activation energy was 47 ± 6 kJ/mol at <75°C and 7.3 ± 1 and 75 ± 40 kJ/mol at >75°C. For specimens aged at 150°C for 24 h, the hyperbolic sine exponent was 4.5 ± 0.4 and the activation energy was 54 ± 7 kJ/mol at < 75°C and 5.9 ± 0.6 and 105 ± 10 kJ/mol at > 75°C. The aged microstructure appeared stable during the creep deformation.

Earlier they [84] also measured the yield stress (40 MPa at – 25°C to 10 MPa at 160°C) at a strain rate of 0.15/h. The elastic moduli (Young's moduli, shear moduli, bulk moduli and Poisson ratios) and the coefficient of thermal expansion were measured also as a function of temperature. Vianco et al. [85] reported the yield stress of Sn4.3Ag0.2Cu, Sn3.9Ag0.6Cu and Sn3.8Ag0.7Cu by compression. The effect of microstructure was noted.

Wiese and Wolter [86] did tensile test on Sn–3.5Ag using a dog-bone type specimen of 3 mm diameter and 117 mm long under a constant load. The liquid solder was solidified in an aluminum mould by cooling the specimen slowly from one end to the other to avoid cavities. They found two stress exponents, 3.5 in the low stress region (below 10 MPa at 343 K and 13 MPa at 293 K) and 11 in the high stress region. The activation energy was 53 kJ/mol in the low stress region and 92 kJ/mol in the high stress region. They also tested Sn–4Ag–0.5Cu in the same way and found stress exponents 3 and 12 and activation energies 35 kJ/mol and 61 kJ/mol. They proposed two simultaneous processes taking place in each case. In addition, they also tested pin through hole and flip chip solder joints of these two materials with varied results.

El-Bahay et al. [87] examined the constant-stress creep of Sn–3.5Ag alloy in the stress range of 4.7–7.0 MPa and the temperature range of 453–493 K. The stress exponent was 3.3–3.5 at low stresses from 4.47 MPa to 5.75 MPa and 6.3–9.3 at high stresses. They concluded that the creep deformation was due to viscous glide at low stresses and due to dislocation climb at high stresses. They observed the scattering of the activation energy, which was 29.5 kJ/mol at 6.6 MPa, 26.6 kJ/mol at 5.8 MPa, and 73.3 kJ/mol at 5.1. The activation energy increased with the increase in stress, which might be created by the presence of Ag_3Sn intermetallics.

Shohji et al. [88] investigated the tensile behavior of Sn3.5Ag, Sn3.5Ag0.75Cu, and Sn3Ag2Bi in the strain rate range of 1.67×10^{-4} to 1.67×10^{-2} s^{-1} and in the temperature range of 233–393 K. They used both the constant strain rate and strain-rate-change tests. The stress exponents were 11–12 for Sn3.5Ag, 7.6–12 for Sn3.5Ag0.75Cu, and 10–13 for Sn3Ag2Bi. The activation energies were 46.6, 47.3 and 61.6 kJ/mol for Sn3.5Ag, Sn3.5Ag0.75Cu and Sn3Ag2Bi, respectively. They concluded that the tensile deformation of all the materials could be related to a slip creep mechanism controlled by pipe diffusion.

Similar to the work of Shohji et al. [88], Lang et al. [89] evaluated the effects of strain rate and temperature on the tensile behavior of Sn3.5Ag solder in the strain rate range of 2.38×10^{-6} to 2.38×10^{-3} s^{-1} and in the temperature range of 223–423 K. The stress exponent was in the range of 12.5–14.9, which increased slightly with temperature. The activation energy was 78 kJ/mol. They concluded that the tensile deformation of Sn3.5Ag was controlled by a pipe-diffusion mechanism.

Lin and Chu [90] studied the constant-load creep rupture of Sn3.5Ag and Sn3.5Ag0.5Cu solders in the temperature of 293–363 K and in the stress range of 4–30.3 MPa. They correlated the minimum creep rate with the applied stress using the Dorn equation and obtained the stress exponents of 5 for Sn3.5Ag and 6 for Sn3.5Ag0.5Cu at temperatures of 333 K and 363 K. Higher stress exponents of 7 for Sn3.5Ag and 9 for Sn3.5Ag0.5Cu at temperature of 293 K were obtained. They suggested that the creep deformation of Sn3.5Ag and Sn3.5Ag0.5Cu were controlled by the lattice-diffusion controlled dislocation climb coupled with a dispersion-strengthening mechanism at high temperature and by a dispersion-strengthening mechanism at room temperature. The activation energies were 41.8–86.4 kJ/mol for Sn3.5Ag and 55.3–123 kJ/mol, respectively, which was stress-dependent.

Mathew et al. [57] recently determined the characteristics of the creep deformation in Sn3.5Ag alloy in the temperature range of 296–473 K and the stress range of 1–30 MPa. The constant-load creep was used in the tests. The stress exponent of 5 and the activation energy of 60.7 kJ/mol were obtained. They suggested that the creep deformation was due to lattice-diffusion-controlled dislocation climb.

4.2 Shear testing

Guo et al. [91–92] studied creep properties of eutectic Sn–3.5Ag solder joints and those reinforced with 15 vol.% mechanically incorporated Cu (6 micron), Ag (4 micron) or Ni (5 micron) particles by using a single shear lap dog-bone-shaped specimen. The metal particles were mixed with the eutectic Sn–3.5Ag solder paste prior to soldering the two half specimens together. Creep tests were conducted at 25, 65 and 105°C. Steady state creep was found under a dead load. The Ni composite showed the best creep resistance. The next best was the Cu composite. The worst were the Ag composite, Sn–3.5Ag eutectic and the Sn–4Ag–0.5Cu alloy. These latter 3 showed similar creep resistance. The activation energy for creep of the Ni composite was 0.64 eV or 62 kJ/mol, somewhat higher than that of Sn–3.5Ag (60 kJ/mol), the Cu composite (54 kJ/mol), the Ag composite (53 kJ/mol) and the Sn–4Ag–0.5Cu alloy (51 kJ/mol).

Jadhav et al. [93] studied stress relaxation on Sn–3.5Ag eutectic solder and a composite with 20 volume%Cu_6Sn_5 using a single shear lap dog-bone-shaped specimen. They found that the percentage of stress relaxed depended on the prior strain and the measured stress exponent was consistent with conventional creep experiments. Later Rhee and Subramanian [94] studied further and found that grain boundary deformation dominated at 150°C while shear banding took place at lower temperatures. The residual stress after a fixed relaxation period decreased with increasing test temperature at a given strain, and increased with increasing prior strain and the rate of prior straining at a given temperature. In a tensile test performed in a RSAIII using the same kind of specimen, Rhee et al. [95] measured creep properties of the eutectic Sn–3.5Ag alloy and found stress exponents between 10 and 12 in the temperature range of 25–150°C and activation energies between 65 kJ/mol and 90 kJ/mol in the stress range of 15–60 MPa. They also reported from microstructual observations that creep in the region 125–150°C was dominated by grain boundary deformation (sliding, relief and decohesion) and at lower temperatures by shear banding. The test was conducted at a constant shear rate of strain. The shear stress reached a maximum and then decreased. The peak shear stress increased with shear strain rate and decreased with increasing temperature.

Wiese et al. [96] designed a micro-shear tester to evaluate the shear creep of SnAg and SnAgCu solders in flip chip joints in the temperature range of 278–323 K and in the stress range of 8–80 MPa. Using the power law relation, they obtained the stress exponents of 11 for Sn3.5Ag and 18 for Sn4Ag0.5Cu. These were much higher than 2 for the Sn37Pb solder in the low stress region. The Sn4Ag0.5Cu alloy showed much lower absolute creep rates than the Sn3.5Ag alloy, which was due to tiny (5–50 nm) precipitates of η-Cu$_6$Sn$_5$. The activation energies were 79.8 kJ/mol and 83.1 kJ/mol for Sn3.5Ag and Sn4Ag0.5Cu, respectively.

Guo et al. [97] studied creep behavior of several Sn–Ag based alloys: (1) eutectic Sn–3.5Ag, (2) Sn–4Ag–0.5Cu, (3) Sn–3.5Ag–0.5Ni and (4) Sn–2Ag–1Cu–1Ni. At room temperature and 10 MPa, the creep resistance increases in the order (4) < (1) < (2) < (3). They used a single shear lap dog-bone-shaped solder joint specimen under deadweight loading. At room temperature and 20 MPa, the order is (2) < (3) < (4) < (1). At 80°C and 6 MPa, the order is (4) < (2) < (1) < (3). At 80°C and 12 MPa, the order is (4) < (3) < (2) < (1). The microstructure consisted of Sn cells surrounded by bands containing Ag$_3$Sn precipitates with other intermetallics dispersed for the alloys such as Cu–Sn and Cu–Ni–Sn. They believe both Ag and Ni are important for creep resistance.

Kerr and Chawla [98] used a microforce testing system to study the shear creep of single solder ball of Sn–Ag/Cu solder at 25, 60, 95 and 130°C. The microstructure consisted Sn rich dendrites surrounded by a eutectic mixture of Sn and Ag$_3$Sn particles which had an average long dimension of 85 nm and a short dimension of 58 nm. The average spacing between particles was 140–200 nm. There appeared a well defined low stress regime and a high stress regime. In the low stress regime, a stress exponent of 6 was obtained at 25°C and 60°C, while at 95°C and 130°C the stress exponent was 4. In the high stress regime, stress exponents of 13–20 were observed, which was suggested to be caused by the resisting stress to dislocation motion from the Ag$_3$Sn particles and was explained by the threshold stress. Applying Eq. (4) for the creep deformation at high stresses, they then obtained the stress exponent of 4–6. The analysis of the threshold stress was supported by the observation of bowed dislocation segments as shown in Fig. 3. The threshold stress in the high stress regime was determined by plotting the strain rate raised to a power of 1/n versus shear stress/shear modulus ratio on a linear scale and extrapolate to zero strain rate. The stress exponent n was adjusted to form a straight line. The same threshold stress/shear modulus ratio was obtained for all four temperatures, 25, 60, 95 and 130°C and the value was consistent with the particle spacing. Both provided strong support for the analysis. At high temperatures the applied stress was lower than the

threshold stress in the low stress regime indicating a climb controlled process bypassing the particles. The activation energy was determined by two methods: (1) single experiments at constant stress, where the temperature was incrementally increased and the strain rate measured, and (2) conventional constant stress, constant temperature experiments. Method (1) gave the activation energy of 58 kJ/mol in the low temperature regime (25–95°C) and 120 kJ/mol at higher temperatures (95–130°C). Method (2) gave 42 kJ/mol at 25–95°C and 135 kJ/mol at 95–130°C. These are consistent with the creep behavior of pure Sn.

More recently Subramanian and Lee [5] used a single shear-lap configuration to study the damage due to thermal excursions. They found internal stresses, increase of electrical resistivity, surface modifications and crack formation.

4.3 Impression and indentation testing

Lucas et al. [99] made nanoindentation studies on Sn–3.5Ag base alloys reinforced with different particles. An attempt to measure the hardness and modulus of FeSn intermetallic compound showed a variation of hardness from 0.37 GPa to 7.9 GPa and that of modulus from 45 GPa to 153 GPa depending on the location of the indent. The most likely values for Ag$_3$Sn platelets were 3 GPa for the hardness and 90 GPa for the modulus. In a creep experiment the stress exponent varied from 11 to 37 due to the difference in microstructure under the indenter.

Dutta et al. [100] used the impression testing to study the creep behavior of Sn3.5Ag, which was rapidly cooled at a cooling rate of 31°C/s between 553 K and 463 K, and ~20 K/s between 463 K and 333 K. The impression creep was performed in the temperature

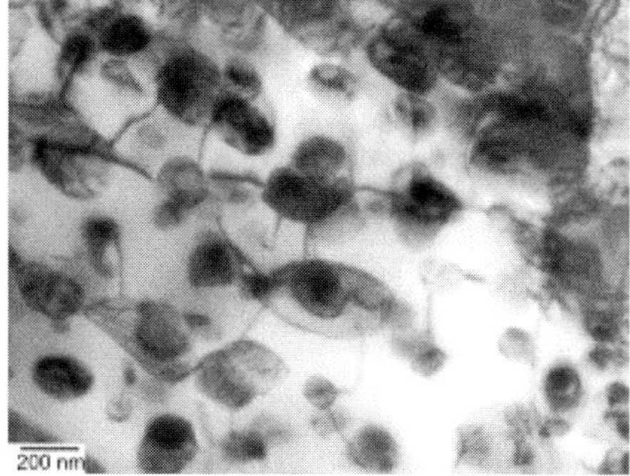

Fig. 3 Dislocations pinned by Ag$_3$Sn particles in the tin matrix after creep deformation, Kerr and Chawla [98]

range of 323–423 K and the punching stress range of 20–50 MPa. They obtained the stress exponent of 2.92–3.04 in the low stress regime and 8.6–9.4 in the high stress regime for temperatures between 323 K and 398 K. At 423 K, they obtained a stress exponent of 4.4 in the low stress regime. The activation energy was 43–46 kJ/mol in the low stress regime and 65–66 kJ/mol in the high stress regime. They concluded that the creep deformation was controlled by two parallel mechanisms, glide-climb with the rate of glide being limited by diffusion of Ag through Sn at low stresses and particle-limited climb by the interaction between dislocations and Ag_3Sn particles with the emergence of a threshold stress at high stresses.

Yang and Peng [101] studied the impression creep of Sn3.5Ag eutectic alloy in the temperature range of 333–453 K and in the punching stress range of 3.4–67.1 MPa. Using a power law between the steady-state impression velocity and the punching stress, the activation energy increased with the punching stress from 44.7 kJ/mol at 6.7 MPa to 79.2 kJ/mol at 46.9 MPa and the stress exponent changed from 1.03 (3.4–13.4 MPa) to 5.9 (20.1–40.2 MPa). However, using a hyperbolic sine function between the steady-state impression velocity and the punching stress, they obtained an activation energy of 51.0 kJ/mol, which was close to the activation energy for grain boundary diffusion of Sn. They suggested that a single mechanism of grain boundary fluid flow is likely the controlling mechanism for the time-dependent plastic flow of Sn3.5Ag eutectic alloy under the testing conditions.

Dutta et al. [102] used the impression testing to evaluate the effect of microstructural coarsening on the creep response of Sn3.5Ag and Sn4Ag0.5Cu. They observed that the creep rate increased proportionately with the size of precipitates at low stresses and the threshold stress for particle-limited creep was altered at high stresses. For both materials, the stress exponent was 6 and the activation energy was 61 kJ/mol.

5 Some other tin alloys

5.1 Sn37Pb eutectic alloy

Yang and Li [103] did impression creep and stress relaxation of Sn37Pb eutectic alloy and used a hyperbolic sine law for the stress dependence of creep rate. By using this law they found a single activation energy of 55 kJ/mol. A mechanism of interfacial viscous shearing between the two eutectic phases was proposed. They also collected many earlier references and mechanisms. More recently Wiese et al. [96] did a micro-shear test on Sn37Pb. They also used a hyperbolic sine stress function with an activation energy of 45 kJ/mol. Plumbridge and Gagg [76] did tensile test on Sn37Pb and found UTS at 22°C, 75 MPa at 0.1/s, 63 MPa at 0.01/s and 32 MPa at 0.001/s. At 0.001/s, the UTS was 72 MPa at − 10°C, 32 MPa at 22°C and 15 MPa at 75°C. The strain rate effect of flow stress showed a stress exponent of 5 at 75°C, 20 at 22°C and 25 at − 10°C. Amagai et al. [77] did a tensile test using specimens of 1 cm diameter and 5 cm gage length over a total length of 14 cm. The strain rate was 1, 10 and 100 mm/min. They used a hyperbolic sine function raised to a power of about 3 to describe the stress dependence of strain rate. They obtained an activation energy of 78 kJ/mol for both Sn37Pb and Sn36Pb2Ag alloys.

5.2 Sn–90Pb alloy

Pan et al. [104] designed and tested a miniaturized impression creep apparatus for small solder balls (about 750 micron diameter) attached to a ball grid array microelectronic packaging substrate. It used a 100 micron diameter cylindrical WC punch with a flat end to be impressed on any one of these solder balls at a temperature from ambient to 423 K. A video imaging system allowed precise alignment between the punch and the ball. The test required no special sample preparation and could produce steady state in about 1–3 h for a total test time of 5–6 h. Multiple creep curves could be generated from different solder balls within the same substrate. For Sn–90Pb alloy they found a stress exponent of about 4 within a punch stress range of 0.002–0.01G (G is shear modulus) for all three temperatures, 323, 343 and 373 K. An activation energy of about 60 kJ/mol was obtained from these tests. This activation energy is consistent with that for dislocation core diffusion, 66 kJ/mol, reported for pure lead [105].

5.3 Sn–Sb alloys

Mahidhara et al. [106] did tensile test on Sn5Sb alloy using a flat specimen of 1 mm thick and 1.24 cm gauge length. The microstructure consisted of 10 μm equiaxed grains with 2–5 micron SnSb intermetallic particles uniformly dispersed. In the low strain rate region, a stress exponent of 3.2 and an activation energy of 18 kJ/mol were obtained. For the high strain rate region, these quantities were 7.0 kJ/mol and 39 kJ/mol.

McCabe and Fine [56] studied creep of Sn–Sb alloys. Their results on pure tin (99.86%) below 101°C are listed in Table 1. When Sb was in solid solution (up to 0.7 wt.%) the stress exponent was similar to that of Sn.

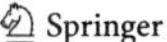

For high concentration alloys (2.9–8.1 wt.%) the SbSn precipitates were in the form of rectangular rods. One kind had its length direction along [001] of tin and other faces parallel to (100) of tin. The typical length was 3–5 μm and width and thickness 126 nm. Another kind had its length along [10$\bar{1}$] of tin and other faces parallel to (101) or (100) of tin. The typical length was 5 μm and width and thickness 205–228 nm. The volume fraction was about equal in all the possible orientations. For such dispersion strengthened alloys the stress exponent and activation energy are usually high as shown in the following table:

Alloy	Low stress		High stress	
	Q kJ/mol	n	Q kJ/mol	n
Sn2.6Sb	101	4.7	117	11.2
Sn5.8Sb	95	5.3	85	11.6
Sn8.1Sb	61	5.6	49	9.8

A common way is to use an internal stress due to the particles such as the Orowan stress needed for the dislocations to bow between particles. This stress is to be subtracted from the applied stress and the difference is the effective stress to be used in the power law equation. With this consideration, the stress exponent in the high stress region was reduced to 8.6 for all 3 alloys and the activation energies were reduced to 77.3, 52.3 and 24.5, respectively. However, the internal stress was found to vary with temperature. Finally the authors combined the internal stress effect with a composite effect in order to describe the data.

Sadykov et al. [107] studied tin babbit (Sb, 10–12 wt.%; Cu 5–6 wt.%). They reviewed some earlier literature. The cast structure consisted of 3 phases, the α phase of solid solution of Sb and Cu in tin having a grain size of 0.5–1 μm, the β phase of SnSb hexagonal crystal and small η phase of Cu_6Sn_5. A change of grain size in the β phase affected both the yield stress and elongation. The rolled alloy appeared to have lower yield stress and better elongation than the case alloy.

Mathew et al. [57] also worked on Sn5Sb alloy. A stress exponent of 5 and an activation energy of 44.7 kJ/mol suggested to them a low temperature viscous flow mechanism. But their alloy should have SnSb precipitates as found by McCabe and Fine [56] and hence dispersion strengthening should be a part of their interpretation.

5.4 Microstructural stability

Yomogita [108] showed the transient effect from one steady state microsturcture to another. He did tensile test of single crystals of Sn of [010] orientation with the slip system of type (a) operating. In a stress relaxation test the strain rate was proportional to the stress rate so the strain rate could relate to the stress. A stress exponent of 11–15 could be deduced. In a strain rate change test, the stress for the second strain rate depended on the duration at the unloading stress. This transient effect depends also on the temperature. In a stress change test, the second strain rate depended also on the duration at the unloading stress but converged to a new steady state value within about 5 min.

Table 1 Activation Energies and Stress Exponents of Creep of β-Tin

Investigator(s) and year	Low temperature(< 150°C)		High temperature(>150°C)	
	Q kJ/mol	n	Q kJ/mol	n
Tyte [1], re-plotted by Breen and Weertman [46]	31			
Breen and Weertman [46]	46	4.8	109	
Weertman and Breen [11]	46 [110]	3.6–5.1[110]	92 [110]	
Weertman [47]	50 [001]		103 [001]	
Bonar and Craig [51]	38			
Chu and Li [3]	37 [001],	4.1–4.4 [001]	105 [001]	4.0 [001]
	42 [100],	3.9–4.5 [100]	109 [100]	3.6 [100]
	39 [110]	4.7–5.0 [110]	109 [110]	4.5 [110]
Suh et al. [53]	55		128	
Suh et al. [54]	59 [100]		128 [100]	
Adeva et al. [55]	35	6		
Nagasaka [44]	48	4.2, 3.8, 3.9	128	3.9
McCabe and Fine [56]	36 (high stress)	6 (high stress)		
	73(Low stress)	8.6 (low stress)		
Ekinci et al. [45]		4.5		
Mathew et al. [57]	60.3	7.6		
Park et al. [59]	42	5		

Mohamed et al. [109] studied the effect of microstructural changes on transient creep and activation energy. They used a power function of time to fit the transient creep curve:

$$\varepsilon_{tr} = Bt^m \quad B = B_0(\dot{\varepsilon}_{st})^\gamma \tag{8}$$

and the constant B was found to be a power function of the steady state creep rate. They calculated the activation energy Q from the following relation:

$$\varepsilon_{tr} = \varepsilon_0 + t^m \exp(-Q_{tr}/RT) \tag{9}$$

The behavior of the following alloys is reported:

(1) Sn0.5 wt.%Zn alloy: The time exponent m changed from 0.52 to 0.8 and the value B changed from 0.011 to 0.045 at a transition temperature of 413 K. Across the same transition temperature the exponent γ changed from 0.26 to 0.42 and the activation energy changed from 25.5 kJ/mol to 41.9 kJ/mol. Based on the binary phase diagram, the Zn rich phase dissolves and disappears completely in the Sn rich phase above 413 K. This provides the cause for the transition.

(2) Sn9 wt.%Zn alloy: The time exponent m changed from 0.53 to 0.87 and the value B changed from 0.0014 to 0.03 at a transition temperature of 423 K. At the same time the exponent γ changed from 0.4 to 1.04 and the activation energy changed from 22.5 kJ/mol to 31.4 kJ/mol across the same transition temperature. Based on the binary phase diagram, the eutectic mixture changed relative proportions across 423 K.

(3) Pb10 wt.%Sn alloy: The time exponent m changed from 0.47 to 0.87 and the value B changed from 0.0016 to 0.0256 at a transition temperature of 423 K. At the same time the exponent γ changes from 0.45 to 0.67 and the activation energy changed from 16 kJ/mol to 27 kJ/mol. Based on the binary phase diagram, the Sn rich phase dissolves and disappears in the Pb rich phase above 423 K.

(4) Pb61.9 wt.%Sn alloy: The time exponent m changed from 0.63 to 0.96 and the value B changed from 0.3×10^{-5} to 4×10^{-5} at a transition temperature of 403 K. At the same time the exponent γ changed from 0.7 to 0.85 and the activation energy changed from 18 kJ/mol to 29.5 kJ/mol. Based on the binary phase diagram, the relative proportion of the two eutectic phases changes across 403 K.

Dutta et al. [102] studied the effect of microstructural coarsening on the creep response of Sn–Ag based solders. They found at low stresses the creep rate increased proportionally with the size of Ag_3Sn particles and at high stresses, precipitate coarsening changed the threshold stress for particle-limited creep. It is possible to have both thermal and mechanically induced coarsening during service. They provided some equations for static and strain enhanced coarsening kinetics for the growth of particles.

6 Summary on plastic deformation

This collection of experiments of Sn and its alloys showed a diversity of results and a variety of proposed mechanisms. Often there were disagreements and inconsistency. The following represents our best assessment of the situation:

(1) Below 1 K, β tin deforms by nucleation of double kinks in a quantum tunneling effect through the Peierls barrier. The flow stress is about 10 MPa independent of temperature.

(2) Between 1 K and 65 K, β tin deforms by thermally activated nucleation of double kinks through a small barrier of about 480 J/mol.

(3) Between 65 K and 291 K, β tin deforms by slip probably through the nucleation of double kinks also. More studies are needed.

(4) Between 291 K and 423 K, β tin and tin-rich alloys deform by slip with an activation energy of 40 kJ/mol and a stress exponent of 4–5.

(5) Between 423 K and 476 K, β tin and tin-rich alloys deform by dislocation climb with an activation energy of 109 kJ/mol and a stress exponent of 3.5–4.5.

(6) Impurities can modify the barrier height and the stress exponent.

(7) When the stress range is large, a single power law cannot fit the whole set of data. This may give a false impression of multiple mechanisms. A hyperbolic sine law should be tried instead. For low stresses Harper–Dorn creep may appear with a dislocation mechanism.

(8) Precipitates can introduce a threshold stress to be subtracted from the applied stress resulting in an effective stress which can be used in the power law or hyperbolic sine law. However, when the temperature is high, dislocation can climb over particles so the threshold stress is reduced or even disappears.

(9)   At high temperatures, intergrain fluidity could be a mechanism for creep when the grain size is small. More studies are needed.

(10) Microstructural changes during service should be monitored and the effect on properties controlled. More studies are needed in this area.

(11) Localized and miniaturized tests are needed for solders. Impression tests are recommended since it is simple and gives steady state information. Pan et al. [104] already developed an apparatus using 100 micron WC cylindrical punch showing steady state creep on 750 - micron solder balls.

The understanding of the fundamental deformation mechanisms will improve our capability of developing new lead-free solders with better mechanical properties. It also provides us the basis to develop reliable and applicable constitutive equations for simulating the evolution of the stress and strain fields in microelectronic interconnections and to improve the design of devices and systems for higher reliability and longer service life.

7 Deformation twinning

The famous tin cry is a mechanical twinning effect. The usual twinning plane is (301) and twinning direction is [$\bar{1}$03], see Clark et al. [110] for some early discrepancies [111, 114]. Another possible twinning mode is in the plane (10$\bar{1}$) along the direction [101], see Lee and Yoo [115].

7.1 Twinning transformation

All atoms in the matrix before twinning will assume new positions after twinning. Similarly any vector in the matrix becomes a new vector in the twin. To figure out the change, one can use the twinning transformation. The following is due to Ishii and Kiho [116]. Any vector (*hkl*) expressed as Miller indices in the bct tin crystal before twinning can be expressed in the Cartessian coordinates (*xyz*) by multiplying with the following matrix:

$$\mathbf{A} = \begin{bmatrix} a & 0 & 0 \\ 0 & a & 0 \\ 0 & 0 & c \end{bmatrix} \tag{10}$$

This vector can now be expressed in the Cartessian coordinates (*x'y'z'*) as shown in Fig. 4 by the following transformation matrix:

$$\mathbf{R} = \begin{bmatrix} -\cos2\theta & 0 & -\sin2\theta \\ 0 & -1 & 0 \\ -\sin2\theta & 0 & \cos2\theta \end{bmatrix} \tag{11}$$

where 2θ is the angle between z and z' so that θ is the angle between z' or z and [$\bar{1}$03]. Hence

$$\tan\theta = a/3c$$

During twinning, the displacement of any point (*x'y'z'*) is equal to the twinning shear (strain) multiplied by the distance between (*x'y'z'*) and the (301) twin boundary. This displacement can be represented by the following transformation:

$$\mathbf{S} = \begin{bmatrix} 1 + e\sin\theta\cos\theta & 0 & e\sin^2\theta \\ 0 & 1 & 0 \\ -e\cos^2\theta & 0 & 1 - e\sin\theta\cos\theta \end{bmatrix} \tag{12}$$

where e is the twinning shear:

$$e = (a^2 - 3c^2)/2ac$$

Finally the coordinates will be changed back to the Miller indices by using $\mathbf{A}^{-1}$. As a result the overall transformation is:

$$\mathbf{T} = \mathbf{A}^{-1}\mathbf{SRA} = -\frac{1}{2}\begin{bmatrix} 1 & 0 & 1 \\ 0 & 2 & 0 \\ 3 & 0 & -1 \end{bmatrix} \tag{13}$$

and its inverse is the same matrix.

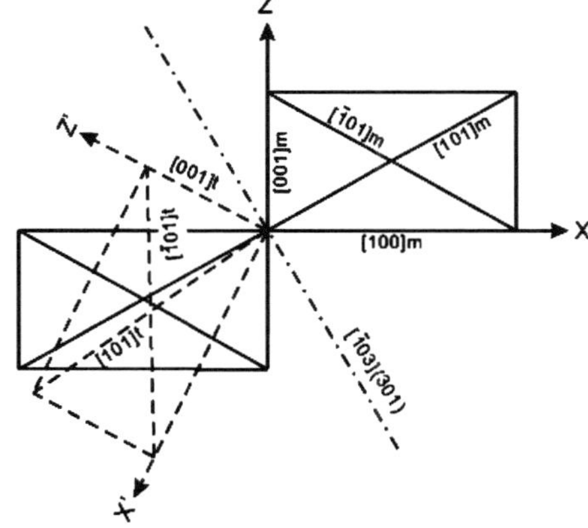

Fig. 4 Coordinates of matrix and twin systems in β-tin, copied from [116]

7.2 Elastic strain energy of twinning

Since twinning is a shear transformation, a twin embedded in the original matrix introduces some strain energy into the system. Lee and Yoo [115] used the Eshelby inclusion theory and calculated such strain energy. For the Eshelby theory see Eshelby [117] or Mura [118]. For an ellipsoidal shaped twin of semiaxes a_1, a_2 and a_3, the strain energy is about $1.18\ \mu\varepsilon^2 V$ for a sphere ($a_1 = a_2 = a_3$) and $0.88\ \mu\varepsilon^2 V$ for a circular cylinder ($a_1 = \infty$, $a_2 = a_3$) where μ is the shear modulus (17.93 GPa for tin at room temperature), ε is the misfit strain (0.0492 for tin) and V is the volume of the twin. The energy is smaller for other shapes. For example, the energy is $1.0\ \mu\varepsilon^2 V$ ($a_1 = a_2 = 2a_3$), $0.8\mu\varepsilon^2 V$ ($a_1 = \infty$, $a_2 = 2a_3$), $0.65\ \mu\varepsilon^2 V$ ($a_1 = a_2 = 5a_3$) and $0.5\mu\varepsilon^2 V$ ($a_1 = \infty$, $a_2 = 5a_3$). Applied stress can lower the total free energy of the system if applied in the right direction. Hence it can enhance or retard deformation twinning.

7.3 Dislocation-twin interactions

Dislocations in bct tin can have [001] and [101] type of Burgers vectors. These dislocations after twinning will change their Burgers vectors according to the transformation just described. The change is summarized in the following table:

Burgers vector in matrix	Burgers vector in twin	Difference
[001]	$[\bar{1}01]/2$	$[101]/2 = b_t$
$[\bar{1}01]$	2[001]	$[\bar{1}0\bar{1}] = -2b_t$
[101]	$[\bar{1}0\bar{1}]$	$[202] = 4b_t$
[011]	$[\bar{1}21]/2 = [0\bar{1}1] + [\bar{1}0\bar{1}]/2$	
$[0\bar{1}1]$	$[\bar{1}21]/2 = [011] + [\bar{1}0\bar{1}]/2$	

The first 3 cases will leave some twinning dislocations at the twin boundary before the dislocation is embedded in the twin. The embedded dislocation is also a slipping dislocation. The last 2 cases will not be transformed into slipping dislocations in the twin. These dislocations may dissociate into other dislocations as shown. Such possibilities of dislocation-twin interactions were compared (Ishii and Kiho, [98]) with experimental observations of Clark and Craig [119], Votava and Hatwell [120] and Fourie et al. [121].

7.4 Twinning mechanisms

Mechanical twinning takes place in three stages [122], a nucleation stage in which a small twin appears due to local stresses, a spreading stage in which the incoherent twin boundaries move to cover the crystal or grain and a thickening stage in which the coherent twin boundaries move perpendicular to the twinning plane, (301). The thickening stage may take place by rotating a partial dislocation around a pole dislocation with a Burgers vector c. Ishii's [122] model of amplitude dependent internal friction was the loss of energy when partial dislocations annihilate each other after one revolution around the pole dislocation. The average spacing between pole dislocations was estimated to be 3 microns and the thickening speed was about 2.3 cm/s. Brunton and Wilson [123] estimated the thickening speed to be about 10 m/s and a broadening (moving of the incoherent boundary) speed of 800 m/s. They observed also that a twin could disappear as fast as it appeared. Mason et al. [124] showed that the twinning dislocations may travel at the speed of sound. Ishii and Kiho [125] provided some discussion on the stresses needed for thickening of twin.

Ishii [126] used impact loading on tin single crystals (a cylindrical specimen with its axis almost perpendicular to [001] impacting along the axis) found the stress for twin initiation scattered between 7 MPa and 25 MPa but more consistently at 8 MPa after indentation. The stress needed for thickening was about 1–2 MPa. When the crystal was oriented for slip, the critical shear stress for slip was about 9 MPa. Maruyama [127] found the shear stress needed for the initiation of twinning in a Sn single crystal is about 5–10 MPa and that needed for thickening is about 1–2 MPa varying not much between 0.025/min and 2.5/min strain rate and $-195°C$ to $20°C$. The thickening stress increases with the thickness of the twin and with the amount of prestrain due to slip before the initiation of twinning. The prestrain increases with deformation temperature. However when the crystal was twinned in advance by impact even at room temperature so the prestrain was very small, the thickening stress had the lowest value of 0.7 MPa. This thickening stress appears consistent with that of Ishii [122] who found that for a twinned single crystal the amplitude dependent internal friction increases at a stress amplitude of about 1 MPa. The initiation stress appears consistent with that found by Jon et al. [128] using ultrasonic deformation, 8–11 MPa.

For Sn single crystal whiskers, Overcash et al. [129] found the shear stress for initiation of twinning to be 74 MPa and that for propagation, 15 MPa.

For thin films, Fourie et al. [121] and Tu and Turnbull [130–131] made interesting observations in the electron microscope. But the stress was unknown and the rate depended on the rate of stress buildup due to a

contamination carbon layer deposited on the specimen surface. As a result quantitative information was absent.

7.5 Effect of straining and twinning on superconductivity and resistivity

Rothberg [132] found that the superconducting transition temperature of [001] tin whisker increased by about 0.5°C at zero magnetic field from 3.5 K to 4 K after 1% elastic strain in tension. In the meantime the critical magnetic field at 0 K increased by about 5 mt (millitesla) from 35 mt to 40 mt. When the elastic strain is relaxed by twinning, the transition became normal, namely returned to the stress-free state unless further elastic straining was applied to the twinned crystal. An [001] whisker twinned to become [101]. The effect of elastic strain on superconducting transition is less for the [101] orientation than the [001] orientation.

Davis et al. [133] measured the superconducting transition temperature and resistivity of tin whiskers of [101], [100], [001] and [111] axial orientations as a function of tensile strain. For the [001] whisker, the transition temperature increased linearly to about 0.5°C for 1.2% axial strain. For a 0.4% axial strain, the resistivity increased 2% at 300 K and 3% at 77 K. For the [100] whisker, the transition temperature increased 0.4°C (purer) or 0.2°C (less pure) for a 1.7% axial strain and the relation is not linear (slower in the beginning). The resistivity of the less pure whisker decreased about 2% at both 300 K and 77 K. These are the extremes. The other orientation of whiskers behaved in between.

8 Concluding remarks

8.1 Sn is an unusual metal

While Sn is a metal, it is not as simple as fcc, bcc or even hcp metals. For example, each atom in fcc metals has 12 nearest neighbors of equal distance with 12 possible equivalent Burgers vectors. Similarly an atom in bcc metals has 8 nearest neighbors with 8 equivalent Burgers vectors. Even in hcp metals, each atom has 6 equivalent neighbors in the basal plane and 6 more in the next plane. However, for Sn, each atom has four nearest neighbors, two next nearest neighbors and 8 third nearest neighbors. The common Burgers vectors are not between the nearest neighbors. There are many more different slip systems in Sn than in any other kind of metals. This contributes to the complexity of plastic deformation of tin.

8.2 Self-diffusion in tin is chaotic

Self-diffusion measurements are confusing as shown in Fig. 2. There are strong short-cut paths such as grain boundaries and dislocations. These fast paths affect creep measurements in a variety of ways. For example, lower temperature climb may take place by dislocation core diffusion with a lower activation energy and higher temperature climb may take place by lattice diffusion with a higher activation energy. For polycrystals and eutectic mixtures, grain boundary or phase boundary fluidity may play an important role in creep deformation. As indicated in the self-diffusion measurements, grain boundary diffusivity may be much faster than dislocation core diffusivity. These provide the sources for complexity in the creep measurements.

8.3 Allotropic forms are uncertain

Even though β-tin can be stable at low temperatures below the transition temperature, 13.2°C, what happens after deformation is uncertain in view of the internal stresses created. Some careful studies of microstructure after low temperature deformation should be made on both cubic and β-tin.

8.4 Future studies

To make an effective Pb-free solder more studies are needed as indicated throughout this review. Due to the complex crystal structure, multipaths of self diffusion and allotropic transformations as just summariezed, Sn is more complicated than a common metal and hence needs more careful studies in order to be applied with confidence in the electronic industry.

Acknowledgements We thank Professor K. N. Subramanian of Michigan State for pointing out several problems related to tin. We also thank Chenny Zhenyu Wang for creating Fig. 4.

References

1. L.C. Tyte, Proc. Roy. Soc. **50**, 193 (1938)
2. K. Löhberg, H. Moustafa, Zeitschrift für Metallkunde **67**(5), 333 (1976)
3. S.N.G. Chu, J.C.M. Li, Mater. Sci. Eng. **39**, 1 (1979)
4. B. Akdim, D.A. Papaconstantopoulos, M.J. Mehl, Phil. Mag. B **82**, 47 (2002)
5. K.N. Subramanian, J.G. Lee, Mat. Sci. Eng. A**421**, 46 (2006)
6. C.S. Barrett in *Metallurgy and Metallurgical Engineering Series*, ed. by R. F. Mehl, 2nd edn. (McGraw-Hill, New York, 1952), pp. 337–379 (see also Metals Handbook, American Society for Metals, Metals Park, OH, 1948, p. 1071)

7. E. Schmid, W. Boas, *Plasticity* (Chapman-Hall, London, 1968)
8. M. Lorenz, Zeitschrift für Metallkunde **59**, 419 (1968)
9. H. Mark, M. Polanyi, Z. Physik **18**, 75 (1923)
10. J. Obinata, E. Schmid, Z. Physik **82**, 224 (1933)
11. J. Weertman, J.E. Breen, J. Appl. Phys. **27**, 1189 (1956)
12. G.I. Kirichenko, V.P. Soldatov, Fizika Metallov I Metallovedenie **54**(3) 560 (1982) Translation, Phys. Met. Metall. **54**, 130 (1982)
13. K. Ojima, T. Hirokawa, Jpn. J. Appl. Phys., Part 1 (Regular Papers & Short Notes), **22**, 46 (1983)
14. M. Nagasaka, Jpn. J. Appl. Phys., Part 1 (Regular Papers & Short Notes), **28**, 446 (1989)
15. R. Fiedler, A.R. Lang, J. Mat. Sci. **7**, 531 (1972)
16. R. Fiedler, I. Vagera, Phys. Status Solidi A**32**, 419 (1975)
17. A.N. Stroh, Phil. Mag. **3**, 625 (1958)
18. B. Duzgun, I. Aytas, Jpn. J. Appl. Phys. Part 1-Regular Papers Short Notes & Review Papers **32**, 3214 (1993)
19. K. Honda, Jpn. J. Appl. Phys. **17**, 33 (1978)
20. K. Honda, Jpn. J. Appl. Phys. **18**, 215 (1979)
21. K. Honda, Jpn. J. Appl. Phys. **26**, 637 (1987)
22. L. Rotherham, A.D.N. Smith, G.B. Greenough, J. Inst. Metals **79**, 439 (1951)
23. E. Fraizier, M.H. Nadal, R. Oltra, Jpn. J. Appl. Phys. **93**, 649 (2003)
24. I. Takamichi, I.L. Roderick, *The Physical Properties of Liquid Metals*, vol. 1 (Clarendon, Oxford, 1988), p. 71, Chapter 3
25. V.M. Mathew, C.S. Menon, K.P. Jayachandran, J. Phys. Chem. Solids **63**, 1835 (2002)
26. W.P. Mason, H.E. Bommel, J. Acoust. Soc. Am. **28**, 930 (1956)
27. P.W. Bridgeman, Proc. Natl Acad. Sci. USA **10**, 411 (1924)
28. K.D. Swartz, W.B. Chua, C. Elbaum, Phys. Rev. B **6**, 426 (1972)
29. R.R. Rao, A. Padmaja, J. Appl. Phys. **62**, 15 (1987)
30. P.J. Fensham, Aust. J. Sci. Res. **3A**, 91 (1950)
31. J.D. Meakin, E. Klokholm, Trans. Met. Soc. AIME **218**, 463 (1960)
32. W. Chomka, J. Andruszkiewicz, Nucleonica **5** 611 (1960) Translation, pp. 521–524
33. C. Coston, N.H. Nachtrieb, J. Phys. Chem. **8**, 2219 (1964)
34. G. Pawlicki, Nukleonica **12**, 1123 (1967), Translation pp. 45–56
35. J.C.M. Li, in *Rate Processes in Plastic Deformation of Materials*, ed. by J.C.M. Li, A.K. Mukherjee (American Soc. Metals, Metals Park, OH, 1975), p. 475
36. K.L. Devries, G.S. Baker, P. Gibbs, J. Appl. Phys. **34**, 2258 (1963)
37. V.D. Natsik, V.P. Soldatov, L.G. Ivanchenko, G.I. Kirichenko, Low Temp. Phys. **30**(3), 253 (2004)
38. V.P. Soldatov, V.D. Natsik, G.I. Kirichenko, Low Temp. Phys. **27**, 1048 (2001)
39. V.D. Natsik, V.P. Soldatov, G.I. Kirichenko, L.G. Ivanchenko, Low Temp. Phys. **29**, 340 (2003)
40. G.I. Kirichenko, V.D. Natsik, V.P. Soldatov, Fizika Nizkikh Temperatur **18**(11), 1270 (Nov. 1992) Transloation: Sov. J. Low Temp. Phys. **18**, 887 (1992)
41. A.N. Diulin, G.I. Kirichenko, V.D. Natsik et al., Low Temp. Phys. **24**, 452 (1998)
42. M. Fujiwara, T. Hirokawa, Jpn. J. Appl. Phys., Part 1 (Regular Papers & Short Notes), **25**, 1598 (1986)
43. M. Fujiwara, Mater. Sci. Eng. A - Struct. Mater. Prop. Microstruct. Proce. **234**, 991 (1997)
44. M. Nagasaka, Jpn. J. Appl. Phys. Part 1-Regular Papers Short Notes & Review Papers **38**(1A), 171 (1999)
45. A.E. Ekinci, N. Ucar, G. Cankaya, B. Duzgun, Indian J. Eng. Mater. Sci. **10**(5), 416 (2003)
46. J.E. Breen, J. Weertman, Trans. Am. Instit. Min. Metall. Eng. **203**(11), 1230 (1955)
47. J. Weertman, J. Appl. Phys. **28**(2), 196 (1957)
48. R.E. Frenkel, O.D. Sherby, J.E. Dorn, Acta Metall. **3**, 470 (1955)
49. K.E. Puttick, R. King, J. Inst. Metals **80**, 537 (1952)
50. C.D. Wiseman, O.D. Sherby, J.E. Dorn, Trans. AIME, J. Metals **9**, 57 (1957)
51. L.G. Bonar, G.B. Craig, Can. J. Phys. **36**, 1445 (1958)
52. O.D. Sherby, J.E. Dorn, Trans. AIME **194**, 959 (1952)
53. S.H. Suh, J.B. Cohen, J. Weertman, Scripta Metall. **15**, 517 (1981)
54. S.H. Suh, J.B. Cohen, J. Weertman, Metall. Trans. A-Phys. Metall. Mater. Sci. **14**, 117 (1983)
55. P. Adeva, G. Caruana, O.A. Ruano et al., Mater. Sci. Eng. A-Struct. Mater. Prop. Microstruct. Proce. **194**, 17 (1995)
56. R.J. McCabe, M.E. Fine, Metall. Mater. Trans. A-Phys. Metall. Mater. Sci. **33**, 1531 (2002)
57. M.D. Mathew, H.H. Yang, S. Movva, K.L. Murty, Metall. Mater. Trans. A-Phys. Metall. Mater. Sci. **36A**, 99 (2005)
58. F. Garofalo, *Fundamentals of Creep and Creep-Rapture in Metals* (Macmillan, NY 1966), p. 101
59. C. Park, X. Long, S. Hoberman, S. Ma, I. Dutta, J. Mat. Sci. to appear (2006)
60. X. Long, D. Pan, I. Dutta, unpublished
61. E.C. Yu, J.C.M. Li, Phil. Mag. **36**, 811 (1977)
62. H.Y. Yu, J.C.M. Li, J. Mat. Sci. **12**, 2214 (1977)
63. J.C.M. Li, Mat. Sci. Eng. A**322**, 23 (2002)
64. M.J. Mayo, W.D. Nix, Acta Metall. **36**, 2183 (1988)
65. V. Raman, R. Berriche, J. Mater. Res. **7**, 627 (1992)
66. M. Fujiwara, M. Otsuka, Mater. Sci. Eng. A - Struct. Mater. Prop. Microstruct. Proce. **319**, 929 (2001)
67. M. Fujiwara, M. Otsuka, J. Jpn. Instit. Metals **63**, 760 (1999)
68. T.S. Kê, Phys. Rev. **71**, 533 (1947)
69. T.S. Kê, Phys. Rev. **72**, 41 (1947)
70. T.S. Kê, Phys. Rev. **73**, 267 (1948)
71. T.S. Kê, J. Appl. Phys. **20**, 274 (1949)
72. F.A. Mohamed, K.L. Murty, J.W. Morris, Metall. Trans. **4**, 935 (1973)
73. W.A. Zama, D.D. Lang, F.R. Brotzen, J. Mech. Phys. Solids **8**, 45 (1960)
74. Japanese Ministry of Health and Welfare Waste Regulation (1998)
75. European Union WEEE Directive, 3rd draft ed. (2000)
76. W.J. Plumbridge, C.R.C.R. Gagg, J. Mat. Sci. Mat. Electr. **10**, 461 (1999)
77. M. Amagai, M. Watanabe, M. Omiya, K. Kishimoto, T. Shibuya, Microelectr. Reliab. **42**, 951 (2002)
78. M.L. Huang, L. Wang, C.M.L. Wu, J. Mat. Res. **17**, 2897 (2002)
79. J. Cadek, *Creep in Metallic Materials* (Amsterdam, Elsevier, The Netherlands, 1988), p. 44
80. J. Yu, D.K. Joo, S.W. Shin, Acta Met. **50**, 4315 (2002)
81. K.S. Kim, S.H. Huh, K. Suganuma, Mat. Sci. Eng. A**333**, 106 (2002)
82. P.T. Vianco, J.A. Rejent, A.C. Kilgo, J. Electr. Mat. **33**, 1389 (2004)
83. P.T. Vianco, J.A. Rejent, A.C. Kilgo, J. Electr. Mat. **33**, 1473 (2004)
84. P.T. Vianco, J.A. Rejent, A.C. Kilgo, J. Electr. Mat. **32**, 142 (2003)
85. P.T. Vianco, J.A. Rejent, J.J. Martin, J. Metals **55**, 50 (2003)
86. S. Wiese, K.J. Wolter, Microelectr. Reliab. **44**, 1923 (2004)

87. M.M. El-Bahay et al., J. Mat. Sci. Electr. Mat. **15**, 519 (2004)
88. I. Shohji, T. Yoshida, T. Takahashi, S. Hioki, Mat. Sci. Eng. A **366**, 50 (2004)
89. F.Q. Lang, H. Tanaka, O. Munegata, T. Taguchi, T. Narita, Mater. Charact. **54**, 223 (2005)
90. C.H. Lin, D.Y. Chu, J. Mat. Sci. Electr. Mat. **16**, 355 (2005)
91. F. Guo, J.P Lucas, K.N. Subramanian, J. Mat. Sci. Mat. Electr. **12**, 27 (2001)
92. F. Guo, J. Lee, J.P. Lucas, K.N. Subramanian, T.R. Bieler, J. Electr. Mat. **30**, 1222 (2001)
93. S.G. Jadhav, T.R. Bieler, K.N. Subramanian, J.P. Lucas, J. Electr. Mat. **30**, 1197 (2001)
94. H. Rhee, K.N. Subramanian, J. Electr. Mat. **32**, 1310 (2003)
95. H. Rhee, K.N. Subramanian, A. Lee, J.G. Lee, Solder. Surface Mount Technol. **15**, 21 (2003)
96. S. Wiese, F. Feustel, E. Meusel, Sens. Actuators A**99**, 188 (2002)
97. F. Guo, S. Choi, K.N. Subramanian, T.R. Bieler, J.P. Lucas, A. Achari, M. Paruchuri, Mat. Sci. Eng. A**351**, 190 (2003)
98. M. Kerr, N. Chawla, Acta Mater. **52**, 4527 (2004)
99. J.P. Lucas, H. Rhee, F. Guo, K.N. Subramanian, J. Electr. Mat. **32**, 1375–1383 (2003)
100. I. Dutta, C. Park, S. Choi, Mat. Sci. Eng. A **379**, 401 (2004)
101. F.Q. Yang, L.L. Peng, Mat. Sci. Eng. A **409**, 87 (2005)
102. I. Dutta, D. Pan, R.A. Marks, S.G. Jadhav, Mat. Sci. Eng. A **410–411**, 48 (2005)
103. F.Q. Yang, J.C.M. Li, Mat. Sci. Eng. A **201**, 40 (1995)
104. D. Pan, R.A. Marks, I. Dutta, R. Mahajan, S.G. Jadhav, Rev. Sci. Instrum. **75**, 5244 (2004)
105. H.J. Frost, M.F. Ashby, *Deformation Mechanism Maps—The Plasticity and Creep of Metals and Ceramics* (Pergaman, Oxford, 1982), p. 21
106. R.K. Mahidhara, S.M.L. Sastry, I. Turlik, K.L. Murty, Scripta Met. Mat. **31**, 1145 (1994)
107. F.A. Sadykov, N.P. Barykin, I.S. Valeev, V.N. Danilenko, J. Mater. Eng. Perform. **12**(1), 29 (2003)
108. K. Yomogita, Jpn. J. Appl. Phys. **9**, 1437 (1970)
109. A.Z. Mohamed, M.S. Saker, A.M.A. Daiem et al., Phys. Status Solidi A-Appl. Res. **133**, 51 (1992)
110. R. Clark, G.B. Craig, B. Chalmers, Acta Crystallogr. **3**(6), 479 (1950)
111. E. Schmid, W. Boas, *Kristallplastizitat* (Springer, Berlin, 1935)
112. C.S. Barrett, *Structure of Metals* (McGraw-Hill, New York, 1943)
113. C.E. Elam, *Distortion of Metal Crystals* (Clarendon Press, Oxford, 1935)
114. B. Chalmers, Proc. Phys. Soc. **47**, 733 (1935)
115. J.K. Lee, M.H. Yoo, Metall. Trans. A (Phys. Metall. Mater. Sci.) **21A**, 2521–2530 (1990)
116. K. Ishii, H. Kiho, J. Phys. Soc. Jpn. **18**, 1122 (1963)
117. J.D. Eshelby, Prog. Solid Mech. **2**, 89 (1961)
118. T. Mura, *Micromechanics of Defects in Solids*, 2nd edn. (Martinus Nijhoff, Dordrecht, The Netherlands, 1987), Chapter 4
119. R. Clark, G.B. Craig, Progr. Metal Phys. **3**, 135 (1952)
120. E. Votava, H. Hatwell, Acta Met. **8**, 874 (1960)
121. J.T. Fourie, F. Weinberg, F.W.C. Boswell, Acta Metall. **8**, 851 (1960)
122. K. Ishii, J. Phys. IV **6**(C8), 269–272 (1996)
123. J.H. Brunton, M.P.W. Wilson, Proce. Roy. Soc. Lond. Ser. A-Mathemat. Phys. Sci. A **309**(#1498), 345 (1969)
124. W.P. Mason, H.J. Mcskimin, W. Shockley, Phys. Rev. **73**, 1213 (1948)
125. K. Ishii, H. Kiho, J. Phys. Soc. Jpn. **18**, 1133 (1963)
126. K. Ishii, J. Phys. Soc. Jpn. **14**, 1315 (1959)
127. S. Maruyama, J. Phys. Soc. Jpn. **15**, 1243 (1960)
128. M.G. Jon, D.N. Beshers, W.P. Mason, J. Appl. Phys. **45**(9), 3716 (1974)
129. D.R. Overcash, E.P. Stillwel, M.J. Skove, J.H. Davis, Philos. Mag. **25**, 1481 (1972)
130. K.N. Tu, D. Turnbull. J. Metals **21**, A18 (1969)
131. K.N. Tu, D. Turnbull, Acta Metall. **17**(10), 1263 (1969)
132. B.D. Rothberg, Philos. Mag. **25**, 1473 (1972)
133. J.H. Davis, M.J. Skove, E.P. Stillwell, Solid State Comm. **4**, 597 (1966)

Mechanical fatigue of Sn-rich Pb-free solder alloys

J. K. Shang · Q. L. Zeng · L. Zhang ·
Q. S. Zhu

Published online: 12 September 2006
© Springer Science+Business Media, LLC 2006

Abstract Recent fatigue studies of Sn-rich Pb-free solder alloys are reviewed to provide an overview of the current understanding of cyclic deformation, cyclic softening, fatigue crack initiation, fatigue crack growth, and fatigue life behavior in these alloys. Because of their low melting temperatures, these alloys demonstrated extensive cyclic creep deformation at room temperature. Limited amount of data have shown that the cyclic creep rate is strongly dependent on stress amplitude, peak stress, stress ratio and cyclic frequency. At constant cyclic strain amplitudes, most Sn-rich alloys exhibit cycle-dependent and cyclic softening. The softening is more pronounced at larger strain amplitudes and higher temperatures, and in fine grain structures. Characteristic of these alloys, fatigue cracks tend to initiate at grain and phase boundaries very early in the fatigue life, involving considerable amount of grain boundary cavitation and sliding. The growth of fatigue cracks in these alloys may follow both transgranular and intergranular paths, depending on the stress ratio and frequency of the cyclic loading. At low stress ratios and high frequencies, fatigue crack growth rate correlates well with the range of stress intensities or J-integrals but the time-dependent C^* integral provides a better correlation with the crack velocity at high stress ratios and low frequencies. The fatigue life of the alloys is a strong function of the strain amplitude, cyclic frequency, temperature, and microstructure. While a few sets of fatigue life data are available, these data, when analyzed in terms of the Coffin–Mason equation, showed large variations, with the fatigue ductility exponent ranging from -0.43 to -1.14 and the fatigue ductility from 0.04 to 20.9. Several approaches have been suggested to explain the differences in the fatigue life behavior, including revision of the Coffin–Mason analysis and use of alternative fatigue life models.

1 Introduction

In most modern electronic devices, a solder connection provides both mechanical and electronic connections between and within the many levels of an electronic package. Notable examples are found in flip-chip and ball grid array packages where solder bumps or balls serve as the necessary leadless interconnections between two separate electronic components. The solder alloys used to make these interconnects are often subject to cyclic loading resulting from repeated mechanical actions or thermal cycling, leading to fatigue failures of the solder alloy and the interconnect. With increasing device miniaturization and power consumption, the solder alloy must withstand larger fatigue loading and higher temperatures so that the fatigue reliability of the solder alloy becomes a critical issue in determining the lifetime of a functional device. Since the fatigue of a solder alloy in an electronic package may involve multiple deformation and

J. K. Shang (✉)
Department of Materials Science and Engineering,
University of Illinois,Urbana-Champaign, Urbana,
IL 61801, USA
e-mail: jkshang@uiuc.edu

Q. L. Zeng · L. Zhang · Q. S. Zhu · J. K. Shang
Shenyang National Laboratory for Materials Science,
Institute of Metal Research, Chinese Academy of Sciences,
110016 Shenyang, China

fracture processes such as atomic diffusion, dislocation motion, grain boundary activities, and development of fatigue cracks, the life prediction for the solder joint or the reliability control of the electronic device remains a serious challenge.

For Pb–Sn solder interconnects, considerable efforts have been made to address their fatigue reliability [1–3], including fatigue testing of real joints [4–17], constitutive modelling of stresses and deformation in the joint [18–20], and bulk property measurements on solder alloys [21–25]. While the fatigue testing of real joints simulates closely the service condition of the joint and is particularly helpful in sorting out a number of critical issues in the failure of the joint, the results of these tests tend to be joint-geometry specific and have poor predictive power. The constitutive modeling, when coupled with experimental measurements of the solder properties, has the predictive potential. However, its proper implementation requires a clear mechanistic understanding of the individual processes involved. Even though the current understanding of those individual time-dependent deformation and fracture processes is still incomplete, the missing gaps for Pb–Sn solder interconnects have been bridged by the vast amount of testing and evaluation experience accumulated over the past several decades. As a result, these experimental and modeling efforts have combined to achieve considerable success in ensuring adequate fatigue reliability of Pb–Sn solder interconnects in many electronic packages.

As Pb–Sn solder interconnects are replaced by the Pb-free solder systems, the differences in their fatigue behavior must be examined and understood before those approaches developed for Pb–Sn systems may be adopted, especially when the design experience with Pb-free solder interconnects is extremely limited. Some of the differences are to be expected because of the inherent differences between the Pb-free and Pb-bearing solder alloys. For example, most of the leading Pb-free candidates are highly rich in Sn and melt at higher melting temperatures than the Pb–Sn eutectic alloy [26–28]. At a given temperature, these Sn-rich alloys are considerably harder than the PbSn eutectic. Under the same strain, they will be subject to higher stresses, therefore prone to stress-induced failures. The low alloying concentration in these high Sn solders also means that the alloy behavior will be largely determined by the Sn-rich phase, and the second phase and phase boundaries generally would not play as significant roles as in the PbSn eutectic.

In this article, the results from the fatigue studies of Sn-rich Pb-free solder alloys are reviewed to gain an overview of the current understanding of the fatigue behavior of these Pb-free alloys. Since fatigue failure results from complex loading conditions and involves a series of complex processes, the review is divided into five major sections, each dealing with one of the fatigue damage processes, namely, cyclic deformation, cyclic softening, fatigue crack initiation, fatigue crack growth, and fatigue lifetimes. In each section, if the data or observations are available, fatigue behavior is related to cyclic stress amplitude, cyclic strain amplitude, the number of fatigue cycles, stress ratio, frequency, temperature, alloy composition and microstructure. When possible, comparisons are made between Pb-free solder alloys and the PbSn eutectic alloy.

2 Cyclic creep deformation

Under a constant cyclic stress amplitude, Sn-rich solder alloys tend to undergo time-dependent creep deformation. As shown in Fig. 1, the creep deformation followed the typical three-stage behavior for most metals, with an extensive steady-state stage. For the same peak stress, the cyclic creep strain is much smaller than the static creep deformation because of repeated unloading of the alloy under cyclic loading. At room temperature, the difference between the static and cyclic creep varies greatly with the peak stress level, stress ratio, and frequency of cyclic loading. The creep rate is generally slower at lower stress ratios and the dependence on the stress ratio seems stronger for larger peak stresses (Fig. 1c). With more frequent unloading, cyclic creep rate is smaller at higher cyclic frequencies. At a stress ratio of zero, the creep rupture strain is notably smaller than that for the static creep.

The stress dependence of the secondary creep is often expressed as a power-law function of the stress for static creep. The same function may be used for cyclic creep of Sn-rich alloys, as illustrated in Fig. 2 for the Sn–3.8Ag–0.7Cu alloy at a stress ratio of zero. For the static loading, the steady-state creep rate, $\dot{\varepsilon}_s$, can be expressed as

$$\dot{\varepsilon}_s = C\sigma^n \tag{1}$$

where σ is the applied stress, and C and n are the fitting constants. For Sn–3.8Ag–0.7Cu alloy, the exponent n is equal to 10.8 at room temperature. The stress exponent does not change much from the static loading to the cyclic loadings at two different frequencies.

Assuming that the dominant creep mechanisms remain the same under the static and cyclic loading, the

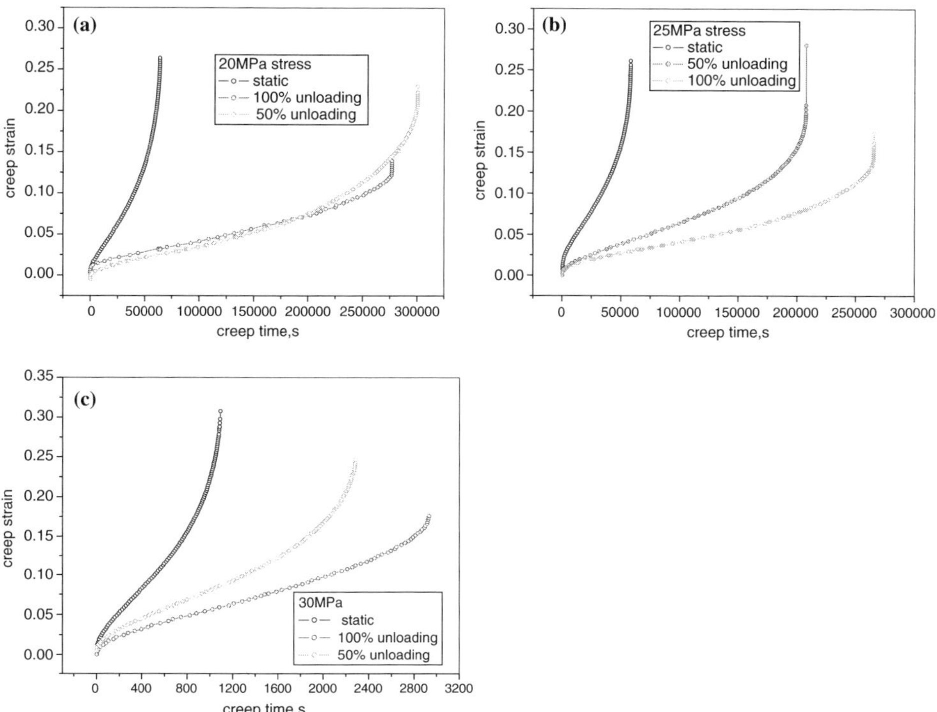

Fig. 1 Cyclic creep curves of Sn–3.8Ag–0.7Cu alloy at 298 K under (**a**) low stress, (**b**) intermediate stress, and (**c**) high stress

cyclic steady-state creep rate, $\dot{\varepsilon}_c$, may be derived from the static creep rate given in Eq. 1. For a triangular waveform, the cyclic stress varies linearly with time, t, as follows:

$$\sigma = 2v\sigma_p t \quad (0<t<1/2v) \tag{2}$$

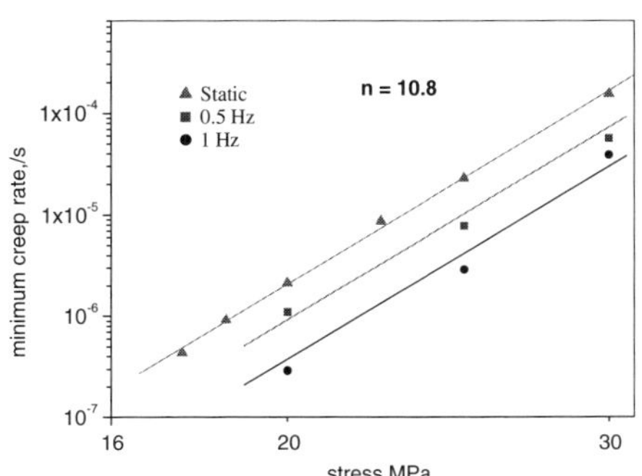

Fig. 2 Dependence of cyclic creep rate on the peak stress in Sn–3.8Ag–0.7Cu alloy

where σ_p is the peak stress and v is the cyclic frequency. Within a half cycle, the total creep strain accumulates to

$$\varepsilon_{tot} = \int_0^{1/2v} C(2v\sigma_p t)^n \mathrm{d}t \tag{3}$$

The average creep rate can then be obtained from the ratio of the total strain in Eq. 3 and the half period, $1/2v$:

$$\dot{\varepsilon}_c = \frac{C\sigma_p^n}{n+1} \tag{4}$$

or

$$\frac{\dot{\varepsilon}_s}{\dot{\varepsilon}_c} = n+1$$

The calculated cyclic creep rate from Eq. 4 is compared to the experimental results in Fig. 3. The calculated creep rate ratio is 2–3 times larger than the experimental values, indicating that the cyclic loading

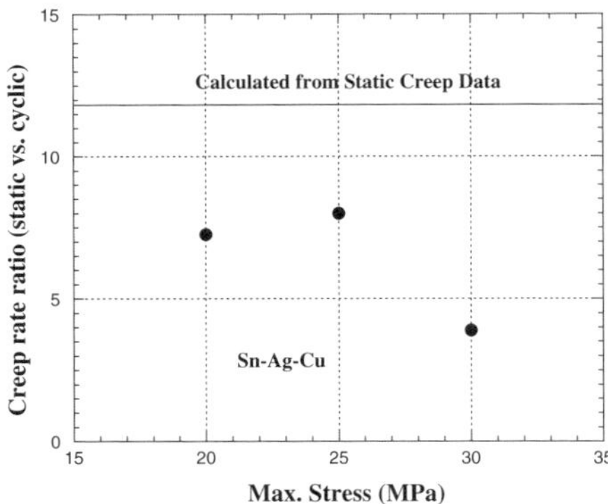

Fig. 3 Comparison between the calculated and experimental cyclic creep rates in SnAgCu Alloy

causes an acceleration of the creep process. The acceleration means that cyclic loading introduces additional mechanisms beyond those in the static creep. Some of the cyclic damage mechanisms are discussed in Sect. 3 below.

3 Cyclic softening

The low cycle fatigue behavior of solder alloys is often examined by conducting fatigue studies at constant strain amplitudes. Under most test conditions, Sn-rich alloys tend to exhibit softening under cyclic loading [29–36]. The softening behavior may appear as cycle-dependent or cycling dependent, as observed in the Sn–Pb eutectic alloy [22, 37].

The low cycle fatigue response of Sn-rich alloys is determined by its microstructure. As shown in Fig. 4, the eutectic or near-eutectic Sn-rich alloys tend to solidify into two equiaxed or dendritic microstructures. The equiaxed microstructures consist of equiaxed Sn-rich grains and a dispersion of a second phase whereas the dendritic microstructures include dendrites of β-Sn surrounded by the fine two-phase eutectic mixture. The dendritic microstructure may evolve into an equiaxed microstructure upon recrystallization. In the dendritic microstructure, cyclic deformation tends to concentrate in the dendritic phase, as shown by the numerous slip bands in Fig. 5a). In equiaxed grain structure, slip bands are also evident in the grains but grain boundary sliding and microcrack develop early (<5% of the life) in the fatigue life along the grain boundaries [36] (Fig. 5b).

With the slip activities confined to the Sn dendrite, dendritic microstructure maintains a steady cyclic stress response under constant strain amplitudes, as

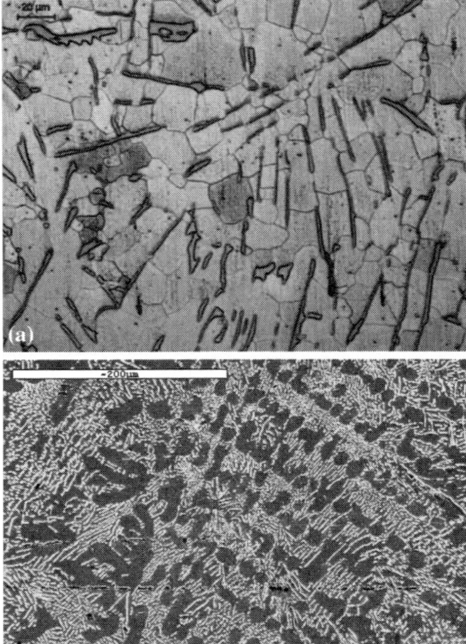

Fig. 4 As-cast microstructure of Sn–3.8Ag–0.7Cu alloy: (**a**) equiaxed and (**b**) dendritic

shown in Fig. 6a). The peak stress stays stable for a great portion of the fatigue life before it falls rapidly later in the fatigue life when fatigue crack growth dominates the cyclic stress response. By comparison, the peak stress in the equiaxed microstructure begins to decay very early in the fatigue life and does not recover to a steady-state, culminating in a cycle-dependent softening behavior (Fig. 6b). Such a cycle-dependent softening behavior is apparently more pronounced in Sn–3.5Ag [31] and Sn–0.7Cu alloys [35] where about $1/4$ drop in the peak stress occurs in the initial five percent of the fatigue life (Fig. 7).

The difference in the peak stress response also extends to the other parts of a fatigue cycle [31, 35, 36]. As shown in Fig. 8, the hysteresis loops reach a stable configuration for the dendritic microstructure in the first 2,000 cycles. For the equiaxed microstructure, the size and shape of the hysteresis loop after 10 cycles already look quite different from the initial configuration. As the number of fatigue cycles increases, the hysteresis loop becomes smaller and smaller (Fig. 8b). The contraction of the hysteresis loop is associated with the reductions both in the plastic flow stress and in the slope of the elastic segment of the stress–strain loop.

The difference in the cyclic behavior between the dendritic and equiaxed microstructures is also evident

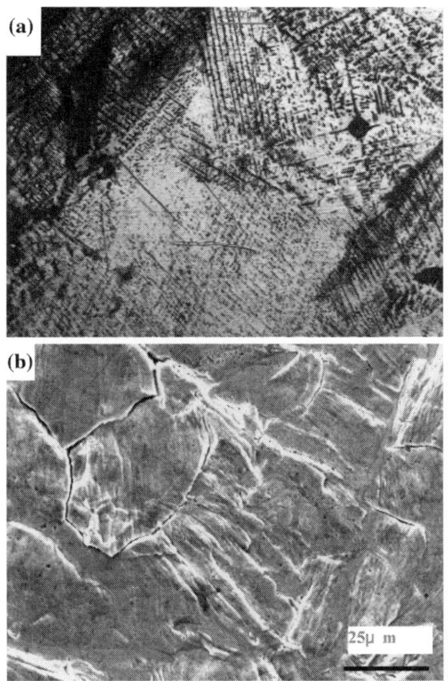

Fig. 5 Cyclic deformation mechanism of Sn–3.8Ag–0.7Cu alloy: (a) slip bands in β-Sn dendrite, and (b) grain boundary cracking in equiaxed microstructure

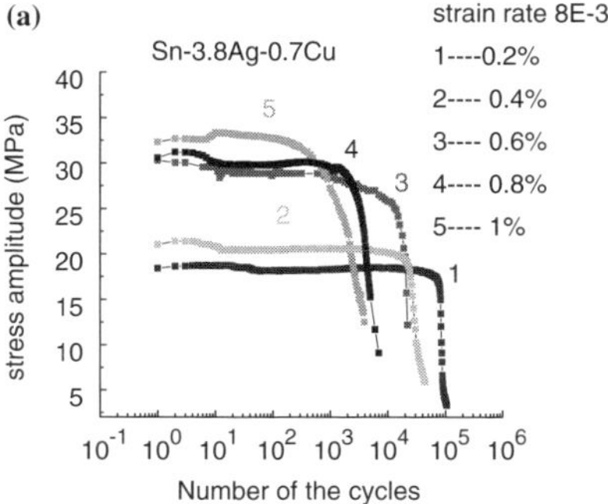

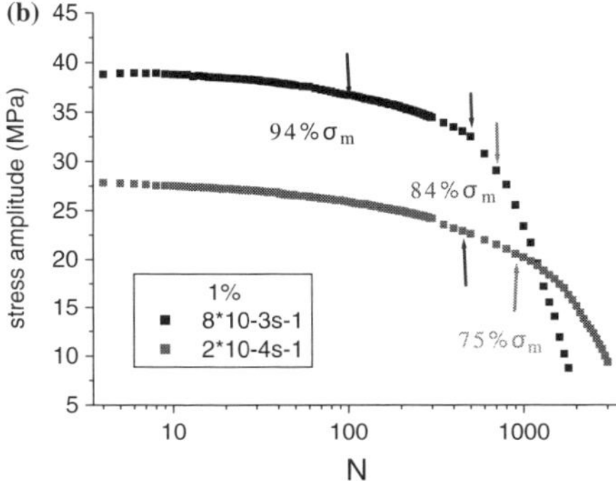

Fig. 6 Cyclic stress behavior of Sn–3.8Ag–0.7Cu alloy under constant strain amplitude loading in: (a) dendritic and (b) equiaxed microstructure

in the cyclic stress–strain curves. As shown in Fig. 9, the cyclic stress–strain curves of the dendritic microstructure appear very similar at 10% and 20% of the fatigue lives [38]. Compared to the initial stress–stress response, the dendritic microstructure showed a slight cyclic softening. The steep rise of the cyclic stress–strain curve at 10% fatigue life indicates that the alloy has a strong tendency to work-harden under cyclic loading. By contrast, the equiaxed microstructure experiences significant cyclic softening; the cyclic stress–strain curve varies greatly with the number of fatigue cycles (Fig. 9b). The much shallower slopes of the cyclic stress–strain curves point to a very weak cyclic work-hardening. The work-hardening rate may decrease further with increasing temperature [39] or decreasing cyclic frequency [40] (Fig. 10).

The cycle-dependent softening has also been observed in ferrous and aluminum alloys [41]. However, the behavior in those alloys is typically transient in that it appears in the initial portion (a few hundred cycles) of fatigue cycling after which the peak stress becomes stable. In contrast, cycle-dependent softening in equiaxed Sn-rich alloys persists over most of the

fatigue life until the final fracture. As a result, equiaxed Sn-rich alloys demonstrate no steady-state cyclic response and no unique relationship between cyclic strain amplitude and cycle stress amplitude under constant strain amplitude cycling. When such a cycle-dependent softening is compared with the time-dependent creep deformation under constant stress amplitude cycling as discussed in Sect. 2, it is clear that the relationship of cyclic stress amplitude vs. cyclic strain amplitude is quite different between stress-controlled and strain controlled fatigue loading.

The cycle-dependent softening rate of Sn-rich alloys is strongly dependent on strain amplitude, temperature and microstructure. As shown in Fig. 11, increasing the strain amplitude leads to a faster softening rate and ultimately a shorter fatigue life. At smaller strain amplitudes (<1.0%), the softening proceeded slowly initially so that the fatigue lifetime may be divided into a slow damage accumulation step and a faster crack growth process. At higher strain amplitudes, the softening is so fast that it is difficult to separate the two fatigue processes. The dependence of the softening

 Springer

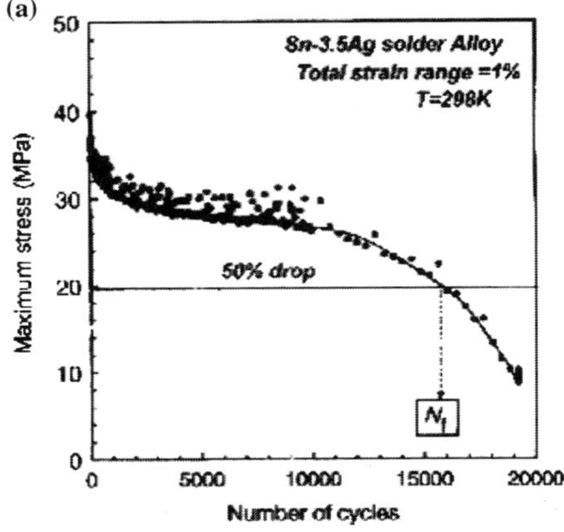

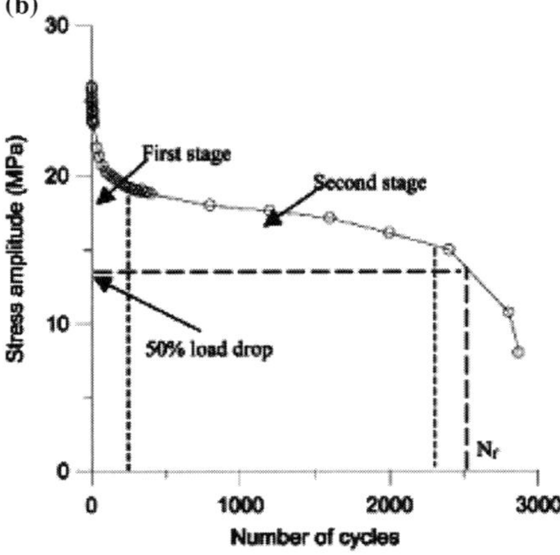

Fig. 7 Cycle-dependent softening of Sn-rich alloys: (**a**) Sn–3.5Ag [31] and (**b**) Sn–0.7Cu alloy [35]

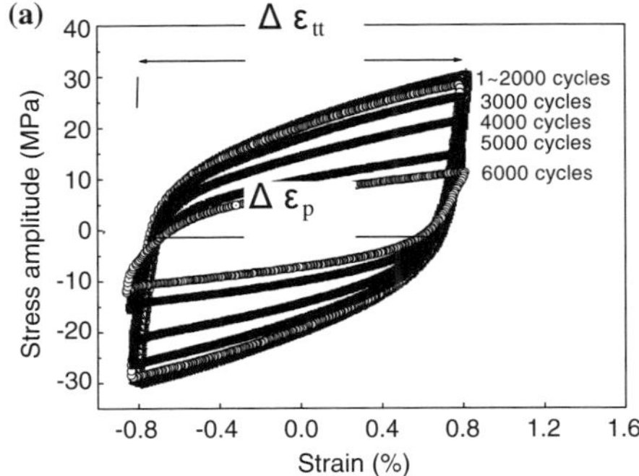

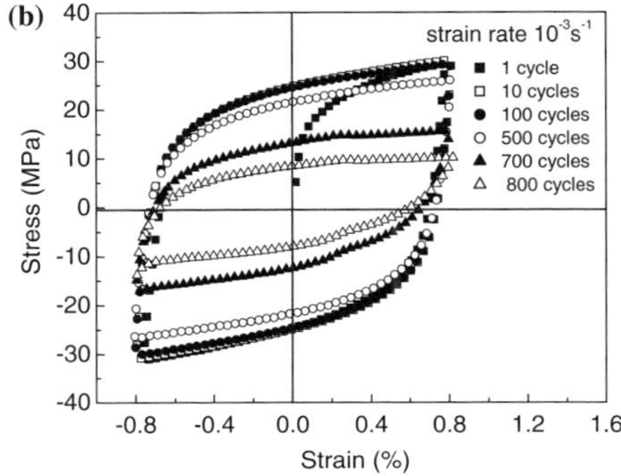

Fig. 8 Hysteresis loops of Sn–3.8Ag–0.7Cu alloy: (**a**) stable loops in dendritic and (**b**) unstable loops in equiaxed microstructure

rate on temperature is presented in Fig. 12, where the softening rate is seen to increase with temperature. Again, the cyclic response at lower temperatures (<60 °C) includes a slow damage accumulation stage whereas no clear separation could be made between the initial damage accumulation and final crack growth at higher temperatures.

The effect of the grain size on cyclic softening is given in Fig. 13, where a comparison is made in the cyclic response between an as-cast and an equal channel angularly pressed (ECAP) Sn–3.8Ag–0.7Cu alloy. In the as-cast condition, the alloy has a coarse-grained microstructure mixed with needle-shaped Ag_3Sn intermetallic phase. The ECAP results in both grain refinement and breakup of needle-shaped Ag_3Sn

intermetallic phase. At the same cyclic strain amplitude, the peak stress of the ECAPed microstructure is much higher initially but falls rapidly with fatigue cycling. Eventually, the fine-grained microstructure reaches a shorter fatigue life.

Although the cyclic deformation of Sn-rich alloys is highly time-dependent as discussed in Sect. 2, Fig. 14 indicates that the cycle-dependent softening in the Sn–3.8Ag–0.7Cu does not bear a clear relation with strain rate within the range of 10^{-4}/s to 10^{-2}/s. At the high strain rates, the softening is slightly faster at a higher strain rates whereas a much lower strain rate appears to accelerate the softening process.

The cycle-dependent behavior has been understood in terms of the grain boundary microcracking model shown in Fig. 15, where an equiaxed grain structure is weakened by a distribution of microcracks along some grain boundaries [36]. At a given microcrack density, ω, the elastic modulus and Poisson's ratio of the solid are reduced to $\bar{E}$ and $\bar{v}$ from the initial values, E_0 and v_0[42]:

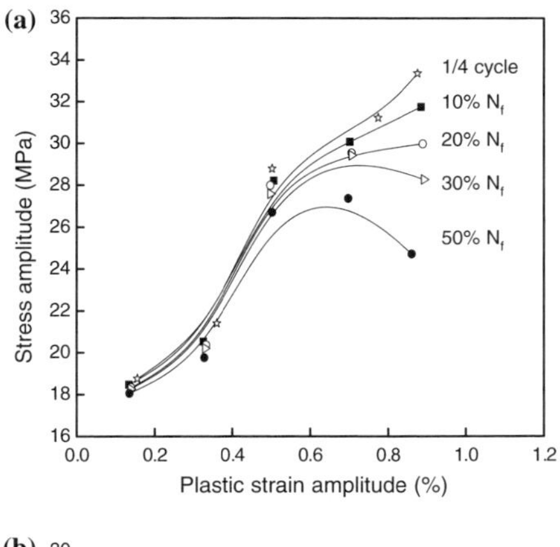

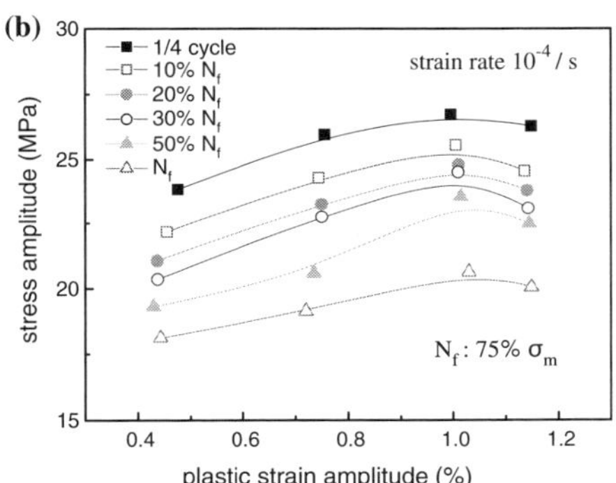

Fig. 9 Cyclic stress–strain curves of Sn–3.8Ag–0.7Cu alloy: (a) dendritic and (b) equiaxed microstructure

$$\bar{E} = E_0 \left[1 - \frac{16(1 - \bar{v}^2)(10 - 3\bar{v})}{45(2 - \bar{v})} \omega \right] \tag{5a}$$

$$\omega = \frac{45(v_0 - \bar{v})(2 - \bar{v})}{16(1 - \bar{v}^2)[10v_0 - \bar{v}(1 + 3v_0)]} \tag{5b}$$

At a constant strain amplitude, the stress amplitude will be reduced to $\bar{\sigma}$ from the initial reading of σ_0 so that

$$\frac{\bar{\sigma}}{\sigma_0} = 1 - \frac{16(1 - \bar{v}^2)(10 - 3\bar{v})}{45(2 - \bar{v})} \omega \tag{6}$$

For a polycrystalline metal, fatigue damage accumulation may be written as a power-law function of the strain amplitude [43], resulting in:

$$\frac{\bar{\sigma}}{\sigma_0} = 1 - \frac{16(1 - \bar{v}^2)(10 - 3\bar{v})}{45(2 - \bar{v})} A\varepsilon_a^\chi N \tag{7}$$

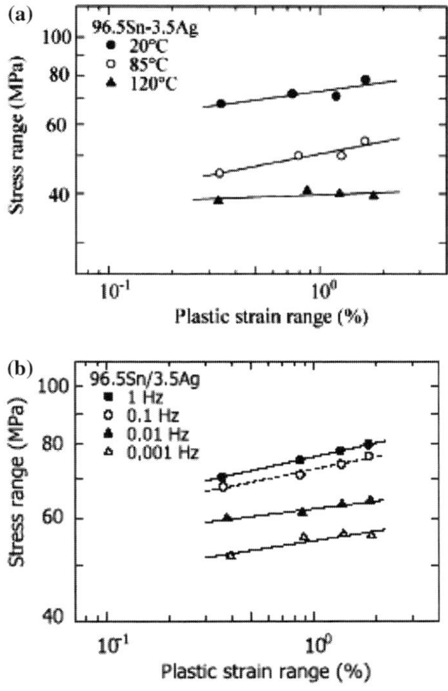

Fig. 10 Cyclic stress–strain response of Sn–3.5Ag alloy: (a) at different temperatures [39] and (b) at different frequencies [40]

where ε_a is the strain amplitude, N, the number of fatigue cycle, A and χ, material constants [36]. A and χ are obtained from experimental measurements by applying the percolation theory to the fatigue life data [36, 44].

The predicted variations of the peak stress with the number of fatigue cycles are shown in Fig. 16 for three

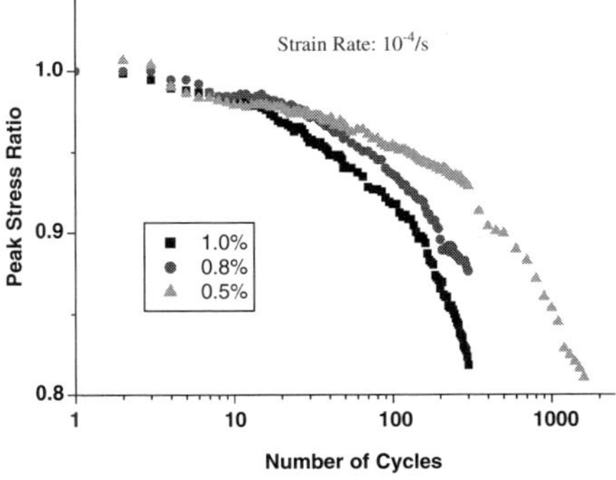

Fig. 11 Cycle-dependent softening of Sn–3.8Ag–0.7Cu alloy at different strain amplitudes

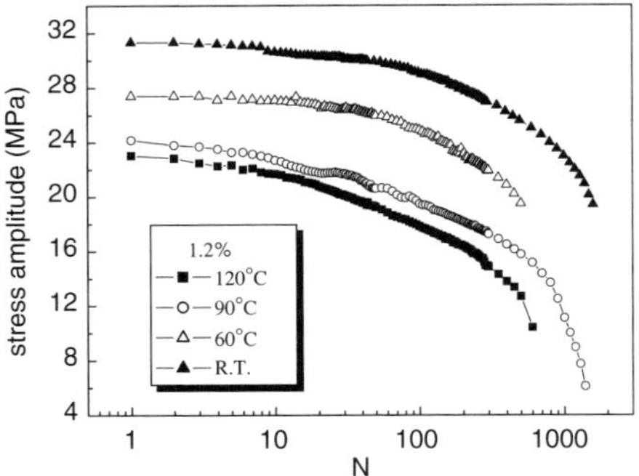

Fig. 12 Cycle-dependent softening of Sn–3.8Ag–0.7Cu alloy at different temperatures

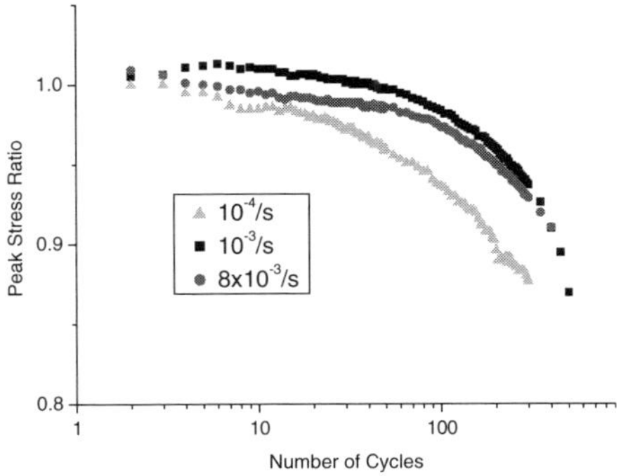

Fig. 14 Cycle-dependent softening of Sn–3.8Ag–0.7Cu alloy at different strain rates (strain amplitude = 0.8%)

different strain amplitudes. Note that the model is valid above peak stress ratio $\bar{\sigma}/\sigma_0 = 0.84$ where the percolation threshold is reached. Within the experimental errors, the model predictions agree closely with the experimental results and thus capture the dependence of the cycle-dependent softening behavior on strain amplitude very well at room temperature.

When the model is applied to different temperatures, the agreements with the experimental results at low temperatures appear quite good (Fig. 17a and b). However, at high temperatures, the model underestimates the softening rate (Fig. 17c), indicating that additional damage process may take place. Further examinations of the high temperature fatigue sample indicate that the alloy undergoes dynamic recrystallization during high temperature fatigue. In the recrystallized regions, a much finer grain structure emerges and many grain boundary cracks develop to weaken the local microstructure.

The microcrack model is also useful understanding the rapid cyclic softening observed in the ECAPed microstructure. As shown in Fig. 18, the application of the microcrack model to the ECAPed Sn–3.8Ag–0.7Cu alloy results in close agreement with experimental results. Therefore, the much faster softening of the ECAPed microstructure shown in Fig. 13 can be explained by the stronger tendency of grain boundary cracking in fine-grained microstructure.

4 Fatigue crack initiation

The initiation of fatigue cracks in Sn-rich alloys bears a close relationship with the alloy microstructure. In dendritic microstructures, accumulation of cyclic

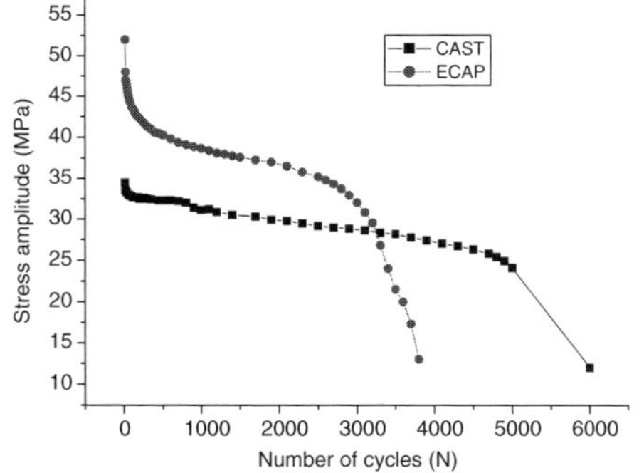

Fig. 13 Comparison of cycle stress behavior in coarse-grained cast and fined-grained ECAP Sn–3.8Ag–0.7Cu alloy

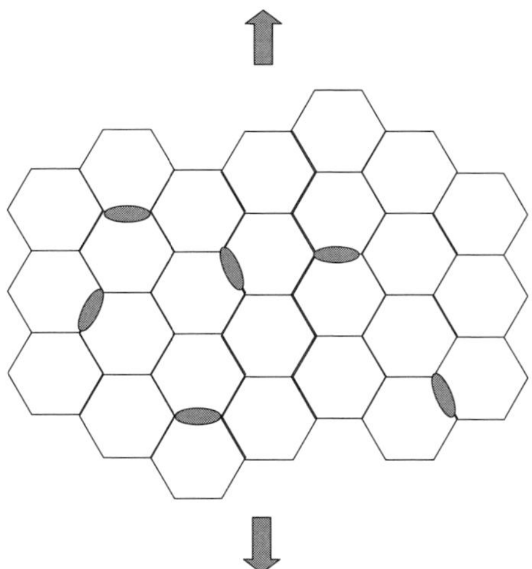

Fig. 15 Schematic illustration of the microcrack model used to analyze cycle-dependent softening behavior in equiaxed microstructure

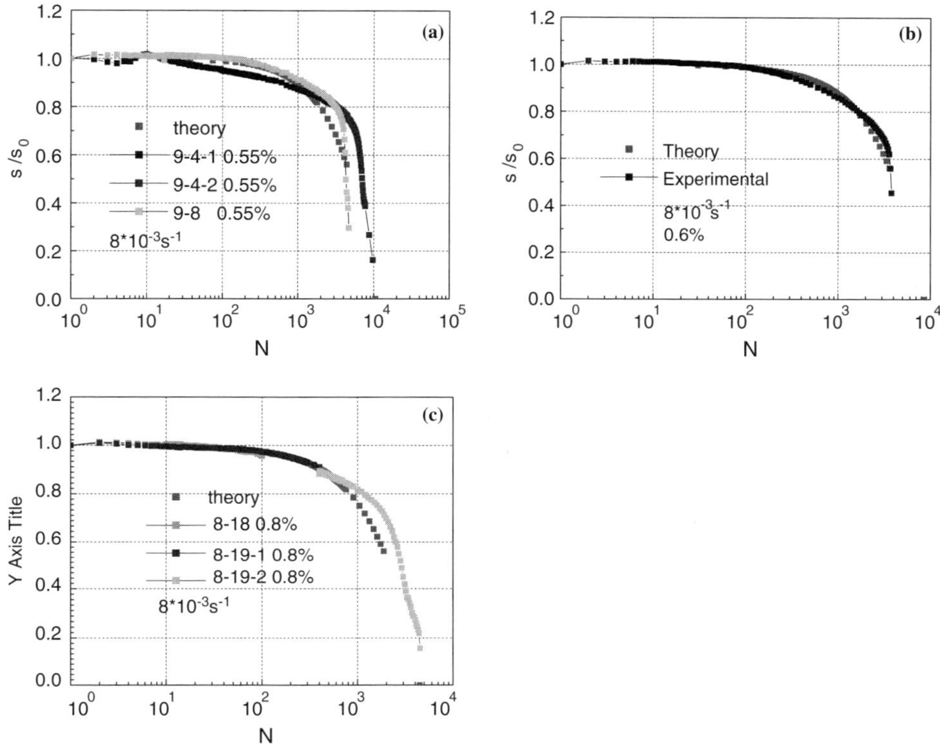

Fig. 16 Comparisons of the microcrack model predictions with experimental results in Sn–3.8Ag–0.7Cu alloy at strain amplitude of (a) 0.55%, (b) 0.6%, and (c) 0.8%

deformation in the β-Sn dendrite leads to fatigue crack initiation along the boundary between the β-Sn dendrite and the eutectic mixture or inside the β-Sn dendrite [38, 40, 45]. At low frequencies, fatigue cracks tend to initiate at the phase boundaries, following the outlines of the β-Sn dendrites (Fig. 19a). In the early stage, the fatigue damage may consist of individual cavities with a dimension of 2–3 μm at the interface (Fig. 19b). The subsequent growth and coalescence of the interfacial cavities culminate in formation of fatigue cracks along the interface. When the dendritic microstructure is subject to high frequency loading, fatigue cracks prefer to initiate in the β-Sn dendrite along the subgrain boundaries [40].

In Sn-rich alloys with equiaxed microstructures, fatigue cracks tend to initiate on the grain boundaries [34, 36, 46] (Fig. 20) and prefer high angle grain boundaries [47]. Early in the fatigue life, the individual microcracks are isolated and widely separated. As the density of microcracks increases with continued cycling and approaches the percolation threshold, individual

microcracks may link together to form a fatigue crack whose subsequent growth leads to eventual fatigue failure of the alloy. As illustrated in Fig. 20, the fatigue crack seems to seek out the recrystallized areas or regions with fine grains or high density of grain boundaries. The presence of elongated second phase can provide an effective barrier to the linkage of the microcracks (Fig. 20).

5 Fatigue crack growth

Fatigue crack growth in Sn-rich alloys depends on a number of mechanical, microstructural, and environmental variables, such as the fracture mechanics driving force, stress ratio, cyclic frequency, and temperature. Under small scale yielding condition, the linear elastic fracture mechanics parameter, the stress intensity range, ΔK, is often used as the driving force for fatigue crack growth. For large scale yielding, the range of J-integral, ΔJ, provides a better correlation with the

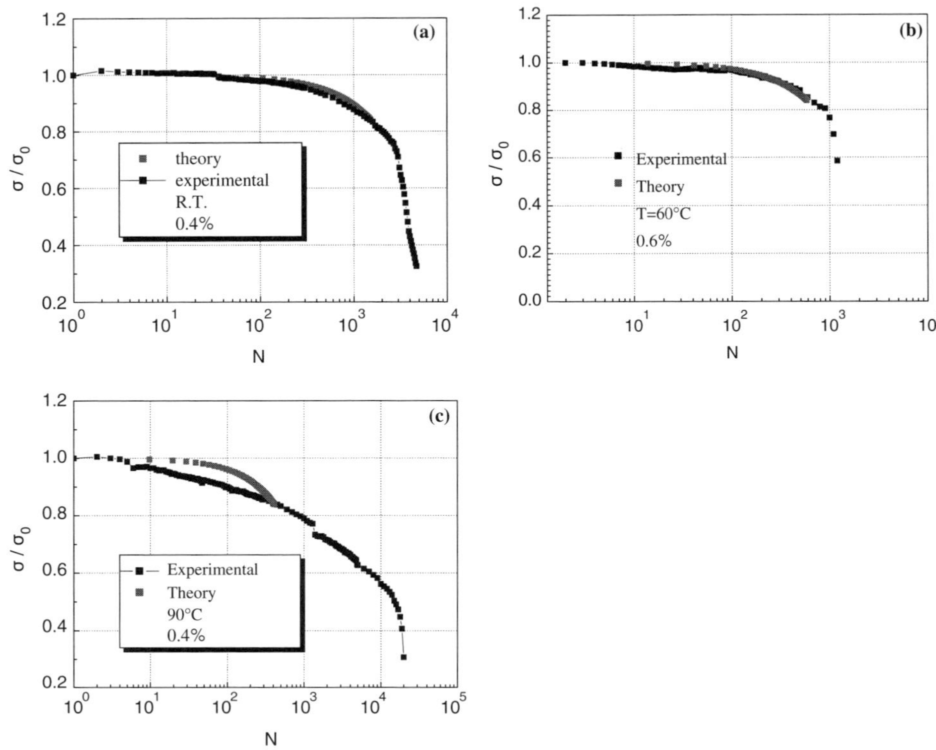

Fig. 17 Comparisons of the microcrack model predictions with experimental results in Sn–3.8Ag–0.7Cu alloy at (**a**) 25°C, (**b**) 60°C, and (**c**) 90 °C

fatigue crack growth. If the time-dependent creep deformation is involved in the fatigue crack growth, time-dependent integral, C^*, may be used.

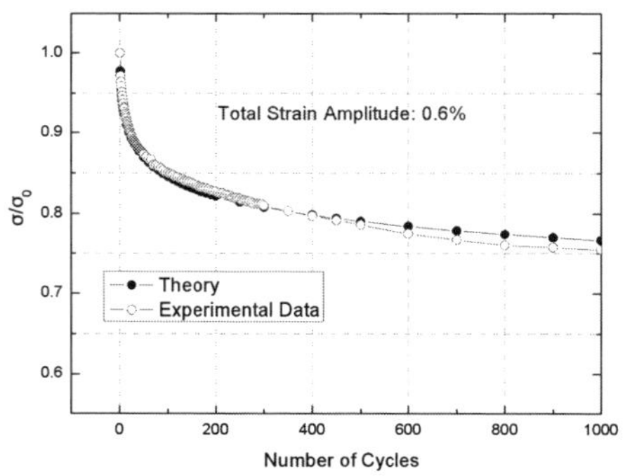

Fig. 18 Comparison of the microcrack model prediction with experimental results in fine-grained ECAP Sn–3.8Ag–0.7Cu alloy

Since Sn-rich alloys are exposed to high homologous temperatures in microelectronic applications, the contribution from the time-dependent deformation must be considered. Zhao et al. [48–51] have shown that the time-dependent processes become dominant at high stress ratios and low frequencies. As illustrated in Fig. 21, on a stress-ratio versus frequency diagram, the fatigue crack growth of Sn-rich alloys may be characterized as either predominantly cycle-dependent or time-dependent [51]. In the cycle-dependent regime, fatigue crack growth is well described by time-independent fracture mechanics parameter, ΔK or ΔJ. On the other hand, the time-dependent parameter, C^*, must be used in the time-dependent regime. The boundary between the two regimes may shift with temperature and strength of the solder alloy. Increasing the testing temperature or reducing the flow strength of the solder alloy expands the regime of the time-dependent crack growth.

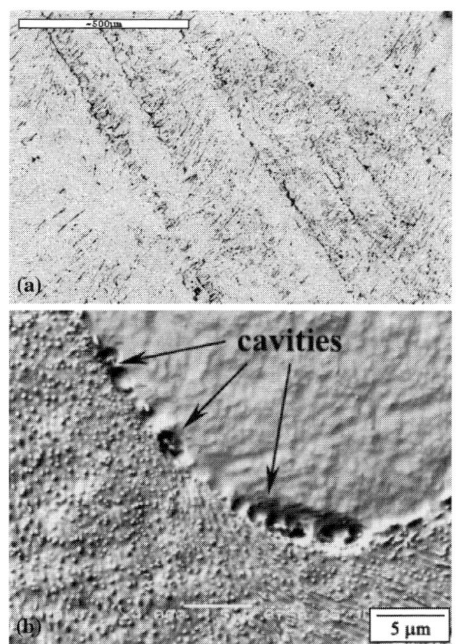

Fig. 19 Fatigue crack initiation sites in dendritic (**a**) Sn–3.8Ag–0.7Cu alloy and (**b**) Sn–3.5Ag alloy [40]

The cycle-dependent fatigue crack growth behavior of common solder alloys is shown in Fig. 22 as a function of the effective stress intensity range, after allowing for the crack closure at low stress ratios. The data from both Pb-bearing and Pb-free solder alloys fall into a narrow band despite the differences in alloy composition and fatigue loading condition. The relationship between the fatigue crack growth rate, d*a*/d*N*, and the effective stress intensity range, ΔK_{eff}, may be expressed as [51]

Fig. 20 Fatigue crack initiation in equiaxed Sn–3.8Ag–0.7Cu alloy. Note that linkage of the grain boundary cracks are blocked by a needle-shaped intermetallic phase

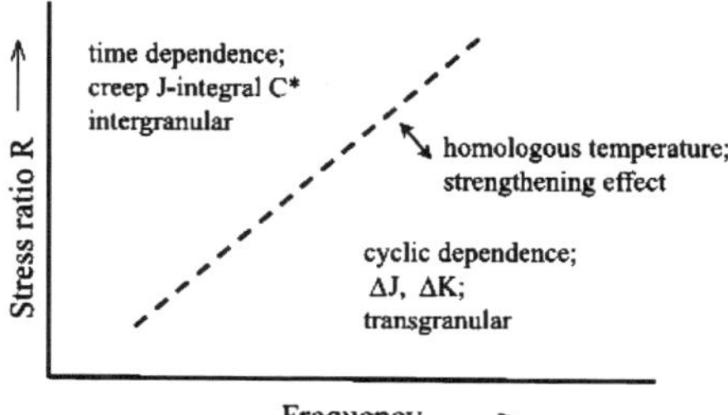

Fig. 8. Schematics of FCG processes in the solders.

Fig. 21 Schematic illustration of the dominant fatigue crack growth process in solder alloys [51]

$$\mathrm{d}a/\mathrm{d}N = 6.83 \times 10^{-2}(\Delta K_{\text{eff}}/E)^{4.05} \tag{8}$$

where E is the Young's modulus of the solder alloy.

For time-dependent fatigue crack growth, the fatigue crack growth behavior is characterized in Fig. 23 in terms of the crack velocity, d*a*/d*t*, as a function of the time-dependent fracture mechanics parameter, C^*. For comparison, the results from creep crack growth (CCG) experiments are also included for the solder alloys. The kinetics of fatigue crack growth appears comparable to that of the creep crack growth. The crack velocity can be expressed as a power-law function of C^* [51]:

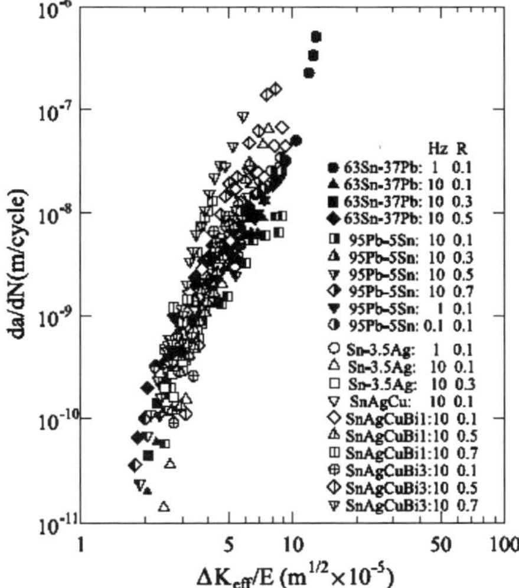

Fig. 22 Cycle-dependent fatigue crack growth in solder alloys [51]

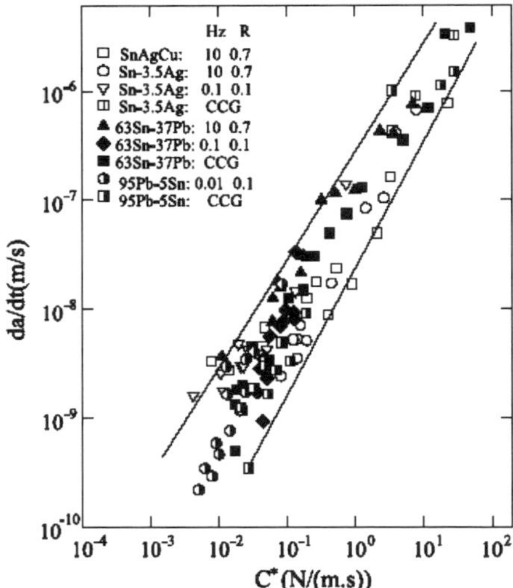

Fig. 23 Time-dependent fatigue crack growth in solder alloys [51]

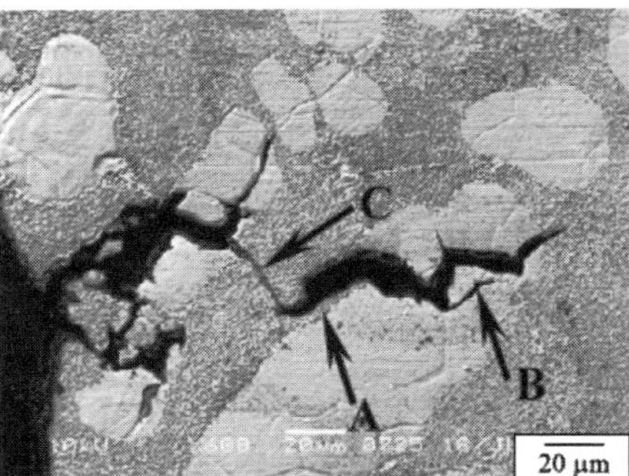

Fig. 24 Fatigue crack growth path in dendritic Sn–Ag alloy along dendrite boundary (arrow A), subgrain boundary (arrow B) and across the eutectic (arrow C) [40]

$$\mathrm{d}a/\mathrm{d}t = 8.34 \times 10^{-8} C^{*1.08} \tag{9}$$

for both Pb-bearing and Pb-free solders tested in the study.

The relationship of fatigue crack growth path to the solder microstructure is shown in Fig. 24 for a dendritic microstructure. As discussed in Sec. 4, fatigue cracks may initiate either inside, or along the phase boundary of, the dendrite phase. They can propagate transgranularly through the stronger eutectic mixture [40] (Fig. 24). In the equiaxed microstructure, the initial microcracks are concentrated on the grain boundaries but the subsequent crack growth may follow a mixture of transgranular and intergranular path. If the fatigue crack growth is predominantly time-dependent, the crack prefers an intergranular path (Fig. 25b). In contrast, more transgranular crack growth is observed for cycle-dependent fatigue crack growth in Sn-rich alloys [48] (Fig. 25a).

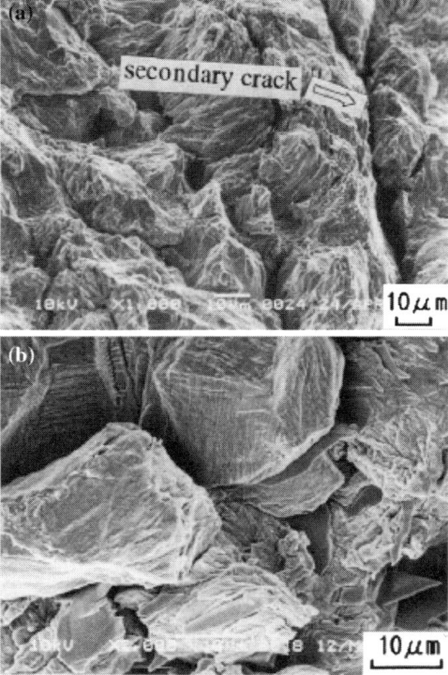

Fig. 25 Fatigue fracture surfaces in Sn–3.5Ag solder alloy [48]: (a) transgranular and (b) intergranular

6 Fatigue Life

Fatigue life behavior of solder alloys is generally characterized in terms of strain vs. life relationship following the Coffin–Mason approach. Since Sn-rich alloys have high ductility at high homologous temperatures, the sample may not break completely apart under constant strain amplitude loading. Accordingly, the fatigue life is often taken as the number of fatigue cycles when the peak stress has fallen to a fraction (e.g., 50% or 75%) of the initial value.

When the results of the low cycle fatigue tests are plotted on a double-log diagram, fatigue life data tend to scatter around a negatively sloped line, as shown in Fig. 26 for the equiaxed Sn–3.8Ag–0.7Cu alloy. The

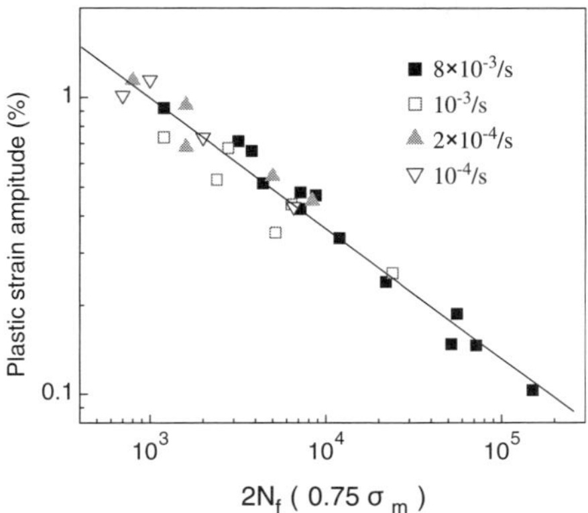

Fig. 26 Fatigue life behavior of Sn–3.8Ag–0.7Cu alloy with an equiaxed grain structure

fatigue life can be written in form of the Coffin–Manson equation,

$$\Delta\varepsilon_p/2 = \varepsilon_f'(2N_f)^c \tag{10}$$

where $\Delta\varepsilon_p$ is the plastic strain range, ε_f' and c are material constants, fatigue ductility coefficient and fatigue ductility exponent, respectively.

Values of the fatigue ductility coefficient and exponent are listed in Table 1 for a number of Sn-rich alloys and SnPb eutectic alloy [32, 34, 35, 52]. Both materials constants vary widely with the alloy composition and testing condition. The exponent, c, ranges from -0.43 to -1.14 and the coefficient, ε_f', from 0.04 to 20.9. The ranges are much greater than the variations for ferrous alloys where c is roughly equal to -0.6 and ε_f' correlates well with the tensile ductility [41].

While some of the variations in c and ε_f' may be related to the effects of alloy microstructure and mechanical variables, considerable uncertainties may have also arisen from lack of sufficient experimental data. For example, it is uncommon to find that the Coffin–Mason parameters have been obtained from a line drawn from only three data points even though fatigue data are well known to have a large scatter. In the low cycle regime, the errors can be quite significant. When the results from three different sources are compared for Sn–3.5Ag alloy, the divergence is quite significant except for the area around the point of 1% strain amplitude (Fig. 27).

Despite those uncertainties, some general trends emerge on the effects of microstructure and fatigue test variables. Compared to the SnPb eutectic alloy, the eutectic Sn–3.5Ag alloy provides a superior fatigue resistance with roughly one order of magnitude longer

Table 1 The Coffin–Mason parameters for Pb-free solder alloys

Alloy	c	ε_f'	Source
Sn–3.5Ag	−0.50	0.6364	[32]
	−0.93	20.8628	[34]
	−0.70	3.1800	[50]
Sn–3.5Ag–0.5Cu	−0.73	3.0685	[34]
	−0.6659	2.2632	[50]
Sn–3.8Ag–0.7Cu	−0.5279	0.9181	[38]
	−0.44	4.84	[36]
	−0.46	5.74	[36]
Sn–3.5Ag–1Cu	−40.43	0.3031	[32]
Sn–3.5Ag–2Cu	−0.44	0.2849	[32]
Sn–3.5Ag–2Bi	−0.55	0.3441	[32]
Sn–3.5Ag–5Bi	−0.60	0.1667	[32]
Sn–3.5Ag–10Bi	−0.54	0.04362	[32]
Sn–3.5Ag–1Zn	−0.55	0.6296	[32]
Sn–3.5Ag–2Zn	−0.49	0.3933	[32]
Sn–3.5Ag–2In	−0.50	0.4879	[32]
Sn–3.5Ag–5In	−0.50	0.2970	[32]
Sn–3.5Ag–0.5Cu–1Bi	−1.14	62.81	[34]
Sn–3.5Ag–0.5Cu–3Bi	−0.96	5.5441	[34]
Sn–0.7Cu	−0.973	20.9051	[35]
	−0.8076	5.1546	[35]
	−0.92	11.8352	[35]
	−0.99	4.4409	[35]
	−0.9022	9.3259	[35]
	−0.7825	2.864	[35]
	−0.7192	1.6298	[35]
SnPb	−0.6083	1.2017	[50]

fatigue lives. The addition of a ternary element tends to degrade the fatigue resistance [31, 32]. At concentrations of up to 2%, Cu, In and Zn only cause a slight decrease in the fatigue life of Sn–3.5Ag% alloy. However, significant reductions in fatigue resistance follow the addition of Bi to the Sn–3.5Ag alloy. The fatigue life decreases as Bi concentration increases so that no more than 2% Bi should be recommended if the Sn–Ag–Bi alloy is to maintain a fatigue resistance comparable to that of the eutectic Sn–Pb alloy. The

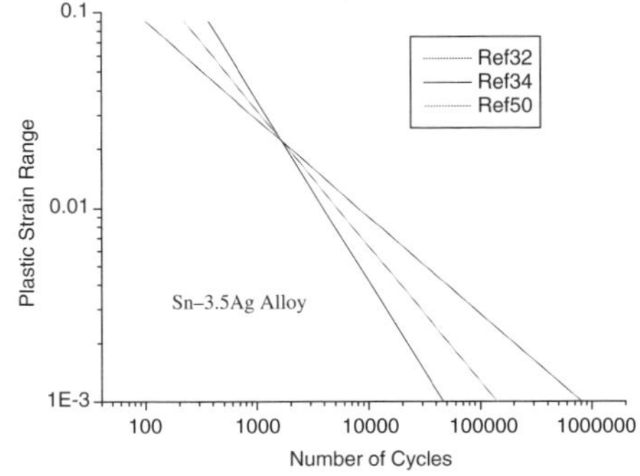

Fig. 27 Comparison of strain-life behavior of Sn–3.5Ag alloy from different sources

detrimental effect of Bi is apparently related to the decrease in the tensile ductility of the alloy with increasing Bi concentration. As shown in Fig. 28, modification of the Coffin–Mason equation by normalizing the strain amplitude with the tensile ductility, D, results in a unique description of the fatigue life behavior of the ternary Sn–Ag–X alloys as following:

$$(\Delta\varepsilon_p/2D \cdot N_f^n = C) \tag{11}$$

where $\alpha = 0.51$ and $C = 0.42$ [32].

For the same alloy, the microstructure also plays a significant role. As shown in Fig. 29 for the eutectic Sn–Ag–Cu alloy, the dendritic microstructure is superior to the equiaxed one, in consistent with the cyclic softening caused by early grain boundary cracking in the equiaxed microstructure. For strain rates between 10^{-4} to 10^{-2}/s,

the room temperature fatigue life of the equiaxed Sn–3.8Ag–0.7Cu is not strongly dependent on strain rate (Fig. 26). However, the dendritic Sn–3.8Ag–0.7Cu and Sn–3.5Ag alloy show strong frequency dependence as shown in Fig. 30, where the fatigue life decreases with decreasing cyclic frequency [40].

The influence of frequency on fatigue life can be described in terms of a frequency-modified Coffin–Manson relationship [35, 41, 53, 55]:

$$(N_f v^{k-1})\alpha\Delta\varepsilon_p = \theta \tag{12}$$

where v is the cyclic frequency, k is frequency exponent, and α and θ are materials constants. For Sn–0.7Cu alloy tested at 398 K, the frequency exponent, k is equal to 0.91 [53]. As shown in Fig. 31 for Sn–0.7Cu [35] and Sn–3.5Ag alloy [40], Eq. 12 provides excellent correlation of fatigue life curves over a wide range of cyclic frequencies.

Alternatively, the frequency dependence of the fatigue life may be analyzed by a frequency-modified Morrow energy model, where the fatigue life is correlated with the plastic strain energy density, W_p [35, 53, 54, 56]. The Morrow model may be further modified by normalizing the plastic strain energy density with the flow stress, σ_f, of the solder alloy as following [53, 54, 56]:

$$(N_f v^{k-1})^m \frac{W_p}{2\sigma_f} = C \tag{13}$$

where m and C are constants. The flow stress is taken as the average between the yield point and the highest stress of the hysteresis loop. Since the flow stress of Sn-rich alloys is highly dependent on strain rate and temperature, in addition to the frequency dependence,

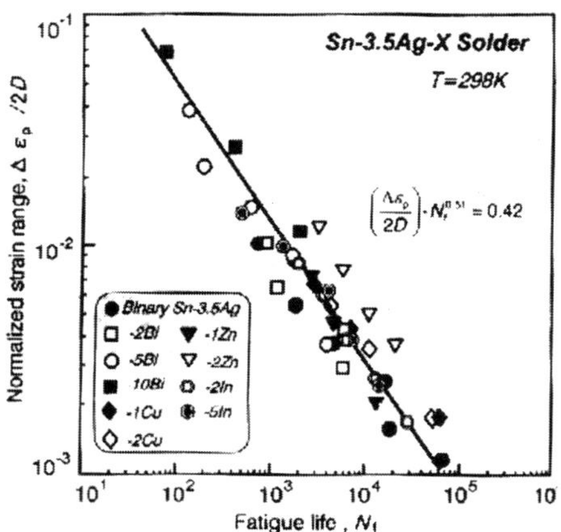

Fig. 28 Effect of alloying additions of fatigue life behavior of Sn–3.5Ag alloy [32]

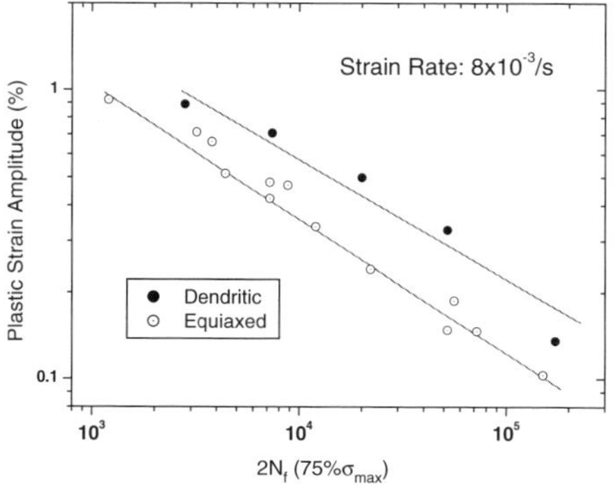

Fig. 29 Effect of microstructure on fatigue life behavior of Sn–3.8Ag–0.7Cu alloy

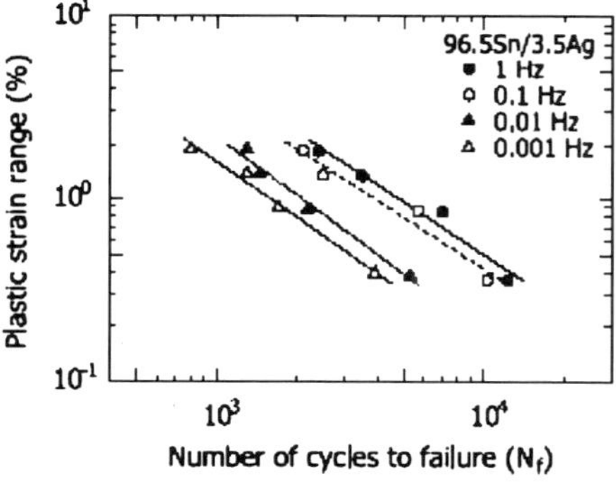

Fig. 30 Effect of frequency on fatigue life behavior of Sn–3.5Ag alloy [40]

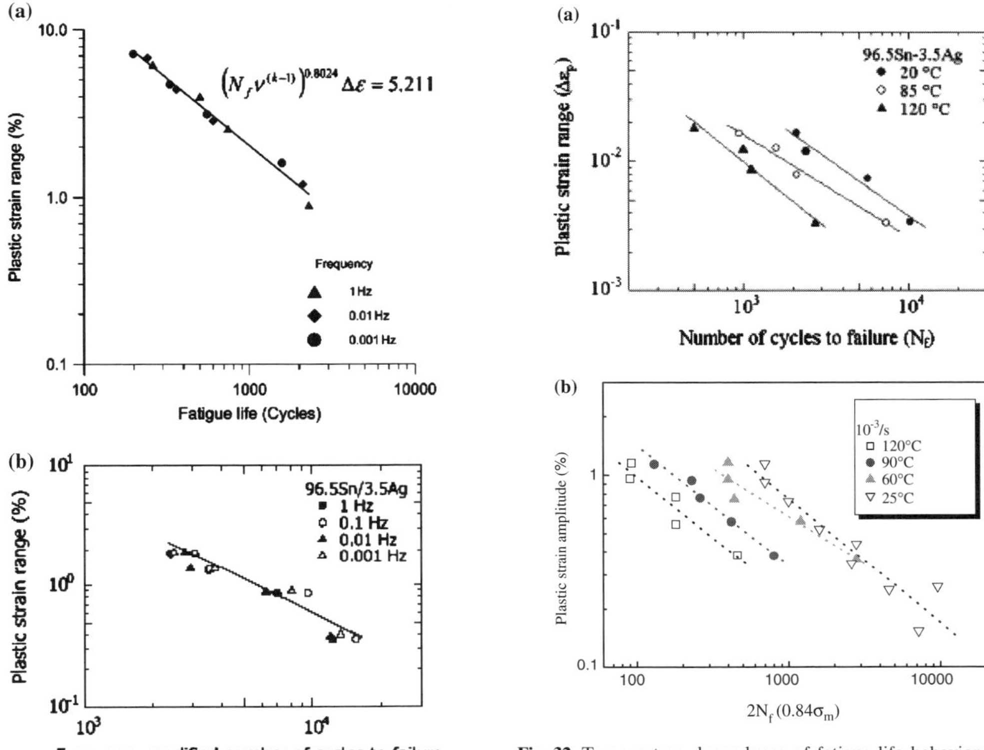

Fig. 31 Correlation of fatigue life behavior by the frequency-modified Coffin–Mason equation: (a) Sn–0.7Cu [35] and (b) Sn–3.5Ag alloy [40]

Fig. 32 Temperature dependence of fatigue life behavior in a) Sn–3.5Ag [39] and b) Sn–3.8Ag–0.7Cu alloy

Eq. 13 also contains the dependence of fatigue life on testing temperature.

The effect of testing temperature on the fatigue life is shown in Fig. 32 for Sn–3.5Ag [39] and Sn–3.8Ag–0.7Cu alloy. As temperature increases, the fatigue life decreases. The slope of the strain-life diagram remains nearly constant from room temperature to 393 K. For the same temperature increase from room temperature to 393 K, fatigue life is shortened nearly by one order of magnitude for the Sn–3.8Ag–0.7Cu alloy and by about half of an order of magnitude in Sn–3.5Ag alloy. As shown by Fig. 33 [54], the temperature dependence in the Sn–Ag alloy may be explained by Eq. 13. However, application of Eq. 13 to the equiaxed Sn–3.8Ag–0.7Cu alloy is problematic because no stable hysteresis can be used to calculate the plastic energy density, W_p, and the peak stress needed for computing the flow stress in eqn. (13) decays rapidly with the number of cycles at elevated temperatures (Fig. 12).

Moreover, examination of the fatigue sample tested at temperatures above 373 K reveals that dynamic recrystallization occurs during fatigue testing. As shown in Fig. 34, the recrystallization results in a much finer grain structure (1~2μm grain size) than the as-cast microstructure and the recrystallized grains contain a very low dislocation density. While the finer grain structure is more susceptible to cyclic softening due to grain boundary cracking, the low dislocation density in the recrystallized grains means that cyclic deformation should be much easier. Therefore, the recrystallization is expected to make a significant contribution to the much shorter fatigue life of the equiaxed Sn–3.8Ag–0.7Cu alloy at high temperatures.

7 Concluding Remarks

The fatigue behavior of Sn-rich alloys has shown a complicated dependence on alloy composition, microstructure, stress amplitude, strain amplitude, stress ratio, cyclic frequency and testing temperature. At

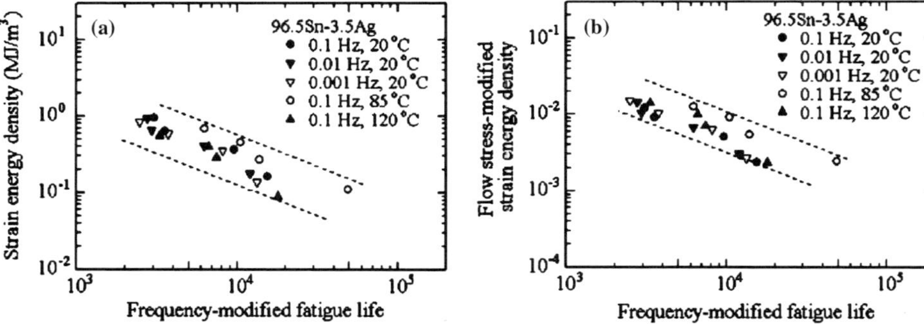

Fig. 33 Correlation of fatigue life behavior in Sn–3.5Ag alloy by (a) frequency and (b) flow-stress modified Morrow energy model [54]

Fig. 34 Recrystallized grain structure in Sn-3.8Ag-0.7Cu alloy after fatigue cycling under constant strain amplitude of 0.8% at 120°C

constant cyclic stress amplitudes, the Sn-rich alloys develop extensive creep deformation. The cyclic creep rate follows a power-law function of the peak stress, with a high stress exponent similar to that under static loading. For a given peak stress, the cyclic creep rate increases with stress ratio but decreases with cyclic frequency. While the creep constitutes a major portion of the cyclic deformation, repeated loading and unloading accelerates the deformation process.

Sn-rich alloys with low alloying concentrations (each element less than 2%) have better fatigue resistance than the eutectic Sn–Pb alloy. However, like Sn–Pb solder alloys, the Pb-free Sn-rich alloys tend to soften cyclically. The softening is more pronounced in equiaxed grain structure where grain boundary microcracking develops early in the fatigue life. The rate of microcrack-induced cyclic softening increases with strain amplitude, testing temperature and grain refinement. In Sn-rich alloys with dendritic microstructure, cyclic deformation is concentrated in the dendrite phase and fatigue cracks initiate around the phase boundary or in the dendrite. Subsequent growth of the fatigue crack occurs intergranually at high stress ratios and low cyclic frequencies. At low stress ratios and high frequencies, fatigue crack growth is predominantly transgranular.

Fatigue crack growth in Sn-rich alloys is cycle-dependent at low stress ratios and high frequencies but time-dependent at high stress ratios and low frequencies. Increasing temperature and reducing the strength of the solder alloy promote time-dependent fatigue crack growth. For cycle-dependent fatigue crack growth, the growth rate follows a unique power-law function of the effective stress intensity range. In the time-dependent fatigue crack growth, crack velocity correlates well with the time-dependent fracture mechanics parameter, C^*, and approaches the creep crack growth rate.

Fatigue life behavior of Sn-rich alloys is well described by the Coffin–Mason equation but the materials constants, fatigue ductility coefficient and exponent, vary widely with alloy composition, microstructure, temperature, cyclic frequency, and source of the data. The effect of minor alloying addition may be rationalized by modifying the Coffin–Mason equation with the tensile ductility of the alloy. The difference between dendritic and equiaxed microstructures is shown to arise from the difference in fatigue damage mechanism, resulting in an inferior fatigue resistance in the equiaxed microstructure. At a given temperature, fatigue life decreases with decreasing cyclic frequency. Such a frequency dependence may be analyzed by either a frequency modified Coffin–Mason equation or Morrow energy model. As temperature increases above room temperature, fatigue lives of Sn-rich alloys decrease and show strong temperature dependence. While part of the temperature dependence may be explained by reduced flow stress, dynamic recrystallization also contributes to the reduction in fatigue life especially at higher temperatures.

Acknowledgements Support for this study was provided by the Chinese Natural Science Foundation under the grant 50228101 and the National Basic Research Program of China, No. 2004CB619306. Discussions with Prof. Zhongguang Wang are greatly appreciated.

References

1. D.R. Frear, W.B. Jones, K.R. Kinsman (eds.), *Solder Mechanics-A State of the Art Assessment* (TMS Publication, Warrendale PA, 1991)
2. J.H. Lau (ed.), *Solder Joint Reliability: Theory and Applications* (Van Nostrand Reinhold, New York, NY, 1991)
3. D.R. Frear, S.N. Burchett, H.S. Morgan, J.H. Lau (eds.), *The Mechanics of Solder Alloy Interconnects* (Van Nostrand Reinhold, New York, NY, 1994)
4. R.N. Wild, Welding J.: Welding Res. Suppl. **51**, 521-s–526-s (1972)
5. E.R. Bangs, R.E. Beal, Welding J.: Welding Res. Suppl. **54**, 377s–383s (1975)
6. D.R. Frear, D. Grivas, J.W. Morris Jr., J. Electron. Mater. **17**, 171–180 (1988)
7. D.R. Frear, D. Grivas, J.W. Morris Jr., J. Metals **40**(6), 18–22 (1988)
8. D.R. Frear, D. Grivas, J.W. Morris Jr., J. Electron. Mater. **18**, 671–680 (1989)
9. D.R. Frear, IEEE Trans. Comp. Hybrids, Manuf. Technol., **12**, 492–501 (1989)
10. D. Tribula, D. Grivas, D.R. Frear, J.W. Morris Jr., ASME J. Electron. Packag. **111**, 83–89 (1989)
11. R. Satoh, K. Arakawa, M. Harada, K. Matsui, IEEE Trans. Comp. Hybrids, Manuf. Technol. **14**, 224–232 (1991)
12. J. Seyyedi, ASME J. Electron. Packag. **115**, 305–311 (1993)
13. N.F. Enke, T.J. Kilinski, S.A. Schroeder, J.R. Lesniak, IEEE Trans. Comp. Hybrids, Manuf. Technol. **12**, 459–468 (1989)
14. T.S.E. Summers, J.W. Morris Jr., ASME J. Electron. Packag. **112**, 94–99 (1990)
15. Z. Mei, J.W. Morris Jr., ASME J. Electron. Packag. **114**, 104–108 (1992)
16. Z. Guo, A. F. Sprecher, H. Conrad, ASME J. Electron. Packag. **114**, 112–117 (1992)
17. Z. Guo, H. Conrad, ASME J. Electron. Packag. **115**, 159–164 (1993)
18. W. Engelmaier, IEEE Trans. Comp. Hybrids, Manuf. Technol. **CHMT-6**, 232–237 (1983)
19. R. Subrahmanyan, J. R. Wilcox, C.-Y. Li, IEEE Trans. Comp. Hybrids, Manuf. Technol. **12**, 480–491 (1989)
20. Y.-H. Pao, IEEE Trans. Comp. Hybrids, Manuf. Technol. **15**, 559–570 (1992)
21. H.D. Solomon, IEEE Trans. Comp. Hybrids, Manuf. Technol. **CHMT-9**, 423–432 (1986)
22. E.C. Cutiongco, S. Waynman, M.E. Fine, D.A. Jeannnotte, ASME J. Electron. Packag. **112**, 110–114 (1990)
23. W.A. Logsdon, P.K. Liaw, M.A. Burke, Eng. Fract. Mech. **36**, 183–218 (1990)
24. P.K. Liaw, M.A. Burke, Scripta Metall. **23**, 747–752 (1989)
25. S.-M. Lee, D.S. Stone, ASME J. Electron. Packag. **114**, 118–121 (1992)
26. K. Suganuma, Curr. Opin. Solid State Mater. Sci. **5**, 55 (2001)
27. M. Abtew, G. Selvaduray, Mater. Sci. Eng. **27**, 95–141 (2000)
28. T. Siewert, S. Liu, D.R. Smith, J.C. Madeni, NIST Report "Database for Solder Properties with Emphasis on New Lead-Free Solders". Sept. 2000
29. S. Vaynman, H. Mavoori, M.E. Fine, Advances in electronic packaging, Proc. international Intersociety electronic packaging Conf.—INTERPAC-95,American society of Mechanical engineers, 135–146 (1995)
30. J. Liang, N. Gollhardt, S.P. Lee, S.A. Schroeder, M.L. Morris, Fatigue Fract. Eng. Mater. Struc. **19**, 1401–1409 (1996)
31. Y. Kariya, M. Otsuka, J. Electron. Mater. **27**, 866 (1998)
32. Y. Kariya, M. Otsuka, J. Electron. Mater. **27**, 1229–1235 (1998)
33. Y. Kariya, T. Morihata, E. Hazawa, M. Otsuka, J. Electron. Mater. **30**, 1184–89 (2001)
34. C. Kanchanomai, Y. Miyashita, Y. Mutoh, J. Electron. Mater. **31**, 456–65 (2002)
35. J.H.L. Pang, B.S. Xiong, T.H. Low, Int. J. Fatigue **26**, 865–872 (2004)
36. Q.L. Zeng, Z.G. Wang, A.P. Xian, J.K. Shang, J. Electron. Mater. **34**, 62–67 (2005)
37. V. Stolkarts, L.M. Keer, M.E. Fine, J. Mech. Phys. Solids **47**, 2451 (1999)
38. Q. Zeng, Z. G. Wang, A.P. Xian, J.K. Shang, Chin. J. Mater. Res. **18**(1), 11–17 (2004)
39. C. Kanchanomai, Y. Mutoh, Mater. Sci. Eng. A **381**, 113–120 (2004)
40. C. Kanchanomai, Y. Miyashita, Y. Mutoh, S.L. Mannan, Mater. Sci. Eng. A **345**, 90–98 (2003)
41. R.W. Hertzberg, *Deformation and Fracture Mechanics of Engineering Materials* (John Wiley & Sons, New York, 1996)
42. B. Budiansky, R.J. O'Connel, Int. J. Solids Struct. **12**, 81 (1976)
43. L.M. Kachanov, *Introduction to Continuum Damage Mechanics.* (Kluwers Academic Publishers, 1986)
44. R. Zallen, *The Physics of Amorphous Solids.* (John Wiley & Sons, New York, 1983)
45. C. Kanchanomai, Y. Miyashita, Y. Mutoh, J. Electron. Mater. **31**, 142–151 (2002)
46. S. Choi, K.N. Subramanian, J.P. Lucas, T.R. Bieler, J. Electron. Mater. **29**, 1249 (2000)
47. M.A. Martin, E.W.C. Coenen, W.P. Vellinga, M.G.D. Geers, Sripta Mater. **53**, 927–932 (2005)
48. J. Zhao, Y. Miyashita, Y. Mutoh, Int. J. Fatigue **23**, 723–31 (2001)
49. Y. Mutoh, J. Zhao, Y. Miyashita, C. Kanchanomai, Soldering Surf. Mount Technol. **14/3**, 37–45 (2002)
50. J. Zhao, Y. Mutoh, Y. Miyashita, S.L.Mannan, J. Electron. Mater. **31**, 879–886 (2002)
51. J. Zhao, Y. Mutoh, Y. Miyashita, L. Wang, Eng. Fract. Mech. **70**, 2187–21 (2003)
52. C. Anderson, Z. Lai, J. Liu, H. Jiang, Y. Yu, Mater. Sci. Eng. A **394**, 20–27 (2005)
53. J.H.L. Pang, B.S. Xiong, T.H. Low, Thin Solid Films **462–463**, 408–12 (2004)
54. C. Kanchanomai, Y. Mutoh, J. Electron. Mater. **33**, 329–333 (2004)
55. X.Q. Shi, H.L.J. Pang, W. Zhou, Z.P. Wang, Int. J. Fatigue **22**, 217 (2000)
56. X.Q. Shi, H.L.J. Pang, W. Zhou, Z.P. Wang, Scripta Mater. **41**, 289 (1999)

Life expectancies of Pb-free SAC solder interconnects in electronic hardware

Michael Osterman · Abhijit Dasgupta

Published online: 23 September 2006
© Springer Science+Business Media, LLC 2006

Abstract The transition from lead (Pb) bearing solder to Pb-free solder has arisen in response to government restrictions on the use of lead (Pb) by the European Union. As a result, electronic manufacturers have sought a material comparable to the conventional 63Sn37Pb solder that has been traditionally used to assemble electronic hardware. Based on extensive review of various solder combination, the majority of electronic manufacturers appear to be adopting a tin–silver–copper (SAC) solder as a popular Pb-free solder replacement. Significant investments have been made by many researchers to characterize the material behavior and durability of this solder system. While the exact composition of the SAC solder is still in question, it now appears that the 96.5Sn3.0Ag0.5Cu (SAC305) solder is gaining wider acceptance as the favored Pb-free replacement, for surface mount assemblies that are going to be subjected predominantly to cyclic thermal environments. This paper presents a review of our current understanding of the life expectancy of Pb-free SAC solder interconnects for electronic hardware. To this end, the paper focuses on material characterization of SAC solder, as well as its temperature cycling and vibration fatigue reliability. From this review, SAC solder interconnects are shown to be suitable for providing adequate life expectancies for temperature cycling in electronic hardware. However, it is clear that there are differences between SAC and the conventional Sn37Pb solder, that need to be understood in order to design reliable electronic hardware.

1 Introduction

As electronic equipment manufacturers transition from the conventional Sn37Pb solder to Pb-free solders, manufacturability and the impact on product reliability have been significant concerns, especially in surface mount assemblies. After reviews by multiple organizations including National Center for Manufacturing Sciences (NCMS), International Electronics Manufacturing Initiative (formally NEMI), and SOLDERTEC, it appears that a tin–silver–copper (SAC) alloy will become the primary replacement for the conventional tin–lead solder used in assembly of electronic hardware. At present, the most likely solder candidate for thermal cycling environments will be Sn3.0Ag0.5Cu (SAC305). The SAC305 alloy has a higher melt point (~217°C) as compared to Sn37Pb (183°C). As a result of this review process and selection, hundreds of studies have been conducted to examine the performance of SAC solders, in comparison to the Sn37Pb solder. In addition, part manufacturers have had to remove any Pb used in their packaging and verify that their packages will handle the higher reflow assembly temperatures.

In electronic hardware, solder usually provides a mechanical and electrical connection. With regards to this mechanical connection, the solder joints are subject to cyclic stress due to temperature excursions,

M. Osterman (✉) · A. Dasgupta
CALCE Electronic Products and Systems Center,
University of Maryland, College Park, MD 20742, USA
e-mail: osterman@eng.umd.edu

vibration, and shock due to equipment operation and use that can eventually degrade the joint and ultimately result in failure. Failure of a joint causes electrical disruptions that may result in loss of function and product failure.

In general, the fatigue life of solder has been modeled to be a function of the applied loading condition. The loading condition is usually characterized by a defined stress (or strain or energy) metric. In addition to the cyclic loading condition, the stress (or strain or energy) metric considers the package geometry and material, the substrate material and geometry, and the solder interconnect material and geometry. As a result, fatigue life is usually plotted on a stress versus number of cycles to failure or S–N curve. A typical fatigue curve for solder is depicted in Fig. 1.

As in most metals the fatigue life of solder falls into low cycle fatigue and high cycle fatigue categories. Low cycle fatigue is usually associated with temperature cycling loads, while high cycle fatigue is typically associated with mechanical or vibration cyclic loads. In both cases, the fatigue life is usually related to the defined cyclic stress metric with a power law equation.

In reviewing the published literature, the majority of Pb-free solder life tests have focused on the reliability of solder joints under temperature cycling with only limited attention paid to vibration and shock. With regards to temperature cycling fatigue models, multiple approaches have been proposed and used [1–6]. Proposed temperature cycle fatigue models have used total strain range, inelastic strain range, total cyclic energy, and partitioned cyclic energies, as the stress metric. Others have simply defined the acceleration factor (ratio of time to failure between test and field conditions) [7].

The establishment of valid models is important in predicting field life expectancies. In developing and validating models, it is necessary to establish the material response of the solder to loading that will occur in test and in use, as well as determining the

reliability of the solder under test and field loads. This paper provides a review of the fatigue life expectancy of Pb-free SAC soldered interconnects in electronic hardware. To this end, a discussion of Pb-free SAC solder as compared to the conventional Sn37Pb solder is provided. The discussion includes information related to material behavior under various mechanical and temperature load conditions. In addition, the temperature cycling and vibration fatigue life testing results are presented.

2 Solder alloy characterization

To understand the fatigue life behavior of SAC solder, it is necessary to first understand how the material system responds to various loading conditions. This behavior, termed constitutive behavior, has been examined and reported by several researchers [8–12]. The constitutive properties include the stress–strain response of the material under monotonic mechanical loads. From the stress–strain response, the elastic and plastic response of the material can be characterized. In addition to the elastic and plastic response of the material, the creep response of solder materials is also important, particularly when examining the material's response to temperature cycle loads. The creep properties are characterized by monitoring the strain history under the action of a constant mechanical stress.

For the stress–strain response, controlled experiments were conducted on samples of SAC (Sn3.9Ag0.6Cu) and Sn37Pb solders [10, 12, 13]. In these tests, a shear loading was applied to a specially designed lap-shear specimen, over a set of temperatures and strain rates. The resulting elastic–plastic creep shear response of Sn37Pb and SAC solder samples are presented in Figs. 2 and 3, respectively. In

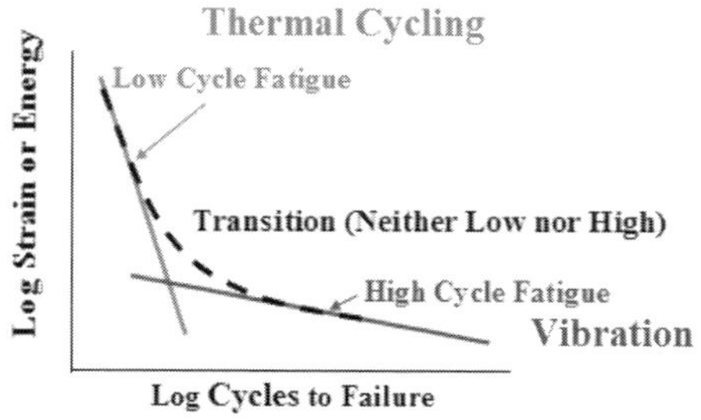

Fig. 1 Typical fatigue curve

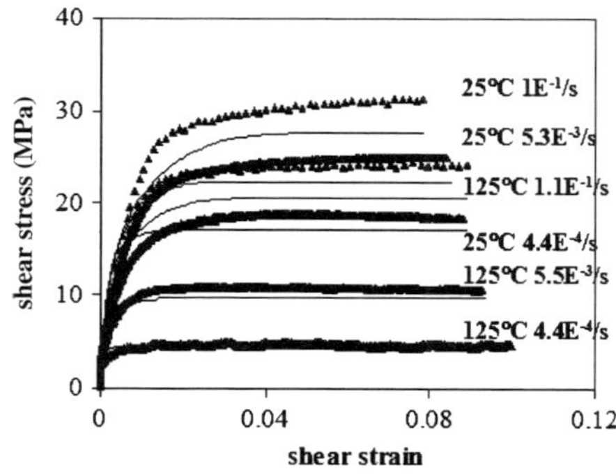

Fig. 2 Shear response of the Sn37Pb solder

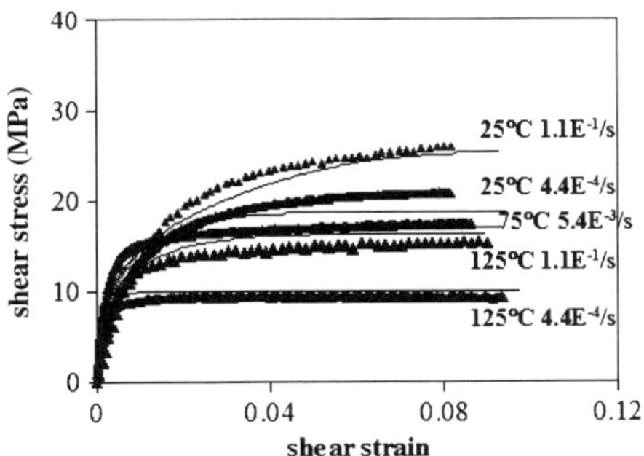

Fig. 3 Strain response of Sn3.9Ag0.7Cu

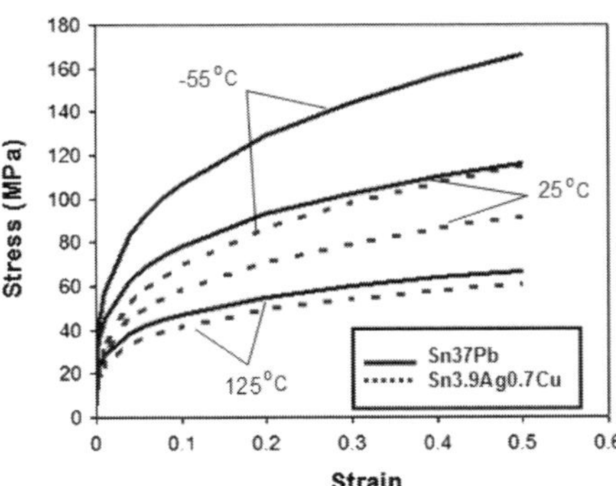

Fig. 4 Plastic shear response of solders (temperature is in degrees Celsius)

these figures, the discrete symbols represent experimental data. The solid lines represent predicted stress–strain curves at defined temperatures and strain rates, based on the constitutive properties given in this paper.

From the shear stress–strain curves, the elastic modulus of both the Sn37Pb and SAC solders was estimated. For a strain rate of 10^{-1}/s, the elastic modulus was found to be [10, 13]:

$$Sn37Pb : E(GPa) = 18.9 - 0.044T(^\circ C)$$

$$Sn3.9Ag0.6Cu : E(GPa) = 18.6 - 0.021T(^\circ C)$$

From Figs. 2 and 3, it can be observed that temperature affects the modulus of Sn37Pb more than that of SAC. This finding is somewhat counter to findings of other researchers that have reported higher modulus values [6, 8]. However, it may in part help explain the finding that SAC Pb-free solders does not increase the likelihood of flex cracking of multiple layer ceramic capacitors (MLCCs) [14]. These elastic moduli, measured under relatively slow strain rates, are found to describe solder deformation quite well for slow thermal cycling loading. However, for rapid loading, as in vibration or mechanical shock, a higher stiffness value is appropriate.

In addition to the elastic response, the plastic behavior is also important in determining the fatigue resistance of a material, especially when the amount of creep deformation is not very high at the test temperatures. This is particularly important in SAC solders which are much more creep resistant than SnPb solder. From the strain history response, the onset of plastic deformation and the change in plastic response is a function of temperature. Figure 4 depicts the plastic shear response of both Sn37Pb and Sn3.9Ag0.6Cu solders. From this plot, it is clear that plastic response of Sn37Pb is more sensitive to temperature than that of Sn3.9Ag0.6Cu. Our measurements reveal that under

shear loading, the SAC family of solders (including several SnAgCu ternaries as well as the Sn3.5Ag and Sn0.7Cu binaries) exhibits lower yield strength, lower hardening, and lower ultimate strength than Sn37Pb solder, under identical loading conditions. This finding agrees with several sources in the literature [15–17] but also differs from some others [18, 19].

In addition to the elastic and plastic response of the solder, the creep behavior plays an important role in determining fatigue resistance. With regards to temperature cycle loading, the primary concern is secondary or steady state creep because the primary transient creep is negligible in comparison, in slow thermal cycling. For solders, the secondary creep can be represented using several different models such as Dorn's model, Norton's model, Weertman's model or Garofalo's model. In this paper we report the model constants for Garofalo's hyperbolic-sine function because the data indicates a clear transition of the creep mechanism within the stress ranges of interest. This transition is manifested as a change in the slope of the creep-rate versus stress curves (on a log–log scale). The form of the equation for modeling steady-state creep is:

$$\frac{d\gamma_{scr}}{dt} = A'[\sinh(\alpha\tau)]^{n'}\exp\left(-\frac{Q}{RT}\right)$$

where: γ_{scr} is shear strain; t is time in seconds; Q is the creep activation energy in J/mol; T is temperature in K; R is the universal gas constant (=8.31451 m^2kg/s^2K mol); τ is the shear stress (MPa), and A' (1/s), α (1/MPa), and n' are model constants (Table 1).

A plot of the secondary creep response is depicted in Fig. 5. The test results were obtained using the same lap-shear test setup discussed in Figs. 2–4.

Based on physical measurement, the Pb-free SAC solder has been found to be significantly more creep

Table 1 Creep model constants

Solder alloy	A'	α	n'	Q(J/Mol)
Sn37Pb	1.15E4	0.20	2.2	5.93E4
Sn3.9Ag0.6Cu	1.50E3	0.19	4.0	7.13E4

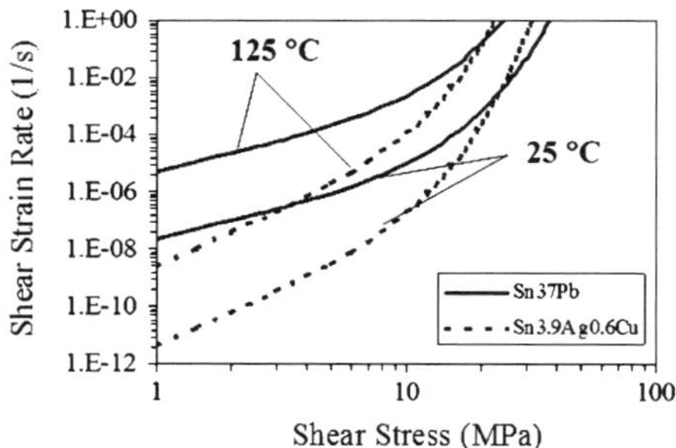

Fig. 5 Creep shear response of solders

resistant than Sn37Pb at low stress levels, (1 MPa), where the creep mechanisms are predominantly diffusion-driven. However, at higher stress levels, (10 MPa), creep mechanisms are predominantly dislocation-driven, and both solders have comparable creep. This finding, in part, explains the longer life observed for the Pb-free SAC solder, in direct comparisons of temperature cycling test results between SAC and Sn37Pb. This result also explains, in part, why acceleration factors for accelerated thermal cycling are higher for SAC solders than for SnPb solder. While the material behavior of solder helps our understanding of solder joint reliability, physical testing is necessary to complete our understanding and ultimately develop predictive reliability models.

3 Temperature cycling fatigue life

With regards to temperature cycling, the majority of reported fatigue failure tests have applied standard temperature cycling test conditions. The standard temperature cycle test conditions are presented in Table 2.

In general, published test results have found the SAC solder to be more fatigue resistant than Sn37Pb

Table 2 Standard temperature cycling test conditions

	Minimum (°C)	Maximum (°C)	Dwell (min)	Ramp (°C/min)
TC1	0	100	5–15	10
TC2	– 40	125	5–15	10
TC3	– 55	125	5–15	10

under temperature cycle testing. In some cases, the fatigue life of SAC solder was found to be 1.4– 2.3 times that of SnPb [20–24]. However, some test results showed the fatigue life of SAC to be lower that SnPb solder [25, 26]. Based on reviewing the test data, an illustrated plot of the experimental test result is presented in Fig. 6.

As depicted in Fig. 6, the fatigue life of SAC is found to be greater than Sn37Pb when the approximated cyclic strain range is lower than a critical value. One observation from reported test results is that SAC was only reported to have a lower fatigue life under temperature cycling when the peak cycle temperature was at to above 125°C and the package architectures generally provided little or no compliance in the solder interconnects. Interestingly, 125°C is above 0.8 melt temperature of SAC (~119°C). Another finding has been that the reduction in fatigue life of SAC solder joints due to longer dwells (hold) times appears to be larger than the reduction observed with Sn37Pb solders [25]. The dwell time result has raised concern over the appropriate dwell time that should be used in qualification tests.

In a recently completed study [27, 28], a designed experiment was conducted to examine the impact of dwell time and the temperature extremes on the temperature cycling fatigue life of Pb-free solders. In this experiment, test specimens were assembled with Sn3.87Ag0.7Cu and Sn37Pb. Each test specimen consists of two 84 IO Ceramic Leadless Chip Carriers (CLCC) and two 68 IO CLCC parts soldered to a 130 × 93 × 2.5 mm high temperature FR4 printed wiring board. A photo of a test specimen is provided in Fig. 7.

To examine the effect of dwell time and the temperature extremes on the solder fatigue life, sets of test coupons were subjected to the temperature cycle conditions identified in Table 3.

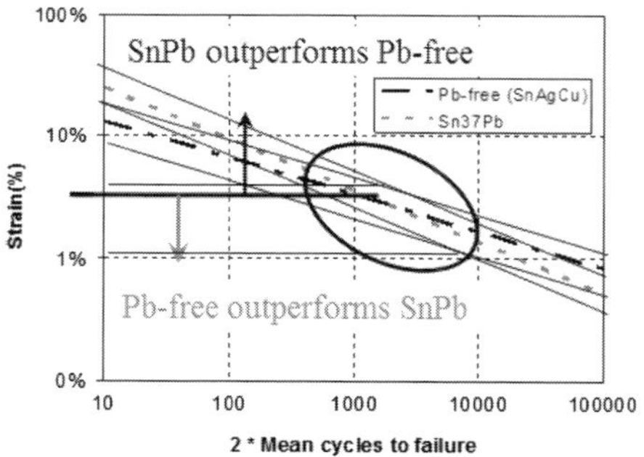

Fig. 6 Illustrated plot of experimental test results

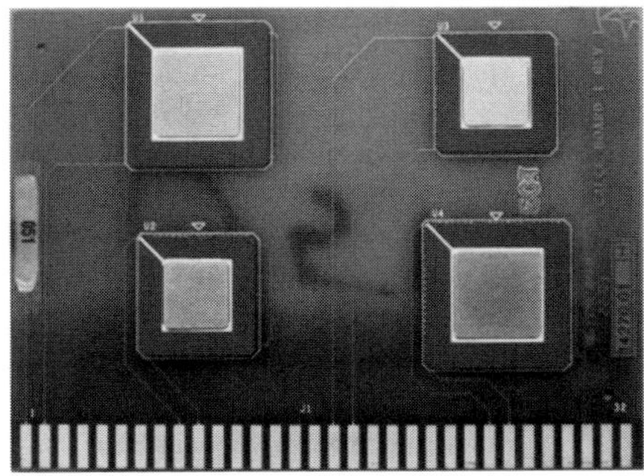

Fig. 7 CLCC test specimen

For each test condition, eight cards for each solder type were used. The parts on each board were electrically monitored continuously during the test. Failure was defined to occur when resistance spikes were observed in ten consecutive temperature cycles. The time of the first resistance spike in the observed failure event was recorded.

After the tests were completed, the failure data was processed and fit to two parameter Weibull distributions. A Weibull plots of the results from the test 1 condition and test 3 conditions are provided in Figs. 8 and 9, respectively.

In these plots, it is interesting to note that the Pb-free solder on the 0–100°C temperature cycle was superior to the Pb-based solder. Only when the temperature cycle was raised to 25–125°C did the Pb-free solder demonstrate a lower reliability. This trend was also observed with the longer 75 min dwell in tests 5 and 6. Based on the above mentioned test results [27, 28], it is unclear that longer dwell times are an efficient approach to establishing solder interconnect reliability. In general, the dwell time at the high temperature has been considered influence fatigue life more that low temperature dwell. However, Lee and Subramanian [29] have reported a larger degree of material degradation with longer low temperature dwell. Further

Table 3 Temperature cycling test matrix

Test	Min. temp. (°C)	Max. temp.(°C)	Temp. range (°C)	Dwell time at max temp* (min)
1	0	100	100	15
2	– 25	75	100	15
3	25	125	100	15
4	– 25	75	100	75
5	25	125	100	75
6	0	100	100	75
7	15	85	70	15
8	15	85	70	75

investigation is needed with regards to low temperature dwells to determine if the material degradation results in earlier electrical failures.

4 Vibration fatigue life

With regards to vibration fatigue studies, only limited testing has been reported in the open literature [30–33] for Pb-free assemblies. Like the temperature cycling testing, most reported vibration tests have generally been a comparison of Pb-free assemblies subjected to currently accepted product qualification test levels. Unfortunately, the testing was time-terminated and not run until a significant portion of part failures occurred. For instance, Tzan and Chu [30] reported no failure on similarly constructed Pb-free and Pb-based test specimens subjected to a 2 g, 20–2,000 Hz, sine sweep test. Similarly, Lee et al. [31] reported no failures on random vibration testing of Pb-based and Pb-free based ball grid array assemblies. Kim et al. [32] used inertia loading to establish a power law relation between fatigue life and shear stress for Pb-based (Sn36Pb2.0Ag) and Pb-free (Sn1.0Ag0.5Cu) solders. For this work, it was reported that Pb-free joints had a lower life than SnPb joints at high stress, but longer life at low stress. The fatigue curve was presented as a relationship between cyclic shear stress range and cycles to failure. This stress-life fatigue curve for Pb-free solders was found to have a smaller slope than Sn37 Pb solder, with a cross-over point at intermediate stress levels. In a recent study, Woodrow reported lower fatigue durability of Pb-free SAC assembled parts, compared to similarly mounted parts assembled with SnPb solder, in random-vibration step-stress tests, conducted as part of the Joint Council on Aging Aircraft/Joint Group on Pollution Prevention (JCAA/JGPP) Pb-free solder study [33]. This study included Ball Grid Arrays (BGAs), Quad Flatpacks (QFPs), Dual In-line Packages (DIPs), Ceramic Leadless Chip Carriers (CLCCs), Thin Small Outline Packages (TSOPs), and leadless resistors.

In quasi-static isothermal mechanical cycling durability studies of Pb-free SAC (Sn3.9Ag0.7Cu) solder, Zhang [12] has reported that the fatigue curve of Pb-free SAC obtained from lap-shear testing, has a smaller slope than Sn37Pb solder. Thus, Sn67Pb was found to be more robust at low load levels (high-cycle-fatigue) while SAC was more robust at high load levels, with a cross-over at intermediate stress levels. This trend in slopes is the opposite of that presented in Kim's stress versus life fatigue plot [32]. It should be noted that unlike Kim's data, Zhang's fatigue data was presented

Fig. 8 Weibull plot of CLCC fatigue life under 0–100°C temperature cycle

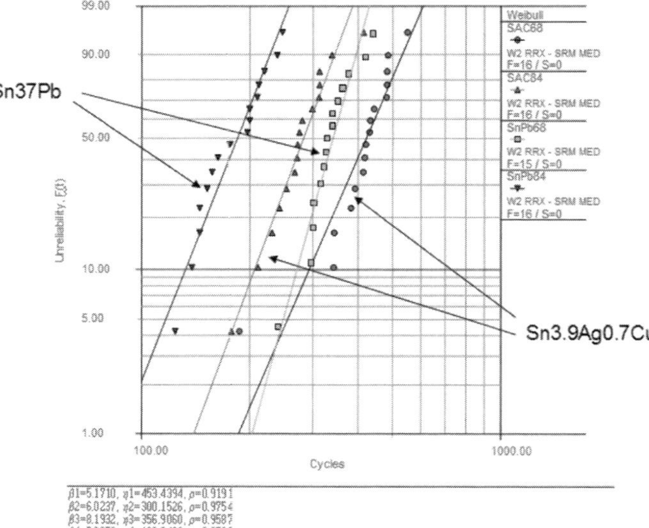

Fig. 9 Weibull plot of CLCC fatigue life under a 25–125°C temperature cycle

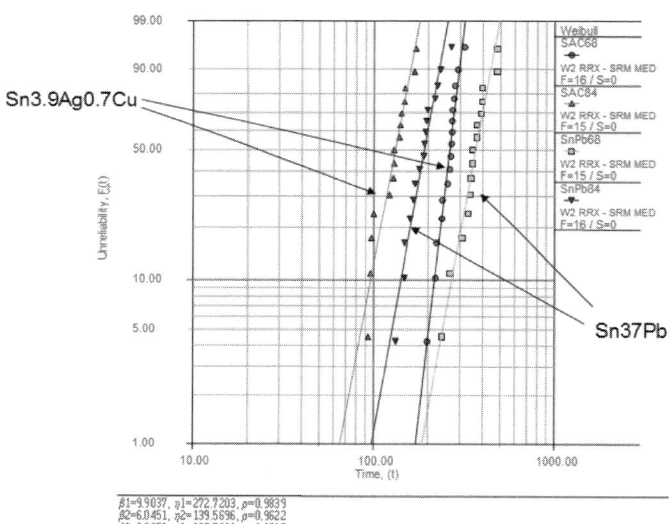

as a plot of the calculated work density versus fatigue life, as shown in Fig. 10. Furthermore, the SAC in these two references have significant differences in composition.

To examine the vibration fatigue reliability of assembled electronic hardware, researchers at CALCE have conducted a mechanical cycling study. The test boards utilized in this study consist of surface mount resistors attached to a printed wiring board constructed of FR4 with dimensions of $140 \times 3.65 \times 1.65$ mm. A sample test board is depicted in Fig. 11. Test boards were assembled with SAC and Sn37Pb solders.

For the vibration loading condition, the test boards were subjected to cyclic four point bending loads. For each test, the strain range was fixed and the loading was applied at a rate between 11 and 17 Hz. The resistance of each resistor was electrically monitored and the time to failure was recorded. With the four

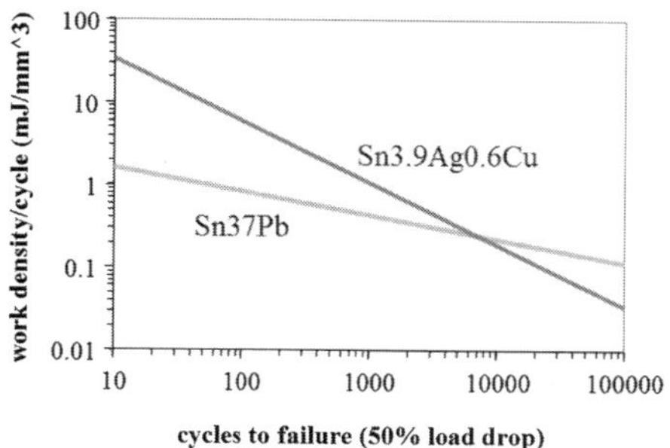

Fig. 10 Isothermal mechanical cycling fatigue life

point bend loading, all parts were assumed to be subjected to the same strain level. As expected, the test to failure resulted in a fair degree of scatter which is not unexpected. A plot of the test results along with the best fit curves is presented in Fig. 12. This plot depicts the mean fatigue life of the surface mount resistors as a function of board curvature for each test condition.

From Fig. 12, the fatigue life of SAC is found to be lower than SnPb at lower stress levels. This result is consistent with the material testing results provided by Zhang [12].

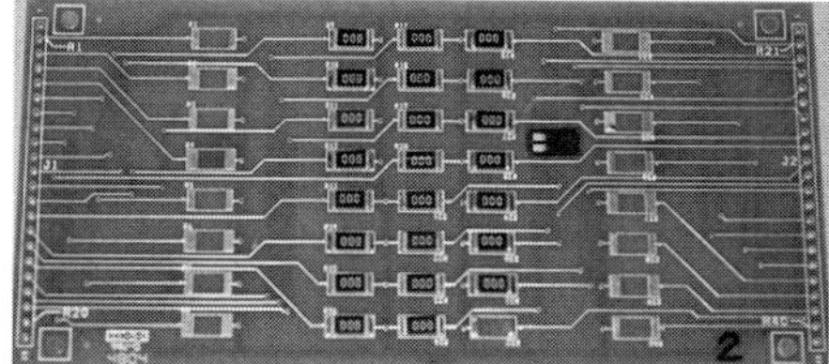

Fig. 11 Vibration test board

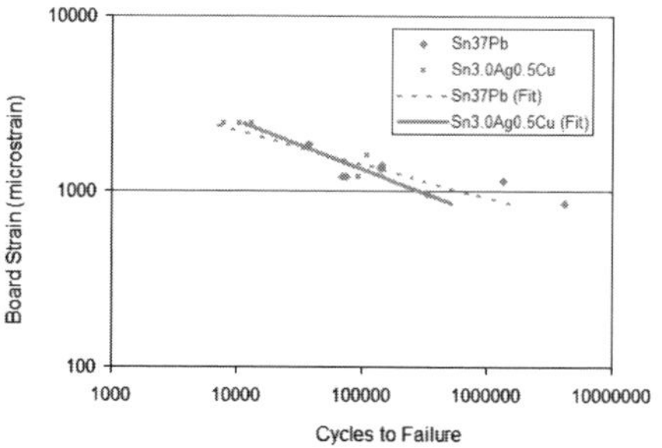

Fig. 12 Vibration fatigue life of BGA. Discrete symbols indicate test data, while solid lines indicate a curve-fit to the data

From the results reported in this paper, it appears that the SAC solder will be less reliable than Sn37Pb under high cycle fatigue due to vibration loading conditions upto moderate loading levels. While this finding may cause some concern for the use of SAC in applications where moderate vibration loading is expected, it should be remembered that vibration loading can usually be mitigated by improved fixturing and electronic enclosure designs. However, more testing is needed to resolve the inconsistency in results reported in the literature and to have a better understanding of SAC vibration fatigue reliability.

5 Summary and conclusions

For more than 20 years, eutectic Pb-based Sn37Pb solder has provided excellent manufacturability and has resulted in reliable electronic hardware. For a successful transition from Pb-based to Pb-free solders, it is critical that the selected Pb-free solder provide equivalent manufacturability and reliability. Based on the research conducted and sponsored by electronic equipment manufacturers, it appears that a ternary tin–silver–copper (SAC) solder alloy will be the most widely used Pb-free replacement for Sn37Pb.

From this review, it has been shown that SAC solder yields at a lower stress and has a higher creep resistance than Sn37Pb. The higher creep resistance helps explain the generally observed longer fatigue life of SAC as compared with Sn37Pb under temperature cycle loading. In the instances where SAC had a lower life than Sn37Pb, the peak cyclic temperature was at 125°C and the package provided little or no compliance in the solder interconnects. With regards to the effect of dwell time during temperature cycling, SAC appears to be slightly more sensitive than Sn37Pb. However, it is unclear whether it is necessary to impose longer dwell time in temperature cycle testing. With regards to vibration, the earlier yield of SAC helps explain its lower fatigue life. As discussed in the vibration fatigue section, more research is need to establish the vibration fatigue life and the low fatigue life may not present a major problem if considered in the design of the mounting points, stiffeners and electronic enclosure. A critical issue to remember when conducting accelerated stress tests using either temperature cycling or vibration, is that the acceleration factors for Pb-free assemblies are going to be quite different from Sn37Pb assemblies because of the differences in the constitutive properties and in the slopes of the fatigue curves. Generally, acceleration factors for thermal cycling tests are larger for Pb-free solders.

In vibration tests, the board and package design will dictate whether the SAC interconnects or Sn37Pb interconnects will have the higher acceleration factor.

From manufacturing and reliability test reports, SAC solder appears to be a reasonable replacement for Sn37Pb. However, it needs to be recognized that SAC behaves differently than Sn37Pb and that our experience in manufacturing and using SAC is still limited. As with Sn37Pb, our understanding of SAC will continue to improve through controlled research and use.

Acknowledgements Much of the information presented in this paper is based on research sponsored by the CALCE Electronic Products and Systems Consortium (EPSC). The authors would like to thank the companies and organizations that sponsored research conducted within CALCE EPSC. In addition, the vibration test information reported this paper is based on the work of Professor Donald Barker and his graduate student Patrice Gregory.

References

1. W. Engelmaier, IEEE Trans CHMT **6**(3), 232 (1983)
2. IPC-SM-785, *Guidelines for Acclerated Reliability Tesing of Surface Mount Solder Attachments* (IPC, Northbrook, IL, 1992)
3. A. Dasgupta, C. Oyan, M. Pecht, D. Barker, Solder Creep-Fatigue Analysis by an Energy-Partitioning Approach, AMSE J Elect Packag, **144** 152 (June 1992). Presented all at Mechanics of Surface Mount Assemblies Symposium, ASME/WAM, GA, Dec. 1–6, 1991
4. J.-P. Clech, in *Proceedings, Surface Mount International Conference* (San Jose, CA, Sept. 8–12, 1996) pp. 136–151
5. A. Syed, *Predicting Solder Joint Reliability for Thermal, Power, and Bend Cycling within 25% Accuracy*, (51st ECTC, 2001), pp. 255–263
6. R. Darveaux, in *Proceedings 50th* (21–24 May 2000) pp. 1048–1058
7. K.C. Norris, A.H. Landzberg, IBM J Res Dev **13**(3), 266 (1969)
8. A. Schubert, R. Dudek, E. Auerswald, A. Gollhardt, B. Michel, H. Reichl, *Fatigue Life Models for SnAgCu and SnPb Solder Joints Evaluated by Experiments and Simulation* (53rd ECTC, 2003), pp. 603–610
9. R. Darveaux, *Shear Deformation of Lead Free Solder Joints*, Electronic Components and Technology, 2005. (ECTC '05. Proceedings, May 31–June 3, 2005) pp. 882–893
10. Q. Zhang, A. Dasgupta, *Constitutive Properties and Durability of Lead-free Solders*, Chapter 3, Lead-free Electronics –2006 edn, ed. by Ganesan, Pecht (Wiley, 2006)
11. S. Wiese, et al., *Microstructural Dependendence of Constitutive Properties of Eutectic SnAg and SnAgCu Solders* (53rd ECTC 2003), pp. 197–206.
12. Q. Zhang, P. Haswell, A. Dasgupta, *Isothermal Mechanical Creep and Fatigue of Pb-free Solders*, International Brazing &Soldering Conference (San Diego, CA, February 16–19, 2003)
13. Q. Zhang, A. Dasgupta, P. Haswell, J Electron Mater **33**(11), 1338 (2004)
14. M. Azarian, *Comparison of Flex Cracking of Multilayer Ceramic Capacitors Assembled with Lead-Free and Eutectic Tin Lead*,CARTS 2006 (April 4–6, 2006)
15. Solder Data Sheet, Welco Castings, 2 Hillyard Street, Hamilton, Ontario, Canada (as reported in Tables 1.13 & 1.14 of NIST Pb-free database on NIST website, http://www.boulder.nist.gov/div853/lead%20free/props01.html)
16. Technical Reports for the Lead Free Solder Project: Properties Reports: Room Temperature Tensile Properties of Lead-Free Solder Alloys; Lead Free Solder Project CD-ROM, National Center for Manufacturing Sciences (NCMS), 1998. (as reported in Table 1.2 of NIST Pb-free database on NIST website, http://www.boulder.nist.gov/div853/lead%20-free/props01.html)
17. J. Madeni, S. Liu, T. Siewert, *Casting of Lead-Free Solder Specimens with Various Solidification Rates*, ASM- International Conference, Indianapolis 2001. (as reported in Table 1.27 of NIST Pb-free database on NIST website, http://www.boulder.nist.gov/div853/lead%20free/props01.html)
18. J. Sigelko, K.N. Subramanian, *Overview of lead-free solders*, Adv. Mat. & Proc., pp. 47–48 (March 2000) (as reported in Table 1.3 of NIST Pb-free database on NIST website, http://www.boulder.nist.gov/div853/lead%20free/props01.html)
19. C. Hernandez, P. Vianco, J. Rejent, *Effect of Interface Microstructure on the Mechanical Properties of Pb-Free Hybrid Microcircuit Solder Joints*, IPC/SMTA Electronics Assembly Expo (1998), p. S19-2-1. (as reported in Table 1.15 of NIST Pb-free database on NIST website, http://www.boulder.nist.gov/div853/lead%20free/props01.html)
20. J.-P. Clech, in *Proc IPC/SMEMA Council APEX 2004 Conf* (Anaheim, CA, Feb. 23–26, 2004), pp. S28-3-1–S28-3-14
21. P. Roubaud, G. Henshall, R. Bulwith, S. Prasad, F. Carson, S. Kamath, E. O'Keeffe, in *SMTA International Proceedings of the Technical Program* (Edina, MN, 2001) pp. 803–809
22. M. Meilunas, A. Primavera, S. Dunford, in *Proc. of the IPC Annual Meeting* (New Orleans, LA, Nov. 3–7, 2002) pp. S08-5–1 to S08-5–14
23. G. Swan, et al. *Proc IPC/SMEMA Council APEX 2001 Conf* (San Diego, CA, Jan. 14–18, 2001) pp. LF2-6-1–LF2-6-6
24. D. Nelson, H. Pallavicini, Q. Zhang, P. Friesen, A. Dasgupta, J SMT **17**(1), 17 (2004)
25. J. Bartelo, et al., in *Proc IPC/SMEMA Council APEX 2001 Conf* (San Diego, CA, Jan. 14–18, 2001), pp. LF2-2-1–LF2-2-12
26. A. Syed, in *Proc. International Symposium on Advanced Packaging Materials Processes, Properties and Interfaces* (Piscataway, NJ, 2001) pp. 143–147
27. S. Yoon, Z. Chen, M. Osterman, B. Han, A. Dasgupta, in *Proceedings, 55th Electronic Components and Technology Conference* (Orlando, FL, May 31–June 3, 2005), pp. 1210–1214
28. M. Osterman, A. Touw, in *Proceeding of IPC/JEDEC 12th International Conference on Lead Free Electronic Components and Assemblies* (March 7–8, 2006)
29. J. Lee, K. Subramanian, J Electron Mater **32**(6), 523 (2003)
30. S.-R. Tzan, S.-L. Chu, *Characterization of lead-free solder by reliability testing*, Electronics Manufacturing Technology Symposium, 2000. Twenty-Sixth IEEE/CPMT International (2–3 Oct 2000) pp. 270–273
31. S.-W.R. Lee, Ben Hoi Wai Lui, Y.H. Kong, B. Baylon, Solder Surf Mount Technol **14**(3), 46 (2002)
32. Y. Kim, H. Noguchi, M. Amagai, in *Proceedings 53rd* (May 27–30, 2003), pp. 891–897
33. T. Woodrow, in *Proceedings, SMTA International Conference* (Chicago, IL, Sept. 25–29, 2005), pp. 771–806

Lead-Free Electronic Solders 237–246

Assessment of factors influencing thermomechanical fatigue behavior of Sn-based solder joints under severe service environments

K. N. Subramanian

Published online: 20 September 2006
© Springer Science+Business Media, LLC 2006

Abstract Reliability of solder joints under thermal excursions encountered in service depends on the solder performance during each stage, its extent in each cycle of temperature excursion, and the cumulative effects of the same under repeated thermal cycling. Extent of such field influence, and the resultant damage, will also be significantly affected by the constraints imposed by the solder joint geometry. Any realistic evaluation of the solder joint behavior should account for all of the above with specimen geometry, and thermal excursion stages that are representative of the actual conditions, encountered in service to arrive at meaningful results. Findings based on studies carried out to evaluate the roles of each of these, with specimens possessing same geometry and prepared under the same conditions, indicate the important contributions of each of the service and material parameters, and the solder alloy composition, for reliability considerations.

1 Introduction

One of the major causes for the failure of electronic components placed in service is due to failure of solder joints from Thermomechanical Fatigue (TMF) that occurs due to stresses that develop due to thermal expansion mismatches that exist between various entities present in the joint [1–5]. Industrial qualification tests for TMF, based on time constraints, usually carry out accelerated thermal cycling (ATC) tests on actual circuit boards [6–12]. Since such studies often use actual components possessing complicated specimen geometry, they are not ideally suited for mechanical or microstructural evaluations of TMF damage accumulation or residual properties. Imposed temperature profiles used in ATC tests, based on time constraints, can not fully represent heating/cooling rates and dwell-times at temperature extremes encountered in actual service [6–12]. As a consequence, analysis of results from such studies can not provide any insight on the damage accumulation or material responses contributing to catastrophic failure of solder joints resulting from TMF. Results obtained from extensive studies carried out to understand the roles of various service and material parameters, by subjecting specimens of realistic dimensions to temperatures excursions representative of severe service environments, are provided in this paper. General trends observed are emphasized, instead of actually measured values to provide an appreciation of issues to be addressed.

All the specimens used in these studies had the same single shear-lap geometry as shown in Fig. 1. These specimens were made by joining long half-dog-bone shaped copper strips with a solder area of 1×1 mm and a thickness of about 100 μm. This joint geometry is highly suitable to carry out shear testing, a method employed in these studies to evaluate the residual shear strength after thermal excursions. All these joints were prepared under identical reflow conditions. Details of the specimen geometry and reflow

K. N. Subramanian (✉)
Department of Chemical Engineering & Materials Science,
Michigan State University, East Lansing, MI 48824-1226,
USA
e-mail: subraman@egr.msu.edu

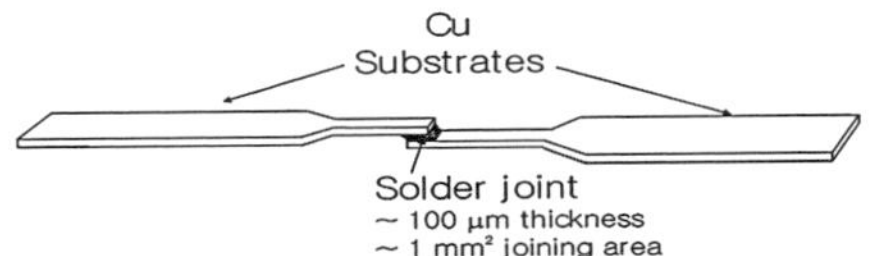

Fig. 1 Schematic of single shear-lap solder joint specimens used in this study

conditions employed are available in several publications [13–17]. Maintaining these fixed through out this study facilitated a valid means to evaluate roles of service and solder material parameters on solder joint reliability. Prior to subjecting the specimens to TMF, one side of the joint region was metallurgically polished so that damage resulting from TMF can be assessed. Metallurgical preparation of specimen surface or sectioning of the specimens after subjecting them to TMF will completely destroy the true damage features since solder is a very soft material.

2 Service parameters

Thermal excursions encountered by the solder joints during service consist of several steps, in addition to imposed temperature difference and the regime of such temperature changes. These features are illustrated in Fig. 2. TMF studies discussed in this paper are based on temperature excursions between − 15°C and 150°C. Due to Coefficient of Thermal Expansion (CTE) mismatches, such a temperature change can

impose a thermal strain of about 0.001 between polycrystalline Cu and polycrystalline Sn, and 0.0–0.002 between adjacent Sn grains due to anisotropy of Body Centered Tetragonal (BCT) Sn [18–20] It should be noted that TMF corresponds to fatigue under constant strain amplitude imposed under varying (increasing or decreasing) temperature. Dwell time issues relate to time duration over which the solder joint is held at the temperature (upper or lower) extremes [21–23]. These dwell times need not be equal at both temperature extremes. For example the dwell at high temperature and low temperature extremes experienced by electronic solder joints in the under-the-hood environment of a long-haul truck or aero-planes during long-distance international flights, will be different from that in an automobile used for commuting to work or in short commuter flights. During the dwell periods stress relaxation under constant strain will take place, resulting in TMF damage repair by recovery/recrystallization/crack healing, or additional crack growth from residual stresses. Rate at which such events can occur will depend on temperature since the mechanisms operative will and could be different. Especially at the low temperature dwell instead of realizing damage repair, crack propagation can proceed from the existing damage [24, 25]. A typical example illustrating such a crack growth during room temperature storage of a joint that had undergone TMF is provided in Fig. 3. In this article for convenience the term "dwell-ratio" will be used to represent ratio of dwell time at high temperature extreme to that at the low temperature extreme.

Supporting studies of isothermal cyclic straining with dwell times at cyclic strain extremes were carried out on eutectic Sn–Ag solder joint so as to understand the dwell time issue [26–32]. Typical stress versus time plot under such conditions illustrating cyclic strain softening is provided in Fig. 4.

In order to differentiate the effects of cyclic straining as compared to that of typical stress relaxation after single strain imposition, studies were carried out by

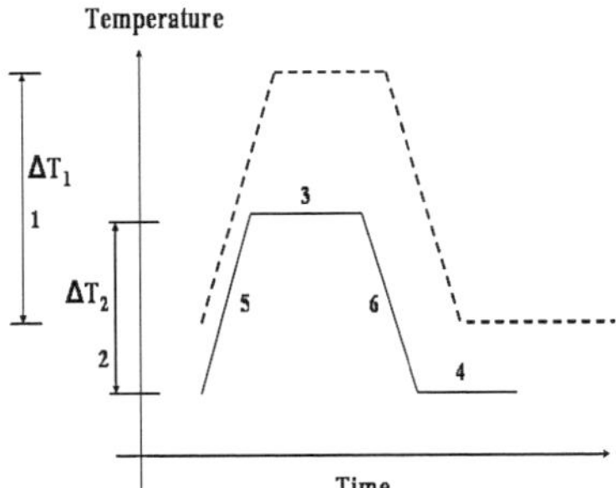

Fig. 2 Steps in TMF temperature profile. ΔT—temperature difference, 1,2—temperature regimes, 3—dwell at high temperature extreme, 4—dwell at low temperature extreme, 5—heating, 6—cooling

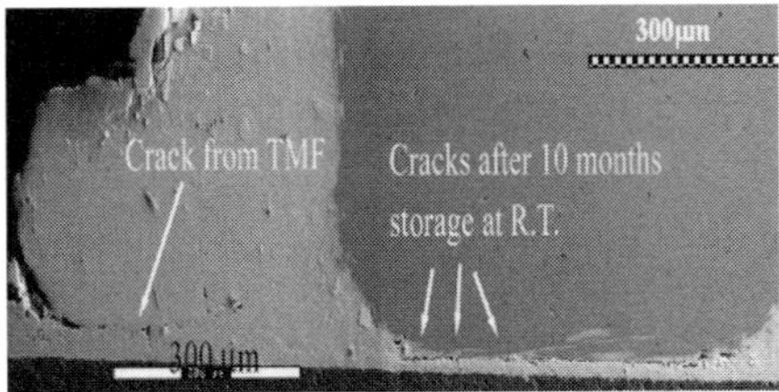

Fig. 3 Crack growth during storage in ambient conditions in a surface mounted joint that had experienced TMF

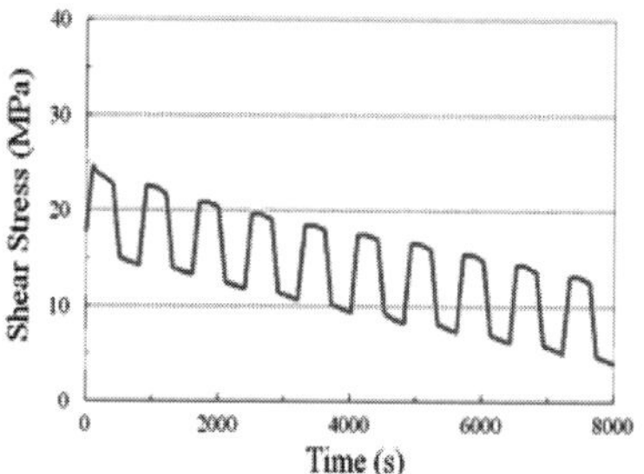

Fig. 4 Shear stress versus time plot of isothermal cyclic straining dwell times at strain extremes

imposing different cyclic strain amplitudes with different strain-rates at different temperatures. Results based on this study are provided in Fig. 5.

These studies indicate that with low cyclic strain amplitude imposed at a low strain-rate at high temperature (150°C) the softening realized is similar to that of stress relaxation after single straining. In other words cyclic straining has no additional influence. This may imply that cyclic straining at high temperatures have a tendency to heal the damage encountered. On the other hand, with large cyclic strain imposed at a high strain-rate at low temperature (room temperature) cyclic straining causes more stress drop than conventional stress relaxation after imposition of a strain. Such an observation suggests that cyclic straining encountered in low temperature regime might cause additional damage resulting in lower load carrying capability since these experiments were strain controlled. The arguments are speculative since the micro-cracks, if present, could not be seen due to resolution capabilities of the available techniques.

Another service parameter, "ramp-rate", representing rate of heating and cooling during temperature excursions is extremely important since they will affect the strain-rate at which the thermal strain is imposed [33]. For example, imposition of a thermal strain of 0.001 over realistic time duration of about 5 minutes (300 s) will correspond to a strain-rate of about 3×10^{-6}/s. However, if one imposes the same thermal strain by thermal shock (TS) like in ATC tests, the strain-rate experienced by the solder joint will be extremely large. Since Sn-based solders exhibit significant strain-rate sensitivity, "ramp-rate" warrants attention for reliability considerations [34, 35]. Supporting deformation studies have shown that for a given imposed shear strain, higher shear stress is realized with higher shear strain-rate as shown in Fig. 6 [26–29]. It is also important to realize that during these ramp periods the thermal strain is being accommodated over a varying temperature and that the solder flow behavior depends very strongly on the temperature. For example, faster heating rate may not facilitate the joint to adapt and reach the imposed temperature quickly, and as a consequence low temperature deformation mode may be more dominant in accommodating the thermal strain. Similarly, in the case of faster cooling-rate the high temperature deformation mode will be dominant as illustrated in Fig. 7.

These studies have also shown that low strain-rate and low temperature accommodate deformation by shear banding, while the high strain-rate and high temperature promote grain boundary sliding, as illustrated in Fig. 8.

Other considerations relate to the temperature difference (ΔT) to be imposed and the regime over which such temperature change exists. Since the temperature difference is directly related to the thermal strain that will be imposed, larger temperature difference will cause more damage and more associated deterioration of properties as shown in Fig. 9.

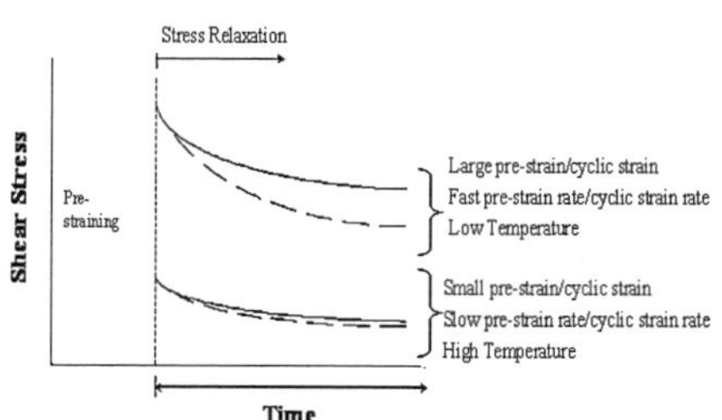

Fig. 5 Effects of temperature, pre- and cyclic-strain amplitudes and rates on stress relaxation. Solid lines correspond to stress relaxation without imposed cyclic straining and dotted lines correspond to those with cyclic straining

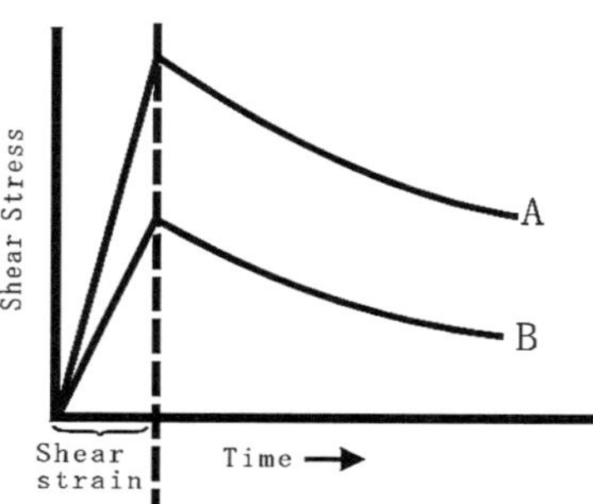

Fig. 6 Effect of strain rate on the peak stress and residual shear stress due to relaxation under isothermal and fixed pre-strain conditions. A: fast strain rate, and B: slow strain rate

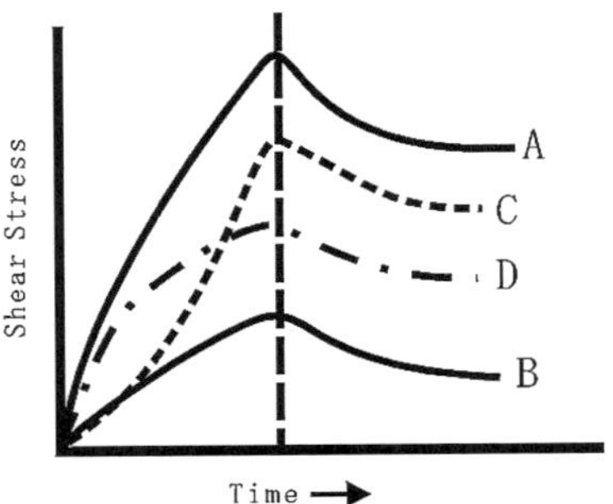

Fig. 7 Stress as a function of time during heating and cooling periods as compared to isothermal conditions at low and high temperature extremes. A: at low temperature extreme, B: at high temperature extreme, C: during cooling, and D: during heating. [Note: Temperature ramp rate during heating and cooling are assumed to be same in this figure for convenience. Such a condition will result in same shear strain-rate in heating and cooling parts of TMF]

However, the effect of the temperature regime in which the imposed temperature difference exists is more complicated since the deformation mode of the solder present in the joint will depend on the temperature range over which the thermal strain is being accommodated. This feature can be noted in Fig. 8. Most of the studies on eutectic Sn–Ag solder joints indicate that the transition between shear banding and grain boundary sliding occurs at a temperature range in the neighborhood of 125°C. In automotive under-the-hood type of applications, which the author of this paper was addressing, the upper temperature extreme imposed was 150°C, and the lower temperature extreme was − 15°C. As a consequence these solder joints were undergoing both deformation modes and the results obtained are based on a mix of both deformation modes.

3 Eutectic Sn–Ag solder joint response under TMF

This section illustrates the damage accumulation and the resultant deterioration of physical properties from TMF in eutectic Sn–Ag solder joint. This solder alloy is used as the basis for comparison in all these studies. Surface manifestation of damage resulting from TMF in solder joints occurs only after several hundred cycles of thermal excursions [36]. However, the physical properties such as residual shear strength and residual electrical conductivity begin to decrease from very early stages of TMF [37, 38]. Hence damage generation and accumulation occurs from very early stages of TMF, although their surface manifestations are not obvious with the currently available microscopic methodologies. Sectioning the specimens and preparing them for microstructure studies to observe the interior damage will completely destroy the true features providing false picture and associated false

Fig. 8 Effect of strain-rate and temperature on deformation mode of eutectic Sn–Ag solder joints

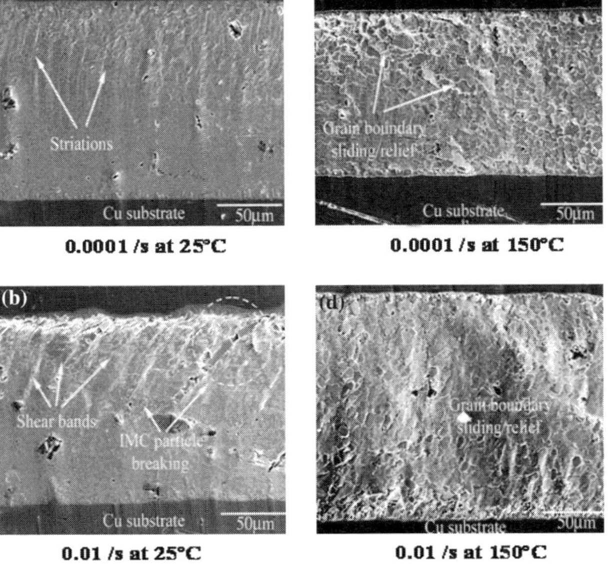

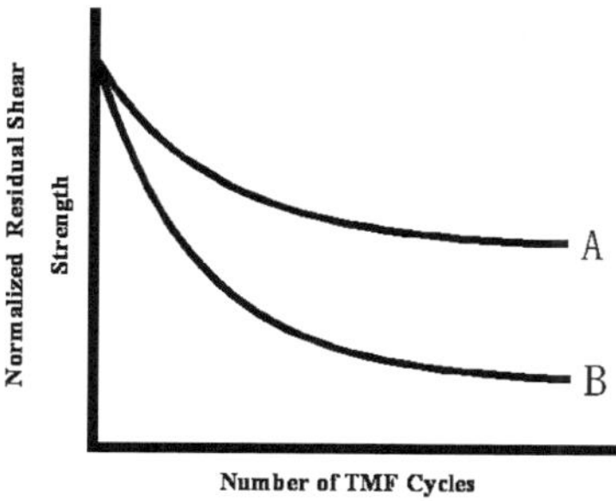

Fig. 9 Effect of the magnitude of imposed temperature difference (ΔT) on the residual strength. Note: A—smaller ΔT, B—larger ΔT

interpretations. Currently available non-destructive methodologies do not possess sufficient sub-micron resolution to observe the interior damage from TMF. As a consequence one has to resort to indirect methods such as residual shear strength and global conductivity changes to characterize the TMF damage [37, 38]. Among these, conductivity changes are influenced by most of the micro-cracks generated within the bulk of the solder [38]. Residual shear strength is influenced by the cracks that are properly oriented to affect the failure propagation during shear testing. Hence, residual conductivity decreases very fast from the very first few cycles of TMF, while the residual shear strength decreases more gradually. However, both stabilize in their values, although the residual electrical conductivity stabilizes within few cycles. Residual shear strength, on the other hand, stabilizes after few hundred cycles.

Schematic in Fig. 10 is provided to discuss the roles of "dwell-ratio" and "heating-rate" issues on the residual properties of eutectic Sn–Ag solder joints.

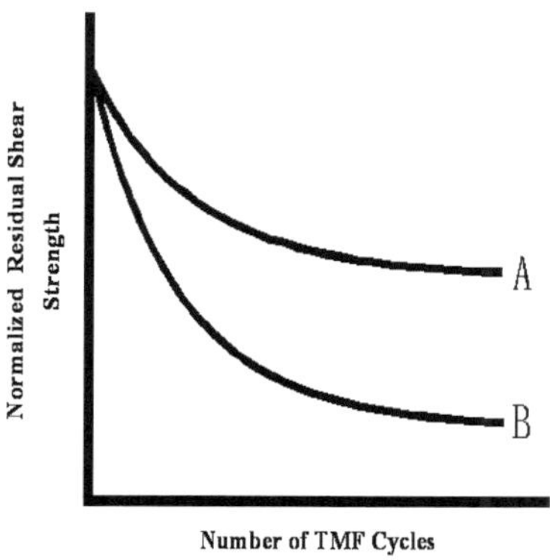

Fig. 10 Effect of TMF parameters during service on residual shear strength. A, B correspond to different service conditions

Plots A and B are used to illustrate the trends. In order to evaluate the effect of either of these two, that parameter has to be varied while the other parameter is maintained constant. For example, A corresponds to higher dwell-ratio, and B corresponds to lower dwell ratio, when ramp rates are maintained fixed. *Decrease in dwell-ratio*, where more dwell occurs at lower temperature extreme (as compared to that high temperature extreme) causes more damage accumulation and more decrease in residual shear strength. Similarly A corresponds to higher heating-rate and B corresponds to slower heating-rate, when the dwell-ratio was maintained constant. Faster heating causes more deterioration of the joint integrity. The experimental set-up used for these studies did not have means to alter the cooling-rate during TMF. It was maintained at about 0.07°C/s in all these studies.

Using the experimental findings from the dwell-ratio and heating rate, a parametric approach to assess solder joint reliability has been proposed [39]. Studies required for evaluation and utilization of such an approach to other solders and joint geometries are yet to be carried out.

4 TMF performance of some other solders

This section compares the residual shear strength after TMF in solder joints made with ternary alloy Sn–Ag–Cu, quaternary alloy Sn–Ag–Cu–Ni, and composite eutectic Sn–Ag reinforced with nano-strctured surface active Si–O cages, with that made using eutectic Sn–Ag [39–43]. The ternary and quaternary solders alloys contained less than 1% Cu and Ni. The nano-structured Si–O cage material used for reinforcement was polyhedral oligomeric silsesquioxanes (POSS) with suitable organic groups [44–47]. Two to three weight percent surface-active POSS tri-silanols were incorporated in the eutectic Sn–Ag solder [40]. Effect of TMF on Normalized Residual Strength in solder joints made with various solders is provided in Fig. 11.

The trends exhibited by the joints made with the above solders as a consequence of increased 'dwell-ratio', and 'heating-rate' during TMF are summarized in Table 1.

Important points to be noted from Fig. 11 and Table 1 are the following:

1. Joints made with Sn–Ag–Cu–Ni solder and POSS reinforced solder do not exhibit significant loss of strength due to TMF. This indicates that the damage accumulation in these solders resulting from TMF is minimal.

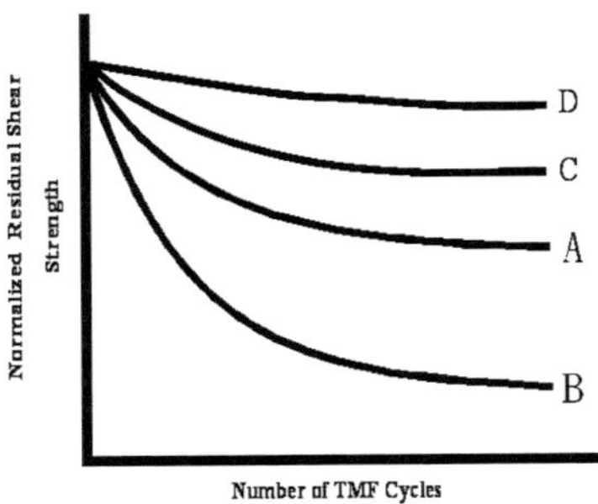

Fig. 11 Effect of TMF on Normalized Residual Shear Strength of joints made with various solders. A—Eutectic Sn–Ag, B—Sn–Ag–Cu, C—Sn–Ag–Cu–Ni, and D—POSS reinforced

2. Joints made with Sn–Ag–Cu solder tend to exhibit more deterioration from TMF as compared to those made with eutectic Sn–Ag solder.
3. Joints made with all the four solders deteriorate more with increased heating-rate during TMF. Among these, the one that deteriorates the most is made using ternary Sn–Ag–Cu solder.
4. Joints made with ternary Sn–Ag–Cu solder deteriorate more with increase in dwell-ratio. Longer high temperature dwell deteriorates this alloy more than the others.
5. Joints made with Sn–Ag–Cu–Ni solder exhibits better TMF performance (with less deterioration) with increase in dwell-ratio. Among the binary, ternary and quaternary alloys investigated this one had the best performance with larger dwell-ratio.
6. Since POSS reinforced solders did not exhibit significant damage from TMF, one can expect that dwell-ratio and heating-rate will not have any significant influences on its TMF performance.

5 Damage progression and failure

Grain boundary sliding is the commonly observed damage from TMF in bulk Sn and bulk eutectic Sn–Ag solder as shown in Fig. 12. Such damage in these bulk specimens results from the severe anisotropy of BCT Sn. Surface topographical observations of the bulk

specimens shown in this figure indicated that grain boundary sliding normal to the surface that resulted from the shear stresses acting on the boundary was about 0.25 µm [38]. As can be noted in this figure, in bulk specimens, normal stresses acting on the boundary has caused cracks resulting in about 10 µm separation between grains meeting at the boundaries [38]. Stresses that develop at the grain boundary regions due to CTE mismatches of grains meeting at the boundary cause such events. Residual property measurements indicate that the deterioration experienced from TMF by bulk Sn and bulk eutectic Sn–Ag solder is about the same [38]. This observation is illustrated in Fig. 13. Such observations indicate that presence of the Ag$_3$Sn IMC in the solder matrix has no obvious influence on the damage accumulation from TMF. Such events occur everywhere with no obvious fracture developing from them.

However, when solder is present in the joint, damage accumulates within the solder in regions very close to solder/substrate interface IMC layer, due to the constraints imposed by the substrate and the interface IMC layer, as shown in Fig. 14.

As TMF progresses such damage becomes intensified in this region and the cracks generated at the grain boundaries join together to form the catastrophic crack that causes inter-granular fracture at about one grain diameter away from the solder/substrate IMC. Under such conditions, grain boundary sliding and crack opening at the boundaries in ranges more than 10 µm have been observed [19].

Damage accumulation in joints made with all the four solders after 1,000 TMF cycles is illustrated with micrographs given in Fig. 15. Better TMF performance of quaternary Sn–Ag–Cu–Ni solder and POSS reinforced solder joints, and worse TMF performance of ternary Sn–Ag–Cu solder joints, as compared to the eutectic Sn–Ag solder joint can be explained using these observations.

Lack of any observable surface damage in the quaternary alloys even after 1,000 cycles provides the reason for its better TMF performance as compared to the eutectic Sn–Ag and the ternary Sn–Ag–Cu solder joints. It can also be noticed in this figure that Sn–Ag–Cu solder joint has experienced more extensive surface

Table 1 Effects of 'dwell-ratio' and 'heating-rate' on *Normalized Residual Strength* in solder joints made with various solders

Service parameter	Eutectic Sn–Ag	Sn–Ag–Cu	Sn–Ag–Cu–Ni
Increased/dwell-ratio	+++	– – – –	+
Increased heating-rate	– – – –	– – – – –	–

Note: "+" indicates that the Normalized strength shows an increasing trend with the increase in the given service parameter, and their number provides a feel for the extent of the change. Similarly " – "corresponds to decreasing trend

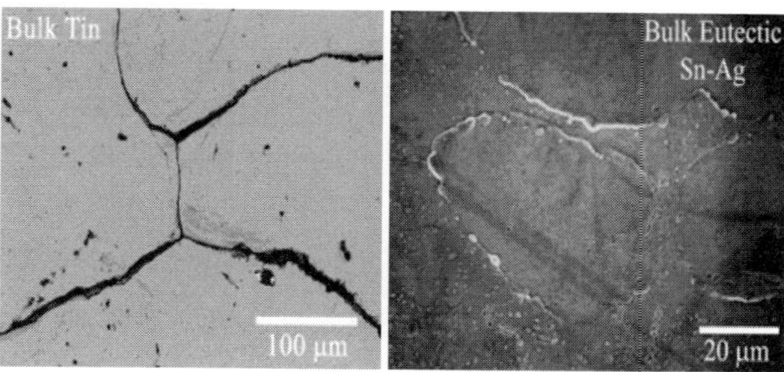

Fig. 12 Grain boundary sliding from TMF observed in bulk Sn and in bulk eutectic Sn–Ag

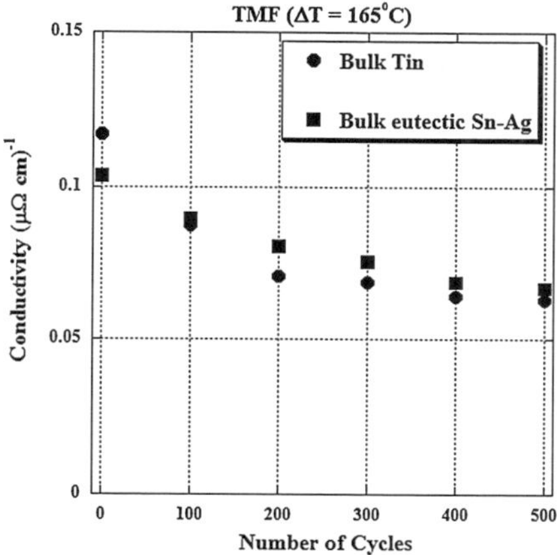

Fig. 13 Electrical conductivity changes resulting from TMF in bulk specimens of Sn and eutectic Sn–Ag

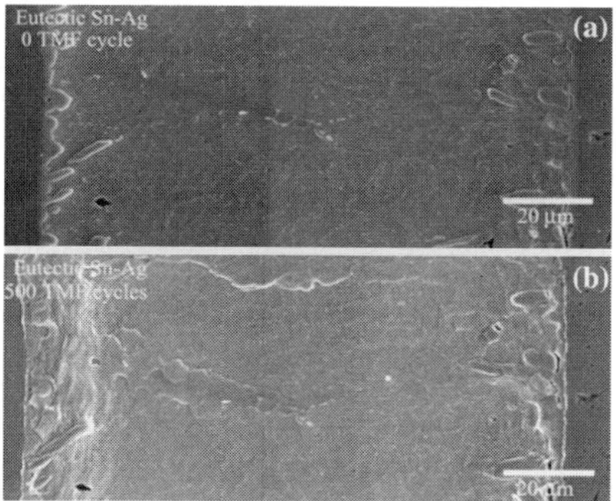

Fig. 14 Damage accumulation within the solder near solder/substrate interface in eutectic Sn–Ag solder joint. Scallop shaped IMC layer at this interface can be clearly seen in (**a**)

damage under similar conditions, thus providing the justification for its poor performance as compared to the eutectic Sn–Ag solder joint. POSS reinforced solder joints exhibit features similar to that of the quaternary Sn–Ag–Cu–Ni solder joints as illustrated in Fig. 16.

Reasons for this observed TMF response of these solders can be understood by noticing the particulates that are present in these solders. In ternary Sn–Ag–Cu solder, Cu_6Sn_5 IMC particles are present at the Sn grain boundaries. In quaternary Sn–Ag–Cu–Ni solders $(CuNi)_6Sn_5$ IMC particles are present at the Sn grain boundaries. In POSS reinforced eutectic Sn–Ag solders, Si–O cages that are strongly bonded to Sn at the Sn grain boundaries are present. Active radicals that were present at the surface of such cages have been utilized to strongly bond the Si–O cage to Sn, especially at the grain boundary locations. Such grain boundary reinforcements improve the strength of the solder.

Dwell times at high temperature extreme can cause significant aging. During 1,000 TMF cycles used in these studies, the solder joint experiences about 2,000 h of aging at 150°C. Effect of such aging on Cu_6Sn_5 present in the Sn–Ag–Cu solder and $(Cu,Ni)_6Sn_5$ are presented in Fig. 17.

As illustrated Cu_6Sn_5 present in Sn–Ag–Cu solders coarsens significantly during TMF, and develops large number of cracks. Such cracking at the grain boundaries will increase the damage accumulation during TMF and deteriorate the residual properties [36, 37] On the other hand $(Cu,Ni)_6 Sn_5$ does not appear to coarsen during this aging process as shown in Fig. 18. Exact reasoning for this is not known at present. A possible explanation can be given on the basis that Cu and Ni are ideally suited for substituting of each other. Cu diffuses very fast, while Ni diffuses very slowly, in Sn. As a consequence, growth of large Cu_6Sn_5 particles at Sn grain boundaries will be easy, while the growth of $(Cu,Ni)_6 Sn_5$ will be relatively slow due to very slow diffusion of Ni. As a consequence, $(Cu,Ni)_6 Sn_5$ remains as an effective reinforcement preventing grain boundary sliding even after 1,000 cycles of TMF. POSS reinforcements are inert in Sn and have no way of

Fig. 15 Damage resulting from 1,000 cycles of TMF in solder joints made with eutectic Sn–Ag, Sn–Ag–Cu, and Sn–Ag–Cu–Ni solders

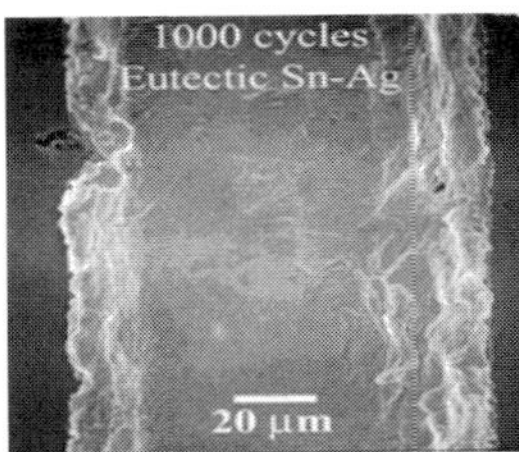

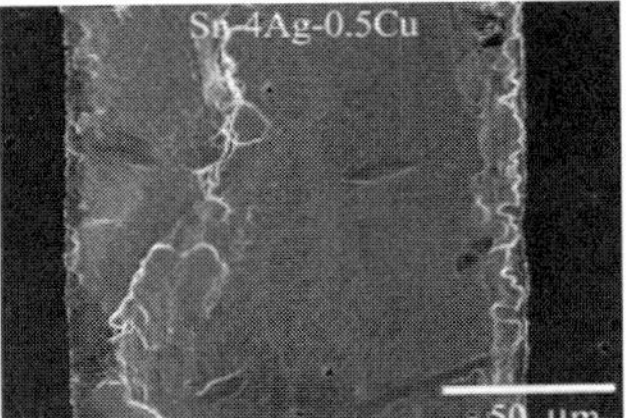

Eutectic Sn-Ag　　　　　**Sn-4Ag-0.5Cu**

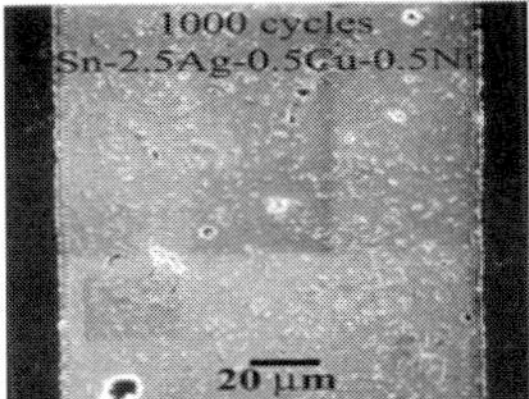

Sn-4Ag-0.5Cu-0.5ni

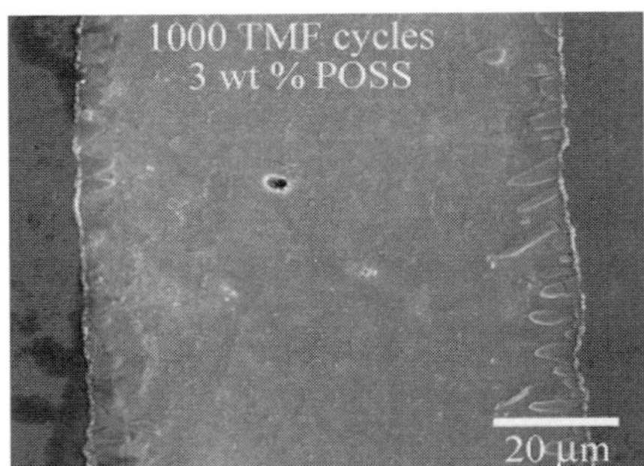

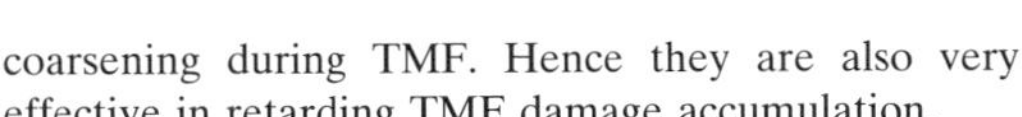

Fig. 16 Lack of any surface damage accumulation in POSS reinforced eutectic Sn–Ag solder joints even after 1,000 TMF cycles

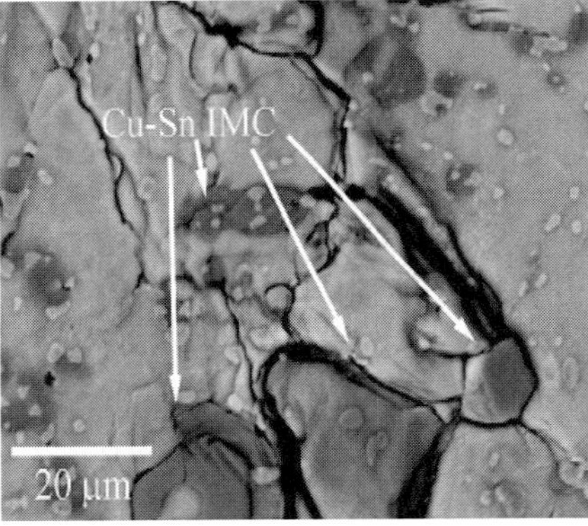

Fig. 17 Coarsening of $Cu_6 Sn_5$ particles during TMF and formation of cracks in such particles

coarsening during TMF. Hence they are also very effective in retarding TMF damage accumulation.

6 Issues that need to be addressed

This article is meant to bring to attention the importance of service and material parameters that need to be incorporated in any TMF reliability modeling. Although this study was focused on using only few solders to address the material response considerations, such responses are highly dependant on the service parameters encountered. For example this study spanned over a severe temperature range where

the mode of deformation of Sn-based solders changes. Such a change appears to be in the neighborhood of 125°C. If the temperature regime employed is low enough to avoid such complications, the findings could be different. Recent trend to carry out tests with a low temperature extreme of − 55°C can create other issues such as β to α transitions. Since α-phase is very brittle, one could expect its influences on the reliability of the solder joint. Also, as has been pointed out, the material parameters and response play very significant roles. Sn–Ag–Cu solder alloys that are being adopted by industry have poor high temperature stability. In other words, one solder alloy that performs extremely well under one set of service conditions may not be suitable

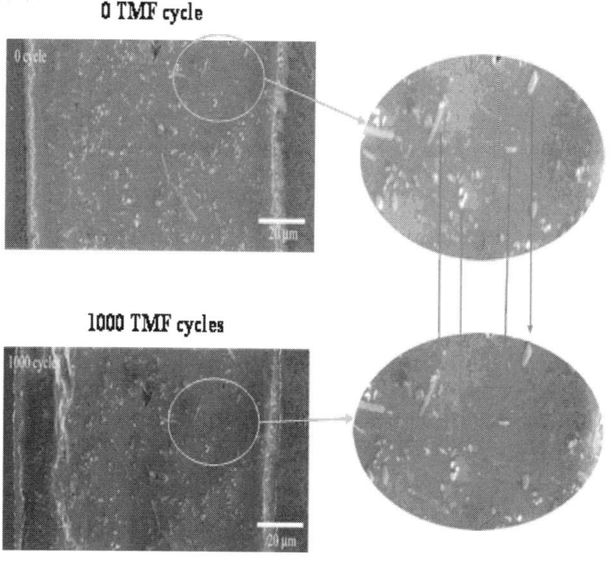

Fig. 18 Absence of coarsening of $(Cu,Ni)_6 Sn_5$ particles during TMF

for some other service environment. Material scientists, who design and develop solders, do not have any control on the service environment and joint geometry. So, it becomes apparent that developing solders in which the damage initiation from the intended service environment is minimal should be the appropriate approach. If damage initiation is minimized or avoided, damage accumulation in any given location of a joint geometry, to cause catastrophic failure, will not become a concern.

Acknowledgements Author thanks Drs. S.L. Choi, J.G. Lee, H. Rhee, F. Guo and A. Lee. Analysis presented in this paper is based on the insight gained from their results. He also thanks D. Choudhuri for help in preparing the manuscript

References

1. S. Choi, K.N. Subramanian, J.P. Lucas, T.R Bieler, J. Electron. Mater. **29**(10), 1249 (2000)
2. J.G. Lee, F. Guo, K.N Subramanian, "Thermomechanical Fatigue Behavior on Pb-Free Automotive Electronic Solders," SAE 2003 World Congress, Detroit, MI, March 3–6, 2003, SAE Paper Number 2003-01-0621
3. J.D. Sigelko, K.N. Subramanian, Adv. Mater. Process. **157/3**, 47 (2000)
4. K.N. Subramanian, A. Lee, S. Choi, P. Sonje, J. Electron. Mater. **30**(4), 372 (2001)
5. W.J. Plumbridge, Solder. Surf. Mt. Technol. **11**(3), 8 (1999)
6. M. Farooq, C. Goldsmith, R. Jackson, G. Martin, J. Electron. Mater. **32**(12), 1421 (2004)
7. C.J. Zhai, Sidharth, R. Blish, IEEE Trans. Device Mater. Reliabil. **3**(4), 207 (2003)
8. B.Z. Hong, L.G. Burrell, IEEE Trans. Comp. Packag. Manufact. Technol. A. **20**(3), 280 (1997)
9. J.C. Suhling, H.S. Gale, R.W. Johnson, M.N. Islam, T. Shete, P. Lall, M.J. Bozack, J.L. Evans, P. Seto, T. Gupta, J.P. Thompson. Sold. Surf. Mt. Technol. **16**(2), 77 (2004)
10. R.H. Raeder, R.W. Jr Messler, L.F Coffin Jr., J. Electron. Mater. **28**(9), 1045 (1999)
11. G.D. Giacomo, S. Oggioni, Microelectron. Reliabil.. **42**(9–11), 1541 (2002)
12. J. Lau, W. Dauksher, J. Smetana, R. Horsley, D. Shangguan, T. Castello, I. Menis, D. Love, B. Sullivan, Solder. Surf. Mt. Technol. **16**(1), 12 (2004)
13. S. Choi, T.R. Bieler, J.P. Lucas, K.N. Subramanian, J. Electron. Mater. **28**(11), 1208 (1999)
14. J. Sigelko, S. Choi, K.N. Subramanian, J.P. Lucas, T.R. Bieler, J. Electron. Mater. **28**(11), 1184 (1999)
15. K.N. Subramanian, T.R. Bieler, J.P. Lucas, J. Electron. Mater. **28**(11), 1176 (1999)
16. J. McDougall, S. Choi, T.R. Bieler, K.N. Subramanian, J.P. Lucas, Mater. Sci. Eng. A **285**, 25 (2000)
17. F. Guo, J.P. Lucas, K.N. Subramanian, J. Mater. Sci. Electron. Mater. **12**, 27 (2001)
18. K.N. Subramanian, J.G. Lee, J. Mater. Sci. Electron. Mater. **15**, 235 (2004)
19. K.N. Subramanian, Fatigue Fract. Eng. Mater. Struct. (in press)
20. J.G. Lee, A. Telang, K.N. Subramanian, T.R. Bieler, J. Electron. Mater. **31**(11), 1152 (2002)
21. J.G. Lee, K.N. Subramanian, J. Electron. Mater. **32**(6), 523 (2003)
22. T. Goswami, H. Hanninen, Mater. Design. **22**(3), 199 (2001)
23. J.H. Huang, Y.H. Jiang, Y.Y. Qian, Microelectron. Reliabil. **34**(5), 897 (1994)
24. K.N. Subramanian, J.G. Lee, Mater. Sci. Eng. A **421**(1–2), 46 (2006)
25. S.G. Jadhav, T.R. Bieler, K.N. Subramanian, J.P. Lucas, J. Electron. Mater. **30**(9), 1197 (2001)

26. H. Rhee, K.N. Subramanian, A. Lee, J.G. Lee, Solder. Surf. Mt. Technol. **15**(3), 21 (2003)
27. H. Rhee, K.N. Subramanian, J. Electron. Mater. **32**(11), 1310 (2003)
28. H. Rhee, K.N. Subramanian, A. Lee, J. Mater. Sci. Electron. Mater. **16**, 169 (2005)
29. H. Rhee, K.N. Subramanian, Solder. Surf. Mt. Technol. **18**(1), 19 (2006)
30. K.C. Chen, A. Telang, J.G. Lee, K.N. Subramanian, J. Electron. Mater. **31**(11), 1181 (2002)
31. J. Howell, A. Telang, J.G. Lee, S. Choi, K.N. Subramanian, J. Mater. Sci. Electron. Mater. **13**(6), 335 (2002)
32. K.J. Ferguson, A. Telang, J.G. Lee, K.N. Subramanian, J. Mater. Sci. Electron. Mater. **14**, 157 (2003)
33. J.G. Lee, K.N. Subramanian, Microelectron. Reliabil. (in press)
34. H. Mavoori, J. Chin, S. Vaynman, B. Moran, L. Keer, M. Fine, J. Electron. Mater. **26**(7), 783 (1997)
35. W.J. Plumbridge, C.R. Gagg, J. Mater. Sci. Electron. Mater. **10**, 461 (1999)
36. S. Choi, J.G. Lee, K.N. Subramanian, J.P. Lucas, T.R. Bieler. J. Electron. Mater. **31**(4), 292 (2002)
37. J.G. Lee, F. Guo, S. Choi, K.N. Subramanian, T.R. Bieler, J.P. Lucas, J. Electron. Mater. **31**(9), 946 (2002)
38. F. Guo, J.G. Lee, T.P. Hogan, K.N. Subramanian, J. Mater. Res. **20**(2), 364 (2005)
39. K.N. Subramanian, J. Electron. Mater. **34**(10), 1313 (2005)
40. A. Lee, K.N. Subramanian, J. Electron. Mater. **34**(11), 1399 (2005)
41. J.G. Lee, K.N. Subramanian, Solder. Surf. Mt. Technol. **17**(1), 33 (2005)
42. A.Z. Miric, A. Grusd, Solder. Surf. Mt. Technol. **10**(1), 19 (1998)
43. W.L. Winterbottom, JOM, **45**, 20 (1993)
44. H.C.L. Abbenhuis, Chem. Eur. J. **6**, 25 (2000)
45. R.A. Mantz, P.F. Jones, K.P. Chaffee, J.D. Lichtenhan, J.W. Gilman, I.M.K. Ismail, M.J. Burmeister, Chem. Mater. **8**, 1250 (1996)
46. J.D. Lichtenhan, J.J. Schwab, W.A. Reinerth, Chem. Innov. **31**(1), 3 (2001)
47. W. Zhang, B.X. Fu, Y. Seo, E. Schrag, B. Hsiao, P.T. Mather, N. Yang, D. Xu, H. Ade, M. Rafailovich, J. Sokolov, Macromolecules **35**, 8029 (2002)

Electromigration statistics and damage evolution for Pb-free solder joints with Cu and Ni UBM in plastic flip-chip packages

Seung-Hyun Chae · Xuefeng Zhang ·
Kuan-Hsun Lu · Huang-Lin Chao · Paul S. Ho ·
Min Ding · Peng Su · Trent Uehling ·
Lakshmi N. Ramanathan

Published online: 22 September 2006
© Springer Science+Business Media, LLC 2006

Abstract A series of electromigration (EM) tests were performed as a function of temperature and current density to investigate lifetime statistics and damage evolution for Pb-free solder joints with Cu and Ni under-bump-metallizations (UBMs). The EM lifetime was found to depend on the failure criterion used, so the results were compared based on the first resistance jump and conventional open-failure criterion. Solder joints with Cu UBM had a longer lifetime than Ni UBM based on the open-failure criterion, but the lifetime with Ni UBM became comparable when the first resistance jump criterion was applied. To determine the temperature in solder joints, the Joule heating effect was investigated with experiments and finite element analysis. The temperature of solder joints was determined to be approximately 15°C higher than that at the Si die surface when 1 A of current was applied. With the appropriate temperature correction, the activation energies and the current density exponents were found to be $Q = 1.11$ eV, $n = 3.75$ and $Q = 0.86$ eV, $n = 2.1$ based on the open-failure criterion for solder joints with Cu and Ni UBM, respectively. Based on the first resistance jump criterion, $Q = 1.05$ eV, $n = 1.45$ for Cu UBM and $Q = 0.94$ eV, $n = 2.2$ for Ni UBM, respectively. For solder joints with Cu UBM, voids were formed initially at the Cu_6Sn_5/solder interface while the final open failure occurred at the Cu_3Sn/Cu_6Sn_5 interface. For Ni UBM, voids were formed initially at the Ni_3Sn_4/solder interface leading to failure at the same interface. The formation of intermetallic compounds (IMCs) was enhanced under current stressing, which followed linear growth kinetics with time. The IMC growth was accompanied by volume shrinkage, which accelerated damage evolution under EM.

1 Introduction

With continuing demands to increase I/O density and power requirement, electromigration (EM) failure of solder joints raises increasing reliability concern for plastic flip-chip packages. It is anticipated that flip-chip solder joints will be subject to a current density in the order of 10^4 A/cm^2 in the near future [1]. Although this current density level is still about two orders of magnitude lower than that for Cu interconnects, EM damage becomes a serious concern for solder joints due to their low current carrying capability [2]. In solder joints, the formation of intermetallic compounds (IMCs) at the interface between the solder and the under-bump-metallization (UBM) plays an important role in controlling EM reliability [2]. Noble or near-noble metals such as Cu or Ni in the UBM can diffuse rapidly in Pb or Sn by an interstitial diffusion mechanism [3–5], and react at a fast rate with Sn to form IMCs [6]. The implementation of eco-friendly Pb-free solders generates further interests in studying the effect of UBM on EM reliability of solder joints. This effect is expected to be more significant for Sn-based Pb-free

S.-H. Chae (✉) · X. Zhang · K.-H. Lu · H.-L. Chao ·
P. S. Ho
Microelectronics Research Center, The University of Texas
at Austin, Austin, Texas 78758-4445, USA
e-mail: chaes@mail.utexas.edu

M. Ding · P. Su · T. Uehling · L. N. Ramanathan
Freescale Semiconductor, Technology Solution
Organization, Austin, Texas 78735, USA

solders. Since Sn is a major constituent of the IMCs, its inexhaustible supply from the solder can greatly enhance the IMC formation to degrade EM reliability [6, 7].

This study investigated the effect of UBM on EM reliability for Pb-free Sn-3.5Ag solder joints. The Cu UBM was compared with the Ni UBM. Due to the rapid reaction rate of Cu with Sn [7], the thick Cu stud in the UBM was of interest. For Ni UBM, the thin Ni film in the UBM was of primary interest due to the slower reaction rate of Ni with Sn [8]. Samples were tested under various temperature and current conditions to obtain EM statistical data. The EM lifetime was found to depend on the failure criterion used, so the results were compared based on the first resistance jump and conventional open-failure criterion. In accelerated solder EM tests, approximately 1 A current is usually applied [6, 7, 9], which can cause substantial Joule heating. Moreover, heat dissipation in plastic packages is much less efficient than in ceramic packages, so the Joule heating effect was taken into account to deduce EM statistics by combining experiments and simulation. EM damage evolution and failure mechanism were investigated by cross-sectional microscopy of solder joints, with emphasis on the morphology changes of solder joints with IMC growth.

2 Experimental

Figure 1 shows a schematic of the cross-sections of Sn-3.5Ag solder joints with two types of UBM in plastic packages. One had a 0.25 μm-TiW/18 μm-Cu UBM and the other a 0.1 μm-Ti/2 μm-Ni UBM. Both Cu and Ni UBM were electroplated after sputter deposition of the seed layers. Thickness variations of Cu and Ni

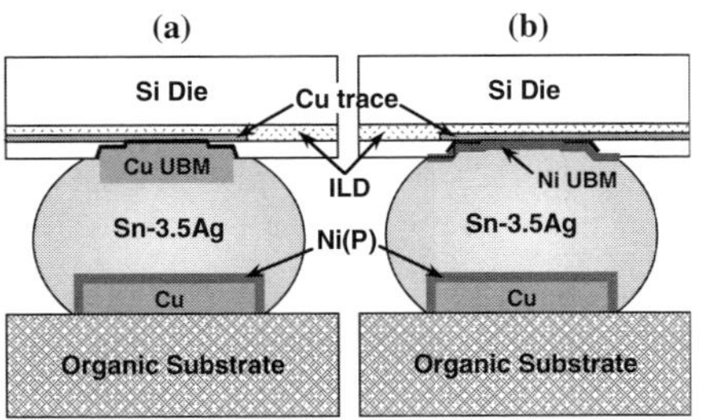

Fig. 1 A schematic diagram of cross-sectioned Sn-3.5Ag solder joints with (**a**) Cu under-bump-metallization (UBM), and (**b**) Ni UBM

UBM were ±4 μm and ±0.5 μm, respectively. The passivation opening of the UBM was 50 μm in diameter. On the substrate side, a 5 μm-Ni(P) was plated electrolessly on the Cu metallization. The Pb-free Sn-3.5Ag solder had a nominal diameter of 130 μm and a height of 80 μm.

The resistance of a solder joint at room temperature was only a few mΩ whereas that of entire circuitry of the test sample was ~1.1 Ω. Thus it was difficult to detect the resistance change of a solder joint before it becomes significantly open [2]. To improve the sensitivity to track the resistance change, this study employed the Wheatstone bridge method. In one leg of the Wheatstone bridge circuit, two pairs of solder joints were connected in series, making the test structure a two-link ($N = 2$) system. Each pair served as one resistor unit of the bridge consisting of a cathode joint and an anode joint. Here, the cathode joint refers to a solder joint where electrons flowed from the substrate side to the die side, and the anode joint refers to one where electrons flowed in the opposite direction. The other leg consisted of a fixed and a variable resistor of the order of 10^3 Ω. In this way, current flowed primarily through the solder joints. To start the EM test, the off-balance voltage V_g in the Wheatstone bridge circuit was set to zero by adjusting the variable resistor. The value of V_g was directly correlated with net resistance changes of solder joints only. Thus the V_g traces enabled us to detect the small resistance changes of solder joints with high sensitivity as damage developed. A detailed description of the Wheatstone bridge method can be found in Ref. [6].

In the EM test oven, a set of samples were arranged in a way that the Si backside of each sample was attached to a Cu plate in order to minimize temperature deviation of each sample and to facilitate Joule heat dissipation in the test sample. Thermocouples were sandwiched between the Si dies and the Cu plate to monitor the Si backside temperature, which was generally higher than the oven temperature by 7–20°C depending on the applied current.

To obtain statistical data, 12–15 samples (i.e. 24–30 pairs of solder joints) were tested in each run. The EM lifetime was measured at 115, 130, 140 and 150°C with 1.01 A of current stressing. The current density dependence was extracted from experiments at 140°C with 0.81, 1.01 and 1.11 A, corresponding to current densities of 4.12, 5.16 and 5.67 in units of 10^4 A/cm², respectively. Current density was calculated based on the area of the passivation opening, and all the above temperatures are referred to the surface temperature at the backside of the Si die.

3 Results and discussion

3.1 Failure criteria

Figure 2 represents typical V_g traces of solder joints with Cu and Ni UBM. Initially, V_g increased slowly. Then an abrupt jump in V_g occurred with approximately 30–40 mV change, which was followed by unstable fluctuations until the solder became electrically open. The V_g increase was more gradual in Ni UBM than Cu UBM. The unstable fluctuations were attributed to simultaneous damage evolution and recovery process in the solder joint, which eventually led to an electrical open failure [10]. At this stage, a solder joint became unstable and vulnerable to failure. The period of unstable fluctuations lasted usually longer for solder joints with Cu UBM than with Ni UBM in this study, which could be related to their thickness difference. Numerical analysis showed that a 30–40 mV change in V_g corresponded to approximately 90–95 % opening of the solder joint, as shown in Fig. 3. This was confirmed by cross-sectional SEM observation and will be discussed in Sect. 3.4. Based on these findings, the first resistance jump was suggested as an alternate failure criterion to supplement the conventional open-failure criterion to evaluate EM lifetime.

3.2 EM lifetime statistics

The cumulative distribution function (CDF) of solder joints with Cu and Ni UBM at various temperatures and currents are plotted in Figs. 4 and 5. The results of

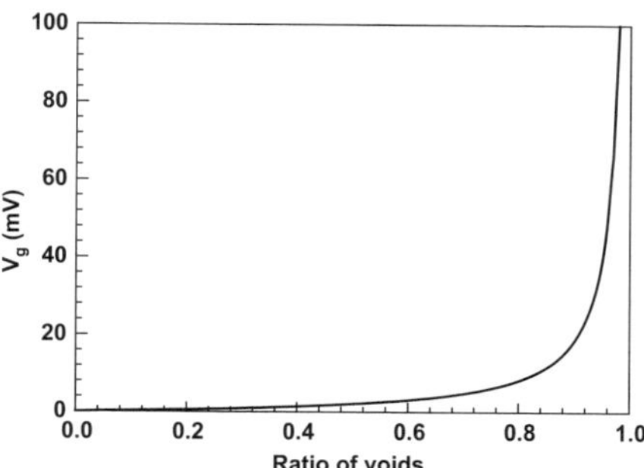

Fig. 3 Calculated off-balance voltage V_g with respect to the void to interface ratio (fraction of open)

EM lifetime statistics are summarized in Table 1. Since each sample was a two-link structure ($N = 2$), CDF was deconvoluted based on the weakest link approximation to deduce the median-time-to-failure (MTTF) t_{50} for single solder joints ($N = 1$) [11]. Based on the electrical open criterion, the lifetime of solder joints with Cu UBM was 1.5–4.0 times longer than that of solder joints with Ni UBM. In contrast, when the first resistance jump criterion was applied the lifetimes of solder joints with Cu and Ni UBM became comparable. This was attributed to the fact that the period of resistance fluctuation in Cu UBM solders amounted to 36–75% of the total lifetime whereas the period in Ni UBM solders amounted to only 7–27%. Considering that the resistance fluctuation already corresponds to significant opening of a solder joint, it is important to select a proper failure criterion to evaluate solder reliability, particularly for a thick Cu UBM.

From the lifetime data the EM activation energy and the current density exponent for solder joints were calculated using the Black's equation [12]:

$$t_{50} = Aj^{-n}\exp\left(\frac{Q}{kT}\right), \tag{1}$$

where t_{50} is the median-time-to-failure (MTTF), A is a constant, j is the current density, n is the current density exponent, Q is the activation energy, k is the Boltzmann constant, and T is the temperature. Based on the open-failure criterion, activation energies of Sn-3.5Ag solder with Cu and Ni UBM were 1.03 and 0.80 eV, with current density exponents 5.1 and 3.3, respectively (Fig. 6). In comparison, based on the first resistance jump criterion, the activation energies of the solder with Cu and Ni UBM were 0.97 and 0.87 eV (Fig. 7a), which are consistent with those from the open-failure criterion. However, the current density exponent was reduced significantly to 2.9 for the solder

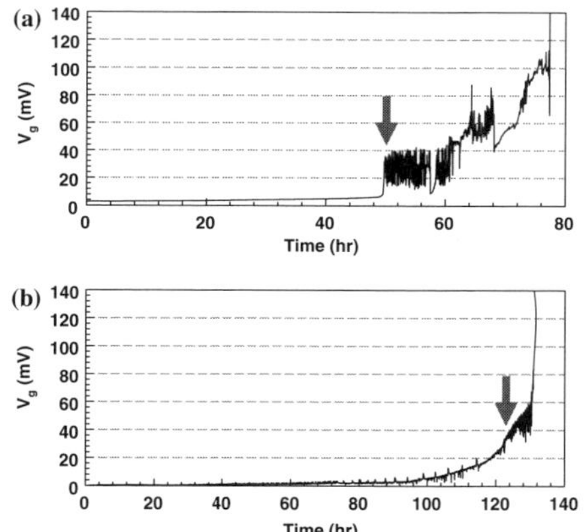

Fig. 2 Typical V_g traces of solder joints with (**a**) Cu under-bump-metallization (UBM), and (**b**) Ni UBM. Arrows indicate where the first resistance jump criterion regards as failure

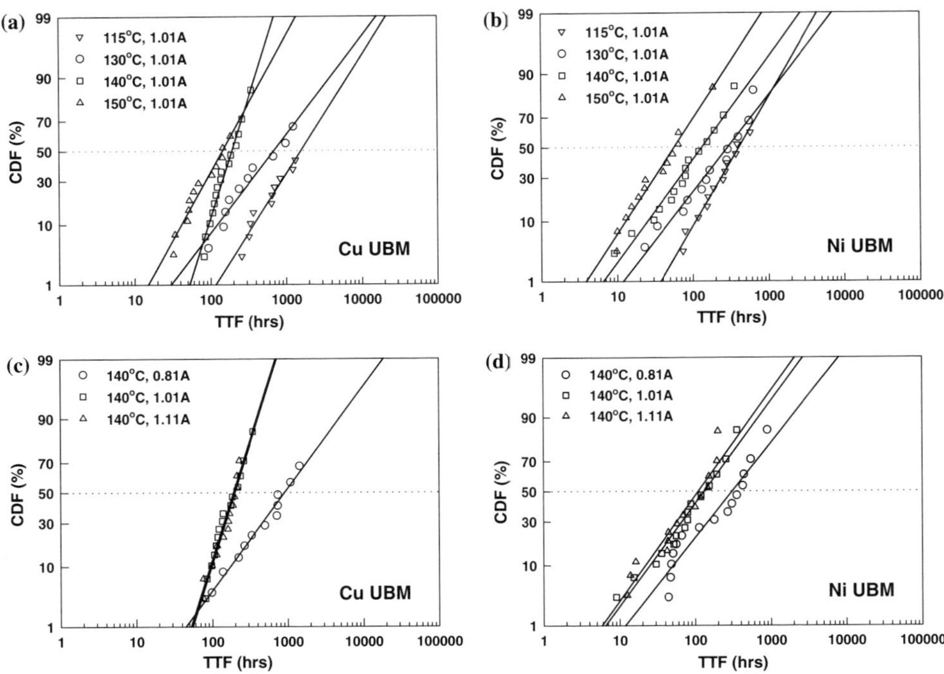

Fig. 4 (**a**)–(**b**) Temperature dependence, and (**c**)–(**d**) current density dependence of electromigration (EM) lifetime statistics of solder joints ($N = 1$) with Cu and Ni under-bump-metallization (UBM), based on the open-failure criterion. (Note: TTF: Time to failure)

with Cu UBM while remaining about the same for the solder with Ni UBM (Fig. 7b). This reflects that the resistance fluctuation period for solders with Cu UBM was substantially longer when the applied current was small. Thus the first resistance jump criterion resulted in more reduction of the EM lifetime at a lower current density. Nevertheless, the current density exponent for solder EM was in general higher than that for Cu interconnects, which is usually between 1 and 2 [13]. This suggests a significant amount of Joule heating during solder EM tests, which has to be properly taken into account to ensure reliable EM parameters.

3.3 Joule heating measurement and simulation

Due to the large applied current (~1 A), the Joule heating will increase the temperature of solder joints under current stressing, although the solder temperature was often assumed to be the same as the temperature at the backside of a Si die [9, 14]. To determine the Joule heating effect, we measured resistance changes as a function of temperature and applied current, and supplemented the experimental results with finite element analysis (FEA). First, the

temperature coefficient of resistance (TCR) was determined by measuring the resistance with increasing temperature at a minimal applied current (50 mA). The TCR was found to be 3.6×10^{-3} $(°C)^{-1}$ as shown in Fig. 8a, which was between those for bulk Cu $(3.9 \times 10^{-3}$ $(°C)^{-1})$ and 0.5 µm wide Cu interconnects $(3.3 \times 10^{-3}$ $(°C)^{-1})$ [15]. Next, the oven temperature was set to 120°C, and resistance changes were measured again with increasing applied current. The relationship between the applied current and the resistance change was found to be $\Delta R/R_0 = 0.158 \times i^{2.165}$ from Fig. 8b. When the oven temperature was set to 140°C the difference in the result was negligible. Combining this relationship with TCR, we found that the average temperature of the test structure was ~30°C higher than the Si backside temperature at 1 A applied current.

To supplement the experimental results, the temperature distribution inside the test structure was calculated using FEA based on the dimensions and configurations of the current path including the solder joints, the Cu trace on the die side and the Cu trace in the substrate. The calculated average temperature in the test structure was force-fitted with the

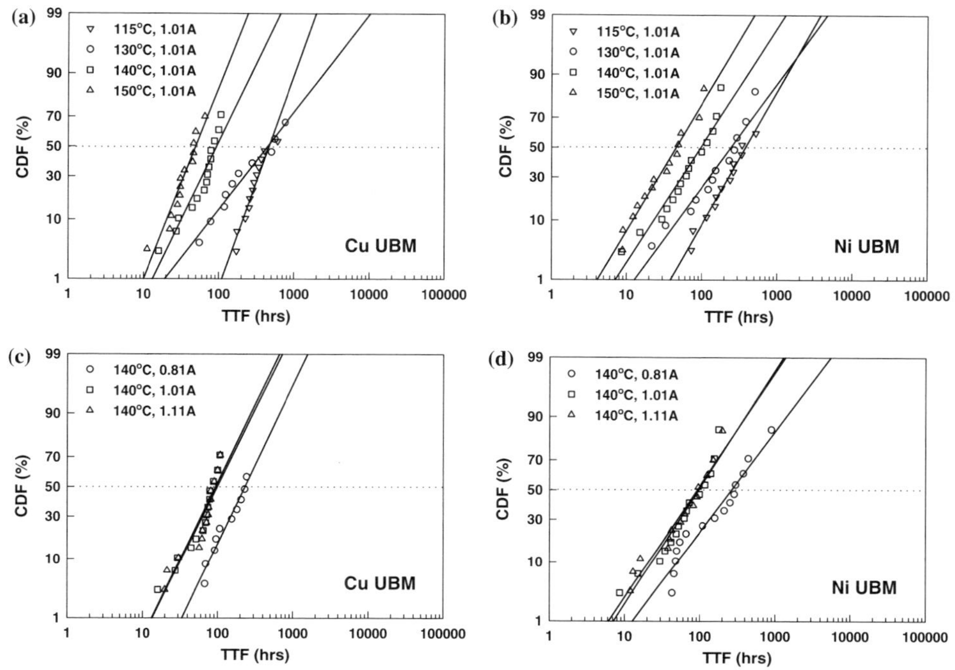

Fig. 5 (**a**)–(**b**) Temperature dependence, and (**c**)–(**d**) current density dependence of electromigration (EM) lifetime statistics of solder joints ($N = 1$) with Cu and Ni under-bump-metalliza-tion (UBM), based on the first resistance jump criterion. (Note: TTF: Time to failure)

experimental result. At 0.81, 1.01 and 1.11 A of the applied current, the solder joint temperature turned out to be higher than that at the Si backside by 10, 15 and 17°C, respectively (Fig. 9). The primary source of Joule heating was the ~33 mm-long Cu trace in the substrate, which resulted in the temperature gradient in the solder as shown in Fig. 9. Accordingly, the solder temperature in the substrate side was ~3°C higher than that in the die side.

Correcting for the actual solder temperature, the activation energies and current density exponents were reanalyzed, and the refined results are summarized in Table 2. The activation energies of EM were then in good agreement with the results obtained from Cu–Sn and Ni–Sn IMC growth in the literature [16–19]. The current density exponents were reduced considerably, although still higher than the typical value for Cu interconnects.

Table 1 Electromigration (EM) lifetime statistics of Pb-free Sn-3.5Ag solder joints with Cu and Ni under-bump-metallization (UBM) based on two kinds of failure criteria at various test conditions

Temperature* (°C)	Applied current (A)	Current density (A/cm²)	Open-failure criterion				First resistance jump criterion			
			Cu UBM		Ni UBM		Cu UBM		Ni UBM	
			t_{50} (h)	σ	t_{50} (h)	σ	t_{50} (h)	σ	t_{50} (h)	σ
115	1.01	5.16×10^4	1570	1.1	407	1.0	464	0.6	379	1.0
130	1.01	5.16×10^4	690	1.3	288	1.4	444	1.3	243	1.3
140	0.81	4.12×10^4	912	1.3	309	1.4	230	0.8	263	1.3
	1.01	5.16×10^4	192	0.6	132	1.3	93	0.8	96	1.1
	1.11	5.67×10^4	200	0.6	110	1.3	98	0.9	92	1.2
150	1.01	5.16×10^4	144	1.0	56	1.2	49	0.7	45	1.0

* Temperature at the backside of the Si die

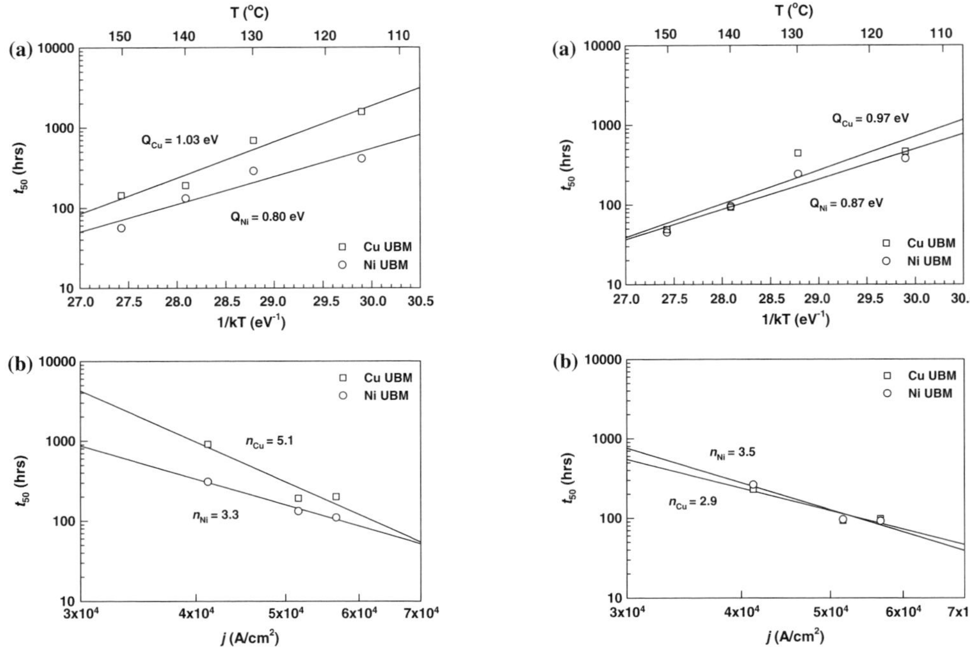

Fig. 6 (**a**) Activation energy, Q, and (**b**) current density exponent, n, deduced from the open-failure criterion

Fig. 7 (**a**) Activation energy, Q, and (**b**) current density exponent, n, deduced from the first resistance jump criterion

3.4 Failure analysis

3.4.1 Damage evolution in solder joints with Cu UBM

Due to the current crowding effect, EM failures were always observed in anode joints where electron flux flowed from the die side to the substrate side, consistent with those reported in other studies [6, 7, 9]. Figure 10b–d show morphological changes evolved in solder joints during EM test for 263 h at 140°C (Si backside temperature). For comparison, a pristine solder joint in the as-received condition was included in Fig. 10a. A small number of Kirkendall voids formed during solder reflow as can be seen between Cu UBM and Cu_6Sn_5. The Cu_3Sn phase should be present between Cu UBM and Cu_6Sn_5 [20] although this layer was too thin to be identified in Fig. 10a.

Although fatal EM damages developed only in the anode joint (Fig. 10d), some microstructural changes were observed in the joint without current stressing and in the cathode joint. Without current stressing (Fig. 10b), the number of Kirkendall voids increased in solder and the scallop-shaped Cu_6Sn_5 phase became flattened during thermal aging. This is because Cu is the dominant diffusing species and the valleys in the scallop-shaped Cu_6Sn_5 phase are the faster diffusion paths for Cu atoms [20]. The increase of voids and flattening process progressed further with longer aging time and increasing temperature.

When an electric current was applied, additional microstructural changes were observed. In the cathode joint (Fig. 10c), $(Cu,Ni)_6Sn_5$ migrated from the substrate side to the die side, where Cu atoms migrated from the initial $(Cu,Ni)_6Sn_5$ phase formed on the Ni(P)

Table 2 Reanalyzed activation energies and current density exponents based on finite element analysis for solder temperature

	Open-failure criterion		First resistance jump criterion	
	Cu UBM	Ni UBM	Cu UBM	Ni UBM
Activation energy (eV)	1.11	0.86	1.05	0.94
Current density exponent	3.75	2.1	1.45	2.2

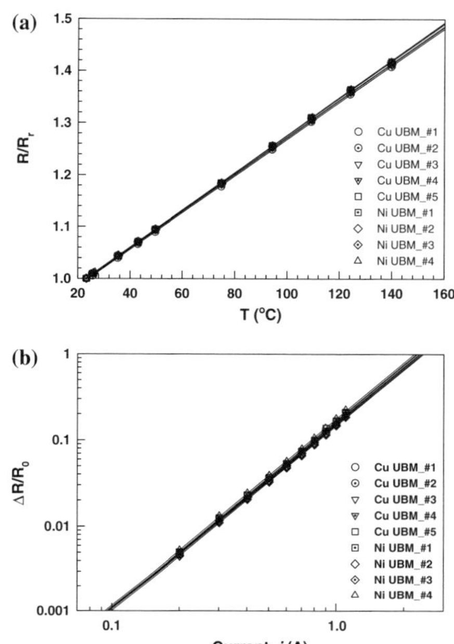

Fig. 8 (**a**) Plot of R/R_r with temperature for the temperature coefficient of resistance (TCR) measurement, and (**b**) plot of the resistance increase with current

substrate finish layer during solder reflow, and Ni atoms migrated from both the $(Cu,Ni)_6Sn_5$ phase and the substrate finish layer. The Ni concentration decreased with distance from the Ni finish layer to the IMC. Meanwhile, at the Cu UBM/Cu_6Sn_5 interface, the Cu_3Sn layer grew thicker under current stressing. The density of Kirkendall voids also increased but the density difference with and without current stressing was not so clear. At this point, most of the Kirkendall voids still located in the Cu_3Sn layer near the UBM/Cu_3Sn interface while the Cu UBM continued to be depleted as shown in Fig. 10c. This indicates that Sn atoms were driven by current through the IMC layers to react with Cu atoms while Cu atoms diffused into solder in the direction opposite to the electron flow. Without Sn diffusion, the additional depletion of Cu UBM cannot be explained as compared with the case without current stressing (Fig. 10b). Without Cu diffusion, Kirkendall voids would have moved toward the substrate side away from the Cu UBM/Cu_3Sn interface. Thus both Cu and Sn atoms interdiffused in the cathode joint, but the diffusion of Cu atoms might be somewhat suppressed due to the electron wind force.

At the anode joint (Fig. 10d), a significant amount of IMC formation and damage evolution was observed. Most of the Cu UBM was dissolved into solder to form IMCs. A large domain of Cu_6Sn_5, enhanced by the electron flow, was formed to develop an extended morphology in concert with the electron flux. In contrast, the Cu_3Sn layer maintained its parallel interfaces, i.e. the Cu UBM/Cu_3Sn interface and the Cu_3Sn/Cu_6Sn_5 interface were almost parallel to each other during Cu_3Sn growth as shown in Fig. 10d. The amount and location of Kirkendall voids in the anode joint were similar to those in the cathode joint. Interestingly, the Kirkendall voids formed during solder reflow had very little to do with EM-induced voids or failure even in the anode joint as shown in Fig. 10d. However, the formation of Kirkendall voids during thermal aging or under current stressing can raise other reliability concerns since they weaken the mechanical strength of solder joints [20].

Figure 11 displays EM damage evolution in the anode joint with Cu UBM. At an early stage, IMC layers grew by consuming Cu UBM (Fig. 11a). The V_g increase was negligible in this stage. Next, EM-induced voids were initiated at the Cu_6Sn_5/solder interface as shown in Fig. 11b. These voids grew independent of Kirkendall voids in the Cu_3Sn layer observed after solder reflow. At this point, V_g was still only a few mV but increased gradually, corresponding to an increase of several mΩ in resistance [10]. The current crowding effect on UBM consumption and void initiation was weak due to the thick Cu UBM structure. Figure 11c depicts a solder joint after an abrupt jump to ~40 mV in V_g. At this point, substantial Cu_6Sn_5 growth was observed, and EM voids were embedded in the Cu_6Sn_5 layer near the Cu_3Sn/Cu_6Sn_5 interface. Before discrete voids were connected to each other (Fig. 11b), the region between voids had a high current density, which further enhanced Cu_6Sn_5 growth with simultaneous void propagation. The resistance increased abruptly when the discrete voids became connected to form a global crack, which could be driven by tensile stress developed during IMC formation. Note that the solder joint was almost open at this stage as was expected based on the result obtained in Fig. 3. At this stage, Cu_6Sn_5 grew preferentially through the region where solder was still connected. When an open failure finally occurred, an extensive formation of Cu_6Sn_5 was observed connecting the IMC on the die side with that on the substrate side as shown in Fig. 11d. The void size increased further while the Cu_3Sn layer grew but at a lower growth rate compared with Cu_6Sn_5. This eventually led to crack that propagated through the Cu_6Sn_5 phase near the Cu_3Sn/Cu_6Sn_5 interface.

 Springer

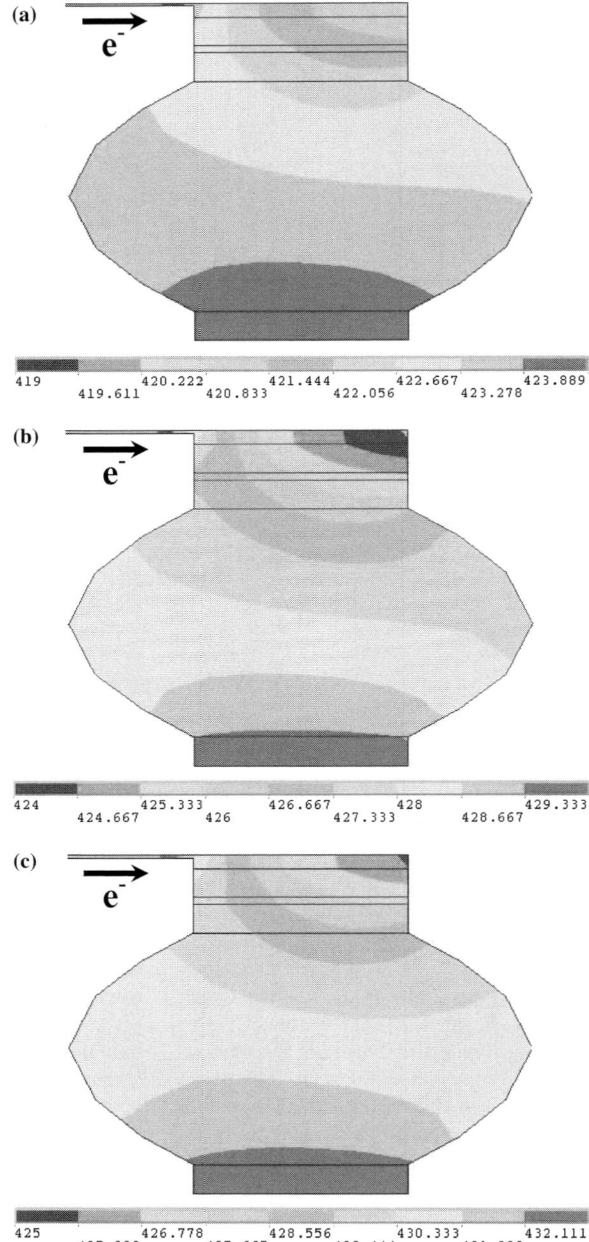

Fig. 9 Simulated temperature (in K) of the anode solder joint under current stressing of (**a**) 0.81 A, (**b**) 1.01 A, and (**c**) 1.11 A. Top: die side; bottom: substrate side

3.4.2 Damage evolution in solder joints with Ni UBM

Figure 12 shows the morphology of cross-sectioned solder joints with Ni UBM after 150 h of an EM test at 130°C (Si backside temperature) with 1.01 A. A thin Ni_3Sn_4 layer had been formed between the Ni UBM and the solder during solder reflow, but little change occurred in the solder without current stressing

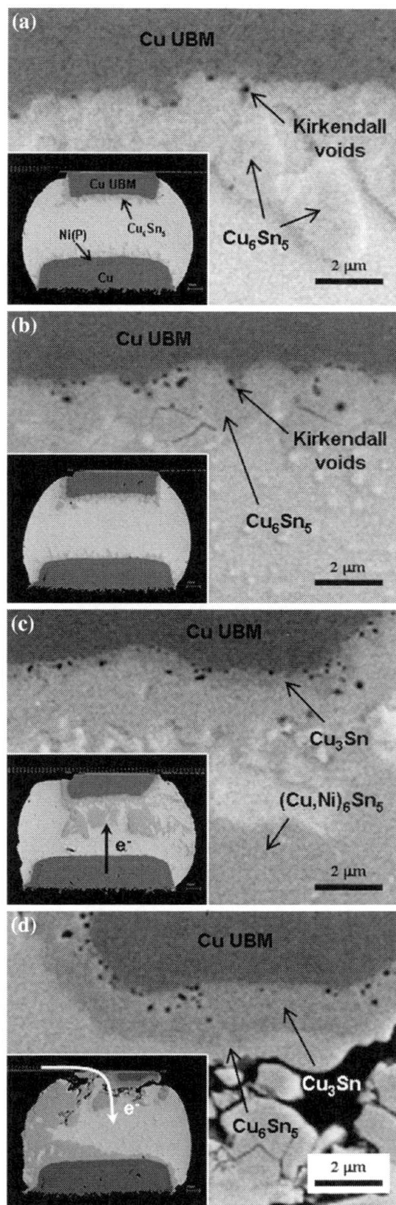

Fig. 10 SEM micrographs of (**a**) the joint with Cu under-bump-metallization (UBM) in a pristine sample, (**b**) the joint without current stressing at 140°C for 263 h, (**c**) and (**d**) the cathode joint and the anode joint from the same sample in (**b**) with 1.01 A current stressing

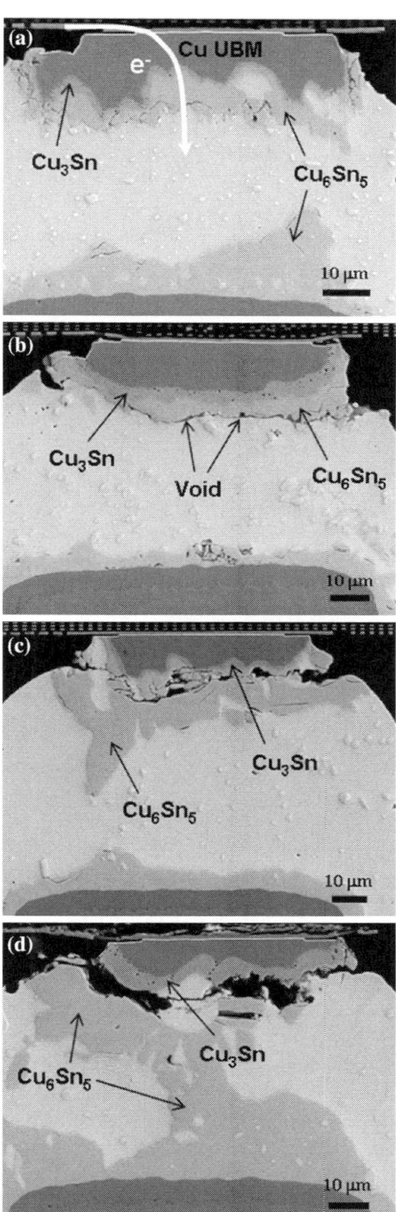

Fig. 11 Electromigration (EM) damage evolution in the anode joint with Cu under-bump-metallization (UBM). (**a**) Initial stage of Cu depletion, (**b**) void initiation prior to the resistance jump, (**c**) crack propagation subsequent to the resistance jump, and (**d**) final open failure

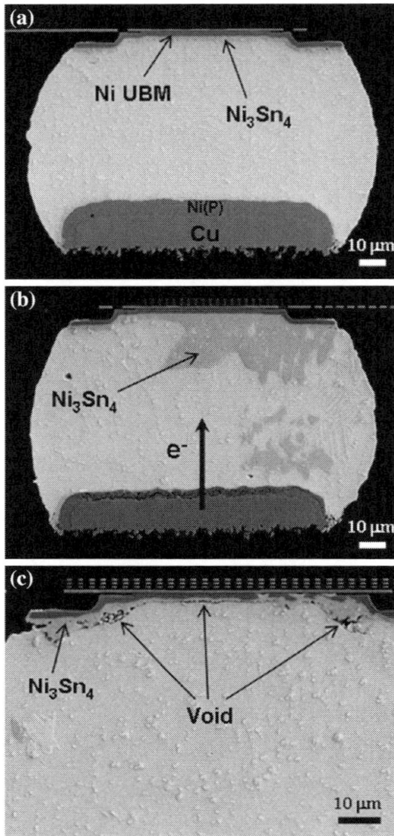

Fig. 12 SEM micrographs of (**a**) the joint with Ni under-bump-metallization (UBM) without current stressing at 130°C for 150 h, and (**b**) and (**c**) the cathode joint and the anode joint from the same sample in (**a**) with 1.01 A current stressing

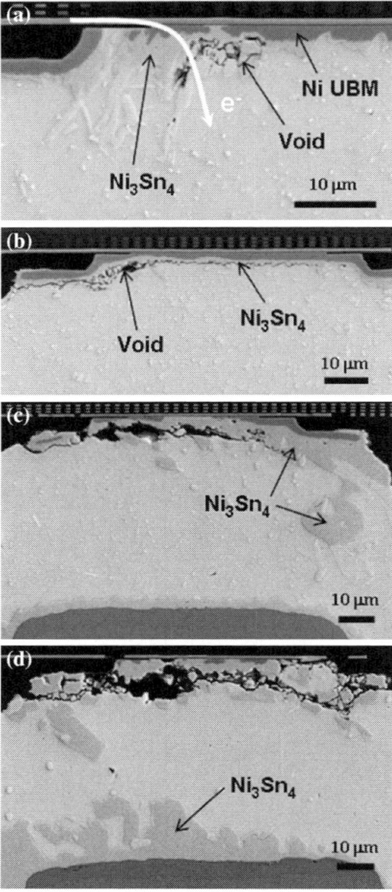

Fig. 13 Electromigration (EM) damage evolution in the anode joint with Ni under-bump-metallization (UBM). (**a**) Ni depletion and void initiation, (**b**) void propagation along intermetallic compound (IMC)/solder interface, (**c**) further IMC and void growth, and (**d**) nearly open

(Fig. 12a). In the cathode joint (Fig. 12b), Ni atoms from the substrate finish layer migrated toward the die side to form Ni_3Sn_4 to sustain the growth of Ni_3Sn_4 on the die side. In contrast, Ni_3Sn_4 growth on the substrate side was enhanced at the beginning but subsequently retarded as Ni in the substrate finish layer was depleted. When Ni was sufficiently depleted, needle-like fissures were found in the Ni(P) finish layer as shown in Fig. 12b. This indicates the migration of Ni instead of Sn during solid state reaction although Chen et al. considered Sn to be the dominant diffusing species in the reaction couple with Ni sandwiched in-between Sn-based solder [21]. In the anode joint (Fig. 12c), EM-induced voids formed at the Ni_3Sn_4/solder interface as a result of enhanced growth of Ni_3Sn_4 by the current flow. The growth of Ni_3Sn_4 was slower than Cu_6Sn_5,

resulting in a slower dissolution of Ni UBM than that of Cu UBM.

Damage evolution in the anode joint with Ni UBM is depicted in Fig. 13. At an early stage where the change in V_g was negligible, initial voids formed at current crowding region as shown in Fig. 13a. Some part of Ni UBM was dissolved into solder to form Ni_3Sn_4. Continuing test led to the growth of voids, which propagated through the Ni_3Sn_4/solder interface (Fig. 13b). At this point, the IMC growth was not substantial with a relatively small volume of void formation. Although voids occupied over a half of the interface, V_g was only a few mV. Compared with Cu

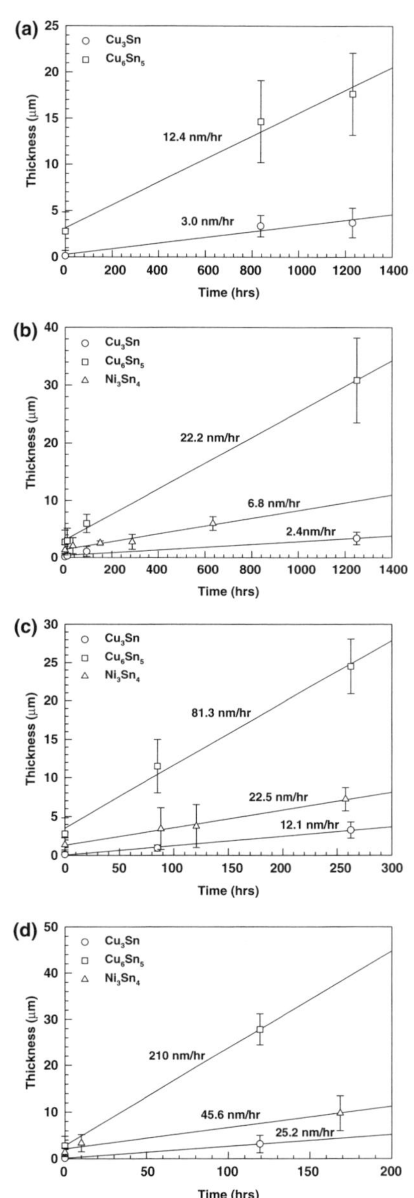

Fig. 14 Nominal intermetallic compound (IMC) thickness with respect to time under 1.01 A current stressing at (**a**) 115°C, (**b**) 130°C, (**c**) 140°C, and (**d**) 150°C of the Si backside temperature

UBM, Ni UBM showed more regular void growth through the Ni_3Sn_4/solder interface, which may account for the more gradual increase of V_g as seen in Fig. 2. Further void growth increased local current density, leading to accelerated IMC and void growth. In Fig. 13c, the amount of Ni_3Sn_4 increased in the region where solder was still connected while voids became much larger. This stage corresponded to ~10 mV of V_g where voids occupied nearly 70% of the IMC/solder interface. Figure 13d shows the EM damage state of a solder joint when unstable resistance fluctuation occurred. At this stage, the solder joint became almost open with a large amount of IMC accumulated on the substrate side. A number of solder joints were melted after this stage due to the increase in Joule heating prior to the final open failure.

3.5 IMC growth rate

Volume change due to IMC formation generated stress within a solder joint. When Cu and Ni atoms react with Sn atoms to form Cu_3Sn, Cu_6Sn_5 and Ni_3Sn_4, the overall atomic volumes decrease by 7.7, 5.1 and 11.4%, respectively. This induces a tensile stress within the solder joint, which can further degrade EM reliability when combined with void formation due to current enhanced IMC formation. Figure 14 plots nominal thickness of each IMC in anode joints as a function of time and temperature. The nominal thickness was calculated by dividing the cross-sectional area of each IMC with the original width of the corresponding UBM. Areas of IMC in both the die side and the substrate side were added since independently grown Cu_6Sn_5 from each side would finally connect as shown in Fig. 11d. Current enhanced IMC growth followed linear growth kinetics as shown in Fig. 14. The growth kinetics of IMC without current stressing had been reported to follow a $t^{1/2}$ or $t^{1/3}$ dependence [17, 19]. With the passage of an electric current, Gan et al. reported parabolic growth kinetics for IMC growth [22], although the time dependency was not reported in another study [21]. Recently we have formulated a kinetic model verifying the linear growth of IMCs when an EM driving force dominates the chemical interdiffusion [23]. In this study, the IMC growth rate was found to follow the order of $Cu_6Sn_5 > Ni_3Sn_4 > Cu_3Sn$. Because Cu_6Sn_5 grew several times faster than the other two IMCs, the overall tensile stress caused by the formation of a large amount of Cu_6Sn_5 could be significant although its molar volume change is the smallest of three IMCs. In contrast, relatively small amount of Ni_3Sn_4 formation would have a large effect on EM damage since both the accompanying molar volume shrinkage and its Young's modulus (133.3 GPa) [24] are the largest of the three IMCs. Continuous growth of IMCs as enhanced by electric current increased the tensile stress with time, leading to significant driving force for crack propagation. Thus IMC growth is one of the key factors controlling EM reliability of Pb-free solder joints.

4 Summary

EM studies were conducted on Pb-free Sn-3.5Ag solder joints with Cu and Ni UBM in plastic flip-chip packages. Si backside temperatures of 115, 130, 140 and 150°C were chosen for tests with 1.01 A current stressing. Two more tests at 140°C were performed with 0.81 and 1.11 A in order to determine the current density exponent. To supplement the open-failure criterion, a first resistance jump failure criterion was used to analyze EM lifetime results. Solder joints with thick Cu UBM had a longer lifetime than those with thin Ni UBM, based on the open-failure criterion; however, lifetime of Ni UBM solders was comparable if the first resistance jump criterion was applied. This was due to the longer resistance fluctuation period of solder joints with thick Cu UBM. Numerical analysis of V_g and failure analysis showed that solder was almost open at the onset of unstable resistance fluctuations. Experiments and finite element analysis were conducted to determine the Joule heating effect and found the solder temperature to be 10, 15 and 17°C higher than the Si backside temperature at 0.81, 1.01 and 1.11 A. Taking this into account, we obtained $Q_{Cu} = 1.11$ eV, $n_{Cu} = 3.75$ and $Q_{Ni} = 0.86$ eV, $n_{Ni} = 2.1$ based on the open-failure criterion. The correction for the Joule heating effect resulted in a reduction of the current density exponents. In solder joints with Cu UBM, EM voids were found to initiate at the Cu_6Sn_5/solder interface but voids for final failure moved near the Cu_3Sn/Cu_6Sn_5 interface. IMCs grew actively as voids grew. The EM-induced voids evolved independently of the initial Kirkendall voids in the Cu_3Sn layer. In solder joints with Ni UBM, the final failure was observed at the Ni_3Sn_4/solder interface where voids initially formed. EM damage evolution was expected to be accelerated with IMC growth, which would induce significant tensile stresses due to decrease in the overall atomic volume. The enhanced IMC growth under current stressing followed linear kinetics with time and

played an important role in controlling EM reliability of Pb-free solder joints.

Acknowledgements This work was supported in part by the Semiconductor Research Corporation and the technical support provided by Freescale Semiconductor Inc. is gratefully acknowledged.

References

1. M. Ding, H. Matsuhashi, G. Wang, P.S. Ho, in *Proceedings of IEEE 54th Electronic Components and Technology Conference* (Las Vegas, NV, 2004), p. 968
2. K.N. Tu, J. Appl. Phys. **94**, 5451 (2003)
3. B.F. Dyson, T. Anthony, D. Turn bull, J. Appl. Phys. **37**, 2370 (1966)
4. W.K. Warburton, D. Turnbull, in *Diffusion in Solids*, eds. by A.S. Nowick, J.J. Burton (Academic, New York, 1975), pp. 171–226
5. D.L. Decker, C.T. Candland, H.B. Vanfleet, Phys. Rev. B **11**, 4885 (1975)
6. M. Ding, G. Wang, B. Chao, P.S. Ho, in *Proceedings of IEEE 43rd Annual International Reliability Physics Symposium* (San Jose, CA, April 2005), p. 518
7. Y.C. Hu, Y.H. Lin, C.R. Kao, K.N. Tu, J. Mater. Res. **18**, 2544 (2003)
8. J.W. Jang, D.R. Frear, T.Y. Lee, K.N. Tu, J. Appl. Phys. **88**, 6359 (2000)
9. J.W. Nah, J.O. Suh, K.N. Tu, J. Appl. Phys. **98**, 013715 (2005)
10. M. Ding, G. Wang, P.S. Ho, Appl. Phys. Lett. (Submitted)
11. M. Gall, PhD Dissertation (The University of Texas at Austin, 1999)
12. J.R. Black, in *Proceedings of IEEE 6th Annual International Reliability Physics Symposium* (Los Angeles, CA, 1967), p. 148
13. A.S. Oates, Appl. Phys. Lett. **66**, 20 (1995)
14. Y.H. Lin, C.M. Tsai, Y.C. Hu, Y.L. Lin, C.R. Kao, J. Electron. Mater. **34**, 27 (2005)
15. K.-D. Lee, PhD Dissertation (The University of Texas at Austin, 2003), pp. 33–34
16. Y.C. Chan, A.C.K. So, J.K.L. Lai, Mat. Sci. Eng. B **55**, 5 (1998)
17. J.-W. Yoon, S.-B. Jung, J. Alloys Compd. **359**, 202 (2003)
18. C.N. Liao, C.T. Wei, J. Electron. Mater. **33** 1137 (2004)
19. Z. Chen, M. He, G. Qi, J. Electron. Mater. **33**, 1465 (2004)
20. K. Zeng, R. Stierman, T.-C. Chiu, D. Edwards, K. Ano, K.N. Tu, J. Appl. Phys. **97**, 024508 (2005)
21. C.-M. Chen, S.-W. Chen, J. Appl. Phys. **90**, 1208 (2001)
22. H. Gan, K.N. Tu, J. Appl. Phys. **97**, 063514 (2005)
23. H.-L. Chao, S.-H. Chae, X. Zhang, K.-H. Lu, J. Im, P.S. Ho, in *Proceedings of IEEE 44th Annual International Reliability Physics Symposium* (San Jose, CA, March 2006), p. 250
24. D.R. Frear, S.N. Burchett, H.S. Morgan, J.H. Lau (eds.), in *The Mechanics of Solder Alloy Interconnects* (Van Nostrand Reinhold, New York, 1994), p. 60

Electromigration issues in lead-free solder joints

Chih Chen · S. W. Liang

Published online: 8 September 2006
© Springer Science+Business Media, LLC 2006

Abstract As the microelectronic industry advances to Pb-free solders due to environmental concerns, electromigration (EM) has become a critical issue for fine-pitch packaging as the diameter of the solder bump continues decreasing and the current that each bump carries keeps rising owing to higher performance requirement of electronic devices. As stated in 2003 International Technology Roadmap for Semiconductors (ITRS), the EM is expected to be the limiting factor for high-density packages. This paper reviews general background of EM, current understanding of EM in solder joints, and technical hurdles to be addressed as well as possible solutions. It is found that the EM lifetimes of Pb-free solder bumps are between the high-Pb and the eutectic composition under the same testing condition. However, our simulation results show that the electrical and thermal characteristics remain essentially almost the same during accelerated EM tests when the Pb-containing solders are replaced by Pb-free solders, suggesting that the melting points of the solders are likely the dominant factor in determining EM lifetimes. The EM behavior in Pb-free solder is a complicated phenomenon as multiple driving forces coexist in the joints and each joint contains more than four elements with distinct susceptibility to each driving force. Therefore, atomic transport due to electrical and thermal driving forces during EM is also investigated. In addition, several approaches are presented to reduce undesirable current crowding and Joule heating effects to improve EM resistance.

C. Chen (✉) · S. W. Liang
Department of Material Science & Engineering, National Chiao Tung University, Hsin-chu 30050, Taiwan, ROC
e-mail: chih@faculty.nctu.edu.tw

1 Overview of the background

Electromigration (EM) has been the most persistent reliability issue in interconnects of microelectronic devices [1]. It is the mass transport of atom driven by combined forces of electric field and charge carriers. As illustrated in Fig. 1, the drifting electrons collide with atoms causing one of the atoms to exchange position with neighboring vacancy during current stressing. The necessary current density to initiate the movement of atom is defined as the threshold current density. Due to the relentless drive for miniaturization of portable devices, the interconnects for those devices are scaling down successively, whereas the required performance continues increasing. As a result, the current density in the interconnect rises continuously with each generation, making the EM a critical reliability issue ever. After stressing for extended time, atoms in interconnects accumulate on the anode end and voids appear on the cathode side, resulting in open failure eventually. In general, the average drift velocity of atom due to EM is given by Huntigton and Grone [2]:

$$v = \frac{J}{C} = BeZ^{*}\rho j = \left(\frac{D_0}{kT}\right)eZ^{*}\rho j \, \exp\left(\frac{-E_a}{kT}\right) \tag{1}$$

where J is the atom flux, C is the density of metal ions, B is the mobility, k is the Boltzmann's constant, T is the absolute temperature, eZ^{*} is the effective charge of the ions, ρ is the metal resistivity, j is the electrical current density, E_a is the activation energy of diffusion, and D_0 is the prefactor of diffusion constant.

Recently, with stringent environmental regulation, lead-free solders have been adopted to replace

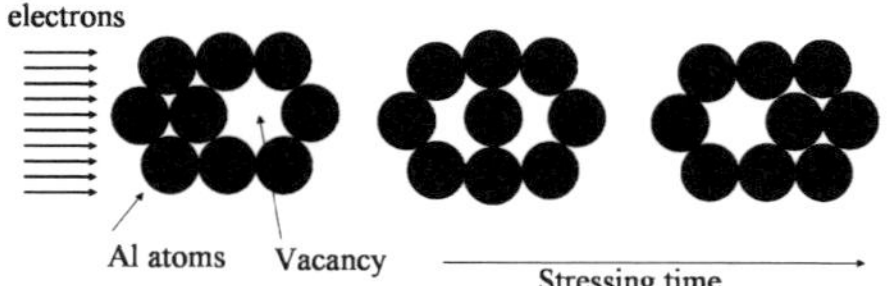

Fig. 1 Schematic diagram showing the diffusion process due to wind force of electrons during EM test. The current density should be high enough to trigger the diffusion process

Pb-containing ones in microelectronics industry. Concurrently, the flip-chip solder joint has assumed the leadership role for high-density packaging in the microelectronic industry as thousands of solder bumps are fabricated into one single chip. To meet higher demand for device performance, the input/output numbers is expected to increase while the dimension of each individual joint is shrunk accordingly. To date, each bump measures at 100 μm or less in diameter. The design rule of packaging dictates that each bump is likely to carry current of 0.2–0.4 A [1], which amounts to current density in the range of 2×10^3 A/cm^2 to 1×10^4 A/cm^2. During device operation, these solder joints frequently reach a temperature as high as 100°C, approximately 77% of the absolute melting temperatures of most Pb-free solder candidate materials including eutectic SnAgCu and SnAg. Understandably, with such high current densities and operation temperatures, facile diffusion of atoms in the lattice is foreseeable. This renders EM a daunting reliability issue for Pb-free implementation [3].

2 Current status

Previous studies on EM of flip chip solder bumps focused mainly on eutectic SnPb solders. In 1998, Brandenburg et al. first reported the failure of eutectic SnPb solder joints under current stressing of 0.625 A at 150°C for 600 h [4]. Tu et al. performed systematic studies and provided insightful reasoning on the EM in Pb-containing solder. They identified that the Sn atoms are the principal diffusion entities at room temperature, whereas Pb atoms dominate at 150°C [5, 6]. In addition, they discovered that the effect for current crowding is more pronounced in the flip-chip solder joints for its unique line-to-bump structure [7]. They performed two-dimensional simulation and the results showed that the local current density at the solder near the entrance of the Al trace was at least 10 times greater than the average value, a number obtained by assuming uniform current spreading in the passivation opening or under-bump-metallization (UBM) opening.

This finding is significant in pinpointing possible location for failure event in EM. Figure 2(a) depicts a three-dimensional (3D) schematic for a solder joint with line-to-bump structure consisting Al trace of 34 μm in width and 1.5 μm in thickness, solder bump with a UBM opening of 120 μm in diameter, as well as Cu line with 80 μm in width and 25 μm in thickness on the substrate side. In this design the cross-section of the Al trace is about 220 times smaller than that of the UBM opening. When the joint was subjected to a current of 0.567 A, the current density in the Al trace reached 1.1×10^6 A/cm^2, whereas the average current density in the UBM opening was only 5.0×10^3 A/cm^2. However, our 3D simulation showed that in thin film UBM configuration, close proximity to the entrance of the Al trace the local current density of the solder could achieve 1.24×10^5 A/cm^2 [8], a value that is 24.8 times greater than the average 5.0×10^3 A/cm^2 one would expect. Among the possible materials used in the joint, solder is considered to exhibit the lowest resistance to EM [1]. Reasonably, we can conclude that solder in this specific location experiences larger electron wind force at a relatively high temperature to its

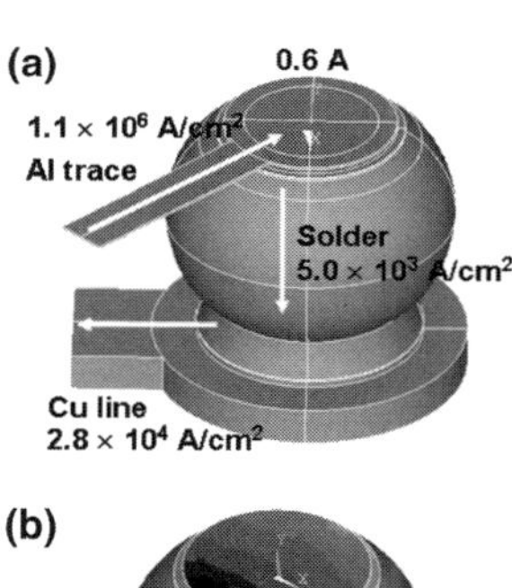

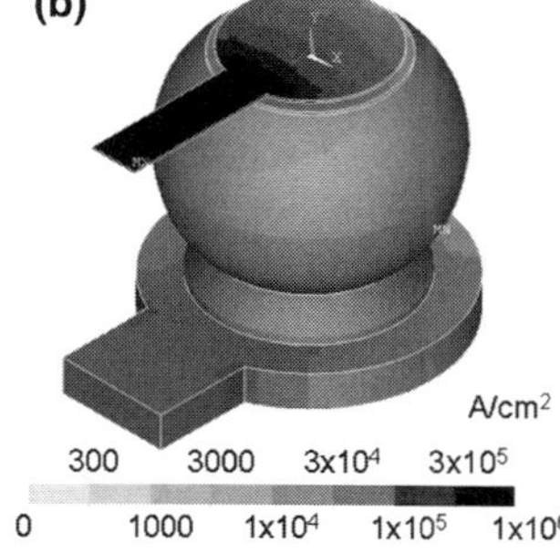

Fig. 2 (a) The 3D schematic for a solder joint with the line-to-bump structure with a 34-μm wide and 1.5-μm thick Al trace. When the joint was applied by 0.567 A, the average current densities in the Al trace, solder bump, and Cu line are shown in the figure. (b) Tilt-view showing the 3D current density distribution in the solder joint when it is powered by 0.57 A. Current crowding occurred very seriously in the junction of the Al trace and the solder bump

melting point. Therefore, flux divergence of mass transport occurs substantially resulting in the formation of catastrophic voids which are directly responsible for interconnect failure.

Besides the current crowding effect, Joule heating also plays a crucial role in the failure mechanism. Recently, findings indicate that a hot-spot exists near the entrance point of the Al trace due to localized Joule heating effect [9]. The resultant temperature difference between the hot-spot and the average temperature in solder can reach 9.4°C under 0.8 A of current flow as shown in Fig. 2(a). In 2003, Ye et al. reported the observation of voids on the chip/anode side [10], and they attributed the void development to thermomigration (TM) as thermal gradient of 1500°C/cm across the solder bump was established during accelerated EM tests. Similarly, Huang et al. found that Sn atom migrates towards the hot side but Pb atom migrates to the cold end [11]. The occurrence of TM during EM complicates the reliability issue. On the other hand, the EM for compositions of the high-Pb and eutectic SnPb solder joints with 5-μm Cu UBM were studied and concluded that the dissolution of the Cu UBM at the current crowding region was primarily responsible for the failure of the joints [12, 13]. The dissolution rate at the current crowding region was accelerated because of higher wind force in combination with elevated temperature for facile diffusion.

Unfortunately, the mean-time-to-failure (MTTF) for Pb-free SnAgCu and SnAg solders joints are shorter than that of the high-Pb solders under the same stressing conditions. Since high-Pb solders are currently adopted for high-density packages, such as microprocessors, EM would be a critical issue for Pb-free implementation. In 2004, Wu et al. demonstrated that the MTTF of SnAg4.0Cu0.5 solder is about five times longer than that of the eutectic SnPb solder, yet is somewhat shorter than that of the high-Pb solder [14]. Gee et al. and Choi et al. also reported that the MTTF of eutectic SnAgCu solder joints was better than that of eutectic SnPb solder joints with the same UBM under the same stressing conditions [15, 16]. To elucidate how the current density and temperature distribute during current stressing, 3D electrothermal-coupled modeling was performed on the solder joints with identical configuration but with different solders materials. They include eutectic SnPb, high-Pb SnPb95 and eutectic SnAg. Figure 3(a)–(c) shows the cross-sectional schematics for these three models. The passivation opening was 85 μm in diameter and the UBM opening was 120 μm in diameter. The dimensions of the Al trace and the Cu line were consistent with those in Fig. 2. Relevant materials characteristics of materi-

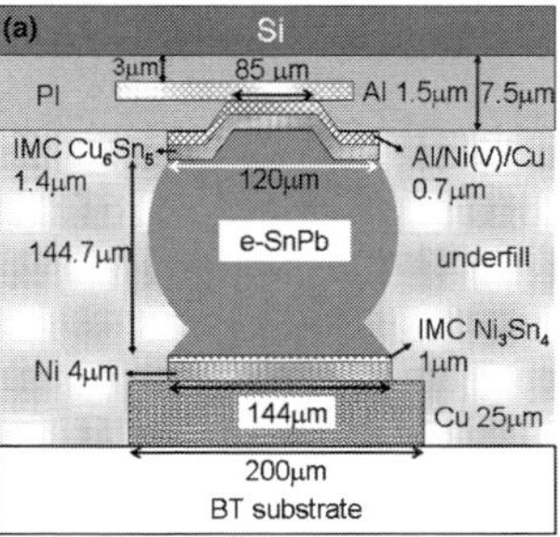

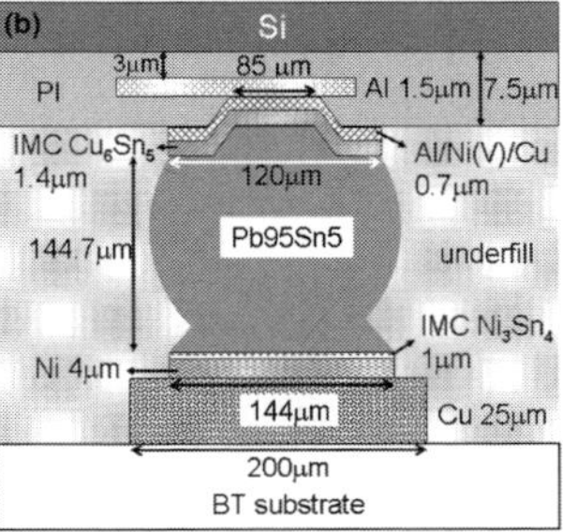

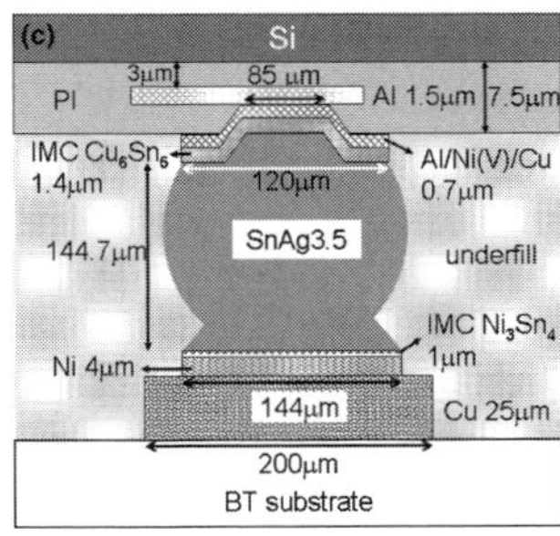

Fig. 3 The cross-sectional schematic showing the three constructed models for electrical–thermal coupled simulation: (a) Eutectic SnPb solder joint; (b) high-Pb solder joint; (c) Eutectic SnAg solder joints. All the features remain are identical except the solder materials

als used in this simulation are provided in Table 1. The dimension of the Si chip was 7.0 mm × 4.8 mm with thickness of 290 μm. The dimension of the bismaleimide triazine (BT) substrate was 5.4 mm in width, 9.0 mm in length, and 480 μm in thickness. The bottom of the BT substrate was maintained at 70°C and the convection coefficient was set at 10 W/m²°C in a 25°C ambient temperature. Constant current of 0.6 A was applied through the Cu lines on the BT substrate [9]. Among these three solders, the Pb-free SnAg possesses the lowest electrical resistivity and thermal conductivity of 12.3 μΩ cm and 33 W/m K respectively. Figure 4(a)–(c) displays the current density

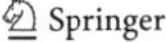

Table 1 The materials properties used in this paper

Materials	Thermal conductivity (W/m K)	Resistivity (mΩ cm)	Temperature coefficient of resistivity (K^{-1})
Al	238	2.7	4.2×10^{-3}
Al/Ni(V)Cu	166.6	29.54	5.6×10^{-3}
Cu6Sn5	34.1	17.5	4.5×10^{-3}
Pb–5Sn	63	19	4.2×10^{-3}
e-SnPb	50	14.6	4.4×10^{-3}
SnAg3.5	33	12.3	4.6×10^{-3}
Ni3Sn4	19.6	28.5	5.5×10^{-3}
Ni	76	6.8	6.8×10^{-3}
Cu	403	1.7	4.3×10^{-3}
Si	147	–	–
BT	0.7	–	–
Underfill	0.55	–	–
PI	0.34	–	–

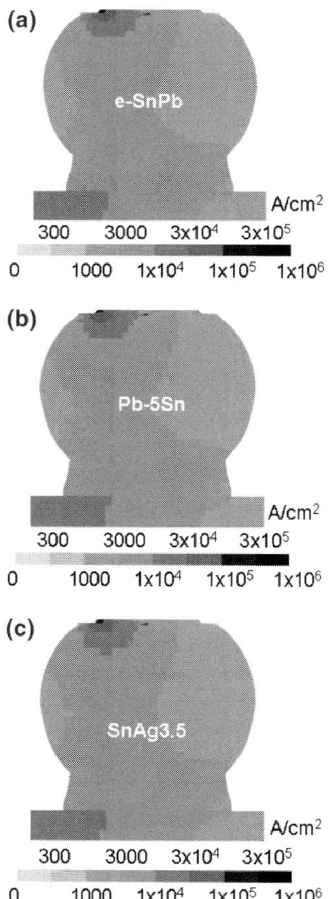

Fig. 4 The simulation results showing the current density distribution under 0.6 A in: (a) Eutectic SnPb solder bump; (b) high-Pb solder bump; (c) Eutectic SnAg solder bump

distribution in the solder joints under the stress current of 0.6 A. The distribution profiles remain essentially the same. The maximum current density were 1.03×10^5 A/cm^2, 9.42×10^4 A/cm^2, 1.11×10^5 A/cm^2 for the eutectic SnPb, high-Pb, and the eutectic SnAg solders, respectively. The Pb-free solder exhibits the highest current crowding effect because of its lowest electrical resistivity. Figure 5(a)–(c) illustrates the temperature distribution in the solder bumps. The solders near the entrance point of the Al trace all show higher temperature than the rest of the solder. Figure 6(a)–(c) shows the cross-sectional views for the temperature distribution. The results indicate the existence of hot-spots in these solder bumps. The hot-spot temperature was 100.0°C, 103.6°C, and 105.4°C, respectively, whereas the average temperature was 95.9°C, 99.2°C, and 98.9°C for the eutectic SnPb, high-Pb, and the eutectic SnAg solder. The Pb-free solder experienced the highest Joule heating effect, which may be due to limited intrinsic capability for heat dissipation and highest current crowding effect. Since the majority of heat source was Al trace [17], lower resistivity of the Pb-free solders did not necessarily render a smaller Joule heating effect. The simulation results are summarized in Table 2.

So far, our data demonstrate that the current crowding and Joule heating effects in Pb-free SnAg solder bump are marginally worse than those in eutectic SnPb solder bump, as shown in Figs. 4 and 5. Nevertheless, the Pb-free solder exhibits far better EM resistance than that of the eutectic SnPb [18]. This surprising improvement may be attributed to the reduced diffusivity for Pb-free solder as its melting point is approximately 50°C higher than that of the eutectic SnPb solder. As a result, the rate of void formation is much lower than that in the eutectic solder. In addition, the highest MTTF for the high-Pb solder may be

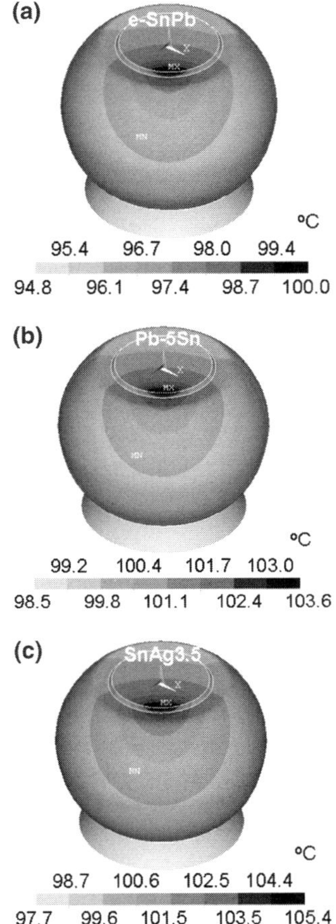

Fig. 5 The simulation results showing the temperature distribution under 0.6 A in: (a) Eutectic SnPb solder bump; (b) high-Pb solder bump; (c) Eutectic SnAg solder bump. A hot-spot exists near the entrance point of the Al trace

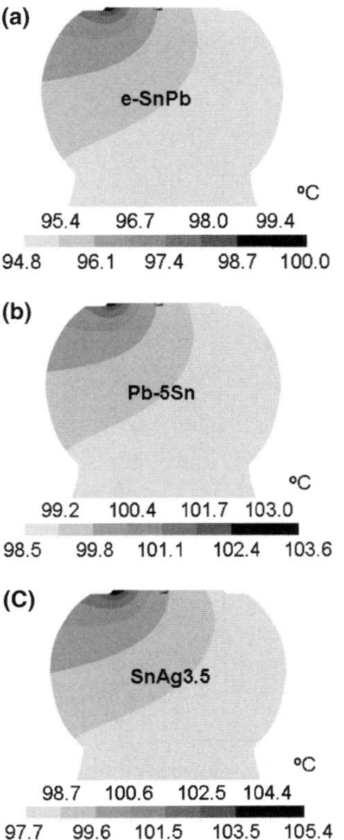

Fig. 6 The cross-sectional view of the results in Fig. 5. (a) Eutectic SnPb solder bump; (b) high-Pb solder bump; (c) Eutectic SnAg solder bump under 0.6 A. The solder on the Si side is hotter than that on the substrate side. Thermal gradient was built across the solder bump

mainly due to its higher liquidus temperature of about 320°C. For example, at stressing temperature of 150°C, it is 93%, 86%, and 71% of the melting points for the eutectic SnPb, eutectic SnAg, and high-Pb solders, respectively. Typically, at melting point metal atoms exhibit a diffusivity of 10^5 cm²/s to 10^7 cm²/s in nature. Therefore, it is prudent to assume that the diffusivity of the Pb-free solder would fall somewhere in between these two Pb-containing solders. This is in accordance to the findings that the EM resistance of Pb-free solder is higher than that of the eutectic SnPb solder, but lower than that of the high-Pb solder.

Likewise, interfacial metallurgical reaction becomes crucial in Pb-free solder joints during current stressing, especially at critical stressing conditions when the solder and the UBM undergo a solid state aging process during EM. It is reported that the Cu–Sn IMC may grow over 13 µm for Pb-free SnAg solder after aging at 170°C for 1500 h [19]. Electron wind force is likely to enhance the dissolution of the UBM materials on the cathode end, and the erosion of the latter may lead to the failure of the solder joints [12, 13]. Thus, the interfacial reaction appears to be the critical factor for the EM of Pb-free solders. Due to the spalling issue for thin-film UBM in Pb-containing and Pb-free solders [3, 20], thick-film Cu or Ni UBM has been adopted for the Pb-free solder joints. This not only solves the spalling problem, but also prolongs the EM lifetime. This point will be discussed later.

The interfacial reactions are accelerated by the stressing current during EM. Chen et al. first investigated the effect of electrical current on interfacial

Table 2 The simulation results on maximum current density, hot-spot and average temperatures, and thermal gradient for the high-Pb, eutectic SnPb, and SnAg solders

	Maximum current density (A/cm^2)	Hot-spot (°C)	Average temperature (°C)	Thermal gradient (°C/cm)
Pb95Sn5	9.42×10^4	103.6	99.2	246.9
e-SnPb	1.03×10^5	100	95.5	259.2
SnAg3.5	1.11×10^5	105.4	98.9	398.7

reactions in solder systems in 1998 [21], and they found that the flow of electrons may enhance or inhibit the growth of the intermetallic compound. For SnPb solder joints, the Pb atoms are the dominant diffusing species when the joints are stressed above 100°C. Figure 7 depicts the schematic for two bumps with opposite current directions. Hence, the Pb atoms move to the chip/anode and substrate/anode sides during current stressing. This phase segregation may enhance the dissolution of the UBM in the chip/cathode end because excess Sn atoms accumulate there. Yet, it prevents the UBM on the chip/anode side from reacting with the solder as the Pb atoms are unlikely to form IMC with Cu or Ni UBM. For Pb-free solders, the interfacial reaction on the chip/anode side becomes noticeable. Extensive IMC formation on the chip/anode during stringent stressing conditions for SnAg solder joints has been reported in literature [22, 23]. Nickel atoms on the substrate/cathode migrated to the chip/anode side and subsequently formed IMC there, a potential failure site. At a reduced stressing current and temperature, the joints failed at chip/cathode side. Unfortunately, the failure modes for thick-film Cu (over 10 μm) or Ni UBM are not yet clear.

Under the electrical (EM) and thermal (TM) driving forces, the diffusion of atoms in the solder joint are rather complicated. Figure 8(a) illustrates the direction of the movement of the atoms under these two driving forces during an accelerated EM test. The simulation assumes a Pb-containing solder in combination with Cu and Ni serving as the UBM material in the chip side and metallization material in the substrate side. In addition, electrons drift upwards in Bump 1, whereas they move downwards in Bump 2. Bump 3 acts as a control sample without passage of current but it undergoes the same thermal history as Bump 1 and 2. During EM testing, the Si die is the hot end due to the excessive Joule heating in the Al trace. For Pb-containing solder bumps, open failure occurs in the chip/cathode of Bump 2 since flux divergence takes place to greater degree at this end. As shown in the Fig. 8(a), due to the apparent current crowding effect in the chip side [8], Sn, Pb, and Cu atoms will migrate away from the chip/cathode end. Furthermore, the built-in thermal gradient also drives the Pb atoms to the substrate side [11]. On the other hand, only an opposite flux of Sn due the thermomigration flows to the chip/cathode end [11]. At testing temperatures above 100°C, Pb atoms are the predominant diffusion species. Thus, both EM and TM forces remove atoms from the chip end to the substrate side in this bump with the net result of void formation in the chip/cathode end. Also, the Pb segregation in the substrate/anode side is quite noticeable after current stressing for extended time [24].

For Bump 1 with upward electron flow, SnPb solder will migrate to the chip/anode side. However, the force

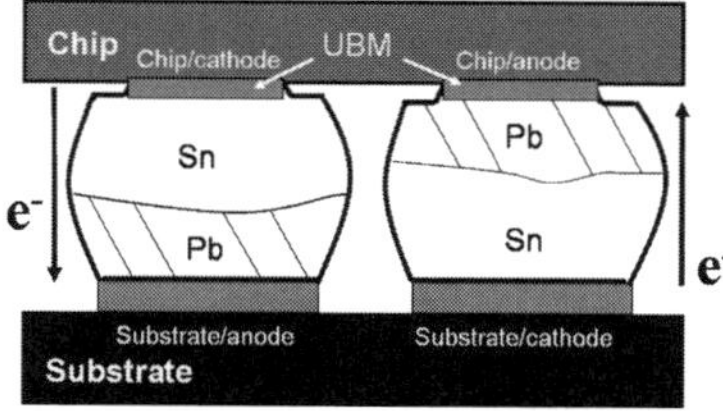

Fig. 7 Schematic drawing showing the polarity effect of during EM testing at temperature higher than 100°C. Lead atoms accumulated on the chip/anode and substrate/anode ends

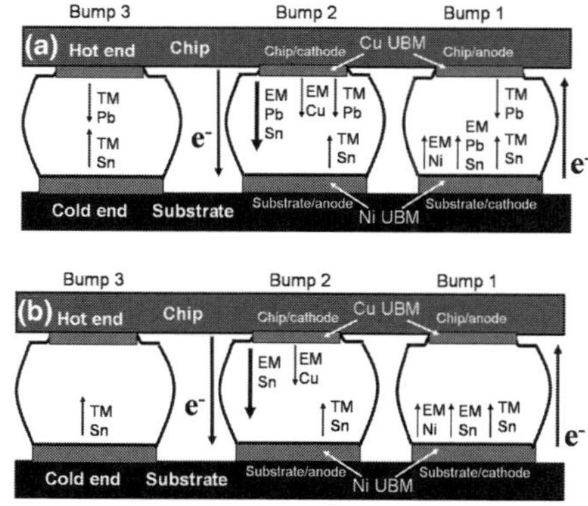

Fig. 8 Diffusion of atoms in solder bumps due to EM and TM forces for: (a) Pb-containing solder bumps; (b) Pb-free solder bumps

is negligible compared with that in Bump 2, since there is almost no current crowding effect in the substrate side due to the large cross-section of the Cu line in the substrate [8]. In addition, Ni atoms in the substrate side will migrate toward chip/anode side due to the EM force [22, 23], and Sn atoms will also diffuse to the chip/anode side under TM force. In contrast, the Pb atoms will move to the substrate/cathode end due to TM force. Thus, the EM and TM forces counteract each other at higher stressing temperatures when the Pb atoms are the dominant diffusion species. Huang have et al. shown that a thermal gradient of 1000°C/cm would provide comparable magnitude of driving force to that of EM force, triggering notable thermomigration in the solder bump [11].

For Bump 3 without passage of any current, the built-in thermal gradient may be close to those in Bump 1 and Bump 2 if the Bump 3 is close to them. Because the heat conduction of Si is appreciable, the Bump 3 may experience similar amount of Joule heating as Bump 1 and 2 do [25]. Hence, the temperature of the solder near the Si die is expected to be higher than that in the substrate side, causing a thermal gradient across the solder joint even without current passage. In addition, since EM force is absent in Bump 3 because there was no current passing through, it offers a great opportunity to decouple the atom migration due to the EM and the TM forces [11]. As depicted in Fig. 8(a), the built-in thermal gradient drives the Pb atoms near the chip side to the substrate side, whereas it triggers the Sn atoms in the substrate side moving to the chip side. The migration of atoms in this bump is mainly attributed to the TM force. It is noteworthy that the current should be sufficiently high to produce a steep thermal gradient across the solder joint. Applied current larger than 1 A was used in those studies that reported thermomigration so far [10, 11]. At lower stressing temperatures, Sn atoms are the predominant diffusion species. For Bump 1, both EM and TM forces propel the Sn atoms towards the chip/anode side. For Bump 2, EM failure may be inhibited to some extent since Sn fluxes due to the EM and TM force are in the favorable direction.

As for Pb-free solder bumps including eutectic SnAgCu and SnAg, Sn atoms are the leading diffusion species during EM. Therefore, the diffusion behaviors driven by EM and TM forces become relatively straightforward, as shown in Fig. 8(b). For Bump 1, both Sn and Ni atoms migrate to the chip/anode end due to EM force, and the TM force also promotes the Sn atoms to that end. As for Bump 2, The EM force drives the Sn and Cu down to substrate/anode end, whereas the TM force prods Sn atom in the opposite

direction. Therefore, the damage caused by EM damage may be limited if the thermal gradient is sufficiently large. In comparison, there is only one force acting on the Bump 3 because it had no electrical current pass through, and Sn atoms are driven toward the chip end. Nevertheless, experimental results are unavailable for thermomigration in Pb-free solders to date.

3 Problems that still need to be addressed and the suggestions for solving them

3.1 Modification of MTTF equation

From engineering point of view, modification of MTTF equation for the use of flip-chip in solder joints is urgently needed. The equation of mean-time-to-failure (MTTF) for Al and Cu interconnects is typically expressed as below [26]:

$$\mathrm{MTTF} = A \frac{1}{j^n} \exp(\frac{Q}{kT}) \tag{1}$$

where A is a constant, j is the average current density, n is a model parameter for current density, Q is the activation energy, k is the Boltzmann's constant, and T is the average bump temperature. As stated above, current crowding and Joule heating effect occurs substantially in flip-chip solder joints, and the failure is usually initiated at the current crowding region in solder, which happens to be the hot-spot. Voids start to form here, or the UBM dissolves quickly at the region. Therefore, the equation should be revised to include current crowding and Joule heating effects for consideration during accelerated EM tests. Tu et al. proposed that the term j^{-n} in the equation needs to be revised to $(cj)^{-n}$ in order to capture the high current crowding effect in the solder joints [16]. Moreover, the temperature factor is later adapted to $(T + \Delta T)$ to account for appreciable Joule heating effect during the accelerated EM test. Further effort is necessary to identify the precise current density term. Since the hot-spot is present, we suggest revising the temperature term to use the hot-spot temperature in predicting the MTTF of the solder joints. We believe that the Joule heating effect needs to be measured with greater accuracy. Otherwise, the MTTF would be underestimated.

3.2 Relieving current crowding and Joule heating effects

Current crowding and Joule heating effects play vital role in the failure of flip-chip solder joints. Hence,

relieving of these undesirable influences is expected to improve the EM resistance significantly. As delineated by Black's equation, reducing the local current density by 50% may extend the MTTF by four times if we take cj term to be the maximum current density and n to be equal to 2. In addition, the MTTF increases exponentially with decreasing bump/hot-spot temperature. Several approaches have been proposed in a previous publication to mitigate these two influences [27]. The approaches will be summarized briefly in following paragraphs.

3.2.1 Keeping the solder away from high current density/hot-spot region

By adopting a thick Cu or Ni UBM, the solder is positioned at far distance from the high current density/hot-spot region. As shown in Figs. 4 and 6, the current density/hot-spot region extends several microns toward solder. By using a Cu or Ni UBM of more than 10 μm thick, the solder will be distant from this region. However, this approach does not reduce the current crowding and hot-spot temperatures in the whole joint. Instead, the current crowding and hot-spot occur in the thick Cu or Ni UBM, and the two metals are much more resistive to EM than the solders are. Our simulation results show that the maximum current density in solder with a 20-μm-thick Cu UBM would be reduced by at least 10 times than that with a thin-film UBM. The hot-spot temperature will be reduced greatly since the local current density in the solder is alleviated and the solder stays away from the Al trace, which is the major Joule heating source. Thus, the MTTF would increase significantly.

3.2.2 Spreading the current uniformly by adding a thin resistive layer

By adding a thin resistive layer between the UBM and the Al trace, the current would be forced to spread uniformly on the passivation opening. This approach will reduce the current crowding effect to a quarter when one adds a thin layer of material with a resistivity of 3000 μΩ cm. Further, this approach could almost eliminate the current crowding effect when the resistivity is increased to above 30000 μΩ cm. Figure 9(a) and (b) shows the current paths schematically with and without this thin resistive layer. Without this layer, majority of the current would drift down near the entrance point of the Al trace, whereas with this layer, it will be forced to drift farther in the Al trace before going down to the solder bump. Thus, the current crowding effect could be greatly reduced with the

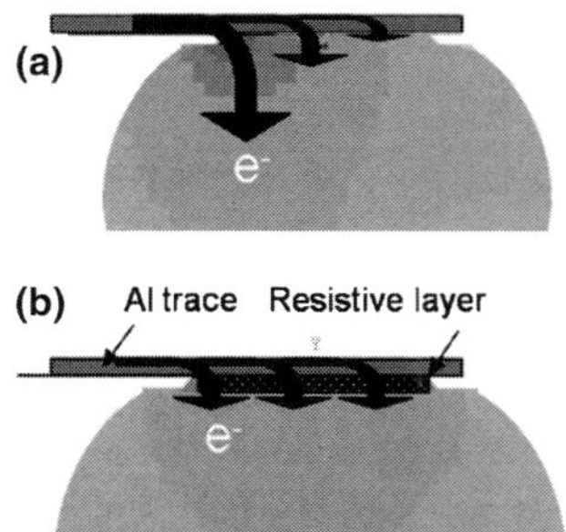

Fig. 9 Schematic drawing showing the current path in the solder joints: (a) without and (b) with a thin resistive layer between the Al trace and the UBM

resistive layer. However, overlaying a resistive layer would inevitably increase the bump resistance, raising the Joule heating effect. Fortunately, the Joule heating effect is not significant at stressing current less than 0.2 A.

3.2.3 Decreasing the passivation opening

With smaller passivation opening, the current is likely to spread out rather uniformly before entering the solder bump, since the cross-section of the Al pad above the passivation opening is larger than that of the Al trace. The current crowding effect can be relieved by decreasing the diameter of the passivation opening. However, the Joule heating effect is not reduced but instead, it increases slightly because the current needs to drift longer in the resistive Al trace.

3.2.4 Enlarging the cross-section of the Al trace

Since the Al trace is the major source for Joule heating during accelerated EM test, enlarging its cross-section would produce a lower resistance. The heating power can be expressed as

$$P = I^2 R = j^2 \rho V \tag{2}$$

where P is the Joule heating power, I is the current, R is the resistance, j is the current density, ρ is the resistivity, and V is the volume. Therefore, the solder joints with a wider or thicker Al trace is likely to have lower Joule heating effect. Also, the current crowding effect will be relieved to some extent. In addition, the solder joint with a shorter Al trace will have lower Joule heating effect. But the current crowding effect remains the same. As the damascene Cu replaces Al metallization, solder joints with Cu trace will be implemented in high-end consumer electronic devices shortly. It follows that the solder joints with the Cu trace would have lower Joule heating effect than that with an Al trace at identical configuration since the resistivity of Cu is only 63% of that of Al [28]. Yet, the current crowding effect remains almost unchanged when the Cu trace replace the Al trace with the same dimension.

Following the rational behind the discussion above and the technology available, the ideal solution for EM in solder joints would be Cu column [29]. Cu column up to 80 μm thick can be fabricated as UBM in solder joint. A thin layer of solder is still required for the joint, as shown schematically in Fig. 10. Due the thick Cu layer, the current spreads out uniformly before reaching the solder. Therefore, there is almost minimum current crowding in solder. In addition, hot-spots in solder may also be eliminated completely due to negligible current crowding effect and superb thermal conductivity of Cu. Furthermore, during current stressing the solder may react with the Cu to form Cu–Sn IMCs. Since the amount of solder is much less than the Cu, the Cu column is unlikely to be consumed completely. Therefore, it may be one of the possible solutions for solder joint EM.

3.3 Investigating failure mechanism for thick-film UBM

Since thick Cu or Ni UBM have been adopted for Pb-free solder joints, the EM behavior and failure mechanism in those joints would be very important. However, only few studies have been performed on this topic [30, 31]. In particular, both electroplated and electroless Ni have been used as UBM materials due to

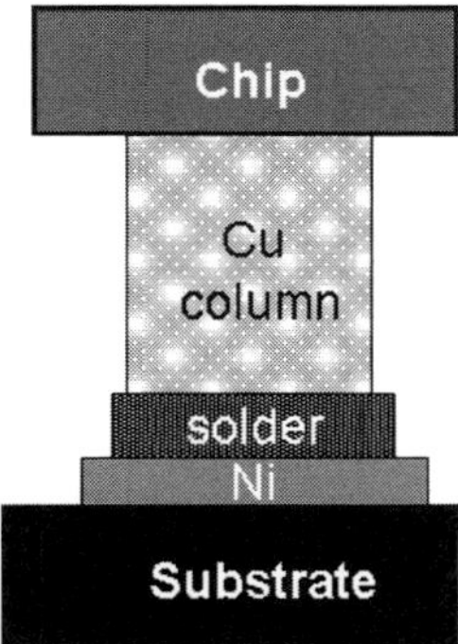

Fig. 10 Schematic drawing showing a solder bump with a Cu column. This structure may be one of the solutions for EM in solder joints

their low reaction rate with solders [32]. Yet, the activation energy for Pb-free solder on the Ni UBMs has not been measured. Since the failure mechanism for Pb-free solder is primarily attributed to the dissolution of the UBM materials, the use of Ni UBM shall be able to inhibit the UBM dissolution rate, and thus to prolong the EM lifetime.

3.4 Thermomigration in Pb-free solders

As stated in previous section, the occurrence of thermomigration in Pb-free solders necessitates further investigations. As depicted in Fig. 8(b), the diffusion of Sn atoms from the substrate/cathode end to chip/anode end is enhanced by TM force for Bump 1. Possible void formation in Bump 1 under thermomigration is an interesting subject in our undergoing study.

3.5 Rotation of solder grains during EM

It has been demonstrated that Sn grains may rotate or grow during current stressing in Sn films, because white tin has anisotropic properties on electrical resistivity [33, 34]. It has a body-center tetragonal crystal structure with lattice parameters $a = b = 0.583$ nm and $c = 0.318$ nm, and its electrical resistivities are 13.25 μΩ cm and 20.27 μΩ cm, respectively. Thus, Sn grains may rotate during EM testing to reduce the total resistance of the Sn stripe. For Pb-fee solders, the matrix consists of Sn grains. However, whether this phenomena would occur in solder bumps is not clear so far, and it needs further investigation.

4 Conclusions

Critical issues of EM in Pb-free solder joints have been reviewed. Our simulation results demonstrated that the distributions of current density and temperature remain almost the same when Pb-containing solders are replaced by Pb-free ones. Many researchers have reported that the EM lifetime for Pb-free solder is higher than that of the eutectic SnPb solders, but shorter than that of high-Pb solders. The underlying reason may be the low melting temperature, as electrical and thermal characteristics of solder bumps during EM testing, are quite similar. Several approaches have been suggested to relieve the current crowding and Joule heating effects in the solder bumps. Among them, thick UBM may be the most effective method to reduce the current crowding effect, whereas the reduction in the resistance of the Al trace would render significantly lower Joule heating. The solder joints with Cu column is likely to be the ideal structures with highest EM resistance.

Acknowledgements The authors would like to thank the National Science Council of R.O.C. for financial support of this study through Grant No. 94-2216-E-009-021. In addition, simulation assistance from the National Center for High-performance Computing (NCHC) in Taiwan is appreciated.

References

1. K.N. Tu, J. Appl. Phys. **94**, 5451 (2003)
2. H.B. Huntigton, A.R. Grone, J. Phys. Chem. Solids **20**, 76 (1961)
3. K. Zeng, K.N. Tu, Mater. Sci. Eng. Rep. **R38**, 55 (2002)
4. S. Brandenburg, S. Yeh, in *Proceedings of Surface Mount International Conference and Exhibition, SM198*, San Jose, CA, 23–27 August 1998, p. 337
5. C.Y. Liu, C. Chen, K.N. Tu, J. Appl. Phys **88**, 5703 (2000)
6. Q.T. Huynh, C.Y. Liu, C. Chen, K.N. Tu, J. Appl. Phys. **89**, 4332 (2001)
7. E.C.C. Yeh, W.J. Choi, K.N. Tu, P. Elenius, H. Balkan, Appl. Phys. Lett. **80**, 580 (2002)
8. T.L. Shao, S.-W. Liang, T.C. Lin, C. Chen, J. Appl. Phys. **98**, 044509 (2005)
9. S. H. Chiu, T.L. Shao, C. Chen, D.J. Yao, C.Y. Hsu, Appl. Phys. Lett. **88**, 022110 (2006)
10. H. Ye, C. Basaran, D. Hopkins, Appl. Phys. Lett. **82**, 7 (2003)
11. A.T. Huang, A.M. Gusak, K.N. Tu, Y.-S. Lai, Appl. Phys. Lett. **88**, 141911 (2006)
12. Y.H. Lin, Y.C. Hu, C.M. Tsai, C.R. Kao, K.N. Tu, Acta Mater. **53**, 2029 (2005)
13. J.W. Nah, K.W. Paik, J.O. Suh, K.N. Tu, J. Appl. Phys. **94**, 7560 (2003)
14. J. D. Wu, C.W. Lee, P. J. Zheng, J. C.B. Lee, S. Li, in *Proceedings of the 54th Electronic Components and Technology Conference*, IEEE Components, Packaging, and Manufacturing Technology Society, Las Vegas, NV, 2004, p. 961
15. S. Gee, N. Kelkar, J. Huang, K. N. Tu, in *Proceedings of IPACK2005, ASME InterPACK 2005*, San Francisco, CA, USA, 2005
16. W.J. Choi, E.C.C. Yeh, K.N. Tu, J. Appl. Phys. **94**, 5665 (2003)
17. T.L. Shao, S.H. Chiu, C. Chen, D.J. Yao, C.Y. Hsu, J. Electron. Mater. **33**(11), 1350 (2004)
18. Y-S. Lai, K.M. Chen, C.W. Lee, C.L. Kao, Y.H. Shao, in *Proceedings of EPTC 2005, 7th Electronics Packaging Technology Conference*, Singapore, 2005, p. 786
19. T.Y. Lee, W.J. Choi, K.N. Tu, J.W. Jang, S.M. Kuo, J.K. Lin, D.R. Frear, K. Zeng, J.K. Kivilahti, J. Mater. Res. **17**(2), 291 (2002)
20. A.A. Liu, H.K. Kim, K.N. Tu, P.A. Totta, J. Appl. Phys. **80**(5), 2774 (1996)
21. S.W. Chen, C.M. Chen, W.C. Liu, J. Electron. Mater. **27**(11), 1193–1198 (1998)
22. T.L. Shao, Y.H. Chen, S.H. Chiu, C. Chen, J. Appl. Phys. **96**(8), 4518 (2004)
23. Y.H. Chen, T.L. Shao, P.C. Liu, C. Chen, T. Chou, J. Mater. Res. **20**(9), 2432–2442 (2005)
24. G.A. Rinne, Microelectron. Reliab. **43**, 1975 (2003)
25. S.H. Chiu, S.W. Liang, C. Chen, D.J. Yao, Y.C. Liu, K.H. Chen, S.H. Lin, in *Proceedings of the 56th Electronic Components and Technology Conference, IEEE Components*, Packaging, and Manufacturing Technology Society, San Diego, CA, 2006
26. J.R. Black, IEEE Trans. Electron. Devices ED **16**(4), 338 (1969)
27. S.W. Liang, T.L. Shao, C. Chen, E.C.C. Yeh, K.N. Tu, J. Mater. Res. **21**(1), 137 (2006)
28. C.Y. Hsu, D.J. Yao, S.W. Liang, C. Chen, J. Electron. Mater. **35**, 947 (2006)
29. J.W. Nah, J.O. Suh, K.N. Tu, S.W. Yoon, C.T. Chong, V. Kripesh, B.R. Su, C. Chen, in *Proceedings of the 56th Electronic Components and Technology Conference*, IEEE Components, Packaging, and Manufacturing Technology Society, San Diego, CA, 2006
30. Y.H. Lin, C.M. Tsai, Y.C. Hu, Y.L. Lin, C.R. Kao, J. Electron. Mater. **34**(1), 27 (2005)
31. J.K. Lin, J.W. Jang, J. White, in *Proceedings of the 53th Electronic Components and Technology Conference*, IEEE Components, Packaging, and Manufacturing Technology Society, New Orleans, USA 2006 p. 816
32. P.G. Kim, J.W. Jang, T.Y. Lee, K.N. Tu, J. Appl. Phys. **86**, 6746–6751 (1999)
33. A.T. Wu, A.M. Gusak, K.N. Tu, C.R. Kao, Appl. Phys. Lett. **86**, 241902 (2005)
34. A.T. Wu, J.R. Lloyd, N. Tamura, B.C. Valek, K.N. Tu, C.R. Kao, Appl. Phys. Lett. **85**(13), 2490 (2004)

Stress analysis of spontaneous Sn whisker growth

K. N. Tu · Chih Chen · Albert T. Wu

Published online: 1 September 2006
© Springer Science+Business Media, LLC 2006

Abstract Spontaneous Sn whisker growth is a surface relief phenomenon of creep, driven by a compressive stress gradient. No externally applied stress is required for the growth, and the compressive stress is generated within, from the chemical reaction between Sn and Cu to form the intermetallic compound Cu_6Sn_5 at room temperature. To obtain the compressive stress gradient, a break of the protective oxide on the Sn surface is required because the free surface of the break is stress-free. Thus, spontaneous Sn whisker growth is unique that stress relaxation accompanies stress generation. One of the whisker challenging issues in understanding and in finding effective methods to prevent spontaneous Sn whisker growth is to develop accelerated tests of whisker growth. Use of electromigration on short Sn stripes can facilitate this. The stress distribution around the vicinity and the root of a whisker can be obtained by using the micro-beam X-ray diffraction utilizing synchrotron radiation. A discussion of how to prevent spontaneous Sn whisker growth by blocking both stress generation and stress relaxation is given.

K. N. Tu (✉)
Department of Materials Science and Engineering, UCLA, Los Angeles, CA 90095-1595, USA
e-mail: kntu@ucla.edu

C. Chen
Department of Materials Science and Engineering, National Chiao Tung University, Hsinchu 30050, Taiwan, ROC

A. T. Wu
Department of Materials and Mineral Resources Engineering, National Taipei University of Technology, Taipei, Taiwan, ROC

1 Introduction

Whisker growth on beta-tin (β-Sn) is a surface relief phenomenon of creep. It is driven by a compressive stress gradient and occurs at ambient. Spontaneous Sn whiskers grow on matte Sn finishes on Cu [1–15]. Today, due to the wide application of Pb-free solders on Cu conductors used in electronic packaging of consumer electronic products, Sn whisker growth has become a serious reliability issue since Pb-free solders are Sn-based and very rich in Sn. Typically, the Cu leadframe used on surface mount technology of electronic packaging are finished with a layer of solder for surface passivation and for enhancing wetting during joining the leadframe to printed circuit boards. When the solder finish is eutectic SnCu or matte Sn, whiskers of Sn are observed. Some whiskers can grow to several hundred microns, which are long enough to become electrical shorts between neighboring legs of a leadframe. The trend in electronic packaging technology is to integrate systems in packaging, so that elements of devices and parts of components are getting closer and closer together and the probability of shorting by whiskers is becoming much greater. Hence, how to suppress Sn whisker growth, and how to perform systematic tests of Sn whisker growth in order to understand the driving force, kinetics, and mechanism of growth are challenging tasks in electronic packaging industry today.

Due to the very limited temperature range of Sn whisker growth, from room temperature to about 60°C, accelerated tests are difficult. If the temperature is too low, there is insufficient kinetics due to slow atomic diffusion and if the temperature is too high, there is not enough driving force because of stress relief.

The whisker growth is spontaneous, indicating that the compressive stress behind the growth is self-generated; no external applied stress is required. Otherwise, one can expect a whisker to slow down and stop when the applied stress that is not applied continuously is exhausted. Therefore it is of interest to ask where is the self-generated driving force coming from, how can the driving force maintain itself to achieve the spontaneous whisker growth, and also how large is the compressive stress gradient needed to grow a whisker?

Spontaneous whisker growth is a unique creep process in which both stress generation and stress relaxation occur simultaneously at room temperature. The three indispensable conditions of Sn whisker growth are (1) The room temperature diffusion in Sn, (2) The room temperature reaction between Sn and Cu to form Cu_6Sn_5 and induce the compressive stress in Sn, and (3) The stable surface oxide on Sn. The last condition is needed in order to produce a stress gradient for creep. When the oxide is broken, the exposed free surface is stress-free, so a compressive stress gradient is developed for creep or the growth of a whisker to relax the stress.

Reviews obtained from examining Sn whiskers by using cross-sectional scanning and transmission electron microscopy of samples prepared by focused ion beam thinning will be discussed in this paper [16]. Use of X-ray micro-diffraction by synchrotron radiation to study the structure and stress distribution around the root of a whisker grown on eutectic SnCu will be discussed later [17].

The growth of a Sn whisker is from the bottom, not from the top, since the morphology of the tip does not change [12]. Many Sn whiskers have been found to grow at room temperature and some of them are long enough to short two neighboring legs of the leadframe as shown in Fig. 1a. It is possible that when there is a high electrical field across the narrow gap between the tip of a whisker and the point of contact on the other leg, just before the tip of the whisker touching the other leg, a spark may ignite fire. The fire may result in failure of the device or a satellite [18–21]. Even when a whisker breaks, it could fall between two neighboring conductors and bridge them.

2 Morphology of spontaneous Sn whisker growth

In Fig. 1b, an enlarged SEM image of a long whisker on the eutectic SnCu finish is shown. The whisker in Fig. 1b is straight and its surface is fluted. The crystal structure of Sn is body-centered tetragonal with the lattice constant "a" = 0.58311 nm and "c" = 0.31817 nm. The whisker growth direction, or the axis

along the length of the whisker, has been found mostly to be the "c" axis, but growth along other axis such as [100] and [311] has also been found.

On the pure or matte Sn finish surface, short whiskers or hillocks were observed as shown in Fig. 1c. The surface of the whisker in Fig. 1c is faceted. Besides the difference in morphology, the rate of whisker growth on the pure Sn finish is much slower than that on the SnCu finish. The direction of growth is more random too.

Comparing the whiskers formed on SnCu and pure Sn, it seems that the Cu in eutectic SnCu enhances Sn whisker growth. Although the composition of eutectic SnCu consists of 98.7 atomic % of Sn and 1.3 atomic % of Cu, the small amount of Cu seems to have caused a very large effect on whisker growth on the eutectic SnCu finish.

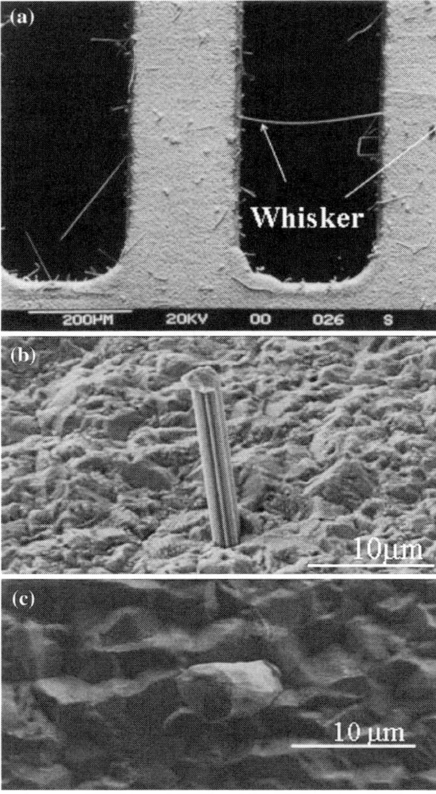

Fig. 1 (a) Many Sn whiskers have been found to grow at room temperature and some of them are long enough to short two neighboring legs of the leadframe. (b) An enlarged SEM image of a long whisker on the eutectic SnCu finish is shown. (c) On the pure or matte Sn finish surface, short whiskers or hillocks were observed

In Fig. 2a, a cross-sectional SEM image of a lead-frame leg with SnCu finish is shown. The rectangular core of Cu is surrounded by an approximate 15 μm thick SnCu finish. A higher magnification image of the interface between the SnCu and Cu layers, prepared by focused ion beam, is shown in Fig. 2b. An irregular layer of Cu_6Sn_5 compound can be seen between the Cu

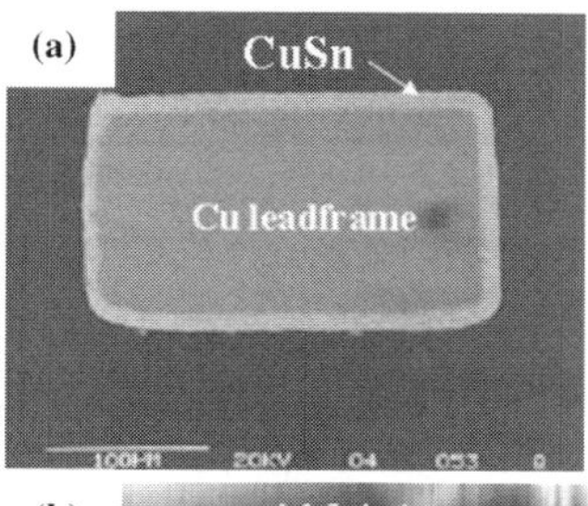

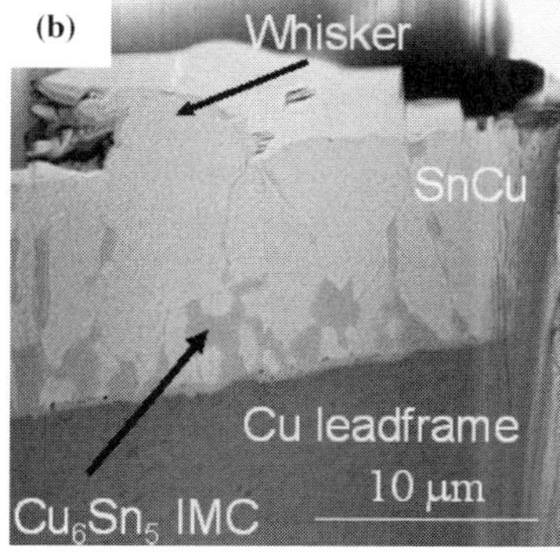

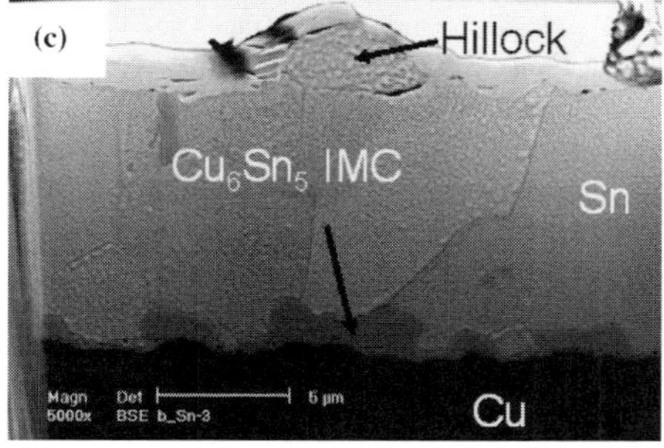

Fig. 2 (**a**) A cross-sectional SEM image of a leadframe leg with SnCu finish is shown. The rectangular core of Cu is surrounded by an approximate 15 μm thick SnCu finish. (**b**) A higher magnification image of the interface between the SnCu and Cu layers, prepared by focused ion beam, is shown. An irregular layer of Cu_6Sn_5 compound can be seen between the Cu and SnCu. No Cu_3Sn was detected at the interface. The grain size in the SnCu finish is about several microns. More importantly, there are Cu_6Sn_5 precipitates in the grain boundaries of SnCu. (**c**) A cross-sectional SEM image, prepared by focused ion beam, of matte Sn finish on Cu leadframe is shown. While the layer of Cu_6Sn_5 compound can be seen between the Cu and Sn, there is much less Cu_6Sn_5 precipitates in the grain boundaries of Sn

and SnCu. No Cu_3Sn was detected at the interface. The grain size in the SnCu finish is about several microns. More importantly, there are Cu_6Sn_5 precipitates in the grain boundaries of SnCu. The grain boundary precipitation of Cu_6Sn_5 is the source of stress generation in the CuSn finish. It provides the driving force of spontaneous Sn whisker growth [21, 22]. The critical stress issue will be addressed later.

In Fig. 2c, a cross-sectional SEM image, prepared by focused ion beam, of matte Sn finish on Cu leadframe is shown. While the layer of Cu_6Sn_5 compound can be seen between the Cu and Sn, there is much less Cu_6Sn_5 precipitates in the grain boundaries of Sn. The grain size in the Sn finish is also about several microns. The lacking of grain boundary Cu_6Sn_5 precipitates is the most important difference between the eutectic SnCu and the pure Sn finish with respect to whisker growth.

TEM images of the cross-section of whiskers, normal to their length, are shown in Fig. 3a, b together with electron diffraction pattern. The growth direction is the c-axis. There are a few spots in the images which might be dislocations.

3 Stress generation (driving force) in Sn whisker growth by Cu–Sn reaction

The origin of the compressive stress can be mechanical, thermal, and chemical. But the mechanical and thermal stresses tend to be finite in magnitude, so they cannot sustain a spontaneous or continuous growth of whiskers for a long time. The chemical force is essential for spontaneous Sn whisker growth, but not obvious. The origin of the chemical force is due to the room temperature reaction between Sn and Cu to form the intermetallic compound (IMC) of Cu_6Sn_5. The chemical reaction provides a sustained driving force for spontaneous growth of whiskers as long as the reaction progresses with unreacted Sn and Cu.

Stress is generated by interstitial diffusion of Cu into Sn and the formation of IMC in the Sn; it generates a compressive stress in the Sn. When the Cu atoms from the leadframe diffuse into the finish to grow the grain boundary IMC, as shown in Fig. 2b, the volume increase due to the IMC growth will exert a compressive stress to the grains on both sides of the grain boundary. As shown in Fig. 4, for a fixed volume V in the Sn finish that contains an IMC precipitate, the growth of the IMC due to the diffusion of a Cu atom into this volume to react with Sn will produce a stress,

$$\sigma = -B\frac{\Omega}{V} \tag{1}$$

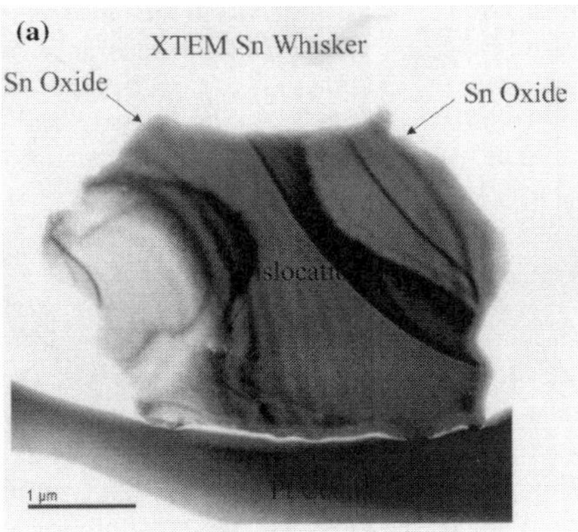

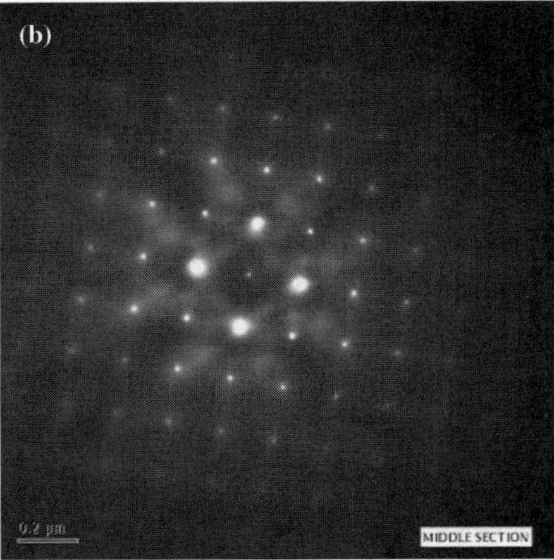

Fig. 3 TEM images of the cross-section of whiskers, normal to their length, are shown in (**a**) and (**b**) together with electron diffraction pattern

where σ is the stress produced, B is bulk modulus, Ω is the partial molecular volume of a Cu atom in Cu_6Sn_5 (the molar volume change of Sn atoms in the reaction for simplicity is ignored). The negative sign indicates that the stress is compressive. In other words, an atomic volume into the fixed volume is added. If the fixed volume cannot expand, a compressive stress will occur. When more and more Cu atoms, say n Cu atoms, diffuse into the volume, V, to form Cu_6Sn_5, the stress in the above equation of σ increases by changing Ω to $n\,\Omega$.

In diffusional processes, such as the classic Kirkendall effect of interdiffusion in a bulk diffusion couple of A and B, the atomic flux of A is not equal to the opposite flux of B. Assuming that A diffuses into B faster than B diffuses into A, a compressive stress in B

will be expected since there are more A atoms diffusing into it than B atoms diffusing out of it. However, in Darken's analysis of interdiffusion, there is no stress generated in either A or B or no analysis of stress was given. Since Darken has made a key assumption that vacancy concentration is in equilibrium everywhere in the sample [23, 24]. To achieve vacancy equilibrium, one must assume that vacancies (or lattice sites) can be created and/or annihilated in both A and B as needed. Hence, provided that the lattice sites in B can be added to accommodate the incoming A atoms, there will be no stress. The addition of a large number of lattice sites implies an increase in lattice planes if one assumes that the mechanism of vacancy creation and/or annihilation is by dislocation climb mechanism. It further implies that lattice plane can migrate, which means Kirkendall shift, in turn it implies marker motion if markers are embedded in the moving lattice planes in the sample. In some cases of interdiffusion in bulk diffusion couples, vacancy may not be in equilibrium everywhere in the sample, so very often Kirkendall void formation has been found due to the existence of excess vacancies [25].

To absorb the added atomic volume due to the in-diffusion of Cu by the fixed volume of V in the finish as considered in Fig. 4, it requires addition of lattice sites in the fixed volume. Furthermore, allowance of Kirkendall shift or addition of lattice plane to migrate is necessary. Otherwise, compressive stress will be generated. Since Sn has a native and protective oxide on the surface, the interface between the oxide and Sn is a poor source and sink for vacancies and furthermore the protective oxide ties down the lattice planes in Sn and prevents them from moving. Therefore, this is the basic mechanism of stress generation. It is worth noting that creep is driven by a stress gradient, not by a stress. Typically creep is defined as a time-dependent deformation under constant load. Actually the driving force

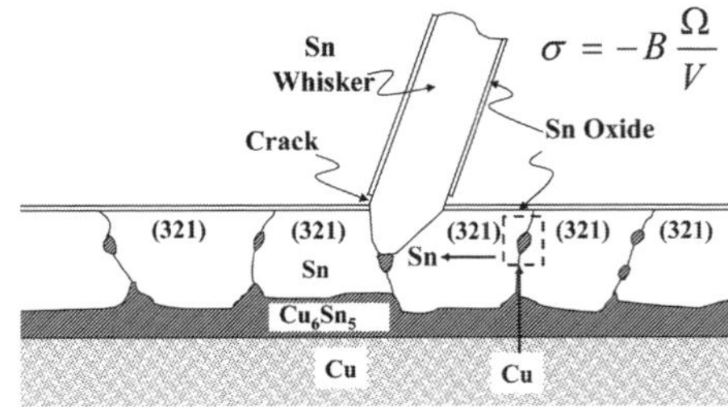

Fig. 4 We consider a fixed volume V in the Sn finish that contains an intermetallic compound (IMC) precipitate, the growth of the IMC due to the diffusion of a Cu atom into this volume to react with Sn will produce a stress

of atomic migration in creep is stress gradient as given in the Nabarro–Herring model. Under a hydrostatic tension or compression, there may be random walk of atoms, but no creep. A uniform compressive stress will not lead to creep, so we need a mechanism to produce a compressive stress gradient in the Sn finish. It will be discussed in next Section.

For the oxide to be effective in tying down lattice plane migration, the finish cannot be too thick. In a very thick finish, say over 100 μm, there are more sinks in the bulk of the finish to absorb the added volume of Cu. Whiskers is a surface relief phenomenon. When bulk relief mechanism occurs, whisker will not grow. There is a dependence of whisker formation on the thickness of finish. Since the average diameter of whiskers is about a few microns, whisker will grow more frequently on a finish having thickness of a few microns to a few times of its diameter.

Sometimes it is puzzling to find that Sn whiskers seem to grow on a tensile region of a Sn finish. For example, when a Cu leadframe surface was plated with SnCu, the initial stress state of the SnCu layer right after plating was tensile, yet whisker growth was observed. Consider the cross-section of a Cu leadframe leg coated with a layer of Sn as shown in Fig. 2a, that experienced a heat-treatment of reflow from room temperature to 250°C and back to room temperature. Since Sn has a higher thermal expansion coefficient than Cu, the Sn should be under tension at room temperature after the reflow cycle. Yet with time, Sn whisker grows, so it seems that Sn whisker grows under tension. Furthermore, if a leg is bent, one side of it will be in tension and the other side in compression. It is surprising to find that whiskers grow on both sides, whether the side is under compression or tension. These phenomena are hard to understand until one recognizes that the thermal stress or the mechanical stress, whether it is tensile or compressive, is finite. It can be relaxed or overcome quickly by atomic diffusion at room temperature. After that, the continuing chemical reaction will develop the compressive stress needed to grow whiskers. So the chemical force is dominant and persistent. In other words, the compressive stress needed for the spontaneous whisker growth on Sn is induced by chemical reaction between Sn (or SnCu) and Cu at room temperature. Room temperature reaction between Sn and Cu was studied by using thin film samples for detection sensitivity [14, 21].

The idea of compressive stress induced by the growth of a grain boundary phase has a few variations. One is the wedge model proposed by Lee et al. [26] that the Cu_6Sn_5 phase between the Cu and Sn has a

wedge shape in growing into the grain boundaries of Sn. The growth of the wedge will exert a compressive stress to the two neighboring Sn grains, same as splitting a piece of wood with a wedge. So far, no such wedge-shape IMC has been observed in XSEM, for example, see Fig. 2b.

4 Effect of surface Sn oxide on stress gradient generation and whisker growth

To discuss the effect of surface oxide, one can refer to the effect on Al hillock growth. In an ultra-high vacuum, no surface hillocks were found to grow on Al surface under compression [27]. Hillocks grow on Al surfaces only when the Al surface is oxidized, and Al surface oxide is known to be protective. Without surface oxide, the free surface of Al is a good source and sink of vacancies, so a compressive stress can be relieved uniformly on the entire surface or the surface of every grain of the Al based on Nabarro–Herring model of lattice creep or Coble model of grain boundary creep. In these models, as shown in Fig. 5, the relaxation can occur in each of the grains by diffusion to the free surface of each grain. The free surfaces are effective source and sink of vacancies. Therefore, the relaxation is uniform over the entire Al film surface; all the grains just thicken slightly. Consequently, no localized growth of hillocks or whiskers will take place.

A whisker or hillock is a localized growth on a surface. To have a localized growth, the surface cannot be free of oxide, and the oxide must be a protective oxide so that it effectively blocks all the vacancy sources and sinks on the surface. Furthermore, a protective oxide also means that it pins down the lattice planes in the matrix of Al (or Sn), so that no lattice plane migration can occur to relax the stress in the volume, V, considered in Fig. 4. Only those metals

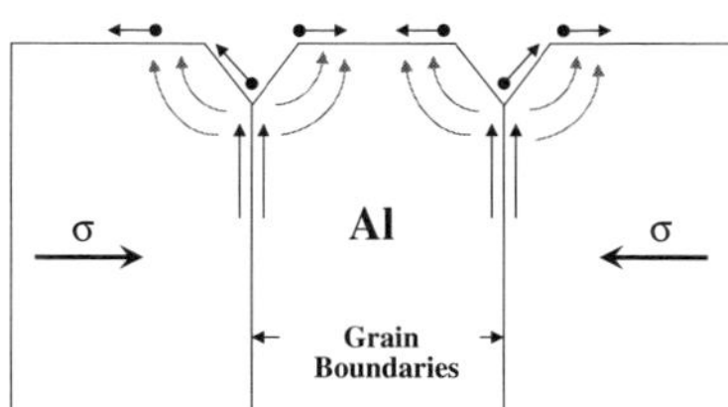

Fig. 5 Nabarro–Herring model of lattice creep or Coble model of grain boundary creep. A schematic diagram to show that when the surface has no oxide, the relaxation of stress can occur in each of the grains by atomic diffusion to the free surface of each grain

which grow protective oxides, such as Al and Sn, are known to have serious hillock or whisker growth. When they are in thin film or thin layer form, the surface oxide can pin down the lattice planes in the near-surface layer easily. On the other hand, it is obvious that if the surface oxide is very thick, it will physically block the growth of any hillock and whisker. No hillocks or whiskers can penetrate a very thick oxide or coating. Thus, a necessary condition of whisker growth is that the protective surface oxide must not be too thick so that it can be broken at certain weak spots on the surface, and from these spots whiskers grow to relieve the stress.

In Fig. 6a, a focused ion beam image of a group of whiskers on the SnCu finish is shown. Figure 6b has the image of the same area after the oxide on a rectangular area of the surface of the finish was sputtered away by using a glancing incidence ion beam to expose the microstructure beneath the oxide. In Fig. 6c, a higher magnification image of the sputtered area is shown, in which the microstructure of Sn grains and grain boundary precipitates of Cu_6Sn_5 are clear. Due to the ion channeling effect, some of the Sn grains appear darker than the others. The Cu_6Sn_5 particles distribute mainly along grain boundaries in the Sn matrix, and they are brighter than the Sn grains due to less ion channeling. The diameter of the whiskers is about a few microns. It is comparable to the grain size in the SnCu finish.

In ambient, one can assume that the surface of the finish and the surface of every whisker are covered with oxide. The growth of a hillock or whisker is an eruption from the oxidized surface. It has to break the oxide. The stress that is needed to break the oxide may be the minimum stress needed to grow whiskers. It seems that the easiest place to break the oxide is at the base of the whisker. Then to maintain the growth, the break must remain oxide-free so that it behaves like a free surface and vacancies can be supplied continuously from the break and can diffuse into the Sn layer to sustain the long range diffusion of the Sn atoms needed to grow the whisker.

Figure 7 is an illustration in which the surface of the whisker is oxidized, except at the base. The surface oxide of the whisker serves the very important purpose of confinement so that the whisker growth is essentially a one-dimensional growth. The surface oxide of the whisker prevents it from growing in lateral direction, thus it grows with a constant cross-section and has the shape of a pencil. Also the oxidized surface may explain why the diameter of a Sn whisker is just a few microns. This is because the gain in strain energy reduction in whisker growth is balanced by the

formation of surface of the whisker. By balancing the strain energy against the surface energy in a unit length of the whisker, $\pi R^2 \varepsilon = 2\pi R \gamma$, providing

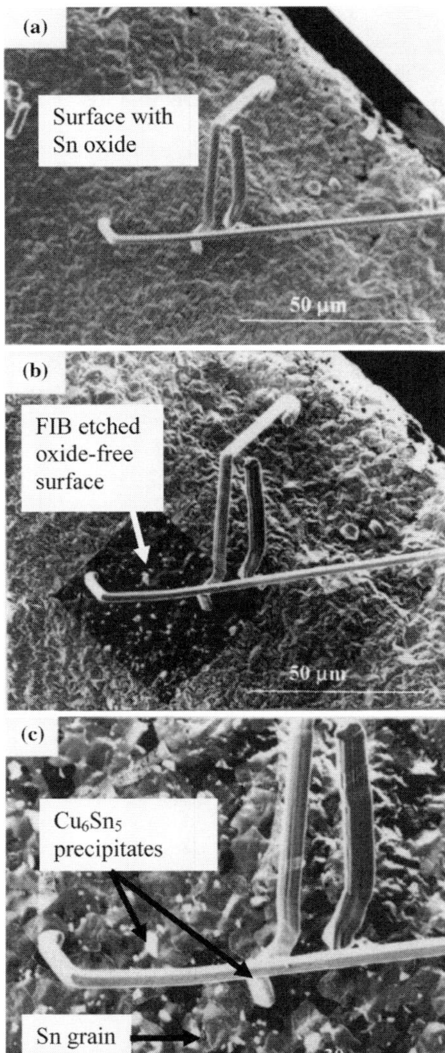

Fig. 6 (**a**) A focused ion beam image of a group of whiskers on the SnCu finish is shown. (**b**) The oxide on a rectangular area of the surface of the finish was sputtered away by using a glancing incidence ion beam to expose the microstructure beneath the oxide. (**c**) A higher magnification image of the sputtered area is shown, in which the microstructure of Sn grains and grain boundary precipitates of Cu_6Sn_5 are clear. The Cu_6Sn_5 precipitates are indicated by the arrows in the figure

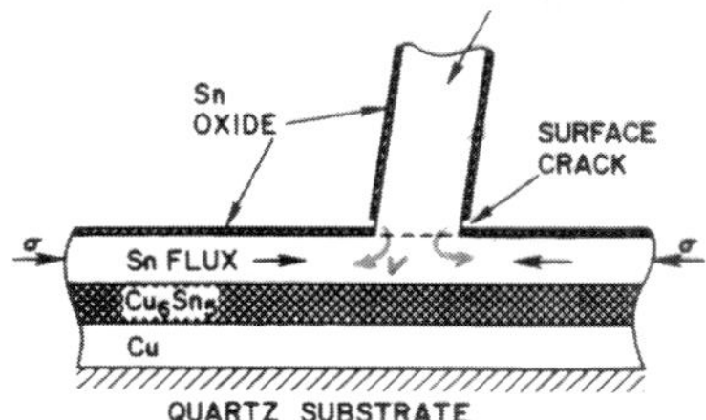

Fig. 7 A schematic diagram depicts that the surface of the whisker is oxidized, except the base

$$R = \frac{2\gamma}{\varepsilon} \qquad (2)$$

where R is radius of the whisker, γ is surface energy per unit area, and ε is strain energy per unit volume. Since strain energy per atom is about four to five orders of magnitude smaller than the chemical bond energy or surface energy per atom of the oxide, the radius or diameter of a whisker is found to be several microns, which are about four orders of magnitude larger than the atomic diameter of Sn. For this reason, it is very difficult to have spontaneous growth of nano-diameter Sn whiskers.

5 Measurement of stress distribution by synchrotron radiation micro-diffraction

The micro-diffraction apparatus in advanced light source (ALS), at Lawrence Berkeley National Laboratory, was used to study Sn whiskers grown on SnCu finish on Cu leadframe at room temperature [28]. The white radiation beam was 0.8–1 μm in diameter and the beam step-scanned over an area of 100 μm by 100 μm at steps of 1 μm. Several areas of the SnCu finish were scanned and those areas were chosen so that in each of them there was a whisker, especially the areas that contained the root of a whisker. During the scan, the whisker, and each grain in the scanned area, can be treated as a single crystal to the beam. This is because the grain size is larger than the beam diameter. At each step of the scan, a Laue pattern of a single crystal is obtained. The crystal orientation and the lattice parameters of the Sn whisker and the grains of SnCu matrix surrounding the root of the whisker were measured by the Laue patterns. The software in ALS is capable of determining the orientation of each of the grains, and displaying the distribution of the major axis of these grains. Using the lattice parameters of the whisker as stress-free internal reference, the strain or stress in the grains in the SnCu matrix can be deter-

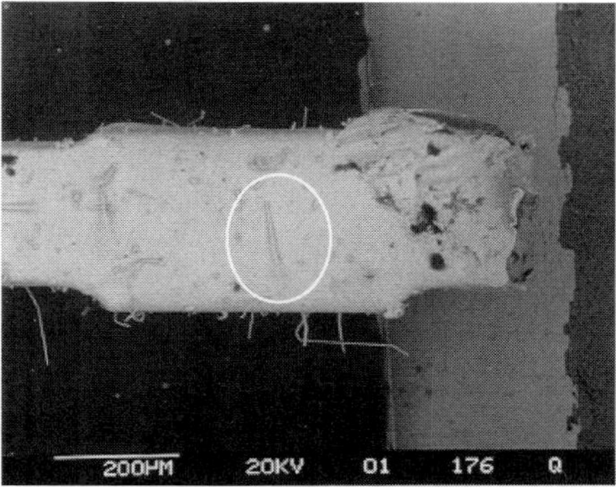

Fig. 8 A low magnification picture of an area of finish wherein a whisker is circled and scanned

mined and displayed. Figure 8 shows a low magnification picture of an area of finish wherein a whisker is circled and scanned.

Figure 9 shows the X-ray micro-diffraction analysis result of stress distribution around whisker. The whisker part is removed in order to observe the stress around the whisker root more clearly. The absolute value of stress in the whisker is higher than that in the surrounding grains. If the whisker is assumed to be stress-free, the surface of SnCu finish will be under compressive stress. The study shows that in a local area of 100 μm × 100 μm the stress is highly inhomogeneous with variations from grain to grain. The finish is therefore under a biaxial stress only on the average. This is because each whisker has relaxed the stress in the region surrounding it. But, the stress gradient around the root of a whisker does not have a radial

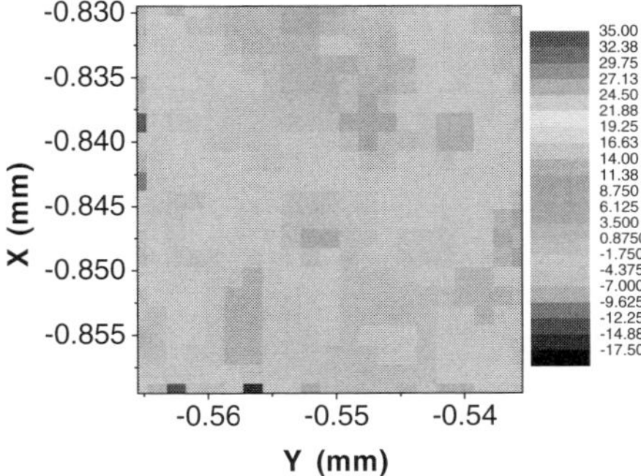

Fig. 9 In the figure, the whisker part is removed in order to observe the stress around the whisker root more clearly. The absolute value of stress in the whisker is higher than that in the surrounding grains. If we assume the whisker to be stress-free, the surface of SnCu finish is under compressive stress

symmetry. The numerical value, and the distribution of stress, are shown in Table 1, where the root of the whisker is at the coordinates of "$x = -0.8415$" and "$y = -0.5475$". Overall, the compressive stress is quite low, of the order of several MPa, however, one can still see the slight stress gradient going from the whisker root area to the surroundings. It means that the stress level just below the whisker is slightly less compressive than the surrounding area. This is because the stress near the whisker has been relaxed by whisker growth. In Table 1, the light-colored arrows indicate the directions of local stress gradient. Some circles next to each other in Table 1 show the similar stress level, which most likely means that they belong to the same grain.

The total strain tensor is equal to the sum of deviatoric strain tensor and the dilatational strain tensor. The latter is measured from energy of Laue spot using monochromatic beam and the former is measured from deviation in crystal Laue pattern using white radiation beam.

$$\varepsilon_{ij} = \varepsilon_{\text{deviatoric}} + \varepsilon_{\text{dilatational}}$$

$$= \begin{vmatrix} \varepsilon'_{11} & \varepsilon_{12} & \varepsilon_{13} \\ \varepsilon_{21} & \varepsilon'_{22} & \varepsilon_{23} \\ \varepsilon_{31} & \varepsilon_{32} & \varepsilon'_{33} \end{vmatrix} + \begin{vmatrix} \delta & 0 & 0 \\ 0 & \delta & 0 \\ 0 & 0 & \delta \end{vmatrix} \quad (3)$$

where the dilatational strain, $\delta = \frac{1}{3}(\varepsilon_{11} + \varepsilon_{22} + \varepsilon_{33})$ and $\varepsilon_{ii} = \varepsilon'_{ii} + \delta$.

The measurements of these two strain tensors can be explained as follows. The deviatoric strain tensor is calculated from the deviation of spot positions in the Laue pattern with respect to their "unstrained" positions. The latter is obtained from an "unstrained"

reference. By assuming that the whisker is strain-free, the Sn whisker itself can be used as the unstrained reference and to calibrate the sample-detector distance and the tilt of detector with respect to the beam. The geometry is fixed. From the Laue spot positions of the strained sample, any deviation of their positions from the calculated positions if the sample has zero strain can be measured. The transformation matrix which relates the unstrained to the strained Laue spot positions is then calculated and the rotational part is taken out. The deviatoric strain can then be computed from this transformation matrix. Presence of more spots in the Laue pattern will facilitate the deviatoric strain tensor determination. The deviatoric strain is related to the change in the shape of the unit cell, but the unit cell volume is assumed to be constant and it consists of five independent components. The sum of the 3 diagonal components should be equal to zero.

To obtain the total strain tensor, the dilatational strain tensor must be added to the deviatoric strain tensor. The dilatational component is related to the change in volume of the unit cell and it consists of a single component of expansion or shrinkage, δ, in the last equation. In principle, if the deviatoric strain tensor is known, only one addition measurement is needed, that is the energy of a single reflection is required, to obtain this single dilatational component. The monochromatic beam can be used to do this. The energy of E_0 for zero dilatational strain can be calculated from the orientation of the crystal and the deviatoric strain for each reflection. The energy can be scanned by rotating the monochromator around this energy E_0 and watch the intensity of the peak of interest on the CCD camera. The energy which maximizes the intensity of the reflection is the actual energy

Table 1 Stress distribution around the root of a whisker, which is at the coordinates of "$x = -0.8415$" and "$y = -0.5475$"

1.5 μm   (Unit : Mpa)

y-axis \ x-axis	-0.5400	-0.5415	-0.5430	-0.5445	-0.5460	-0.5475	-0.5490	-0.5505	-0.5520	-0.5535	-0.5550
-0.8340	-2.82	-3.21	-2.26	0.93	0.93	-0.23	-8.17	2.22	1.49	1.6	-0.03
-0.8355	-2.26	-2.64	-2.64	-1.04	1.37	1.37	-1.31	0.87	0.87	0.87	-0.7
-0.8370	-2.53	-3.21	-3.21	-2.64	-1.04	3.61	0.75	0.87	0.7	0.7	-0.19
-0.8385	-7.37	-9.62	-6.57	-2.64	3.61	4.52	3.61	0.29	-1.31	0	-4.79
-0.8400	-7.37	-8.22	-6.57	-1.18	0.75	4.23	0.75	-2.25	-2.27	-2.91	-6.91
-0.8415	-4.17	-4.84	-4.17	-1.81	-0.67	0.00	-1.96	-1.96	-3.74	[illegible]	[illegible]
-0.8430	-4.17	-4.17	-3.63	-1.81	-1.81	[illegible]	-2.29	-1.96	-1.96	-3.27	-3.27
-0.8445	-4.14	-4.17	-3.86	-3.63	-2.79	-4.64	[illegible]	-0.84	-1.4	-1.49	-3.27
-0.8460	-3.14	-3.63	-3.86	-3.63	-3.13	[illegible]	[illegible]	0.04	0.04	-1.41	-2.33
-0.8475	-4.14	-4.9	-4.9	-4.64	-3.86	-6.04	-1.72	3.55	3.55	-0.41	-2.33
-0.8490	-3.33	-5.67	-6.29	-6.29	-2.66	-2.08	[illegible]	-1.79	0	-1.79	-3.73

Whisker

of the reflection. The difference in the observed energy and the E_0 gives the dilatational strain.

Since, $\sigma'_{xx} + \sigma'_{yy} + \sigma'_{zz} = 0$ by definition, $-\sigma'_{zz}$ is a measure of the in-plane stress (Note that for a blanket film, with free or passivated surface, on the average the total normal stress $\sigma_{zz} = 0$), from that σ_{b} (biaxial stress)$=(\sigma_{xx} + \sigma_{yy})/2 = (\sigma'_{xx} + \sigma'_{yy})/2 - \sigma'_{zz} = -3\sigma'_{zz}/2$. This relation is always true on the average. A positive value of $-\sigma'_{zz}$ indicates a overall tensile stress whereas a negative value indicates an overall compressive stress. However, the measured stress values, corresponding to a strain of less than 0.01%, are only slightly larger than the strain/stress sensitivity of the white beam Laue technique (sensitivity of the technique is 0.005% strain). No long range stress gradient has been observed around the root of a whisker, indicating that the growth of a whisker has released most of the local compressive stress in the distance of several surrounding grains.

6 Stress relaxation (kinetic process) in Sn whisker growth by creep

Whisker growth is a unique creep phenomenon in which stress generation and stress relaxation occurs simultaneously. Therefore, one must consider two kinetic processes of stress generation and stress relaxation and their coupling by irreversible processes [29]. About the two processes in whisker growth, the first is the diffusion of Cu from the leadframe into the Sn finish to form grain boundary precipitates of Cu_6Sn_5. This kinetic process generates the compressive stress in the finish. The second is the diffusion of Sn from the stressed region to the root of a whisker which is stress-free to relieve the stress. The distance of diffusion in the second process is much longer than the first and also the diffusivity in the second process is slower too, so the second process tends to control the rate in the growth of whiskers.

Since the reaction of Sn and Cu occurs at room temperature, the reaction continues as long as there are unreacted Sn and Cu. The stress in the Sn will increase with the growth of Cu_6Sn_5 in it. Yet the stress cannot build up forever; and it must be relaxed. Either the added lattice planes in the volume, V, in Fig. 4 must migrate out of the volume, or some Sn atoms will have to diffuse out from the volume to a stress-free region.

Since room temperature is a relatively high homologous temperature for Sn, which melts at 232°C, the self-diffusion of Sn along Sn grain boundaries is fast at room temperature. Therefore the compressive stress in the Sn induced by the chemical reaction at room temperature can also be relaxed at room temperature by atomic rearrangement via self grain boundary diffusion. The relaxation occurs by the removal of atomic layers of Sn normal to the stress, and these Sn atoms can diffuse along grain boundaries to the root of a stress-free whisker to feed its growth.

7 Accelerated test of Sn whisker growth

One of the most annoying behaviors of Sn whisker growth is that it does not grow when it is needed but grows when it is not needed. In order to predict the lifetime of Pb-free solder finish without whisker growth, one should conduct accelerated tests as in most reliability problems. An accelerated test can be conducted at larger driving force or faster kinetics, provided that the mechanism of failure remains the same. Typically, tests at higher temperatures are performed to obtain the activation energy of the rate controlling process, which will enable the extrapolation of the life time at the device operation temperature. For Sn whisker growth, while it is possible to conduct the tests up to 60°C, the rate of whisker growth is still quite slow due to slow atomic diffusion. When the temperature approaches 100°C, the diffusion is faster, yet the stress will be relieved due to fast atomic diffusion. Hence, competition between driving force and kinetics determines the rate of whisker growth. Although Cu can be added to Sn to have a faster whisker growth as in eutectic SnCu solder, the rate is still not fast enough. Besides, the effect of Cu on whisker growth needs to be isolated.

In 1954, Fisher et al. developed a method for accelerated growth of Sn whiskers by using a metallographic clamp [30], and they reported the maximum growth rates at clamping pressures of 7500 psi to be about 10,000 Å/s. Compared with the spontaneous growth rate of about few Å/s without external pressure [31], the growth rate for accelerated growth is much faster and the mechanism of growth might be different. Lee et al. reported that Sn whiskers grew on the Sn film plated with Cu concentration above 2% after high acceleration stress test for 300 h. Arnold published a paper in 1956 on investigating the methods to repress the growth of whiskers [32]. He found that neutron bombardment enhanced the growth of Sn whiskers. When the plated Sn film was bombarded at a neutron flux density of 10^{12} cm^{-2} s^{-1} for 30 days, dense whiskers were found in the film. He also found that electrical and magnetic fields had no effect on acceleration effect.

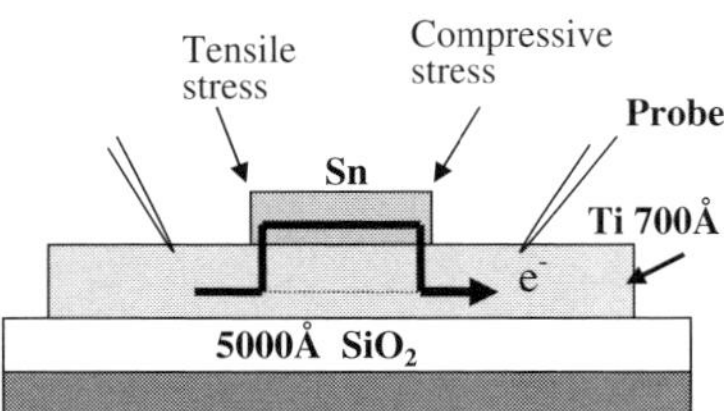

Fig. 10 The cross-sectional schematic drawing showing the structure of the Blech Sn film. Most of the electrons drift in the Sn film, as indicated by the arrow in the figure

Liu et al. demonstrated that the growth of Sn whiskers can be accelerated by electromigration [33]. When electric current is applied to leadframes or continuous Sn films, it produces no back stress. The sample that Liu et al. used was 5000 Å pure Sn film on a Ti conducting layer, which is also known as Blech specimen [34], as shown schematically in Fig. 10. The resistance of the Ti layer is much higher than the Sn film. Thus most of the electrons take a detour into the Sn film on the cathode end (left side) of the Sn stripe, and exit near the anode end (right side) of the Sn stripe. Since the Ti layer has higher electromigration resistance, the Ti atoms almost do not migrate during the current stressing. In electromigration, Sn atoms migrated from the left to the right, resulting in the formation of voids on the cathode end and hillocks on the anode end.

Due to current crowding induced flux divergence at the lower corner of both ends of the stripe, a compressive stress built up on the anode end and tensile stress on the cathode end during current stressing, as indicated in Fig. 10. At the anode, due to current crowding, electrons exit at the lower corner, where the compressive stress is the highest because Sn atoms will be driven there. However, at the upper corner of the anode, the compressive stress is low because of low current density. Hence a vertical compressive stress gradient is created and it will drive atoms to move up.

When the oxide at the upper corner is broken, the free surface is stress-free and it enhances whisker or hillock growth. In these samples, there were no reacting layers under the Sn film, such as Cu or Ni. Thus, the driving force for the whisker growth originated mainly from the current stressing.

Figure 11 shows the plan-view SEM image for a fabricated Sn stripe after the storage at room temperature for two months. No obvious growth of Sn whiskers was observed on the Sn film. However, accelerated growth of Sn whiskers has taken place when it was applied by a high current density, as shown in Fig. 12a–f. The Sn stripe was 5000 Å thick on a Ti layer of 700 Å thick. Current density of 1.5×10^5 A/cm^2 was applied at room temperature. After 10 h of the current stressing, a whisker started to form on the anode end, as seen in Fig. 12b. As the stressing time increased, more whiskers appeared and they grew longer, as illustrated in Fig. 12c–f. The growth rate was measured to be about 3 Å/s, which was higher than that of the spontaneous growth (about 0.1–1.0 Å/s) [35]. Yet, it was much slower than that by using a metallographic clamp. However, this technique of electromigration appears to be more controllable. Sn whiskers grow on the anode end, and the growth rate can be controlled by varying applied current (driving force) and temperature (kinetics). The grow rate increased as the applied current increased, and it also increased to 7.7 Å/s at 50°C under the same current density. In addition, the growth rate was almost linear with the applied current, which is similar to that obtained by Fisher et al. using a metallographic clamp [30].

In using electromigration to conduct accelerated tests of whisker growth and by measuring the growth rate and the diameter of the whisker, the volume change per unit time of the whisker can be obtained, $V = JA\mathrm{d}t\ \Omega$, where J is the electromigration flux in units of number of atoms/cm^2-s, A is the cross-section area of the whisker, t is unit time, and Ω is atomic volume. On this basis

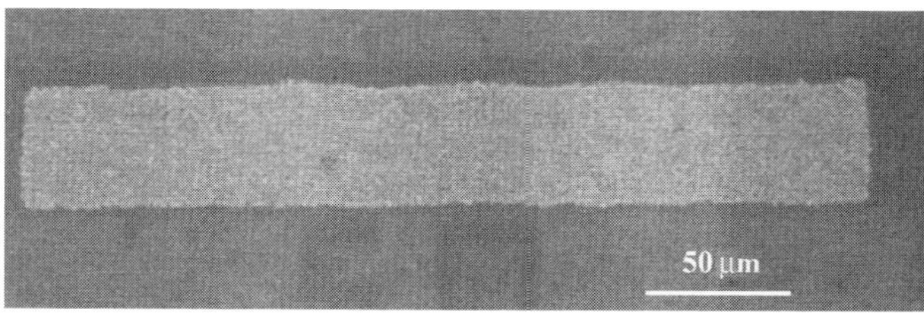

Fig. 11 Plan-view SEM image of a pure-Sn Blech sample which was stored under room temperature without current stressing for over 2 months. No obvious Sn whiskers or hillocks were found

Fig. 12 SEM images showing the growth of the Sn whiskers driven by electric current. The applied current was 40 mA at room temperature, resulting in the current density of 1.5×10^5 A/cm^2 in the Sn film. (**a**) 0 h, (**b**) 10 h, (**c**) 30 h, (**d**) 70 h, (**e**) 140 h, (**f**) 260 h

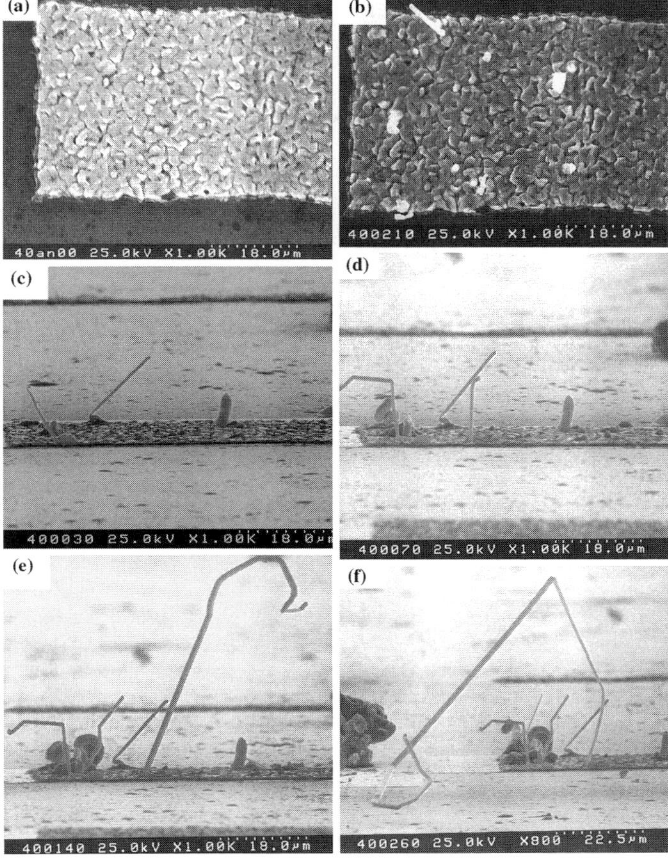

$$J = C \frac{D}{kT} \left(\frac{d\sigma\Omega}{dx} + Z^* ej\rho \right) \tag{4}$$

where $C = 1/\Omega$ in pure Sn, D is diffusivity, kT is thermal energy, σ is stress at the anode, $d\sigma/dx$ is the stress gradient along the short stripe of Sn of length of dx, Z^* is the effective charge number of the diffusing Sn atoms in electromigration, e is electron charge, j is current density, and ρ is resistivity of Sn at the test temperature. The stress at the cathode can be assumed to be zero. σ can be evaluated from the last equation. By keeping the stripe dimension and the applied current density unchanged, the activation energy of whisker growth when the growth as a function of temperature can be obtained.

In-situ stress distribution near the anode end of the stripe can be obtained by micro-beam X-ray diffraction in synchrotron radiation [36, 37]. The sample can be scanned by the micro X-ray beam while a high current density is passing through the sample. Furthermore, using focused ion beam, the cross-section of the sample can be prepared and studied by the micro X-ray beam diffraction.

Figure 13 shows SEM image of whiskers at the anode ends of a set of stripes of eutectic SnPb driven by electromigration. The stripes were cut by focused ion beam, and it was found that the cut has enhanced whisker growth at the anode ends. Sn whiskers grew even at temperature as high as 100°C. Thus accelerated test at higher temperature can be performed using the Blech specimens.

To control the diameter of a whisker growth at the anode, a thin coating of quartz can be sputtered over the entire Sn stripe with an etched hole of given diameter at the anode, as depicted in Fig. 14. Driven by an applied current density, a whisker can be pushed out from the hole at the anode. The growth rate and the volume of the whisker as a function of current

Fig. 13 SEM image showing the growth of the Sn whiskers driven by electric current in SnPb solder stripes. The applied current was 61 mA at 100°C for 65 h

density, temperature, and time in a controlled manner can be measured with this technique. At the same test condition, the Sn stripe can be replaced by an eutectic SnCu stripe or a stripe of a bilayer of Sn/Cu for comparison. However, the accelerated test may not be meaningful without a confirmation that the whisker driven by electromigration has the same growth behavior and mechanism as the whisker grown spontaneously on the Pb-free finish.

8 Prevention of spontaneous Sn whisker growth

On the basis of the analysis presented in the previous Sections, there are three indispensable conditions of spontaneous whisker growth; they are (1) The room temperature grain boundary diffusion of Sn in Sn, (2) The room temperature reaction between Sn and Cu to form Cu_6Sn_5, which provides the compressive stress or the driving force needed for whisker growth, and (3) The stable and protective surface Sn oxide. A break of the surface oxide produces the stress gradient needed for whisker growth. If we remove any one of them, in

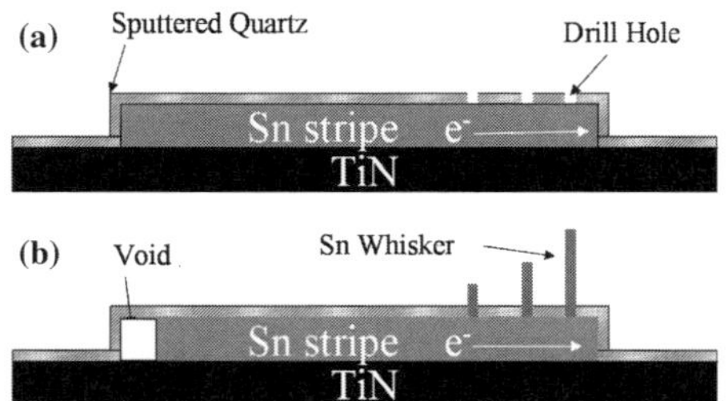

Fig. 14 Schematic drawing showing the controlled growth of Sn whiskers from the drilled holes in the anode end

 Springer

principle there will be no whisker growth. However, synchrotron radiation study has shown that it takes only a very small stress gradient to grow Sn whiskers, hence it is difficult to prevent whisker growth. National Electronics Manufacturing Initiative (NEMI) has recommended a solution to remove the condition (2) by preventing Cu from reacting with Sn. To remove the condition (3), no oxide or a very thick and unbreakable oxide or coating are necessary, such a condition is unrealistic. We propose here to remove the condition (1) by blocking the grain boundary diffusion of Sn. Furthermore, if one can remove both conditions (1) and (2), it is even better.

Since Sn whisker growth is an irreversible process which couples stress generation and stress relaxation, it is essential to uncouple them in order to prevent Sn whisker growth. In other words, both stress generation and stress relaxation should be removed. Stress generation can be removed by blocking the diffusion of Cu into Sn. The NEMI solution is to stop the diffusion of Cu into Sn by electroplating a layer of Ni between the Cu and the Sn solder finish. Ni serves as a diffusion barrier to prevent the diffusion of Cu into Sn. One can also use Cu–Sn intermetallic compound, instead of Ni, as the diffusion barrier. An annealing of the plated leadframe above 60°C will lead to the formation of Cu_3Sn, the Cu_3Sn formed between the Cu leadframe and the Sn finish can serve as diffusion barrier.

However, up to now, no solution to remove stress relaxation is given. In other words, how to prevent the creep process or the diffusion of Sn atoms to the whiskers is unknown. This may be accomplished by using another kind of diffusion barrier to stop the diffusion of Sn. To block the diffusion of Sn atoms from every grain of Sn in the finish is non-trivial. This may be accomplished by adding several percentage of Cu or another element into the matte Sn or the eutectic SnCu solder. We recall that the Cu concentration in the eutectic SnCu is only 1.3 atomic % or 0.7 wt%. We shall add about several (3–7) wt% of Cu. We recall the cross-sectional microstructure of the eutectic SnCu as shown in Fig. 2 that Cu_6Sn_5 forms as grain boundary precipitates. The reason to add a few more percentage of Cu is to have enough precipitation of Cu_6Sn_5 in all the grain boundaries in the finish, so that every grain of Sn in the finish will be coated by a layer of Cu_6Sn_5. Thus, the grain boundary coating becomes a diffusion barrier to prevent the Sn atoms from leaving each of the grains. When there is no diffusion of Sn, there is no growth of Sn whisker since the supply of Sn is cut. Figure 15 depicts a layered structure of a Sn–Cu finish on a Ni diffusion barrier on a Cu leadframe. What is the optimal concentration of Cu in the finish requires

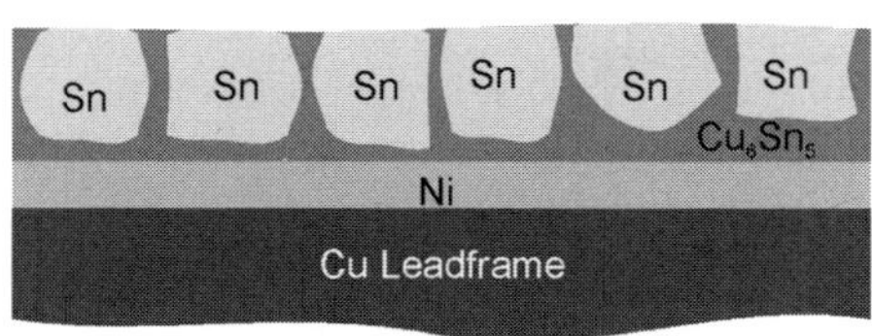

Fig. 15 A layered structure of a Sn–Cu finish on a Ni diffusion barrier on a Cu leadframe

more study. Cross-sectional SEM and focused ion beam images of the samples should be obtained to investigate the microstructure of electroplated Sn–Cu finish, especially the distribution of grain boundary phase of Cu_6Sn_5 in Sn as a function of Cu concentration from 3 to 7 wt%.

There are two key reasons of the selection of Cu (or another element) to form grain boundary precipitate in Sn. The first is that when the Sn has so much supersaturated Cu, it will not take more Cu from the leadframe. The second is that the adding of Cu will not affect strongly the wetting property of the surface of the finish. This is an important consideration since without a good property of wetting, it cannot be used as finish on the leadframe. To plate the Sn finish with a few percent of Cu is much easier than the plating of eutectic SnCu which has only 0.7 wt% Cu. It is difficult to control the concentration less than 1% in electroplating.

Whether the addition of several percent of Cu is effective and whether there are other problems must be studied. For example, whether the grain boundary precipitate is brittle or not should be investigate. Perhaps, it is the idea of how to prevent grain boundary diffusion of Sn should be explored. Whether there are other elements better than Cu for the purpose of whisker prevention also remain to be studied. It is known that the addition of several percent of Pb will prevent Sn whisker growth since Pb is soft and it tends to reduce the local stress gradients in Sn. Also because Sn–Pb is a eutectic system, the eutectic microstructure consists of two separated and intermixing phases, and they block each other in terms of long range diffusion. So, adding several percent of the other soft elements that have eutectic phase diagram with Sn, such as Bi, In, and Zn, may be good choice. If there is no Sn diffusion, we expect no Sn whisker growth.

References

1. C. Herring, J.K. Galt, Phys. Rev. **85**, 1060 (1952)
2. G.W. Sears, Acta Metall **3**, 367 (1955)
3. A.P. Levitt, in *Whisker Technology* (Wiley-Interscience, New York, 1970).
4. U. Lindborg, Metall. Trans. **6A**(8), 1581 (1975)
5. T. Nagai, K. Natori, T. Furusawa, J. Jpn. Inst. Met. **53**, 303 (1989)
6. I.A. Blech, P.M. Petroff, K.L. Tai, V. Kumar, J. Cryst. Growth **32**(2), 161 (1975)
7. N. Furuta, K. Hamamura, Jpn. J. Appl. Phys. **8**(12), 1404 (1969)
8. J.D. Eshelby, Phys. Rev. **91**, 755 (1953)
9. F.C. Frank, Phil. Mag. **44**, 854 (1953)
10. S. Amelinckx, W. Bontinck, W. Dekeyser, F. Seitz, Phil. Mag. **2**, 355 (1957)
11. W.C. Ellis, D.F. Gibbons, R.C. Treuting, in *Growth and Perfection of Crystals*, eds. by R.H. Doremus, B.W. Roberts, D. Turnbull (John Wiley, New York, 1958), pp. 102.
12. R. Kawanaka, K. Fujiwara, S. Nango, T. Hasegawa, Jpn. J. Appl. Phys. Part I **22**(6), 917 (1983)
13. U. Lindborg, Acta Metall. **24**(2), 181 (1976)
14. K.N. Tu, Acta Metall. **21**(4), 347 (1973)
15. W.J. Boettinger, C.E. Johnson, L.A. Bendersky, K.-W. Moon, M.E. Williams, G.R. Stafford, Acta Materialia **53**, 5033 (2005)
16. G.T.T. Sheng, C.F. Hu, W.J. Choi, K.N. Tu, Y.Y. Bong, L. Nguyen, J. Appl. Phys. **92**, 64 (2002)
17. W.J. Choi, G. Galyon, K.N. Tu, T.Y. Lee, in *Handbook of Lead-free Solder Technology for Microelectronic Assemblies*, eds. by K.J. Puttlitz, K.A. Stalter (Marcel Dekker, New York, 2004).
18. Ivan Amato, Fortune magazine **151**(1), 27 (2005)
19. Rob Spiegel, Electronic News 03/17/2005
20. http://www.nemi.org/projects/ese/tin_whisker.html
21. K.N. Tu, R.D. Thompson, Acta Met. **30**, 947 (1982)
22. K.N. Tu, Phys. Rev. **B49**, 2030 (1994)
23. P.G. Shewmon, *Diffusion in Solids* (McGraw-Hill, New York, 1963)
24. D.A. Porter, K.E. Easterling, *Phase Transformations in Metals and Alloys* (Chapman and Hall, London, 1992)
25. K. Zeng, R. Stierman, T.-C. Chiu, D. Edwards, K. Ano, K.N. Tu, J. Appl. Phys. **97**, 024508-1 (2005)
26. B.-Z. Lee, D.N. Lee, **46**(10), 3701 (1998)
27. C.Y. Chang, R.W. Vook, Thin Solid Films **228**, 205 (1993)
28. W.J. Choi, T.Y. Lee, K.N. Tu, N. Tamura, R.S. Celestre, A.A. MacDowell, Y.Y. Bong, L. Nguyen, Acta Mat. **51**, 6253 (2003)
29. K.N. Tu, Phys. Rev. **B49**, 2030 (1994)
30. R.M. Fisher, L.S. Darken, K.G. Carroll, Acta Metallurgica **2**, 368 (1954)
31. V.K. Glazunova, N.T. Kudryavtsev, Translated from *Zhurnal Prikladnoi Khimii*, **36**(3), 543 (March 1963)
32. S.M. Arnold, *Proc. 43rd Annual Convention of the American Electroplater's Soc.*, vol. 43, (1956), pp. 26–31
33. S.H. Liu, Chih Chen, P.C. Liu, T. Chou, J. Appl. Phys. **95**(12), 7742 (2004)
34. I.A. Blech, J. Appl. Phys. **47**, 1203 (1976)
35. George T. Galyon, Annotated Tin Whisker Bibliography
36. A.T. Wu, K.N. Tu, J.R. Lloyd, N. Tamura, B.C. Valek, C.R. Kao, Appl. Phys. Lett. **85**, 2490 (2004)
37. A.T. Wu, A.M. Gusak, K.N. Tu, C.R. Kao, Appl. Phys. Lett. **86**, 241902 (2005)

Sn-whiskers: truths and myths

J. W. Osenbach · J. M. DeLucca · B. D. Potteiger ·
A. Amin · F. A. Baiocchi

Published online: 23 September 2006

Abstract Whisker growth from the solid state has been documented in the public domain for more than 60 years. Unfortunately, a basic understanding upon which a quantitative model, one which would provide a means for predicting whisker growth and risk and ultimate prevention strategies, remains elusive. This paper is an attempt to critically review (hopefully with minimal bias) the state of Sn-whisker growth and mitigation. This is done by: (i) examining the existing experimental data and the limitations of collecting such data, (ii) analyzing the proposed driving forces, mechanism and models for whisker growth, and (iii) carefully evaluating the proposed mitigation strategies and how subsequent assembly processes and device applications may or may not impact these strategies. In each area, the validity of the model and mitigation strategy is examined by comparing it to existing experimental data. Areas where the experimental data is insufficient to adequately test the theory or predict risk are identified. In addition, the areas where experimental difficulties are found that could potentially negatively impact data collection and analysis are reviewed. The authors hope that at a minimum this review will provide a starting point for discussion between the users of the final products, the producers of the final products and the electronic components used in the final products on the topic of Sn-whisker growth, and its mitigation and risk management. In the best case scenario it could provide directions for the advancement of both experimental work and theoretical understanding of Sn-whisker growth and mitigation.

1 Introduction

Overmolded lead frame integrated circuit packages have traditionally used tin–lead (Sn/Pb) alloys as the surface finish of choice. However, the European Restriction of Hazardous Substances (RoHS) legislative requires the elimination of Pb from electronic devices by July 2006 [1]. As such Pb must be removed from lead frame finishes. The most non-disruptive and economical way of doing so is to replace the Sn–Pb finish with a pure Sn finish [2]. Unfortunately, Sn-whiskers have been found on pure Sn-plated Cu lead frames [3–15]. Figure 1 shows an example of one of the types of problems that can occur with pure Sn finishes. The device is a 176 lead thin quad flat pack with 400 μm pitch leads. As shown in the inset, there is a high density of whiskers that are 10–25 μm long and 4–8 μm in diameter. There is no observable corrosion anywhere along either lead. In addition there are three "needle-like" whiskers, two growing from the right hand lead and one growing from the left hand lead, that have grown across the gap toward the adjacent lead. The three needle-like whiskers have diameters between 0.5 and 1.0 μm and lengths between 100 to 150 μm. In this case, two whiskers from adjacent leads are in contact with each other creating an electrical short. The small diameter and long length make this type of whisker susceptible to fracture. If fracture occurs,

J. W. Osenbach (✉) · J. M. DeLucca · B. D. Potteiger ·
A. Amin · F. A. Baiocchi
Agere Systems, 555 Union Blvd, Allentown, PA 18109-3286,
USA
e-mail: osenbach@agere.com

Fig. 1 Example of excessive whisker growth on Sn on 400 μm pitch electronic component. The inset is a higher magnification micrograph of the section of the lead identified with the box that shows a large density of 4–8 μm diameter, 10–25 μm long whiskers

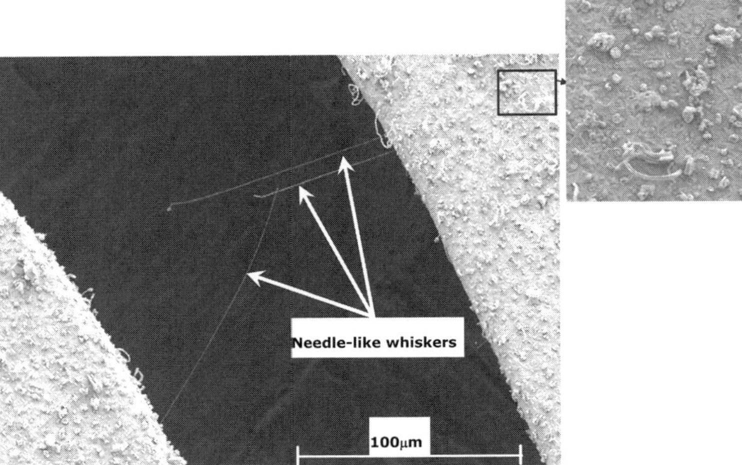

then the whisker could interfere with the operation of optical and mechanical devices. Clearly these types of whiskers can lead to devices failure in a variety of ways including but not limited to those discussed above.

The potential for whisker growth on pure Sn-finishes establishes the importance of developing and implementing a whisker inspection/acceptance procedure for pure Sn plated devices. The inspection/acceptance procedures have been developed and are defined in two JEDEC standards, JEDEC22-A121 and JEDEC-201 [16, 17]. In addition many system users have placed additional inspection/acceptance criteria on component suppliers. These inspection/acceptance procedures require intensive-time consuming optical microscopy inspection of many leads on many devices. The optical inspection is often supplemented with scanning electron microscopy (SEM) inspection of some of the leads on some of the devices. Thus although the manufacturing infrastructure (capital equipment, plating process, and process flow), for both the component and board assembly is essentially unchanged with the transition from Pb/Sn to Sn, additional costs are incurred with this transition because of the extensive inspection/acceptance required to insure the plated films are in fact "whisker-resistant". Sn films that do not show signs of whisker growth during the acceptance testing are termed as "whisker-resistant" not "whisker-free" because, to paraphrase Brusse and Leidecker [18], "the absence of evidence of whiskers is not necessarily evidence of the absence of whiskers". In fact, as shown in latter sections of this paper, the statistical variations in the incubation time and the density of whiskers measured on devices from different Sn-plating lots made with ostensibly the same "whisker-resistant"-Sn bath and plating process grown on the same substrate metal alloy, are sufficient to produce one lot of material that passes the criteria and a subsequent lot that fails the criteria.

Whisker formation and growth is not unique to the Sn–Cu system. Whiskers have been observed in other metal systems including but not limited to Ag, Au, Zn, and Cd. Both metallic and non-metallic whiskers can and have been grown via vapor phase nucleation and growth on metal and metal-oxide surfaces. Although there may be similarities in nucleation and growth mechanisms that control solid state whisker growth and vapor phase whisker growth, as well as solid state whisker growth in different metal systems, only solid state whisker growth in the Sn–Cu and Sn–Ni systems is considered in this paper. In particular, consideration is given only to Sn-whiskers found on devices that are operated under the typical environmental conditions for which active electronic components are subjected to. These environments typically do not include the application of excessive external mechanical forces or abrasion and wear. Furthermore, high vapor pressures of Sn and or Cu are not usually present under typical operating conditions.

2 General overview

A brief review of the literature to provide the basis for the balance of the paper is given in this section. A detailed review of the literature published on this topic from the first publication by Cobb in 1946 on Cd whiskers though that published in 2004 has recently been published by Galyon [19]. The reader is referred to this review and the articles referenced in the review for more details on both the experimental and theoretical work done on this topic. Although Sn-whiskers are typically single crystal metallic Sn-filaments, polycrystalline whiskers have also been reported. There has been a report of a single whisker with both single crystal regions and polycrystalline regions [20]. Typically, whiskers are found to spontaneously grow out of electroplated Sn films on various Cu based substrates. Here spontaneous is used to describe the phenomena were at time, t, there are no indications of whisker nuclei or whiskers, but at some later time, $(t + \Delta t)$, whiskers are observed. The incubation time, $\{t + (\Delta t - \partial t)\}$, could be a few tens of hours or tens of thousands of hours. Similarly, the length and density of whiskers could vary by orders of magnitude from one device to another. Although there does not yet exist a quantitative model that defines the mechanistics of whisker nucleation and growth, it is known that the incubation time depends upon bath chemistry, plating process, base alloy, post-plate heat treatments, and ambient conditions the device is exposed to.

Although Sn-whiskers have been known for more than six decades, are technologically important and scientifically intriguing, there is relatively limited literature in the public domain on Sn-whisker formation and growth at least as compared to literature on other technologically important topics pertaining to Pb-free solder alloys and assembly of Pb-free electronics. For example, the 2005 and 2006 IEEE Electronic Components Technology Conference, ECTC, had more than 200 papers each year dealing with Pb-free solders in electronic packaging. However, including the 1 day Sn-whisker workshops held each year, there were less than 25 combined Sn-whiskers papers given at these two conferences. Similarly, the Pb-free sessions and workshop of the 2006 Annual meeting of The Materials, Minerals and Metals Society (TMS) had more than 200 papers dealing with different aspects of Pb-free solders in electronics, but only four papers on Sn-whiskers. In fact, during the past 10 years thousands of papers dealing with Pb-free solders in electronics have been published, and at least as many presentations in conferences. In contrast, during the entire period of about 60 years there have been a few hundred publications and talks on whisker growth in the public domain. Furthermore, there are a significant number of conflicting and sometimes contradictory results reported in these whisker papers and talks, causing various researchers to arrive at diametrically opposed conclusions.

In spite of the limited data set, the apparent conflicting data, and lack of agreement on the detailed mechanisms that are at the root of Sn-whisker formation and growth, it is generally agreed that whisker formation and growth is influenced by a number of variables including: (1) the intrinsic stress and stress distribution in the as-plated Sn-film; (2) the extrinsic stress arising from chemical reactions between the Sn-film and the substrate alloy; (3) the purity of the Sn plate; (4) the grain size and grain crystallographic orientation of the Sn plate; (5) the lead frame surface finish; (6) the pre-plate chemical treatment of the lead; (7) the process used to form the lead, (8) the bend radii of the lead; (9) the Sn-plating thickness; (10) the post-plate thermal processing; and (11) the storage and operating ambient. Unfortunately, a quantitative relationship between these variables and whisker growth does not yet exist.

One of the reasons for the relatively small number of publications and the lack of a quantitative understanding of whisker growth is undoubtedly related to the fact that a simple solution to alleviate this problem, namely the addition of Pb, was available and widely used for more than five decades. With the implementation of the RoHS legislation in July 2006, this solution is however, no longer available. The RoHS legislation is most probably the reason for the increased number of publications and work by government laboratories, industry, academia, and standards groups on Sn-whiskers over the past decade. Based on the authors own experience, another prime reason for the lack of published work is the experimental difficulties in carrying out the time-consuming long-range experimental studies. The difficulties are a result of both technical and practical limitations. Excluding the difficulties associated with the identification of whiskers by optical and scanning electron microscopy inspection, four categories of limitations which the authors have directly encountered in their work are discussed below: (1) incubation time; (2) statistical variations; (3) limited acceleration rates; and (4) water condensation induced corrosion.

2.1 Incubation time

Figure 2 is a plot of the maximum whisker length versus time at 60°C/93%RH for a number of different Sn films on Cu-alloy CDA 7025. As shown the non-whisker resistant film (6 μm-NA) begins to whisker almost immediately after it is put on test, however, the "whisker-resistant" films do not begin to whisker until they are subjected to 60°C/93%RH for 10–25 weeks (3–6 months). Other films are even more whisker

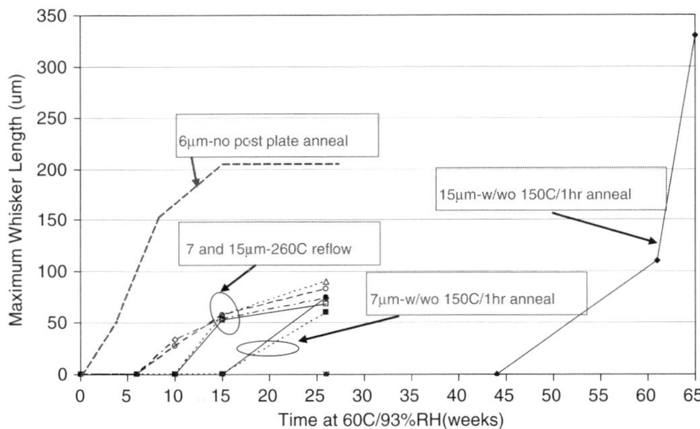

Fig. 2 Incubation time for whisker growth. The film thickness is represented by the number of microns. The post-plate anneal appears to improve whisker resistance on some films but not others. The 15 μm thick films exposed to the 150°C/1 h anneal became more whisker prone after reflow

resistant (15 μm A) and do not begin to whiskers for as many as 45–60 weeks (approximately 1 year). The whisker novice might interpret this result as an indication that incubation time is a good measure of whisker resistance of a particular film. Experience of the authors of this paper indicates that this is true for "whisker-prone" films, i.e. films that readily grow many whiskers quickly such as the 6 μm-NA film shown in the figure. However, for the "whisker-resistance" films this is not necessarily true. For example, the non-reflowed film that did not whisker for close to a year, when reflowed in component form began to whisker in 10 weeks. In this case, reflow at the component level reduced the whisker resistance of the Sn-films [21]. Furthermore, the 10–25 weeks variation in incubation time observed for the whisker resistant films appears to be related to an incubation time distribution effect and may not be representative of a difference in the whisker immunity, as illustrated in Fig. 3. This figure is a plot of the whisker incubation time at 60°C/93%RH for device taken from five different Sn-plating lots. All of the films were plated in the same in-line plating system with the same commercially available known whisker resistant MSA Sn-plating bath. These five lots were plated at roughly 4 month intervals over a 1.5 year period of time. All these lots were plated on CDA-7025 Cu-alloy. The devices were either 132-lead or 176-lead quad flat pack. After Sn-plating, the devices were subjected to a 1 h, 150°C anneal in dry air to stabilize the Sn film and Sn–Cu interface. The devices were then loaded onto JEDEC trays, baked in dry air at 125°C for 8 h after which the trays and devices were sealed in dry bags. After 1–3 months of storage at room temperature, the devices were removed from the dry bags. Some of the devices were subjected to a simulated board attach

Pb-free reflow process at a peak temperature of 260°C in a dry nitrogen ambient. Others were not reflowed. Trays of both reflowed and non-reflowed devices were then placed into an environmental chamber pre-set at 60°C/93%RH. Periodically the devices were removed from the chamber and inspected with an optical microscope for whiskers. The optical inspection was augmented with scanning electron microscope inspections. The details of the experimental procedure used are given in reference [21]. As shown, the incubation time for the reflowed devices ranged from 10 to 25 weeks, whereas for the non-reflowed devices it ranged from 33 to 62 weeks. Note that there is no correlation between the incubation time after reflow and that observed on non-reflowed devices. Within the measurement capability of the monitoring techniques used to characterize the bath chemistry, plating process, bake and bagging process, reflow process,

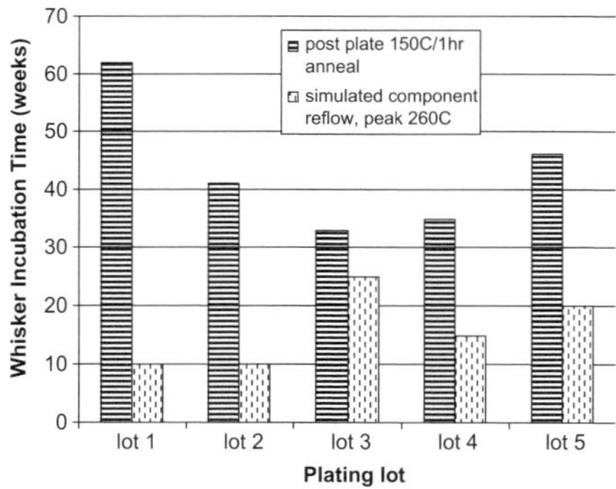

Fig. 3 Incubation time for whisker growth at 60°C/93%RH as a function of plating lot. All of the Sn-plating was done in the same equipment set using the same process and process chemistry

stress test, and inspection procedures used, there were no obvious differences between the conditions employed for these lots. Thus the stress test results indicate that the whisker-incubation time can vary by a factor of approximately 2–3 for films that are apparently the same when they are exposed to 60°C/93%RH for extended periods of time. In other words comparison of an incubation time taken from a single lot of "whisker-resistant" Sn-films to another lot of "whisker-resistant" Sn-films made by the same or different processes, chemistries or suppliers is not necessarily a measure of the level of whisker immunity of the particular Sn-film. In fact, this data indicates that a variation in incubation time for whisker growth in 60°C/93%RH for two different Sn-films of less than approximately 2–3 × means the different Sn-films may have identical whiskers resistance.

2.2 Statistical variations

Figure 4 is a plot of the number of devices per total number of packages in the cell that had whiskers versus variables encountered in plating namely, vendor, plating-bath chemistry, plating-bath supplier with plating lot, and exposure time as the parameters. Included in this plot are data taken from devices that were initially coated with a 15 and 7 μm thick film as denoted by the numbers in the individual bars. Note that all of these devices were subjected to a simulated board attach

reflow at 260°C peak temperature prior to subjecting them to the accelerated aging test. Similarly, Fig. 5 is a plot of the total number of leads with whiskers per total number inspected versus plating-vendor, plating-bath chemistry, and plating-bath supplier with plating lot, and exposure time as the parameters. It should be noted that 25–30% of the leads that had whiskers at the 15 week inspection point did not have whiskers at the 26 week inspection point. It is likely that the missing whiskers were fractured when the devices were handled.

The data in Figs. 4, 5 show significant variations in the whisker response for devices plated in different plating lots. The largest variation in whisker response occurred between different lots of devices plated by one supplier using the same plating bath and obsessively the same process. The implication of this result is that the variation in whisker response of reflowed parts may not fundamentally be controlled by the plating chemistry, plating bath supplier, lead frame plating supplier, or lead frame Cu-alloy but by some other yet to be determined variable or variables. However, independent of the mechanism/s responsible for the statistical variations in whisker response, whiskers were observed on at least one lead on at least one device in the each and every plating lot of devices subjected to a simulated solder reflow with a peak temperature of 260°C, and extended exposure to 60°C/93%RH. The results indicate that the number of

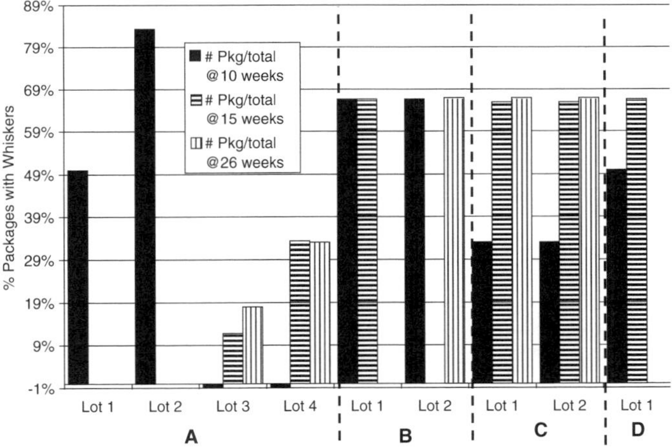

Fig. 4 Percent of the packages per aging cell with whiskers after exposure to 60°C/93%RH with aging time as the parameter. The nominal Sn-film on these packages was 15 μm thick. Following plating the devices were subjected to a 150°C/1 h anneal. They were sealed and then stored in dry bags for approximately 3 months and subsequently subjected to a simulated board attach reflow at a peak temperature of either 260 or 245°C. (**A**) Lead frame plating supplier A using bath chemistry 1 from bath supplier 1; (**B**) lead frame plating supplier B using bath chemistry 1 from bath supplier 1; (**C**) lead frame plating supplier B using bath chemistry 2 from bath supplier 2; (**D**) lead frame plating supplier C using bath chemistry 1 from bath supplier 3

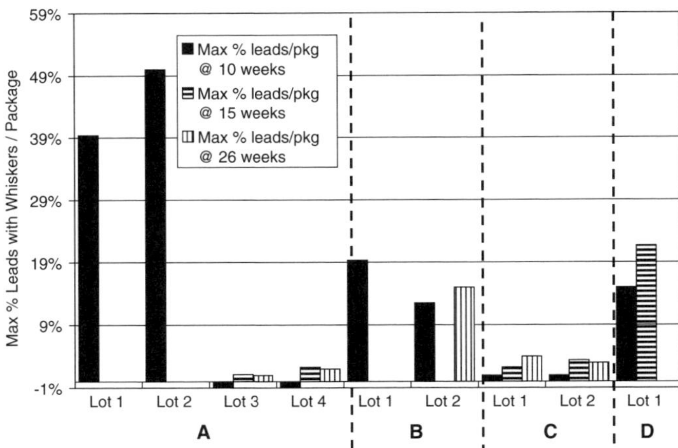

Fig. 5 Percent of the leads per aging cell with whiskers after exposure to 60°C/93%RH with aging time as the parameter. The nominal Sn-film on these packages was 15 μm thick. Following plating the devices were subjected to a 150°C/1 h anneal. They were sealed and then stored in dry bags for approximately 3 months and subsequently subjected to a simulated board attach reflow at a peak temperature of either 260 or 245°C. (**A**) Lead frame plating supplier A using bath chemistry 1 from bath supplier 1; (**B**) lead frame plating supplier B using bath chemistry 1 from bath supplier 1; (**C**) lead frame plating supplier B using bath chemistry 2 from bath supplier 2; (**D**) lead frame plating supplier C using bath chemistry 1 from bath supplier 3

leads and possibly number of devices that must be inspected to insure that a "no-whiskers" found signature is truly statistically meaningful should be sufficiently large as to take care of the type of statistical variations shown in Figs. 4, 5. This is especially true as processes and chemistries are developed that lead to improvements in the fundamental "whisker-resistance" of the Sn film.

2.3 Difficulty in accelerating whisker growth in non-condensing conditions

Figure 6 is a plot of the maximum whisker length versus accelerated aging condition for a 6 μm-thick Sn-films. With the exception of one sample (TF/NA) at 60°C/93%RH the whisker growth behavior appears to be similar for all devices in all tests even after aging for very long periods of time under stress. For all of the acceleration tests used, except the 60°C/93%RH condition, the whisker lengths are of the order of 40–60 μm. For thicker Sn films, the whisker lengths in these tests tend to be less than 40 μm [21], which at present is not considered problematic for most field use conditions [17].

It is interesting and somewhat troubling that aging at the highest acceleration condition, 85°C/85%RH for 0.5 years did not result in whiskers whereas whiskers were found after 0.5 years at 90°C/93%RH or 1 year at room ambient. This result clearly shows the difficulty in accelerating whisker growth. It also shows that in a highly humid ambient at 85°C whisker growth is suppressed relative to that found at 60°C. This is consistent with the literature which indicates 52°C may be the optimal whisker nucleation and growth temperature [22]. The most probable reason for the reduction in growth at higher temperatures is a change in mechanism by which excessive system energy is reduced. In particular, it appears that at low temperatures, energy reduction is achieved by mechanisms that lead to whisker growth, whereas at higher temperatures, energy reduction is achieved by mechanisms that do not lead to whisker growth but by those that lead to more conventional grain growth and recrystallization.

This leads to a fundamental problem in designing accelerated aging qualification/reliability tests for whiskers. As shown, the maximum temperature where whisker growth was found was 60°C. But 60°C is not much different than the temperature that many devices routinely experience during normal operation in the field. Therefore, when a device is stressed at 60°C/93%RH, the relative humidity is the only accelerant. Unfortunately there are a number of issues associated with high humidity testing. These limitations will be discussed in the next section.

2.4 Water induced corrosion

To the authors' knowledge, corrosion-assisted whisker growth was first reported (at least implied) by Arnold in 1966 [23], followed by Britton in 1974 [5]. Recently

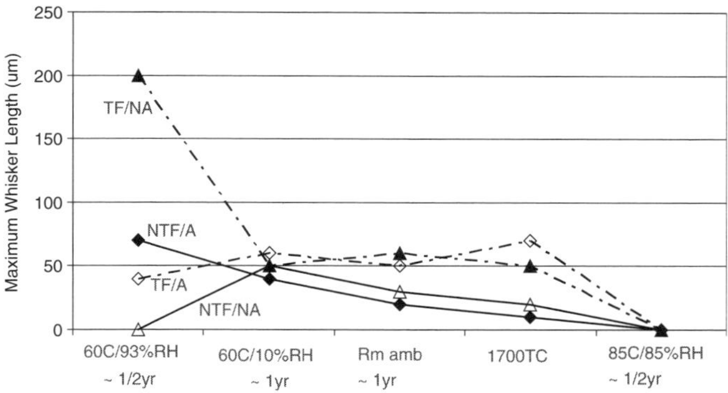

Fig. 6 The maximum whisker length as a function of whisker stress test. In all cases the plated-Sn film was 6 μm thick: TF/NA means trim and formed leads, non-post-plate annealed film; NTF/NA means non-trim and formed leads; TF/A means trim and formed leads, post-plate 150°C/1 h anneal; NTF/A means non-trim and formed leads, post-plate 150°C/1 h anneal

Barsom et al. have proposed a Sn-oxidation model to explain this effect [24]. The authors have reported an increase in whisker propensity when Sn-films were exposed to high humidity in the absence of condensed water [21]. Furthermore, the authors have reported excessive whisker growth when samples with and without Ni-underplate were exposed to a 60°C/93%RH and intentional or unintentional water condensation [20]. Figure 7 shows an example of these two effects. Sample 7A was exposed to 60°C/93%RH for 6,550 h. In this case, care was taken to avoid water condensation during the test. Localized corrosion is clearly observed as the dark spots (dark spot corrosion) on the Sn finish on all the leads of this device, however, no whiskers were observed on any of the leads at this test point. Note that whiskers were observed when the device was subjected to an additional 6 weeks, total of 7,550 h, of aging at 60°C/93%RH. In general, for 12–15 μm thick Sn-films dark spot corrosion has been found to occur after 5,000–9,000 h of exposure to 60°C/93%RH when water condensation is avoided. Non-condensing black spot corrosion can occur in 2,000–4,000 h at 60°C/93%RH on thinner Sn-films. The device shown in Fig. 7B was exposed to a 60°C/93%RH ambient for 2,500 h after intentional water contamination. In this case both dark spot corrosion and whiskers in and around the dark spot corrosion can be observed. This result clearly indicates that water condensation enhances localized corrosion and whisker growth. This is consistent with recent work by Oberndorff et al. [25], and Su and Chopin [26], who reported whisker growth on corroded regions of samples when exposed to 60°C/93%RH for less than 2,000 h. Because water condensation leads to rapid whisker growth, it is important to prevent water condensation from occurring in the field as well as in the accelerated whisker testing.

Figure 8 is a plot of the local relative humidity as a function of temperature change at the device surface relative to the ambient temperature the device is exposed to with ambient relative humidity as the parameter. In this case the ambient temperature is fixed at 60°C. The fit is made assuming constant pressure [27]. The plot is read as follows: the ambient humidity is that which occurs at $(\Delta T) = 0$. For those situations were the device is at a temperature below ambient temperature, (ΔT) is negative. For those situations where the temperature of the device is greater than ambient temperature, (ΔT) is positive. In those situations where the (ΔT) is sufficient to drive the local relative humidity to 100%, condensation will probably occur somewhere on the device. This could be the case for example when a portable electronic device (cell phone, lab top computer, etc.) is taken from a cold outdoor setting to a warm indoor setting. Thermal mass, gaseous diffusion, and thermal conductivity effects, could also create a condensation situation when a portable electronic device is taken from a warm humid setting such as an office building to a cold outdoor setting. The time for the device to reach the new ambient condition is dependent upon many factors including the thermal mass, air flow and geometry. Here the most important geometry effects are those related to thermal conductivity and gas exchange rates.

For accelerated testing the issue of condensation is related to chamber control and device handling procedures. Note that when the chamber ambient is 60°C/93%RH a change in temperature somewhere in the chamber of only −1.3°C causes the local relative humidity in that region of the chamber to reach 100%. Therefore, the temperature in every location in the chamber must be within −1.3°C of the 60°C/93%RH set point to insure water condensation does not occur. For an ambient setting of 60°C/80%RH a temperature

Fig. 7 Corroded regions on matte Sn on Cu–lead frame found after 60°C/93%RH storage: (**A**) alloy-EFTEC-64T after exposure to 60°C/93%RH for 6,550 h. In this case, care was taken to avoid water condensation during the test. Localized corrosion is clearly observed as the dark spots (dark spot corrosion) on the Sn finish on all the leads of this device, however, no whiskers were observed on any of the leads at this test point; (**B**) alloy-7025 after exposure intentional water contamination and a 60°C/93%RH ambient for 2,500 h. In this case both dark spot corrosion and whiskers in and around the dark spot corrosion can be observed

change of less than approximately –4°C insures water condensation does not occur. In many humidity chambers the temperature control is of order ±2–2.5°C, making it impossible to prevent condensation everywhere in the chamber when the relative humidity is approximately 85%. Fortunately, the largest variation in the chamber temperature usually occurs near

the walls and door of the chamber. As such, condensation can be prevented from reaching the device surface by placing a moisture–permeable–condensed-water–physical barrier between the devices and the chamber walls. Careful experimental studies indicate that a barrier is needed to prevent condensation from occurring on devices placed in a humidity chamber set at a relative humidity greater than 85%RH. In addition, condensation occurs when devices that are at room temperature are inserted into a humidity chamber pre-set at a higher temperature and humidity significantly higher than the room ambient conditions. In this case condensation occurs on the devices themselves as well as along the walls and exposed surfaces when the chamber is opened. These two effects can be avoided by loading the device into the humidity chamber that is set at a temperature between 22 and 30°C and relative humidity less than 40–45%RH. After the devices are inserted, the temperature can be ramped to 60°C followed by a slow ramp of the humidity to the level of interest. In this case the humidity ramp has to be controlled to prevent overshooting. The procedure is then reversed when the devices are removed from the chamber for inspection. Implementation of such a procedure requires 6–8 h to reach the elevated temperature and humidity conditions and 6–8 h to ramp down to < 30°C/ < 50%RH. Although this prevents condensation, there are often times and equipment constraints that do not allow for the implementation of such a process. Details of an alternative procedure for minimizing the probability of condensation on the devices is given in the following section. Devices can be loaded into JEDEC trays and covered with a moisture–permeable–condensed-water–physical barrier shroud. The shrouded JEDEC tray can be placed in contact with a plate that has a large thermal mass such as a large piece of brass or steel (the mass of the plate should be at least 5 kg). Pre-heat the

Fig. 8 Local relative humidity as a function of change in device temperature with humidity as the parameter. In this case the system temperature is set to 60°C

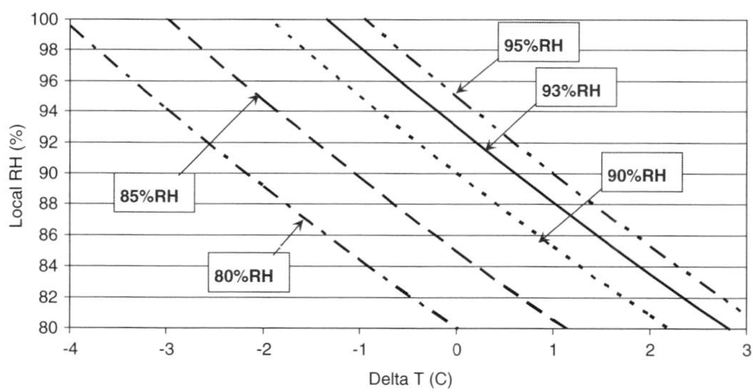

entire assembly at 60°C/ < 10%RH and transfer it to the humidity chamber that is preset at 60°C and the humidity of interest. If the transfer takes less than 30 s, condensation is essentially prevented. Similarly to avoid condensation when the devices are removed from the chamber, the removed devices should be placed into a 60°C/,10%RH chamber for at least 1 h within 30 s of removal from the 60°C/93%RH chamber.

Before concluding this section, it is appropriate to discuss contamination enhanced corrosion and condensation. It is well recognized that atmospheric pollutants can either be adsorbed, in the case of gaseous pollutant, or deposited and ionized, in the case of particle pollutants, onto metal surface. Typical airborne gaseous pollutants include: SO_2; NO_2; H_2S; NH_3; and Cl_2. Typical airborne particle pollutants contain: sodium-, potassium-, calcium-, magnesium-, and ammonium-based sulfates; nitrates; and chlorides. In the presence of moisture these ionic contaminants increase the metal corrosion rate [28–30]. Although, outdoor environments contain significantly higher concentrations of pollutants than indoor environments, the contamination levels in indoor environments are often high enough to cause increases in the corrosion rate of metals. For some metals, such as Ag, there is essentially no difference in the corrosion rate measured in an outdoor environment as compared to that measured in an indoor environment. Rice et al. [28] have defined the corrosion rate of metals in the presence of pollutants as the integrated effects of water vapor, temperature, and pollutant concentration,

$$\text{Corrosion rate} = Af(\text{RH})f(T)f(C)\partial\text{RH}\partial T\partial C,$$

where, $f(\text{RH})$ is the relative humidity dependence of the corrosion rate, $f(T)$ is the temperature dependence of the corrosion rate, and $f(C)$ is the pollution, contamination dependence of the corrosion rate. The functions $f(\text{RH})$, $f(T)$, and $f(C)$ have been experimentally determined for a number of pollutants for a number of metal systems including Cu, Ni, Ag, and Fe, unfortunately they are not available at the present time for Sn.

In addition to enhancing the corrosion rate, ionic substances that dissolve in water have a critical relative humidity (CRH) that is below 100%. The CRH can be defined as the equilibrium humidity above a saturated water solution of the substance. Thus when an ionic substance is exposed to an ambient humidity that is equal to or greater than the CRH for that substance, water will be absorbed until a saturated solution is formed, i.e. a water droplet containing ionic species is formed. For example, in high SO_2 environments, nickel sulfate is formed which has been reported to have a CRH of approximately 68%RH at room temperature [28]. Above this CRH there is a marked increase in its corrosion rate, presumably due to water condensation. In contrast, when copper is exposed to high SO_2 environments there is no evidence of a CRH. The nickel and copper examples were used here because both are ubiquitous in the electronic industry as is the presence SO_2 in our environment (a byproduct of fossil fuels) to illustrate the complex nature of pollution-induced corrosion of metals. Based on an early study, the authors of this paper have reported accelerated growth of clusters of whiskers called flowers, shown in Fig. 9. In the locations where flowers were observed, a high concentration of sulfur was also noted along with evidence of water condensation. It is possible that condensation from the chamber walls fell onto the exposed Sn-plated leads and that the chamber condensation lead to corrosion and accelerated whisker growth. Other possible root causes include contamination of the lead by sulfur bearing particles with a CRH below 93%RH or the development of a Sn–sulfate complex on the Sn surface followed by water condensation. Although sulfur is a common air born pollutant, sulfur bearing moieties are contained in most commercially available matte–Sn plating baths. Therefore, it is possible that the plating process and chemistry could play a role in whisker formation in high humidity via a segregation or ion exchange/corrosion mechanisms.

As discussed, Sn-corrosion in general, and condensation induced corrosion in particular, significantly

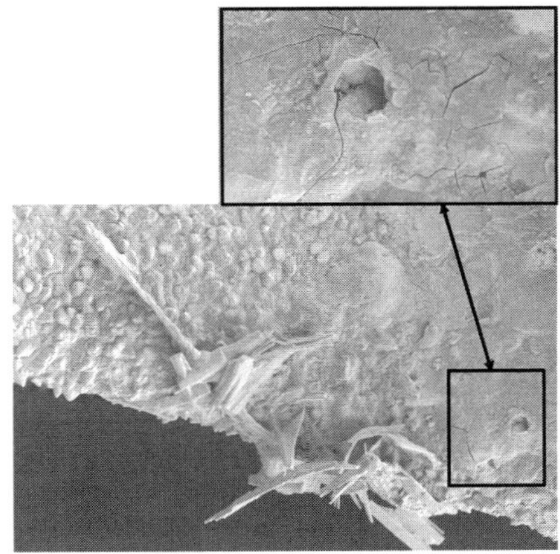

Fig. 9 Flower clusters of Sn whiskers on two different leads of a device after 672 h of exposure to 60°C/93%RH. In this case there was evidence of water condensation sometime during the test. Note that EDAX detected S in and around the cluster

enhances Sn-whisker formation and growth. Therefore, it is important to consider possible atmospheric pollution and plating contamination effects when Sn-whisker risk assessments are made. Unfortunately, to the authors' knowledge, there is no literature in the public domain on this topic.

3 Driving force and mechanisms

In spite of conflicting data and lack of agreement on the detailed mechanisms that are at the root of Sn-whisker formation and growth, it is generally agreed that whiskers are a form of abnormal recrystallization or crystal growth, not deformation and flow. It is interesting that there have been and continues to be disagreement about the role or lack there of dislocations in whisker growth. Perhaps the best example of this is in the work of Ellis [31, 32]; his 1958 paper argued against dislocation mechanisms, whereas his 1966 a paper argued for dislocation mechanisms. The early work on whiskers that ruled out dislocation mechanisms was based on examination of the whisker growth directions and angular bends in whiskers showing that not all directions were low-indice glide planes. More recent work clearly shows many, if not all, whiskers do not grow by the addition of material to the top of the whiskers, but rather by addition of material to the base of the whisker, Fig. 10. This figure shows a whisker with a segment on the top that has a microstructure that is clearly different from the rest of the whisker. The top has a larger diameter than the bulk of whisker as evidenced by the over-hang. The top has a distinctly different microstructure than the rest of

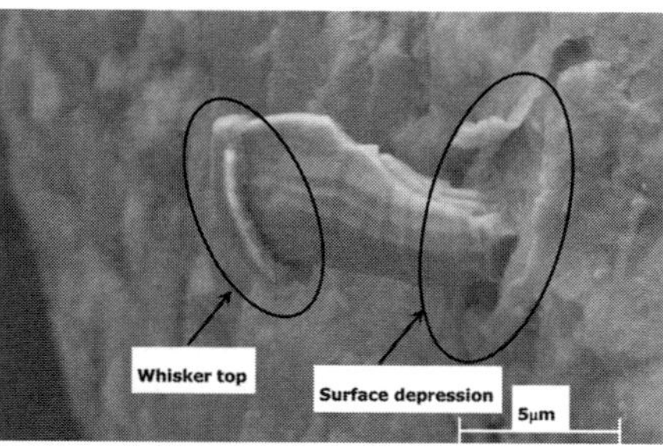

Fig. 10 Whisker found after 1,000 temperature cycles. The Sn-film was 15 µm thick. A 150°C/1 h post-plate anneal was applied to the device within 12 h of Sn-plating. Note that the whisker top appears to match almost perfectly the surface depression surrounding the whisker. Furthermore, there appears to be a tear in the surface layer of the Sn film near the edge of the depression

the whisker. The shape of the top appears to be a good match to the shape of the surface depression in the Sn film that is surrounding the whisker were it exits the Sn-film surface. The surface of the Sn film appears to be mechanically altered as evidenced by the apparent surface tear. This type of structure is consistent with whisker growth from its base towards the surface of the Sn film. Such findings have been used to rule out most dislocation mechanisms. This interpretation of growth by dislocation mechanisms might be flawed as discussed below. There are three theories for the driving force of whiskers: (i) surface energy effects [33, 34]; (ii) stored strain energy [31, 35, 36]; and (iii) internal mechanical stress [9, 12]. Independently, Eshelby [33] and Frank [34] suggested that negative surface energy was responsible for whisker growth. The localized negative energy drives the formation of Frank–Read emission of dislocation loops. The loops glide to the surface where they create a whisker. In this model material is added to the base of the whisker not the tip of the whisker. Eshelby developed an expression for the whisker growth rate that is proportional to κD, where κ depends upon the stress field and D is the Sn-self diffusion coefficient. Using Eshelby's estimate for the constants in this whisker growth rate expression a maximum whisker growth rate of less than 0.001 mm/year can be calculated. This rate is significantly less than 1 mm/year that has been measured [37]. Thus the experimental findings do not support this model assuming Eshelby's estimates for the parameters used in the negative surface energy model are correct. Furuta and Hamamura [35] advanced the original model proposed by Ellis [31] and developed an expressions for the growth rate of whiskers based on a balance of the excess strain energy per unit volume of Sn with the excess surface energy per surface area of Sn-whisker. Using Furuta and Hamamura's estimates for Sn diffusion and excess strain energy, the model predicts growth rates as high as 1 mm/year which is consistent with the experimental data. Boguslavsky and Bush [36] suggested that excess stain energy could lead to whisker growth via a dislocation mechanism originally proposed by Lindborg [38]. Lindborg proposed a two stage model; in the first stage a Bardeen–Herring source generates dislocation loops, and in the second stage the dislocation loops glide towards the surface creating whisker. Lindborg specifically states that growth occurs by the pushing the whisker up one atomic step at a time in the direction of the Burgers vector, i.e. addition of material to the base not the tip of the whisker. In this model, the whisker growth rate is bounded by the generation rate of dislocation loops and rate that these loops glide to the surface. Lindborg derived a whisker growth rate expression of the form, rate P σ^n, for this model where $n = 1$ if dislocation loop generation is rate limiting effect and $n > 10$ if dislocation glide is rate limiting. Clearly this model is consistent both with incubation time effects as well as discrete periods of time where rapid whisker growth occurs.

The first direct evidence for stress induced whisker growth can be found in the work of Fisher, Darken and Carroll published more than five decades ago [37]. The whiskers grown in this study as a result of the mechanical stress had many of the same characteristics that whiskers grown on a "non-externally mechanically" stressed Sn film have; single crystal growths not extrusions, a variable stress dependent incubation time, a variable stress dependent growth rate, and the possibility of a threshold stress below which the whisker growth rate becomes zero. Since this landmark publication there have been a number of other publications that clearly show that stress does drive whisker growth [39–42]. Although stress is the physical quantity that is often related to whisker growth, thermodynamically it is the excess energy that is fundamental to whisker growth. For the majority of the cases where Sn-whiskers occur, the excess "stress" energy originates from the application of strain not stress to the system. The strain is resultant from things such as coefficient of thermal expansion mismatches, reactions between the base alloy and the Sn-film, and Sn-oxidation. Other sources of excess energy are (i) grain boundaries, (ii) dislocations, (iii) plating contamination such as hydrogen, carbon, oxygen, etc., and (iv) porosity. Although the driving force for whisker formation and growth is excess energy, excess energy itself is not the mechanism of whisker growth, i.e. stress is necessary but not sufficient for whisker growth to be realized. At least two additional constraints must be placed on the system for solid state whisker growth to be realized: (i) mechanisms that prevent the more commonly observed energy relaxation mechanisms of grain growth and recrystallization must be present, i.e. grain boundary pinning mechanisms [37]; (ii) mechanisms that lead to both whisker nucleation and growth must also be present in the system.

The grain boundary pinning mechanism was first proposed by Ellis [31]. In Ellis's words "Substantial grain boundary immobility is required to affect a decrease in free energy through the growth of a whisker, rather than by migration of a grain boundary". He proposed two possible pinning mechanisms: (1) impurity segregation at the grain boundaries, and (2) surface tarnish (oxidation). Boguslavsky and Bush [36] pointed out that normal grain growth is inhibited or even stopped when the grain size approaches the thickness of the film. Note that a similar pinning constraint occurs in the Al system when the film deposition conditions are not sufficiently oxygen free to prevent oxide growth at the grain boundaries of the film in the Al system [43]. Boettinger et al. [44] have found that high purity water was required in the Sn plating bath to avoid whisker growth, and that the addition of Cu to the Sn bath lead to whisker growth even in the presence of high purity water. Both of these results are consistent with the concept of impurity pinning of the grain boundaries. Kehrer and Kadereit [45] measured the whisker propensity of evaporated Sn films on Cu. The films that were evaporated at 10^{-4} torr background pressure in the evaporation chamber whiskered whereas those deposited at 10^{-6} torr background pressure did not. This result is consistent with the concept of tarnish induced grain boundary pinning originally proposed by Ellis.

In 1994, 36 years after Ellis's original publication, Tu [12] further developed the tarnish theory and added for the first time the concept that the non-planar topology in the Cu–Sn intermetallic layer formed at low temperatures as the primary source of excess stress. Tu proposed that the reaction kinetics were such that the CuSn–IMC would grow fairly rapidly even at room temperature. Because of the exaggerated growth along the grain boundaries, IMC growth leads to the development of a non-uniform compressive stress in the Sn film. Note that although Tu defined the physical situation as compressive stress, the stress was developed as a result of differential strain not stress. The strain induced excess energy in the system could not be relieved by "normal" recrystallization mechanisms because the system was pinned. However, Tu recognized that the excess energy in the system could be reduced if and when a source of vacancies became available. He proposed that this could occur when localized "weak" spots in the surface oxide (tarnish) are breached. When such breaching occurs, "un-pinning" can occur via ambient introduced vacancies at these oxide fracture points. This theory is often referred to as the "cracked oxide theory". In the original formulation Tu proposed that the crack would occur at those locations were there was a "weakness" in the oxide. Tu derived a whisker growth rate expression of the form, where whisker growth rate is proportional to σD, where, σ is the stress level in the film and, D is the Sn diffusion coefficient. Boettinger et al. [44], have developed a model that indicates the stress is mainly related to stresses in the as-deposited film plus that due to precipitation of supersaturated impurities, with only minor changes resulting from the Cu–Sn IMC growth

over time. Galyon and Palmer [46] have argued that the stress is mainly from the growth on IMC's over time. These issues are discussed in more detail later in this paper.

Osenbach et al. [21] modified Tu's model slightly when they proposed differential oxidation of differently oriented grains on opposite sides of the grain boundaries, not local "weak spots in the oxide", could be responsible for oxide fracture. Osenbach et al. recognized other possible mechanisms could lead to oxide fracture including, defects as per Tu, externally applied stress, mechanical damage and other internally driven stresses and strains. The two main sources of external stresses are temperature cycling and externally applied mechanical forces (e.g., such as connector matting). Because Sn is anisotropic its thermo-mechanical properties are orientation dependent [47]. The directionally dependent thermo-mechanical properties can induce a differential stress response within the polycrystalline Sn-film when it is subjected to thermal excursions. This differential stress could be sufficient to cause the oxide to fracture at grain boundaries of differently oriented grains. Additional stress would be applied via the differential CTE's of the substrate,

IMC, and surface oxide relative to the Sn film itself. Scratch-induced mechanical deformation adds compressive stress and a localized high radius of curvature regions in the Sn. Scratches could also introduce contaminants into the system. These contaminates could locally alter the oxidation kinetics and/or provide stress risers in themselves or their Sn-reaction products. Finally, mechanical damage may degrade any protective barriers, such as a Ni underlayer, that may be present in the system. Thus mechanical damage is expected and sometimes found to be problematic from a whisker growth perspective.

This model can be represented by an Arrhenius flux/temperature plot, Fig. 11. The fundamental assumptions are: (1) Sn diffuses by a vacancy mechanism; (2) Sn diffusion is required to reduce the excess energy present in the system; and (3) the grain boundaries and surfaces of the Sn film are pinned. The intrinsic Sn flux as a function of temperature, assuming it is not vacancy availability limited, is shown in this figure. In the context of this paper flux is defined as the effective mass transport coefficient, i.e. diffusion coefficient. The approximate temperature ranges where the dominant diffusion mechanism, surface, grain boundary, and

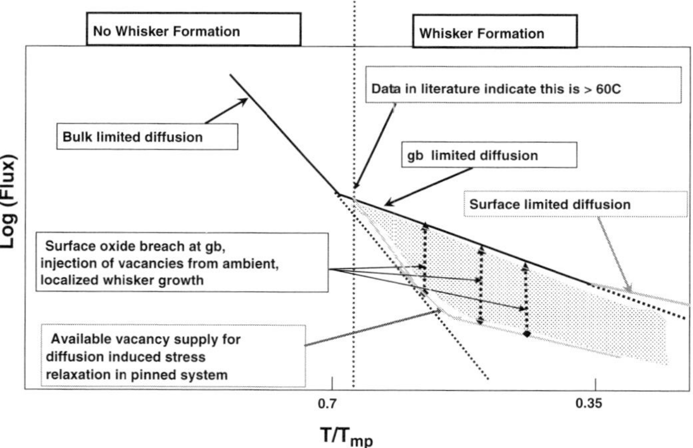

Fig. 11 Schematic Arrhenius plot of the Sn and vacancy flux versus normalized temperature in a system that is pinned. In this plot T_{mp} is the melting point of Sn. Here it is assumed that the limited vacancy supply is in part responsible for the pinning of the system. The approximate temperature ranges were lattice (bulk), grain boundary (gb) and surface diffusion dominate the mass transport in the system are schematically shown. Also shown is the "pinned" vacancy flux in the film itself. Note that when the oxide that is present on the surface of the Sn-film is breached, the air injects vacancies into the system until the surface in this location is re-oxidized. It is assumed that these vacancies can diffuse along the grain boundaries as a counter flux to the excess Sn that is present in the system as a result of compressive stresses. In this way the excess Sn can re-distribute and relieve the stress. The Sn diffuses to a lower energy site in the film were it can begin the whisker nucleation and growth process. Each time the oxide is breached the air vacancy supply is re-activated and vacancies are injected. This leads to addition Sn mass transport that provides more stress relaxation and additional whisker growth. The process continues until the stress in the system is relaxed. Note that whisker growth is only possible when the system is pinned, when the temperature is high enough the system becomes unpinned and other stress relaxation processes begin to relieve the system stresses

lattice, occur are shown. Because the system is pinned, the actual vacancy flux may be many orders of magnitude lower than the intrinsic Sn flux at all temperatures where whisker growth occurs. Hence, the system is pinned because the available vacancy supply is significantly lower than that required to sustain intrinsic Sn diffusion at that temperature. The pinning constraint is partially removed when vacancy supply is no longer the rate controlling step in the stress relaxation process. The dotted vertical lines represent the instantaneous supply of vacancies that would be injected into the system when the pinning constraint is removed such as when an oxide fracture occurs. When this occurs, Sn diffusion along these high diffusivity paths is no longer constrained by the supply of vacancies, and Sn diffusion along the high diffusivity paths can and does occur in such a way as to relieve the excess energy in the system. Because the grain boundaries are pinned, when the vacancy supply is injected into the system, energy reduction occurs by whisker growth instead of by normal grain growth. Note that one expects a similar oxide fracture mechanism in the Pb/Sn system. However, the vacancy supply and grain boundary mobility in the Pb/Sn film, even at room temperature, appears to be sufficiently high for stress reduction as evidenced by the observation of grain growth in Pb/Sn systems at room temperature. Presumably this is related to the low surface tension of Pb. Thus during isothermal stressing, the vacancy supply that is available when the oxide is fractured apparently is not required for stress relaxation in the Pb/Sn–Cu system. As such whisker growth in the Pb/Sn system is not expected to occur except in those locations and or stress conditions where there is an insufficient supply of vacancies or grain boundary mobility. One such case would be where the Pb is distributed in

such a way as to form a non-homogeneous film with regions of Sn-only and other regions with both Pb and Sn. The rapid application of differential stress such as might occur during temperature cycling could also lead to whisker formation in the Pb/Sn system. In this case the strain-rate could exceed the time constant associated with energy reduction by the intrinsic vacancy flux and grain boundary motion. Clearly this type of effect would diminish as the strain-rate and magnitude of the stress decreases or as the average temperature in the system increases.

Figure 12 shows SEM micrographs of two whiskers. Examination of these whiskers reveals circular striations at discrete points along the length of the whisker along with the longitudinal striations. Careful examination of these whiskers, and literally hundreds of others, indicated that in addition to the longitudinal surface texture/striations, circular striations are also present on most whiskers. This type of structure is consistent with the model according to which whisker growth could occur at discrete points in time between oxide fracture and re-oxidation of the surface of the Sn-film. This overall model appears to be consistent with many of the observations of Sn-whisker growth including; (1) it helps to elucidate why large variations in whisker incubation times are possible; (2) why whisker growth is found in both Sn on Cu and Pb–Sn on Cu in temperature cycling tests but only in Sn on Cu in isothermal tests (time constant driven); (3) why humidity is an accelerant; (4) why whisker growth propensity can be dependent on the Sn-film thickness; (5) why whisker growth is suppressed at higher temperatures; (6) why the post-plate anneal may increase the incubation time but not necessarily provide whisker immunity; (7) why whiskers are often found to have features that appear to be consistent with an

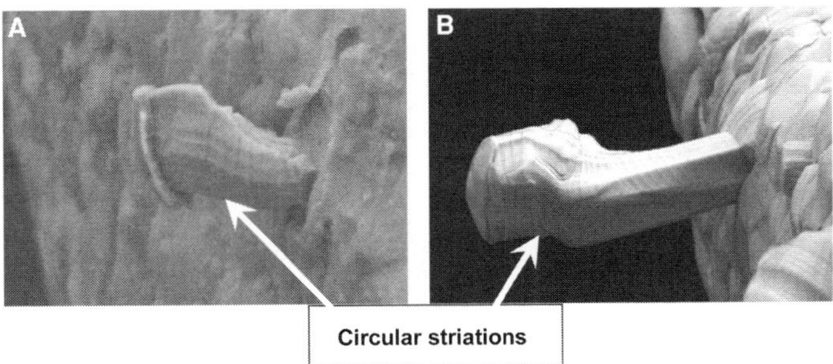

Fig. 12 (**A**) Whisker found after 1,000 temperature cycles on a MSA-based 15 μm thick Sn on Cu plus post-plate 150°C/1 h anneal. (**B**) Whisker found after 1,000 temperature cycles on a mixed acid-based 7 μm Sn on Cu plus no post-plate 150°C/1 h anneal. Note the circular striations along the length of the two whiskers. The striations appear to be perpendicular to the growth direction

extrusion like process; (8) why smaller grain size films are more susceptible to whisker growth than larger grain size films; (9) why contamination in the film can significantly degrade whisker immunity; (10) why certain barrier layers and/or processes are found to provide significant, if not complete, whisker growth immunity, and (11) why mechanical damage typically reduces the film's resistance to whisker growth independent of base metal. Furthermore, it does not prohibit the possibility of whisker growth by creep and or dislocations.

4 Mitigation strategies

Although there have been a significant number of investigations directed at identifying and eliminating the driving force for whisker growth, relatively minimal attempts are being made to directly identify the mechanisms responsible for grain boundary pinning or those required for nucleation and growth of whiskers. The prevailing, but not universal, view is that stress is the dominant driving force for whisker growth. Thus identification and elimination of the sources of stress has received most of the attention. The stress state of the film is influenced by a number of sources including: (1) pre-plate chemical treatment of the lead frame; (2) plating process and bath chemistry; (3) film thickness, grain size and grain crystallographic orientations; (4) externally applied loads such as those that are imparted during the trim and form process; (5) post-plate heat treatments; (6) Cu–Sn IMC growth; (7) the material Sn is plated onto (Cu, Ni, Ag, brass, alloy 42), (8) CTE mismatches in the system; (9) board assembly; and (10) oxidation and corrosion. Rosen documented the influence of the plating chemistry on the propensity for Sn-whisker growth in 1968 [48]. Zhang et al. [49] evaluated the whisker propensity of films created from five different Sn-plating solutions. They found that films containing >0.2 wt.% carbon grew longer and a higher density of whiskers than those containing <0.2 wt.% carbon. They also reported difficulty in reflowing the higher carbon content films because of discoloration and de-wetting. Although low carbon, matte–Sn, has generally been found to have a lower propensity for whisker growth than higher carbon content, bright-Sn, Hilty has recently reported results that indicate the opposite for connector applications [50]. Moon et al. [51] showed that small amount of Cu (0.8 ppm by mass) in the plating bath would enhance the propensity of whisker growth and result in Cu_6Sn_5 precipitates in the Sn-grain boundaries of the plated Sn film. They also noted that their results are in conflict

with commercial Sn-plating practice where electrolytes containing up to 300 ppm Cu are considered acceptable. Tu and Zeng evaluated the whisker propensity of Cu–Sn eutectic plating on Cu. They also found Cu_6Sn_5 precipitation in the grain boundaries of the Sn plate. They concluded the Cu_6Sn_5 grain boundary precipitates enhanced whisker growth. In contrast, Schetty et al. [52] found 6.4 wt.% Cu in the bath had a beneficial effect of whiskers. Most of the commercially available Sn-plating baths are formulated such that dissolution of Cu and other metals such as Fe, Zn, etc. is possible. Many of these metals are incorporated in the growing Sn film during plating. This incorporation tends to minimize the probability of excessive build up of these contaminants in the plating bath. Thus care must be taken to control the impurity concentration in the plating baths during the plating process itself to minimize the chance of "doping" the plated film with unwanted unknown impurities. This is especially true for those elements that can be incorporated into interstitial sites in the Sn lattice. In the authors' opinion dissolution of the substrate and or hardware used in the plating system may be one of the causes for conflicting whisker results. It is possible that the lot to lot and lead to lead statistical variations in whisker performance discussed previously is related to unwanted uncontrolled localized bath contamination. These results and others clearly show whisker mitigation critically depends on control of the impurity level (intentional and un-intentional) in the plating bath.

Glazunova and Kudryavtsev [53] studied the whisker propensity as a function of Sn-thickness on a variety of substrate materials including copper, brass and steel. They found that whiskers did not grow on films less than 0.5 µm. On Cu, the whisker growth rate was maximum when the Sn-thickness was 2–5 µm, whereas on steel it was maximum when the Sn-thickness was 5–10 µm. For Sn-thicknesses greater than 20 µm, whiskers were suppressed on copper and steel, but not on brass. They argued that reduction in the whisker growth rate with increasing thickness was probably caused by a reduction in the internal stress of the film although no experimental data was given. Others have reported increased thickness of the Sn plate reduced the propensity of whisker growth. The literature is clear that thicker Sn reduced the propensity for whisker formation, or at least increase the incubation time, such that whiskers are often not observed in the standard whisker tests.

In 1962 Glazunova [54] reported the beneficial effects of post-plate anneal on whisker propensity of plated Sn films. In 1963 Glazunova and Kudryavtsev reported the beneficial effects of post-plate heat treatment at temperatures between 100 and 180°C for times ranging from 1 to 24 h on reducing the propensity of whisker growth [53]. All of the heat treatments improved the whisker resistance of the films. They concluded that the role of heat treatment was one of reduction of the internal stress in the film. Similar results were reported by Rozen [48] in 1968. Rozen also noted that annealing changed the Cu/Sn IMC making it more resistant to hydrochloric acid and increasing its grain size. Brittan [5] found that annealing Sn on Cu at 200°C eliminated whiskers for up to 5 years at room ambient. Britton also stated that a 150°C, 1 h anneal had no beneficial effects on whisker growth. As pointed out by Galyon, careful analysis of Britton's data indicate the 150°C, 1 h anneal delayed the onset of whisker growth, by increasing the incubation time [19]. Ishii et al. [55] reported that a 2 h 150°C post-plate anneal mitigated whiskers in ultra-fine lead frames. Ditte et al. [56] reported that a 1 h post-plate anneal at 150°C was an effective whisker mitigation countermeasure even for thin, 1.5 μm, films. They proposed that the anneal led to a more planar Sn to IMC interface and a reduction in the compressive stress of the film. Figure 13 is a focused ion beam (FIB) micrograph of a Sn-film on Cu-alloy 7025 that was post-plate annealed within 2 h after plating at 150°C for 1 h. The sample was then subjected to 2,500 h at 60°C/93%RH. As shown, the IMC layer has significant surface topology at Sn-grain boundaries and at the edge of the lead frame. This is clear direct evidence that the post Sn-plate 150°C/1 h anneal does not completely eliminate localized thickness variations in the Cu/Sn IMC. Osenbach et al. [21] reported improvements in the whisker immunity for non-reflowed 1.5, 6, 10 and 15 μm thick Sn-films when they were subjected to a post-plate 150°C, 1 h anneal. However, whiskers were

found on all but the thickest non-reflowed Sn films, when exposed to 1,000 temperature cycles or thousands of hours at 60°C/93%RH. All of the data to date indicate that a post-plate anneal improves whisker performance, however, it appears that the improvement is resultant from an increase in incubation time not necessarily complete immunity.

Although most results indicate that the stress state of the system changes over time, to the authors' knowledge except for Boettinger et al. [44], all literature on this topic states that the stress in the system becomes more compressive with time due to the growth Cu–Sn intermetallics. Boettinger et al. developed and applied a time varying four layer model (Sn, Sn_6Cu_5, Cu_3Sn, Cu) to their stress measurements and found that the stress in the Sn film was not significantly dependent upon the growing IMC layer. Here the time dependence used was the IMC growth rate. Since it is generally considered that whiskers are driven by excess compressive stress in the Sn film, in the authors' opinion, a similar multilayer analysis of all existing cantilever beam-time dependent bow measurements is needed to gain a better understanding of the fundamentals of stress versus time in the Cu/Sn system. Such an analysis could help determine if there is a universal stress dependence in the Sn–Cu system as the IMC grows.

Galyon and Palmer [46] have proposed a model for stress generation consistent with Tu [12] and Ditte [56] proposal that the growing IMC is responsible for increased compressive stress in the film. Based on the movement W inert markers imbedded in a Sn/Cu diffusion couple, Tu and Thomson [57] reported that Cu is the primary diffuser in a Sn/Cu diffusion couple when exposed to room ambient. This is the basis for Galyon and Palmer's model. Galyon and Palmer

Fig. 13 Focused ion beam (FIB) and transmission electron microscope (TEM) micrographs of a Sn-film on Cu-alloy 7025 that was post-plate annealed within 2 h after plating at 150°C for 1 h. The sample was then subjected to 2,500 h at 60°C/93%RH

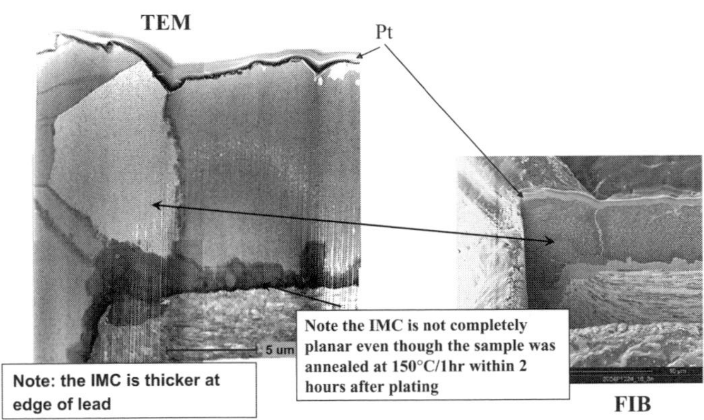

proposed that flux of Cu through the IMC is greater than the flux of Sn though the IMC. As a result there is a net loss of Cu in the Cu near the Cu/IMC-interface. Galyon and Palmer further stipulate that the process requires 6 Cu atoms to diffuse into the space occupied by 11 Sn atoms or for a higher degree of reaction 12 Cu atoms into the space occupied by 11 Sn atoms, leading to a net volume expansion near the Sn/IMC interface of 21 and 43%, respectively. Their final stipulation is that the intermetallic region expansion will be limited by the restraining forces of the overlying Sn and a shrinkage zone in the Cu immediately adjacent to the IMC interface. The Cu shrinkage zone is a direct result of the Kirkendall effect. This model is qualitatively consistent to the macroscopic stress measurements for Sn on Cu and Sn on Ni bimetallic strips as well as whisker propensity data.

Further evidence for the validity of this model are the FIB micrographs taken by Galyon and Palmer of porosity on bright Sn on Cu samples aged for 3–5 years at room temperature immediately below whiskers [46]. This porosity is identify as Kirkendall voiding. In investigations by the authors, void formation at the Cu side of the joint in Sn/Cu–lead frame-alloy couples after extended periods of time at room temperature and 60°C has not been found. Figure 14 shows FIB cross-sections of a Sn on Cu sample after exposure to 60°C for 2 years. No voids are observed at the Cu/IMC interface even after 2 years of aging at 60°C. The authors are not aware of any other direct observation of Kirkendall voids on plated, non-reflowed, Cu/Sn samples aged at low temperatures, <70°C. However, it should be pointed out that the absence of such findings

in the investigations by the authors or others does not disprove the Galyon–Palmer theory. It does however indicate that at a minimum at least one additional constraint must be placed on the system for Kirkendall voiding to occur. This would be consistent with Galyon and Palmer's observations that voids are only observed 10–20% of the time. The question then is, what is this additional constraint/s that when applied provides the opportunity to directly observe Kirkendall diffusion induced voiding. Often quoted as further evidence of the Kirendall effect, not necessarily correctly so, is the voiding that has been observed on solder balls connected to Cu pads. In most, but not all studies the solder ball data was collected at temperatures in excess of 100°C. There have been a number of publications that show Sn is the primary diffuser in the Cu–Sn system at temperatures greater than 125°C [58–64]. Thus at the temperatures where most of the solder balls ageing data was collected, one would expect void formation on the Sn side of the joint not in the IMC/Cu side of the joint if Kirkendall effects were playing a role in void formation. This is in fact the opposite of what is observed. The solder ball aging data is thus ruled out as direct evidence of Kirkendall diffusion induced voids in the Cu side of Sn–Cu joints formed at low temperature. It is interesting that studies carried out at high temperatures indicate Sn to be the primary diffuser [58–64], whereas low temperature investigations indicate Cu to be the primary diffuser in the Sn/Cu system [57]. This may indicate that there is a change in dominant diffusion mechanism between room temperature and 125°C. This is surprising since reaction-rate data from 25 to 225°C appears to be well

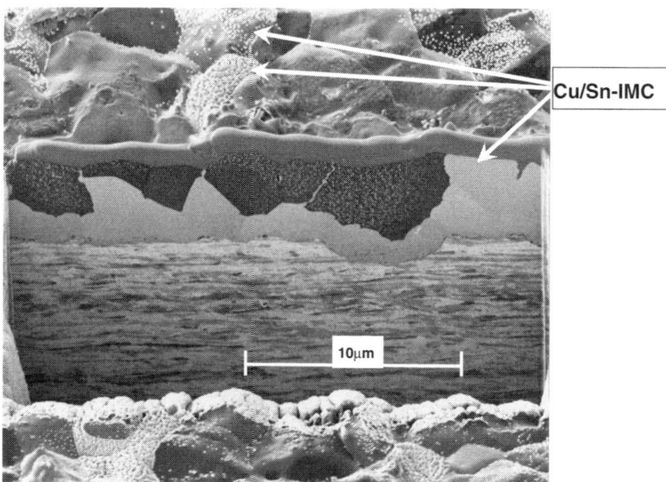

Fig. 14 FIB cross-section of a 15 μm Sn over Cu-alloy CDA-7025 after exposure to 2 years of 60°C/10%RH. There are no void present in this sample. In some regions the entire thickness of the Sn-film has been converted into IMC, whereas in other regions less than 30% of the Sn film has been converted

represented by one activation energy [21]. This is clearly an area where there is a need for further work on identifying the dominant diffusive species in the Cu/Sn system from room temperature to 200°C.

5 Effectiveness of Sn on Cu mitigation strategies over the product life cycle

This section mainly considers devices that are exposed to board reflow. However, it will be obvious to the reader how similar arguments could be applied to non-reflowed parts. As discussed in great detail previously, the as-plated Sn on Cu film is most probably not in a stress-free (lowest energy) state. The stress state depends on all of the materials and processes used to create the Sn-plated Cu device. For simplicity, one can assume the magnitude of the as-plated stress is non-zero, either compressive or tensile, since these assumptions have minimal, if any, effect on the analysis to follow. The built-in excess energy can be reduced by a post-plate anneal. The final stress state depends upon the starting stress state, the anneal temperature and time, the ramp-rates used for the anneal, the anneal atmosphere (moisture content, oxygen content, etc.), and possibly the anneal time-lag between plate and annealing furnace temperature. During the anneal, stress reduction occurs by standard grain growth and recrystallization mechanisms. Thus the anneal can reduce or eliminate stresses due to non-planar IMC growth, volume changes associated with IMC growth, contaminants in the plating, grain orientation and grain size, gas entrapment, oxidation, etc. Here it is assumed that there is some combination of temperature, time, ramp-rates, post-plate delay time, and anneal ambient condition that could be found for each and every plated Sn on Cu film that would yield a stress-free Sn/Cu system post-anneal. This anneal process may or may not be the 150°C/1 h anneal process commonly used by the industry for whisker mitigation. It is the authors' opinion that it would be highly unlikely that one anneal condition would produce stress-free material from all suppliers given all of the variations in plating chemistry, Cu-alloys, pre-plate etch-time and etch-chemistry, plating equipment, and plating process (time–temperature–pH, deposition-rate, etc.) that is currently being used by the industry. But to show that the stress state of the device prior to board attachment does not necessarily translate to the stress state of the device after board assembly and field use, it is assumed that to first order 150°C, 1 h air anneal is sufficient to produce a nearly stress-free material from all suppliers. If this condition could be maintained for the entire product life cycle, then one would conclude, as has been suggested by some in the literature, that the application of a 150°C/1 h post-plate anneal to a Sn on Cu plated film created from a "whisker-resistant" plating bath and process would be *whisker free* for the entire life of the product. Unfortunately, as is discussed below, even for components starting life in stress free Sn state, subsequent processing and field exposure will cause a change in the stress state of the Sn film which could lead to the nucleation and growth of whiskers.

Up until recently, the prevailing thought was after reflow the Sn-film would be in an stress-free, equilibrium, state after reflow. And because stress, or more properly excess energy, is required for whisker growth, reflowed Sn would not whisker. As such, relatively little attention was paid to reflow effects on Sn whiskers in recent years. The authors of this paper were among the first to publish experimental data taken on Sn-films produced with the new "whisker-resistant" plating bath chemistries that questioned this reflow theory [21]. Since this publication, a number of other publications and presentations have shown similar effects, as did some older studies on films made from less optimal plating chemistries [41, 65–67]. In fact there is no physical basis for the belief that reflow produces a stress-free, equilibrium state in the Sn film. In the liquid state the film is most probably stress free, but when it solidifies it is typically not. At temperatures above the melting point of Sn (232°C) the dissolution rate of Cu is at least 0.1 μm/s [68]. The solubility of Cu in Sn at temperatures greater than 232°C is at least 0.7 wt.% and it increases with increasing temperature [69]. The maximum Cu solubility will be influenced by the solder paste alloy. Assuming uniform Cu dissolution into a 15 μm thick Sn film (typically the Sn film thicknesses on electronic components is less than 15 μm), the equivalent Cu thickness required to reach the saturation limit at 232°C is approximately 0.1 μm, and at 260°C approximately 0.2 μm. If thinner Sn films are used, the Cu thickness required to reach saturation is less. Thus, the Sn film will always contain a Cu concentration that is at least the equilibrium saturation at the maximum temperature the device is exposed to during reflow. In addition to Cu doping, additional phenomena occur during reflow: incorporation of other elements into molten Sn film, changes in the IMC thickness and planarity, surface tension induced film flow, solder wetting and flow, and Sn-oxidation. The additional elements could come from the flux, the solder paste, the board metallization, and board or mold compound out-gassing. There are no requirements on concentration or spatial distribution of these elements within the Sn-film except those related to

equilibrium solubility limits. It is likely that some regions of the Sn film will contain solder alloy elements whereas other may not. Furthermore, it is possible that the spatial distribution of the additional elements will vary within one lead as well as between leads, and from device to device. The spatial distribution will be dependent upon the reflow process, the lead and pad geometry, the board thermal mass and mass distribution, and the solder paste volume. The IMC thickness will increase and the interface between it and the Sn will develop a scalloped-non-planar structure. Depending on the reflow conditions, temperature–time–ambient and flux, the solder paste thickness, and the device geometry, the Sn film thickness can remain the same, decrease, or increase during reflow. It is also likely that there will be rather large spatial variations in Sn-thickness within and between the leads of different devices. Given that thicker Sn films tend to be more resistant to whisker growth than thinner Sn films one needs to concentrate on those cases were the film thickness decreases.

After the Sn and solder are melted, the device temperature is reduced and the solder begins to solidify. Because the cooling rate is typically significantly faster than that required to maintain equilibrium, super saturation of Cu and possibly other elements is expected to occur. The exact concentration of excess Cu trapped in the Sn lattice and its spatial distribution is dependent upon the cooling rate which is dependent upon the flow dynamics of the reflow oven, geometry of the board and device, and the thermal mass of the device and the board. Thus there will be design-to-design, device-to-device, and even lead-to-lead variations in the Cu super saturation concentration. It will also be dependent on the reflow oven and ambient used. In general a slower ramp down will lead to less super saturation, however, in all cases some degree of super saturation is to be expected. When the film solidifies, thin regions within the film, if they occur, will occur at high radius of curvature regions of the lead such as convex bends and edges. In addition there may be thin regions where the lead enters the molded body of a lead frame device. Furthermore, if the solder paste volume is insufficient or wetting is poor, then there will be a thin solder fillet near the tip of the lead.

Over time, the super saturated Cu atoms will migrate by diffusion to a lower energy state and eventually nucleate and grow Cu_6Sn_5 intermetallics within the Sn film. According to Boettinger et al. [44], Sn super saturated with Cu occupies a smaller volume than does a binary mixture of Sn and Cu_6Sn_5. Thus when the supersaturated Cu begins to precipitate out in the form of Cu_6Sn_5 the film will experience a volume increase. This leads to a localized compressive stress in the Sn film in the vicinity of the precipitate. This will tend to increase the propensity for whiskering in the film after reflow as compared to that observed prior to reflow. This is in fact the argument put forth by the authors in their original publication that for the first time reported data showing an increased whisker propensity on Sn films that were reflowed as compared to those that were not reflowed. Experimental evidence supporting the discussion above is given in Fig. 15. Figures 15A, B show the change in the Sn thickness that occurs when a trim and formed lead frame is exposed to simulated board attach reflow at a peak temperature of 260°C. The reflow has resulted in a

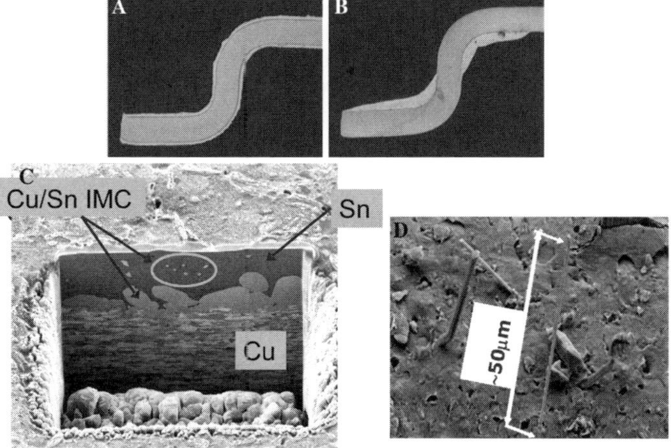

Fig. 15 (**A**, **B**) Optical micrographs taken from mechanical cross-sections of as-plated and post-simulated component reflow at a peak temperature of 260°C. (**C**) FIB cross-section of a lead taken from the device shown in (**B**) showing Cu/Sn IMC formation in the Sn film. (**D**) SEM of whisker found on samples whose cross-sections are shown in (**B**, **C**) after exposure to 10 weeks of 60°C/93%RH

significant redistribution of the Sn film. The thinnest region of the film after reflow is approximately 3 µm and occurs at the convex region of the lead. Figure 15C is a FIB cross-section of a lead on a device stored at room temperature for approximately 1–2 months after exposure to a simulated board attach reflow at a peak temperature of 260°C. The Cu_6Sn_5 precipitates are readily visible in the Sn film. Figure 15D is a micrograph taken with a scanning electron microscope of a lead on a device after reflow and exposure to 2,500 h of 60°C/93%RH. Whiskers are clearly visible on the lead in and around the thinned region of the film. It should be noted that there was no evidence of corrosion on the Sn film indicating that corrosion was not the root cause of whisker. Henshall [67] has recently reported whisker growth on some devices but not others that were reflowed to boards prior to an exposure to an 60°C/87%RH ambient for multiple thousands of hours. The whiskers were found in regions of the lead that had thin Sn, approximately 3 µm. In contrast, the devices that did not whisker had thick Sn, at least 16 µm, along the entire length of the lead.

Using the same logic that Boettinger et al. [44] applied to the Cu–Sn system, it is reasonable to conclude that the volume of a Sn film super saturated with other metal elements, Me, that are interstitial diffusers in Sn, such as Ni, is smaller than a binary mixture of Sn and the MeSn–IMC phase. Thus, if incorporated as a super saturated species in a Sn film, metal interstitial diffusers will eventually precipitate out in such a way as to create a localized volume increase within the Sn film. The super saturation concentration of any element is dependent upon the solubility of the element at the peak reflow temperature, the kinetics of dissolution, and the reflow process. Given that Ni is expected to behave similarly to Cu when incorporated as a super saturated element, the question one is left with is, why does Ni provide better whisker immunity post reflow than Cu [21]. At 260°C the solubility of Ni in molten Sn is significantly less than 0.1 wt.% closer to 0.01 wt.% [70], 10 to 100 times less than that the Cu solubility limit in Sn at the temperatures of interest. Thus the probability of creating excessive stress in the film as a result of Ni dissolution, super saturation, and precipitation is significantly lower in the Ni–Sn system than in the Cu–Sn. Figure 16 is a FIB cross-section of a Sn on Ni on Cu–lead on a device stored at room temperature for approximately 1–2 months after exposure to simulated board attach reflow at a peak temperature of 260°C. The white haze covering most of the Sn is an artifact of the FIB cross-sectioning process. As shown, no Ni_3Sn_4 precipitates are observed in the Sn film. Furthermore, the Ni_3Sn_4 IMC thickness is significantly less and more planar than the Cu–Sn IMC after reflow, as can be observed in Fig. 15. Finally, the solid state reaction rate of Ni and Sn is at least one order of magnitude lower than that of Cu and Sn. Therefore independent of which element is the primary diffuser in the IMC, increased stress as a result of IMC growth during field use of the device will be significantly lower

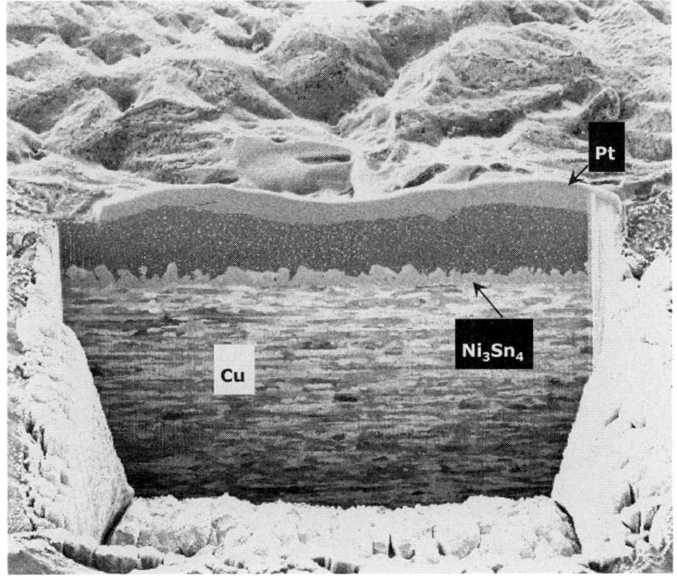

Fig. 16 FIB cross-section of a lead taken from the device with Ni underplate after exposure to simulated component reflow at a peak temperature of 260°C. Note there are no Ni/Sn IMC particles in the Sn film. The Pt layer is deposited in the FIB system prior to the FUB cut to pressure the original surface

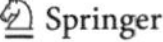

in the Ni–Sn system than in the Cu–Sn system. The whisker data is consistent with this analysis in that the whisker propensity was not found to be different in reflowed and non-reflowed Ni–Sn devices [21].

This model is helpful in understanding why some hot tin dipped devices are more susceptible to whisker growth than others. The data and model discussed above imply that hot-Sn dipping might lead to differing whisker responses depending upon the volume of Sn in the solder-pot, the age of the pot, shape and exposed surface area of the lead, the temperature of the pot, and the details of the dipping process used. This is because every time metal containing Cu or other metals that dissolve into the liquid Sn is dipped into the pot some of the dissolved metal will be left behind in the liquid Sn. Thus the concentration of foreign metals increases with every dip. In the limit the metal concentration in the liquid Sn will reach its solubility limit. Thus over time, the pot becomes less and less like an infinite sink for the dissolved metal and more like a mixture of Sn and the dissolved metal/s. When this occurs the foreign metal is likely to be incorporated into the solidified film in the form of a super saturated solution. If this occurs, then the whisker propensity increases.

After board attach the devices are then subjected to field use. Environmental exposure to temperature, relative humidity, cyclic temperature, air born contamination, etc. will change the stress state of the film. Figure 17 is an Arrhenius plot of the Cu diffusion

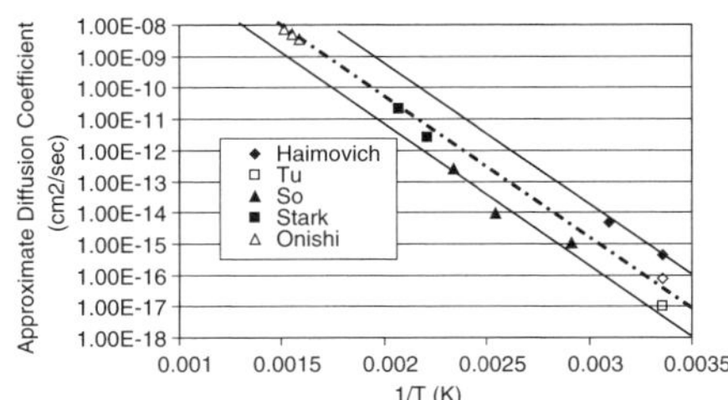

Fig. 17 Arrhenius plot of the estimated Cu diffusion coefficient through the Cu/Sn IMC as per the referenced data sets. The dashed dotted line is the least square regression fit to all of the data sets. The dotted lines bound the range of expected diffusion coefficients at any temperature of interest

coefficients taken from or estimated by fitting a Fickian diffusion model to reaction rate data in the literature [71–76]. The data are identified as per the reference they were taken from. Note that the data were taken at temperatures between 25 and 404°C. The data above 170°C were taken from diffusion couples of Cu and Cu_6Sn_5. The dash-dot line is the least square regression fit to all of the data. As shown, when the data is treated as a "black box", it is well represented by a single line $[D = 0.0169\exp(-0.86\ eV/kT)]$ with an R-square of approximately 0.96. The best fit line can be used to estimate the time constant required for Cu to diffuse through either a single phase or multiphase IMC layer at typical device operating temperatures. The dotted lines represent the authors attempt to bound the

Fig. 18 FIB cross-section of Sn/Cu couple after 1 h/150°C anneal. Note that silver like layer on top of the Sn film is Pt. The Pt layer was deposited in the FIB chamber prior to the FIB cut. It is used to identify the original surface of the sample

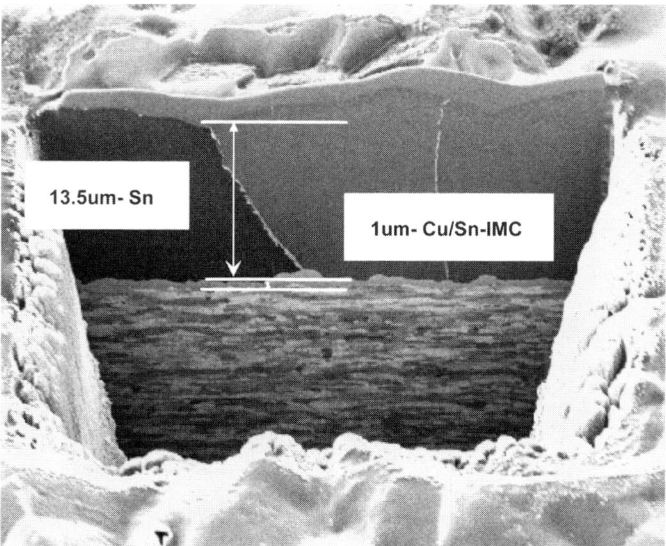

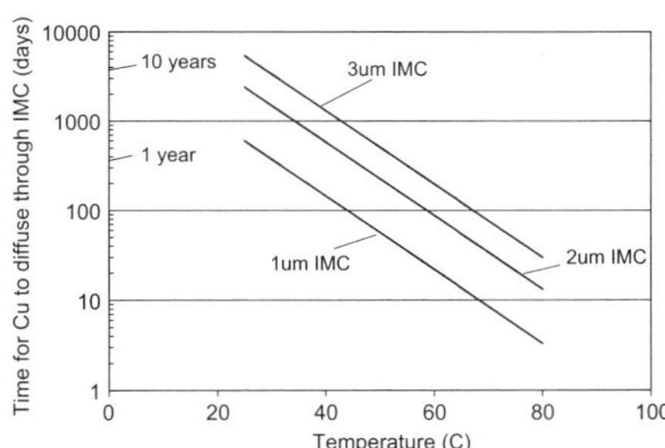

Fig. 19 Estimated whisker incubation time improvements for three different thicknesses of IMC

spread in the mass transport data. The two lines represent a constant slope, $E_a = 0.86$ eV, applied to the outer most data in the figure. This presumably provides a means for placing an upper and lower bounds on incubation time improvement that any post-plate annealing process might provide. The validity of this empirical curve fitting procedure can be checked by comparing the estimated reaction rates to those measured. Figure 18 is a micrograph showing a FIB cross-section of a matte–Sn over Cu sample subjected to a 150°C/1 h post-plate anneal. The anneal was carried out within 2 h after plating and the lead frame alloy was 7025. In this case the measured IMC thickness is approximately 1 μm. The calculated thickness based on the best fit line in Fig. 16 is approximately 0.85 μm, indicating the procedure is valid.

The estimated incubation time improvements that a 150°C/1 h anneal provides, assuming Fickian diffusion, is shown in Fig. 19. Figure 19 is an Arrhenius plot of the estimated time constants for Cu diffusion through a 1 μm thick IMC layer. Also shown in Fig. 19 are the estimated time constants for Cu diffusion through a 2 and 3 μm thick Cu/Sn IMC. The estimated diffusion coefficient at any temperature are within approximately one order magnitude of that given by the best fit linear regression analysis, provided in Fig. 17. Therefore, the time constant for Cu diffusion through the IMC is within a factor of $\pm$ $\sqrt{10}$ of that shown in Fig. 19. Thus even taking into account the possibility of a decrease in mass transport of one order of magnitude, this analysis indicates that if the 150°C/1 h anneal does in fact provide a whisker-free solution for devices whose required lifetimes exceeds 5 years, then the anneal must do more than just improve planarity and reduced excess of Sn at the Sn–IMC interface.

The authors have recently reported an increase in whisker propensity when Sn-films were exposed to high humidity in the absence of condensed water [77]. Figure 20 is a plot of the maximum whisker length vs. time with stress condition as the parameter. The data clearly show whisker growth is strongly temperature and humidity dependent between room temperature and 60°C, however at 85°C/85%RH no whiskers were observed. The equation is the best fit to all the data taken between room temperature and 60°C. Note in this case the samples were as plated, not annealed or reflowed, furthermore corrosion was not observed on any of the samples. This results clearly shows that long-term exposure to a non-condensing ambient can increase the whisker propensity of an as-made "stress"-free film.

There have been a number of publications over the last decade demonstrating whisker growth under temperature cycling. The density and length of whiskers appear to be dependent upon the magnitude of the difference in the coefficient of thermal expansion (CTE) of the base metal and the Sn plate [42]. For example the length is lowest for Sn on Cu (ΔCTE~6 ppm/C), intermediate for Sn on Ni (ΔCTE ~11 ppm/C), and highest for Sn on alloy 42 (ΔCTE ~19 ppm/C). Fortunately, in the Sn on Cu and Sn on Ni

Fig. 20 Dependence of whisker length on time with stress conditions as the parameter

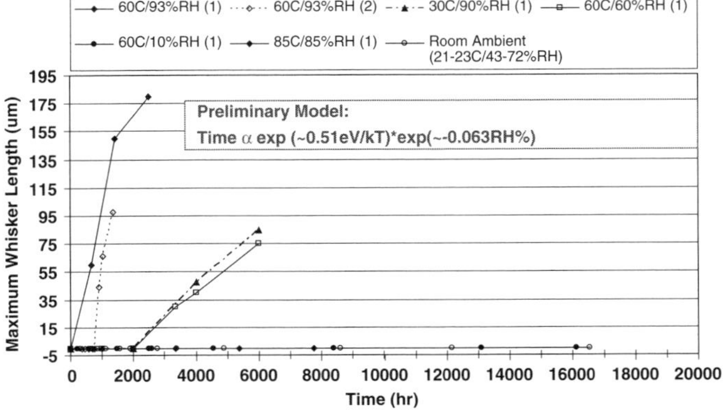

system, temperature-cycle induced whiskers are typically less than 50 µm in length with a diameter of 5 µm or more. In addition their length appears to saturate after 1,000–2,000 cycles between −55 and +85°C. Thus for most applications temperature cycling induced whiskers do not pose great danger to the operation of the device.

6 Summary

Whisker growth from the solid state has been documented in the public domain for more than 60 years. Unfortunately, a basic understanding upon which a quantitative model, one which would provide a means for predicting whisker growth and risk and ultimately prevention strategies, remains elusive. This paper is an attempt to critically review the state of Sn-whisker growth and mitigation. This is provided by: (i) examining the existing experimental data and the limitations of collecting such data, (ii) analyzing the proposed driving forces, mechanism and models for whisker growth, and (iii) carefully evaluating the proposed mitigation strategies and how subsequent assembly processes and device applications may or may not impact these strategies. The models for the most part are extensions of models proposed at least 3–4 decades ago. Likewise, with the exception of improved plating chemistry and possibly pre-plate base alloy etch depth, the mitigation strategies have been known for at least three decades. In each area the validity of the model and mitigation strategy is examined by comparing it to existing experimental data. Areas where the experimental data is insufficient to adequately test the theory or predict risk are identified. In addition, the areas where experimental difficulties are found that could potentially negatively impact data collection and analysis are reviewed. The authors hope that at a minimum this review provides a starting point for discussion between the user of the final product, the producer of the final product, and the producer of electronic components used in the final product on the topic of Sn-whisker growth, mitigation, and risk management. In the best case, it should also provides direction for the advancement of both experimental work and theoretical understanding of Sn-whisker growth and mitigation.

References

1. M. Warwick, J. SMT **12**, 1 (1999)
2. Y. Zang, Circuit Tree **94** (2001)
3. V.K. Glaxunova, N.T. Nudryavstev, J. Appl. Chem. USSR **36**, 519 (1963)
4. P.L. Key, in *Proc. Electronics Components Conference* (1970), p. 155
5. S.C. Britton, Trans IMF **52**, 95 (1974)
6. B.D. Dunn, ESA Sci. Tech. Rev. **2**, 1 (1976)
7. R.F. Dielh, N.A. Cifaldi, in *Proc. 8th Annual Connector Symposium* (1975), p. 328
8. S.M. Arnold, Plating PLATA **53**, 96 (1966)
9. G.T.T. Sheng, C.F. Hu, W.J. Choi, K.N. Tu, Y.Y. Bong, L. Nguyen, J. Appl. Phys. **92**, 64 (2002)
10. W.J. Choi, T.Y. Lee, K.N. Tu, N. Tamura, R.S. Celestre, A.A. MacDowell, Y.Y. Bong, L. Nguyen, G.T.T. Sheng, in *Proc. 53rd Electronic Components and Technology Conference* (2002), p. 628
11. B.Z. Lee, D.N. Lee, Acta Mater. **49**, 3701 (1998)
12. K.N. Tu, Phys. Rev. B **49**, 2030 (1994)
13. Tin Whisker Formation in Electronic Circuits, Aerospace Report No. TR-92, Dec. 1 (1992)
14. T. Kakeshita, R. Kawanaka, T. Hasegawa, J. Mater. Sci. **17**, 2560 (1982)
15. Y. Zhang, J.A. Abys, Circuit World **25**, 30 (1998)
16. JEDEC22-A121
17. JEDEC-201
18. H. Leidecker, J. Brusse, Tin Whiskers: A History of Documented Electrical System Failures. Briefing for NASA Space Shuttle Program, April 2006 (personal communication)
19. G.T. Galyon, IEEE Trans. Electron. Packag. Manufact. **28**, 94 (2005)
20. J.W.Osenbach, J.M. DeLucca, B.D. Potteiger, A. Amin, R.L. Shook, F.A. Baiocchi, Presented at ECTC Sn-Whisker workshop, June, 2005, accepted for publication in IEEE Trans. Electron. Packag. Manufact
21. J.W. Osenbach, R.L. Shook, B.T. Vaccaro, B.D. Potteiger, A.N. Amin, K.N. Hooghan, S. Suratkar, P. Ruengsinsub, IEEE Trans. Electron. Packag. Manufact. **28**, 36 (2005)
22. N.A.J. Sabbagh, H.J. McQueen, Metal Finishing, March, 21(1975)
23. S.M. Arnold, Plating PLATA **53**, 96 (1966)
24. M.W. Barsoum, E.N. Hoffman, R.D. Doherty, S. Gupta, A. Zavalianos, Phys. Rev. Lett. **93**, 206104-1 (2005)
25. P. Oberndorff, M. Dittes, P. Crema, S. Chopin, in *Proc. of the 55th Electronic Components and Technology Conference* (2005), pp. 429–433
26. P. Su, M. Ding, S. Chopin, in *Proc. of the 55th Electronic Components and Technology Conference* (2005), pp. 434–440
27. See for example, J.W. Osenbach, Semicond. Sci. Technol. **11**, 155 (1996)
28. D.W. Rice, P.B.P. Phipps, J. Electrochem. Soc. **127**, 563(1980)
29. D.W. Rice, R.J. Cappell, W. Kinsolving, J.J. Laskowski, J. Electrochem. Soc. **127**, 891 (1980)
30. J.D. Sinclair, L.A. Psota-Kelly, C.J. Weschler, Atmos. Environ. **19**, 315 (1985)
31. W.C. Ellis, D.F. Gibbon, R.G. Treuting, in *Growth and Perfection of Crystals*, ed. by R.H. Doremuss, B.W. Roberts, D. Turnbull (John Wiley and Sons, New York, 1958), pp. 102–120
32. W.C. Ellis, Trans. Met. Soc. AIME **236**, 872 (1966)
33. J.D. Eshelby, J. Appl. Phys. **91**, 755 (1953)
34. F.C. Frank, Phil. Mag. **44**, 854 (1953)
35. N. Furuta, K. Hamamura, J. Appl. Phys. **8**, 1404 (1969)
36. I. Boguslavsky, P. Bush, in *APEX Conf.*, 2003
37. R.M. Fisher, L.S. Darken, K.G. Carrol, Acta Metall. **2**, 368 (1954)
38. U. Lindborg, Met. Trans. **6A**, 1581 (1975)
39. B.Z. Lee, D.N. Lee, Acta Mater. **49**, 3701 (1998)

40. J. Liang, X. Li, Z. Xu, D. Shangguan, in *IPC/JEDEC 8th International Conference on Lead Free Electronic Components and Assemblies*, San Jose, CA, 18–20 April 2005
41. C.H. Pitt, R.G. Henning, J. Appl. Phys. **35**, 460 (1964)
42. Y. Zhang, C. Fan, C. Xu, O. Khaselev, J.A. Abys, in *Proc. IPC SMEMA APEX Conf.* (2002)
43. C.Y. Chang, R.W. Vook, Thin Solid Films **288**, 205 (1993)
44. W.J. Boettinger, C.E. Johnson, L.A. Bendersky, K.-W. Moon, M.E. Williams, G.R. Stafford, Acta Mater. **53**, 5033 (2005)
45. H.P. Kehrer, H.G. Kadereit, Appl. Phys. Lett. **16**, 411 (1970)
46. G.T. Galyon, L. Palmer, IEEE Trans. Electron. Packag. Manufact. **28**, 17 (2005)
47. J.A. Rayne, B.S. Chandrasekhar, Phys. Rev. **120**, 1658 (1960)
48. M. Rozen, Plating **55**, 1155 (1968)
49. Y. Zhang, C. Xu, C. Fan, J. Abys, J. Surf. Mount Tech. **13**, 1 (200)
50. R.D. Hilty, iNEMI meeting at Herndon VA, (2005) (communication)
51. K.W. Moon, M.E. Williams, C.W. Johnson, G.R. Stafford, C.A. Hanweker, M.J. Boettinger, in *Proc. Pacific Rim Inter. Conf. Advanced Materials and Processing* (2001), pp. 1115–1118
52. R. Schetty, N. Brown, A. Egli, J. Heber, A. Vinckler, in *Proc. AESF SUR/FIN Conf.* (2001), pp. 1–5
53. V.K. Glazunova, Kristallografiya **7**, 761 (1962)
54. V.K. Glazunova, N.T. Kudryavtsev, Zh. Prikladnoi Khim. **36**, 543 (1963)
55. M. Ishii, T. Kataoka, H. Kruihara, in *Proc. 12th Eur. Microelectronics and Packaging Conference* (1999), pp. 379–385
56. M. Ditte, P. Oberndorff, L. Petit, in *Proc. 53rd Electronic Components and Technology Conference* (2003), pp. 822–830
57. K.N. Tu, R.D. Thompson, Acta Met. **30**, 947 (1982)
58. A. Paul, A.A. Kodentsov, F.J.J. van Loo, Z. Metallkd. **95**, 913 (2004)
59. M. Oh, Ph. D. Thesis, Lehigh Univ. (1994) p. 60, 99
60. M. Onishi, H. Fujibuchi, JIM **16**, 539 (1975)
61. H.C. Bhedwar, K.K. Ray, S.D. Kulkarni, V. Balasubramanian Script Metall. **6**, 919 (1972)
62. L.C. Correa da Silva, R.F. Mehl, Trans. AIME **191**, 155 (1951)
63. H. Oikawa, A. Hosoi, Scripta Metall. **9**, 823 (1975)
64. S.-J. Kim, K.-S. Bae, in *Proc. Electronic Packaging Conference* (2000), pp. 81–85
65. K. Cunningham, M. Donahue, in *Proc. 4th International SAMPE Electronics Conf.* (1990)
66. G. Henshall, Presented at the 3rd annual ECTC Sn-whisker workshop, 2006
67. H. Renolds, Presented at the 3rd annual ECTC Sn-whisker workshop, 2006
68. W.G. Bader, Welding J. Res. Supl. **48**, 551 (1969)
69. See for example, M. Hansen, K. Anderko (eds.), *Constitution of Binary Alloys*, 2nd edn. (McGraw-Hill, New York, 1958), pp. 633–637
70. See for example, M. Hansen, K. Anderko (eds.), *Constitution of Binary Alloys*, 2nd edn. (McGraw-Hill, New York, 1958), pp. 1041–1047
71. K.N. Tu, Mater. Chem. Phys. **42**, 217 (1996)
72. D. Unsworth, in *Modern Solder Technology for Competitive Electronic Manufacturing*, ed. by J. Hwang (McGraw Hill, New York, 1996), pp. 397–400
73. A.C.K. So, Y.C. Chan, J.K.L. Lai, IEEE Comp. Pkg. Manuf. Technol, part B, 20 (1997)
74. M. Onishi, H. Fujibuchi, Trans. JIM **16**, 539 (1999)
75. E. Starke, H. Wever, Z. Merallk **155**, 108 (1964)
76. J. Haimovich data, in NEMI user group web site. http://www.nemi.org
77. J. W. Osenbach, in NEMI user group web site. http://www.nemi.org

Tin pest issues in lead-free electronic solders

W. J. Plumbridge

Published online: 27 September 2006
© Springer Science+Business Media, LLC 2006

Abstract Tin pest is the product of the $\beta \rightarrow \alpha$ allotropic transition at 13.2°C in pure tin. It is a brittle crumbly material, often responsible for the total disintegration of the sample. The transformation involves nucleation and growth, with an incubation period requiring months or years for completion. Experimental observations reveal a substantial inconsistency and an incomplete understanding of the process. Some alloy additions promote tin pest by reducing the incubation time, whereas others retard or inhibit its formation. Traditional solder alloys have generally been immune to tin pest in service due to the presence of lead, and bismuth and antimony as common impurities. However, the new generation of lead-free solders are more dilute—closely resembling tin. A much debated question is the susceptibility of these alloys to tin pest. Bulk samples of tin-0.5 copper solder undergo the transition at − 18°C although not at − 40°C after five years exposure. Other lead-free alloys (Sn–3.5Ag, Sn–3.8Cu–0.7Cu and Sn–Zn–Bi) are immune from tin pest after a similar period. Large scale model joints exhibit tin pest but it appears that actual joints may be resistant due to the limited free solder surface available and the constraint of intermetallic compounds and components. It seems likely that impurities are essential protection against tin pest, but for long term applications there is no certainty that tin pest and joint deterioration will never occur.

W. J. Plumbridge (✉)
The Open University, Milton Keynes, UK
e-mail: w.plumbridge@open.ac.uk

1 Introduction

The World's largest industrial sector, Electronics, is currently facing two major challenges. The first concerns the demand for continuing miniaturisation of equipment and the associated increase in density of component packing and in functional efficiency. In this case, maintenance of overall structural integrity and avoidance of electrical contact (shorting) between adjacent component parts are key criteria to be achieved. The nature of the materials involved is of secondary concern. In contrast, the second challenge is material-centred, and relates to existing and proposed legislation to remove toxic lead from solder alloys. The damage to health from lead in items, such as paint and piping, is well known and its presence has been eliminated many years ago. Considerably more debate has surrounded the possible hazards associated with the disposal of goods containing lead in their electronics systems, largely because of the much reduced volumes involved. Nevertheless, the Environmental Lobby has prevailed and, with the additional driver of potential commercial profit, adoption of lead-free solder technology is well underway [1]. For example, in the European Community, after 1st July 2006, lead-containing solders were prohibited in the vast majority of applications. This ban involved purchasers as well as producers, so the ramifications for a global industry, such as Electronics, are clear.

Soldering is the preferred method of joining components to printed circuit boards (PCBs) in electronics. It is generally a relatively low temperature (< 250°C) process, so avoiding thermal damage to the polymeric materials of the PCB or the component. The process involves the formation of an intermetallic compound (IMC) between the metallic materials being joined. Unlike welding, no melting occurs. The fundamentals and practicalities are described briefly in Ref. [2].

Since antiquity, the most common forms of solder have been based upon alloys of tin and lead—notably Sn–37 mass per cent Pb, which is the eutectic composition and possesses the lowest melting point of the tin–lead system. Further advantages associated with this alloy are good fluidity, wettability, mechanical properties and a pleasant shiny appearance. It is an almost perfect alloy for its purpose, and it was no surprise that it assumed a dominant role in the production of interconnections in electronics. Some two decades ago, the toxicity problem surfaced, but the removal of lead was resisted by commercial/industrial intransigence and the highly satisfactory nature of the current Sn–37Pb alloy. Irrespective of this, the search for new solder alloys without lead had been triggered. Details of the investigations are reviewed elsewhere [3, 4], but the principal outcome was that there was no direct (or drop-in) replacement of Sn–37Pb. The favourites were again tin-based systems with a eutectic, and containing silver, copper or zinc. More precisely, Sn–3.5Ag, Sn–0.5Cu, Sn–3.8Ag–0.7Cu, Sn–8Zn–3Bi were popular amongst those proposed. Small variations in composition from those cited above may be found due to patent restrictions or producer preference. While these new alloys have been found to have properties at least equal to those of Sn–37Pb [5, 6], all of them suffer the disadvantage of a higher melting point (i.e. up to 223°C as compared with 183°C for the Sn–37Pb alloy). This can impose significant constraint on processing and PCB assembly. Another area of potential vulnerability of lead-free solders is the possibility of forming tin pest on prolonged exposure to low temperatures. The present review examines the available information on tin pest in pure tin and its alloys, with particular emphasis upon the new generation of lead-free solder alloys and the likelihood of its appearance on actual joints in service.

2 Tin and its alloys

For an engineering metal, tin has an uncommon crystal structure which varies according to the temperature. Between its melting point (232°C) and 161°C, γ Sn occurs with a rhombic crystalline configuration. Below 161°C and at room temperature, it exists in a body centred tetragonal (bct) form as β, or 'white' tin. However, it undergoes an allotropic transition at 13.2°C to a diamond cubic structure, known as α, or 'grey' tin which is a non-ductile semiconductor (Fig. 1). The lattice parameters are: for β Sn, a = 5.831 and c = 3.181 × 10^{-10}m, and for α Sn, a = 6.489 × 10^{-10} m [7], and a considerable volume change (~ 27 percent) is associated with the transition. Consequently, localised rupture occurs which is normally observed on surfaces or corners of samples where constraint is a minimum. The surface spots or eruptions (warts) are known as tin pest or tin plague. The transition occurs by a process of 'nucleation and growth', and the incubation period may be prolonged, ranging from a few months to several years [8].

Tin pest was first reported more than 150 years ago [9] and the vast majority of work on the $\beta \rightarrow \alpha$ transition was performed over 50 years ago. In most studies, the long and uncertain period of incubation was substantially eliminated by *inoculation*, or seeding, with grey tin particles, simply pressed into the white tin surface. For example, by this approach it was possible to achieve 100 per cent transformation in a few days, as compared with the decades required in the original

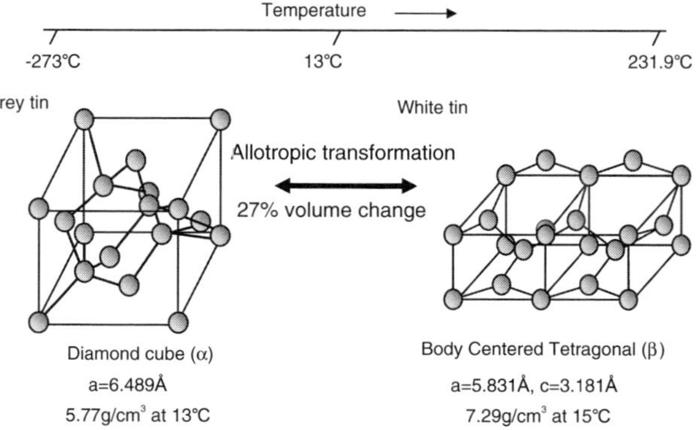

Fig. 1 The allotropic transformation in tin

condition [10]. Consequently, the majority of the available literature on tin pest relates to the *growth* phase of grey tin following artificial nucleation. Its relevance to the present day service situation is therefore somewhat limited.

Once nucleated, subsequent growth is relatively rapid [8, 11]. It was considered to be a diffusionless transformation but the observational methods available at the time were insufficiently sensitive to confirm this [8]. The maximum rate of transformation is reported to be at around − 40°C, associated with a maximum linear growth rate of the grey tin phase of 0.06 mm per day [10]. At − 18°C, the growth rate is about one third of this [8]. The process rarely goes to completion [12, 13] and partial reversal can be achieved by heating to about 60°C or above. A potentially worrying scenario to the designer of equipment likely to operate in conditions conducive to tin pest formation is that of *'self inoculation'* i.e. during thermal cycles in service, if tin pest is nucleated during the cold phase of one cycle, and the process is incompletely reversed during the following warm phase, the requirement for a prolonged incubation period in the next cold phase will disappear. The tin is said to be *'activated'* and significantly more tin pest will form—and so on. No such event in service has been reported to date, but as with most disasters, it is not impossible that a combination of circumstances could contrive to set up such conditions.

A key point emerging from a survey of the literature from that period is the lack of consistency in the findings—both in terms of the occurrence, or not, of the transformation and the rate at which it proceeds when it does. In particular, any quantitative observation is entirely specific to the conditions of the experiment being described. This is in no way a criticism, but rather a recognition of the enormous challenges involved—the precise roles of composition, impurity levels, strain effects and microstructure are not clearly understood. This is well illustrated by an extensive study at Battelle Laboratories [14, 15] on non-inoculated tin samples stored at − 40°C. Twenty one varieties of a commercially pure tin, with small amounts of Pb, Sb, Bi, Cu, As, Fe as impurities, were exposed for ten years and examined periodically. Only in one batch did the majority (22 out of 24) cast samples transform in periods of between 5 and 114 months, but following cold rolling, 11 out of 12 of identical samples from that batch did not transform after 10 years exposure. Similar erratic behaviour was exhibited by four other brands when less than 20 per cent of identical samples transformed within the 10 year period. No tin pest formed at all in any of the remaining brands. Chemical

analysis failed to resolve the matter. Further, subsequent work indicates that cold work promotes rather than retards the transformation [14, 16, 17].

It has been suggested that stress relaxation in the white tin ahead of the interface plays an important role in the growth of tin pest [18]. The high stresses that are generated by the large volume change form a zone of plasticity in front of the β/α interface, the rate of migration of which is controlled largely by the average stress level in its vicinity. Relaxation via point defect creation, dislocation motion, twinning or crack formation reduces these stresses and more rapid growth occurs. This hypothesis can account for an observed thickness effect on tin pest growth. With reduced thickness, below 500μm, the plastic zone size is less but the dislocation density within it is higher. This hinders relaxation, maintains a higher local stress level and favours slow growth of the interface. For foil thicknesses above 500μm, stress relaxation occurs quite rapidly and the thickness effect is not apparent (Fig. 2). The controlling plastic zone size is generally much smaller that the grain dimensions, so no grain size effect is observed. A steady state balance between work hardening and recovery is envisaged in the zone ahead of the interface, and the activation energy for growth is most closely associated with diffusion along dislocation cores. Naturally, the balance between core and bulk diffusion varies with temperature, and it is not unusual for either mechanism to dominate according to the temperature.

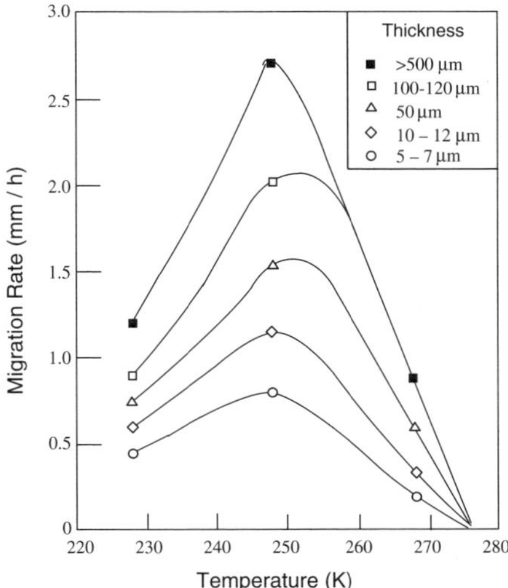

Fig. 2 Effect of sample thickness on migration rate of the β/α interface in pure tin (18, modified)

TEM studies of the transformation are limited due to the brittle nature of grey tin and the large volume change involved. Some success has been achieved by performing the transformation in samples preprepared for TEM examination. Observations of the $\beta \rightarrow \alpha$ transition in pure tin reveal a preferred orientational relationship between the two phases [19, 20]. The (111) plane of grey tin is parallel to the (001) plane of white tin, and the [211] direction is nearly parallel to the [010] of white tin. The absence of grey tin in foils thinner than 130 nm was attributed to surface energy effects [19]. Overall, the available evidence—the curved nature of the β/α interface, the lack of a three-dimensional crystallographic relationship between the phases and the retention of Xenon cavities in both phases [21] —suggested that the transformation was massive [22] rather than martensitic [19]. It is suggested that the transformation occurs partly by mass transformation and partly by atom motion that is massive in nature. There is a hysteresis effect between the directions of the reaction, with the $\alpha \rightarrow \beta$ transformation occurring about 20°C above the $\beta \rightarrow \alpha$ transition [23].

3 Effect of alloying additions to tin

The difference between an impurity and an alloy addition is a matter for academic debate, but there is strong evidence that the presence of elements soluble in tin, such as Pb, Bi, Sb suppresses the $\beta \rightarrow \alpha$ transition by raising the transition temperature [10, 12, 16, 24] and that Cd, Au and Ag retard it [16, 25–27]. Extremely low levels of solute can be effective—as low as 0.0035 mass per cent in the case of bismuth [28]. In contrast, insoluble elements, such as Zn, Al, Mg and Mn accelerate the transformation by lowering the temperature at which it occurs [16, 29] while Cu, Fe and Ni are reported to have little influence [16]. However, later work suggests that Cu and Zn promote grey tin formation [27, 30] and it was suggested that the influence of any alloy addition could be changed by the presence of other elements [27] i.e. an element could either promote or inhibit the transformation according to the other elements present. From the application perspective, consideration of the amounts of second element addition reported in the literature indicates that an overlap with commercial forms of 'tin' is possible (commercially pure tin contains typically up to 0.2 mass per cent impurities). This range of compositions suggests one possible explanation for the inconsistent and sometimes conflicting findings reported.

The presence of copper (0.8 mass per cent) and heavy cold work (up to 90 per cent) promotes tin pest formation at $-30°C$ [17]. Incubation times are reduced to less than 10 days and the transformation is complete after 40 days. The effect was more marked as the degree of cold work increased (Fig. 3). Tensile pre-strain, rather than compressive, has been found to promote tin pest on tin-coated steel [27]. Even in the absence of prior deformation, as-cast samples transformed in about six months—which is much more rapid than generally observed. The composition of that alloy was (mass per cent);

Cu, 0.8; Pb, 0.001; Fe 0.0003; Sb, 0.0002, As 0.0002; Bi 0.0001.

In dilute Sn–Ge alloys, the temperature, T_c, of the reverse transformation from grey to white tin is raised by increasing the Ge content, according to the expression [31]

$$T_c = 32.0 + 51.8C^{0.33}$$

where C is the amount of Ge in atomic per cent. For example, the transition temperature for a Sn–0.5Ge alloy is about 75°C. This effect is attributed to the reduced mobility of the β/α interface. In Sn–0.6Si alloys, grey tin remains stable on heating to 90°C [32].

The role of solute (or impurity) atoms was investigated using germanium (0.1–0.3 at per cent) [18]. Within this range, the migration rate of the β/α interface slowed with increasing Ge content (Fig. 4). This was attributed to the solute atoms impeding dislocation climb—the less mobile the solute, the stronger the effect. Hence, stress relaxation was restricted and the advance of tin pest retarded.

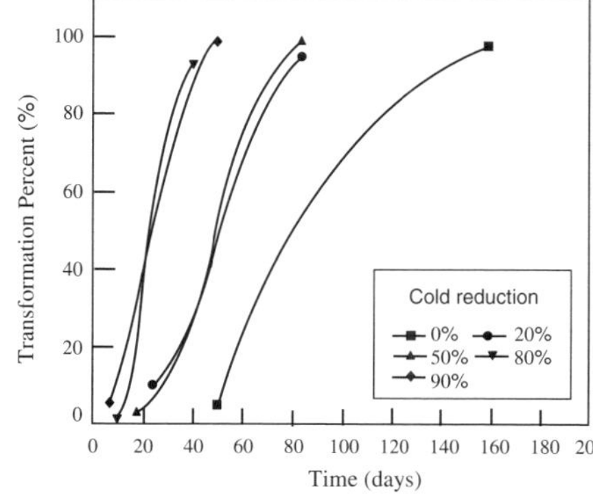

Fig. 3 Influence of cold work on the transformation in a Sn-0.8 mass per cent Cu alloy at $-30°C$ (17, modified)

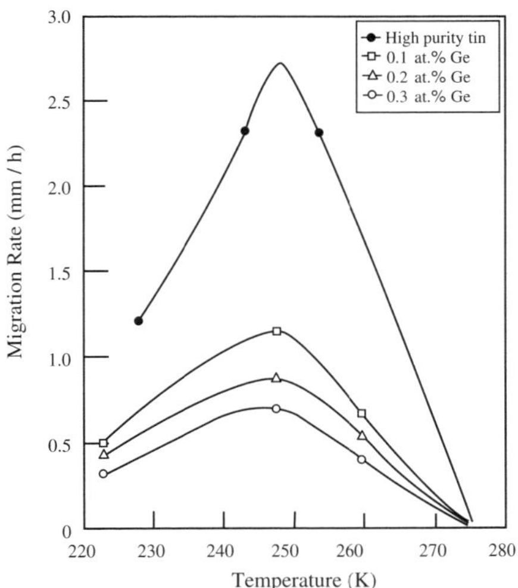

Fig. 4 Effect of solute concentration on the migration rate of the β/α interface in dilute Sn–Ge alloys (18, modified)

External pressure has a significant effect on the temperature of transformation and the rate at which it occurs [23, 33, 34]. For example, in dilute Sn–Ge alloys (0.1 to 0.5 mass per cent) the transition temperature is reduced by 50 K per kbar.

An extensive and systematic study by Boremann [35] confirmed the variability in the duration of the incubation period for pure tin (99.9944 mass per cent) at – 100 F (– 73.3°C). One of four samples in the non-inoculated condition transformed after six months exposure, but the remainder were unaffected after four years. Cold worked pure alloys, after inoculation, exhibited tin pest in Sn, Sn–40Pb and Sn–50Pb but this could usually be prevented for at least four years by the addition of 0.1 Bi or Sb. Dilute binary and ternary alloys (Sn + Pb, Sb and Bi), without inoculation were also studied. Unfortunately, their initial condition differed from that employed previously (a 2 h homogenisation anneal was applied after cold working). However, it did indicate the critical amounts of solute necessary to prevent tin pest formation after 14 months at exposure at – 100 F. These were 0.18, 0.047 and 0.27 mass per cent for Pb, Bi and Sb respectively. The existence of intermediate ranges of solute content below these levels in which the transformation did occur highlights the difficulties associated with this topic but nevertheless the study provides an valuable source of experimental detail. It was concluded that Pb, in the amounts generally found in solders, will not prevent the transformation from occurring. Small amounts of Bi or Sb are effective inhibitors—more so when they are in combination.

It seems that the strength of the β Sn phase plays a significant role, as the α Sn warts expand into it [18]. With increase in strength, tin pest growth is slowed. For similar amounts, it is likely that elements soluble in tin will make a greater contribution to solid solution strengthening than insoluble atoms will enhance particle strengthening. This suggestion is in line with the experimental observations. Elements soluble in tin, such as Pb, Bi and Sb, suppress the transition by producing a stronger β matrix around the tin pest warts. Conversely, insoluble elements, such as Zn, Al, Mg and Mn, have little effect on strength and appear to promote the transformation. The role of prior deformation is less clear, and this is probably due to it having opposing effects on the initiation and growth stages. In Sn–0.8Cu at – 30°C, it has been clearly demonstrated that cold work reduces the incubation period by up to an order of magnitude [17] whereas the transformation rated is increased by around threefold. With the exception of the Battelle study [14, 15], most previous investigations concur that the transition is promoted by cold work.

Tin pest has also been observed by Ogden (R. Odgen, Private Communication) in a reputedly Sn–3.0Cu solder alloy after low temperature storage in a domestic freezer (typical temperature a – 18°C approx) for 21 months. None was apparent after 15 months.

4 Tin pest in modern electronics?

There are numerous applications in which the electronics systems enter the domain of tin pest formation. From the IPC categories of operating conditions [37] aeronautical, aerospace and automobile are obvious candidates. However, because lead and impurities, such as Bi and Sb, suppress the $\beta \rightarrow \alpha$ transition and because the vast majority of traditional solders to date contain these elements, tin pest has not been a problem. But the emerging generation of lead-free solders are rather different; they are much more dilute—up to a 70 × in the case of Sn–0.5Cu. Microstructurally, they comprise small intermetallic particles, as opposed to substantial proportions of second phases in the conventional Sn–Pb solder alloys, together with a tin-rich solid solution. In other words, they resemble pure tin more closely than traditional lead-containing solders, so the key question is whether tin pest is more likely to form in the new lead-free alloys?

Studies by the Solder Research Group at the Open University, UK, were commenced about a decade ago as a spin off from their evaluation of the mechanical behaviour of solder alloys [38]. Due to their low

melting point, room temperature represents a significant homologous temperature (~0.65) for solders which are microstructurally unstable under ambient conditions. Common practice to preserve stability was to store the as-cast test pieces in a freezer at $-18°C$ prior to testing. After extended periods, tin pest warts were observed on certain samples, and these findings sparked more systematic investigations which are now described [39, 40].

A range of lead-free alloys, including Sn–3.5Ag, Sn–0.5Cu, Sn–3.8Ag–0.7Cu, and Sn–8Zn–3Bi, with Sn–37Pb as a comparator, has been exposed to sub-critical temperatures, (between $-18°C$ and $-40°C$) for periods up to 10 years. Details of their compositions are given in Table 1. Most samples were in the form of tensile test pieces (diameter 11.1, gauge length 60 mm) in the as-cast condition and cooled by water quenching, air or furnace cooling (cooling rates 20, 0.2 and 0.02°C s^{-1} respectively). They were removed from storage and examined periodically. Some specimens were aged or subjected to monotonic or cyclic strain prior to low temperature storage.

Of the alloys investigated, only Sn–0.5Cu aged at $-18°C$ regularly exhibited tin pest formation. Surface spotting occurred after a few months, followed by surface eruptions, the development and growth warts, severe cracking and eventually total disintegration of the sample. Figures 5 to 8 show this sequence. Both the individual warts and the transformation interface are clearly visible (Fig. 5). The effect of surface strain, induced in this case by machining the specimen heads for gripping purposes, in promoting the transition is clear. The head regions of the sample are virtually covered with tin pest, whereas the specimen gauge length which retains its original cast surface exhibits isolated warts (Fig. 6). Spread of tin pest is largely a surface event, although penetration into the interior of the sample does gradually occur (Fig. 7). Electron Backscatter Diffraction Analysis (ESBD) confirmed that growth of tin pest occurred at the interface of the α/β phases and also the absence of grey tin in the interior of the sample (Fig. 8). Complete disintegration of the sample eventually occurs as shown in Fig. 9. Eventually, the sample fragments would become

powder. It is important to note that the above observations pertain to a single sample; while a substantial proportion of Sn–0.5Cu specimens revealed tin pest formation, they did so at different rates under identical conditions. Moreover, in other samples of this alloy, no tin pest was observed although their history was identical. This observation applied to all three cooling rates. Tin pest was not formed after ageing at $-40°C$ in the Sn–0.5Cu alloy, nor in any of the other alloys examined, irrespective of prior treatment, after storage at either temperature.

5 Discussion

The allotropic transformation from β to α at 13.2°C in pure tin is well established, and there is substantial evidence that a similar transition occurs in many dilute tin alloys. Also mirroring the findings on tin, is the inconsistency and variability of the process in the alloy, with nucleation as the critical event. While acknowledging that the exact mechanism of the transition is unclear, the salient questions from the electronics viewpoint are; (1) *will tin pest form in other lead-free solders?* and (2) *will tin pest form in actual solder joints?* These possibilities are now considered further.

5.1 Formation of tin pest in other solder alloys

The efficacy of solute additions in inhibiting or promoting the $\beta \rightarrow \alpha$ transition in tin was described earlier. In terms of *global composition*, Pb, Bi and Sb suppress the allotropic transformation whereas elements insoluble in tin promote it. Much of the published work involved very low solute concentrations which could be effective in their actions. From this, two aspects merit further consideration; the level and type of impurities, and compositional variations i.e. *local composition* effects.

A wide variation exists in the purity levels of commercial lead-free solder alloys and few agreed limits apply. Typical ranges for elements commonly found are, in mass percent, R. Bilham (Private Communication);

Table 1 Composition of SRG alloys

Sn	Pb	Ag	Cu	As	Bi	Fe	Al	Sb	Zn	Others
Balance	37	–	<0.08	<0.02	0.25	<0.02	0.005	0.2	0.005	0.08
Balance	0.014	0.008	0.55	<0.001	<0.001	0.001	<0.0002	0.012	<0.001	<0.003
Balance	0.04	3.47	0.003	<0.001	<0.001	0.007	<0.0002	0.025	<0.001	<0.003
Balance	<0.1	3.8	0.7	<0.03	<0.1	<0.02	<0.001	<0.1	<0.001	<0.2
Balance	0.081	–	0.006	0.001	2.92	0.002	–	0.005	7.89	–

Fig. 5 Early stages of tin pest formation in a Sn–0.5Cu alloy during storage at − 18°C

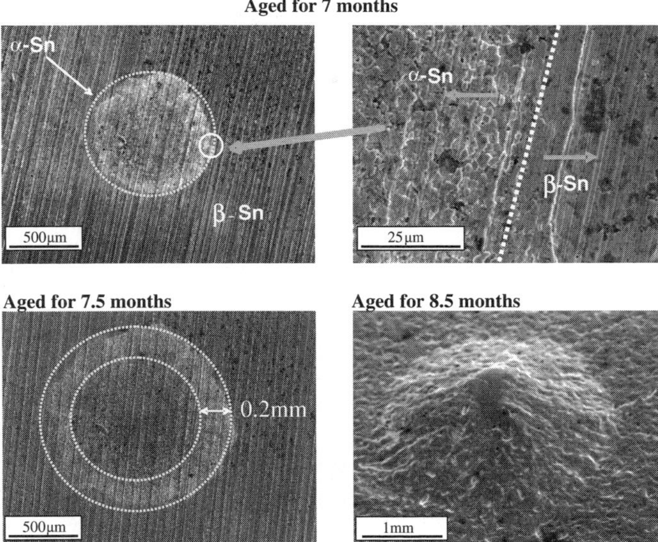

Fig. 6 Development of tin pest over a test piece, exposed at 18°C, showing a higher concentration of warts on machined areas

Pb (0.03–0.07); Bi(0.003–0.030); Cu(<0.01)

Zn, Cd, Al($\leq$ 0.001); Fe, As, Ni(0.003–0.01).

The extreme values are roughly 0.04 to 0.12 mass per cent when alloys have either a minimum amount or a maximum amount of each impurity. These levels overlap with compositions of the intentional alloys made up to study the $\beta \rightarrow \alpha$ transition. For comparison, in the classic Batelle study [14, 15] of 'pure' tin, the impurity levels of individual batches ranged from 0.002 to 0.246 mass per cent. The composition of the Sn–0.5 Cu alloy described previously (Table 1) coincides well with the above bands, and in its entirety would be regarded as a relatively pure alloy (total impurity content ~ 0.04 mass per cent). Notably, the lead content appears low. The significant point is the similarity in levels of intentional addition in the published work on

tin pest in 'pure' tin and typical amounts of impurity likely to be encountered in commercial lead-free alloys. So, in present day commercial alloys, the impurity levels could play a vital role in determining the susceptibility to tin pest formation.

A dichotomy arises between the scientific requirement for total understanding and the engineering need to evaluate the probability of tin pest formation in service. The studies described previously on Sn–0.5Cu involved a commercial alloy, the composition of which was given in Table 1. Despite the presence of three established tin pest inhibitors (Pb, Bi and Sb) the transition occurred eventually in the majority of samples. Detailed chemical analysis of individual specimens has yet to be performed.

Of all the candidate lead-free solder alloys, Sn–0.5Cu is the most dilute, i.e. it should resemble pure

Fig. 7 Penetration of α tin (tin pest) into the interior of the gauge head, and the interface region between the α and β phases

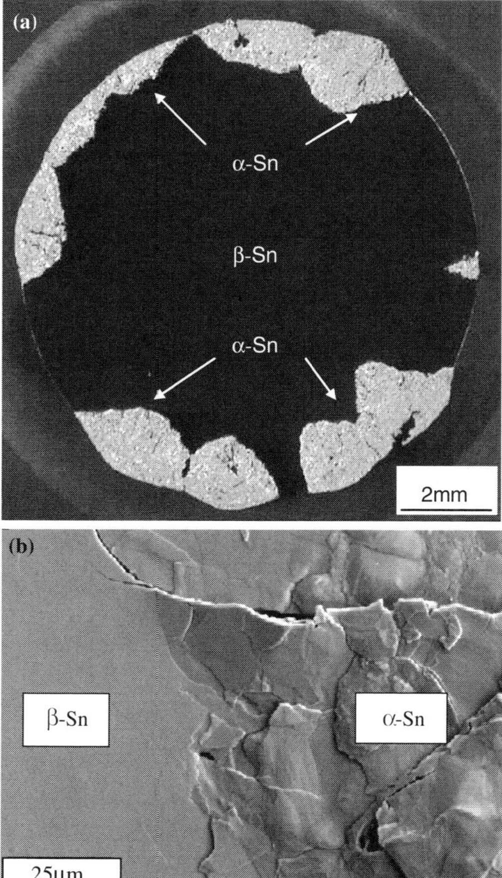

Fig. 8 Electron backscatter diffraction image of α/β interface and the region ahead of it

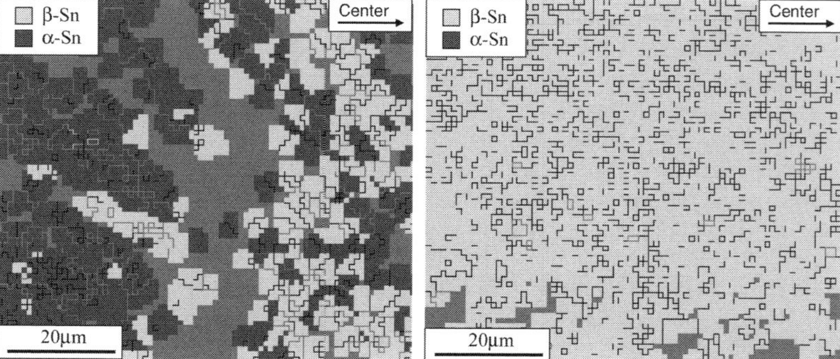

tin more closely than other selected systems. This is borne out by its propensity for tin pest formation. Approximately, one atom in every 100 in Sn–0.5Cu is a copper atom; in the ternary Sn–3.8Ag–0.7Cu, one in twenty is a solute atom, while in Sn–37Pb, one atom in five is lead. In the case of tin pest formation in alloys, two criteria are worthy of consideration; the proximity of the solid solution composition to pure tin (*the*

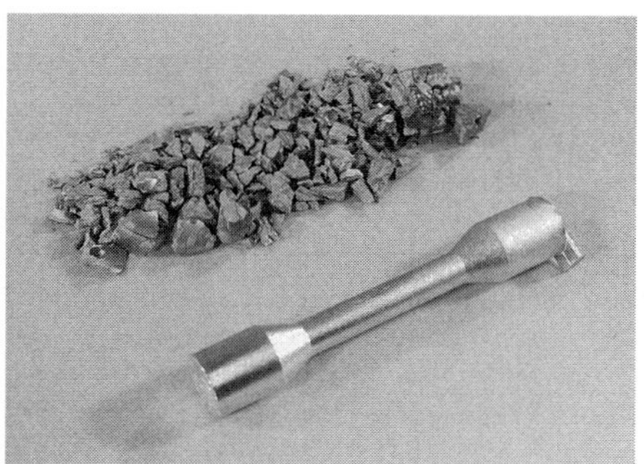

Fig. 9 Total disintegration of a Sn–0.5Cu test piece after prolonged exposure at – 18°C

proximity hypothesis), and the *excess solute* situation. From the available evidence, additions that suppress tin pest are soluble, thus affecting the solid solution with little or no excess solute atoms. Elements insoluble in tin are likely to be associated with excess solute atoms which may cluster or form precipitates (IMCs) and the conditions of instability conducive to the $\beta \rightarrow \alpha$ transition. The solid solution will contain a limited amount of solute, will approximate more closely to pure tin and hence be more susceptible to tin pest formation.

The discussion so far has largely focussed upon global composition considerations, but the possibility of local variations should also be considered. There will be regions in the microstructure that more closely resemble pure tin than indicated by the overall composition. Microstructure and local composition depend upon initial cooling rate and any subsequent heat treatment. For solders, room temperature represents a significantly high homologous temperature for thermal instability and continued changes in microstructure and local composition—so storage periods should be included. The equilibrium solid solubilities of Ag and Cu in tin at the respective eutectic temperatures are 0.04 mass per cent (0.042 atomic per cent and 0.0063 mass per cent (0.012 atomic per cent) respectively [42]. Considerably higher values have been found in a commercial version of the ternary Sn–3.8Ag–0.5Cu alloy (0.1 mass per cent Ag and 0.16 mass per cent Cu) [43], and this has been attributed to supersaturation during non-equilibrium cooling during the casting process [44]. This level of supersaturation was retained subsequently, even after prolonged annealing at 150°C and above [44]. In addition to the extent of solute atoms in solution, the volume fraction of 'tin' dendrites is likely to be sensitive to the cooling rate, if the findings for the family of Sn–3.8Ag–0.7Cu [45] apply to

Sn–0.5Cu. With the eutectic composition, the volume fraction of 'tin' regions (containing less than 0.1 mass per cent Ag and 0.006 mass per cent Cu) varied from 5 per cent under slow cooling ($1°C\ s^{-1}$) to 65 per cent after rapid cooling ($100°C\ s^{-1}$). The situation was further complicated by sensitivity to the Ag and Cu content—for hypoeutectic concentrations, an increase in cooling rate increased the volume fraction of 'tin' dendrites, whereas the opposite occurred for hypereutectic amounts of Cu and Ag. However, no effect of cooling rate was observed in the study of tin pest in Sn–0.5Cu described earlier.

The absence of tin pest on samples of Sn–0.5Cu aged at – 40°C for up to five years is in contrast to the earlier work on pure tin which suggested that the maximum rate of formation was around this temperature. While the thermodynamic driving force will undoubtedly be greater at the lower temperature, irrespective of any change in transition temperature induced by copper, the associated kinetics of the transformation could be slowing the process to time-scales beyond those in the present investigation. Alternatively, this finding might be attributable to differences between 'pure' tin and the Sn–0.5Cu alloy, but since the latter comprises only one copper atom in a hundred, such a key difference is not immediately apparent. The absence of tin pest after exposure at very low temperatures, when it forms at higher temperatures, has been reported for Cd and Bi additions to tinplate [27].

6 Tin pest in solder joints

The possibility if tin pest formation in solder joints was suggested by Bornemann [35] and by Williams [30] about half a century ago. Model joints, comprising samples of copper soldered with pure tin, a high purity Sn–Pb solder or a commercial solder, were exposed at – 40°C *after inoculation*. No tin pest was observed in the sample made using the commercial solder alloy—a fact attributed to the Bi an Sb impurities present.

In addition to the compositional effects which themselves may be distorted by conditions within the joint, the dimensions, geometry and scale of an actual interconnection may play a role in the formation of tin pest. From a geometrical and mechanical perspective, a soldered joint can be regarded a very thin layer of soft material (the solder) covered on its upper and lower surfaces by much harder materials (the intermetallic layer and the component/substrate surfaces). As miniaturisation proceeds, the solder layer is becoming almost a plane, and the significance of the IMC layer increases. Unlike the bulk solder samples described earlier, the proportion of free surface area in a joint is considerably limited.

Although tin pest forming on actual joints in service is unknown to the Author, it has been observed on relatively large *model* Cu/Sn/Cu joints in which the degree of constraint is much less (Y. Kariya, Private Communication). Figure 10 again demonstrates the propensity for tin pest to form at free surfaces. A soldered interconnection could contain several, if not all, classes of material (metal, ceramic, polymer and composite) each possessing quite disparate properties [47]. The mechanical response is then determined by the entire structure rather than simply by the behaviour of a single component in it. The constraint emanating from the surface layers is likely to restrict the large volume expansions associated with tin pest formation. Some of the initial studies on pure tin showed that the growth rate of the transformation product was halved at high pressure (90 atmospheres) [33, 34], suggesting perhaps that investigations of tin pest for aerospace applications should involve a vacuum? Simple physical constraint has been shown to inhibit the transformation in regular-shaped samples of tin [48].

Returning to chemical factors, regions throughout a soldered joint may experience quite different cooling rates, either between different joints or within a single unit. This will result in both microstructural and local compositional variations as discussed previously, but in a more complex manner. Similarly, initial or service-induced stress or strain (e.g. thermomechanical cycling) can also produce compositional changes which might affect the propensity for tin pest formation.

With the transition to lead-free solders, it should not be forgotten that other sources of lead (e.g. finishes and terminations) exist and may constitute over half of the lead available [49]. Implementation of lead-free technology involves more than simply changing the solder alloys.

It is hoped that the scenario envisaged by Lasky [50] that 'purification' of solder alloys (i.e. reducing Sb, Bi and Pb levels) might promote tin pest formation with dire consequences will never take place. However, this approach might be a valuable exercise in 'closing the door *before* the horse has bolted!

7 Areas for future investigation

Studies of phase transformations, such as the white-to-grey in tin, are generally associated with the heyday of X-ray crystallography and dilatometry. More than 70 years have elapsed since the fundamentals of the α/β

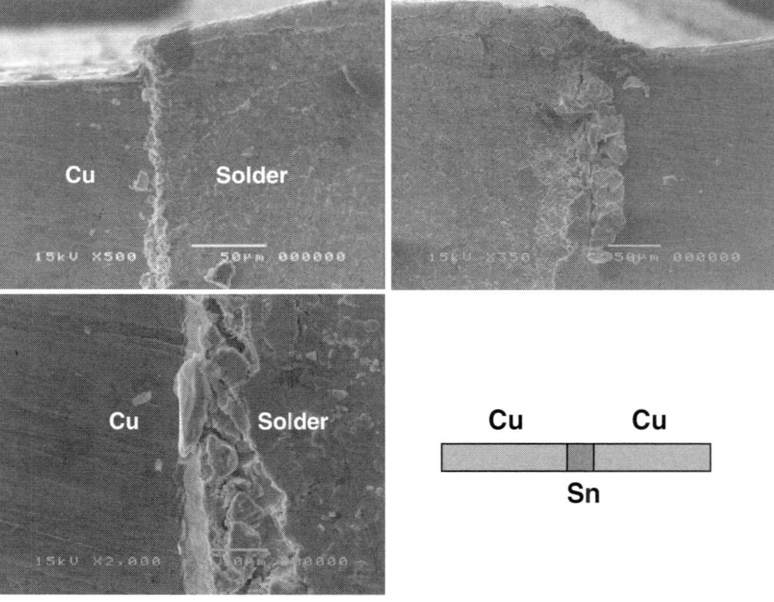

Fig. 10 Tin pest formation in a model Cu/Sn/Cu joint[46]

transition in tin, described earlier in this paper, were identified. During that period, tin pest was never a problem in solders because of the presence of lead and other inhibitors in solder alloys. However, with the advent of 'lead-free' solders and the findings of the most recent investigations, a technological driving force has been created and the subject merits re-investigation. The variety and capacity of observational methods now available (SEM, TEM, neutron diffraction, EPMA and so on) would have been unimaginable in the early days of the investigations. With this in mind, the following avenues seem worthy of further study;

- Illucidation of the precise mechanism of the transformation, particularly the nucleation stage and the criteria for its onset. Detailed crystallography, orientation relationships etc.
- Clarification of the roles of additional elements to tin—and why some retard the transition and others promote it. Considerable experimental evidence exists on this but the underlying reasons for the observed effects are unclear. Measurements of the transformation temperature.
- Examination of exposed samples with identical histories, only a proportion of which may exhibit tin pest. Identification of chemical and microstructural differences both globally and locally. Apart from near-surface regions, determination of preferable sites for nucleation.
- Evaluation of the kinetics of growth and its activation energy.
- With respect to tin pest formation in actual joints, determination of local microstructure and composition that will differ from that in bulk samples, and to some extent will be joint specific.
- The role of mechanical constraint warrants further investigation, and offers opportunity for modelling and stress analysis.

While these aspects themselves represent a formidable challenge, it is a salutary exercise to consider the potential variables involved (Table 2). Even inserting a minimal number in to the matrix produces a dauntingly large number of combinations.

8 Conclusions

A survey of tin pest formation in pure tin and tin-based alloys has indicated the following points relevant to this re-emerging and sometimes controversial topic:

- Like the pure metal, bulk samples of several dilute tin alloys experience the $\beta \rightarrow \alpha$ transition forming tin pest after prolonged exposure at low temperatures. Initial solidification rate appears to have little significance. The process involves nucleation and growth, with the former reaction being inconsistent in terms of time and its actual occurrence. Such sporadic behaviour is also found in pure tin. Due to this, the initiation stage has often been circumvented by seeding, and subsequent studies have been focused upon growth.
- A large volume change is associated with the $\beta \rightarrow \alpha$ transition, and tin pest is formed on the sample surface where constraint is a minimum. This causes eruptions (warts), cracking and eventual disintegration.
- The factors governing the allotropic change are uncertain. Elements soluble in tin, such as Pb, Sb and Bi, inhibit tin pest appearance, while insoluble additions such as Zn, Al and Mg promote its formation. Very small amounts appear to be influential, and it could be that impurities are unknowingly beneficial in suppressing tin pest. In the Sn–0.5Cu system, the addition of copper, at the level of one atom in a hundred, overrides the presence of three tin pest inhibitors, albeit in much smaller quantities.
- The exact criteria for initiation and the mechanisms by which growth occurs are not well understood. Both shear and massive transformations have been proposed, although the latter is currently in favour. Many of the experimental observations can be explained in terms of the resistance to the volume expansion emanating from either the strength of the white tin phase or the presence of a mechanical constraint.
- As well as global composition, local composition and microstructure may be important, and these are affected by cooling rate and the overall composition with respect to the associated eutectic.

Table 2 Factors affecting tin pest formation

Composition	Intended		Impurities	
Condition	As cast (Cooling Rate)	Cold worked (% CW)	CW+ Annealed (Annealing time, temp)	
Exposure	Inoculated	Non-Inoculated	Temperature	Time

- Tin pest formation in actual solder joints as they become even smaller is likely to be limited by the lack of free solder surface and the constraint imposed by the intermetallic layer and the adjacent component/substrate. The situation is further complicated by the possibility of variable cooling rates throughout the joint volume.
- An extensive investigation of the behaviour of lead-free solder alloys (Sn–3.5Ag, Sn–0.5Cu, Sn–3.8Ag–0.7Cu, Sn–8Zn–3Bi, with Sn–37Pb as a comparator) at − 18 and − 40°C, for periods of up to ten years, has shown that only the Sn–0.5Cu solder is vulnerable to the appearance of tin pest. To date, no tin pest has been observed in this alloy after storing at − 40°C for periods up to five years.
- The timescales over which experimental data are available are often shorter than required in some applications. Acceleration of the complete nucleation and growth process is difficult without resorting to artificial inoculation. So, perhaps the salient enquiry should not be *'Does tin pest occur...'*, but *'When does tin pest occur...'*.

References

1. 'Engineering the Future', Open University Course, T173, Block 3, Part 5, 'Lead-free solders', (2001)
2. W.J. Plumbridge, R.J. Matela, A. Westwater, *Structural Integrity and Reliability in the Electronics – Enhancing Performance in a Lead-Free Environment* (Kluwer, Dordrecht, Boston, London, 2003) Chapter 1
3. M.R. Harrison, J.H. Vincent, H.A.H. Steen, Solder. Surf. M. Technol. **13**, 21 (2001)
4. B. Richards, C.L. Levogner, C.P. Hunt, K. Nimmo, S. Peters, P. Cusack, *Lead-Free Soldering – An Analysis of the Current Status of Lead-Free Soldering* (Dept. Trade and Ind.,1999)
5. W.J. Plumbridge, C.R. Gagg, J. Mater. Sci.- Mater El. **10**, 461 (1999)
6. W.J. Plumbridge, C.R. Gagg, S. Peters, J. Electron. Mater. **30**, 1178 (2001)
7. 'The Properties Of Tin', (Tin Research Inst., Publication 218, 1954)
8. W.G. Burgers, L.J. Groen, Faraday Soc. Discussions **23**, 183 (1957)
9. O.L. Erdmann, J. Prakt. Chem. **52**, 428 (1851)
10. G.V. Raynor, R.W. Smith, Proc. Roy. Soc. **244A**, 101 (1958)
11. F. Vnuk, Metal Congress **45**, 175 (1975)
12. G. Tammann, K.L. Dreyer, Z. Anorg. Chem. **199**, 97 (1931)
13. E.S. Hedges, *Tin and its Alloys* (Edward Arnold Ltd, London, 1960)
14. R.M. Macintosh, Tin Uses **6**, 72 (1966)
15. J.L. Gissy, J.G. Kuva, Battelle Memorial Inst. Columbus Ohio Report (1960)
16. E. Cohen, A.K.W.A. van Lieshout, Proc. K. Akad. Wet., Amsterdam **39**, 1174 (1936)
17. Y.J. Joo, T. Takemoto, Mater. Lett. **3678**, 793 (2002)
18. A.A. Matvienko, A.A. Sidelnikov, J. Alloys Comp. **252**, 172 (1997)
19. D.R.G. Mitchell, S.E. Donnelly, Philos Mag. A **63**, 747 (1991)
20. K. Ojima, A. Takasaki (1993). Philos Mag. Lett. A **68**, 237
21. N. Blake, R.W. Smith, J. Mater. Sci. Lett. **5**, 103 (1986)
22. M. Kaya, F. Vnuk, R.W. Smith, Proc. Conf. on Phase Transformations, (Cambridge, Inst. of Metals, 1988), p. 647
23. F. Vnuk, A. De Monte, R.W. Smith, J. Appl. Phys. **55**, 4171 (1984)
24. E. Cohen, W.A.T. Cohen de Meester, J. Landsman, Proc. K. Akad. Wet., Amsterdam **40**, 746 (1937)
25. E. Cohen, A.K.W.A. van Lieshout, Proc. K. Akad. Wet., Amsterdam **39**, 352 (1936)
26. E. Cohen, A.K.W.A. van Lieshout, W.A.T. Cohen der Meester, Z. Phys. Chem. **178**, 221 (1937)
27. R.R. Rogers, J.F. Fydell, J. Electrochem. Soc. **100**, 383 (1953)
28. C.W. Mason, W.D. Forgeng, Metals Alloys **6**, 87 (1935)
29. E. Cohen, W.A.T. Cohen der Meester, Proc. K. AKAD. Wet. Amsterdam **51**, 860 (1938)
30. W.L. Williams, *Symposium on Solder* (ASTM STP 189, Philadelphia, 1956), p. 149
31. F. Vnuk, A. De Monte, R.W. Smith, Mater. Lett. **2**, 67 (1983)
32. W.M.T. Gallerneault, F. Vnuk, R.W. Smith, J. Appl. Phys. **54**, 4200 (1983)
33. G. Tammann, R. Kohihass, Z. Anorg. Chem. **199**, 209 (1931)
34. E. Cohen, A.K.W.A. van Lieshout, Proc. K. Akad. Wet., Amsterdam **39**, 596 (1936)
35. A. Bornemann, *Symposium on Solder*, (ASTM STP 189, Philadelphia, 1956), p. 129
37. Performance Test Methods and Qualification Requirements for Surface Mount Attachments, (IPC – 9701, January 2002)
38. The Solder Research Group at the Open University, Update Report (http://www.materials.open.ac.uk/srg/srg-index.html) (2005)
39. Y. Kariya, C.R. Gagg, W.J. Plumbridge, Solder. Surf. M. Technol. **13**, 39 (2000)
40. Y. Kariya, N. Williams, C.R. Gagg, W.J. Plumbridge, J. Mater. **53**, 39 (2001)
42. C.E. Homer, H. Plummer, J. Inst. Met. **64**, 169 (1939)
43. K.W. Moon, W.J. Boettinger, U.R. Kattnewv, F.S. Biancaniello, C.A. Handwerker, J. Electron. Mater. **29**, 1122 (2000)
44. L. Snugovsky, C. Cermignani, D.D. Perovic, J.W. Rutter, J. Electron. Mater. **33**, 1123 (2004)
45. L. Snugovsky, P. Snugovsky, D.D. Perovic, J.W. Rutter, Mater. Sci. Technol. **21**, 61 (2005)
47. W.J. Plumbridge, R.J. Matela, A. Westwater, *Structural Integrity and Reliability in the Electronics – Enhancing Performance in a Lead-Free Environment*, (Kluwer, Dordrecht, Boston, London, 2003) Chapter 2
48. E.S. Hedges, J.Y. Higgs, Nature **169**, 621 (1952)
49. A. Brewin, Sixth European Surface Mount Conference, Brighton Workshop 6, November, (2004)
50. R. Lasky, Proc. Surface Mount Technology Assoc. Int. Conf., Chicago, Sept, (2004)

Issues related to the implementation of Pb-free electronic solders in consumer electronics

D. R. Frear

Published online: 8 September 2006
© Springer Science+Business Media, LLC 2006

Abstract Consumer electronic applications are the primary target of the Pb-free initiative and package assembly and performance is affected by the move from eutectic Sn–Pb to Pb-free solder alloys. This paper outlines the key issues and mitigation possibilities for package assembly using Pb-free solders: High temperature reflow, Interfacial reactions, and Reliability. At the high temperatures required to reflow Pb-free alloys, moisture absorbed into the package can result in delamination and failure. The reaction of the Pb-free solder with Ni and Cu metallizations results in interfacial intermetallics that are not significantly thicker than with Sn–Pb but provide a path for fracture under mechanical loading due to the increased strength of the Pb-free alloys. The reliability issues discussed include thermomechanical fatigue, mechanical shock, electromigration and whiskering. The Pb-free alloys tend to improve thermomechanical fatigue and electromigration performance but are detrimental to mechanical shock and whiskering. Design trade-offs must be made to successfully implement Pb-free alloys into consumer applications.

1 Introduction

Consumer electronic applications are now driving packaging technology much in the same way computers enabled new packages for the past 25 years. Consumer

D. R. Frear (✉)
Freescale Semiconductor, Temple, AZ, USA
e-mail: darrel.frear@freescale.com

electronic applications consist of typically hand-held devices and home use products. Examples include: cell phones, game controllers, radios, MP3's and large applications like televisions and smart appliances. The trend of these devices is smaller, faster, less expensive and built in high volume. The applications are also becoming more diverse with mixed technologies that include digital microprocessors (embedded and stand-alone), memory, RF wireless and along features. In many cases, especially RF applications, the package is part of the electronic system and defines the performance. The packaging trend for these applications is moving to modules that are simpler to integrate and create an electrical system but add to the packaging complexity (i.e., multiple die and passive devices into a single package). This results in an increasingly complex package that is critical in system integration and cost. Figure 1 is an example of the complexity of this packaging in an integrated Radio Frequency (RF) module.

The interconnect material of choice for the integrated consumer electronic system is solder. The solder is used to join discrete packages or modules to boards and is increasingly used to interconnect the semiconductor die to the package through either die attach or as a flip chip interconnect. The solder acts not only as an electrical conduit but also as a thermal path and mechanically holds the parts in place. Solder has excellent electrical and thermal properties carrying current and heat very well and has been used from the very beginning in electronic systems.

The traditional solder material is a Sn–Pb alloy. Sn–Pb solder is used because it has a relatively low melting point, good wetting behavior, good electrical conductivity and can be used in hierarchical soldering. Hierarchical soldering is the utilization of a solder that

Fig. 1 Integrated RF module with solder die attached die and discrete passives

has a lower melting temperature than all other solder interconnects that precede it. The lower temperature solder must have a working temperature sufficiently low so that it does not melt a higher temperature interconnect. The melting temperature range for Sn–Pb solders is from 310°C for Sn–97Pb to 183°C for eutectic Sn–37Pb. The basic requirement of the solder interconnect is to form a reliable electrical and mechanical connection that retains integrity through subsequent manufacturing processes and service conditions. The joints are also required to have the capacity to dissipate strains generated as a result of coefficient of thermal expansion mismatches under service conditions over the lifetime of the assembly.

Recently, however, Pb has come under increasing scrutiny as a heavy metal toxin that can damage the kidneys, liver, blood, and central nervous system. The voiced concern is that as the volume of electronics increases in the waste stream that the Pb can leach from landfills into the water table forming a potential source of long-term contamination of soil and ground water. International laws have recently been proposed to expand control laws to limit or ban the use of Pb in manufactured electronics products. The most aggressive and well known effort is the European Union's Waste in Electrical and Electronic Equipment (WEEE) and Restriction of Hazardous Substances (RoHS) and the China RoHS directives propose a ban on Pb in electronics by mid year 2006. The products impacted most heavily by these laws are consumer products (because they have a relatively short lifetime and the volume of that will end up in landfill) are

significant. A comprehensive review of the status of Pb-free solders can be found in this current publication and in the literature [1–3].

Pb-free solders for electronic applications are based on Sn-rich compounds that fall into a melting temperature range 40°C or more above eutectic Pb–Sn solder alloys (183°C). These include eutectic Sn–3.5Ag with alloying elements of Bi, Cu, Sb, In, or Zn. Other alloys based on the Sn–Cu, Sn–In, Sn–Sb, Sn–Bi and Sn–Zn systems have also been proposed. A small two-phase region (temperature difference between liquidus and solidus) is desired because it prevents the joint from moving and becoming disturbed during solidification. Binary or ternary near-eutectic alloys are also desired because simpler alloys reduce the potential for compositional variations that affect the behavior of the solder joint. A chart of the melting temperatures of potential solder alloys is shown in Fig. 2. The Pb-free solder alloys Sn–0.7Cu, Sn–3.5Ag, and Sn–3.8Ag–0.7Cu are among the most promising based on the criteria given above and Sn–Ag–Cu has the highest degree of acceptance in the electronics industry. The following section outlines the physical metallurgy of Pb-free solders that forms the basis for how these solders perform physically and mechanically.

Along with the increase in melting temperature, the Pb-free alloys have differences in soldering performance, intermetallic formation and mechanical behavior compared to eutectic Sn–Pb. These differences contribute to a number of key challenges that must be addressed for Pb-free solder to be successfully implemented into consumer electronics applications. These are listed below. The following sections of this

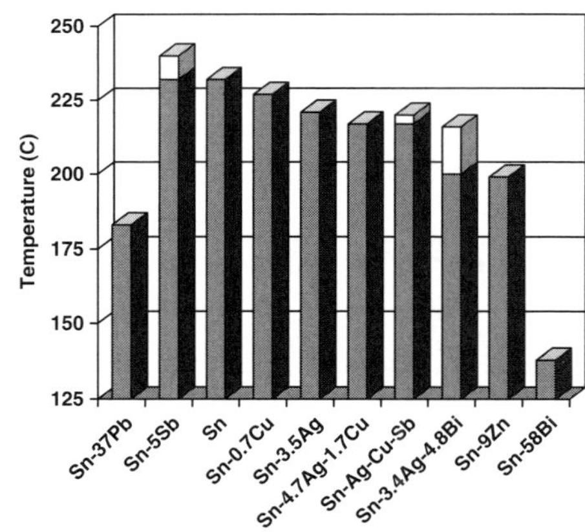

Fig. 2 Chart of Pb-free solder temperatures (dark=solidus, light=liquidus)

 Springer

paper go into the metallurgical details of each of these challenges and how they can be addressed.

1.1 Key issues in consumer electronics applications

- Higher temperature reflow requirement
- Interfacial reactions (intermetallic formation)
- Reliability (thermomechanical fatigue, mechanical shock (drop), electromigration, and whiskering)

2 Metallurgy of Pb-free solders

The microstructure of eutectic Sn–Pb and Pb-free solders are shown in Figs. 3–6. The microstructure defines the behavior of the solder alloy. The Sn–Pb eutectic has a structure of Sn-rich and Pb-rich phases that form as lamella, Fig. 3. Similarly oriented lamella form cells, or colonies, separated by slightly coarsened cell boundaries. This structure is known to be susceptible to heterogeneous microstructural evolution that concentrates strain at the cell boundaries and causes them to further coarsen and, eventually, be the site of failure [4].

Figure 4 shows the Sn–0.7Cu microstructure. This solder is composed of large Sn-rich grains with a fine dispersion of Cu_6Sn_5 intermetallics. The solder grains form and grow out from the bond pad interfaces. The grains are large; on the order of 20–50 µm in size.

The Sn–3.5Ag microstructure consists of a fine structure of alternating Sn-rich/Ag_3Sn intermetallic lamella, Fig. 5. Grain colonies also form in this microstructure but the boundaries are not coarsened. In addition to the fine Ag_3Sn intermetallics, large needles of Ag_3Sn are present and are typically attached to one of the bump pad interfaces. The Sn–3.8Ag–0.7Cu

Fig. 4 Optical micrograph of the microstructure of Sn–0.7Cu solder on Cu

solder microstructure is shown in Fig. 6 and is very similar to that of the Sn–3.5Ag with the addition of discrete Cu_6Sn_5 intermetallics throughout the bulk of the solder.

3 Key issues in consumer electronics applications: reflow temperature issues

The melting temperature of the Pb-free solder is higher than the 183°C of eutectic Sn–Pb (see Fig. 2) and the packages must reach the melting temperature plus at least 20°C to achieve reflow. For Sn–Pb this reflow temperature is typically 220°C. For Pb-free solders, the reflow temperature is typically between 240°C and 260°C. The added temperature for Pb-free reflow is needed because the solders do not wet as well as Sn–Pb and the higher temperature is used for compensation. This higher reflow temperature has a potentially significant impact on commercial electronics.

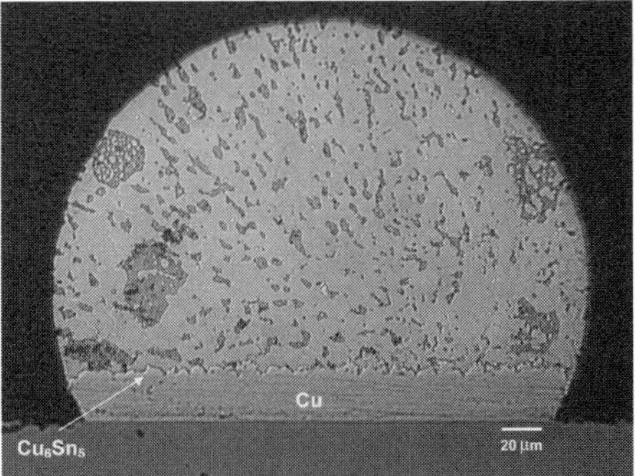

Fig. 3 Optical micrograph of the microstructure of Sn–37Pb solder on Cu

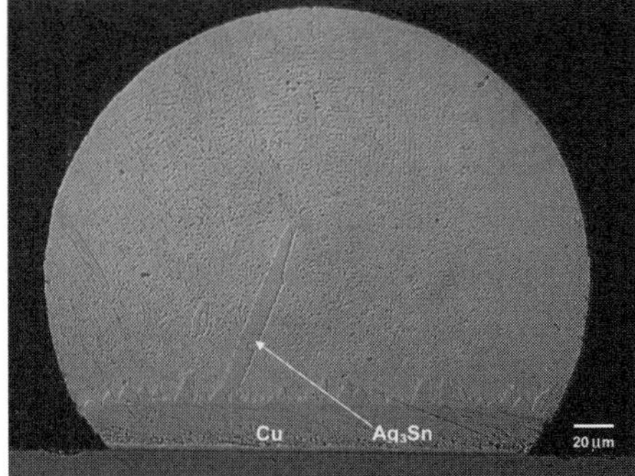

Fig. 5 Optical micrograph of the microstructure of Sn–3.5Ag solder in a flip chip bump on a Cu UBM

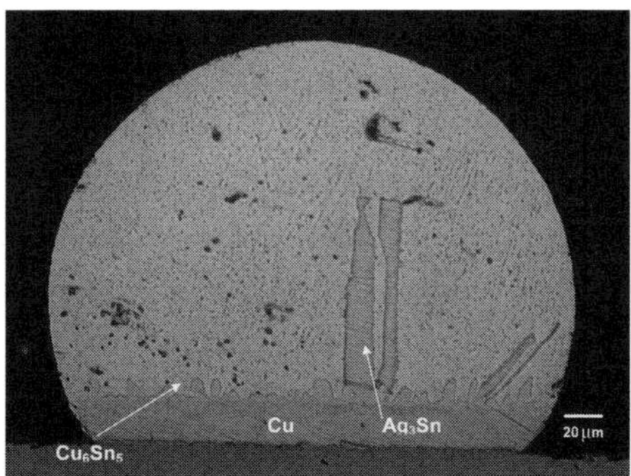

Fig. 6 Optical micrograph of the microstructure of Sn–3.8Ag–0.7Cu solder in a flip chip bump on a Cu UBM

The electronic packages for commercial applications consist of many materials, and an increasing number of polymer materials, and interfaces between these materials. These polymers include glass/epoxy substrate, organic die attach, solder masks, and mold compounds (epoxy/silica composite). Figure 7 is an illustration of a cross section of a module package that shows these many interfaces. This is an issue because polymers absorb moisture as can interfaces. Before reflow, the packages are exposed to the environment and will absorb moisture from the air. As the packages undergo reflow, the moisture turns into steam and expands creating an internal stress in the package. This can result in interfacial delamination in the package and electrical failure if interconnect lines are broken or

warpage of the package due to differential stresses. This situation is exasperated by the higher reflow temperatures of Pb-free solders. Figure 8 shows a scanning acoustic microscope image of a module package showing delamination (light areas in the image). This is not as serious an issue for Sn–Pb eutectic applications because the materials set in the lower temperature reflow conditions were optimized for Sn–Pb applications. The increase of 20–40°C for Pb-free solder reflow has created this problem.

There is no simple solution to this issue. The higher temperature reflow is required. The problem must be addressed by improved adhesion of the materials in the package and decreasing the water permeability of the polymers. Increased adhesion can be achieved by adding chemical bonding agents to the polymer materials and by increasing surface roughness of the materials to be joined thereby increasing the mechanical adhesion. New mold compounds are needed with lower water uptake and higher modulus (higher fill content), with no change in manufacturability, to limit the potential for steam generation in the package and decrease the potential for warpage by making the package stiffer (higher modulus).

4 Key issues in consumer electronics applications: interfacial reactions

The transformation of solder-wettable coatings into intermetallics by solid-state reactions can also degrade

Fig. 7 Illustration of a package with a die solder die-attached. Note the many polymers and interfaces in the package

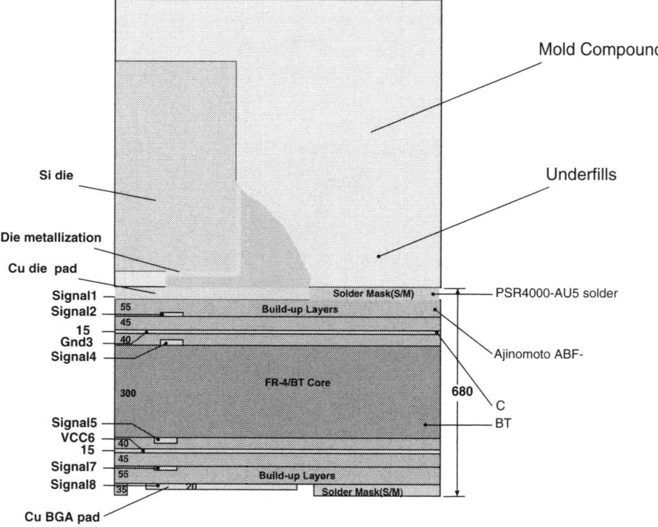

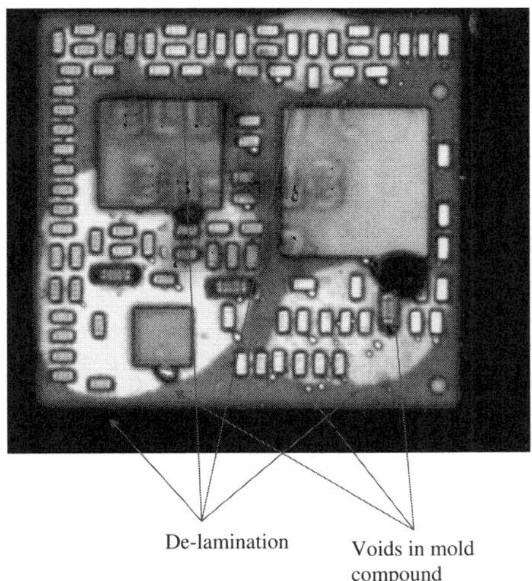

Fig. 8 Scanning Acoustic image of a module package showing evidence of delamination caused by water uptake then reflow

mechanical properties. The interfacial intermetallics are brittle due to their complex crystal structures with few crystallographic planes available to accommodate stress and may fracture when strain is imposed, especially if the strain is tensile in nature. There is also a concern that the growth of the intermetallics will be accelerated in high Sn-content Pb-free solders. For commercial electronics applications the interfacial intermetallic structure can have a significant impact on performance and this must be understood.

When the molten solders come into contact with the Ni or Cu surfaces they wet and react to form interfacial intermetallics. The intermetallics grow out into the solder as rods, or plates, and continue to grow when the solder is in the solid state. Even though all the solders studied are Sn-rich, the morphology and reaction kinetics differ between alloys. As will be discussed later in this paper, the interfacial intermetallics play a role in the reliability and mechanical behavior of the solder interconnect. If the intermetallics grow too thick they can consume the underlying metallization and fall off. If too thick, the intermetallics can also be the weak link in the joint and the fracture path under fast loading conditions.

On Cu, Sn–37Pb forms a two-phase intermetallic of Cu_6Sn_5 adjacent to the solder and Cu_3Sn adjacent to the Cu. The Cu_3Sn is planar with a columnar grain structure and the Cu_6Sn_5 consists of elongated nodules. On Ni, eutectic Sn–Pb forms irregularly shaped Ni_3Sn_4. The formation and growth of the interfacial intermetallics between Cu, Ni and eutectic Sn–Pb solder are well known [5, 6]. The intermetallics follow parabolic growth kinetics and do not extensively spall off into the solder.

The intermetallic formation for the Pb-free solders on Ni is shown in Fig. 9. Figure 9 shows the structure of the interface on electroless Ni–P, but the same observations were made for the electrolytic Ni UBM. The intermetallic that forms is Ni_3Sn_4. The Ni_3Sn_4 intermetallic between Sn–0.7Cu and Ni is thin and regular and is the most uniform of the Pb-free alloys. The Sn–3.5Ag solder on Ni has a different intermetallic

Fig. 9 SEM micrographs of the Pb-free solders on an electroless Ni UBM showing the morphology of the interfacial intermetallics

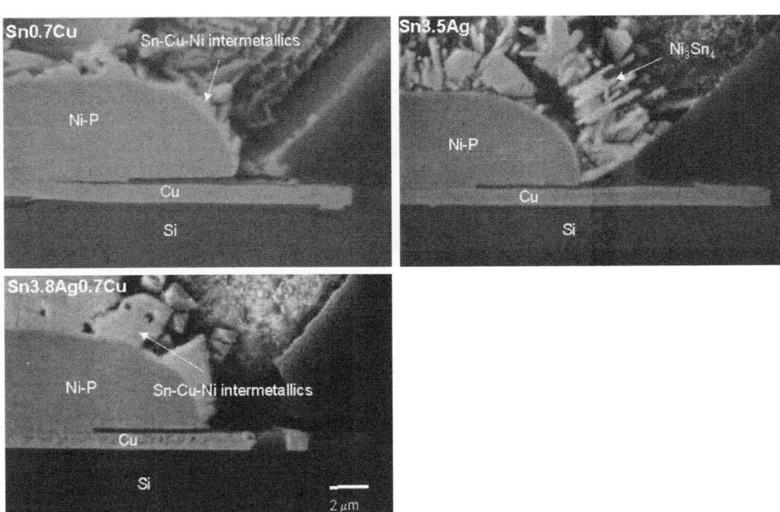

morphology that consists of nodules and chunks of Ni_3Sn_4 that spall off into the solder. This morphology has been attributed to the lack of Cu in the solder. It is hypothesized that the Cu acts to saturate the solder with respect to the Ni and inhibits dissolution and spalling of the intermetallic into the solder [7]. However, this mechanism remains to be fully understood.

The growth of interfacial intermetallics while the solder is in the solid-state is of concern because if the intermetallic layer coarsens significantly, it can consume the solderable metallization and cause the joint to dewet at the layer beneath the UBM. Also, the intermetallic is brittle and if it becomes a significant fraction of the solder joint it can act as a site for crack initiation and propagation when the joint is deformed. The consumption of the Ni layer by the formation of Ni_3Sn_4 intermetallic during solid-state aging for each of solder alloy is shown in Fig. 10 for 1000 h aging at 150° and 170°C. Figure 10 shows results on an electroless Ni–P/Au UBM (similar results were observed on the electrolytic Ni UBM). For these solders, less than 2 µm transformed into Ni_3Sn_4. The Ni reacts slowly with Sn-based solders and is the reason why it is often preferred as the metallization for Sn-rich solders. The Pb-free solders consume more Ni and form more intermetallic than Sn–Pb eutectic solder but this increase is relatively small. The Ag_3Sn intermetallic plates are attached to the Ni_3Sn_4 interfacial intermetallics, similar to that observed on the Cu UBM.

Figure 11 shows SEM micrographs of the solders on a Cu. The intermetallic that forms is Cu_6Sn_5, no Cu_3Sn was found but it may have been too thin to be observed. The interfacial intermetallic has the same morphology for all the Pb-free alloys as for Sn–Pb eutectic. The intermetallic consists of regularly spaced nodules of Cu_6Sn_5. The Ag-containing solders all have large Ag_3Sn intermetallics that are attached to the Cu_6Sn_5 interface. All three Pb-free alloys also have small discrete particles of Cu_6Sn_5 present in the bulk of the solder. For Sn–0.7Cu, the Cu is present in the

solder before joining to the UBM. For the Sn–3.5Ag solder the Cu_6Sn_5 is present due to the dissolution of some of the UBM into the solder. Spalling of the interfacial intermetallics into the molten solder was not observed and this is believed to be due to the presence of Cu in each of the solders during reflow. The Cu inhibits growth and spalling of the intermetallic because it saturates the solder [8]. Of the three Pb-free alloys on Cu, the Sn–0.7Cu solder structure is the most uniform, and has the thinnest intermetallic structure.

One of the concerns of using Sn-rich Pb-free solders is the reaction of the Cu with the solders is feared to be so fast that the UBM will dissolve during reflow. A plot of the Cu consumed after 2 reflows for Sn–Pb and the Pb-free solders is shown in Fig. 12. The Pb-free solders consume only 10–20% more Cu than Sn–Pb and this is less than 2 µm after 2 reflows. A plot of consumed Cu during solid-state aging is shown in Fig. 11 for the solders on Cu at 150°C for 500 and 1000 h. In the solid state, the Cu was consumed at a slower rate in Pb-free solders than for eutectic Sn–Pb. The Pb appears to play a role in enhancing intermetallic growth, perhaps by enhancing Sn diffusion to the intermetallic/solder interface. Sn–0.7Cu had the slowest consumption rate of the Cu UBM.

5 Key issues in consumer electronics applications: reliability

The reliability of the commercial electronic system includes everything that can happen to the package after it is manufactured. The environmental stresses that drive degradation of the package, such as temperature, thermal cycling and passing electrical current (electromigration) are typically externally imposed and can not be controlled in commercial applications for cost reasons. The following describes the issues associated with thermal cycles, mechanical shock/drop, electromigration and whiskers that can affect consumer product reliability.

5.1 Thermal cycling

The solder joint is a mechanical interconnect that must have sufficient strength to hold the joined component in place but must also deform sufficiently so that imposed strain is not translated to the joined components. The shear strength of the Pb-free solder alloys is shown in Fig. 13 as compared to eutectic Sn–Pb. Sn–37Pb and Sn–0.7Cu solder bumps have similar values of shear strength. The Sn–3.5Ag solder alloy has the greatest shear strength at 25% greater than

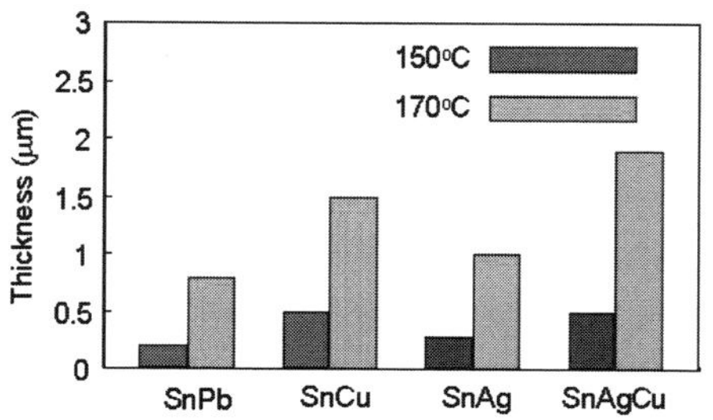

Fig. 10 Thickness of consumed Cu for the solders on electroless Ni–P/Au UBM after 1000 h aging at 150 °C and 170 °C

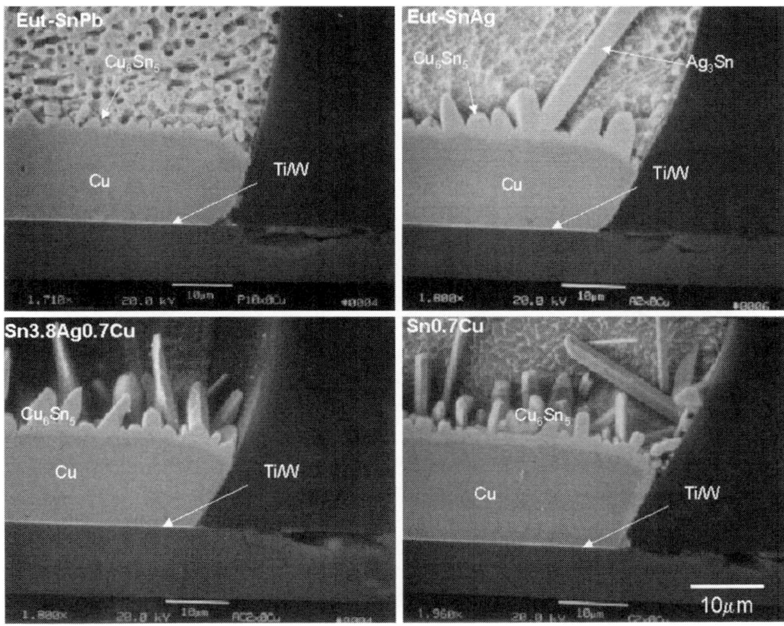

Fig. 11 SEM micrographs of the Pb-free solders on Cu UBM

Sn–0.7Cu with the Sn–3.8Ag–0.7Cu with a slightly lower strength than Sn–3.5Ag. Eutectic Sn–3.5Ag has shown susceptibility to brittle interfacial delamination in surface mount interconnects [5].

Creep behavior is important for solder interconnects because the solders deform to relax stress over time when held at a constant strain. The creep rate of a solder must be sufficiently fast so that the strain is minimized in joined bulk components. However, the creep rate must not be so fast that the components move over time. The creep behavior of solders can be summarized empirically using one of two equations:

$$\mathrm{d}\gamma/\mathrm{d}t = A\sigma^{n}\mathrm{e}^{-Q/RT} \tag{1a}$$

$$\mathrm{d}\gamma/\mathrm{d}t = A\sinh(\alpha\sigma)^{n}\mathrm{e}^{-Q/RT} \tag{1b}$$

where $\mathrm{d}\gamma/\mathrm{d}t$ is the creep rate in shear, A is a constant, α is the stress constant, σ is the flow stress, n is the stress exponent, and Q is the creep activation energy. Equation (1a) works well for creep mechanisms that remain constant over all test temperature. Equation (1b) is the Garofalo, or $\sinh$, creep relation that captures up to two different creep mechanisms in a single formulation. Table 1 represents the fitted

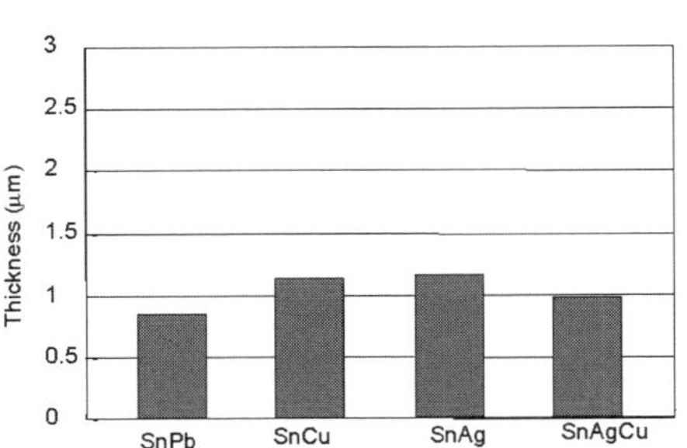

Fig. 12 Consumed Cu thickness after 2 reflows for the solders on a Cu UBM

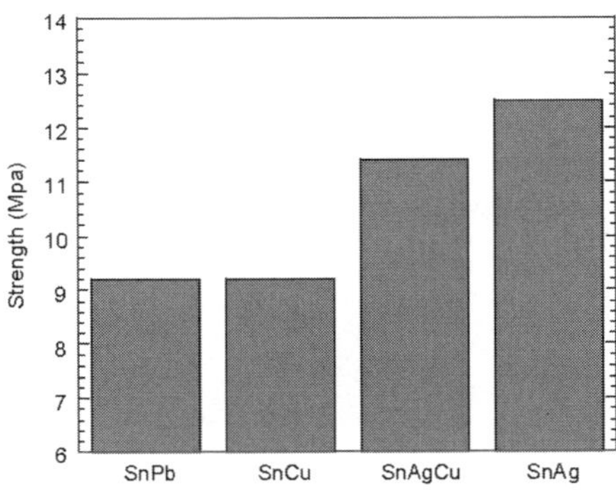

Fig. 13 Shear strength of solder alloys

Table 1 Constitutive creep relations

Alloy	A (s^{-1})	α	n	Q (kJ/mol)	Ref
Sn-40Pb	2.48×10^4	0.0793	3.04	56.9	9
Sn-40Pb	1.1×10^{-12}		6.3	20	10
Sn-3.5Ag	9.3×10^{-5}		6.05	61.2	11
Sn-3.8Ag-0.7Cu	2.6×10^{-5}		3.69	36	11
Sn-1Cu	1.41×10^{-8}		8.1	79.4	12

results of creep tests on Pb-free solders using these equations. In general, the creep rate of Sn–Ag–X solders is significantly slower than Sn–Pb. A faster creep rate is desired because damage can be accommodated by the solder, rather than the more brittle joined components. Furthermore, a faster creep rate often translates into a longer thermomechanical fatigue lifetime.

Temperature variations encountered during use conditions, combined with the materials of differing coefficients of thermal expansion in the electronic package, result in cyclic temperature and strain on the solder joints (thermomechanical fatigue). A diagram of thermal fatigue life versus strain range for the Pb-free solder alloys is shown in Fig. 14, The thermal fatigue data from Sn–0.7Cu is the longest followed by Sn–Ag–Cu, Sn–Pb and then Sn–Ag.

The Sn–37Pb flip chip interconnects fail by crack formation and propagation through heterogeneous coarsened bands near the UBM/solder interface as shown in Fig. 15. The heterogeneous coarsened region forms as a result of the strain concentrating at the weak eutectic cell boundaries and cracks eventually form at these boundaries [8, 13–14]. The Sn–0.7Cu solder deforms in thermomechanical fatigue by grain boundary sliding. The cracks were observed to propagate at the grain boundaries significantly removed from the

UBM/bump interface, near the center of the joint. This solder is the most compliant in thermal fatigue and undergoes massive deformation before failing by crack propagation. The failure behavior of Sn–3.5Ag solder is similar to the Sn–3.8Ag–0.7Cu alloy with crack initiation and propagation through the intermetallics and at the intermetallics/solder interface. The Sn–3.5Ag solder has a shorter thermal fatigue lifetime than the other Pb-free alloys. It is hypothesized that the large Ag$_3$Sn intermetallic plates strengthen the joint and further reduce the compliance of the solder thereby shortening the thermal fatigue life.

In summary, solder joint strength and creep behavior has a direct correlation to fatigue performance. The weaker, more compliant, and faster steady-state creep rate solder (Sn–0.7Cu) has a longer fatigue lifetime. Eutectic Sn–Pb has the same low shear strength as Sn–0.7Cu but has a shorter thermal fatigue lifetime due to heterogeneous coarsening. The Sn–0.7Cu alloy undergoes massive deformation during thermal cycling. This deformation protects the semiconductor device from damage due to imposed strain. Interestingly, even with significant surface deformation, the crack formation takes longer in Sn–0.7Cu than the other solders that had no surface deformation. The Sn–0.7Cu accommodates the thermal strain by grain boundary sliding and rotation. The self-diffusion of Sn to accommodate the sliding and rotation is sufficient to delay the formation of cracks. An additional positive aspect of the Sn–0.7Cu solder is that failure is only observed in the solder joint away from the brittle intermetallic/solder interface.

5.2 Mechanical shock/drop

For consumer electronic applications, one of the greatest challenges for the package assembly is to survive a very challenging use environment that includes being dropped. For example, a cellular phone or MP3 player is expected to work if it is accidentally dropped while in use (e.g., surviving a fall of up to 2 m onto a hard surface). This drop is a mechanical shock loading condition and is increasingly recognized as a significant reliability concern. Portable consumer electronics are more likely to be dropped than affected

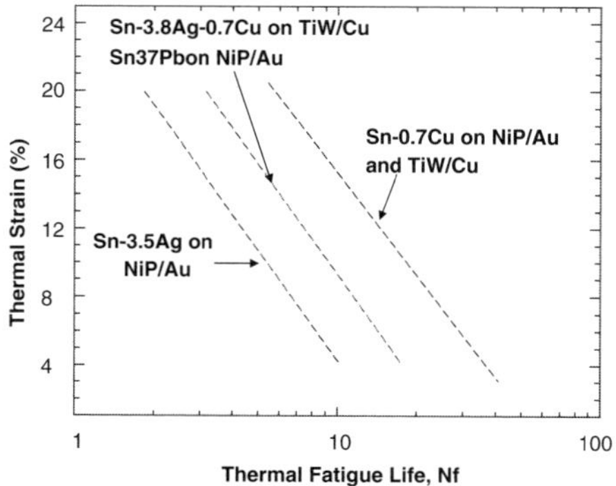

Fig. 14 Plot of thermal fatigue life as a function of applied thermal strain for the Sn–3.5Ag, Sn–37Pb, Sn–3.8Ag–0.7Cu and Sn–0.7Cu

Fig. 15 Optical micrographs
of solder joint cross sections
of solder joints on Cu UBM/
Cu pads on organic substrates
after thermal cycling 0 °C to
100 °C

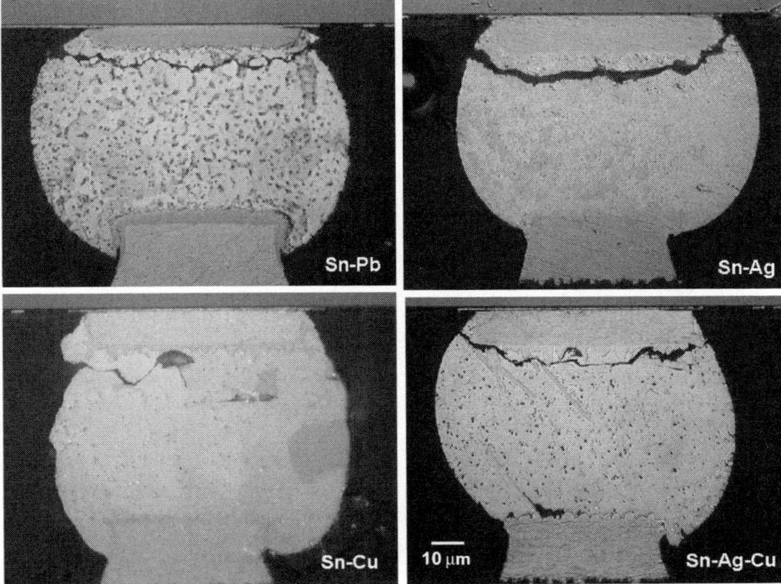

by changes in the temperature of the ambient environment, reliability requirements have added mechanical shock loading to the traditional moisture sensitivity level, thermal cycling and power cycling tests [15].

The literature on mechanical shock loading of electronic packages and assemblies is starting to grow [16–20]. The weak link in the package is the board level solder interconnect between the package and the printed wiring board. It has also been noted that Pb-free solder interconnects are more prone to failure in drop conditions than Sn–Pb solder.

At the time of impact, the mechanical shock is translated into a rapid displacement in the board in the form of a decaying sine wave. The board bends with this wave rapidly putting the components alternatively into tension or compression. Clearly, the peak strain in tension is where the solder joints are most likely to fail. The strain levels imposed on solder joints are complex and are dependent upon the configuration of the board, the board material and its thickness, the angle at which the board hits the surface, the location of the component on the board, the density of the solder joints on the component and the height of the solder joints. For example, a densely populated board would be stiffer and have less strain imposed, a thicker solder joint would experience more strain than a thinner joint, etc. However, a typical shock of 1500 g would last 1 ms at a drop height of 1.5 m. Depending upon the board,

component placement and angle of contact this translates into a strain of 2000–4000 µ strain and a strain rate of 10^2. There is very limited materials data at this strain rate and strain range, particularly for solder materials.

There are a variety of proposals on how to evaluate drop performance of portable consumer electronics. European cell phone manufacturers have a drop test that involves dropping the finished cell phone 1.8 m onto a linoleum covered concrete floor five times in each of four different orientations and then verifying the absence of cracking. The existing JEDEC drop test method (JESD22-B111) is only a relative performance tool. Both of these methods are pass/fail and provide no fundamental data on solder joint performance. Dynamic 3- and 4-point bend testing is currently preferred to characterize solder interconnects in drop [21]. This testing involves a soldered component on a board that is placed in a 3- or 4-point bend fixture (component down) and a weight or a load is dropped on the board. The height at which the weight is dropped or the rate of load displacement determines the level of strain.

The failure of board-level solder interconnects in drop tends to be more extensive in ball grid array packages than land grid array because the joint is thicker and more dynamic strain is imposed. Pb-free solder joints fail at a lower level of strain than eutectic Sn–Pb. The Sn–Ag and Sn–Ag–Cu solders have a higher strength than Sn–Pb. Rather than the solder deforming during mechanical shock, the strain is translated to the interface of the solder joint and cracks form through the interfacial intermetallics. Figure 16 shows a SnAgCu solder joint after drop testing and a crack has propagated through the interfacial intermetallic resulting in electrical failure of the joint.

There are a few ways to improve the drop behavior of solder interconnects. The major issue with Pb-free solders is that the strength is significantly increased over Sn–Pb eutectic and this forces failure through the more brittle interfacial intermetallics. This can be alleviated by decreasing the strength of the solder (perhaps by reducing the Ag and Cu content). The issue with changing the solder alloy is that the higher strength SnAgCu solders tend to have longer thermo-mechanical fatigue lifetimes than the more compliant solders so there is a trade-off in performance that would have to be optimized. The other method to improve drop performance is to minimize the strain imposed on the interconnects by stiffening the board. This can be done by adding more solder joints on the component or adding "dummy" joints in the high strain regions. From a system architecture point of view, minimizing strain can also be accomplished by designing the board layout to be stiffer (more soldered components) in high strain regions.

5.3 Electromigration

In both Pb–Sn and Pb-free flip chip applications, the size of the solder interconnect is decreasing to enable higher i/o count. The small ball size has raised reliability concerns, especially electromigration. Electromigration is not an issue for board-level interconnects, they are too large, but for flip chip this is becoming increasingly significant. Current circuit design rules require that each interconnect carry a current of up to 0.2 amp with an increase to 0.4 amp in the near future. A current of 0.2 amp through an interconnect 50 μm in diameter results in a current density of 1×10^4 a/cm^2. This is less than that found in Al or Cu interconnects on semiconductor die (1×10^5 a/cm^2 or more) but the solder has a much lower homologous temperature and, therefore, has a greater atomic diffusivity at the device operating temperature of ~100°C. The high current density and low melting temperature—can result in the electromigration of flip chip solder joints. Examples of published work on electromigration of solders can be found in the literature [22, 23].

For Sn–Pb solder, hillocks form on the anode side of the solder ball and voids formed on the cathode side. The anode becomes Pb-rich indicating Pb is the dominant diffusing species. Microstructural evolution was extensive in the solder induced by electromigration where both Sn and Pb phases (and grains) coarsen significantly. In Pb-free solders the electromigration mechanism is different. Figure 17 shows a typical high Sn solder bumps (Sn–3.5Ag) both anode and cathode on Cu pads. The electromigration enhances the formation of Cu–Sn intermetallics that react and across the bump. Failure occurs when the Cu metallization is consumed where electrons enter the interconnect and voids form and joint together.

The electromigration of Pb-free solders is still being investigated and the fundamental relationships are being established. However, there is strong evidence that Pb-free solders have a much longer mean time to failure in electromigration than do Sn–Pb eutectic alloys (on the order of twice the lifetime). This could be due to the difference in damage accumulation in the solders (Sn–Pb undergoes coarsening and failure in the solder itself, Sn-rich solders fail by the dissolution of the base metallization and intermetallic growth).

5.4 Whiskers

Solderable surfaces in an electronic package must have metallization (surface finish) that is compatible with Pb-free solders and can easily be wet by the solder. Many current surface finishes are Sn–Pb based and must be replaced with Pb-free alternatives. The two surface finishes that have found the most widespread use are Ni/Pd/Au and pure Sn. The Ni/Pd/Au is a higher cost alternative and is more difficult to wet than

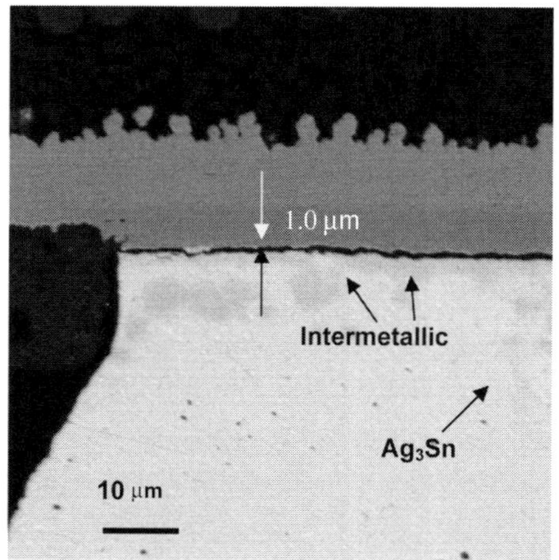

Fig. 16 SEM of a SnAgCu solder after drop test showing interfacial cracking through the intermetallic

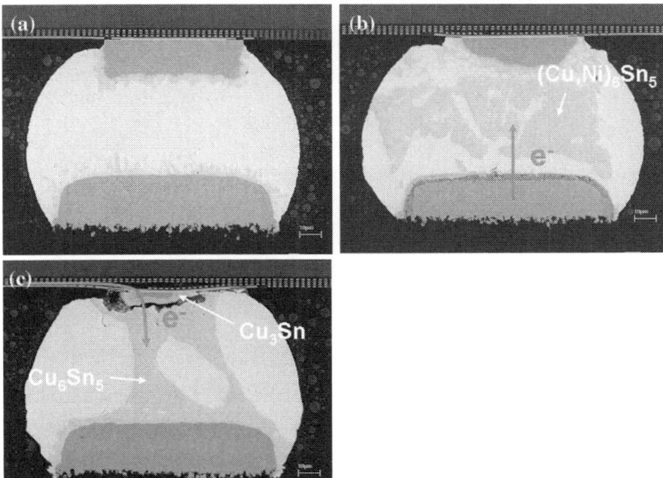

Fig. 17 SEM micrographs of Sn–3.5Ag solder bumps after 836 h at 115 °C. (**a**) No current bump, (**b**) cathode bump with 5.1×10^4 A/cm^2, (**c**) anode bump with 5.1×10^4 A/cm^2

Sn–Pb or Sn metallization so it is not applicable for consumer applications. The Sn metallization is low cost but can experience whiskering. Whiskers of Sn are metallic crystals that grow out of the surface of the metallization very rapidly and can be very long (several mm). If the whiskers join across conducting contacts it can result in an electrical short. A scanning electron micrograph of a Sn surface with whiskers is shown in Fig. 18. The mechanism of whisker growth is a time dependent (creep) phenomenon and the driving force is compressive stress in the Sn film [24, 25]. The internal compressive stress in the film is generated from the formation of the Cu$_6$Sn$_5$ intermetallic that occurs at room temperature after the Sn is plated. The whisker formation is creep in the film that occurs due to the presence of a stress gradient that allows the stress to be relaxed. In the Sn plating this gradient is a crack in the Sn surface oxide. The whisker then grows out of this discontinuity and the Sn diffuses via the grain boundaries in the film.

Tu et al. [24, 25] propose that minimization of the whiskering can be achieved by removing the internal stress in the Sn film, reducing the internal diffusivity of the Sn or eliminating the stress gradient. Eliminating the stress gradient is exceptionally difficult because only a very small gradient is required and modifying the Sn oxide is difficult. Removing the internal stress can be accomplished by stopping the formation of Cu$_6$Sn$_5$ intermetallics by plating Ni on the Cu leads prior to Sn plating. Alternatively, annealing the plated film at 160°C allows Cu$_3$Sn intermetallic to form (it does not form at room temperature) and this will also reduce the stress. Reducing the internal diffusivity of Sn has not been demonstrated but could be achieved by adding binary elements to the Sn film (such as Cu) that could coat the boundaries and reduce Sn grain boundary diffusion.

6 Summary

There is a strong effect on the performance of commercial electronics caused by the move to Pb-free solders. This is particularly true for thermomechanical fatigue behavior, electromigration, mechanical shock (drop) and whiskering. The Pb-free alloys are Sn-rich and change the microstructure and mechanical behavior of the interconnects. The Pb-free solder consist of a Sn-rich phase with dispersed intermetallics of SnAg and CuSn that result in a higher strength alloy than

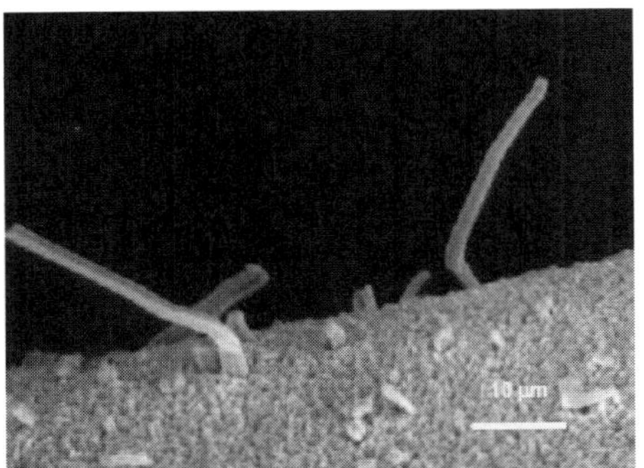

Fig. 18 SEM image of Sn whiskers forming on a Sn-plated metallization

eutectic Sn–Pb. This improves thermomechanical fatigue because there is no heterogeneous coarsening that shortens the life of eutectic Sn–Pb. It also improves electromigration performance because there is no coarsening and lifetime is dictated by interfacial intermetallic growth. However, the strength of the solder combined with the interfacial intermetallics detrimentally affects mechanical shock (or drop) performance. (Note: of all the properties discussed in this paper, mechanical shock is the one that must clearly be addressed for consumer (portable) applications.) The Pb-free solders also require a Pb-free surface finish that is typically Sn-plating that can form whiskers that can result in failures due to electrical shorts. The mitigation of these effects through metallurgical changes in the solder result in trade-offs between properties (e.g., improving drop performance by decreasing the strength of the solder can degrade thermomechanical fatigue performance). The best mitigation incorporates the overall architecture of the package/board design for optimal solutions.

Acknowledgements I would like to acknowledge the excellent technical support of the following people in Freescale Semiconductor. Pb-free thermomechanical fatigue: Jong-Kai Lin, Jin-Wook Jang, Mechanical Shock: Norm Owens, Electromigration: Lakshmi Ramanathan.

References

1. J. Glazer, Int. Mater. Rev. **40**, 65 (1995)
2. J. Glazer, J. Electron. Mater. **23**, 693 (1994)
3. D.R. Frear, J.W. Jang, J.K. Lin, C. Zhang, JOM **53**, 28 (2001)
4. D.R. Frear et al., *Solder Mechanics: A State of the Art Assessment* (TMS Publications, Warrendale, PA, 1990)
5. D.R. Frear, P.T. Vianco, Metal. Trans. **25A**, 1509 (1994)
6. W.J. Boettinger et al., in The Mechanics of Solder Alloy Wetting and Spreading, ch. 4 (Van Nostrand Reinhold, New York, 1993)
7. H.K. Kim, K.N. Tu, P.A. Totta, Appl. Phys. Lett. **68**, 2204 (1996)
8. D.R. Frear, D. Grivas, J.W. Morris Jr., J. Electron. Mater. **17**, 171 (1988)
9. J.J. Stephens, D.R. Frear, Metall. Trans **30A**, 1301 (1999)
10. Z. Mei, J.W. Morris, J. Electron. Mater. **21**, 599 (1992)
11. D.R. Frear, Sandia National Labs Report # SAND96-0037 (1997)
12. J. Liang et al., Mat. Res. Soc. Symp. Proc. **445**, 307 (1997)
13. D.R. Frear, D. Grivas, J.W. Morris Jr., J. Electron. Mater. **18**, 671 (1989)
14. D.R. Frear, in *The Metal Science of Joining*, ed. M.J. Cieslak, M.E. Glicksman, S. Kang, J.H. Perepezko (TMS Publications, Warrendale, PA, 1992) p. 191
15. V. Puligandla, S. Dunford, J. K. Kivilahti, in *Handbook of Lead (Pb)-Free Technology for Microelectronics Assemblies*, ed. by K. Puttliz (Marcel Dekker Inc., New York, 2004) p. 769
16. T.T. Mattila, J.K. Kivilahti, J. Electron. Mater. **35**, 250 (2006)
17. K. Mishiro, S. Ishikawa, M. Abe, T. Kumai, Y. Higashiguchi, K. Tsebone, Microelectron. Reliability **42**, 77 (2002)
18. K. Kujala, M. Kulojärvi, in *Proc. IMAPS Nordic Annual Conference*, 2003, p. 21
19. T.Y. Tee, H.S. Ng, C.T. Lim, E. Pck, Z. Zhong, in *Proc. 53th IEEE-ECTC Conf.*, 2003, p. 121
20. M. Alajoki, L. Nguyen, J. Kivilahti, in *Proc. IEEE-ECTC Conf.*, 2005, p. 637
21. R. Darveaux, A. Syed, in *Proceedings of the SMTA*, 2000
22. J.W. Nah, K.W. Paik, J.O. Suh, K.N. Tu, J. Appl. Phys. **94**, 7560 (2003)
23. Y.C. Hsu, C.K. Chou, P. C. Liu, C. Chen, D.J. Yao, T. Chou, K.N. Tu, J. Appl. Phys. **98** (2005) 033523
24. K.N. Tu, J.O. Suh, A.T.C. Wu, N. Tamura, C.H. Tung, Mater. Trans. **46**, 2300 (2005)
25. K. N. Tu, C. Chen, Albert T. Wu, J. Mater. Sci.: Materials in Electronics (to be published 2006)

Impact of the ROHS directive on high-performance electronic systems

Part I: need for lead utilization in exempt systems

Karl J. Puttlitz · George T. Galyon

Published online: 22 September 2006

Abstract The European Union enacted legislation, the ROHS Directive, that bans the use of lead (Pb) and several other substances in electronic products commencing July 1, 2006. The legislation recognized that in some situations no viable alternative Pb-free substitute materials are known at this time, and so provided exemptions for those cases. It was also recognized that certain electronic products, specifically servers, storage and storage array systems, network infrastructure equipment and network management for telecommunication equipment referred to as high-performance electronic products, perform tasks so important to modern society that their operational integrity had to be maintained. The introduction of new and unproven materials posed a significant potential reliability risk. Accordingly, the European Commission (EC) granted an exemption permitting the continued use of Pb in solders, independent of concentration, for high-performance (H-P) equipment applications. This exemption was primarily aimed at assuring that the reliability of solder joints, particularly flip-chip solder joints is preserved. Flip-chip solder joints experience the most severe operating conditions in comparison to other applications that utilize Pb in electronic equipment. This paper briefly describes the solder-exempted H-P electronic products, their capabilities, and some typical tasks they perform. Also discussed are the major attributes that differentiate H-P electronic equipment from consumer electronics, particularly in relation to their operational and reliability requirements. Interestingly, other than the special solder exemption accorded to H-P electronic equipment, these products must meet all the other requirements for ROHS compliancy. The EC was aware that issues would surface after the legislation was enacted, so it created the Technical Advisory Committee (TAC) to review industry-generated requests for exemptions. The paper discusses three exemption requests granted by the EC that are particularly relevant to H-P electronic products. The exemptions allow the continued use of lead-bearing solder materials.

K. J. Puttlitz (✉)
Puttlitz Engineering Consultancy, LLC, Wappingers Falls, NY 12590, USA
e-mail: KJPuttlitz@aol.com

G. T. Galyon
IBM eSystems and Technology Group, Poughkeepsie, NY 12601-5400, USA
e-mail: Galyon@us.ibm.com

1 Introduction

The toxic effect of lead (Pb) is well documented and widely reported to be related to numerous health risks. Therefore, in the US the use of lead (Pb) has been banned in such products as paint, solder for plumbing, and as an additive in gasoline among others for sometime. But its continued use in electronic assemblies has led to a global movement seeking to establish a lead-free environment. In particular, it has been noted that so-called electronic waste, consisting mostly of consumer electronic products, is primarily disposed of in landfills where it poses a serious potential health risk. The concern is related to the possibility that rainwater will leach lead (Pb) from the solder on printed circuit boards (PCBs) and other parts containing Pb, and that the leachate will seep into the groundwater system and contaminate municipal water supplies. Based on these concerns the European Parliament and the Council of the European Union (EU) passed legislation, "Restriction of the Use of Certain Hazardous Substances (ROHS) in Electrical and Electronic Equipment," Directive 2002/05/EC effective July 1, 2006. The ROHS Directive defines the requirements that each EU Member State must embody in separate laws enacted and enforced by each member country. Among these requirements is that a maximum defined concentration of four metals (mercury, lead, cadmium, and hexavalent chromium) and two organic-based fire retardant substances not be exceeded for certain products. This paper contains references to "lead-free" solder alloys, which has become an industry-standard term meaning the solder alloy should contain less than the ROHS Directive limit of 0.1 wt% (1,000 ppm) lead. It does not mean the solder alloy will necessarily contain no lead at all. The scope of the legislation essentially covers all electrical and electronic equipment including consumer electronics, information technology equipment (e.g. laptop, desktop computers), electric light bulbs, etc. There are some product categories that are outside the scope of the legislation: medical devices, military products and monitoring/control devices (under review). Also, the ROHS Directive does not apply to spare parts for the repair of electrical and electronic equipment (EEE) placed on the market before July 1, 2006; or the reuse of EEE placed on the market before that date. Lead is pervasively used in electronic products of all types. Among the allowable usages of lead (Pb) are: glass in monitors, brass and aluminum hardware, free-machining steel members, certain electronic components (e.g. capacitors, piezo electric devices), lead-acid batteries for operation or backup, cable sheathing and solder for components and assembly to PCBs. The quest to replace Pb-bearing solders is a daunting task and is the largest and most costly effort the electronics industry has ever undertaken. Lead (Pb)-based solders have been utilized to assemble circuit boards in electrical products for more then 50 years. The conversion to Pb-free technology has raised some serious reliability concerns within the industry. The European Commission (EC) recognized that the introduction of a new material set in electronic products posed risks, particularly as it relates to reliability. Accordingly, some exemptions were granted because Pb was used in some applications for which there are no known suitable substitutes. For example, the EC allowed the use of Pb in the glass screens of cathode ray tube monitors to provide protection from X-rays. An exemption was also granted for the use of Pb in solder materials, but it only applied to

high-performance systems (servers, storage, network infrastructure/telecommunication systems). These electronic products are markedly different from consumer products in almost every way. This paper briefly describes those products that received the special solder exemption and their capabilities because they perform mission-critical operations, a necessary element of modern society. Reliability is the single most important attribute that distinguishes so-called high-performance (H-P) electrical equipment from consumer products. It is imperative that H-P electronic equipment possess the capability to operate successfully and uninterrupted for years on end to perform their important tasks, as contrasted with the occasional use of consumer electronics that provide entertainment and convenience.

The European Commission (EC) further recognized that issues would arise during the lead (Pb)-free implementation process. Therefore, the EC created the Technical Advisory Committee (TAC) to review industry requests for exemptions (i.e. allow the use of Pb) for specific applications. The requests were to cover applications where there currently is no suitable replacement for lead (Pb), or where the replacement substance(s) posed a greater potential risk to human health or the environment than lead. Several important exemption requests were granted by the EC that have particular relevance to high-performance products. The exemptions allow the use of mixed solder joints to attach chips; and the use of Pb for compliant pins, and part of a thermal-dissipation subsystem. These exemptions are all discussed in detail.

2 High-performance electronic systems

The European Union's ROHS legislation expressly grants an exemption for the use of lead (Pb) in solders, regardless of composition in high-performance, high-reliability systems specifically, "servers, storage and storage array systems, network infrastructure equipment for switching, signaling, transmission as well as network management for telecommunications" [1]. But the intention is to phase out the exemption, and conduct progress reviews every few years with the initial review scheduled for 2008. These products have several characteristics and requirements in common that differentiate them from consumer-electronic products. The most important characteristic is that they all require a very-high degree of reliability since they are relied upon to carry out mission-critical operations and must virtually perform flawlessly over their operational lifetime. Equipment failures could have severe societal or human life consequences. A brief description of the high-performance systems granted a special solder exemption and their reliability requirements follows.

2.1 Solder-exempted systems and their functions

2.1.1 Server systems

Servers are often classified into three categories. (a) Entry-level (Fig. 1 b, d) systems, which possess a very significant processing capability and are utilized for a variety of applications: management and integration of small-to-large business operations, and optimized to integrate network and distributed computing. They are also used for specialized applications such as patient monitors; and for (control) equipment in hospitals, industry, etc. In some cases sophisticated desk-top personal computer (PC) systems may possess similar capabilities, but even so-called entry-level servers are designed and tested to provide a very high level of reliability not characteristic of PCs. (b) There are several types of mid-range systems, some are architected to provide optimum management and integration of business operations (Fig. 1f); while others are designed to maximize computing power for business, government, and scientific applications with the ability to model complex systems, perform detailed data mining, etc. by reliably performing massive scalar-integration operations (Fig. 1e). (c) High-end server systems, often referred to as mainframe or enterprise systems (Fig. 1c) are configured to provide unmatched computing performance, data integrity, security, scalability, and interconnection capability. Because of these unique attributes they are used to provide massive, reliable computing power for such applications as

insurance, international banking, and airline reservations in the business arena; are the basis of national security systems; perform operations at all levels of government and military; research; development; etc. [2].

2.1.2 Storage Systems

The storage function is an integral part of complex and sophisticated information systems that store, provide, and direct the flow of information to server systems. Disk libraries (Fig. 2a) are large stand-alone systems used to access frequently needed information. These systems are faster than tape systems and typically consist of more than 125 drives, where each drive provides storage of more than 70 giga bits (GB) of information. Tape systems (Fig. 2b) are also stand-alone units with a capacity of about 6,000 tape cartridges, and with tape drives that can read and write tape media. They provide long-term storage and access libraries. Storage Area Network (SAN, Fig. 2c)

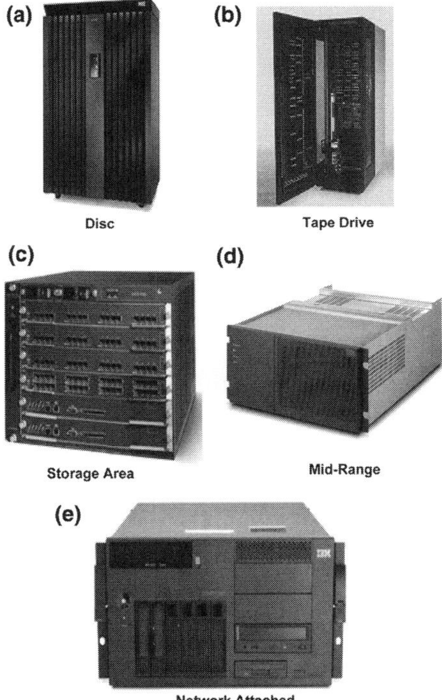

Fig. 2 The photograph depicts a range of storage systems. (**a**) Disc library. (**b**) Tape drive system. (**c**) Storage Area Network (SAN). (**d**) Mid-range system. (**e**) Network Attached Storage (NAS) (after Ref [2]) (courtesy of IBM Corporation)

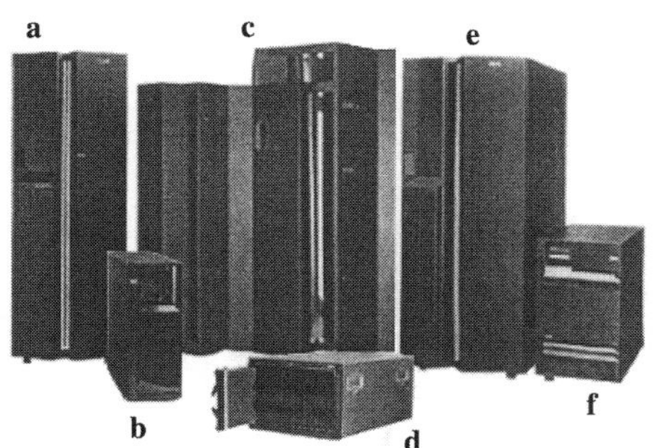

Fig. 1 The photograph depicts a range of IBM server systems. (**a**) pSeries, intermediate high-end. (**b**) iSeries, low-end. (**c**) zSeries, high-end. (**d**) xSeries, configurable blade server. (**e**) pSeries, high-end server. (**f**) iSeries, mid-range server (courtesy of M. Hoffmeyer, IBM Corporation)

systems serve as a switch or a director of information among servers and/or storage units. These units are capable of very rapid data rates, e.g. 10 GB/s or more. Mid-range Systems (Fig. 2d), so-called because they are often utilized to support mid-sized businesses, are rack-mounted units. They consist of a small number (15 or so) of disc drives that provide a continuous flow of data across storage networks. Network Attached Storage (NAS, Fig. 2e) systems are also rack-mounted units that contain a small number of disc drives and configured to serve as an interface between a LAN and SAN [2, 3].

2.1.3 Network infrastructure and telecommunications equipment

The solder exemption granted for network infrastructure equipment specifically relates to routing, switching, signaling, transmission, network management or security and telecommunications. Network infrastructure equipment consists of integral and complex elements of information systems that provide mission-critical support for communication networks.

2.2 High-performance systems vs. consumer electronics

The requirements and characteristics of high-performance electronic products are very different from consumer electronics (Table 1). The PCBs are very complex compared to consumer products and typically consist of a large number and high mix of attached components, with thousands of solder joints (Fig. 3). High-end equipment is expensive, low volume, and usually owned by corporations. The equipment is

Table 1 Typical characteristics and requirements of high-performance systems vs. Consumer electronics

Characteristic/ Requirement	High-performance systems	Consumer electronics
Complexity	High to very high	Low
Development time	2–5 years	1–3 mos.
Owners	Corporations	Individual end users
Product life	Servers/Storage: 5–15 years Network Infra. Equip: 15–20 years Telecommunications: up to 25 years	1–5 years
Utilization	24 h/day, 7 days/week over product life time	Intermittent over life time
Volumes	Very low	Very high

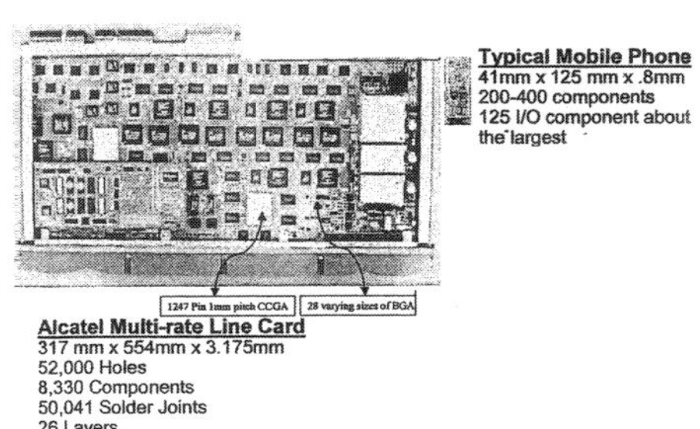

Fig. 3 Photographs that illustrate the difference in complexity between a telecommunications infrastructure product (high-performance) PCB and a mobile phone PCB (consumer product). The photographs are approximately to scale (after Ref. [4]) (courtesy of Alcatel)

sophisticated, requiring long development cycles to assure the demanding functional and reliability requirements are met. These products experience continuous utilization over their operational life time, with virtually no allowed down time. In some applications such as satellite systems there is no opportunity for repair; while others must operate under harsh environmental conditions, experiencing temperature extremes between − 40 and 70°C [4]. The successful operation of high-performance systems is absolutely necessary, given their potential societal and life/death consequences in the event of failure, compared to annoyances of consumer-product failures. Changes that have the potential of compromising reliability such as the introduction of Pb-free solder joints cannot be incorporated into high-performance systems until they are demonstrated to pose a very low risk.

2.3 Reliability and operational requirements

The reliability and operational requirements of the high-performance systems granted a solder exemption, far exceed those of any other electronic equipment in use today. For example, these systems are expected to be capable of continuous operation 24 h per day, 7 days per week, for at least 10 years in most cases [2, 3] and up to 25 years for some telecommunications equipment. The allowed "downtime" for high-performance systems varies from an average of 5.3 min per year for network infrastructure equipment [5] to essentially no allowable failures for mainframe computer systems (Table 2) and the storage systems that support them. For example, based on an average of 100 installed units, a mainframe computer typically only experiences one incident repair action (IRA) in over 10 years of operation. However, the system is still fully or partially functional so repairs can be made at the

Table 2 Typical mainframe computer operational and reliability capabilities

Feature/Aspect	Capability
Typical field life	90,000–100,000 h
Lifetime operating percentage	99.9954%
Average total lifetime downtime	Less than 4 h
Frequency of an Incident Repair Action (IRA)*	> 10 years
Frequency of an Unscheduled Incident Repair Action (UIRA)*	> 30 years

*Average based per 100 installed units in the field (after Ref. [3])

user's convenience. Sometimes no failures are involved, but modifications made to update the system such as installing a microcode patch, change a channel card, etc. An actual system crash or so-called unscheduled incident repair action (UIRA) typically only occurs about once every 30 years or more, also an average based on every 100 systems installed in the field. The requirements for high-performance systems far exceed those of consumer electronics, which are only used occasionally and whose operational lifetime is normally 1–5 years. Also, there are no particular reliability norms associated with consumer-electronic products.

2.4 Use of lead in high-performance electronic systems

High-performance electronic systems typically are complex. For example, a large-scale electronic processor consists of 10,000 or more individual parts that are combined into subassemblies that are themselves combined into functional assemblies or systems, e.g. power, memory, logic, cooling, etc. Many of these contain lead (Pb). Individual parts, subassemblies, and assemblies are categorized into subclasses to efficiently manage parts purchasing, the vendor supply line, and production. Although there are no industry standards, some typical subclass categories that utilize lead and common to many high-performance products are listed in Table 3. Lead (Pb) is an ingredient in a variety of materials utilized in electronic products including cathode ray tubes, (CRTs) for shielding radiation, in cables as a PVC stabilizer, in certain ceramic components (e.g. capacitors, piezo electric devices), but predominantly in solders. It is estimated that over 200,000 electronic components contain lead-bearing solders. Solders are literally the "glue" that hold electronic products together and also possess the necessary electrical, mechanical and thermal properties. Solder is utilized as a coating material on the traces (i.e. copper circuitry) of PCBs and component lead frames. Solder

Table 3 Some lead (Pb) containing hardware categories common to many high-performance electronic products

Part/Assembly category	Lead (Pb) utilization
Electronic functions	Card/component lead finishes
	Solder joints
Memory functions	Component lead finishes
Logic functions	Component lead finishes
Storage finishes	Component leads/card
	Solder joints
Power supply	Components/solder/card/finishes
Voltage Reg. Modules	Components/solder/card/finishes
I/O & system cards	Surface finishes
Actives/Optics/Passives	Lead finishes
Cables/Connectors	Thermal stabilizer in cables
	Connector finishes
Mechanical parts	Metal joining (solder)
Black/Grey boxes	Hard drives
	Monitors
	Power supplies

is used for flip-chip solder bumps, and solder-ball terminated components (BGAs, CSPs). It is also used to attach all the circuit elements to PCBs (components, connectors, passives, etc.), to form hermetic seals for some components, etc. Eutectic (63Sn–37Pb) solder has been universally used for the assembly of cards and boards, but both low-Pb solders (85Sn–15Pb) and high-Pb solders (97Pb–3Sn) are utilized in electronic products as well [6].

3 Exemptions important for high-performance electrical systems

The ROHS legislation notes a variety of applications (Table 4) where the use of lead (Pb) is permitted regardless of the electronic product category since no viable alternative materials are known. It is also stated

Table 4 ROHS Lead (Pb) exemptions for any electronic product

Exemptions
Non-solder related
Piezo electric devices
All glass products
Cathode ray tubes, fluorescent tubes, filter glass
Lead utilized as an alloy with certain metals
0.35% in steel
0.4% in aluminum
4.0% in copper
Electronic ceramic components
Lead-acid storage batteries
Solder related
High-lead content solders, > 85% Pb
Two-element solders, > 80% Pb, but < 85% Pb

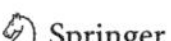 Springer

in the Annex of the ROHS legislation that the use of lead in solders is permitted (i.e. with no concentration limitations) in servers, storage systems, and network infrastructure/telecommunications equipment. It should be noted however, that the use of lead in these high-performance systems is also banned, except for solder materials. The European Commission (EC) never issued any guidance documents that defined the term "solder", which has caused some controversy and confusion in the industry. For many applications there is broad agreement, while in others it is not clear if the exemption applies. In any event, the general solder exemption (> 85% Pb), and the solder exemption for high-performance systems were granted to assure the reliability of solder joints, which will be discussed in detail later in this paper.

The EC realized that as the electronics industry pushed forward in implementing Pb-free technology new issues were likely to arise. Accordingly, a procedure was put in place whereby the EC's Technical Advisory Committee (TAC) reviews industry-generated exemption requests. Three exemption requests granted by the EC that are of particular importance to high-performance systems are: flip-chip mixed solder joints, the use of Pb for compliant-pins, and for a C-ring utilized in a cooling system.

4 Lead (Pb) use in solder exemptions for microelectronic packages applications

4.1 Components with wirebonded chips

The ROHS Directive makes it necessary to eliminate lead (Pb) (i.e. < 1,000 ppm level) in microelectronic components utilized in electronic products that do not qualify for an exemption such as personal computers (PCs), personal data assistants (PDAs), camcorders, digital cameras, and other consumer electronics products. Although lead (Pb) is used internally for some microelectronic components, it is most often used for component terminations (i.e. lead-frame finish, or solder ball/bump) and the solder paste to reflow attach components to the next level of assembly, typically a printed circuit board (PCB). The use of Pb for some major microelectronic package types is illustrated in Fig. 4. A quad flat pack (QFP) consists of a chip that is adhesively back-bonded to the chip carrier. Thin gold or aluminum wires are bonded to pads located around the periphery of the chip that provide the electrical path between the chip and plastic chip carrier. Lead–tin coated metal fingers, called leads, extend beyond the molded plastic body and are used to attach the package to the PCB. A Pb-free lead finish (e.g. pure Sn) is used to render these components ROHS compliant. This is also true for solder-plated pads on quad flat non-leaded (QFN) and bumped chip carrier (BCC) packages. They also utilize adhesively-mounted wirebonded chips (Fig. 4b, c). Plastic ball grid array (PBGA) components and their enhanced versions EPBGA or HPBGA with wirebonded chips (Fig. 4d, e) consist of an array of solder balls on the under side of the plastic chip carrier, used for attachment to a PCB. The standard solder ball composition is typically eutectic Sn–Pb or eutectic Sn–Pb with 2% Ag to help reduce leaching Cu from terminal pads during reflow operations. Both compositions are compatible for assembly with Sn–Pb solder paste. ROHS-compliant versions are created by utilizing Pb-free solder balls, usually a Sn–Ag–Cu (SAC) solder alloy.

4.2 Components with flip-chips

4.2.1 Attributes of flip-chips

Flip-chips provide for large numbers (several thousand) of tiny (about twice the diameter of a human

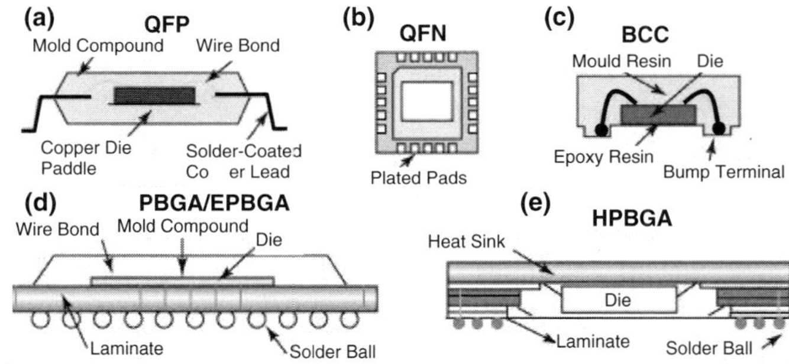

Fig. 4 Illustration depicting several microelectronic package types: (**a**) Quad flat pack (QFP). (**b**) Quad flat non-leaded (QFN). (**c**) Bumped chip carrier (BCC). (**d**) Plastic ball grid array (PBGA)/enhanced plastic ball grid array (EPBGA). (**e**) High-performance ball grid array (HPBGA) (after Ref. [7])

hair), densely-spaced solder bumps spread over the entire active surface of the device. The solder bumps are located below the chip allowing close chip-to-chip spacings (typically referred to as brickwalling). Close chip proximity is important since it allows for the efficient use of chip-carrier real estate, but even more important, it reduces the chip-to-chip distance, thereby significantly increasing the circuit speed, i.e. performance. Flip-chip interconnections also provide superior electrical performance (i.e. low inductance and resistance) since they are very short, large-diameter conductors compared to the wire interconnections utilized for wirebonded chips. Flip-chip solder joints have demonstrated a very high degree of reliability in the field, the highest of any chip interconnection technology. Their very-high interconnection density capability, excellent electrical characteristics (even for high frequency and multiple simultaneous switching conditions), and high reliability are some of the major reasons that flip-chips are so widely used in high-performance, high-reliability electrical equipment. It is also thought to be the reason the EC provided a general exemption for solders with > 85% Pb, i.e. to assure the integrity and desirable mechanical properties of solder joints, and those of flip-chips in particular.

4.2.2 Flip-chip solder joint requirements

Flip-chip solder joints serve several important purposes: they provide both an electrical and thermal path between the chip and next level of assembly, as well as a mechanical attachment. Wirebonding only provides an electrical path. Flip-chip solder joints must, therefore, maintain adequate mechanical properties, particularly the ability to withstand thermal fatigue. Thermal fatigue is the result of a mismatch in the coefficient of thermal expansion (CTE) between the materials attached to the opposite ends of the solder joint (i.e. the silicon chip and next level of assembly material). Solder joints experience shear stresses when exposed to temperature cycles that result from electrical equipment being turned "on" and "off". Lead (Pb) is almost unique in its ability to withstand failure upon being bent back and forth during the course of up to several thousand thermal cycles. That is, high-Pb solders have the ability to relieve stress through plastic deformation, in contrast to high-Sn solders where the ability to undergo plastic deformation is considerably more difficult. The ability to undergo plastic deformation (i.e. be ductile) without undergoing failure is an important characteristic for flip-chip solder joints. High-Pb solder joints accommodate shear stresses

internal to the solder joint instead of "passing" the stress on to the fragile, stiff and brittle silicon chip.

Finally, high-Pb, high-melting solders such as 97Pb–Sn (MP = 322°C) often used for flip-chip solder bumps, have the ability to withstand high current densities over long periods without undergoing a degradation process referred to as electromigration, discussed in Part II.

4.2.3 Flip-chips mounted in BGA packages

Converting BGA microelectronic packages with wirebonded chips to ROHS-compliant (i.e. Pb-free) versions requires changing the solder balls to a Pb-free solder. But the situation is more complicated in the case of packages with flip-chip mounted semiconductor devices. Consider for example a flip-chip with 97Pb–Sn solder bumps reflow attached to a ceramic BGA chip carrier, (Fig. 5a). Although the BGA solder balls are 90Pb–10Sn (Fig. 5b), the attached component may or may not be ROHS compliant. The composition of the flip-chip solder bumps are compliant. But the composition of the BGA solder joints may not be compliant depending on the fraction of eutectic solder volume comprising the joints. The component consists of eutectic Sn–Pb solder fillets that attach the 90Pb–10Sn solder balls to the CBGA. Depending on the eutectic

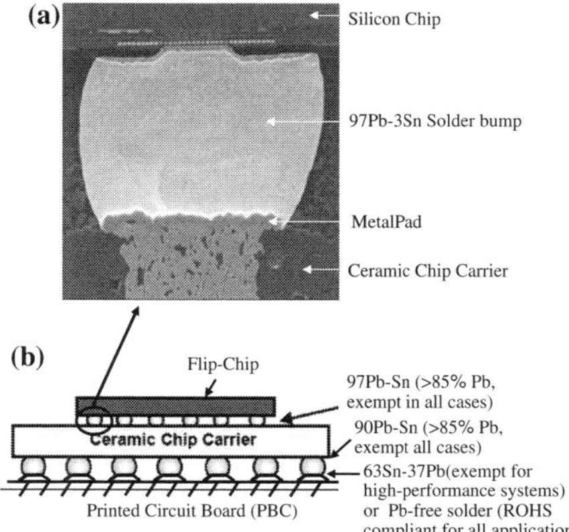

Fig. 5 A flip-chip with high-Pb (97Pb–3Sn) solder bumps is attached to a ceramic chip carrier that has high-Pb (90Pb–10Sn) solder balls. (**a**) Photomicrograph of a vertical cross section of a flip-chip solder bump directly reflow-attached to a metallized pad on the ceramic chip carrier (after Ref. [8], courtesy of IBM Corporation). (**b**) Sketch depicting two ways the ceramic BGA can be attached to the PCB: with eutectic Sn–Pb (exempt for high-performance systems) or Pb-free solder, which is ROHS compliant but creates a mixed solder joint that is not forward compatible

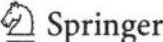

Sn–Pb solder volume utilized to surface-mount attach the component to the board determines if the lead content in the CBGA solder joints are > 85% in order to be ROHS compliant. That is, if the amount of eutectic Sn–Pb solder paste used to SMT-attach the CBGA component is too great, the resulting lead (Pb) concentration in the CBGA solder joints will be < 85%; and therefore not ROHS compliant. Exempted high-performance equipment (i.e. servers, storage and storage array systems, etc.) are not affected by the eutectic Sn–Pb solder volume issue since they can utilize any solder independent of Pb content. Note that if the BGA component with 90Pb–Sn solder balls is attached to the PCB with a Pb-free solder, a mixed solder-joint condition is created. Although ROHS compliant, it is considered a reliability risk (i.e. not forward compatible) as discussed in Part II.

The use of plastic BGA chip carriers has become widespread owing to their reduced weight and lower cost compared to ceramic BGAs. The flip-chip solder bumps of some BGA part numbers may only be available with high-Pb solders (i.e. 97Pb–Sn) that require a 340°C or higher reflow temperature which is much higher than plastic chip-carrier materials can withstand. The chips are therefore attached with a low-melting-point solder, usually eutectic Sn–Pb to circumvent the problem. The use of a low-melt solder to attach flip chips with high-melt solder bumps (Fig. 6a) has recently been approved by the European Commission. Although this combination consists of two different solder compositions, they are both of the same Pb–Sn solder family. The degree of high-Pb solder dissolution into the low-melting eutectic Sn–Pb solder is normally quite limited under standard assembly reflow conditions. This solder combination typically results in high-reliability solder joints because most of the solder joint is high-Pb which characteristically provides superior fatigue resistance. Flip-chips with eutectic Sn–Pb solder bumps can be directly reflowed to plastic chip carriers without the concern of thermally degrading the carriers. These solder joints are ROHS-compliant even though the Pb content is less than 85%. The inherent thermal-fatigue resistance of all-eutectic Sn–Pb flip-chip solder joints is less than their dual-solder counterparts, i.e. 97Pb–3Sn solder bumps mounted to a PBGA component with eutectic Sn–Pb solder paste. However, the thermal fatigue resistance of all-eutectic flip-chips with an underfill mounted on PBGA chip carriers is equal to or exceeds the fatigue resistance of dual-solder flip-chips. The use of an underfill material becomes an issue in situations where chip replacement (i.e. rework) may be necessary, such as multichip modules, where it may become

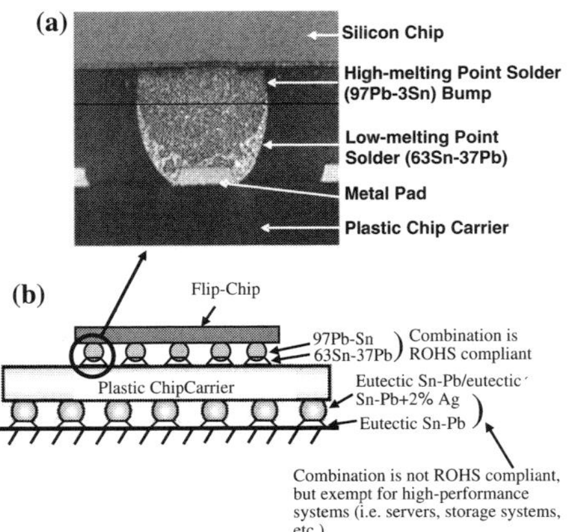

Fig. 6 A flip chip with a high-Pb (97Pb–3Sn) solder bumps is attached to a plastic chip carrier that has low-melt solder balls (eutectic Sn–Pb). (**a**) Photomicrograph of a vertical cross section of a flip-chip solder bump attached to a metallized pad on the plastic carrier with a low-melting point solder, eutectic Sn–Pb (after Ref. [8]) (courtesy of IBM Corporation). (**b**) Sketch depicting the attachment of the plastic chip carrier with eutectic Sn–Pb solder paste forming an all-eutectic Sn–Pb solder joint that is not ROHS compliant, but exempt for high-performance systems

necessary to replace one or more chips. Replacing chips with underfills is time consuming, and a reliability concern. It is not a recommended procedure for H-P equipment. Plastic BGA components normally have eutectic Sn–Pb (some contain 2% Ag) solder balls (Fig. 6b) and are attached to the PCB with eutectic Sn–Pb solder paste. This lead-containing BGA solder joint is not ROHS compliant but is exempted (i.e. allowed) for high-performance systems. It was the standard surface-mount (SM) assembly process in use prior to the ROHS legislation going into force.

5 Exemption for the use of Sn–Pb coated compliant pin (push pin) connectors

Compliant-pin refers to an interconnection technology that mechanically and electrically joins a connector to a printed circuit board (PCB) without the use of a solder process [9]. This technology has been widely utilized in applications requiring high reliability such as the telecommunications and defense industry for more than 20 years [10, 11]. The most common designs are illustrated in Fig. 7. Initially, solid pressed-fit pins were utilized where the pin diameter was slightly greater than the plated-through-hole (PTH) diameter, requiring some deformation of the pin and hole material

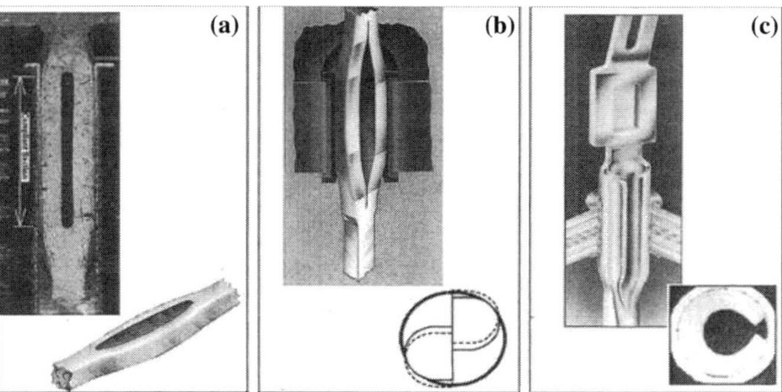

Fig. 7 Illustration depicting several compliant-pin designs. (**a**) Eye of the needle. (**b**) Tyco Electronics patented "Actin Pin". (**c**) Winchester Electronics patented "C-Press" (after Ref. [12])

when the pin was inserted into the hole. Very tight tolerances of both pins and PTHs were required to avoid damage upon pin insertion or extraction. Since compliant pins compress, the allowed tolerances of both pin and hole diameters is much greater than for solid pins, and the insertion and extraction forces much less compared to solid pins [13]. Compliant-pin connections are widely used in high-performance systems because they allow for very high-density and high reliability designs (Fig. 8).

5.1 Advantages of compliant pins over soldered connections

Compliant-pin connections have several important advantages over soldered connections. In high-performance systems it is often necessary to locate active components on both sides of a PCB, thus installing wave-soldered, through-hole components becomes problematic since many active components cannot withstand wave soldering temperatures. The situation becomes even more aggravated in the case of high-melting point Pb-free solders. Compliant-pin connections allow the attachment of large, high-thermal-mass components, connectors or daughter cards to backplanes without the complications inherent to solder processes. Additionally, compliant-pin connections provide the capability for very-high packaging density due to the ability for connectors to share the same contacts and stagger components or connectors on the top and bottom of a PCB (Fig. 7a, b). Since there are no minimum space limitations, as there are for soldering processes, smaller plated-through-holes (PTHs) can be utilized with compliant-pin systems which allows for a higher signal line density between PTHs. Signal integrity is also better in compliant-pin systems

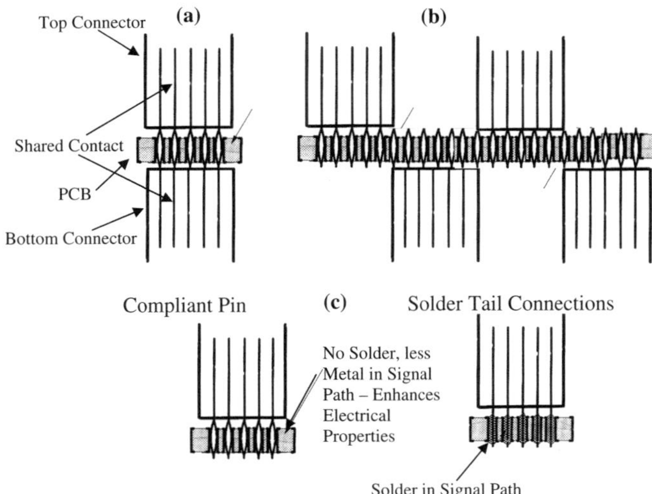

Fig. 8 Illustration depicting several advantages of compliant-pin connections over soldered connections. (**a**), (**b**) higher packaging density, thus enhanced electrical performance. (**c**) Enhanced signal integrity due to interconnection features (after Ref. [14], (courtesy of P. Isaacs, IBM Corp.)

compared to the configuration of soldered PTHs (Fig. 7c). All of these factors contribute to provide a superior electrical performance capability. Accordingly, compliant-pin terminations have become the predominant method of attachment for backplane systems; providing reliable connections with high-performance capability, i.e. signal integrity at multi-megabit transmission rates [15, 16]. High-density, compliant-pin connections are becoming even more important because of greatly increased network bandwidth and speed requirements. It is now a key consideration given the ever increasing data-transfer rates between a mother board and daughter cards in high-performance systems [17].

5.2 Role of Sn–Pb solder coating

Compliant-pin connector systems depend on a very thin (1.27–1.78 µm thick) tin–lead, solder-plated coating with the lead content ranging from 3 to 10%. The lead (Pb) acts as a lubricant to reduce the force necessary for inserting components into a PCB. This lubrication effect ensures that the PTHs in the PCB are not damaged when inserting or removing high-pin-count components. A typical insertion force ranges between 4 and 10 lbs. per pin [18]. The Sn–Pb coating aids in removal as well. However, the retention force must be sufficient to maintain the integrity of the connection [18]. Typical compliant-pin retention forces vary between 2 and 10 lbs. per pin. The need for a sufficient lead (Pb) content in the Sn–Pb solder coating is illustrated in Fig. 9. In the case of high-pin count, high-density boards utilized in mainframe computers, a nominal 90Sn–10Pb composition is utilized (50–70 µ thick), with a minimum of 3% Pb as noted earlier. If the Pb concentration is too low (approximately 2% or less) the PTHs are observed to become damaged when inserting or extracting components (Fig. 9b). When the insertion and removal forces are excessive the plating and hole shape of PTHs is damaged and so distorted that reliable connections cannot be made [11]. Lead–tin coatings also possess the necessary contact point wiping characteristics. Surface oxides and other debris are wiped away at both the pin and PTH surfaces when the pin is inserted. The clean, fresh surfaces make physical contact with each other, and provide the optimum conditions to create an air-tight electrical connection (Fig. 10).

5.3 Search for Pb-free solder coatings

Several studies were conducted investigating the effect of various combinations of Pb-free solders utilized (Table 5) with PTHs and eye-of-the-needle design compliant pins [19–21]. It was determined that all compliant pin and PTH combinations tested created less than the maximum allowable hole distortion, and left more than the minimum allowable plated thickness remaining in the PTHs. Also, no cracks were observed in the plating of the through-holes as required by IEC 60352-5 specifications [22] in an initial study [19]. A subsequent re-evaluation [20] confirmed that all the test combinations also conformed with a more stringent PTH deformation requirement as set forth in R4–10 of the Network Equipment Building System (NEBS) Telcordia Specification, GR-78-CORE. Another test was conducted [21] with Au over Ni-coated pins in addition to the Pb-free combinations listed in Table 5 evaluated in the earlier tests. Based on the combined results of these studies the effect of the

Fig. 9 Cross section of a board (**a**) with a compliant-pin connector that met the solder-coating requirements, therefore there is no damage to the plated through hole both during pin insertion (top) and after extraction (bottom). (**b**) A compliant-pin connector whose Pb concentration in the Sn–Pb coating was below requirements resulted in damage to some PTHs after the compliant-pin connector was removed (courtesy of IBM and Molex Corporations)

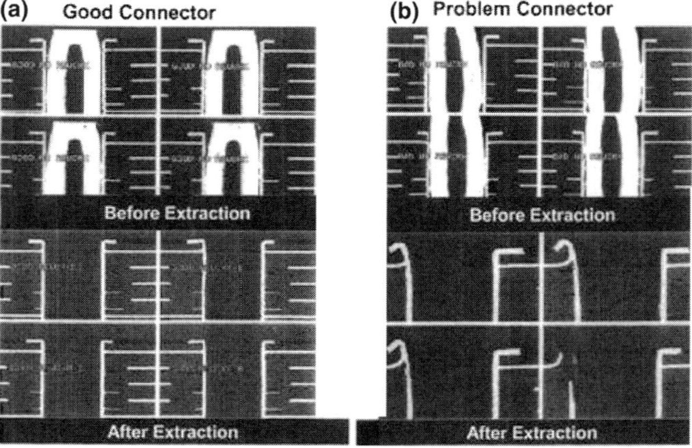

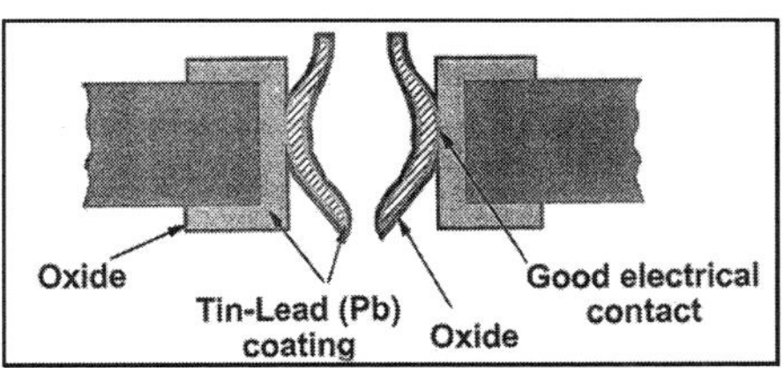

Fig. 10 Illustration depicting a compliant-pin with debris and oxide wiped from the pin and PTH surfaces that are in contact forming an optimum air-tight electrical connection (after Ref. [12])

Table 5 Potential Lead-free alternatives evaluated

Compliant-pin coating	Printed Circuit Board Material (i.e. PTH)
Sn–Pb (benchmark)	Electroplated Ni/Au
Pure Sn, matte	Copper/OSP
Pure Sn, bright	Immersion Tin (Sn)
	Immersion Ni/Au
	Immersion Ag

various pin coatings on insertion force are ranked as: matte Sn > gold > bright Sn–Pb > bright Sn. In general, the use of Pb-free compliant-pin finishes causes the insertion force to increase by approximately 15% compared to a Sn–Pb coating [23]. For high-pin-count components, connectors, etc., this translates to a significant increase in the total force applied to a card or board. The impact of this increase must be well-understood before Pb-free solders and coatings can be utilized for compliant pins in high-performance, high-reliability systems. Similarly the retention forces for the pin finishes are ranked as: matte Sn > bright Sn–Pb > bright Sn > gold. To determine the effect of environmental factors the various Pb-free pin and PIH combinations were exposed to three conditions: room ambient (23°C, 30–50% RH/5,000 h), heat/humidity (60°C/93% RH/5,000, 6,000 h), and thermal cycling (– 40 to 85°C/air-to-air, 10 min dwell/2,500 cycles) [21]. Tin whiskers were observed to nucleate and grow on all three pin coatings even though a Ni undercoat was used to mitigate their formation (Table 6). The temperature/humidity condition had the greatest effect in accelerating Sn-whisker formation and growth. Based on the Sn-whisker sizes observed, both the matte and bright-Sn pin coatings met the Class 2 acceptance criterion of the iNEMI Sn-whisker test, i.e. < 40-μm size whiskers. The level of risk introduced by whiskers of this size may be reasonable for some applications, but not high-performance equipment without a thorough investigation including

Table 6 Tin-whisker growth observed during testing

Compliant-pin finish	Sn-Whisker formation, Size (μm)
Ni/bright Sn–Pb	Yes, 20–30
Ni/matte Sn	Yes, 20–37
Ni/bright Sn	Yes, ~20

extensive qualification testing. This also includes the use of gold finishes for compliant pins. Although Au finishes eliminate the Sn-whisker issue, they nevertheless result in unacceptably high insertion forces, low retention forces, and a very significant increase in cost. The trade-off of adverse mechanical performance for the elimination of Sn whiskers is also not acceptable. The EC granted an exemption to allow the use of Pb for compliant-pins, and in consideration of the issues noted, for good reason. However, studies are continuing in an effort to resolve the outstanding concerns.

6 Use of Lead (Pb)-coated C-rings: an important test case

6.1 Single-user exemption request

All the exemptions granted when the ROHS legislation was enacted (Feb 13, 2002) pertained to practices in common usage across the electronics industry. Exemption requests were generally made through an industry association and carried the unanimous consensus of the association's membership. The matter of an exemption request that only affected the product(s) of a single company first surfaced with an IBM request for the use of lead rather than a lead (Pb)-free solder for an application specific to IBM mainframe computers.

6.2 Cooling requirements

All server products are designed to perform massive computing tasks. Significant quantities of heat are generated due to a high degree of integration for numerous high-power devices. Each server has its own unique heat-removal requirements. Even servers manufactured by the same company, like IBM, often utilize different hardware design concepts to achieve the desired performance and reliability objectives through custom designs that efficiently remove heat from a particular platform. The architecture of IBM's very-high performance mainframe computers (i.e. zSeries), results in the generation of significant quantities of heat, the source of which is a unique processor module referred to as a thermal conduction module (TCM). This technology allows a large number of chips to be placed in close proximity to each other on a single ceramic chip carrier, which greatly enhances both performance and reliability, but also localizes the heat that the chips generate. IBM introduced the TCM technology in its mainframe systems starting with the 3081 Processors (1981). At that time a combination of 118 closely-spaced logic and memory devices were mounted on a 90-mm multilayer ceramic (MCM) chip carrier. The TCM assembly was designed to maintain the chip-junction temperature below 81°C with a maximum thermal load of 4 watts per chip, and a cooling capacity of up to 300 watts. Cooling was achieved by a separate spring-loaded piston contacting the back side of each chip (Fig. 11). The array of pistons was contained in a housing that in turn contacted a water-cooled cold plate that removed the heat. A key aspect of the TCM's design in meeting its heat-removal requirements was charging the assembly with helium gas to a pressure of 0.16 MPa (1.6 atm.) to minimize the thermal resistances at the chip-to-piston and piston-to-housing interfaces [25]. But the TCM assembly's ability to meet its functional cooling requirements over its operating lifetime depends on maintaining the integrity of a seal created by a C-ring (Fig. 12).

The C-ring is clamped between the cooling housing and a flange brazed to the ceramic chip carrier. Since the computing power of mainframe computers has steadily increased, correspondingly the TCM's heat-removal requirements dramatically increased as well. By 1991, TCMs utilized in the IBM Enterprise System 9000™ family of processors had a heat dissipation capability of 600 watts [26]. The number of chips (typically about 16) mounted on TCMs utilized in IBM's current mainframe systems (i.e. eServer z990) is much reduced due to very large scale integration (VLSI) technology. However, these very complex chips each possesses 10K or more I/O solder bumps and dissipates 180–200 watts. Currently TCMs are required to dissipate between 800 and 1,200 watts depending on the particular system. The current, more compact and simplified TCM design (Fig. 13) utilizes a thermal interface material (TIM) to provide a more direct heat-removal path between a chip and the hat to which an air-cooled heat sink is attached. Helium is also charged into these TCM assemblies to provide a benign internal atmosphere. The C-ring must prevent oxygen or other gaseous species from entering the TCM and degrading the TIM or chemically attack the flip-chip solder joints.

6.3 C-Ring requirements

The C-ring core material is Inconel 718, 0.38μm thick and has a 2.5μm non-compressed height. It is electroplated with pure Pb, 63–115 microns thick, resulting in a deposited average weight of 3.1 g. The lead (Pb) is overcoated with a very thin coat of a multi-component, blended wax [27] (Fig. 14). The Pb coating exhibits excellent sealing characteristics to both ceramic and metal surfaces, to maintain near hermetic conditions (i.e. a helium leak rate of 10^{-8} atm-cc/sec) over the anticipated 10-year life of a TCM. The seal experiences 500 lbs. of force per inch of the seal. The Pb seal also

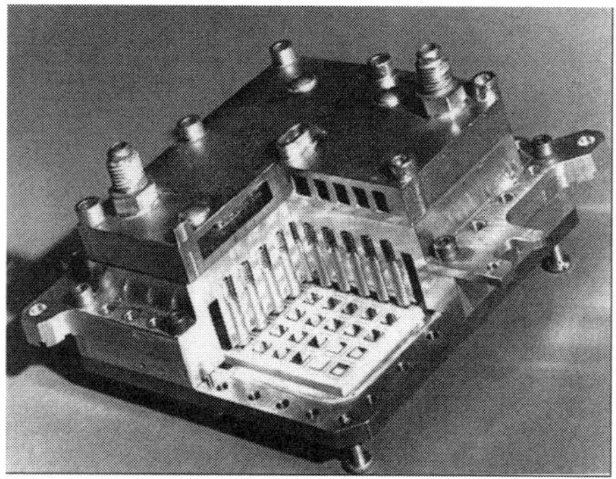

Fig. 11 Oblique-view photograph of the TCM assembly utilized in IBM 3081 processors, with a cut-out section to make some pistons and mounted chips visible (after Ref. [24])

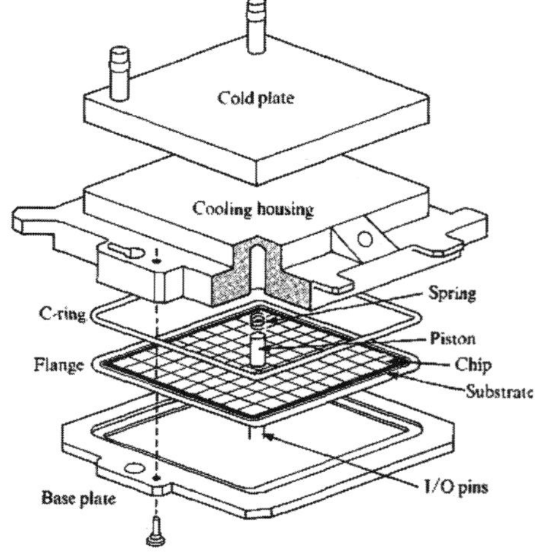

Fig. 12 Illustration depicting the major elements comprising a TCM assembly used in IBM 3081 processors (after Ref [25])

Fig. 13 Major elements comprising TCMs utilized in IBM eServer z990 mainframe computers. (**a**) Typical top surface floor plan of chips and decoupling capacitors mounted on a 93 mm × 93 mm ceramic chip carrier. (**b**) Side-view schematic depicting a TCM assembly (after Ref. [27])

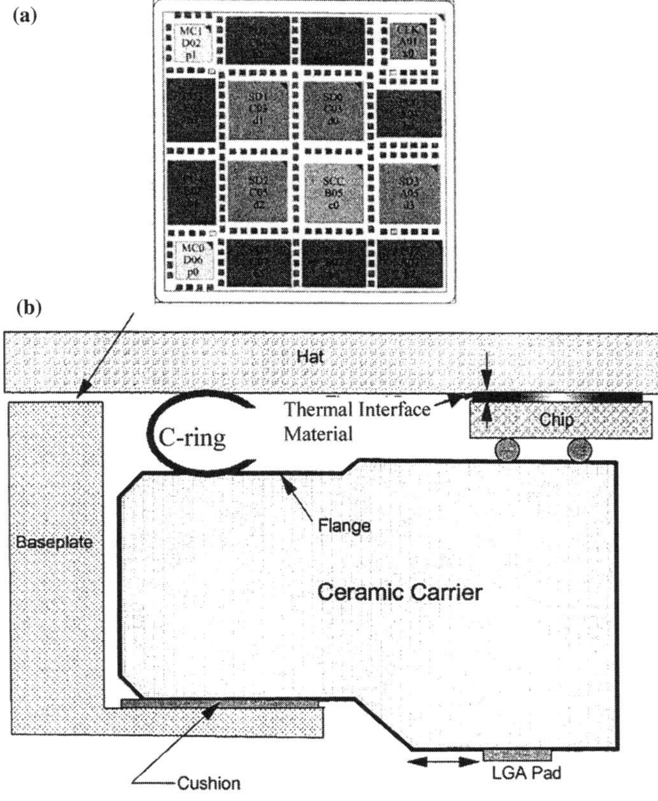

provides excellent wear characteristics when exposed to accelerated thermal cycle (ATC) stress testing, eliminating the effect of C-ring motion resulting from the coefficient of thermal expansion (CTE) mismatches between the C-ring and the materials it is in contact with [27].

6.4 Lead-free alternatives

A number of solutions were evaluated, among them: elastomer and metal O-rings; elastomer, metal, and composite gaskets; and plated elastomers. Other plated coatings were evaluated on the Inconel core in place of Pb, among them: Au, Ag, Pb–Sn, and In. All those potential solutions failed to meet the requirements. Among the reasons were: high gas permeation rate, high wear rate causing an unacceptable loss of normal force, and poor ATC test performance [27]. Lead (Pb) was the only material found suitable to meet the performance requirements. In early 2004 the European Commission (EC) granted an exemption for this application.

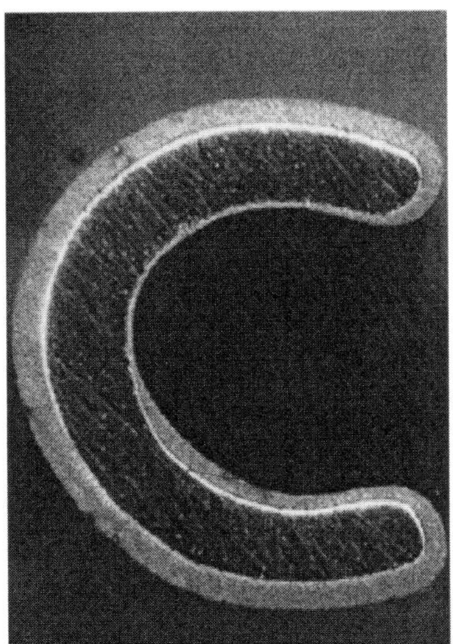

Fig. 14 Photomicrograph of a C-ring in cross-section, consisting of a 0.015-in. thick Inconel core, electroplated with 0.002–0.004 in. of lead (Pb) (after Ref. [28])

This request served as an important test case. If viable solutions exist to implement Pb-free technology, it is generally understood that the EC requires that those solutions be adopted, with cost only a minor consideration from an EC perspective. However, this case involves an important principal. Other mainframe manufacturers have solutions that do not involve the TCM approach, but have completely different architectures and methods of cooling. What is at issue here is the impact that forcing ROHS compliance could have on innovation, product competitiveness, and the pursuit for optimum product quality and performance. That is, IBM and its competitors each adopted a system architecture and design in the belief it provided the best approach to be competitive and provide the highest performance products over time. These are typically long-term strategies, as shown in IBM's case. If the EC had insisted that IBM abandon its TCM-based approach, and acted similarly with other entities in instances requiring a major design departure from their products, these actions among other things would only serve to have a serious adverse impact on technical innovation and progress, clearly not objectives of the ROHS legislation. This unintended negative impact was avoided by the granting of IBM's exemption request.

7 Summary/Conclusions

This paper has discussed the most important differences between consumer electronic products and those high-performance products that the EC permitted the continued use of lead (Pb) in its solder exemption. In addition, the paper discussed three specific industry exemption requests that have a particular importance and relevance to high-performance equipment applications. It was concluded that:

(1) High-performance equipment differed markedly from consumer electronics in several important respects.

    (a) High performance equipment is relatively expensive and low volume. It has a 10-to-25-year operational life, is often refurbished, and those portions that are ultimately discarded only represent a very small fraction of the electronic waste in landfills.

    (b) The complexity of high-performance equipment is often orders of magnitude greater than that of consumer electronics. Introduction of new Pb-free assembly materials makes the product assembly task more difficult and that increased difficulty directly translates into the creation of defects and a corresponding reduction in reliability.

    (c) The most important factors that differentiate high-performance from consumer electronic equipment are a demanding set of reliability requirements. Among these are the ability to operate continuously for 10–25 years with virtually no downtime due to failures. Some telecommunication equipment must meet these requirements in harsh outdoor environments or satellite systems in space environments with little or no opportunity for repair.

(2) Lead is an ingredient in a variety of materials utilized in electronic products including batteries, glass in monitors, cable sheathing, some ceramic components, and in solders. Solders are used as a coating material for PCBs, component lead frames and solder bump/ball terminations; and are universally utilized to attach components, connectors, and other circuit elements to printed circuit boards (PCBs). A main board in a high-performance electronic product may have 10,000 or more attachments, resulting in the need to create thousands of sound and reliable solder joints.

(3) A variety of leaded chip and component termination combinations have been exempted (i.e. allowed to be used) by the EC for high-performance systems. It is permitted to attach a flip-chip with high-Pb solder bumps to the next level of assembly (> 85% Pb) with eutectic Sn–Pb solder (< 85% Pb). These high-Pb flip chip/eutectic tin–lead solder joints typically exhibit good fatigue properties since both solders are of the same solder family (Pb–Sn), and also because most of the high-Pb bumps remain undissolved in the eutectic Sn–Pb solder. This was an important exemption since the use of plastic over ceramic for BGAs is increasing rapidly. A eutectic Sn–Pb attachment provides a sufficiently low-reflow temperature to preserve the integrity of plastic chip carriers. Being ROHS compliant does not necessarily qualify a solder joint for high-performance applications. For example, BGA components with high-Pb solder balls (> 85% Pb) reflow-attached to PCBs with Pb-free solder paste, creating a joint consisting of a solder mixture.

(4) A close proximity of components is necessary in high-performance systems to reduce the delay

time in transmitting information point-to-point. Densification therefore plays a large part in enhancing performance. The use of compliant-pins is the preferred method of interconnection in high-performance systems because they provide a much higher level of densification than possible with soldered pin-in-hole (PIH) connectors. It also avoids the processing problems introduced by locally heating large and complex assemblies. A Pb–Sn coating, with 3–10% Pb has been the only known material that provides the requisite combination of characteristics (i.e. contact wipe, low insertion forces, and sufficient retention force). All Pb-free materials, including gold, were found not to be adequate for high-performance systems. Very recent work utilizing tin over a nickel underlay on top of a copper substrate appears to provide an acceptable combination of properties and is being implemented by many of the major compliant-pin manufacturers.

(5) IBM was granted a request to use lead (Pb) as a coating for a C-ring that creates a seal that prevents the oxidation of flip-chip solder joints. It also prevents the degradation of a material that provides the heat-removal path for chips contained in an assembly central to the performance of IBM's mainframe computers. Lead (Pb) is the only known metal that satisfies all the requirements, although many others have been investigated. Aside from the fact that the request involved a significant percentage of the world's mainframe computing capacity, the request served as an important test case. By granting the request, the EC indicated it would consider the needs of even individual manufacturers, particularly in cases where forced compliance clearly would result in impeding progress and innovation.

In summary, it was absolutely necessary for the EC to recognize that the integrity and reliability of high-performance systems had to be preserved. Granting a solder exemption was an important step in helping to assure the reliability of H-P systems. But since even high-performance systems are required to be ROHS compliant, so it is likely that unanticipated issues will continue to surface during the implementation process and into the future. Three already have, as discussed in this paper and the EC is already considering others.

Acknowledgements The authors gratefully acknowledge the efforts of Terri Pinto, Hudson Presentations, in the preparation of the manuscript and graphics generated for this paper; and to Marie Cole for her very helpful comments and suggestions.

References

1. "Directive 2002/95/EC of the European Parliament and the Council on the Restriction of the Use of Certain Hazardous Substances in Electrical and Electronic Equipments," Official Journal of the European Union, Jan. 27, 2003
2. G. Galyon, Presentation to ERA representing the European Commission, Leatherhead, Surrey, UK, Aug. (2004)
3. K.J. Puttlitz, in *Proc. International Seminar on Electronics: Lead-free technologies and recycling*, Nov. 10–11, (Beijing, China, 2003), pp. 147–156
4. M. Wang, J. Zhu, J. Smetana, in *Proc. International Seminar on Electronics: Lead-free technologies and recycling*, Nov. 10–11, (Beijing, China 2003), pp. 173–178
5. Vicki Chin, in *Proc. International Seminar on Electronics: Lead-free technologies and recycling*, Nov. 10–11, (Beijing, China, 2003), pp. 162–166
6. David Towne, in *Proc. International Seminar on Electronics: Lead-free technologies and recycling*, Nov. 10–11, (Beijing, China, 2003), pp. 59–64
7. M.S. Cole, K.J. Puttlitz, Power Architecture Community Newsletter, http://www-128.ibm.com/developerworks/library/pa-ni15-community/ March 15, 2005
8. M.S. Cole, *Presentation to ERA representing the European Commission*, Leatherhead, Surrey, UK, Aug. 2004
9. R.D. Hilty, *Tyco Electronics Communication to European Commission*, Brussels, Belgium, July 1, 2004
10. R. Spiegel, *Design News*, Aug. 29, 2005
11. Naval Avionics Facility Indianapolis (NAFI), Interconnection Systems, http://www.albacom.co.uk/web/site/def_products/def_nafi.asp
12. K.J. Puttlitz, in *Proc. International Seminar on Electronics, Lead-free technologies and recycling*, Nov. 10–11, (Beijing, China, 2003), pp. 221–226
13. "Press-fit Technology" http://www.opt.de/english/press-fit_technology/overview.htm
14. P. Isaacs, in *Presentation to ERA representing the European Commission*, Leatherhead, Surrey, UK, Aug. 2004
15. International Engineering Consortium, On-Line Education, http://www.iec.org/online/tutorials/signal_integrity/
16. R. Holt, June 21, 2005, http://www.connectorsupplier.com/tech_updates_backplanes_6-21-05.htm
17. S. Nagarajan, "High-density Compliant Pin Connectors—Assembly Process Issues and Solutions, Electronics Manufacturing Research and Services (EMRS) Masters Abstracts, 2002, http://www.emrs.binghamton.edu/thesis_ma02_nagas.html
18. N. Nguyen, G. Robertson, R. Schlak, T. Do, Printed Circuit Design, Nov. 2002, 12–13
19. G.J.S. Chou, R.D. Hilty, in *Proc. of the IPC Anneal Meeting, Minneapolis*, MN, Sept. 28–Oct. 2, 2003, 507–2-1 to 10
20. G.J.S. Chou, Tyco Electronics, Publication 503–2, Rev. A, Nov. 16, 2004
21. G.J.S. Chou, R.O. Hilty, in *Proc. SMTA Inter. Conf.*, Chicago, IL, Sept 25–29, 2005, pp. 104–119
22. IEC 60352–5 Specification, "Solderless Connections—Part 5: Press-In Connections—General Requirements, Test Methods, and Practice Guidance," 2nd Edn, 2001
23. D. Parenti, Lead-Free Magazine, Nov. 9, 2005, http://www.leadfreemagazine.com/pages/vol005/press_fit_components_vol5_1.html

24. IBM Archives, http://www.03.ibm.com/ibm/history/exhibits/vintage/vintage_4506-2137.html
25. A.J. Blodgett, D.R. Barbour, IBM J. Res. Develop.**26**(1), 30 (2001)
26. J. Knickerbocker, G. Leung, W. Miller, S. Young, S. Sands, R. Indyk, IBM J. Res. Develop. **35**(3), 330 (1991)
27. M. Hoffmeyer, Presentation to ERA representing the European Commission, Leatherhead, Surrey, UK, Aug. 2004
28. K.J. Puttlitz, in *Proc. International Seminar on Electronics: Lead-free technologies and recycling*, Nov. 10–11, (Beijing, China, 2003), pp. 207–214

Impact of the ROHS Directive on high-performance electronic systems

Part II: key reliability issues preventing the implementation of lead-free solders

Karl J. Puttlitz · George T. Galyon

Published online: 22 September 2006
© Springer Science+Business Media, LLC 2006

Abstract There are important logistical and techno-logical issues that confront the electronics industry as a result of implementing lead (Pb)-free technology to comply with the European Union's (EU) ROHS Directive effective July 1, 2006. This paper focuses on the technological matters that pose the greatest potential risk to the reliability of so-called high-performance (H-P) systems (i.e., servers, storage, network infrastructure/telecommunication systems). The European Commission (EC) specifically granted a special use of lead (Pb) in solder exemptions, independent of concentration, for applications in H-P systems. The intent was to preserve the reliability of solder joints, particularly flip-chip solder joints. H-P systems perform mission-critical operations, so it is imperative that they maintain continuous and flawless operation over their lifetime. H-P systems are expected to experience virtually no downtime due to system failures. This paper discusses several major technological issues that impede the implementation of Pb-free solders in H-P systems. The topics discussed include solder compatibility and the reliability risks of mixed solder joints due to component availability problems. The potential effects of microstructural factors, such as the presence of Ag_3Sn platelets, and ways to eliminate them are described. Also discussed are the effects that stress-test parameters have on the thermal-fatigue life of several Pb-free solders compared to eutectic Sn–Pb. Another issue discussed involves the effect that tin's body-center tetragonal (BCT) crystal structure has on sol-der-joint reliability, and the complete loss of structural integrity associated with an allotropic phase transfor-mation at 13.2°C, referred to as tin pest. Yet another important consideration discussed is the tendency of pure tin or tin-rich finishes to grow "whiskers" that can cause electrical shorts and other problems. Finally, the paper notes the potential for electromigration failures in Pb-free, flip-chip solder joints. Based on the current status of the issues discussed, it appears likely that the exemption allowing the use of lead (Pb) in solder will need to be extended by the EC when scheduled for review in 2008, and perhaps well into the future.

K. J. Puttlitz (✉)
Puttlitz Engineering Consultancy, LLC, Wappingers Falls,
NY 12590, USA
e-mail: KJPuttlitz@aol.com

G. T. Galyon
IBM eSystems and Technology Group, Poughkeepsie, NY
12601-5400, USA
e-mail: Galyon@us.ibm.com

1 Introduction

The European Union (EU) legislation, "Restriction of the Use of Certain Hazardous Substances (ROHS) in Electrical and Electronic Equipment" (Directive 2002/95/EC) effectively bans the use of lead (Pb) and several other substances in electrical products. This paper contains references to "lead-free" solder alloys which has become an industry-standard term meaning that the solder should contain less than the ROHS Direc-tive limit of 0.1 wt% (1,000 ppm) for lead (Pb). It does not mean that the solder alloy will necessarily contain no lead (Pb) at all. The worldwide effort to eliminate the banned substances, particularly Pb, has surfaced some significant logistical and technical challenges to the electronics industry. How to disposition Pb-bearing inventory? How to easily distinguish and maintain separation of converted and Pb-bearing parts to avoid confusion and costly errors on the manufacturing floor? What due diligence procedures should be followed to assure compliance and avoid potentially severe legal repercussions? How to establish the necessary supply-line partnerships and controls to assure the continuous availability of truly compliant parts? While these logistical matters are important it is the technology-based issues that pose the greatest concern due to their potential impact on reliability. No widely accepted "drop-in" Pb-free solder replacement for eutectic Sn–Pb has been identified as yet. Only sparse laboratory test data was initially available for the several Pb-free solder candidates being evaluated across the global electronics industry, and the melting point for most of the lead-free solders is about 30°C higher than the melting point of eutectic Sn–Pb. In addition, it has not been possible to infer reliability performance for the lead-free solders from the existing Sn–Pb reliability data base because the metallurgical characteristics of high-lead content and eutectic Sn–Pb solders are significantly different compared to Pb-free solders. Based on these considerations the EC provided several solder-usage exemptions when the legislation was initially enacted and since then has granted several additional exemptions requested by the electronics industry. From the outset the EC recognized that there was a special class of electronic products (e.g., servers, storage, network infrastructure/telecommunications equipment) used in mission-critical operations whose operational integrity and reliability had to be preserved. Reliability greatly depends upon individual solder joint integrity. Of particular concern are flip-chip solder joints owing to their importance in high-performance circuits and the harsh mechanical conditions they must endure: e.g., exposure to repeated shear stress fluctuations resulting from the coefficient of thermal expansion (CTE) mismatches between the materials on either side of the solder joint. The EC did not provide a permanent exemption for the use of Pb in solder for high-performance products, but did agree to review the matter every 4 years starting in 2008. Another reliability concern is associated with the implementation of lead-free solder attachments involves the mixing of leaded and lead-free solders. During the initial transition phase it will be inevitable that electrical components with leaded (Pb) solders on their attachment points will be used in lead-free, solder-attachment processes. Mixed-solder joints can result in severe reliability problems. Several other significant technical issues existed at the time the

legislation was enacted that also helped prompt the EC to grant the continued use of lead (Pb) in solder exemption. While progress has been made in some cases, in others the problems are inherent and must be sufficiently characterized in order to determine if a "safe" operating condition or window can be established. For example, the wettability of all the most popular Pb-free solder candidates is less than Sn–Pb. There are solidification peculiarities associated with Sn–Ag–Cu or SAC alloys that can result in the presence of large Ag_3Sn platelets randomly oriented throughout the microstructure under conditions similar to those often encountered during manufacturing. The thermo-mechanical fatigue characteristics for SAC solders have been found to be very different compared to Sn–Pb solders. There are several other lead-free solder issues that did not exist with Pb–Sn solders which are directly related to tin's crystal structure. Pure tin undergoes a phase change at about 13°C that results in a condition called "tin pest" that literally causes the tin to disintegrate structurally into a dust-like powder. Still another very serious reliability concern is the tendency for tin electroplated finishes to form filamentary growths called whiskers. These whiskers are conductive and have caused electrical shorting in several mission critical applications involving heart pacemakers, space capsules, and missile control systems. For many decades lead has been added to electroplated tin finishes to mitigate or reduce the tendency to form whiskers. The elimination of lead will, unless otherwise mitigated, significantly increase the risk of whisker shorting failures. Finally, solder-joint electromigration resistance is becoming increasingly more important due to increased electrical current levels associated with high-performance electrical circuits.

2 Mixed solder joints

2.1 Reliability implications due to parts availability

The ROHS lead elimination directive will have a significant impact even for those manufacturers (e.g. medical, military, server, telecommunication, storage) whose products are permitted continued usage of leaded (Pb) solders. Manufacturers who purposely choose not to be compliant will also be impacted. The high volumes associated with parts utilized in certain consumer-electronic products will dictate that a particular component is solely manufactured as a Pb-free version [1, 2]. The demand for most Pb-free part numbers will be so great in relation to their leaded counterparts

 Springer

that many suppliers will elect not to support both lead (Pb)-bearing and lead-free versions of the same parts. Some suppliers (e.g., Infineon) have stated plans to continue providing leaded products to customers whose markets allow them to use lead-bearing parts [3]. Shortages of some Pb-free part numbers are anticipated early on presenting consumer-electronic manufacturers with product development and market introduction challenges that will eventually dissipate [4]. More serious will be the unavailability of some key Pb-bearing parts by manufacturers of high-performance products who plan to invoke the exemptions provided to assure the reliability of those products as intended, or entities that simply wish to remain non-compliant. Some degree of mixed (i.e., hybrid) metallurgical solder joints raises the potential for serious reliability problems. This situation is particularly ironic because the reason for granting the lead-usage exemptions was because of the uncertainty associated with the reliability of Pb-free solder joints [5]. The currently available reliability data base for hybrid solder joints is even more scant than that for lead-free solders.

2.2 Component and process compatibility

There are eight solder joint types that can be created during solder reflow operations (Table 1) as a result of solder composition variations between the component attachment point (i.e., lead-frame finish or solder ball) and the solder paste.

No compositional issues are introduced in cases involving component lead finishes or solder balls and solder paste of the same composition assuming the solder joints are formed utilizing an adequate reflow-temperature profile. A lead (Pb)-free component is said to be backward compatible if it can be reliably joined to a PCB utilizing eutectic Sn–Pb solder with a standard Sn–Pb reflow-temperature profile. Similarly, a Sn–Pb component is said to be forward compatible if it can be reliably reflowed to a PCB using a Pb-free solder paste and a Pb-free reflow temperature profile. Both forward and backward compatibility can be achieved for both Sn–Pb and Pb-free components

(Table 1) and such processes are practiced by many major manufacturers [6, 7].

2.3 Forward compatibility

The forward-compatible lead-frame products with Pb-based finishes are virtual drop-in replacements for their Pb-free counterparts. But some caution must be exercised. Pb-based components are designed to withstand Sn–Pb reflow process temperatures, typically with a maximum 230°C rating. Component manufacturers have made appropriate material changes such that, Pb-free components are capable of withstanding the significantly higher Pb-free process temperatures. So care must be exercised in assuring that the reliability of a Pb-bearing component is not compromised when subjected to the higher temperature (235–260°C) utilized for a Pb-free reflow process. Care must also be exercised in selecting a Pb-free solder paste that is metallurgically compatible with lead (Pb) [8]. For example, Pb-free solders containing bismuth (Bi) can form a low-melting compound in the presence of Pb that very significantly reduces the maximum-allowable application temperature. Bismuth-containing solders are successfully utilized by limiting the bismuth content to about 5% maximum. Some manufacturers who favor bismuth solders due to their relatively low melting points in comparison to the more commonly used SAC solders. Lead-free solder paste compositions vary considerably with respect to the various elements combined with Sn, Cu, Ag, Bi, In, Bi and others.

2.4 Backward compatibility

2.4.1 Lead-frame components

Generally speaking manufacturers have found that most lead-free surface finishes on lead-frame components are fully backward compatible with Pb–Sn solder and will not cause soldering or reliability problems [9] in spite of their higher melting points. The thin plating dissolves into the molten Pb–Sn solder volume, providing backward compatibility. However, depending on the joint solder volume, it is sometimes necessary

Table 1 Solder joint variations

Note: The reflow temperature range for eutectic Sn–Pb is typically between 220 to 235°C and 235 to 260°C for Pb-free solder palstes

Component		Solder paste and reflow utilized	
Termination	Solder	Lead (Pb) free	Tin–lead
Lead finish	Sn–Pb	Hybrid (forward compatible)	Homogenous
Lead finish	Pb-free	Homogeneous	Hybrid (backward compatible)
Ball	Sn–Pb	Typically not forward compatible	Homogeneous
Ball	Pb-free	Homogeneous	Typically not backward compatible

 Springer

to slightly increase the peak reflow temperature to achieve good-quality solder joints [7]. Backward compatibility covers a variety of lead-frame components: TQ, PQ, UQ, SO, VO, SOIC, PLCC, QFP, etc [10, 11]. It offers a valuable solution to manufacturers utilizing a Sn–Pb assembly process who must incorporate one or more lead-free components. They can do so with no adverse affect on reliability with only minor changes in the assembly process [10].

2.4.2 Solder ball terminals (ball grid arrays or BGAs)

As noted, the hybrid solder joints resulting from plated lead-free lead-frame components are generally considered a low reliability risk. But in the case of Pb-free ball grid array (BGA) terminations the joint reliability depends upon the homogeneity of the microstructure as a result of the reflow conditions. The issue here is that the melt temperature of the Pb-free solder balls is significantly greater than that for eutectic Sn–Pb solder paste. For example, Sn–Ag–Cu (SAC 305) melts at approximately 217°C vs. 183°C for eutectic Sn–Pb. As a result, the solder spheres either partially melt or do not melt at all during the Sn–Pb reflow process [2]. In general, it has been demonstrated that if the reflow results in an inhomogeneous solder joint due to only a partial melting of the solder ball, the solder joints exhibit a much-reduced reliability compared to standard eutectic Sn–Pb BGA solder joints [12–15]. Incomplete melting of the solder ball can cause poor self-alignment, open solder joints, and latent defects [15]. Conversely it was demonstrated for a wide range of components with Sn–Ag–Cu solder balls, if the peak reflow temperature was > 217°C and the time above liquidus (183°C) was sufficiently long, complete melting and mixing was observed in the solder joints of CSPs [12, 13], PBGA-196, 256 [15], and CBGA-937 components [14, 15]. These studies have shown that the reliability of some fully mixed and homogeneous solder joints are equivalent to or better than comparable Pb–Sn joints. Although some joints were observed to be statistically worse than their Pb–Sn counterparts, they were judged to not pose a severe reliability risk compared to the high reliability risk of joints with partially melted Sn–Ag–Cu solder balls [13]. Interestingly, the failure modes of both fully and partially reflowed BGA solder joints in accelerated thermal cycle (ATC) tests were observed to be the same. Cracks initiated at the component/solder interface and propagated through the region with coarsened grains, i.e., areas of high stress concentration. The lower fatigue life of partially mixed solder joints was attributed to a reduced ability of an inhomogeneous microstructure to dissipate the stress [15]. Unfortunately, to achieve consistently homogeneous solder joints requires utilizing either a lead-free or higher than normal tin–lead reflow profile. The higher temperature itself has its own potential for reliability problems associated with degraded component and PCB materials.

Adjustments in reflow parameters may be necessary based on ball size and composition since they play a significant role in mixing kinetics. Large BGA balls require longer times and higher temperatures to adequately mix with Sn–Pb solder. SAC305 solder balls require a 3–5°C higher reflow temperature than SAC405 solder alloys. Under the same reflow parameters (peak temperature, time above liquidus, and solder paste/solder joint ratio) the mixed portion of the solder joint is smaller in size for SAC305 compared to SAC405 [15]. In addition, excessive voiding has been observed in BGA solder joints based both on size and composition. For example, voids were observed to occur within 1.0-mm and finer pitch BGA balls, but to only occur minimally with larger pitches (e.g., 1.27 mm). The cause of the voids is not understood given that soak times of 60–120 s are normally ample to allow flux volatiles to escape prior to solder reflow [9]. The incidence of voiding was noted to be least for BGA solder joints constructed from SAC405 solder balls and paste, increasing with a mixture of SAC405 balls/ SAC305 paste, and worst for a mixture of SAC405 balls/Sn–Pb paste. The pure Pb-free joints were reflowed at 235–245°C, while the mixed solder joints were reflowed at a 220°C peak temperature [14].

2.5 Processing

2.5.1 Pb-free finished lead frame components

As indicated in Fig. 1, a standard Sn–Pb profile can be utilized to reflow attach Pb-free finished lead-frame components. The optimum profile depends upon additional factors such as component mass and distribution (i.e., layout) on a board, board size, geometry, etc. Semiconductor suppliers (e.g., National Semiconductor) are recommending that board-level assemblers adopt J-STD-020C to establish an industry-wide reflow standard [16]. Owing to their backward compatibility Pb-free finished lead-frame components can be utilized along with Pb-based components without the need to alter the standard Sn–Pb assembly process [8].

Fig. 1 Schematic depicting temperature profiles for reflow attaching Pb-free finished lead-frame components with either Sn–Pb or Pb-free solders (after Ref. [10])

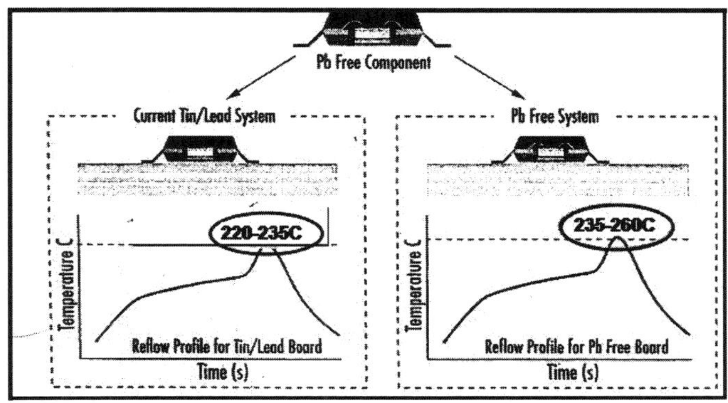

2.5.2 Pb-free solder ball (BGA) terminated components

The predominant position in the industry has been not to create mixed BGA solder joints consisting of Pb-free solder balls and eutectic Sn–Pb solder paste, because as has been noted earlier, the resulting inhomogeneous solder joint created by standard Sn–Pb reflow conditions exhibits unacceptable fatigue characteristics. However, the realities are that in many instances it will be almost impossible to avoid Pb-free BGA mixed solder joints in an assembly. Although the options are limited, these joints can be made to be backward compatible. The most desirable solution, although rarely available, is to reball the components with Sn–Pb solder balls [9]. In the absence of that capability, steps must be taken to determine the profile conditions (peak temperature, dwell time) that have the least detrimental effects on the assembly materials and still capable of producing fully mixed and homogeneous solder joints. Acceptable ATC test results must confirm the adequacy of the mixed solder joints.

High-performance systems manufacturers will in the main continue to use eutectic Sn–Pb to assemble their PCBs. It will be inevitable that some components are only available with lead-free terminations. Lead-free, lead-frame components generally are not a concern since they can be assembled utilizing standard Sn–Pb processes (i.e., they are backward compatible). But lead-free BGA components will be a source for concern since they are not backward compatible. In some instances it will be possible to achieve compatibility through process and material modifications, while in other cases it may not be viable to do so.

3 Reduced wetting characteristics of Pb-free solders

A variety of Pb-free solders have been investigated as replacements for eutectic Sn–Pb solder. None of them can be considered as a "drop-in" replacement for eutectic Sn–Pb. For reflow applications the consensus lead-free solder choice is a near-ternary eutectic Sn–Ag–Cu alloy with, in some cases, relatively minor additions of other elements. A comprehensive review of Sn–Ag, Sn–Cu, Sn–Ag–Cu, Sn–Ag–Cu–X, and Sn–Ag–X solders and solder joints is available in the literature [17, 18]. Some unique Sn–Pb solder attributes have made the search for a Pb-free system with similar characteristics difficult. Sn–Pb is a binary eutectic system with only one chemically active constituent, tin (Sn). All the chemical bonds that Pb–Sn solders form do so by the creation of intermetallic compounds (IMCs) containing only Sn. However, lead (Pb) does play an important role by providing molten Sn–Pb solders with a low surface tension value that is a major contributing factor to the excellent solderability (i.e., wetting and spreading) characteristics exhibited by Sn–Pb solders [19]. For example, the surface tension of molten Pb is 450 dynes/cm (at $T_{\text{melt}} = 326°C$), and molten Sn is 550 dynes/cm (at $T_{\text{melt}} = 232°C$). The wetting characteristics of eutectic Sn–Pb exceed those of all Pb-free solders of current interest when soldered to the base metals typically utilized in electronic assemblies (Figs. 2, 3). Minor elemental additions of Ag, Bi, Cu, In, Sb, Zn, etc. are used to lower the melt temperature and to enhance the mechanical properties of many lead-free solders. However, these alloy additions also significantly degrade solderability characteristics. Consider for example the binary solder Sn–58Bi utilized for applications requiring good wettability and a low-melt temperature. As shown in Fig. 4, a 1% addition by each of several alloying elements results in a moderate-to-significant loss in wettability (i.e., higher contact angle). The reliability implications related to reduced wetting characteristics is an unknown. This is just another example from a long list of technology concerns associated with the implementation of lead-free solders for high-performance systems assembly operations.

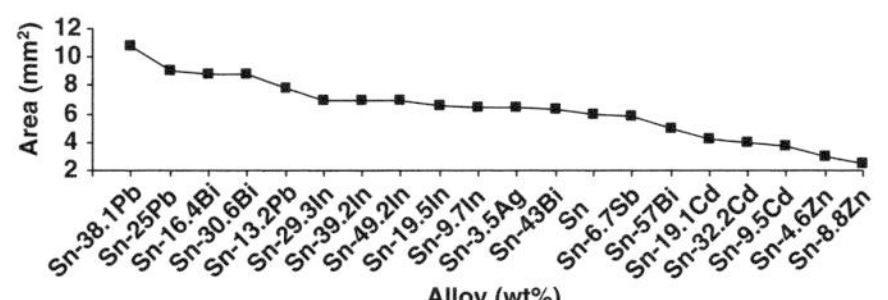

Fig. 2 Comparative spreading area measurements for several lead-bearing and lead-free solder alloys on copper utilizing a resin flux. Near-eutectic Sn–Pb exhibited better wetting (i.e., greatest spreading area) compared to all Pb-free alloys tested (after Ref. [20]) (courtesy of CRC Press)

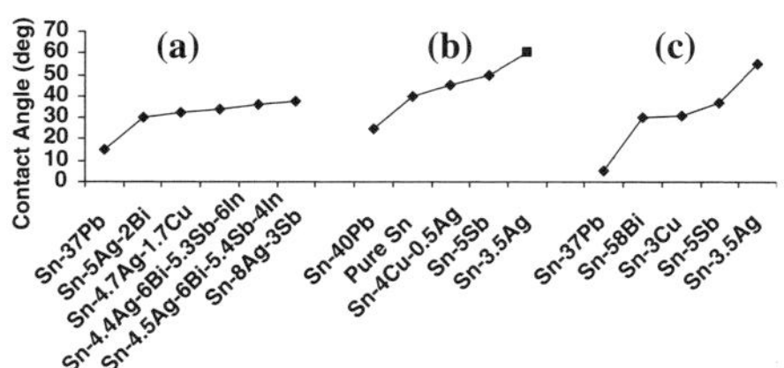

Fig. 3 Equilibrium wetting contact angles of several Pb-free solder alloys under various reflow conditions. (**a**) Cu substrate with RMA (GF-1235) flux (from Drewin et al. [21]). (**b**) Cu substrate with RMA (Alpha 611) flux ~40°C superheat over the liquidus temperature (from Vianco et al. [22]). (**c**) Cu substrate with RMA flux at ~30°C superheat over the liquidus temperature where the liquid alloy is dispensed onto a heated substrate (from Pan et al. [23]). Eutectic or near-eutectic Sn–Pb exhibits better wetting (i.e., lowest contact angle) compared to all Pb-free solders tested, under all test conditions (courtesy of CRC Press)

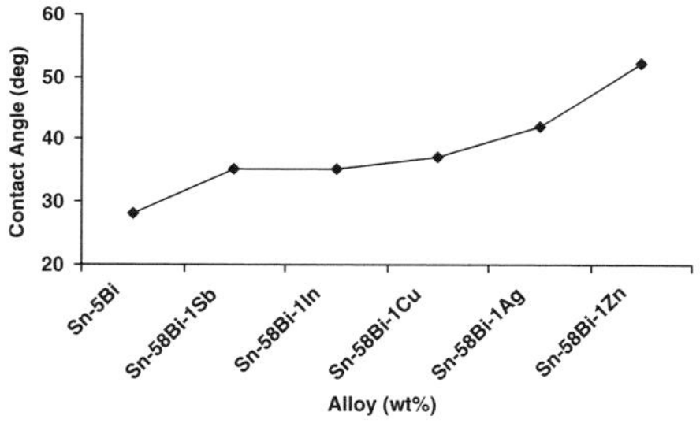

Fig. 4 Comparison of the contact angles of eutectic Sn–Bi and individual 1% ternary element additions when reflowed on Cu at a 200°C peak reflow temperature and using Kester 197 rosin flux. In each case a 1% ternary addition resulted in reduced wettability (i.e., increased contact angle) compared to the binary Sn–58Bi solder alloy (from Ref. [24], courtesy of CRC Press)

4 Solidification and microstructure of near-eutectic Sn–Ag–Cu alloys

4.1 Highly non-equilibrium structure

The Sn–Ag–Cu eutectic composition is thought to be Sn–3.5Ag–0.9Cu with a eutectic melting point at 217°C

[25, 26]. Some commercially available, near-eutectic alloys are: Sn–4.0Ag–0.5Cu and Sn–3.8Ag–0.7Cu. These near eutectic Sn–Ag–Cu alloys consist of three phases (β-Sn, Ag_3Sn and Cu_6Sn_5) that form immediately upon solidification from the liquid state. Both the Ag_3Sn plates and Cu_6Sn_5 rods formed are intermetallic compound phases. However, none of the phases are present in the amounts predicted by the equilibrium phase diagram since they are not formed as an equilibrium microstructure. The intermetallic compound (IMC) phases nucleate and grow with minimal undercooling. These pro-eutectic phases may undergo excessive growth under slow-cooling conditions due to the fact that they continue their growth from the liquid state until the molten solder has been sufficiently undercooled by 15–30°C. Undercooling is required to nucleate β-Sn crystals from the liquid phase. By comparison tin–lead based solders only require a small amount of undercooling to initiate solidification. Once β-Sn nucleates solidification occurs very rapidly with a dendritic morphology. A typical Sn–Ag–Cu solder joint substantially solidifies in less than a second after β-Sn nucleation commences and with β-Sn dendrites forming more than 90 wt% of the solder joint [27]. The eutectic intermetallic phases (Ag_3Sn and Cu_6Sn_5) nucleate simultaneously from the liquid located between impinging β-Sn dendrites. These IMC phases are manifested as fine-precipitate particles that outline the dendritic β-Sn grain's structure. The microstructure of a slow-cooled BGA solder ball depicted in Fig. 5 is typical of the highly non-equilibrium structure formed as a result of the events described. The dendritic β-Sn structure noted is very evident in Fig. 6. It was observed that as-solidified BGA solder joints approximately 900μm (0.035 in.) in diameter are typically composed of 1–10β-Sn grains [28], with an average of about eight grains. The small number of grains, together with the anisotropy of β-Sn, are reasons to anticipate mechanical property variations for Pb-free solder joints. The reliability impact from these mechanical property variations is, as yet, unknown.

4.2 Ag$_3$Sn plates

4.2.1 Observed effects

The presence of Ag_3Sn platelets were reported to have adverse effects on the plastic deformation properties of solidified solder joints [29]. Henderson et al. [28] subjected BGA solder joints to thermo-mechanical testing (0–100°C) and observed strain localization and large strains due to grain boundary sliding at the boundary between the Ag_3Sn and β-Sn phases. The orientation

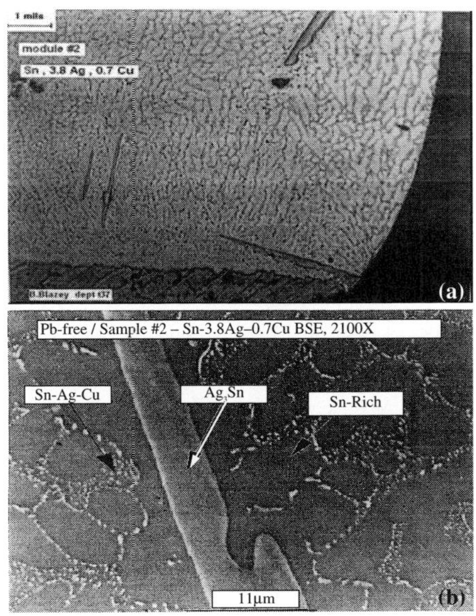

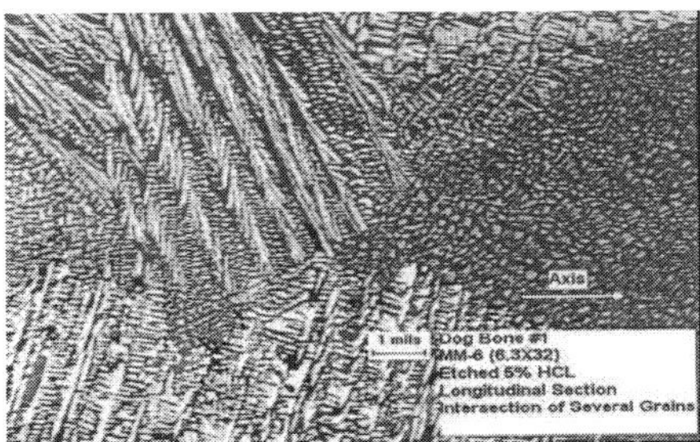

Fig. 5 SEM micrograph showing the typical microstructure of a slow-cooled Sn–3.8Ag–0.7Cu alloy in a BGA solder ball at two magnifications. The tin dendritic arms (light appearing phase) in (**a**) are surrounded by Ag_3Sn and Cu_6Sn_5 particulates as is clearly evident in (**b**) that decorate the β-Sn phase (after Ref. [27])

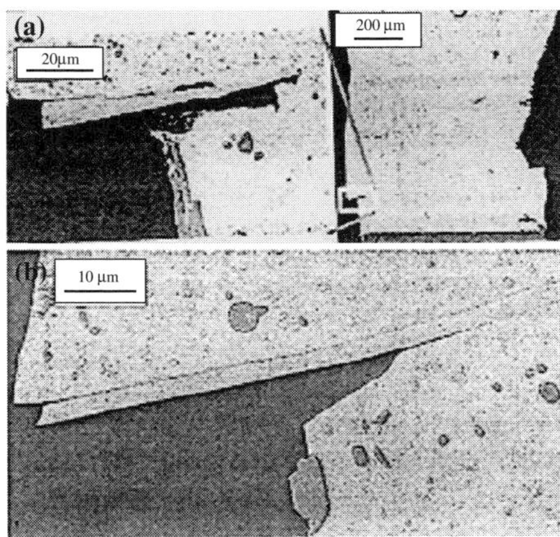

Fig. 6 Scanning electron microscope (SEM) image of a slow-cooled test sample of Sn–3.8Ag–0.7Cu solder alloy lightly etched to reveal the dendritic growth structure of the β-Sn phase (after Ref. [27])

of some plates were conducive to causing the top portion of some solder joints to shift relative to the lower portion as shown in Fig. 7. Strain localization at the β-Sn/Ag_3Sn boundary was also observed earlier by Frear et al. [30].

4.2.2 Factors that control Ag_3Sn plate formation

Extensive studies were conducted [28, 31, 32] to determine the controlling factors for Ag_3Sn platelet

Fig. 7 Vertical cross-section view of two solder joints that exhibit large strains localized at the boundary between a Ag_3Sn plate and β-Sn phase subsequent to an accelerated thermal cycle (ATC) test conducted at 0–100°C. The top portion of the solder joint has shifted in relation to the bottom portion along the Ag_3Sn/β-Sn boundary in each of these joints (after Ref. [28])

formation in near-eutectic Sn–Ag–Cu solder alloys. Of interest was the relationship between typical joint assembly cooling rates (0.2–4.0°C/s) and Ag_3Sn platelet formation. Results showed that cooling rate had a significant effect on Ag_3Sn platelet formation utilizing 0.88 mm diameter Sn–3.8Ag–0.7Cu BGA solder joints mounted on electroless Ni/immersion Au-plated pads on an organic chip carrier (Fig. 8). At low cooling rates (< 1°C/s) Ag_3Sn platelets grow large enough to span the entire BGA cross-section. For cooling rates between 1.5 and 2.0°C/s the platelet size is much smaller than the solder joint dimension, and at very fast cooling rates (> 3.0°C/s) platelet formation can almost be completely suppressed [28]. By conducting similar experiments with a hypoeutectic alloy, Sn–2.5Ag–0.9Cu (Fig. 9), it was determined that Ag content has a significant effect on Ag_3Sn plate formation. Results with this lower Ag content solder alloy showed that Ag_3Sn platelets rarely form, even at slow cooling rates; but that large Cu_6Sn_5 IMC rods appear [28]. Copper content was found to have only a minor effect on Ag_3Sn plate formation [31]. But lowering the Cu content is beneficial in reducing the so-called "pasty range"; the temperature range between the liquidus and solidus temperatures, where the equilibrium structure is a mixture of liquid and solid phases. Copper concentrations above the eutectic composition of 0.9 wt% in SAC alloys increases the pasty range [32]. Reducing the pasty range typically results in fewer solder joint defects, such as pad lifting of plated through holes (PTHs). A reduction in silver (Ag)

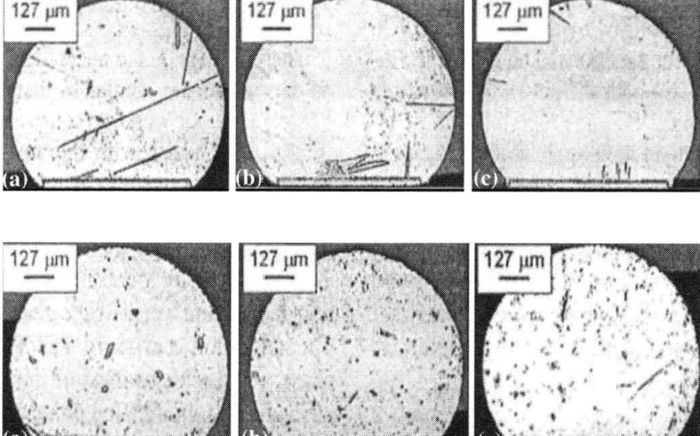

Fig. 8 Microstructure of Sn–3.8Ag–0.7Cu BGA solder balls cooled from a 240°C peak reflow temperature at various cooling rates. (**a**) 0.2°C/s, (**b**) 1.2°C/s, (**c**) 3.0°C/s (after Ref. [28])

Fig. 9 Microstructure of Sn–2.5Ag–0.9Cu solder balls cooled from a 240°C peak reflow temperature at various cooling rates. (**a**) 0.2°C/s, (**b**) 1.2°C/s, (**c**) 3.0°C/s (after Ref [28])

content is the current preferred method for eliminating or minimizing Ag_3Sn platelet formation. Rapid cooling may not be possible or practical in a manufacturing environment due to such things as large variations in the size and thermal mass of components that result in the introduction of excessive stresses in some assemblies and their solder joints.

5 Fatigue properties of Pb-free solders

5.1 Shear stresses due to CTE mismatch

Thermal–mechanical fatigue resistance is among the most critical reliability concerns with the introduction of lead-free solder joints. The ability to withstand repeated deformations due to stresses generated by the CTE mismatches between materials at the opposite ends of solder joints is critical to the reliability of H-P electronic equipment. Flip-chip and BGA solder joints are particularly susceptible to thermal–mechanical fatigue failure because there are some large CTE differences amongst the materials that comprise these assemblies (Table 2). For example, flip-chip solder joints can experience moderate-to-severe thermally induced shear stresses depending on whether they are mounted on ceramic or organic (plastic) chip carriers. The situation is similar for BGA solder joints where the thermally induced stress levels also depend upon whether the carrier material is ceramic or organic. Thermal excursions result from machine power "on/off" cycling and from operating between full and reduced (i.e., sleep or rest) power modes utilized to conserve power.

5.2 Parameters that affect fatigue life

Numerous fatigue studies have been reported covering Pb-free solder joints for several types of mounted components exposed to a variety of stress conditions. Some studies show that Pb-free, solder-joint fatigue life far exceeds the fatigue life for standard eutectic Sn–Pb solder joints [33]. However, other studies show that eutectic Sn–Pb has superior fatigue life in comparison to lead-(Pb)-free solder joints. For example, the Sn–Ag–Cu solder joints of a 48-I/O TSOP and 2512-sized resistor cycled between 0–100°C, −40–125°C, and −55–125°C [34] had a fatigue life significantly less than similarly tested eutectic Sn–Pb joints. Some of these seemingly contradictory results in the literature may be due to differences in the thermal-fatigue behavior of near-eutectic Sn–Ag–Cu solder alloys compared to eutectic Sn–Pb under various stress conditions. Such differences were very apparent in a fundamental study conducted by Bartelo et al.[35]. A comparison was made between the thermo-mechanical fatigue life of both Sn–3.8Ag–0.7Cu (SAC) and Sn–3.5Ag–3.0Bi (SAB) compared to eutectic Sn–Pb solder that served as a benchmark, using ceramic BGA modules attached to an organic card as a test vehicle. The initial test cycle investigated was 0–100°C and three cycle times: 30, 60, and 120 min. The accumulative failures in each case were plotted lognormally as a function of their thermal-cycle life, and then replotted to normalize all the data relative to the N_{50} value of the Pb–Sn solder joints, made to equal 1 for ease of comparison purposes (Fig. 10a). N_{50} represents the number of thermal cycles at which 50 percent of the test population failed, and was used as the comparison value among the three

Table 2 Coefficients of thermal expansion (CTE) values for materials of mounted BGA components

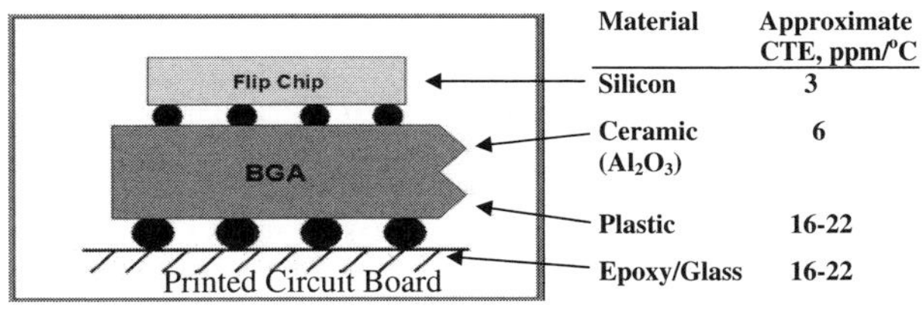

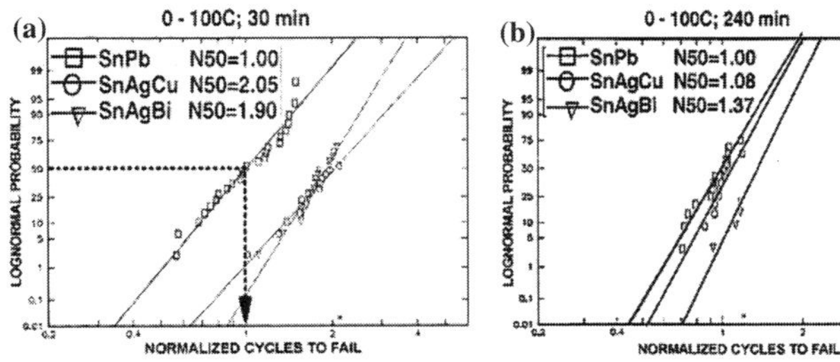

Fig. 10 Accelerated thermal cycle (ATC) test results comparing the thermo-mechanical fatigue life of two popular lead (Pb)-free solders (Sn–3.8Ag–0.7Cu or SAC, and Sn–3.5Ag–3.0Bi or SAB) in relation to eutectic Sn–Pb whose N_{50} was normalized to equal 1. The test conditions: 0–100°C cycle, cycle time (**a**) = 30 min, (**b**) = 240 min. (after Ref. [35])

solder compositions. For the 0–100°C/30-min. test condition, both the SAC and SAB alloys exhibited a fatigue life approximately two times better than Sn–Pb (2.05 and 1.9, respectively). For the longer 240-min. cycle the fatigue life of both Pb-free solders was markedly reduced with SAC nearly equivalent to Sn–Pb, and SAB only marginally better (Fig. 10b). Subsequently, the temperature cycle was altered to –40°C–125°C (a ΔT = 165°C instead of 100°C). Both Pb-free alloys experienced further reductions in thermal fatigue life relative to Sn–Pb due to the increased temperature deltas and become worse with increased cycle time (see Table 3). For example, the comparative fatigue life of both Pb-free solders for the 240-min cycle time was SAC-73% and SAB-96%, with 100% representing equivalence with Sn–Pb. A

final test was conducted to determine if an increased temperature range and/or an increased peak temperature was responsible for the loss in fatigue life, since both stress-test parameters were increased in the prior test. In the final test all test conditions utilized a constant cycle time = 30 min, and a constant temperature range (ΔT) = 100°C, but the peak temperature varied (60, 100 and 125°C). As noted on the right side of Table 3 both the SAC and SAB solders suffer a dramatic reduction in fatigue life relative to eutectic Sn–Pb with an increase in peak temperature.

5.3 Effect of cooling rate

Microstructure plays a significant role in determining solder joint mechanical properties. As previously

Table 3 Effect of temperature range, peak temperature, and cycle time on the fatigue life of two lead-free solders relative to Sn–Pb

Solder	Fatigue life relative to Sn–Pb				
	– 40°C to 125°C, ΔT = 165°C		30 min, ΔT = 100°C		
	4 Min	240 min	–40–60°C	0–100°C	25–125°C
Sn–Ag–Cu (SAC)	0.84	0.73	2.32	2.05	1.25
Sn–Ag–Bi (SAB)	1.21	0.96	2.93	1.90	1.12

Based on Ref. [35]

discussed, the solidification cooling rate has a large effect on the near-eutectic Sn–Ag–Cu solder microstructure. A study [36] addressed the effects these microstructural changes have on the fatigue life of SAC alloys. Ceramic BGA modules were mounted to organic cards with Sn–4.0Ag–0.5Cu solder paste. The Sn–Ag–Cu solder ball compositions ranged in Ag content from 2.1 to 3.8%. Parts from all compositions were cooled from a 240°C peak temperature at two rates, 0.5°C/s (slow) and 1.7°C/s (fast). The actual Ag composition of the reflowed solder joints was calculated to be slightly higher (by 0.15–0.2 wt%) than the BGA solder ball composition owing to the higher Ag content in the solder paste. Three ATC test conditions (0–100°C for 30 and 120-min. cycle times; –40–125°C for 42 min) were used. As a practical matter, the results of the 0–100°C ATC temperature range condition was considered to be of particular significance because it is the most widely used application range in computer, storage and network infrastructure/telecommunications systems which are the focus of this paper. For the 0–100°C/30-min. cycle, the slow-cooled 3.8Ag alloy exhibited the longest fatigue life, followed by the slow-cooled 2.1 and 2.5Ag alloys. The fast-cooled 2.1Ag joints exhibited the shortest fatigue life among the conditions tested (see Table 4). For the 120-min. cycle the trend was reversed, with the slow-cooled 2.1%-Ag joints exhibiting the longest fatigue life, followed by slow-cooled 2.5%-Ag and, fast-cooled 2.3Ag (and 0.2%-Bi) exhibiting the shortest fatigue life. In all cases, the number of cycles to failure decreased upon increasing the cycle time from 30 to 120 min, consistent with the prior study. This points to the fact that near-eutectic Sn–Ag solders require a much longer cycle time (i.e., stress relaxation period) in order for the maximum damage to occur, which is much longer than required for Sn–Pb solder joints. Accordingly, if the stress cycle used during ATC testing is too short, the results are likely to be more optimistic than those actually achieved under field conditions. It also implies that longer test-cycle times are necessary for meaningful Pb-free solder ATC testing in comparison to Pb–Sn solders. The Pb-free solder fatigue life values were even more drastically reduced for the – 40–125°C, 42-min stress conditions (Table 4) due to the larger temperature range ($\Delta T = 165$°C compared to 100°C) and higher peak temperature ($T = 125$°C). In essence, under these severe test conditions, the fatigue life of each test cell (i.e., solder alloy at a specific cooling condition) was observed to be only about 1/3 of its 0–100°C, 30-min condition fatigue life. This study further demonstrates the complexities involving the reliability of Pb-free solders, particularly the fatigue characteristics of near-eutectic Sn–Ag–Cu alloys. For example, based on the ATC data low-Ag content solder joints provide a superior thermal fatigue life in the case of 0–100°C, long-cycle time (120 min.) stress conditions, but not for short-cycle time, nor the – 40 to 125°C condition, for which the slow-cooled, high-Sn (3.8%) alloy was noted to perform best. This is in spite of the fact that the slow-cooled high-Sn joints were observed to contain large Ag_3Sn plates oriented in random directions, suggesting that other factors, such as the stress conditions may be more important than the presence of Ag_3Sn plates in determining solder-joint thermal fatigue life. Nevertheless steps to remove Ag_3Sn plates from a solder joint microstructure should be taken since their presence is believed to pose a reliability risk.

5.4 Benefits of low Ag

The benefit of reduced Ag content to avoid the formation of Ag_3Sn plates remains unclear. However, what is clear is that low-Ag joints have a much reduced fatigue life sensitivity to increasing cycle times. It

Table 4 Average fatigue life (N_{50}) determined from ATC failure data of Sn–Ag–Cu BGA solder joints and their ranking

Alloy composition (cooling rate)	ATC stress conditions					
	0–100°C		0–100°C		–40–125°C	
	(30 min)	Rank	(120 min)	Rank	(42 min)	Rank
Sn–3.8Ag–0.7Cu (Slow)	1,408	1	1,012	4	452	1
Sn–3.8Ag–0.7Cu (Fast)	1,164	5	982	5	392	4
Sn–2.5Ag–0.9Cu (Slow)	1,212	3	1,108	2	446	2
Sn–2.5Ag–0.9Cu (Fast)	1,200	4	1,054	3	384	5
Sn–2.5Ag–0.5Cu–0.2Bi (Slow)	1,212	3	953	6	407	3
Sn–2.5Ag–0.5Cu–0.2Bi (Fast)	1,130	6	934	7	376	6
Sn–2.1Ag–0.9Cu (Slow)	1,224	2	1,188	1	372	7
Sn–2.5Ag–0.9Cu (Fast)	1,064	7	1,012	4	384	5

After Ref. [36]

was observed that low-Ag solder joint fatigue life decreased less with increasing ATC cycle time in comparison to joints with higher Ag contents [36]. Low Ag content does provide a microstructure considered to be more favorable to enhancing fatigue life. Near-eutectic Sn–Ag–Cu alloys with reduced Ag contents have a lower volume fraction of eutectic constituents, a higher volume fraction of the β-Sn dendrite phase, and an increased Sn-dendrite size. These microstructural changes have been observed to result in decreased microhardness when Ag content is decreased [36]. The creep deformation of near-eutectic Sn–Ag–Cu alloys is understood to be mainly associated with the β-Sn phase rather than the eutectic constituents [37]. A coarse dendritic structure resulting from a reduced Ag content would be expected to exhibit an enhanced resistance to the creep component of thermal-cycle tests relative to similarly tested high-Ag joints.

In addition to being insensitive to cycle time, the thermal-fatigue life of low-Ag solder joints was also observed to be quite insensitive to cooling rate variations. This is a distinct advantage because they therefore are also less sensitive to variations in reflow process conditions during manufacturing. High-Ag content alloys such as Sn–3.8Ag–0.7Cu, exhibit high sensitivity to cooling rate variations [36].

The above fatigue studies demonstrate that Pb-free solder thermal fatigue behavior is very dependant on the test stress conditions and much more so than similarly tested Pb–Sn solders.

6 Tin crystal structure related anomalous behavior

There are fundamental crystalline structure differences between Pb and Sn that directly effect physical properties. At room temperature Sn exists as a body-centered tetragonal (BCT) crystal structure, referred to as white tin (β-Sn), compared to lead's face-centered cubic (FCC) structure.

The BCT Sn crystal structure is less symmetric than the face-centered cubic Pb structure and is, therefore, unable to as easily deform under stress. The BCT asymmetry also results in anisotropic physical properties like the CTE. Differences in CTEs between adjacent tin grains create grain-boundary stress when thermally cycled that can cause cracks to initiate at grain boundaries and may result in solder joint failure.

Yet another difference between Pb-bearing and pure Sn or some tin-rich solders is that the Sn-rich alloys undergo a phase change at 13.2°C which is in the middle of most ATC temperature extremes. When the temperature dips below 13.2°C tin changes from BCT

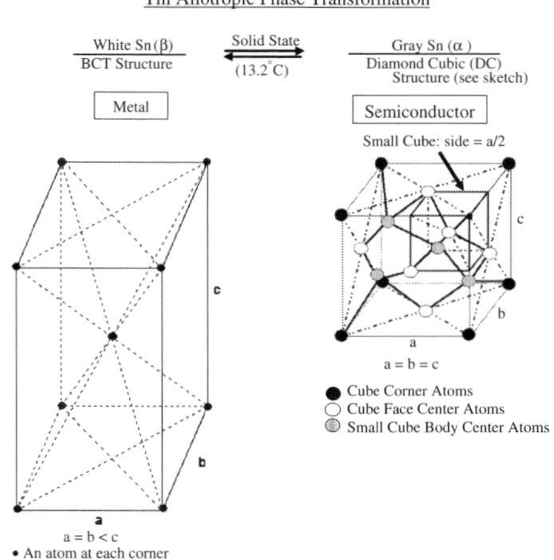

Fig. 11 Illustration depicting an allotropic transformation from a body-centered tetragonal (BCT) to diamond cubic (DC) crystal structure at 13.2°C

white-tin (β-Sn), with essentially metallic properties, to a diamond cubic (DC) structure gray tin (α-Sn) with essentially semiconductor characteristics (see Fig. 11). Because of the very brittle nature of gray Sn and the 21% increase in volume that accompanies the transformation, the material suffers a severe loss in structural integrity, and loss of thermo-mechanical fatigue resistance due to the build-up of internal stresses that result in fractures [38]. However, an extended period below 13.2°C is necessary to initiate the tin pest transformation. Consequently rapid temperature cycling does not result in transforming white Sn to gray Sn. But the transformation can occur in areas such as airplane storage areas where temperatures typically range between – 40 and 60°C. The surface has a rough appearance and contains many cracks (see Fig. 12). The β-Sn to α-Sn transformation also results in an approximately 25 times increase in electrical resistivity which is sufficient to cause failure in many sensitive high-performance electronic circuits. The adverse physical changes that accompany white tin's allotropic phase change to gray tin is referred to as tin pest. Tin pest has not been observed for all Pb-free solder candidates and finishes (e.g., the Sn–Ag–Cu alloys) but it has been observed in Sn–0.7Cu [38]. Although Sn-rich solders are prone to exhibit tin pest, minor additions of certain alloying elements, particularly Bi and Sb, have been proven to be effective in retarding the tin-pest reaction. Tin pest is one of the new concerns brought about by the use of Pb-free alloys that did not exist

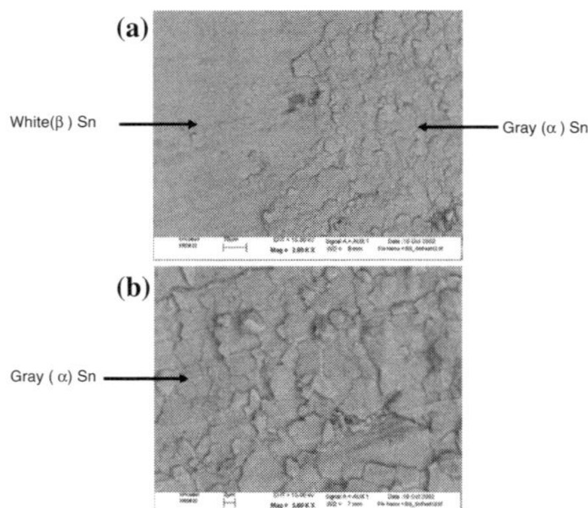

Fig. 12 Scanning electron micrographs of the surface of a Sn–Cu sample that has partially transformed from white (β) tin to gray (α) tin. (**a**) Right side of the photograph shows the rough surface of the transformed region (i.e., α-Sn) as contrasted with the smooth surface of the untransformed β-Sn on the left side. (**b**) An enlargement of the right side of (**a**), i.e., transformed to α-Sn (courtesy of IBM Corporation)

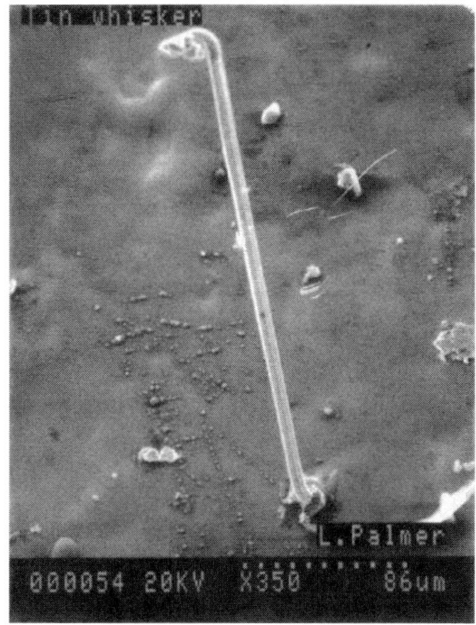

Fig. 13 A scanning electron microscope (SEM) image of a whisker that grew from a bright-tin plated film (after Ref. [39])

with the use of Sn–Pb solders in electronic products, because Pb is among those elements that resist the tinpest transformation.

7 Tin-whisker issues and concerns

Metal whiskers are thin, hair-like single crystal filaments that erupt from a metal surface (Fig. 13) and can be straight, kinked and even curved. Some have a fluted surface, while others do not. Most, if not all nonnoble metals grow whiskers under certain ambient or near-ambient conditions. However, the greatest focus is on whiskers that typically grow from tin-plated finishes due to their pervasive use in electronic assemblies. Whiskers grow in length over time, and can become several millimeters long. Because they are conductive, they can cause a number of failure conditions. In high-impedance, low-current circuits a whisker can cause a permanent short circuit which may require as much as 50 mA to "burn out". But whiskers can also cause temporary, or intermittent, shorts prior to being vaporized. Whisker vaporization can also cause serious problems. For example, in some space applications a whisker's vaporization can create a plasma with a current-carrying capacity of hundreds of amperes. Tin whiskers can also have an effect even when not causing short circuits. For example, in high-frequency RF (> 6 GHz) or in fast digital circuits (rise time < 59 ps), whiskers can act as a conductive stub (i.e., antennae) affecting circuit impedance and causing

reflections. In electromechanical systems, such as disk drives, whiskers may break off and cause head crashes or contaminate bearing surfaces. Similarly they may contaminate the optics in electro-optical systems and reduce the effective isolation for high-voltage components in power-supply systems [40].

7.1 Tin plating types

There are several types of pure-tin plated coatings or finishes (Table 5). Each tin finish type has characteristics that result from plating conditions and the chemicals added to the plating bath. So-called bright tins are highly reflective due to their smooth texture and small grain size. They are often selected for connector applications because they possess good abrasion resistance and a lower coefficient of friction relative to other tin coating types. However, bright tins have a relatively high residual film stress and are more prone to whisker growth so they tend to be long (up to 5000 microns in length). Matte tins have a dull appearance because of their rough surface texture and larger grain size (typically > 5 μm). Matte tins are often utilized for PCB and component lead finish applications. Matte tins are much less prone to whisker growth than bright tins and rarely grow greater than 500 microns in length. Satinbright films are intermediate in their appearance

Table 5 Common types of pure tin finishes and their characteristics

Sn plating	Appearance	Comments/characteristics
Matte tin	Dull appearance due to rough surface texture	Relatively large grains Typically μm Only Sn plating up to mid-1960s Large polygonal grains Usually extend through thickness of the film Achieved by organic additions to the bath Residual stress < bright Sn Grows whiskers but < density and smaller compared to bright Sn
Bright tin	Bright, highly reflective due to a smooth texture	Relatively small grain size Typically < 1 μm To achieve brightness add organic agents to plating bath Residual stress > matte Sn Whisker growth more prolific than matte Sn, longest whiskers
Satin bright tin	Brightness intermediate between matte and bright Sn	Stress is intended to be as close to zero as possible

between matte and bright tins and are designed to have a near-zero film stress. Tin plating has been the plating of choice throughout the electronics industry since the early 1950s. Around 1960 it became common practice to add lead (Pb) to electroplated tin films to enhance solderability and to mitigate whisker formation.

7.2 General areas of agreement

The European Union's ROHS regulations require that the lead (Pb) content be drastically reduced (to less than 1,000 ppm) which exposes the end user to whisker-related reliability failures due to whisker shorting. These reliability concerns have caused a flurry of research investigating whisker formation starting around 2000 and still in progress today. Whiskers have been a topic of scientific interest since about 1950 and there are a significant number of whisker-related studies described in the scientific literature [41]. The great majority of the published research concludes that compressive stress is the driving force for whisker formation and that whisker formation is one of several stress relieving mechanisms. The more recent research does not contradict these historical findings relative to compressive stress and the iNEMI (International Electrical Manufacturer Initiative) whisker modeling committee has taken the position that compressive stress is, beyond any reasonable doubt, the driving force for whisker formation and growth [42]. All published works agree that diffusion is the mechanism by which tin atoms move towards the whisker from adjacent regions, and most researchers believe that grain-boundary diffusion is the specific diffusion mechanism

responsible for whisker growth. There is a (minority) body of opinion that holds surface diffusion to be at least partially responsible for some movement of tin atoms towards the whisker-growth site. All published researchers, without exception, acknowledge that whiskers grow from the addition of tin atoms to the base of the whisker and not from the addition of tin atoms to the growth tip [43]. There is little or no consensus on other aspects of whisker growth including recrystallization, creep, grain boundary slip, and dislocation mechanisms. Nevertheless, the basic tenet that whisker growth is one form of stress relaxation is generally accepted by the published research.

7.3 The role of recrystallization

Tin deformation has long been known [44] to accelerate whisker formation, and if a deformation stress is maintained through a clamping action the whisker growth rate is accelerated by several orders of magnitude. A pure clamping action without deformation does not appear to accelerate whisker formation [45]. When the clamping action results in plastic deformation it is observed that whiskers form at an accelerated rate from the deformed material. Focused ion beam (FIB) examinations of whiskers growing from deformed material has shown that extensive recrystallization occurs within the mass of the deformed material and that whiskers grow from a grain (i.e., a whisker grain) which is itself unique in morphology with respect to both other recrystallized grains and the parent grain structure of the as-deposited film [46]. This unique morphology can best be described as a set of oblique (i.e., at an angle), relatively straight-sided, grain boundaries that intersect the surface at angles between 30 and 60°. In cases involving deformed tin, it is highly likely that whisker grains result from a recrystallization event that itself was the result of tin deformation. These straight-sided, oblique-angle whisker grains are always observed with whiskers no matter what the circumstances. However, for matte and satin-bright tins without a clear deformation event the oblique-angled whisker grains are not associated with an obvious recrystallization event. Instead, the matte/satin-bright whisker grains appear to form directly from the columnar-grain structure of the as-deposited tin. Most researchers acknowledge that the deformation events and the whiskers that form from very-bright tins involve a recrystallization event, but there is no agreement on whether recrystallization events play a role in whisker formation from matte and satin-bright tin films.

7.4 Problematic conditions

A variety of practices and environmental conditions have been identified that are known to exacerbate whisker formation. For example, copper readily co-deposits with electrolytic tin and forms Cu_6Sn_5 intermetallic particles that are dispersed throughout the tin matrix. It has long been known that the introduction of copper to a tin film in amounts ranging upwards of approximately 100 ppm greatly increases the tendency to nucleate and grow tin whiskers. It appears that copper alloying increases the built-in compressive stress state of the film and, thereby, increases the probability of whisker formation [47]. Unfortunately, copper may be present in a tin film even when it is not specifically requested. One of the authors (G. Galyon) has dissected many supposedly pure-tin films and found that they contain significant amounts of copper. Invariably, these copper-containing tin films were observed to have tin-whisker formation problems. Temperature cycling is also known to add stress to tin films and accelerates whisker formation. In fact, many whisker researchers utilize temperature cycling to induce whisker formation. The difference in the CTE between the deposited-tin film and the base metal upon which it is deposited (i.e. substrate) is the direct cause for the stress buildup that occurs during temperature cycling. Temperature-cycling induced stresses are particularly problematic for tin films deposited on Alloy42 (Fe42Ni) substrates due to the large CTE differences between film and substrate (23 vs. 10 ppm, respectively). High-humidity environments (typically > 85%

RH) accelerate whisker formation. The iNEMI group investigating Sn-whisker formation capitalized on the concept by utilizing a combination of temperature cycling and high humidity environments [48] to induce whisker formation. Their analysis showed that high-humidity environments not only increase the average surface oxide thickness, but that the conversion to tin oxide can, in some localized areas, extend from the film surface down to the film/substrate boundary. Oxide formation should, and apparently does (although no direct data exists confirming this fact), add to the compressive stress state and, thereby, accelerates whisker formation. Interestingly, tin deposited on copper or copper-alloy base metal surfaces that experience temperature cycling or temperature storage in the absence of high humidity results in the formation of pedestals. Pedestals are individual grains that are converted entirely to a copper–tin intermetallic compound (Cu_6Sn_5). Pedestal grains are observed to be scattered throughout the matrix of pure-tin, columnar-like grains. The pedestals form because for some unknown reason(s) certain tin grains appear to have anomalously high bulk diffusion parameters for Zn, Cu, and oxygen so they are readily oxidized or diffused. Consequently these "special" grains form what are termed as pedestals [46, 49].

Brass (Cu–Zn) surfaces are the most prone to tin whisker formation. Zinc is an extremely fast diffuser in tin and diffuses from the brass substrate to the tin-film surface in significant quantities very soon after deposition. A common mitigation practice for tin-plated brass substrates is the use of a copper underlay > 1.0 micron in thickness. However, the basis for the effectiveness of a copper underlay is not understood since there are no known studies that have addressed this matter.

Tin-plated steel for cabinets and drawers is relatively rare today. Zinc coatings are currently the finish of choice for the steel sheet-metal industry. Tin whisker formation is accelerated when steel substrates are not adequately cleaned. The belief is that a rusty (i.e., oxidized) steel surface induces stresses within the deposited tin film to promote whisker formation.

7.5 Mitigation practices

Whisker prevention (mitigation) in tin and tin-alloy films is largely based on stress reduction. There are no absolute preventative techniques. The objective is to mitigate, or lessen the probability of whisker growth. Several mitigation techniques that have been practiced are noted in Table 6.

 Springer

Table 6 Tin-whisker mitigation practices

Practice	Comments
Alloying	
With Pb	Banned by EU as of 7-1-06
With Ag, Bi	Rarely practiced, supported by iNEMI
Fusing	Melt the Sn film, slowly cool to relieve stresses
	Rarely practiced
Annealing	Typically heat to 150°C, 1 h
	Widely practiced, supported by iNEMI
Use of underlay films	Thin film between the base metal and Sn-film layer
Nickel (Ni)	Widely practiced by connector companies, supported by iNEMI
Silver (Ag)	Rarely practiced
Copper (Cu)	Rarely practiced
	Utilized for Sn films deposited on brass base-metal surfaces
Pure tin plated films	
Matte Sn	Widely practiced
	Not supported by iNEMI, controversial
Satin bright tin	Rarely practiced
	A lower stress variant of bright Sn

These mitigation techniques are intended to modify the tin stress state and so reduce and/or eliminate tin-whisker formation. Each technique has some supporting data. However, it is also true that not all the experimental data supports the effectiveness of each mitigation technique noted. Each mitigation technique has its strength and weakness. The effectiveness of several of these techniques is very dependent on proper procedure and practice. Consider, for example, underlays (whether copper, nickel, or silver) which must be thick enough to provide good coverage of the underlying base material. If the underlay is too thin the tin film will come into direct contact with the copper base metal at point locations, thereby negating the mitigating effect of the underlay material. Typically underlay thickness of > 1.0 micro meter are necessary to be effective, a greater thickness may be necessary in cases of surfaces with a high surface roughness.

The potential for tin whisker formation on tin-rich finishes and the reliability risk they pose is too great to recommend using these finishes in life-threatening applications such as heart pacemakers or military avionics. However, the reality is that manufacturers of these and high-performance electronic products will sometimes not have the ability to procure components with non-tin based finishes. In these cases the authors recommend that mitigation practices be adopted that best minimizes the risk of whisker formation and growth.

8 Electromigration concerns

There are several physical processes that can cause solder joint mechanical and electrical properties to degrade. One of these physical processes is electromigration. Electromigration is a mass transfer phenomenon resulting from the momentum-transfer between moving electrons and the constituent atoms in the conducting material. Electromigration is of particular concern for flip-chip solder joints because of the strong demand for continually increasing power levels and ever-smaller solder joints. Current densities in flip-chip solder joints are anticipated to soon reach $\sim 2.74 \times 10^3$ A/cm^2, a level at which there is concern for failures due to electromigration [50].

Conducting electrons are believed to collide with diffusing metal atoms that are then pushed in the direction of the electron flow as a result of those collisions. The phenomenon can occur with any current carrying metal but it is usually negligible and can safely be ignored. It becomes a consideration in cases where the diffusion rates are relatively rapid and the current density is high; conditions not unusual for advanced semiconductor devices and their flip-chip terminations as utilized in high-performance electronic equipment. "Effective charge" is a quantity that relates how diffusing atoms interact with or absorb the impact of conducting electrons, which itself is a complex function of a metal's atomic structure and is directly related to the driving force for electromigration. For solder materials, including tin, the effective charge is relatively high [51]. Other important electromigration factors are resistivity and current density. Conductor lines or terminations consisting of material with a combination of a high effective charge, one or more fast diffusing elements, and experiencing a high current density is a candidate for electromigration effects and potential catastrophic failure.

8.1 Fast diffusers

Copper (Cu), silver (Ag), gold, (Au), and nickel (Ni) are rapid diffusion elements in both Pb and Sn. These elements are routinely utilized in electronic assemblies for integrated circuits, printed circuit boards (PCBs), and solder joint terminations including the under-bump metallization (UBM) of flip-chip solder joints. They also frequently form IMCs with Sn. Fast diffusers occupy interstitial sites in the host-metal lattice. Lead

(Pb) has a close packed, FCC lattice, whereas tin's body-centered tetragonal lattice (BCT) is not close packed. Both Pb and Sn are large atoms and consequently contain relatively large interstitial sites that easily accommodate the much smaller noble and near-noble metal atoms. Accordingly, it has been determined that these smaller atoms diffuse interstitially in Pb and Sn and do not require vacancies in order to jump (i.e., diffuse) [51].

8.2 Electromigration failure in flip-chip solder joints

Electromigration effects in flip-chip solder joints usually result from the mass transport of a major solder species (i.e., Pb) or alloying element. It is also possible to have electromigration failures due to the mass transport of a minor, fast-diffusing species.

8.2.1 Solder-related failure

Flip-chip, solder-joint electromigration failures where the mass transport involves a major constituent are usually the result of void formation. Large voids cause a significant increase in electrical resistance with an associated increase in temperature eventually leading to an open circuit. Surface extrusions can also be cre-

ated by electromigration that cause short circuits [52–54]. A shallow void is often formed with a lenticular shape (Fig. 14), located at the interface of the solder and IMC associated with the UBM terminal metal. The geometry of flip-chip solder joints at the capture pad and narrow chip-via region (Fig. 15) causes current crowding, resulting in high local current densities, about 50 times greater than the solder bump [55, 56]. The condition also causes the generation of heat (i.e., increased temperature) that further exacerbates the situation. This is the mode of failure often observed when the direction of electrical current flow is from the chip to the chip carrier (Fig. 16a). But if the current density is sufficiently high when the direction of current flow is from the chip carrier to the chip, failure can occur at the chip-carrier side of the solder joint as well (Fig. 16b).

8.2.2 UBM-related failure

Electromigration failures can also involve degradation of the underbump metallurgy (UBM) structure for flip-chip solder joints; particularly if fast-diffuser elements are incorporated in the structure. Fast diffusers result in UBM dissolution and stress generation due to excessive IMC formation that causes interfacial fracture, and ultimately circuit failure [55, 57].

8.3 Electromigration in lead (Pb)-free solders

Electromigration data related to Pb-free solders is sparse. It is known that Sn acts much like Pb as a host for fast diffusers. Metals that diffuse rapidly in Pb also diffuse rapidly Sn. But there are some important differences between Pb and Sn with respect to diffusion. One of the most important differences is due to tin's BCT lattice structure which results in anisotropic diffusion properties. For lead (Pb) the diffusion

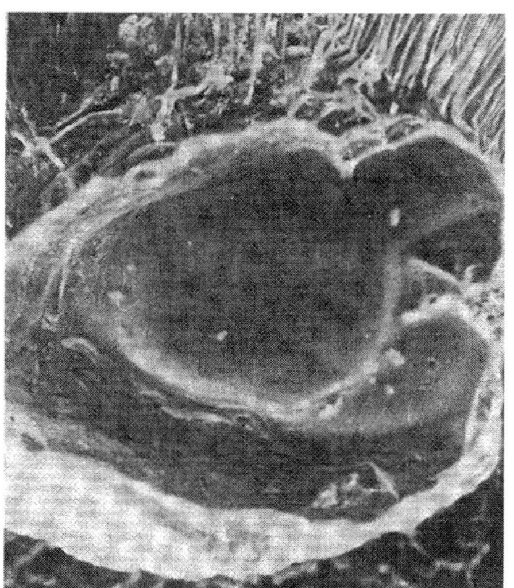

Fig. 14 Scanning electron microscope (SEM) image that shows the smooth surface characteristics of a lenticular-shaped void that often involves the entire cross section of the chip via in electromigration failures of flip-chip solder joints (after Ref. [51])

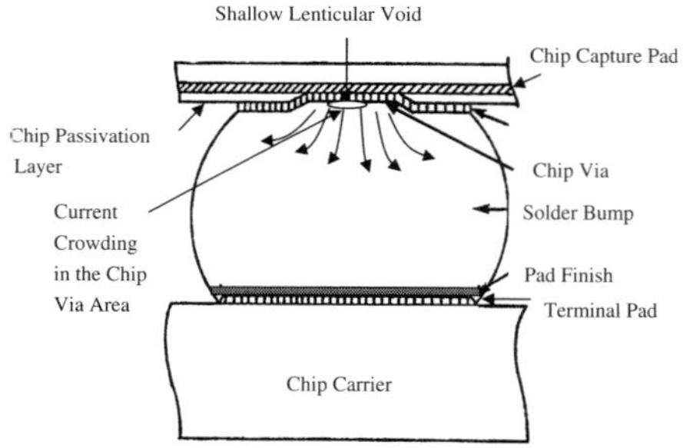

Fig. 15 Illustration depicting some aspects often associated with electromigration failures in flip-chip solder joints

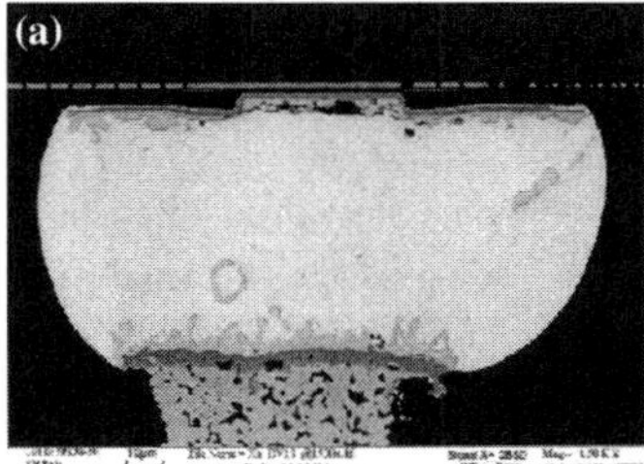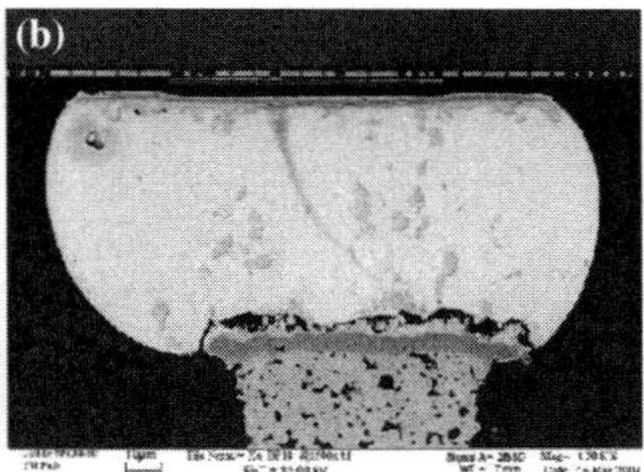

Fig. 16 Photomicrographs of vertical cross sections of lead-free flip-chip solder joints that have failed due to thermomigration during testing. (**a**) Failure occurred near the chip via, the direction of electrical current flow was from the chip to the chip carrier. (**b**) Failure occurred near the chip carrier, the direction of electrical current flow was from the chip carrier to the chip. (Courtesy of H. Longworth, IBM Corporation)

coefficient is independent of orientation, but not so in tin where the diffusion coefficients are very different for directions parallel ("*a*" direction) and perpendicular ("*c*" direction) to the basal plane. This is also true for self-diffusion. In the case of fast diffusers in tin, the ratio of the *a/c* diffusion coefficient varies by a factor of 30–50 at 200°C [51]. A greater variation in electromigration susceptibility may, therefore, be anticipated among Pb-free solder joints due to these diffusion coefficient orientation dependencies and the fact that flip-chip solder joints typically consist of very small numbers of individual grains (usually < 10). Electromigration test results for Pb-free flip-chip solder joints have been quite favorable in comparison to eutectic Sn–Pb. For example, in a test consisting of commercially available electroplated Sn–Ag solder bumps with a Ti/Cu/Ni UBM the data showed an electromigration lifetime one order of magnitude greater in comparison to eutectic Sn–Pb bumps at constant currents up to 800 mA and temperatures up to 165°C [58]. High-performance electronic circuits utilize high-Pb (i.e., 97Pb–3Sn) solder to benefit from both the superior electromigration and fatigue resistances exhibited by this material [59]. Another study was conducted [50] comparing the electromigration resistance of Sn–3.5Ag and 90Pb–10Sn flip-chip solder joints at similar current density levels (3.5 and 4.1×10^4 A/cm^2). However, a different set of temperatures were selected close to the melting temperature of the test solders in an attempt to hasten the failure process (185 and 202°C for high Pb; 122 and 155°C for Pb-free). The high-Pb solder joints exhibited an electromigration lifetime approximately four times longer than the Pb-free joints in spite of being tested at a much higher temperature. The Pb-free solder joints performed better than the eutectic Sn–Pb solder joints investigated in the study cited earlier [59]. Flip-chip solder joint dimensions are steadily decreasing in response to the need for higher

input–output (I/O) densities. At the same time, flip-chip solder joints used in high-performance equipment are experiencing ever higher current levels which are soon expected to be 0.3–0.5 A or more per solder joint [57, 60]. Although the available reliability data is meager, what is available raises disturbing concerns as to whether Pb-free solders possess the necessary electromigration (EM) resistance. Even at present current-density levels it appears the reliability risk for lead-free solders is too great. These reliability issues require a much better understanding before Pb-free solders can be relied upon to provide a safe level of EM resistance.

9 Summary/conclusions

This paper has discussed the major technological issues impeding implementation of Pb-free solder materials for high-performance electronic systems (i.e., servers, storage, network infrastructure/telecommunications systems) due to their potential negative reliability impact. It was concluded that:

(1) The solder wetting characteristics of all the major Pb-free solders currently under consideration are worse than eutectic Sn–Pb; some of the Pb-free solders are significantly worse than eutectic SnPb. The actual reliability impact of lead-free solder implementation is not well understood, but it is generally agreed that sound solder joints can be made with Pb-free solder. However, the process window for the formation of acceptable Pb-free solder joints is more restricted given the higher melting points for most of these materials.

(2) The large degree of undercooling required to nucleate the β-Sn phase in Sn–Ag–Cu alloys gives rise to a highly non-equilibrium structure in relation to that predicted by the phase diagram. In addition, under slow cooling conditions from

the melt, very large randomly oriented Ag$_3$Sn platelets are formed throughout the solder joint. There is some controversy as to whether these platelets have an adverse affect on reliability. It is generally felt that the Ag3Sn platelets are a reliability risk and their formation can be greatly reduced and even eliminated by reducing the Ag content to approximately 3% or less.

(3) The fatigue characteristics of the major Pb-free solder candidates (SAC, SAB) are very dependant upon test or application conditions (i.e., peak temperature, temperature range), and the cooling rate during the solder attachment process. Under some conditions the fatigue resistance of SAB and SAC is superior to eutectic Sn–Pb; for others it is approximately equivalent or markedly worse. The stress-relaxation time constants for SAC and SAB solders are far longer than those for eutectic Sn–Pb. As a result, ATC tests whose cycle times are too short will predict an overly optimistic field life. It should be apparent from these discussions that longer test times will be required to adequately characterize Pb-free solders. Consequently, due to these additional complications much more needs to be done to fully understand the very important issue of thermomechanical fatigue resistance in Pb-free solder joints.

(4) Tin's BCT crystal structure gives rise to anisotropic properties including the CTE, which is expected to have a negative impact on flip-chip solder joint reliability because they consist of few grains (~10). The BCT structure is also less capable of dissipating stress compared to lead's FCC structure and so maintains higher stress levels for longer periods of time. Tin also transforms to a very brittle DC structure when cooled below 13.2°C and this phase change results in a complete loss of structural integrity and a large increase in electrical resistivity. The condition is known as tin pest and is one that pure tin and tin-rich solders are particularly susceptible to, but not Pb–Sn solders.

(5) Fine, hair-like filaments referred to as tin whiskers that can form on pure Sn or Sn-rich finishes and often utilized on component lead frames can cause electrical shorts and numerous other problems. Although the industry has defined several mitigation practices that all serve to reduce risk, none of them is an absolute guarantee to halt whisker formation. Until one is determined, tin-whisker formation will continue to be a major concern for high-performance systems.

(6) With the continuous reduction in solder joint cross sections, particularly those of flip-chip solder joints, and the need for them to support higher electrical current levels, resulting in higher current densities, it is questionable whether Pb-free solders are capable of reliably performing their current-carrying task without degradation due to electromigration. It is not sufficient that they appear to possess better thermomigration characteristics compared to eutectic Sn–Pb. Only high-Pb solders appear to possess both the necessary thermal fatigue and thermomigration properties necessary to meet flip-chip solder joint requirements.

In consideration of the findings in relation to all the key issues discussed, it is concluded that the solder exemption accorded to high-performance systems was both necessary when granted by the EC and most likely will require an extension when reviewed in 2008, and perhaps well into the future.

References

1. Military and Obsolete Components—ROHS Directive, Dionics, Dec. 29, 2005. http://www.high-rel.com/rohs.htm
2. R.D. Hilty, ECN, Nov. 1, 2005. http://www.ecnmag.com/article/CA6279217.html
3. J.F. Mason, Electronic News, Dec. 22, 2005. http://www.read-electronics.com/electronicnews/index.asp?layout=articleprint&articleID
4. S. Deffree, Electronic News, Dec. 23, 2005. http://www.read-electronics.com/electronicnews/index.asp?layout=articlePrint&articleID
5. B. Dastmaichi and R. Vermeij, Transition to ROHS: the seven common pitfalls to avoid, Nov. 11, 2005
6. R.D. Hilty and M.K. Myers, Tyco electronics publication 503–1001, 2004, Revision 0
7. Implementation of Lead-free Packages, Spansion, 2005, http://www.spansion.com/support/lead_free_flash/Lead_Free_Packages.html
8. Backward and Forward Compatibility, Dallas Semiconductor, http://www.maxim-ic.com/emmi/backward_forwards_compatibility.cfm
9. R. Rowland, Surface Mount Technology, Nov. 2005, http://www.surfacemount.printthis.clickability.com/pt/cpt?action=cpt&title=surface+Mount+TEC
10. A. Maheshwari, Xcall Journal, Summer 2004, abhay.maheshwani@xilinx.com
11. Lead Free and ROHS Data, Freescale Semiconductor, 2004/2005, http://www.freescale.com/wehap/sps/site/overview.jsp?nodeId=061R7NMSSR
12. F. Hua, R. Aspandiar, T. Rothman, C. Anderson, G. Clemons, M. KLIER, in *Proc. Surface Mount Technol. Assoc. (SMTA) International Conf.*, Sept. 22 (2002)
13. F. Hua, R. Aspandiar, C. Anderson, G. Clemons, C. Chung, M. Faizul, in *Proc. Surface Mount Technology Assoc. (SMTA) International Conf.*, Sept 21, (2003), pp. 246–252

14. S. Bagheri, P. Snugovsky, Z. Bagheri, M. Romansky, M. Kelly, M. Cole, M. Interrante, G. Martin, C. Bergeron, in *Proc. CMAP International Conf. on Lead-free Soldering*, (May, 2006), Toronto, Canada
15. P. Snugovsky, M. Kelly, A. Zberzhy, M. Romansky, in *Proc. CMAP International Conf. on Lead-free Soldering*, May, 2005, Toronto, Canada
16. Impact to Customers: Forward/Backward Compatibility, National Semiconductor, http://www.national.com/packaging/leadfree/impact.html
17. K.J. Puttlitz, in *Handbook of Lead-Free Solder Technology for Microelectronic Assemblies*, eds. by K.J. Puttlitz, K.A. Stalter (Marcel Dekker, Inc, New York, 2004), pp. 239–280
18. S.K Kang, Ibid, 281–300
19. P.T. Vianco, Ibid, 167–210
20. W. Feng, C. Wang, M. Morinaga, J. Electron Mater. **3** 185 (2002)
21. C.A. Drewin, F.G. Yost, S.J. Sackinger, J. Kern et al. Sandia report, SAND 95–0196. UL-704, Sandia National Laboratories, Feb. 1995
22. P.T. Vianco, F.M. Hosking, J.A. Regent, in *Proc. NEPCON West 1992* (Des Plaines, IL), pp. 1730–1738
23. J.M. Tsung-Yu Pan, H.D. Nicholson, R.H. Blair et al., in *Proc. 7th International SAMPLE Electronics Conf.* (Parsippany, NJ, June 1994), pp. 343–354
24. T.J. Singler, S.J. Meschter, J. Spalik, in *Handbook of Lead-free Solder Technology for Microelectronic Assemblies*, eds. by K.J. Puttlitz, K.A. Stalter (Marcel Dekker, Inc., New York, 2004), pp. 331–429
25. K.W. Moon, W.J. Boettinger, U.R. Kitten, F.S. Biancaniello and C.A. Handwerker, J. Electron. Mater. **29**(10), 1122 (2000)
26. M.E. Loomans, M.E. Fine, Metall. Mater. Trans. A, Phys. Metall. Mater. Sci., **31A**(4), 1155 (April 2000)
27. D.W. Henderson, J.J. Woods, T.A. Gosselin, J. Bartelo et al., J. Mat. Res. **19**(6), 1608 (2004)
28. D.W. Henderson, T. Gosselin, A. Sarkhel, S.K. Kang, W.K. Choi, D.Y. ShIH, C. Goldsmith, K. Puttlitz, J. Mater. Res. **17**(11), 2775 (2002)
29. K.S. Kim, S.H. Huh, K. Suganuma, Mater. Sci. Eng. A **333**, 106 (2002)
30. D.R. Frear, J.W. Jang, J.K. Lin, C. Zhang, J. Minerals, Metals, Mat. Sci. (JOM) **53**(6), 28 (2001)
31. S.K. Kang, W.K. Choi, D.Y. Shih, D.W. Henderson, T. Gosselin, A. Sarknel, C. Goldsmith and K.J. Puttlitz, in *Proc. of 53rd Electronic Components and Technology Conference* (New Orleans, LA, May 2003), pp. 64–70
32. S.K. Kang, P.A. Lauro, D.Y. Shih, D.W. Henderson, K.J. Puttlitz, IBM J. Res. Dev. **415**, 607 (2005)
33. A. Woosley, G. Swan, T.S. Chong, L. Matsushita, T. Koschimieder, K. Simmons, in *Proc. Electronics Goes Green: Fraunhoffer Institute, Berlin, Germany* (Sept. 2000)
34. S. Prasad, F. Carson, G.S. Kim, L.S. Lee, A. Garcia, P. Rouband, G. Henshall, S. Kamath, R. Herber, R. Bulwith, in *Proc. SMTA Internat. Conf.* (Chicago, IL, Sept. 2000), pp. 272–276
35. J. Bartelo, S. Cain, D. Caletka, K. Darbha, T. Gosselin, D.W. Henderson, D. King, K. Knadle, A. Sharkel, G. Thiel, C. WoychiK, D.Y. Shih, S.K. Kang, K.J. Puttlitz, J. Woods, in *Proc. of APEX 2001* (San Diego, CA, Jan, 2001), 14–18, LF2–2, 1–12
36. S.K. Kang, P. Lauro, D.Y. Shih, D.W. Henderson, J. Bartelo, T. Gosselin, S.R. Cain, C. Goldsmith, K.J. Puttlitz, T.K. Hwang, W.K. Choi, Mater. Trans. (Jpn. Inst. Metals) **45**(3), 695 (2004)
37. J.W. Morris, H.G. Song, F. Hua, in *Proc. 53rd Electronic Components Technology Conf.*, (New Orleans, LA, May 2003), pp. 54–57
38. M.J. Sullivan, S.J. Kilpatrick, in *Handbook of Lead-free Solder Technology for Microelectronic Assemblies*, eds. by K.J. Puttlitz, K.A. Stalter (Marcel Dekker, Inc., New York, 2004), pp. 915–978
39. W.J. Choi, G. Galyon, K.N. Tu, T.Y. Lee, IBID, 851–913
40. R.A. Quinnell, Test and Measurement World, 9–1-2005, http://www.tmworld.com
41. G.T. Galyon, IEEE Trans. Electr. Packag. Manufact. **1**, 94 (2005)
42. Meeting, International Electrical Manufacturers' Initiative (iNEMI) Whisker Modeling Committee, Herndon, VA, November 2005
43. S.E. Koonce, S.M. Arnold, J. Appl. Phys. **3**, 365 (1953)
44. R.M. Fisher, L.S. Darken, K.G. Carroll, Acta Metal. **3**, 368 (1954)
45. B.D. Dunn, European Space Agency (ESA) Report, STR-223, September 1987
46. G. Galyon, L. Palmer, IEEE Trans. Electr. Packag. Manufact. **1**, 17 (2005)
47. W.J. Boettinger, C.E. Johnson, L.A. Bendersky, K.W. Moon, M.E. Williams, G.R. Stafford, Acta Mater. **11**, 5033 (2005)
48. N. Vo, M. Kwoka, P. Bush, IEEE Trans. Electr. Pkg. Mfg., **1**, 7 (2005)
49. S.C. Britton, M. Clarke, in *Proc. 6th Internat. Metal Finishing Conf.*, May 1964, pp. 205–211
50. P. Su, M. Ding, T. Uehling, D. Wonton, P.S. HO, in *Proc. 55th Electronic Components & Technology Conf.*, (Lake Buena Vista, FL, May 31–June 3, 2005), pp. 1431–1436
51. J.R. Lloyd, K.N. Tu, J. Jaspal, in *Handbook of Lead-free Solder Technology for Microelectronic Assemblies*, eds. K.J. Puttlitz, K.A. Stalter (Marcel Dekker Inc., New York, 2004), pp. 827–850
52. C.T. Liu, C. Chen, K.N. Tu, J. Appl. Phys. **88**, 5703 (2000)
53. S. Brandonburg, S. Yeh, in *Proc. Surface Mount Internat. Conf. and Exposition* (San Jose, CA, 1998), pp. 337–344
54. T.Y. Lee, K.N. Tu, S.M. Kuo, D.R. Frear, J. Appl. Phys. **89**, 3189 (2000)
55. T.Y. Lee, K.N. Tu, D.R. Frear, J. Appl. Phys. **90**, 4502 (2001)
56. S.Y. Yang, J. Wolf, W.S. Kwon, K.W. Park, in *Proc. 52nd Electronic Components and Technology. Conf.* (San Diego, CA, 2002), pp. 1213–1230
57. M. Ding et al., in *Proc. Electronic Components and Technology Conf.* (Las Vegas, NV, June 1–4, 2004), pp. 968–973
58. B. Ebersberger, R. Bauer, L. Alexa, IBID, 683–691
59. J.D. Wu, P.J. Zheng, C.W. Lee, S.C. Hung, J.J. Lee, in *Proc. 41st IEEE Internat. Rel. Physics Symp.* (Dallas, TX, 2003), p. 132
60. Y.C. Hu et al., J. Mat. Res. **11** 2544 (2003)

Index